A Guide to
**The Culture of
Science, Technology, and Medicine**

A Guide to
The Culture of
Science, Technology, and Medicine

General Editor
Paul T. Durbin
University of Delaware

Area Editors

History of Science
Jerome R. Ravetz
University of Leeds

History of Technology
Melvin Kranzberg
Georgia Institute of Technology

History of Medicine
Saul Benison
University of Cincinnati

Sociology of Science and Medicine
Bernard Barber
Barnard College, Columbia University

Philosophy of Science
Peter D. Asquith
Michigan State University

Philosophy of Technology
Paul T. Durbin
University of Delaware

Philosophy of Medicine
Warren T. Reich
Georgetown University

Science Policy Studies
Derek de Solla Price
Yale University

 THE FREE PRESS
A Division of Macmillan Publishing Co., Inc.
NEW YORK

Collier Macmillan Publishers
LONDON

The project on which this volume is based was funded jointly by the
National Endowment for the Humanities and the National Science
Foundation.

The Free Press
A Division of Macmillan Publishing Co., Inc.
866 Third Avenue, New York, N.Y. 10022

Collier Macmillan Canada, Ltd.

Library of Congress Catalog Card Number: 79–7582

Printed in the United States of America

printing number

   3   4   5   6   7   8   9   10

```
Library of Congress Cataloging in Publication Data
Main entry under title:

A Guide to the culture of science, technology, and
    medicine.

    Includes bibliographies.
    1. Science--Social aspects.  2. Technology--Social
aspects.  3. Medicine--Social aspects.  I.  Durbin,
Paul T.
Q175.5.G84          301.24'3          79-7582
ISBN 0-02-907820-2
```

Contents

Analytical Table of Contents

Chapter 4.   Philosophy of Science: Historical, Social, and
             Value Aspects   197

Preface

The heart of *A Guide to the Culture of Science, Technology, and Medicine* is a set of state-of-the-field surveys of nine academic disciplines that take as their object science, technology, and medicine. They are integrated in this volume by a focus on value issues in the sciences and biomedicine. The surveys are a product of close collaboration between the authors and the area editors. The editors were also responsible for overall coordination; Arnold Thackray, who served on the board in addition to authoring a chapter, was particularly helpful in this respect.

Possible users of this volume, it was assumed from the outset, should include scholars in neighboring fields—philosophers of science, for instance, who may want to learn more about the history or sociology of science. However, members of the technical community and the interested educated general public ought also to be interested.

More than anything else, the volume should be useful to teachers of interdisciplinary courses in technology and values or science and society. Such teachers have often been faced with a difficult task of collating materials from innumerable and disparate sources. We have tried to gather as many of these as we could, discussing them explicitly, commenting on them in bibliographic essays, or, at the very least, listing them in bibliographies that attempt to balance comprehensiveness with utility.

Another recent source, *Science, Technology and Society: A Cross-Disciplinary Perspective* (1977), edited by Ina Spiegel-Rösing and Derek de Solla Price, focuses more on the social sciences. With the *Guide* and *Science, Technology and Society,* teachers in these interdisciplinary courses now have useful guides to both social science and the humanities.

The idea behind *A Guide to the Culture of Science, Technology, and Medicine* is due mainly to Dr. Richard Hedrich, director of science values projects at the National Endowment for the Humanities. When he was approached by Carl Mitcham for support of ongoing bibliographic efforts in the philosophy of technology, the request set him thinking about applications he had been receiving from a number of similar fields. Ought they not be related? Why not try to find some integration among them? The idea was discussed with two of our editorial board members, Melvin Kranzberg and Derek Price. Kranzberg then undertook the effort to publicize the venture in scholarly circles. It fell to the general editor to collect a wide variety of inputs and give the project its final form. In due course the project was funded by NEH along with the National Science Foundation.

The able assistant to the general editor for the project was Elise B. Harvey; her warmth brought a human quality to the sometimes exasperating interchanges between editors and authors which was appreciated by all. Fran Durbin contributed generously to this process and put up with much, especially during final editing. Final typing was done by Betty Dickinson, with a whole host of secretaries at the authors' various home institutions handling earlier versions of chapters and other chores.

Introduction

Two basic problems are addressed in this volume. One is the continuing split between the sciences and the humanities—once infamous as the "two-cultures" gap.[1] The other problem is less familiar. It has to do with fragmentation. Each of the humanistic and social science disciplines focusing on science and technology—the history of science or technology, for instance, or philosophy of science or science policy studies—goes its own way. And all these disciplines, for the most part, steer clear of explicit application to ethical and other value issues associated with science, technology, and medicine.

The volume has a dual aim: to attempt to establish an interdisciplinary whole by surveying nine currently fragmented humanistic and social fields dealing with science, technology, or medicine; and to bring about this integration by focusing explicitly on the help these disciplines might afford to those attempting to deal with major contemporary value questions in science and technology.

This objective is achieved in two steps. The general introduction suggests one way in which the nine fields surveyed here might be integrated —under the heading "culture of science disciplines." The second step is the remainder of the volume, comprising surveys of the fields, with extensive bibliographies. Whenever possible, the state-of-the-field surveys refer to other fields and to the overall integrative focus around value issues.

PART I. GENERAL INTRODUCTION

The idea behind calling this volume *A Guide to the Culture of Science, Technology, and Medicine* is similar to the one that motivated others to title educational programs "Science, Technology, and Values" or "Human Dimensions of Science." Indeed, another satisfactory title might have been "Value Issues in Science, Technology, and Medicine." One reason for preferring "culture" to "values" in the title is that the volume addresses not only value issues *in* the sciences but also the value *of* science—the place of science, technology, and medicine in contemporary culture.

Approaches to Value Questions

"Axiology" or "the general theory of value(s)" has, it is claimed, "provided unification for the study of a variety of questions—economic, moral, aesthetic, and even logical—that had often been considered in relative isolation." [2] Philosopher Ralph Barton Perry, often said to have provided, in his *General Theory of Value* (1926), the *magnum opus* of twentieth-century treatments of values, equates "value" with "any object of any interest." [3] Later, in *Realms of Value* (1954),[4] he touched upon eight such values: morality, religion, art, science, economics, politics, law, and custom. Most authors would probably agree readily with Perry that morality, art, and custom belong among the "realms of value." Others would find little difficulty in adding law and religion. The inclusion of economics can be explained historically: the term "value" was widely used in economics before being generalized, mainly in the twentieth century, to other areas. What about science as a "realm of value"? This is the central issue with which we are concerned in *A Guide to the Culture of Science, Technology, and Medicine.*

The Meaning of "Values"

Values can be considered, at risk of oversimplification, in two ways. The social sciences, from psychology to anthropology, from economics to political science, attempt to *describe* values, value preferences, and value systems. Sometimes the descriptions are embedded in cross-cultural theories, personality theories, or general economic theories; more often values are presented purely descriptively. Philosophers, on the other hand, have traditionally attempted to evaluate values—to define what *ought* to be, rather than simply what is, valued.[5]

In this volume, such an oversimplified distinction, although helpful as a first approximation, will not do the whole job. Historians of science and technology, for instance, do not fit neatly into either category. They sometimes describe the values of science or the cultural values that lead to advances in technology; but at other times they also make value judgments with respect to particular periods or groups in the history of science. Similarly, some sociologists of science or medicine venture into "normative sociology," taking stands on issues facing the technical community. This is also clearly true in science policy studies. Even the philosophers reported on in this volume are divided, some feeling freer to talk about what *ought* to be the case, and others trying to be merely descriptive.

Most generally, at least since the time of Plato, philosophers have discussed the question whether something is valuable (good, etc.) because it is valued, or whether it is valued because it is (intrinsically, objectively, intuitively) valuable.[6] Again most generally and at the risk of oversimplification, it can be said that there are four main types of philosophical value theorists: There are "subjective absolutists" (for instance, G. W. F. Hegel and other Absolute Idealists) and "objective absolutists" (e.g., natural law theorists); there are also "objective relativists" (utilitarians, for example) and "subjective relativists" (existentialists).

A fifth category perhaps ought to be added: "critical theorists" who claim to do no more than criticize the statements or views of the other four types. Immanuel Kant's "critical" philosophy may be viewed as the first of such modern ethical theories. More recent "metaethicists" (the "emotivist" A. J. Ayer, or the "prescriptivist" R. M. Hare, for instance) clearly fall under this heading.

It is useful to make some distinctions among key value terms:

1. "Value," "good," "goal," "end," "motivation" are all value terms with somewhat different connotations. Sometimes, as in Perry's theory, all such value terms are taken to be equivalents of "interest." At other times, an author will reserve "value" for something concrete—e.g., economic values—and leave "good" to the philosophers. The other terms usually appear in specialized contexts.

2. For "value," the basic distinction is between "inherent" and "instrumental."

3. "Goal" is generally used only in making a long-range/short-range distinction.

4. "End" is frequently used as a synonym for "goal" or "good" or even "value" where the means–end distinction is important.

5. "Motivation" is often a substitute for other value terms in social science literature.

6. Finally, we must add "policies" to the list, meaning choices—usually social—among goals or ends. There are long-term policies and short-term policies, and "strategic" and "tactical" choices *among* policies, as well. Going even farther afield, such terms as "ideology," which have more specialized uses, have a value component.

This wide divergence among usages of value terms is one thing that has made a general theory of value(s) seem attractive.

A second set of distinctions bears on differences between science, applied science, and technology. Some maintain that, at least in the ideal, pure science is different from applied science and technology in that its only value is truth for its own sake. Applied science is assumed to add the dimension of practicality, and technology to bring in a whole host of social, cultural, and even ideological values. Medicine is an even more complex matter, involving, as it does in its current Western mode, science, technology, and the "art of healing."

All these distinctions, however clear they may seem, are subject to a wide variety of interpretations and to much dispute, as is clear in the chapters that follow.

Social Science Theories

There is no neat schematization of social science treatments of values that closely parallels the above list of general philosophical theories. Anthropologists have almost always been concerned with cross-cultural comparisons of value systems. Sociologists, especially of the methodology-oriented American school, for a long time avoided problems of value, usually getting at values only indirectly through the study of organizations or groups, though in more recent years, especially with the rise of "normative sociology," this has been less true; however, value neutrality still remains a goal of many, especially behaviorist-oriented, sociologists.[7] (Sociologists of science for the most part followed the patterns of sociologists generally, though early on there was an attempt to discover empirically the peculiar value system of the scientific community.[8]) Social psychology is a field sharply divided between behaviorists and more "subjective" theorists (for example, symbolic interactionists), with the latter much more likely to focus on values and value systems, typically those of small groups. Historians and political scientists have recently come to be considered partly within the social science category, especially when they make use of quasieconometric models. Traditionally they would have been thought of as closer to philosophers in saying not only what is, but what ought to be, valued. Among practitioners of the disciplines surveyed here, science policy students have very often attempted to use econometric or "systems theory" models.

Some social scientists believe that it is possible to measure value preferences and to differentiate between various groups—national cultures, ethnic groups, occupational groups including scientists—and so on—in quantitative terms.[9] They even believe that such a venture can lead to significant theoretical unification among the social science disciplines, though it is currently too early to tell what other social scientists will make of this bold claim.

In short, the discussion of values by social scientists (where they discuss them at all, without reducing valuation behavior to other forms of behavior) takes on a bewildering variety of forms. In the *Guide,* a permissive attitude has been taken: our surveyors of social-science-of-science fields have been given *carte blanche* to report any sort of value focus in the writings surveyed.

Values, Science, and Technology

This area encompasses three possible concerns: the alleged value neutrality of science and/or technology; the value *of* science (the place of science, technology, and medicine in contemporary culture); and value issues *in* science. These will be taken up in turn, the first quite briefly.

Science, Technology, and Value Neutrality. Especially by scientists and other members of the technical community, but also among the general public, it is widely held that science is "objective" or "value free," or at least "value neutral." A similar claim is often made that "technology is a neutral tool, bad or good only insofar as it is used for some purpose." Neither claim will stand up to criticism, though there is clearly *something* to be said for them.

First, technology: Although the "neutrality" claim is often used in a defensive manner by members of the technical community when technology is under attack, it is hardly consistent with another claim that technologists have few qualms about making—namely, that technology is methodologically distinct from science insofar as it embodies practical applications and social goals. In fact, no particular technology, construed as a technological object, gadget, process, or system, or even an isolated bit of technological knowledge or know-how, can be morally neutral. It was designed or conceived for some purpose, and any such purpose is subject to moral or ethical evaluation. What defenders of the neutrality of technology probably have in mind is that a particular technology designed for one purpose (presumably good) may be used for other purposes and that the designer or producer of an object should not be held responsible for its misuse by someone else. This defense, however, does not get one very far. It is conceivable that someone could design a bit of technology for destructive purposes, and most people would consider a designer at least callous if he or she did not care whether a particular, perhaps technically perfect, product was to be used for good or evil purposes. Defenders of "technological neutrality" seem to intend the phrase to have a positive connotation. They are thus defending technology not as neutral but as good, whatever bad uses others may make of it.

The value neutrality of science is perhaps a slightly trickier question, though the predominant view among philosophers of science—who worry most about such issues—may have swung around recently to the argument that science is not value neutral either.

One version of the "science is value neutral" thesis can be found in views of Arthur Kantrowitz, chief advocate in the United States of the "science court" idea. Kantrowitz believes that the "separation of facts from values can always be made," that "scientists take great pains to separate the factual statements they make from any of their personal values." [10] To this claim Alex Michalos (an author in this volume) has replied: "It is plainly logically impossible to say anything good or bad about anything without saying something that is value laden. It is just literally absurd to pooh-pooh values in the interests of *good* science." Michalos probably speaks for most philosophers of science today when he concludes: "The most one can hope to do," whether by way of science courts or through other means, "is get general agreement about this or that evaluation,"[11] whether of facts or of values.

All this is not to say that science is not "objective" in some sense, that there is no ideal of objectivity toward which scientists strive.[12] It is only to say that objectivity, as thus pursued, is itself a value and must be weighed against other values.

The Place of Science and Technology in Contemporary Culture. Our next concern is the value *of* science, technology, and medicine, especially in a modern society or culture based on an industrial economy. The concerns of this *Guide to the Culture of Science, Technology, and Medicine* cannot be divorced from the broader social environment of the late twentieth century. One large concern in that context is what might be called the "questioning of science" movement. In the early 1960s there was heated discussion in intellectual circles over the "two-cultures" controversy, especially as brought to public attention in 1958 by C. P. Snow. The late 1960s saw this concern transformed into a pro-and-con debate over the value of technology; in the U.S. this debate may be said to have culminated in Robert Pirsig's *Zen and the Art of Motorcycle Maintenance* (1974).[13] Where Snow had been an open advocate of science and technology, Pirsig may be labelled "protechnology," but only in a very limited sense. Many members of the scientific-technical community feel that the 1970s have seen still further erosion in the public attitude toward science and technology, even though opinion polls do not support this view.[14] Facts alleged in support of the alarmist view include: a sharp decline in the rate of increase of U.S. federal spending for research; signs that other countries are rapidly catching up with the U.S. in terms of technological innovation; and increased expression of alarmist views in the mass media by antitechnology opinion leaders.

There are indicators of negative public sentiment about the scientific-technological community: public perceptions of scientists, engineers, and biomedical researchers are often couched in negative terms.[15] Some even maintain that "technique" dominates modern consciousness to the exclusion of every other kind of thinking in a "one-dimensional society."[16] Problems affecting science and technology have been explained as simply a function of "bigness."[17] Others warn that, as science gets bigger, its problems smaller, and competition fiercer, sharp practices and dishonesty can be expected;[18] some claim that they see this trend exemplified in recent celebrated cases of falsification of scientific data. Still others bemoan the fact that "big science" has achieved its current size as a result of military research funding.[19]

Another problem is the apparent failure of professional codes of ethics, especially in medicine and engineering, to protect the public adequately. Medical and psychological experimentation on human subjects is mentioned below in our list of specific value issues facing the sciences. This remains a problem in spite of the fact that worldwide codes of ethics for medical practitioners and researchers have undergone revision after revision. Within the sphere of the various engineering societies, it has been claimed that codes of ethics are ineffective safeguards—that at times, indeed, they have served as obstacles to protection of the public.[20]

Perhaps the best known among the outspoken critics of the scientific-technological community are authors associated wtih the "Frankfurt school" in Germany. Most influential in the U.S. is the Marxist-revisionist philosopher, Herbert Marcuse; in Europe, Jürgen Habermas is probably more influential. Marcuse's basic critique, most clearly stated in *One-Dimensional Man: Studies in the Ideology of Advanced Industrial Society*,[21] has remained substantially the same through numerous revisions.[22] It is primarily a criticism of capitalist technology and of misuses of technology under bureaucratic socialism. Marcuse's fundamental thesis is that technology, a tool in the hands of the ruling class, helps guarantee the enslavement of the masses by its totally alienating rational objectivity. It is his focus on reason-rationality as the culprit that distinguishes Marcuse's thought. The idea of technological rationality—that "technological controls appear to be the very embodiment of Reason for the benefit of all social groups and interests— occurs to such an extent that all contradiction seems irrational and all counteraction impossible."[23] Some of Marcuse's favorite targets are behaviorist psychology as a tool of thought control, and analytical philosophy as the enforcer of the linguistic status quo. The "one-dimensionality" of his title refers to the flattening out of experience in contemporary culture; Marcuse would call it "pseudo-culture."

Marcuse's explanation of contemporary ills is a sort of "devil theory." It is not unlike Jacques Ellul's equally famous attack on "technique" as the enslaving spirit of our age[24] or related attacks on the spirit of the age by existentialist philosophers.[25] Marcuse's Marxist thought became more influential, and more threatening to the science establishment, than the ideas of other intellectuals because it was adopted by "New Left" radicals of the 1960s bent on disrupting, among other things, scientific professional meetings.

The ideas and impact of another critic, Lewis Mumford, have been much more genteel. Although Mumford has been lumped with four others as "the leading antitechnologists" by one protechnology author,[26] this accusation oversimplifies Mumford's complex and subtle thesis. With great erudition, especially in the two-volume study *The Myth of the Machine* (1967) and *The Pentagon of Power* (1970),[27] Mumford mounts an attack on the theory of tool-using as the evolutionarily distinctive character of *Homo sapiens*. Mumford claims to have discovered such a myth recurring throughout history used by powerful rulers who have been willing to organize their subjects into vast machine-like organizations for the efficient attainment of their own goals. Mumford's analogy between the organization of manpower for the building of the pyramids and that of technical experts involved in getting men to the moon is widely viewed among literary intellectuals as a penetrating insight. Mumford's overall thesis is that this "myth-of-the-machine" organization is usually oriented toward the selfish aims of rulers and, as often as not, toward military victory. Indeed, the other major image that Mumford presents as a description of the dangers inherent in contemporary technology is that of "the pentagon of power." In this respect, his critique shares some characteristics with the views of those who have attacked the "military-industrial-technological complex."

As mentioned, such criticisms have not filtered down to the masses in technological society. Popular culture remains deferential toward technical experts even while it exhibits a receptive attitude toward negative stereotypes of those experts. It is not too much to say that the culture-versus-science/technology debate remains significant today.

Value Issues in Science and Technology. Recall some of the value distinctions made earlier: between values, "the good," and motivations; between long and short-range goals; between tactical and strategic policies. These distinctions suggest some of the concrete value issues that arise in science, technology, and biomedicine today.

First and most obvious of the value issues are the moral issues. Some of the most widely publicized cases involving such questions are those associated with bioethics: experimentation on human subjects, both generally (questions of informed consent) and with respect to particular subjects (prisoners, hyperactive children, fetuses); turning off exotic life-preserving devices or allocating them among needy patients; the right to health care; controversies over medical malpractice; and so on.

Engineering ethics may soon catch up with biomedical ethics as a matter of public debate. To date, the most widely discussed ethical problem in engineering has been that of "whistle-blowing": the effort of particular engineers (or applied scientists) to call public attention to what they see as unethical practices on the part of their employers. Other problems that have received attention include the lack of institutional support (for instance on the part of professional societies) for engineering whistleblowers and others who demand support for what they view as ethical behavior; or alleged unethical behavior of particular engineers—for instance in the design process, or in kickbacks to public officials for contracts to do public work projects.[28] In addition, some questions have been raised—particularly by social critics of technology and certain literary figures—about the narrowness of the value system of engineers in an age demanding social responsibility from all the professions.

Another, and entirely different, sort of value concern in the sciences is the vast area of policy decisions with respect to science, technology, and biomedical research. Every time a policy decision must be made there is a value issue; often there is also a debate over alternatives, or at least the opening for such a debate. Whether the issue is the funding level for the sciences generally; the funding balance between basic and applied science; relative priorities among the physical and life sciences and medical research; questions about the economic payoff of research and development—in all these cases conflicting values are at stake. In each case, some perspective or bit of information from the science-humanities or social-science-of-science disciplines might lead to more enlightened and reasonable discussion.

These are the more obvious value issues in the sciences. There are others as well. Some social science and historical studies in particular have focused on determining the value system(s) of the general scientific-technical community or of particular segments of it. Others have focused on the significant differences between various general social value systems or cultures as those differences impinge on science and technology.

Another less obvious but no less important set of issues revolves around medicine. In some senses medicine is the most "human" of the sciences surveyed here, and it is, of course, an art as much as it is a science. Some of the issues involved in medical science include the values certain cultures place on health; the degree to which the art or practice of medicine ought to be science based; and the relative weight that ought to be assigned to preventive medicine, health care, clinical research, and so-called "basic" biomedical research.

Last among value issues *within* the sciences, there are broad (often tendentious) questions raised about the relationship between science or technology and ideology. There are issues of an "environmental ethic." And there are a number of general social issues to whose solution many people feel science and technology should contribute: world hunger, energy problems, and the like. Direct guidance on all these issues—even help on how to be reasonable with respect to disagreements about them—will not be found in this volume. Still, it is hoped that enough of these issues will have been touched upon so that helpful analogies or extrapolation may be made that carry over to issues not specifically raised. One hoped-for outcome of *A Guide to the Culture of Science, Technology, and Medicine* is a sizable list of researchable topics for future discussion.

Some Value Issues Not Taken Up in This Volume

Contemporary art and architecture have clearly been influenced by technology and, to a lesser extent, by science. R. Buckminster Fuller is only one—admittedly an extreme example—of many architects influenced by technology, just as Marcel Duchamp is only one of many painters who have used science and technology as explicit themes. Most contemporary abstract painting and sculpture could not have appeared except in an era of high technology. Certainly an article on the relationship between science and art would seem to belong in a guide to the culture of science. Our task, however, has been to focus large bodies of existing literature on the value aspects of science, and the available material on science and art either is r.ot ample enough or would not blend into the structure of this volume.

Although a study of science and literature would be meaningful for the cultural values aspect of contemporary science, technology, and medicine, that too has been omitted. In this area there is an abundance of material, including not only criticism of science and technology by literary figures and popular social commentators, but also science fiction.

An excellent critical overview of science fiction is provided by Robert Scholes and Eric S. Rabkin in *Science Fiction: History, Science, Vision* (1977).[29] Although Scholes and Rabkin recognize prototypes of science fiction from the time of Galileo on, they agree with most historians in designating Mary Shelley's Frankenstein (1818) as the true beginning of the genre. The major works between that date and the present that they survey include Jules Verne, *20,000 Leagues under the Sea* (1870); H. G. Wells, *The Time Machine* (1895); Yevgeny Zamyatin, *We* (1920); David Lindsay, *A Voyage to Arcturus* (1920); Olaf Stapledon, *Star Maker* (1937); Arthur C. Clarke, *Childhood's End* (1953); Walter M. Miller, Jr., *A Canticle for Leibowitz* (1959); Ursula K. Leguin, *The Left Hand of Darkness* (1969); and John Brunner, *The Shockwave Rider* (1976). Also included are the "Golden Age" authors: A. E. van Vogt, Robert A. Heinlein, Isaac Asimov, and Theodore Sturgeon. In passing, Scholes and Rabkin touch upon science fiction in media other than literature—for example, Stanley Kubrick's film "A Clockwork Orange" (based on Anthony Burgess' novel) and TV's "Star Trek." The authors conclude that "when the most important American novel of the past several years, Thomas Pynchon's *Gravity's Rainbow* (1973), is also a case of straddling the old dividing line between science fiction and general or 'serious' fiction, then it is obvious . . . that the line itself has virtually disappeared." They also make a prediction about the future: "Science fiction will not exist. But the whole shape of literature will have changed." Scholes and Rabkin's overall judgment is this: "The premise upon which this book is based is that a sufficient number of works of genuine merit have been produced in this field to justify its study as an aspect of modern literature as well as an important feature of contemporary culture."[30]

Along with works of science fiction, which are basically optimistic about science and technology, indeed pushing the technological imagination in fantastic directions, there has been from the beginning a steady influence of modern science on literature. A recent bibliographic survey includes items as early as Boccaccio's *Decameron* (protoscience; fourteenth century); numerous nineteenth-century works, including poems by Elizabeth Barrett Browning, Robert Browning, Ralph Waldo Emerson, and Edgar Allen Poe, and essays by Matthew Arnold and T. H. Huxley; as well as recent works such as Boris Pasternak's *Doctor Zhivago* and Robert M. Pirsig's *Zen and the Art of Motorcycle Maintenance*.[31] Valuable also are Jacob Bronowski's explicit discussions of science and the literary imagination in *William Blake and the Age of Revolution* and *The Identity of Man*.[32]

There is a long history of antiscience and antitechnology literature. Names here include Goethe (though he is not totally antiscience, even in *Faust*), Mary Shelley (again for *Frankenstein*), the popular trio of Aldous Huxley (*Brave New World*, 1932), George Orwell (*1984*, 1949), and Anthony Burgess (*A Clockwork Orange*, 1962), and—perhaps the most popular in the U.S. in recent years—Kurt Vonnegut (*Player Piano*, 1952, and *Cat's Cradle*, 1963), among others.

Contemporary science and technology have also begun to affect writing in general history and philosophy. Some examples are the excellent coverage of science and technology in Daniel Boorstin's *The Americans*,[33] and John Passmore's *Man's Responsibility for Nature: Ecological Problems and Western Traditions*.[34]

Passmore's work should remind us that one of the most significant issues in contemporary culture involving the impact of science and technology is that of ecology and the environment. In this volume a choice had to be made between covering the topic incidentally, noting where the issues have been raised in the various disciplines, and attempting to be comprehensive in a separate article. The decision was to cover environmental issues incidentally, within the state-of-the-field surveys. Perhaps the seed will have been planted for someone else to attempt an encyclopedic survey of science, technology, and the environment.

Two other areas remain that one might expect to find covered here: science, technology, and law; and science and religion.

A number of law-and-environment issues were excluded by the decision not to treat science and the environment comprehensively. With respect to law, science, and technology more generally, quite a few studies have appeared.[35] In 1970, for example, the *Denver Law Journal* published the proceedings of a symposium entitled *The Implications of Science-Technology for the Legal Process*, and a number of other forums and symposia have been conducted. One particular concern has been product liability.[36] With the exception of the well-established field of patent law, this is a new and rapidly changing area that seemed inappropriate for inclusion in this volume, however central the law may be in the culture of a society.

Science-and-religion controversies have had a long history. For centuries, proscience literature made much of the conflict between Galileo and the Church. The conflict was a major subtheme in much Enlightenment literature throughout the West. The nineteenth century ushered in a heyday of science–religion controversies, notably the Huxley-Wilberforce debates on Darwinism and Robert G. Ingersoll's militant atheistic crusade. All these events and more are surveyed in Andrew Dickson White's *History of the Warfare of Science with Theology in Christendom* (1897).[37] Although these debates continued into the twentieth century, notably in the famous Scopes trial in Tennessee in 1925, much of the sting of the issue had subsided. Perhaps the last major voice in the debate was that of Bertrand Russell, for example in *Religion and Science* (1935).[38] More balanced treatments have been the norm in recent decades, especially the 1970s.[39] Although religion is and is likely to remain a significant part of man's cultural life a survey of the often needlessly emotional science–religion debate seems out of place in the present volume, though some attention is paid to it in chapter 1. A number of religious critiques of technology and technological culture are also included in chapter 5.

Finally, why is there no survey of insiders' views of the value issues facing science and technology? From the beginning, in the planning of this volume, there has been a minority viewpoint insisting that what was to be done could not be done without including representation from the scientific-technical community. The view that prevailed was that the views of scientists, engineers, and biomedical researchers would be included only incidentally—in the state-of-the-field surveys and at the discretion of individual authors. The success of this method must be left to the judgment of the readers of this volume. It has been our hope from the outset that members of the scientific-technological community would be among the users of the volume—not least those who have courses to teach with such titles as "Science and Values," or "Science, Technology, and Society." Perhaps some users will feel the need to supplement the resources here with firsthand statements of scientists, engineers, or biomedical researchers, in order to achieve a balanced view on substantive issues.

PART II. INTRODUCTIONS TO THE FIELD SURVEYS

The authors of *A Guide to the Culture of Science, Technology, and Medicine* were given two mandates. One was to survey the state of the art in their field. The second was to attempt some integration with the other fields, especially by focusing on value issues in science, technology, and biomedicine. The latter task was ambiguous from the outset. It can be interpreted as involving: a section or sections on value questions, which might then be discussed either directly or in an encylopedic-descriptive fashion; discussions of various sorts scattered throughout a chapter; or even a treatment in terms of the second-level value choices influencing the treatment (or neglect) of particular value issues in certain culture of science fields. Almost predictably, our authors interpreted their instructions in different ways, with all the possibilities mentioned and variations thereon. A comment will be made on these choices in each of the chapter introductions.

A word needs to be said in each case, as well, on the degree of integration with respect to the other fields—the extent to which the author(s) succeeded in demonstrating possibilities for integration, as well as the objective possibilities that existed for him or her to work with.

Finally, something should be said about the concern that underlay the assembling of this *Guide* from its inception. There is a feeling, for example, that if the history of science is worth anything at all, over and above its own inherent antiquarian value, it may be in helping someone work toward solutions to value issues facing the scientific community. (At various times it has also been valued for other external purposes, including as an introduction to science for nonscience college students.) This view is a variation on the theme that those who are ignorant of history are doomed to repeat its mistakes. In other areas, it is not the value of historical relevance that is emphasized, but the advantages of looking at issues from the height of philosophical generality or, where possible, analyzing value issues empirically, utilizing the methods of the social sciences. In this introduction to the fields, this issue of "relevance"—possibly the primary one for teachers of courses in technology and values, or science and society—will be taken up largely in the single chapter in Part IV, Policy Studies.

History of Science

Arnold Thackray in his survey of the history of science takes as his principal focus the way in which external social and cultural values have decisively influenced historical studies of science in different ways at different times. His account moves briskly through the "prehistory" of the history of science; his survey proper begins with the nineteenth century.

As Thackray sees the history of science, it was the French, in terms of Auguste Comte and his "positivist" disciples, who supplied the bridge from nineteenth-century historians to the twentieth-century academic profession of the history of science. And the bridge was the work, for the most part, of one man—George Sarton. Thackray surveys the work of Sarton, that of the 1930s Marxists, and what he labels "The Idealist Program" following World War II. In the last category his main focus is the work of Alexandre Koyré, which he says provided the first true "paradigm"—the way of doing history of science for a generation of scholars—in the history of the history of science.

Thackray ends this section with a quick overview of the wide diversity of current professional approaches to the history of science, calling this "the new eclecticism." Up to this point, then, Thackray's account is a values-oriented externalist history of the history of science movement.

He then devotes the second half of his chapter to "some central domains in the history of science," ranging from the social roots of science, to ancient and medieval science studies, to science and religion, and to relations with other fields surveyed in the *Guide*. The bibliography is, relatively speaking, concise; however, it is a judicious selection that should prove helpful to students and scholars in other fields.

History of Technology

Carroll Pursell's survey of the history of technology emphasizes two things: the variety of people working in the field—economic historians, social historians, scholars in American Studies programs, "industrial archaeologists," as well as "professional" historians of technology—and the tendency of professional historians of technology to differentiate themselves from their predecessors by focusing on the social as well as the mechanical aspects of technology. According to his outline, the field came into being as a result of the interests of students of industrial history in Britain (the Newcomen Society for the Study of the History of Engineering and Technology, founded in 1919) and of engineering educators in the United States (the Society for the History of Technology, founded in 1958).

In this section Pursell emphasizes the diversity of history of technology approaches—internal history of technology, work stressing business strategies and economic change, the biographies of famous inventors or entrepreneurs combined with histories of individual industries; and at least two types of social history of technology, institutional and intellectual.

The second major portion of Pursell's chapter is devoted to problem areas of special interest in the history of technology: medieval technology, the professionalization of engineering, and women and technology, among others. For the most part, this section is reportorial and bibliographic.

In the final third of his survey, Pursell discovers the relationship of history of technology to other fields—history of science, general history, and so on—and, in a brief section, the relation of history of technology to value issues. Here Pursell suggests that *all* of history of technology is inherently value-oriented because of the nonneutrality of technology.

History of Medicine

Gert Brieger's survey of the history of medicine, like Pursell's on the history of technology, is an insider's survey. It is punctuated occasionally with flashes of Brieger's wry wit or the gentle needling of his colleagues. The "insider" aspect shows up in the authoritative judgments of someone who has been at the intellectual center producing much of the history of medicine scholarship on which he reports.

Brieger moves meticulously through the entire range of approaches to history of medicine, old and new. There are particularly good treatments of the social history of medicine in general, the social history of hospitals, social history work on psychiatry, and oral history. Brieger is also illuminating on the relationship of the history of medicine to other fields, especially the history of science, women's studies, sociology and anthropology, and art history.

On the history of medicine and bioethics, Brieger comments, "Unfortunately, with few exceptions, those who have written in the broad field of bioethics have not generally approached their subjects historically"; and his treatment of public policy and medicine is the most extensive and definitive in the *Guide*. Finally, a unique feature of Brieger's chapter—alone among the surveys in the *Guide*—is his final section on teaching in the history of medicine. The bibliography is as comprehensive as possible, given the literature in the field.

Philosophy of Science

Chapter 4 on the philosophy of science, by Alex Michalos, is a somewhat uneasy amalgam of two approaches. The first part includes two encyclopedic surveys, one of the history of philosophy of science, the other of work done in the twentieth century outside the United States. The second part focuses on Anglo-American professional analytic philosophy of science.

Michalos' survey of analytical philosophy of science touches on some principal illustrative problems—scientific significance, theory and observation, discovery and growth, explanation, induction and probability—and combines the simplicity of an introductory textbook with the subtlety of ongoing debates in the philosophy of science. Michalos does not hesitate to include argument for what he sees as the best solution to problems currently being debated in the field. But his most intriguing contribution in this part of his survey is to discover value issues inherent in the philosophical positions taken on these issues. In a field noted for the abstractness and value-eschewing intentions of its practitioners, this is no mean feat. Michalos' bibliography is extensive, though in this field today no bibliography can really be comprehensive.

Philosophy of Technology

In his survey of the philosophy of technology, Carl Mitcham deftly inserts conceptual analyses of major thinkers in the field into his synthesis. The synthesis is divided into three parts: the historical development of treatments of technology broad enough to be called philosophical; a survey of metaphysical-epistemological-definitional analyses; and a broad overview of principal ethical problems associated with technology or the technological world.

Mitcham's historical survey focuses primarily on three broad schools of thought: West European, Anglo-American, and East European. He bases what he calls his "metaphysical" survey on a conceptual scheme that divides treatments of technology according to their focus on technological objects, processes, knowledge, or volition, or on some combination of these. He chooses eight representative ethical problems associated with technology and work, war, culture, the environment, and social-political-economic development.

The three main conceptual essays focus on Jacques Ellul, Martin Heidegger, and Friedrich Dessauer. Two somewhat shorter conceptual essays discuss a representative East European collective volume, *Man, Science, and Technology* (1973), and the work of Lewis Mumford. What Mitcham manages in these essays is to combine some actual philosophy-of-technology analysis with his encyclopedic overview of the field.

Philosophy of Medicine

This chapter, by H. Tristram Engelhardt, Jr., with Edmund L. Erde, is really two chapters in one. An historical introduction is followed by two parts, one covering basic epistemological issues in the philosophy of medicine and the other bioethics.

In the section on biomedical ethics, Engelhardt and Erde emphasize the rights and duties of patients as well as physicians and other health providers; concepts of the person and human nature; and the just distribution of biomedical goods and harms. A final section acknowledges the contribution of religious perspectives to bioethics. There is some overlap with chapter 8, but it provides differing perspectives on some of the same issues.

The historical introduction clearly demonstrates the perennial character of the kinds of issues, both epistemological-ontological and ethical, raised in the chapter and in recent philosophy of medicine literature.

The most original portion of the chapter is the epistemological part, which includes extensive discussions of recent analytically oriented literature on the language of medicine, theory and explanation in medicine, clinical judgment, and the philosophy of mind in medicine. This section should go far toward demonstrating the legitimacy of a philosophy of medicine that goes beyond just biomedical ethics.

Sociology of Science and Technology

Jerry Gaston's survey of the sociology of science goes farther than any other chapter in the *Guide* in arguing the case for a paradigmatic approach to research in a field. According to Gaston, the dominant paradigms for testable, empirical sociology of science are those of Robert Merton and Thomas Kuhn. Gaston presses this claim so hard that some are likely to accuse him of overlooking other social perspectives that may not be so narrowly oriented toward the empirically testable.

Gaston surveys the field in seven sections: social and intellectual contexts of the sociology of science and technology; models of science; the scientific community; growth and change; the sociology of technology; sociology of science and technology outside the United States; and the problem of values.

As Gaston sees the matter, the sociology of science focuses on five areas: role performance and social stratification in science; social processes in science; social control in science; sociological studies of the origins and early development of science; and studies of disciplines, specialties, and research areas. He admits that although the sociology of technology has as yet to emerge as a specialized scientific subdiscipline, with a recognizable set of practitioners, as the sociology of science has developed, there is an abundance of material on technology. The opening for the development of a specialty is there, and he thinks the prospects are promising.

In his section on value issues in the sociology of science and technology, Gaston lists issues that may, and perhaps should, be explored. Nevertheless, he suspects that they are more likely to be explored by humanists than by professional sociologists bent on testing theories against hard data.

Medical Sociology and Science and Technology in Medicine

The chapter by Linda Aiken and Howard Freeman surveys medical sociology, or the sociology of medicine, but the survey is relatively brief and serves principally as background for a separate, original essay. The latter deals with what has and has not been done by sociologists and other social scientists on the impact of science and technology on contemporary Western medicine. There is also a fairly detailed discussion of issues relevant to the redirection of medical ethics, turning it into bioethics, as a result of the impact of science and technology.

The brief survey of medical sociology concentrates on the development of the field, institutional roles of medical sociologists, and the work they have done on health behavior, human resources, the organization of medical care, and the delivery of health services, as well as program evaluation.

Noting that contemporary medicine and health care systems are dominated by scientific and technological developments, Aiken and Freeman take as one of their principal aims solving the puzzle of why medical sociologists have so largely neglected what has been done along these lines. There is, in fact, a fairly large body of work done by health research administrators and policymakers on scientific and technological impacts on health care. Their essay on science and technology in medicine covers work that has and has not been done on medical and health research and researchers; the organization, delivery, and costs of medical care; and clinical practice. Bioethics concerns are clustered in the first and third of these sections, focusing on human subject experimentation and issues in clinical practice, such as death and dying, euthanasia, and the use and distribution of large medical-technological equipment.

Science and Technology Policy

Diana Crane's chapter on science and technology policy is a straightforward survey of a widely scattered field that has so far stubbornly resisted unification and may continue to do so indefinitely. Crane reports the efforts of those attempting to integrate the various science policy approaches into a superdiscipline, most notably Ina Spiegel-Rösing and Derek de Solla Price in their edited work *Science, Technology and Society: A Cross-Disciplinary Perspective* (1977). But Crane's emphasis falls on the diversity of the field(s). According to her, there are too many science policy specialists using too many different methodologies—policymakers and policy analysts, sociologists and political scientists, science reporters and R&D administrators, et cetera, et cetera, et cetera—for science and technology policy studies to be systematically unified.

For the most part, Crane's survey is a literature review. She makes occasional comments that indirectly reveal her slant on the issues, but for the most part she dispassionately reports the findings of a disparate group of authors. The chapter is divided into four parts, focusing on the nature and determination of policies for science and technology; national policies for controlling the impact of science and technology; value issues—especially those involving impact assessment and scientific-technological expertise; and science, technology, and international relations.

In the longest section, that on science policy in specific sectors, the focus is on "technoscience" or "research and development systems," which Crane defines as "the whole spectrum of basic and applied research associated with the production of a specific type of technology." She discusses five R&D sectors—agricultural, military, academic, civilian, and social (each is called a "technology" for short)—while omitting three others: space technology, biomedical technology, and technology for the service sector.

As Crane defines science and technology policy, it is a post-World War II phenomenon—that being the period when science administrators began to make conscious policy for R&D, and also when academic and other researchers consciously took up the study of the process. The absence of pre-World War II emphases suggests reading the chapter in conjunction with others: those on the history of science, the history of technology, and the history of medicine, and perhaps also the chapter on sociology of science and technology.

APPLICATIONS OF *GUIDE* SURVEYS TO POLICY ISSUES

A useful comparison can be made between Crane's treatment of value issues and that in other chapters. Here more than anywhere in the *Guide,* it is possible to examine one of the hypotheses on the basis of which the volume was put together—namely that humanistic and social science studies of science, technology, and medicine may be helpful in addressing contemporary value issues that arise in and in connection with those fields.

Crane's survey of value issues focuses on three points at which cultural values, ideologies, and ethics are particularly significant: (1) the selection of priorities for the allocation of research funds; (2) the role of the scientist as expert and political advisor; and (3) the assessment of the impact of science and technology on society.

On criteria for the selection of research priorities—her first general issue, Crane reports a discussion involving Michael Polanyi (science needs no external justification), Alvin Weinberg (technological applications and social utility are relevant), C. F. Carter (economics alone determines), and Stephen Toulmin (all the above, depending on circumstances). In chapter 4 Alex Michalos discusses the same general issue, though in different terms and with a different set of discussants in mind. Michalos focuses on two things: good reasons for accepting scientific hypotheses; and the possibility of determining in advance the advantages of a particular research area or research program. In either case, the point at issue is whether it is true, as Michalos puts it, that "moral, political, esthetic, religious, economic, and social values are irrelevant to the scientific enterprise." The philosophical discussions that Michalos reports are cast in terms that are extremely abstract, but those discussions are directly relevant to the sort of issue Crane raises.

A second value issue Crane discusses is whether or not there are adequate guidelines for the scientific researcher offering advice to people outside science. In passing, Crane mentions the controversy over whether or not there are even any guidelines within science. Jerry Gaston's chapter on the sociology of science details the controversy that Crane can only mention, and he offers one framework within which it can be understood. There is a generally close tie between these two chapters; Crane even skips a section needed to fill out her extremely orderly, meticulous outline (the subject is the organization of basic research, mainly in academia), referring the reader to Gaston's chapter. More generally, the sociology of science offers the only well-developed theoretical framework for strictly scientific empirical research on research in all the fields surveyed in the *Guide*.

Crane's last topic in her survey of values and science-technology policy concerns the sources of opposition to science and technology. At one point she says there is a "small literature" (she mentions Jacques Ellul, Jürgen Habermas, Herbert Marcuse, and Theodore Roszak) of opposition to science, and especially to technology, "which appears to be widely read in certain social groups," as if she were puzzled at the fact. A great deal of Carl Mitcham's survey of the philosophy of technology is devoted to critiques of technology from every quarter. If they do not read a fair amount of the literature Mitcham surveys, many members of the scientific-technological community are going to continue to be puzzled by the fairly widespread opposition to technology, and sometimes to science as well.

The antitechnology critical literature suggests a final example. Crane discusses a series of issues—she calls them "political and moral dilemmas"—

associated with the expert advisory role often played by scientists and technologists. If one chooses a government advisory role, one is bound for the duration to adhere to government policy; to refuse means loss of the opportunity to influence policy. Advice, when given, is often distorted or abused. Furthermore, there is the question whether advice should be given only on means, or whether one ought to serve also as an advocate for particular ends. Carroll Pursell, as noted, concludes his chapter on the history of technology with an argument against the common conception of technology as somehow neutral; consequently, for him, the history of technology must be nonneutral as well.

This impossibility of maintaining a neutral posture brings up, in turn, a delicate question about the applicability of the culture of science disciplines to science-technology policy and to other value issues within scientific or technological research. Many (by no means all) historians of technology—as well as philosophers, sociologists, and others—would agree with Pursell. Yet many scientists, engineers, and other members of the technical community express grave reservations about letting humanistic value judgments enter into their decisions unless the judgments are somehow made "objective." What is more, some scientists are extremely resentful of humanists and social scientists who doubt the possibility of scientific objectivity. What we may have here is a paradox not unlike the one which Crane reports (and which echoes all the way back to Plato): If humanities scholars wish to influence engineers, for example, their tone cannot represent a total stumbling block to communication; yet if they blunt their criticisms, they may feel untrue to their calling. In short, even if the culture of science and technology disciplines could aid in science values decisions, the help might not always be welcome.

SOME HINTS ON HOW TO READ THIS *GUIDE*

A guide of this sort, like an encyclopedia, is not meant to be read straight through from beginning to end. Some discussion of the way it may be most effectively used is therefore in order.

Its primary usefulness, in the minds of the editors and authors who assembled the *Guide,* has always been thought of as scholarly. For whatever purposes, scholars in one field often need a handy guide to another field. What we have tried to do is make readily available eight proximate fields for anyone working in one of the nine fields surveyed. Most often a scholar will look to a classic in a neighboring field for a quick overview, and to identify that classic will seek advice from a friendly professor in the appropriate department. We have tried to play that professor's role for a wider array of colleagues, hoping that they will find us appropriately friendly and helpful. "Classics" are listed in all the fields—differently of course in each, and they are placed in a disciplinary history context.

A second expected user is the graduate student, primarily in particular fields, but also, perhaps, in cases where a student might wish to combine two fields. Here the bibliographies should be especially helpful; moreover, most chapters suggest significant researchable topics still available in the fields.

To be realistic, most users are likely to be teachers—particularly those teaching interdisciplinary courses who are perhaps uncomfortable in unfamiliar fields. Some further guidance for such people, and for the general reader, is in order.

Several "cuts" through the *Guide* may be appropriate. Some may want to look at a particular set of topics from different disciplinary perspectives. In that case one may wish to use together the surveys on: history and philosophy of science and sociology of science and technology; or the three medicine-biomedicine chapters; or the chapters on history of technology, philosophy of technology, and science policy studies. Alternatively, it is easy to see how someone may want to use the three history chapters together, or the three philosophy chapters, or the two sociology chapters. Once started, the potential variations are numerous indeed.

One final note on bibliography. In each chapter all references are to the bibliography in that chapter unless otherwise noted. Each chapter has a section labeled "Bibliographic Introduction," but there are marked differences among these sections. In them the authors chose, and were allowed, to do a number of different things: list classics, annotate entries (something that could not be done in the main bibliographies), and so on. The bibliographies proper are intended to include, as far as possible, *only* works relevant to each field, and among those a selection that would be as comprehensive as possible, yet without overwhelming potential users of the volume.

NOTES

1. See C. P. Snow, *The Two Cultures and the Scientific Revolution,* 11th ed. (1959; New York: Cambridge University Press, 1963) and Gerald Holton, ed., *Science and Culture* (Boston: Houghton Mifflin, 1965).
2. *Encyclopaedia Britannica,* 15th ed., Micropaedia, s.v. "Axiology"; see also William K. Frankena, "Value and Valuation," in *The Encyclopedia of Philosophy,* edited by Paul Edwards (New York: Free Press, 1967).
3. Ralph Barton Perry, *General Theory of Value: Its Meaning and Basic Principles Construed in Terms of Interest* (Cambridge, Mass.: Harvard University Press, 1926).
4. Ralph Barton Perry, *Realms of Value: A Critique of Human Civilization* (Westport, Conn.: Greenwood Press, 1954; reprint 1968).
5. *Encyclopaedia Britannica,* "Axiology"; and *Encyclopedia of Philosophy,* "Value and Valuation" (see footnote 2, above).

6. *Encyclopaedia Britannica,* 15th ed., Macropaedia, s.v. "Ethics," by Alan Gewirth.

7. Albert J. Reiss, Jr., "Sociology: The Field," in *International Encyclopedia of Social Sciences,* vol. 15, edited by David L. Sills (New York: Free Press, 1968).

8. Warren Hagstrom, "Science: Sociology of Science," in *International Encyclopedia of Social Sciences,* vol. 14.

9. Milton Rokeach, *The Nature of Human Values* (New York: Free Press, 1973).

10. Arthur Kantrowitz, "The Science Court Experiments: Criticism and Responses," *Bulletin of the Atomic Scientists* 33 (April 1977): 44.

11. Alex C. Michalos, "A Science Court: Objections and Replies"; paper read at the second University of Delaware Philosophy and Technology Conference, June 1977; to appear in *Research in Philosophy & Technology,* vol. 3 (Greenwich, Conn.: JAI Press, forthcoming 1980).

12. Karl R. Popper, *Objective Knowledge: An Evolutionary Approach* (London: Oxford University Press, 1972).

13. Robert M. Pirsig, *Zen and the Art of Motorcycle Maintenance* (New York: William Morrow, 1974).

14. National Science Board, *Science Indicators 1974: Report of the National Science Board 1975* (Washington, D.C.: The National Science Board, National Science Foundation, 1976).

15. George Basalla, "Pop Science: The Depiction of Science in Popular Culture," in *Science and Its Public,* edited by G. Holton and W. A. Blanpied (Dordrecht, Holland: D. Reidel, 1976), pp. 261–278.

16. Herbert Marcuse, *One-Dimensional Man: Studies in the Ideology of Advanced Industrial Society* (Boston: Beacon Press, 1964), p. 9; and Jacques Ellul, *Technological Society* (New York: Alfred A. Knopf, Random House, 1964).

17. Derek de Solla Price, *Little Science, Big Science* (New York: Columbia University Press, 1963).

18. Robert K. Merton, "Priorities in Scientific Discovery," in *The Sociology of Science,* edited by Norman W. Storer (Chicago: University of Chicago Press, 1973), p. 309.

19. Carroll W. Pursell, Jr., *The Military-Industrial Complex* (New York: Harper & Row, 1972).

20. Edwin T. Layton, *The Revolt of the Engineers: Social Responsibility and the American Engineering Profession* (Cleveland: Press of Case-Western Reserve University, 1971); and an unpublished MS by Layton, "Historical Perspectives on Problems of Engineering Ethics," presented as part of the symposium "Ethical Problems in Engineering and Applied Science," University of Delaware, June 1977.

21. See footnote 16, above.

22. Herbert Marcuse, *Reason and Revolution: Hegel and the Rise of Social Theory* (Boston: Beacon Press, 1960; reprint ed., Atlantic Highlands, N.J.: Humanities Press, 1968); *An Essay on Liberation* (Boston: Beacon Press, 1969); *Negations: Essays in Critical Theory*, tr. by Jeremy Shapiro (Boston: Beacon Press, 1969); *Five Lectures* (Boston: Beacon Press, 1970); *Counterrevolution and Revolt* (Boston: Beacon Press, 1972); *Studies in Critical Philosophy*, tr. by Joris De Bres (Boston: Beacon Press, 1973).

23. Herbert Marcuse, *One-Dimensional Man*, p. 9.

24. Jacques Ellul, *Technological Society;* see footnote 16, above.

25. Perhaps especially Gabriel Marcel, *Man Against Mass Society*, tr. by G. S. Fraser (Chicago: Henry Regnery, 1952.)

26. Samuel C. Florman, *The Existential Pleasures of Engineering* (New York: St. Martin's Press, 1976), p. 91.

27. Lewis Mumford, *The Myth of the Machine: Technics and Human Development* (New York: Harcourt Brace & World, 1967); and *The Myth of the Machine: The Pentagon of Power* (New York: Harcourt Brace Jovanovich, 1970). Note that the references in the text use the popular titles.

28. See Robert J. Baum and Albert Flores, eds., *Ethical Problems in Engineering* (Troy, N.Y.: Center for the Study of the Human Dimensions of Science and Technology; Rennselaer Polytechnic Institute, 1978).

29. Robert Scholes and Eric S. Rabkin, *Science Fiction: History, Science, Vision* (New York: Oxford University Press, 1977).

30. *Ibid.,* pp. 99, vii.

31. Joanne Trautmann and Carol Pollard, *Literature and Medicine* (Philadelphia: Society for Health and Human Values, 1975).

32. Jacob Bronowski, *William Blake and the Age of Revolution* (New York: Harper & Row, 1965; reprint 1969); and *The Identity of Man* (Garden City, N.Y.: Natural History Press, Doubleday, 1965; rev. ed., 1972).

33. Daniel Boorstin, *The Americans: The Democratic Experience* (New York: Random House, 1973).

34. John Passmore, *Man's Responsibility for Nature* (New York: Charles Scribner's Sons, 1974).

35. Morris L. Cohen; Jan Stepan; and Naomi Ronen, compilers, *Law & Science: A Selected Bibliography.* Edited by V. Shelanski and M. La Follette (Cambridge, Mass.: Science, Technology, & Human Values; Harvard University, 1978).

36. See, for instance, *Duquesne Law Review* 12:3 (Spring 1974), "Product Liability: An Interaction of Law and Technology"; and Henry R. Piehler et al., "Product Liability and the Technical Expert," *Science* 186 (20 December 1974).

37. Andrew Dickson White, *History of the Warfare of Science with Technology in Christendom* (New York: Dover; reprint of work originally published in 1896; reprinted again, with an introduction by Bruce Mazlish, New York: Free Press, 1965).

38. Bertrand Russell, *Religion and Science* (New York: Oxford University Press, 1935; reprint 1965).

39. Ian G. Barbour, *Issues in Science and Religion* (New York: Harper & Row, 1968); *Science and Secularity: The Ethics of Technology* (New York: Harper & Row, 1970); and *Myths, Models, and Paradigms: A Comparative Study in Science and Religion* (New York: Harper & Row, 1974).

Part I
Historical Disciplines

Chapter 1
History of Science

Arnold Thackray, UNIVERSITY OF PENNSYLVANIA

INTRODUCTION

A. The Uses of Nature

The pattern of attributes assigned to nature by members of any culture is part of a moral vision. Depending upon the character of this vision, the natural world may be approached in a variety of ways. Nature may be deemed lawlike and discoverable or capricious and unapproachable by naked reason. Its very substance may be thought of as enduring (witness the various "conservation laws" of modern physics) or as forever subject to waste and change, while its forms may be construed as matter, or as spirit, or as some combination of the two, or even as intermediate agencies of different kinds (as in alchemy). Nature may be susceptible of control by man. On the other hand it may itself control man (as in various astrological systems). Choices of belief and of interpretative stance in these matters are but an aspect of the moral vision prevailing within a culture. Accordingly, the work of the historian of science belongs within—and offers one contribution to—a wider cultural investigation.

Choices of belief and of interpretation are also connected to the more immediate social interests of the community by subtle, complex ties. For instance, it has been argued that the categories into which nature is cast may serve as homologies of ordinary social experience. Again, the relations between man and nature may mirror the character and extent of social power, and symbolize the canons of authority, decorum, and meaning appropriate to a community. It is anthropologists who have provided the most intriguing texts. The *locus classicus* is Durkheim and Mauss (1901–1902), while Geertz (1973) and Douglas (1970 and 1975) are influential recent statements. The thesis is simple of statement, but dark and controversial

3

of investigation: how, say, has the evolving character of Western science been shaped by (and in turn shaped) the socioeconomic transformations of mercantile capitalism or of industrialism, of urbanization or liberal democracy?

An allied theme is less controversial but no less complex in its implications. Propositions about the ways in which science has developed, and about the conditions that have impeded or facilitated its growth, have themselves been part of the structure by which symbolic meaning is attributed to nature. These propositions about science have increased in importance in the last several decades. The new visibility of the scientific enterprise; its greater number of practitioners and its differentiation into dozens of domains; its expanding array of possible technical implications, from cloning to intergalactic communication; above all the increasing moral and social burdens placed upon our symbolisms of man in nature: severally and together these factors have deepened the need for mediation between the many groups with interests in the social construction of scientific knowledge. The history of science provides one means for such mediation. It is for this reason, among others, that the subject has enjoyed much saliency in recent years.

B.   Interests, Values, and the Past of Science

Modern science is an enterprise of massive scope and complexity. It includes extensive repertoires of subtly nuanced concepts, theories, hypotheses, and laws. It makes recurrent use of certain interlocking ideas, which may themselves be only partially articulated (Holton, 1973). It embraces a great variety of methods (mathematical abstraction, laboratory experimentation, field observation, systematic data collection, etc.). It often depends on the use of sophisticated instruments and special techniques (e.g., radio telescopes, particle accelerators, chromatography, carbon dating, the focussed interview, the use of control groups). It encompasses millions of facts. Yet to the dismay of logicians, its theoretical forms are underdetermined by those facts (Hesse, 1974).

Modern science is also a patterned structure of values, embedded in an array of supporting institutions and their social organization of cultural assumptions (Merton, 1973). In this sense science inheres in learned societies, symposia, and conferences; in "invisible colleges" and informal networks; in textbooks, examinations, apprenticeships, and Ph.D. programs; in supporting cadres of technical assistants, disciplinary organizers, and publicists; and in journals, prizes, and funding agencies. Modern science may also be considered in its international aspects, or viewed within the confines of particular nations. It may be seen as one "unified science," or as arranged in certain broad divisions (behavioral sciences; natural sciences; physical sciences; social sciences), or as a collocation of many imperfectly connected specialties, from experimental psychology, through synthetic organic chemistry, to X-ray crystallography.

No single definition of science will be attempted or adhered to in this essay. As an activity, as a body of knowledge, as a methodology, and as a set of institutions, science is a complex construct with ramified roots. The history of science is therefore a multifaceted subject. The same implication flows from the role of the history of science in articulating, affirming, and giving currency to particular symbolisms of man-in-nature, and in establishing the canons of legitimate knowledge.

Different groups of practitioners with different values, agendas, assumptions, and interests have been drawn to one or another facet of the history of science at different times. Changes in the social character or intellectual program of these groups have often been caused by events remote from those that might be derived from any internal logic of science or history. Only very recently, and partially, has the history of science become the recognizable property of a coherent, continuing group—that of the academic practitioners of the history of science as an organized discipline. The late appearance of the strong intellectual constraints associated with academic professionalism is a positive advantage for the purpose of this essay. In the absence of such forms, paradigms of explanation were more plastic and more open to contextual influences. The paradigms may tell us directly about the values of the scientist and of the historian alike.

For instance, the early Victorian construction of a *History of the Inductive Sciences* (Whewell, 1837) was part of a concerted effort to redraw the boundaries of knowledge in ways appropriate to a new social order. A century later, a recognizable strain of secular optimism was to run through all of George Sarton's organizing work, occasionally breaking surface in statements such as, "The history of science is the history of mankind's unity, of its sublime purpose, of its gradual redemption" (Sarton, 1927–1948, I, p. 32). A very different genre of optimism was able to make a boldly original view of Newton's *Principia* part of the argument for the dawning world order of Soviet communism (Hessen, 1931). In contrast, the suggestion of those norms that have since been widely held to make up the ethos of science came partly in response to the Nazi attack on the intellectual autonomy of science. "Science and Technology in a Democratic Order" incorporated the notion that scientists "recognize their dependence on particular types of social structure" in the wider community (Merton, 1942, p. 115). More recently, the espousal of "history of science as rational reconstruction" was closely linked to the protection of the universities from student revolt (Lakatos, 1974, p. 236). The point also emerges with admirable clarity at the start of C. C. Gillispie's *Edge of Objectivity*. That "Essay in the History of Scientific Ideas" expertly encapsulated and convincingly confirmed the values of an era: "The hard trial will begin when the instruments of power created by the West come fully into the hands of men not of the West. . . . And what will the day hold when China wields the bomb? And Egypt? Will Aurora light a rosy-fingered dawn out of the East? Or will Nemesis?" (Gillispie, 1960, p. 9).

As the general argument indicates, and the quickly limned illustrations display, moral judgments and social commitments pervade the field. The role of values in the history of science may therefore best be treated otherwise than by a search for explicit discussions of "value issues in science." Instead, it may be more fruitful to examine historical enquiry directly. That examination will reveal a great variation among implicit assumptions about how the interrogation of nature has been conducted, and about what that interrogation signifies. Different assumptions have gained resonance and credibility in historical enquiry at various times. This resonance must always be understood within its social context.

C. The Structure of This Essay

What follows is not meant as a survey of received understandings of the development of science. To provide one such survey within four large volumes required heroic feats of compression and necessitated major omissions (see Taton, 1963–1966). No more does this essay offer encyclopedic treatment of past or current literature. That literature is simply too large and too variously specialized. For instance, a recent annual volume of the *Isis Critical Bibliography* takes 181 pages simply to *list* 2850 new books and articles in the history of science, using over 100 subcategories. Instead, this essay offers a brief *guide* to the historical study of science (cf. Sarton, 1952).

The pathways by which the subject has developed will first be traced out. If those pathways are known, it is possible to see why certain domains in the history of science have experienced greater attention at particular times, as values have changed or argument and understanding have developed. Some of these domains will be examined at length in the second half of the essay. Armed with this knowledge, one can appreciate the texture of the historical literature, its strengths and limitations. A concluding section offers further signposts to the geography of the discipline. By way of postscript there is a listing of works cited.

Because this essay is only a *guide,* it is partial and incomplete. Some classic treatments, standard works, and influential monographs have almost surely been omitted. Many important areas of research and writing in the history of science have perforce been ignored or given only scant attention, in the interest of brevity. A deliberate bias in favor of books, as against research reported only in scholarly articles, has been necessary to bring the work within bounds. The medical sciences, "applied science," and the relations between science, economics, and public policy have been treated less fully than they deserve because of separate chapters in this volume on the history of technology (chapter 2), the history of medicine (chapter 3), and on science policy (chapter 9). What is offered is not so much a map of the whole territory of the history of science, as a compass and some initial instructions. With their aid, readers may navigate to the destination they desire, locating the necessary history as the journey unfolds.

I. THE HISTORIOGRAPHY OF SCIENCE

A. Its Prehistory

One way to legitimate a new activity is to present it as old. The desire of practitioners of unseasoned arts to seek historical precedents for their work is deep rooted in Western civilization. As scientific enquiry gained momentum in the Renaissance, historical argument was brought into play in aid of the novel venture. Then as later, the intellectual and social needs of scientific practitioners were powerful forces in the shaping of patterns of historical explanation.

The Paracelsian and alchemical compilers of chemistry texts regularly discussed the great antiquity of, and traced the line of progress in, chemical understanding (Debus, 1962). To point to an example of a rather different kind, Francis Bacon in the early seventeenth century was at pains not simply to call for a "novum organum" but to urge the importance of a new tracing of the development of natural knowledge up to his time (Rossi, 1957). A legitimating account of still a third, institutional kind may be seen in Thomas Sprat's defense of unfamiliar social arrangements in the form of a *History of the Royal Society* (London, 1667).

By the eighteenth century, natural knowledge was well established as a distinct cultural mode. That knowledge was increasingly pursued in formal societies, modelled on the Royal foundations in London and Paris. History was brought into play, both as a consequence of and as an aid to this development. The magisterial *éloges* of its deceased members, which were delivered before and published by the Paris Academy of Sciences, constitute the most outstanding example of a form of history—heroic biography—in the service of specific institutional interests (Delorme, 1961). This form was later adapted to varied ends (e.g., Arago, 1857). In the present day (e.g., in the *Biographical Memoirs of the National Academy of Sciences* and the *Biographical Memoirs of the Fellows of the Royal Society*) it offers important accounts of many otherwise unstudied episodes in the history of recent science.

Quite different in character but also widely influential in defining the style and the boundaries of natural knowledge were various systematic treatises. In England a Unitarian Divine wrote classic histories of electricity (Priestley, 1767) and optics (Priestley, 1772). An account of the scientific progress of the eighteenth century was even produced within the United States (Miller, 1803). In France a minor Royal functionary devoted his life to the history of mathematics (Montucla, 1799–1802; see Sarton, 1936), while a professor of "physique" produced a two-volume philosophical history of his subject (Libes, 1810–1813). In Germany, matters were more ordered. Scholars associated with the flourishing school of universal history at Göttingen labored on mathematics (Kästner, 1796–1800), physics (Fischer, 1801–1808), chemistry (Gmelin, 1797–1799), and technology (Beckmann, 1784–1805), while another professor in Erfurt also gave his attention to chemistry (Trommsdorff, 1806). Their encyclopedic histories stood within, and brought to a fresh stage of sophistication, the developing genre through which devotees of scientific knowledge presented the work of their own generation as the inevitable, latest stage in the unravelling of nature's single truth.

As these examples suggest, the volume of materials dealing with man's relationship to the natural world was already considerable by the time of the Napoleonic wars. Later work was to draw heavily on the facts, anecdotes, and categories established by these writers. However, this early activity was stronger on chronicle and compilation than on historical or philosophical analysis. It was but feebly informed by the continuity of problematics and communication between practitioners that together characterize a socially mature intellectual discipline. In this latter sense the history of science, like history itself, saw its real development in nineteenth-century Europe.

B. Nineteenth-Century Europe

The emergence of industrial society brought with it new cognitive forms, social relations, and educational structures. The need for fresh intellectual arrangements was felt throughout the West: a pattern of changes may be traced not only in such central states as Britain, France, the Low Countries, and Germany, but also in Scandinavia and the United States, and—to a lesser extent—in such industrially peripheral areas as the Austro-Hungarian Empire, Italy, Russia, and Canada. For all these countries the development of novel social organizations of science, fresh cognitive orderings of the map of knowledge, and unaccustomed rationalizing functions for history and for nature were part of the process by which "modern" culture emerged (Shils, 1972).

The variety in the political forms and institutional categories of these different countries meant that natural knowledge was severally accommodated in Royal academies, in bureaucratically regulated institutes, in military schools, in museums, in voluntary societies, in private colleges, in state universities, in industrial firms and other proprietary organizations, in religious orders, and in avowedly secularist groupings. (For a classic discussion of different national "styles" in nineteenth-century science, see Merz, 1896.) One result is that it becomes impossible to outline—in a book, let alone a chapter—the variety of forms assumed by and subjects treated in the history of science in the nineteenth century. Instead, this discussion will focus on developments in the English-speaking world, as a matter of expedience. Other facets of the subject (principally various developments in France and, to a lesser extent, Germany and Russia) will be touched on only as they impinge on interests and actors on the narrower stage.

In the wake of the Industrial Revolution, scientific practitioners in the British Isles became more self-conscious about their status and functions. The British Association for the Advancement of Science, founded in 1831, gave expression to their needs. The association's restrictive and preemptive use of the word "science" to describe its carefully circumscribed activities was one signal of the state of affairs. So too was the coining of the word "scientist" by William Whewell in the context of the association's foundation (see Ross, 1962). History was also called into play. One of the association's moving spirits composed a classic of Victorian biography, in which Isaac Newton was duly shorn of his psychological, theological, and intellectual obsessions in order that he might be presented as the exemplar of scientific rationality (Brewster, 1831 and 1855). Other members produced extended works of scholarship in which particular individuals were rescued from oblivion (e.g., Baily, 1835, on Flamsteed, the first Astronomer Royal) or the whole developmental pattern of the sciences was reassessed (e.g., Powell, 1834; Whewell, 1837). An early, short-lived section of the British Association was devoted to the history of science. It was soon succeeded by an equally ephemeral "Historical Society of Science," which sought to reprint classic texts and foster attitudes appropriate to the new dignity of science and the position of the scientist (see Hornberger, 1949).

Other notable developments of the period included Thomas Morell's *Elements of the History of Philosophy and Science* (London, 1827) and the historical supplements to the *Encyclopaedia Britannica*. These eventually consisted of a massive quartet of philosophically inclined and sagely optimistic disquisitions on the intellectual progress of physical and moral science, written by members of the flourishing professoriate of Edinburgh University (Stewart et al., 1815–1835). Not to be outdone, Glasgow University's foundation professor of the discipline of chemistry produced the first English monograph on *The History of Chemistry* (Thomson, 1830–1831). In London, Charles Lyell extensively rewrote the history of his subject as a strategic introduction to the *Principles of Geology* (1830).

These early, legitimating activities of the growing community of academically oriented gentlemen of science in Britain did not lead on to any systematic ventures in the history of science. But in the *Reports* of the B.A.A.S., the Anniversary Discourses of the Royal Society, the popular effusions of such writers as Mrs. Arabella Buckley, and the profusion of heroic biographies religiously compiled to celebrate the lives both of earlier pioneers and of leading Victorian men of science, materials were steadily accumulated to explain and justify the role of science in the national life. Those materials had greater philosophic depth and richer factual and documentary support than had earlier investigations.

What was new, however, was not so much the character of the materials as their very number and their variety. What was uniquely British was the diversity of writers, themes, and audiences, a situation reflecting the decentralized, voluntarist nature of British science in this era. The works produced included meticulous enquiries into episodes in the internal history of the increasingly self-confident physical sciences (e.g., Rigaud, 1838), massive compilations on institutional developments (e.g., Weld, 1848, on the Royal Society), exemplary editions of scientific correspondence (e.g., Rigaud, 1841; and Edleston, 1850), and technically exacting histories of the intellectual development of particular disciplines (e.g., Grant, 1852, on physical astronomy). Later in the century, as the sciences continued to proliferate and subdivide, there came learned histories of new research specialties (e.g., Schorlemmer, 1879, on organic chemistry), popular treatises on recent developments (e.g., Clerke, 1885, on astronomy) and, eventually, specialist monographs on the technical histories of particular problem areas (e.g., Freund, 1904, on chemical composition; Whittaker, 1910, on the aether).

Despite these considerable British contributions, it was in Germany that works of scholarship in the field were most zealously nourished in the nineteenth century. The extensive development of scientific departments in the German universities; the clear differentiation of disciplines and specialties; the creation of research institutes—each served to foster an interest in the production of historical works that would explain, elucidate, justify, and guide the development of scientific understanding. That interest reached one culmination in an ambitious, multiauthor, multivolume *Geschichte der Wissenschaften in Deutschland* (see Carus, 1872; Kobell, 1864; Sachs, 1875). The *Geschichte* also profited from the search for national identity and the respect accorded to learning. Another milestone was the foundation of the Deutsche Gesellschaft für Geschichte der Medizin und der Naturwissenschaften, in Hamburg in 1901.

Among the classic works that German scholars devoted to the histories of particular disciplines were Kopp, 1843–1847 (chemistry); Kobell, 1864 (mineralogy); Carus, 1872 (zoology); Sachs, 1875 (botany); Cantor, 1880–1908 (mathematics); Heller, 1882–1884, and Rosenberger, 1882–1890 (physics); and Zittel, 1899 (geology). Encyclopedic listings of scientists and of discoveries, owing something to the spirit of German historical scholarship, were also produced (e.g., Poggendorff, 1863, and Darmstädter and DuBois-Reymond, 1904). Finally, there arose in the latter part of the century a sophisticated school of writings upon the neoKantian problems posed by recent developments in the sciences (in, e.g., energetics, electricity, and evolutionary theory). Two lines of argument were to generate widespread interest and to have an important effect on developments both in science and in its historiography throughout the Western world. The one was monistic and metaphysical and found its classic expression in the corpus of writings of Wilhelm Ostwald. Ostwald was averse to atomism and mechanism. He sought an energy-based union of the natural and social sciences with the humanities, that would hasten the triumph of pacifism, internationalism, and a common world language. His ambitions motivated brilliant scientific, historical, and philosophical investigations (Ostwald, 1896; see also Lasswitz, 1890, as well as the well known works of Ernst Haeckel, not cited in the bibliography). The second line of argument was critical and positivistic and was brilliantly exemplified in the writings of Ernst Mach, especially his *Die Mechanik in Ihrer Entwicklung: Historisch-Kritisch Dargestellt* (Leipzig, 1883). This latter work characteristically and unashamedly displayed its interest in placing history in the service of contemporary philosophical interests by stating that "its aim is to clear up ideas, expose the real significance of the matter, and get rid of metaphysical obscurities" (Mach, [1907] 1883, p. ix).

The strivings of a growing and self-confident professoriate powered much British and most German work on the history of the sciences, in the Victorian era. Philosophical ideas about the unity of those sciences often receded before the pressure of proliferating institutions and fragmenting research. The developmental path was quite different in France. There the history of science was to be given its first mandate not simply as the expression of an "interest" but as the vehicle of a secular faith. That faith—positivism—was articulated by Auguste Comte (1798–1857), in a by then traditional French context of animosity between advanced intellectuals and the Roman Catholic Church. Positivism placed its trust in the analytical and experimental methods of science. The necessary progress of humanity toward love and peace would find its full expression and its agent in a completed science of humanity itself. Correspondingly heavy programmatic ambitions were placed upon the history of science. As in the later case of Marxism, the ambitions were to remain unfulfilled.

Comte's ideas found a sympathetic echo in the English Utilitarian tradition. Some of his work was translated and another strand added to the web of reasons for studying the development of scientific knowledge (see, e.g., Lewes, 1864). In France Comte's influence, if still slight, was more significant and more direct. It led not only to the founding of the first chair to be devoted to the "histoire générale des sciences" at the Collège de France in 1892 (see Sarton, 1947, and Paul, 1976) but also to the nourishing of sustained, significant scholarship. By the close of the nineteenth century there had emerged in France a cluster of writers committed to studying the history of the sciences from a generalizing, philosophical point of view. These writers were indebted to Comtean ideas and Comtean programs, though they were often in revolt against them. Among their number were Henri and Lucien Poincaré, Arthur Hannequin, Léon Brunschvicg, Pierre Duhem, and Paul Tannery. The enduring intellectual monuments to their work include the ten volumes of Duhem (1913–1959) and the daunting array of historical and other enquiries collected together in Tannery (1912–1950). The organizational landmark was the first international gathering of historians of science, in Paris and under Tannery's leadership, in 1900 (see Guerlac, 1950).

The Comtean tradition of analysis and the philosophical ferment in France were profoundly to influence a young Belgian scientist and gentleman-scholar, George Sarton (1884–1956). Imbued with a secularist and progressivist faith in science, Sarton in 1913 published the first number of *Isis*, his "revue consacrée à l'histoire des sciences." *Isis* was to emulate and apply in Sarton's chosen discipline, the ambitious synthetic program by which French scholars such as Henri Berr hoped to achieve a philosophical synthesis based upon historical understanding. Sarton's plans found many sympathizers, including Charles Singer in England and the progressive Italian scientist Aldo Mieli, who had studied physical chemistry under Wilhelm Ostwald. In due course Mieli was to edit the first Italian journal of the history of science (*Archivio di Storia della Scienza,* founded in 1919 and renamed *Archeion* in 1927) before exiling himself to Paris in 1928. Sarton's own editorial work was more quickly interrupted. The outbreak of World War I spelled an end to Comtean hopes in Europe. It also, adventitiously, provided the context in which Sarton's program was transferred to the more fruitful soil of the United States (Thackray and Merton, 1972). In like manner, the looming shadow of World War II was to cause Mieli to transfer himself and his journal to Argentina, which proved a less propitious environment.

C. The United States

The experiences of industrialism, urbanization, and liberal democracy were to find their fullest expression in the United States. North America was also to provide a receptive home for modern science and for the scholarly study of its history. Social developments that were slow or incomplete elsewhere in the Western world, taking up two centuries to reach a matured form (as industrialism in France, or an adequate mass higher education system in England), proceeded with vigor and completeness in the United States. The period from the Centennial Exhibition to the First World War saw a great transition in American culture. In this era the research university emerged, flourished, and came to early maturity. The natural sciences established themselves as proper components of that university. The Ph.D. machine, the research professoriate, the organizational plan of departments and subject specialties—all found acceptance as cognitive expressions of social realities in the new, industrialized America.

It was in this context that the history of science burgeoned. The subject was believed to be a useful means of socializing students into science, of integrating the increasingly differentiated scientific specialties and sub-specialties, and of demonstrating the centrality of science in Western civilization. It also provided a common mode of discourse to the emerging class of science organizers—to foundation and university presidents, deans, department chairmen, heads of observatories, directors of laboratories and museums—and, outside the field, to publishers and politicians. By the eve of World War I the subject was widely introduced in American colleges and universities (Brasch, 1915). Appropriate textbooks were written by American professors for their student audiences. Those texts dealt with both the "general history of science" (e.g., Sedgwick and Tyler, 1917; Libby, 1917) and the histories of particular sciences (e.g., Cajori, 1899, on physics; D. E. Smith, 1906, on mathematics; Moore, 1918, on chemistry). A first stirring of original scholarship was also discernible among scientists eager to use historical enquiry in clarifying contemporary problems, or to stimulate pride in American achievements, or to ensure fidelity to European disciplinary canons of performance (e.g., Venable, 1896; E. F. Smith, 1914; Cajori, 1919).

The experience of World War I and the subsequent wave of concern for the future among certain liberal intellectuals together formed the context in which, in 1924, the History of Science Society came into existence. That society embraced a variety of interests and several contradictory programs. Rich and influential patrons and elder statesmen of science favored a discriminating bibliophilia. In this they were at one with a cognate group of physicians who were active in establishing the American Association for the History of Medicine in the same period. Some scientists remained strongly influenced by their exposure to German thought in their student days. The historico-philosophical programs of Mach and Ostwald were their particular passion. On the other hand, James Harvey Robinson and his associates saw the history of science and technology as but one facet of the "new history" that would provide the themes appropriate to a democratic mass culture (H. Barnes, 1919). Encouraged by this belief, some of Robinson's students worked on scientific societies, magic and science, and the place of science in intellectual history (respectively, Ornstein, 1913; Thorndike, 1923–1958; Preserved Smith, 1930–1934). Robinson's revolutionary program did not have science as its exclusive focus. Neither did it capture any significant number of members of the historical fraternity. Their main interests remained with other constituencies. The analyses of the "new historians" were more acceptable to those professional scientists and administrators who were central to the History of Science Society and were the main patrons of the history of science within universities.

These patrons might agree with Robinson and the "new historians" on the importance of the history of science. Even so, they found that the more imperialistic positivism of Sarton's program—science as *the* leading thread of history—made better sense. Domination of science over, not integration of science with, the humanities was the implicit message. (See chapter 3 in this volume for an account of Sarton's clash with Henry Sigerist over the relative positions of the histories of science and medicine.) A stress on remote periods, esoteric subjects, critical bibliography, and heroic biography were also agreeable to the patrons. Sarton was thus able to exercise an increasing influence over what gradually came to be seen as "his" discipline. The encyclopedic vision and the learned antiquarianism that he favored, as the developments of the 1930s dealt heavy blows to his progressivist faith, were not such as to command excitement or disciples. While Sarton's loosely-defined association with Harvard University served him well in his roles as organizer, propagandist, and exemplar of scholarship in the history of science, he was unable to give the cognitive direction that would channel recruits to the field. Instead, other intellectual programs took center stage as the actual and symbolic relations between man and nature underwent the profound strains associated with the Depression and World War II.

D. The 1930s Marxists

Auguste Comte possessed a powerful vision of the place of science in society and the corresponding importance of the history of science. So, too, did his near-contemporary Karl Marx (1818–1883). Marx's views, as developed by Engels and Lenin, were to become the official, dogmatic basis of one major institutional and international program in the history of science, that of the Soviet Union and its sympathizers. More recently, those same views have also become dogma in China. The development of modern Soviet and Chinese historiography cannot be pursued in this essay, though it should be noted that the Academies of Science in the U.S.S.R. and the People's Republic of China sponsor specialist research through their respective Institutes of the History of Science and Technology. According to one Soviet publication, "Since 1953 the Institute has prepared and published more than 614 monographs, including 23 works on the history of mathematics, 3 on astronomy, . . . 52 on chemistry, . . . 57 on biology, . . . and 15 bibliographical reference books and dictionaries" (Nauka, 1977, p. 7). The barriers of language have helped to limit the impact of these works outside the U.S.S.R. However, an earlier piece of Soviet historiography was destined to have an important influence in the West and to serve as a reference point for the writings of the "1930s Marxists."

 With high drama and a fitting sense of occasion, a Soviet delegation flew to the Second International Congress of the History of Science (London, 1931) to announce Russian progress and plans in science and to contrast the optimistic future of the Socialist countries with the impending doom of the Depression-ridden system of capitalism. The symbolism of man and nature was pressed into deliberate service, most notably in Hessen's paradigm-setting analysis of the "Social and Economic Roots of Newton's

Principia." The message was that "the method of production of material existence conditions the social, political, and intellectual process of the life of society." Hence "the source of Newton's creative genius, . . . the content and the direction of his activities," could be laid bare by "applying the method of dialectical materialism and the conception of this historical process which Marx created." "The roots of all ideas without exception"— even the commanding abstract ideas of Newtonian mechanics—were to be found in "the state of the material productive forces" (Hessen, 1931, pp. 151–152).

Hessen's audacious example encouraged a remarkable and variegated group of leftward-leaning scientists based in Cambridge, England. They set out to articulate appropriate histories (see Werskey, 1979). Most prolific was J. G. Crowther, whose journalistic essays were to be a familiar feature of the Western scene through the next four decades (e.g., Crowther, 1935, 1936, 1937, and 1970). Crowther made seven separate journeys to the Soviet Union in the early 1930s. His extended visit to Harvard in 1937 (he was being appraised for his suitability as assistant and complement to George Sarton) helped to spread knowledge of Hessen's work in North America. The polymathic crystallographer J. D. Bernal was equally engaged with questions of science policy and the history of science, though with less time to devote to them (see Bernal, 1939 and 1954). More peripherally, C. P. Snow, J. B. S. Haldane, P. M. S. Blackett, Julian Huxley, and Lancelot Hogben were all to make use of Marxist historical insights in their popular writings and in their policy-making work on the place of science in modern civilization. Even "ordinary" historians were moved to contribute to the debate on science and social welfare (e.g., Clark, 1937). The most deeply rooted influence of Marxist thought was to come to maturation in the profuse and learned writings of Joseph Needham. A Christian Socialist perspective also contributed to his investigations of the development of Chinese science (Needham, 1954—; see Teich and Young, 1973; Nakayama and Sivin, 1973).

As Needham's publications illustrate, Marxist ideas never died away entirely in the English-speaking world. The same ideas influenced the widely popular *History of the Sciences* produced by S. F. Mason (1953) and may be seen in a collection of scholarly essays published in the journal *Centaurus* (Lilley, 1953). The Dutch *emigré* mathematician and historian of science Dirk Struik continually sought to stimulate Marxist history of science in the United States from his position at M.I.T. (Struik, 1948; see Cohen et al., 1974).

E.   The Idealist Program

An upsurge of high-minded sentiment in certain sections of the populace following World War I had helped foster the institutionalization of history of science in North America. A similar emotional wave after World War II in both Britain and the United States was congruent with changed appreciations of the place of man in nature. The explosion of the atomic bomb seemed to herald an unprecedented, worrisome era. The conservative, almost escapist assumptions characteristic of this period were profoundly at odds with the radical, activist orientations of the 1930s Marxists. Instead, these assumptions derived from earlier intellectual traditions that were tinged with idealism in the technical philosophic sense. The classic statement of the new position with respect to the history of science was provided not by a leftward-leaning scientist, but by a Cambridge historian. Herbert Butterfield's *The Origins of Modern Science* (1949) was to capture the imagination of many, and to show how science could be understood historically as abstract, imaginative thought.

The idealist program was based on precedents in the history of philosophy, especially in the work of such Americans as E. A. Burtt, A. J. Snow, and A. O. Lovejoy, together with the American work of A. N. Whitehead, the ideas of R. G. Collingwood, and appropriate translations of the writings of Ernst Cassirer (see Burtt, 1925; Whitehead, 1925; Snow, 1926; Cassirer, 1932; Lovejoy, 1936). Its proximate origins lay quite elsewhere. Science was one important element in the French historical tradition of philosophical enquiry that, via Tannery and Duhem, had come to a fresh flowering between the two World Wars in the work of Émile Meyerson, Pierre Brunet, and Hélène Metzger (see, e.g., Meyerson, 1930 and 1931; Brunet, 1931, Metzger, 1923 and 1938). Just as Sarton had brought an earlier version of French thought to American shores, so another *émigré*—the "white" Russian Alexandre Koyré—was to bring this alternative tradition to the attention of American scholars. Koyré's own early writings were in French (e.g., Koyré, 1939). Numerous translations and fresh productions in English appeared in the 1950s and 1960s (see, e.g., Koyré, 1957 and 1968). He himself was a frequent visitor to the Institute for Advanced Study in Princeton and to other strategic centers. The Koyré paradigm—the principle that science "is essentially *theoria,* a search for the truth" and that this search has "an inherent and autonomous" development (Crombie, 1963, p. 856)—became the vehicle of a fresh idealism in the history of science. The formulations of that idealism proved compelling to an Anglo-American generation recoiling from the implications of an age of nuclear weapons. If contemporary events showed science inextricably entangled in state secrecy, politics, funding battles, and national security, history would redeem the purity of the intellectual pursuit of nature (Gillispie, 1973; Toulmin, 1977).

The quarter century following World War II was to see a triumphant growth in the size of American academe, and the rapid expansion of government provision for, and general public support of, the natural sciences. Radar, the proximity fuse, and the atomic bomb were only some obvious examples of the power that was held to flow from science. The unexpected launching of a space satellite by Russia (Sputnik) served to crystallize an already present American determination that the natural sciences must be more fully supported and more vigorously deployed (Greenberg, 1967). In a context of exuberant growth, of further differentiation, and of increased specialization within science, a need for academic study of the history of science was widely accepted. To the older desires to socialize students into science, to reintegrate the sciences, and to assert their cultural hegemony, were now added newer hopes. It was believed that the "bridge" subject of the history of science might unite the "two cultures" of science and humanistic scholarship. There was also the recognition that the field constituted a legitimate professional pursuit in its own right, in a world of learning minutely partitioned among specialties (see Stimson et al., 1959; Crombie and Hoskin, 1963).

The major patrons of the subject were academic natural scientists. The first generation of full-time practitioners was drawn mainly from the ranks of fledgling men of science. The Koyré paradigm offered a useful program in terms of which the practitioners could organize their professional interests. Analyzing scientific ideas utilized prior scientific training, while the results of such analyses were accessible to scientific patrons and scientific audiences. The international flavor, implicit in the French roots and Anglo-American fruits of the program, was also widely acceptable. In the hands of such practitioners as Marshall Clagett, I. Bernard Cohen, Charles C. Gillispie, Henry Guerlac, Thomas S. Kuhn, and A. Rupert Hall, the method of conceptual analysis opened up a powerful vision of the intellectual grandeur and austere challenge of (physical) scientific thought (see, e.g., Clagett, 1959a and 1964–1978; Cohen, 1956 and 1971; Gillispie, 1951 and 1971; Guerlac, 1961 and 1977; Hall, 1952; Hall and Hall, 1962; and Kuhn, 1957). The method was communicated to younger recruits and to the first group of graduate students the field had ever known. They in their turn refined and extended the technique of conceptual analysis, with impressive results (e.g., Boas, 1958; Greene, 1959; Woolf, 1959; Hesse, 1961; Hoskin, 1963; Coleman, 1964; Mendelsohn, 1964; Debus, 1965). The utilities of the technique were confirmed by a pragmatic division of labor. "Internalist" historians of science explored the filiations and intellectual contexts of successful ideas (the "high agenda" of science). The smaller, but growing, number of "externalist" sociologists—following a quite separate but compatible paradigm developed by Merton—traced out the contemporary structure and functions of a massively expanded scientific community (see Hall, 1963; Young, 1973; Cole and Zuckerman, 1975).

Such a neat division of labor could last only while each group of practitioners was preoccupied with establishing its own identity, and the relationship of science to society was untroubled. The political and technological contexts in which the relationship of science to society would command attention have become a part of familiar experience in the years since the Vietnam war. In fact, a brilliant incursion into sociological territory by the physicist-turned-historian-of-ideas, Thomas S. Kuhn, took place as early as 1959. Kuhn's historical theses about the philosophical and sociological character of the scientific community would take on major importance in the late 1960s and the 1970s (Kuhn, 1962; see Hollinger, 1973; Kuhn, 1977; Merton, 1977; and Wade, 1977; see also the discussions in chapters 4 and 7 of this *Guide*.)

F. The New Eclecticism

Things were never so tidy as the foregoing sketch would suggest. Even in the heyday of the Koyré paradigm, it was not difficult to find English-language historians pursuing other approaches, in addressing widely varied constituencies and interests. For instance, Richard Shryock was the leader among those Americanists (including Arthur Schlesinger, Sr., and Merle Curti) who began the serious historical exploration of the place of science in American society. In so doing, he and his pupils drew on categories as remote from Koyré as from Marx (Shryock, 1936 and 1948; Bell, 1955 and 1965; Hindle, 1956; Reingold, 1964). In both Britain and the United States, the economic, political, administrative, and applied aspects of science continued to command their own audiences—audiences that could be interested in historical investigation on occasion. Thus, Hunter Dupree meticulously explored *Science in the Federal Government: A History of Policies and Activities to 1940* (1957), while D. S. L. Cardwell provided a study focused on the nineteenth-century *Organization of Science in England* (1957). Official historians in each country began the major task of writing contemporary histories of the atomic bomb projects (Hewlett and Anderson, 1962; Gowing, 1964; see also Hiebert, 1961). There was no cessation in the labors of dedicated natural scientists with a learned or antiquarian interest in the history of their several disciplines; Partington (1961–1970) was one characteristic result. The Department of History and Methods of Science at University College, London (launched by Charles Singer, as early as 1923), proved especially hospitable to this "scientific" approach to history (see, e.g., McKie and Heathcote, 1935; Crosland, 1962).

The idealist position was always most closely associated with the study of parts of the physical sciences. Students of the history of the biological and medical sciences found themselves drawn to subjects and categories that held little in common with rationalistic Platonism. Joseph Needham and Walter Pagel, erstwhile collaborators (Needham and Pagel, 1938), offered two alternative visions. Much of the now widely admired and quoted work of Pagel dates from the 1950s and 1960s (Debus, 1972).

With all these qualifications admitted, it remains true that the Koyré paradigm provided a point of reference to the history of science when the disciplinary interests of its professional practitioners were undergoing their first articulation. The paradigm has since been discarded, but has not been replaced by any new agreed synthesis. The reasons why deserve discussion, for they illuminate the contemporary forms and utilities of the history of science.

The first reason has to do with the sheer size of the academic discipline, most especially within the United States but also in the other English-speaking countries and throughout the West. As recently as 1950, graduate training in history of science was offered in only three or four American institutions, and there were less than twenty professors of the subject in the whole world (Guerlac, 1950, p. 206). By the late 1960s some thirty-odd universities in the United States together enrolled several hundred graduate students and employed around 150 professional historians of science, according to one estimate (Price, 1969). Expansion has continued since that time, though at a decreasing rate. The community of discourse has been further swelled by the development of sizeable professional cadres sharing a common language in Great Britain, Canada, and Australia. Today, there are more than seventy graduate programs in the English-speaking world. The list of theses recently completed or in progress in British universities alone, has over 400 entries (Morrell, 1978). In the United States, hundreds of historians of science are in academic employment, while increasing numbers work in museums, in archives, on editorial projects, for federal and state agencies, and for "R&D contractors." Some thirty-odd English-language journals are at least partly devoted to publishing learned articles on one or more aspects of the history of science (*Isis*, 1980). Easier travel and communication heighten the effect, while the barriers between the Anglo-American community and French, German, and Scandinavian scholars are semipermeable.

The increase in the number of avowed historians of science is only one small facet of the overall growth of the professional classes within society. With this latter growth has come a multiplication of the interests served by historians of science. New specialties, subspecialties, and research clusters have formed around these interests. For instance, the history of physics has its own (American) center supported by the physics discipline, together with specialist archives, journals, conference mechanisms, and research programs (Weiner, 1972). Learned societies have been established to foster subjects as varied as the history of the behavioral sciences and social studies of science—respectively "Cheiron" and "4S." The latter society alone has a membership as large as that enjoyed by the History of Science Society only forty years ago. New methodologies—quantitative, structural, prosopographical, and hermeneutical—compete for attention. (See, e.g., Forman et al., 1975; Hannaway, 1975; Shapin and Thackray, 1974). Enlarged resources are devoted to the locating, preserving, cataloging, and rendering accessible of basic archival materials. The *Sources for the History of Quantum Physics* were the object of a pioneering *Inventory and Report* (Kuhn et al., 1967). Today, an office in Oxford, England, concentrates on securing permanent homes for, and providing working inventories of, the papers of leading twentieth-century British scientists; another in Philadelphia, Pennsylvania, focuses on sources for the history of biochemistry and molecular biology; a third, in Berkeley, California, is concerned largely with the electronics industry. Each employs its own staff; the first two issue occasional newsletters. (See Symposium 7, "Problems of Source Materials," in Forbes, 1978.) Fresh efforts are being made to provide a partial, statistical accounting of the size and direction of past scientific activity (Price, 1963; Menard, 1971; Elkana et al., 1978). At the same time historians of science are recruited from more diverse backgrounds (physics, philosophy, history, sociology; see French and Gross, 1973, p. 163) and trained in programs with quite different styles and foci (liberal studies in science, history, history of science, history of science and medicine, history and philosophy of science and technology, history and sociology of science, science studies, etc.).

The history of science discipline now contains within it variegated congeries of practitioners of more or less connected specialisms. There are those who visualize the subject as a series of enquiries, each closely linked with its parent scientific subject; those who, following Sarton, see the subject as a unitary and synthetic discipline, at least self-sufficient if not central to the progress of knowledge; those who emphasize the place of science in historical studies and accordingly focus on particular periods (the ancient world, the twentieth century) or on national cultures (American science, Chinese science); those who stress the links with cognate fields such as philosophy or sociology or economic history, and define their problematics accordingly; those who seek a closer alliance with the history of medicine or the history of technology, which, more recent in their growth as self-confident disciplines, have often been seen as appropriate auxiliaries to the history of science; and to conclude, though not to exhaust the possibilities, those who are concerned with science policy and with the ethical implications of modern scientific knowledge.

It should be apparent why no fresh consensus is about to dominate the field. At the same time, certain broader cultural concerns seem likely to have an enduring impact on the history of science. A discernible public disenchantment with higher education—and with modern science, its most visible offspring—has helped to fuel both defensive and self-questioning moods in the scholarly community (*Daedalus,* 1974 and 1978). A shift in the "center of gravity" of historical study from medieval and early modern to nineteenth- and twentieth-century science may be connected to this change of mood. That shift is possibly related in some way to the emergence of the "big science" that goes with industrial research, military applications, and modern medical concerns (see Price, 1963; Ravetz, 1971; and Gowing, 1974). The by now unavoidable comparison of Western with Oriental and "third world" cultures may also help to shape the agenda of the field (Nakayama and Sivin, 1973; preface to Sivin, 1977). Can logical or historical distinctions between science and magic be sustained (Horton, 1967; Elkana, 1977)? Is science culture free in any but a trivial sense (Barnes, 1974)? Is progress or truth a useful concept around which to organize historical work? How are the patterns of growth and decay in scientific activity and scientific creativity to be conceptualized and measured (Thackray, 1978)? Is science essentially an ideology or form of oppression (Marcuse, 1964; Habermas, 1971)?

There is currently no agreement on any of these questions. Historians of science are skeptical as to whether nature itself can legitimately be understood as setting any definite, narrow limits to the ways in which scientific knowledge may develop. The hope for a true "world science" of which modern (Western) science is but the first foreshadowing appears only a little less delusive than the hope for world government based on a (Western) parliamentary model. Perhaps more realizable is the wish for a coherent understanding of how different scientific systems reflect the cultures that sustain them (Rudwick, 1975). Such an anthropology of science remains far distant (Barnes and Shapin, 1979). For now, the ways in which the relations of man to nature are shaped by the moral and social realities of a given culture remain a matter for allusion, allegory, and speculative probing. More accessible and more concrete are the ways in which the central domains of enquiry in the history of science have developed in response to the values and interests of the subject's practitioners, patrons, and audiences.

II. SOME CENTRAL DOMAINS IN THE HISTORY OF SCIENCE

Certain periods, problems and people have been central to enquiry and debate in the history of science over comparatively long stretches of time. The character of the argument has shifted as new groupings of practitioners have brought their voices to the discussion. These shifts may be explained, though not predicted, in the light of the values, interests, and concerns of the various participants. Again certain subjects have enjoyed great prominence, only to enter on long periods of desuetude or perhaps to experience oblivion. This review of central questions will be personal, partial, and prejudiced. It may, however, serve as a guide to current understandings and previous practice. It may also show in greater detail how values and interests inform and shape the examination of scientific activity.

An obvious place to begin is with the social history of science and with the associated question of the social *roots* of scientific activity. This discussion points toward a major preoccupation of historians of science in the recent past—the idea of the "Scientific Revolution." Studies of that revolution are connected to work on ancient and medieval science and, less directly, to the newly important question of science in nonwestern cultures. The study of science in national settings and the examination of "discipline histories" are two other central domains of enquiry. Also central, but less dominant in the recent past, is the issue of science and religion. Certain related fields that are not themselves major domains but that also deserve attention are the connections of science with medicine and technology, and the philosophy, psychology, and sociology of science. Finally, the domain of "great man studies" incorporates aspects of the other fields and takes us full circle to the historiography of science.

A. The Social Roots and Social History of Science

The question of what are the social roots of science may be understood in two quite separate ways. On the one hand, science may be viewed primarily as a system of knowledge and attempts made to find social causes for the emergence of that knowledge (and, most recently, for its form and content in widely varied historical contexts). On the other hand, science may be viewed in terms of the emergence of certain procedures (experimentation, "the scientific method") and institutions (the Paris Academy of Sciences, the Royal Society of London). Explanations may then be sought for those procedures and institutions in such things as the union of intellectual and craft traditions of enquiry, or the development of the bourgeoisie. The second set of concerns is really a subset of the first.

Neither version of the social roots of science was the subject of concern while the relation of man to nature was undisturbed by rapid technological change, and while science was almost entirely a small-scale activity pursued by lone practitioners. The issues were not addressed in any eighteenth-century or earlier historical work. This is not to deny that many writers, from Bacon on, found reasons for a new stress on natural knowledge in the decline of the authority of the church, the end of the intellectual rule of Aristotle, and so on. The explanations given were, however, far from being social ones, or from consisting of developed arguments on social causes. No more did those causes feature in the early nineteenth-century analyses of Whewell or Comte, though in their different ways both authors were aware of the transformations of science and society through which they were living.

The question of the social roots of science was first formally posed, in abstract form, in the writings of Karl Marx and Friedrich Engels. Marx's 1847 declaration is characteristic: "Social relations are closely bound up with productive forces. . . . The hand mill gives you society with the feudal lord; the steam mill, society with the industrial capitalist. The same men who establish their social relations in conformity with their material productivity, produce also principles, ideas, and categories, in conformity with their social relations. Thus these ideas, these categories are as little eternal as the relations they express. They are historical and transitory products" (1847, p. 109). This general attitude was not translated into the beginnings of a theory of the social roots of science until the appearance in 1882 of Engel's *Dialectics of Nature*. That work contains many statements of the form that: "If, after the dark night of the Middle Ages was over, the sciences suddenly arose anew with undreamt-of force, developing at a miraculous rate, once again we owe this miracle to—production" (Engels, 1940 [1882], pp. 214–215). Neither in Marx nor in later writings of Engels are these insights developed enough for it to be clear whether they are to be considered as explanations of the social roots of scientific *knowledge*, or of *scientific methods and institutions*, or both.

Engels wrote against the background of industrial England. A similar interest in and ambiguity about the social roots of science—but quite different intellectual views—may be seen in other theorists of the late nineteenth and early twentieth century, in places as far apart as Germany and the United States (e.g., Weber, 1904–1905; Veblen, 1906). It was to take a combination of growth in the scale of the scientific enterprise and the common experience throughout the Western world during the 1930s Depression to translate fragmentary interest into a coherent program of research. The clarion call was sounded by Boris Hessen. His essay on the "Social and Economic Roots of Newton's *Principia*" was clear in its concern with the determinants of scientific thought.

The same concern with the social roots of science was to motivate Robert K. Merton's seminal study *Science, Technology and Society in Seventeenth Century England* (1938), in which statistical and prosopographical techniques were marshalled in service of insights Weber had broached and Hessen and others made newly salient. Merton was to focus primarily, though not exclusively, on the social causation of scientific work. An interest in the institutions and methods of science rather than the content of scientific ideas was also to motivate a brilliant series of essays in sociological history published by Edgar Zilsel in the final, North American stage of his career (e.g., Zilsel, 1942; see Keller, 1950). This focus on the social roots of scientific activity also characterized the more traditional historical monograph of G. N. Clark. Stimulated by Hessen's polemic, he offered a penetrating examination of *Science and Social Welfare in the Age of Newton*. In surveying the science of the period, Clark found six different groups of interests at work: "those from economic life, from war, from medicine, from the arts, and from religion . . . and from the disinterested love of truth" (Clark, 1937, pp. 87–89).

In the altered climate of scholarship following World War II, it was the last of Clark's six interests that received prime attention. The shift in emphasis may be seen reflected in A. R. Hall's 1948 Ph.D. dissertation on *Ballistics in the Seventeenth Century* (published as Hall, 1952). In striking refutation of Hessen's proposition that the practical concerns of gunners helped mold the high theory of a Newton, Hall sought to establish that the *theory* of even ballistics itself was independent of practical concerns or practical value. The point was well made, though Hall may not have appreciated sufficiently that failure in application does not entail irrelevance in motivation. The developing argument about the primacy of intellectual over social factors was elaborated in Hall's 1954 textbook on *The Scientific Revolution* and in his critique of "the Merton thesis." In contrast to the moderate earlier position of G. N. Clark, Hall was prepared to argue that "the intellectual change of the seventeenth century is one whose explanation must be sought in the history of the intellect; to this extent . . . the history of science is strictly analogous to the history of philosophy" (Hall, 1963, p. 11).

The dominance of such idealist, philosophical history through much of the 1950s and 1960s naturally directed attention away from *any* discussion of the social roots of science. Where that discussion did occur, it took place in carefully demarcated territory and was promoted by sociologists rather than historians of science. The discussion accepted the idealist proposition that "the succession of ideas is [to be] explained as a result of the discovery of logical flaws within the models [of nature] or bad fits between the models and the natural events they were supposed to explain" (Ben-David, 1971, p. 1). The task of the historian of science was limited to elucidation of the logical and historical filiation of those ideas. Many historically inclined sociologists also accepted the view that scientific thought was not amenable to social analysis. They therefore contented themselves with examining "the conditions that determined the level of scientific activity and shaped the roles and careers of scientists and the organization of science in different countries at different times" (Ben-David, 1971, p. 14).

The tidy division of labor implicit in these remarks began to break down in the late 1960s (see Kuhn, 1968). The last several years have witnessed a surge of works devoted to exploring the social roots of science. Influenced by anthropological traditions and concerns, and harking back to Durkheim and Weber as much as to Marx, these works have taken not only the social forms but also the cognitive content of science to lie within their purview. The theoretical problems that arise when scientific knowledge is viewed in this relativistic way are considered in Hesse (1974) and Bloor (1976). Actual historical investigations include Rattansi (1963), Webster (1975), and Jacob (1976) on the familiar topic of seventeenth-century English science, and Shapin (1975) and Berman (1978) on the less familiar topics of the social roots of science in nineteenth-century Edinburgh and London. For explorations of the situation in America and in Weimar Germany, see Rosenberg (1976) and Forman (1971). For such an approach applied to the social forms and cognitive content of a particular scientific discipline (geology), see Porter (1977).

More broadly, it is true to say that of all genres of the history of science, it is social history that has caught the imagination of scholars in the past decade. The wider, historical concern to study movements and ideologies "from the bottom up" coincided with moods of public disenchantment over higher education and its most visible offspring—science—in the late 1960s. Disenchantment implied a fresh and more critical analysis (Ravetz, 1971). At the same time, study and research in the history of science was expanding rapidly. New generations of graduate students were coming forward, including in them many with some historical training but little formal knowledge of scientific ideas or of foreign languages, classical or modern. Undergraduate lectures and public discussions on science and society became popular (creating a demand for teachers, texts, and "experts"), and the problems and allure of a social explanation of scientific ideas remained intellectually enticing. The shifting dialogue between Western nations and the third world, and the difficulties encountered by missionary attempts to implant "modern science" in traditional cultures, reinforced the awareness that science was as much a social and cultural as an intellectual phenomenon. Historians were quick to perceive that the formal institutions of science—societies, laboratories, university departments, prizes, journals, funding agencies, and so on—would lend themselves to social analysis using both documentary and sociological techniques. The widely influential writings of Thomas S. Kuhn persuasively argued that "scientific knowledge, like language, is intrinsically the common property of a group or else nothing at all" (Kuhn, 1970, p. 210). For these among other reasons, the social history of science has developed rapidly in the last several years.

The genre includes within it studies of the social roots of science and much of national science studies and discipline histories. Thus many works noted below in sections E and F also belong under this head, such as Miller (1970); Rossiter (1975); Allen (1976); Kevles (1978); Geison (1978); and so on. A rich tradition of in-house histories serving the interests of scientific organizations continues to thicken and fructify. (See, Weld, 1848, or Lyons, 1944, on the Royal Society of London, as older examples; Burns, 1977, on the Electrochemical Society, or Schrock, 1977, on the geology department at M.I.T., for recent works.) This tradition has been joined by a steadily rising stream of work by professional historians (e.g., Crosland, 1967, on the Society of Arcueil; Hahn, 1971, on the Paris Academy of Sciences; Kohlstedt, 1976, on the American Association for the Advancement of Science; Russell et al., 1977, on the Royal Institute of Chemistry; and Oleson and Brown, 1976, and Oleson, 1978, on learned societies in America). There have also been works on Nobel laureates (Zuckerman, 1977), on journals (Kronick, 1976), and on technical education (Artz, 1966, for France; Sanderson, 1972, for Britain). In addition, studies of the place of science in popular culture, of the social meanings and intellectual character of the "fringe" sciences (phrenology, astrology, mesmerism, etc.), of long-run statistical trends in scientific activity, and of the politics of science, currently exist only in the form of Ph.D. dissertations or journal articles. The *"4S" Newsletter, Minerva,* and *Social Studies of Science* are three periodicals where this work may be followed.

The new interest in social history has led to a rapprochement between historians of science, economic historians, contemporary historians, and policy planners. Peter Mathias, an economic historian, has edited a set of essays on *Science and Society, 1600–1900* (Cambridge, 1973). Roy M. MacLeod has written many fine articles on policy and the administration of British science in the Victorian era (e.g., MacLeod, 1971). Two historians of science are among the co-editors of a volume on the utilities of science indicators (Elkana et al., 1978). Works like *France in the Age of the Scientific State* (Gilpin, 1968) and *The Scientific Estate* (Don Price, 1965) contribute substantially to contemporary history, from the perspective of the political scientist. Among the memoirs of scientists engaged with intelligence and policy issues, Kistiakowsky (1976) and Jones (1978) are important, while Clark (1965) offers a biography of Sir Henry Tizard who was influential in British war work. York (1975), Smith (1965), and Stern (1969) deal with weapons research, scientists' reaction to the atomic bomb and the Oppenheimer affair, respectively. Official historians have also offered much—as in Gowing (1964 and 1974), in Hewlett and Anderson (1962) and Hewlett and Duncan (1969), and in the studies of the space program (e.g., Hall, 1977) and of the wartime Office of Scientific Research and Development (Baxter, 1946). The interrelations of social history with these other fields are pursued further in chapter 2 of this *Guide* and in the invaluable review essay by Roy MacLeod in Spiegel-Rösing and Price (1977).

B. The Scientific Revolution

Historical investigations of the "Scientific Revolution" are analytically distinct from, but in practice closely allied to, the discussions of the social roots of science treated above. The Scientific Revolution has become the *locus classicus* of debate within the history of science. The reasons are not hard to discern: this "revolution" offers simple yet profound ideas well suited to the idealist method of conceptual analysis. It gives common ground to historians and philosophers of science in the study of a Bacon, a Galileo, or a Descartes, or of, say, Newton and Locke. It is liberally studded with great men, each in himself worthy of a lifetime's scholarship. It gives pride of place to those "hard" mathematical sciences most dominant in the recent past. It provides obvious populations small enough for the historical sociologist to handle by individual effort, and it can usefully be linked with central themes in English and in American colonial history.

It is therefore salutary to be reminded that the concept of "the Scientific Revolution" was foreign to the eighteenth-century authors discussed earlier, and unknown to Whewell, Comte, and Marx (but see Cohen,

1976). Indeed the idea *as an organizing principle* is conspicuous by its absence from the writings of such a major recent historian as George Sarton or from his English counterpart, Charles Singer. Its first sustained use appears to have taken place in the United States early this century, in those essays in which James Harvey Robinson celebrated the future of democratic society with his "new history." Robinson's student Martha Ornstein refers to the first half of the seventeenth century as accomplishing "through the work of a few men a revolution in the established habits of thought and enquiry, compared to which most revolutions registered in history seem insignificant" (Ornstein, 1913, p. 21). This perception was given fuller expression by Preserved Smith. He devoted much of *History of Modern Culture* (1930–1934) to "The Scientific Revolution," giving that title to a crucial chapter.

Hessen's essay helped feed the concentration on this revolution, as did the writings of Clark and Merton. In the years after World War II, many able scholars concentrated their attention on the leading figures in astronomy and mechanics, from Copernicus and Kepler to Descartes and Newton. Using the tools of conceptual analysis most skillfully wielded by Koyré, they undertook to find the origins of modern science in the triumph of certain "rational" ideas of matter, motion, and mathematics. A picture was thus built up of "The Establishment of the Mechanical Philosophy" (Boas, 1952), of the movement *From the Closed World to the Infinite Universe* (Koyré, 1957) and of *The Mechanization of the World Picture* (Dijksterhuis, 1961). Herbert Butterfield's lucid textbook on *The Origins of Modern Science* (London, 1949) helped make the Scientific Revolution a familiar concept throughout the English-speaking world. Butterfield for one was not afraid to make the extraordinary claim that this revolution "outshines everything since the rise of Christianity and reduces the Renaissance and Reformation to the rank of mere episodes, mere internal displacements, within the system of medieval Christendom" (Butterfield, 1949, p. viii). The popular studies of Cohen (1960), Gillispie (1960), Hall (1954), and Kuhn (1957) all emphasized the new intellectual beginnings and the development of objectivity and scientific method, bound into one or another facet of the Scientific Revolution.

Although the 1950s and 1960s saw a massive entrenchment of the idea of such a revolution, the concept was not without its difficulties. The fact that Hall's textbook *The Scientific Revolution* was subtitled *1500–1800* indicates one such problem. The same difficulty appears if one juxtaposes Kuhn's exploration of the "Copernican Revolution" with Butterfield's belief that a "postponed scientific revolution in chemistry" was taking place more than two centuries later. If these dates were taken seriously, revolution in science became a somewhat leisurely affair! One possible response was to move the locus of revolution back to an earlier era altogether (as in Crombie, 1953). Another was to generalize the notion and have not one but many revolutions (as in Kuhn, 1962). Still a third was to subject particular facets of the Scientific Revolution to a far more sustained enquiry.

Walter Pagel has laid stress on the medical and the neo-Platonic aspects of that revolution through detailed examination of such figures as Paracelsus and William Harvey (Pagel, 1958 and 1967). Hermetic and "irrational" influences have been heavily emphasized by Frances Yates (e.g., 1964 and 1972) and increasingly recognized by other scholars. (See, e.g., the exchange between Hesse, 1973, and Rattansi, 1973; the essays in Bonelli and Shea, 1975; and Debus, 1977.) The continuing role of Aristotelian and other "traditional" influences has been stressed (Schmitt, 1971). The importance of "unsuccessful" sciences and unfulfilled hopes for reform has been the subject of masterly examination (Webster, 1975). With each passing year it becomes more difficult to believe in the existence or coherence of a single unique "Scientific Revolution." Even so, the "Scientific Revolution of the seventeenth century" remains a central heuristic device in the field and the subject of myriad textbooks and courses.

C. Ancient and Medieval Science

Enquiries into ancient science are themselves of great antiquity, reflecting the centrality of the classics within the Western tradition until very recent times. The scholarship and sophistication that was early lavished on facets of ancient science may be seen in J. S. Bailly's *Histoire de l'Astronomie Ancienne* (Paris, 1775) or in Delambre's later two-volume work of the same title (Delambre, 1817). Other notable nineteenth-century studies include Lenz (1856, on zoology), Berthelot (1888, on alchemy) and Zeuthen (1896, on mathematics).

Works of this kind lay within the tradition of "discipline histories" of particular sciences. But late in the nineteenth century, sustained efforts were made by Tannery and others to write integrated histories of the sciences, or of science, in the ancient world. (See, e.g., Tannery, 1887, and Gunther and Windelband, 1888). This latter ambition also lay behind George Sarton's heroic, unfinished endeavor to produce an *Introduction to the History of Science* (1927–1948). More recently, attempts have turned toward placing ancient science within the context of a wider classical scholarship, as in Farrington (1944), or in Geoffrey Lloyd's two volumes on *Early Greek Science: Thales to Aristotle* (1970) and *Greek Science after Aristotle* (1973); see also Edelstein (1952) and Stahl (1962, on Roman science).

In contrast, medieval science was for long unstudied and unsung. While intellectuals had to contend with entrenched ecclesiastical authority, the "Middle Ages" was viewed as a wasteland of the mind. From fifteenth-century humanists to nineteenth-century positivists, the verdict was unanimously negative. This verdict appeared fully corroborated by the negligible evidence of achievements in science for over a millenium of European history. Defenders of one or another aspect of medieval culture could always be found among clerics, traditionalists, or romantics, but the paradoxical task of promoting medieval *science* waited until early in the present century. Perhaps it required a very paradoxical champion: Pierre Duhem was a devout Catholic, yet a friend of Modernists and Dreyfusards, and anti-thomist, indeed rather Pascalian, in his philosophy of science. Duhem was experienced as a scientist and skilled in the use of historical-philosophical polemics. He discovered early, embryonic versions of modern statics among fourteenth-century remains and eventually satisfied himself, both on a continuous thread of phenomenalist philosophy and on real precursors of later developments.

Subsequent studies have indeed shown much solid medieval work in the mathematical sciences (see, e.g., Maier, 1949–1958, and Clagett, 1964, 1976, and 1978). Attempts to locate the methodological roots of later work within the Middle Ages (e.g., Crombie, 1953) have proved less enduring. The medieval period, more than any other epoch in Europe, raises anthropological questions as to the nature and function of science in a culture (Southern, 1963). There is a real rationale in Thorndike's assimilation of magic and experimental science in his encyclopedic compilation (Thorndike, 1923–1958). The conjunction of what appear to us as strange bedfellows serves to remind us of the bias of our own perspectives.

The massive expansion of students and publications in the last few years has had its effect on medieval science. Grant (1974) is an exemplary sourcebook. Monographic studies of scrupulous exactitude, dealing with particular texts or individuals or scientific problems, have begun to appear (e.g., Grant, 1971; North, 1976; Lindberg, 1976). With the emergence of a more self-confident, autonomous grouping of scholars, there has also come a renunciation of "the pox of precursoritis." Instead, historians of medieval science have begun to argue that their work should be placed within the wider context of medieval culture (see, e.g., Murdoch and Sylla, 1975). This ambition is not the less important, given that even a general survey of *Science in the Middle Ages* now requires the combined efforts of sixteen scholars (Lindberg, 1978).

Whatever their particular interests, students of ancient and medieval science have always agreed on the rewards to be gained by moving the locus of enquiry in the history of science backwards in time, within the Western tradition. Other scholars—considerable numbers of them in recent decades—have seen greater profit in moving the locus of enquiry outwards and, sometimes but not necessarily, backwards.

D. Science in Non-Western Cultures

Ancient science has most often been understood as Greek science. The importance of Babylonian contributions to the emerging structure of

Western science has also long been recognized. The nineteenth-century succession of learned Orientalists certainly had its effect on George Sarton. His emphasis on the cumulative character of scientific knowledge, his humanism, and his desire for encyclopedic history made him aware of and sympathetically disposed to Islamic, Babylonian, and Eastern contributions (see, e.g., Sarton, 1927–1948). The nineteenth-century development of Assyriology also made many scholars newly conscious of Babylonian science.

Astronomical observation in particular depends on measurements repeated at long historical intervals. A tradition of interest in past observations has thus always been implicit in astronomy. That tradition has been used to especial advantage in work on ancient astronomy. Babylonian, Egyptian, and Greek contributions have been meticulously explored by Willy Hartner and Olaf Pedersen in Europe, and by Otto Neugebauer and the scholars associated with his institute at Brown University. (See, e.g., Pedersen and Pihl, 1974; Neugebauer, 1952 and 1975; Maeyama and Saltzer, 1977.) More recently, the science of medieval Islam has become a renewed focus of attention, in connection with the burgeoning ethnic pride and economic power of the oil-rich states of the Middle East; the newly launched *Journal of the History of Arabic Science* (1977–   ) and its home institute at Aleppo in Syria give promise of important future scholarship. For the present, European scholars wanting a synthetic account of Islamic science must rely on Aldo Mieli's pioneering study of *La Science Arabe* (Leiden, 1938), which may be supplemented by Nasr (1968).

If medieval, Islamic, and Babylonian science were discoveries of the nineteenth century, Chinese and Japanese scientific efforts have entered the consciousness of Western historians only in the most recent decades, and African science is barely recognized as a subject for serious research. Interest in these fields owes much to the "postcolonial" situation that has prevailed since World War II. It was Joseph Needham's wartime exposure to China that, combined with his earlier Christian-Marxist interest in science and society, led to the decision to embark on his monumental study, *Science and Civilization in China* (Needham, 1954–   ; see volume 1 for a discussion of the historiography of Far Eastern science). Other scholars (e.g., Sivin, 1968 and 1977) have joined this enterprise, and there is now an American journal titled *Chinese Science*. Japanese science has also been the subject of increasing scrutiny in Western languages (e.g., Nakayama et al., 1974). There have been, of course, well established groups of historians of science in China and Japan for decades. The question of "African science" is more complex (see Horton, 1967). Indeed, any sustained enquiry into African beliefs about nature raises the question of whether there is, and what might be the essence of, a "demarcation criterion" between science and nonscience. The decay of the confidence of Western philosophers in the existence of any such principle (see chapter 4 in this volume) has been nicely matched by the growing awareness of Western anthropologists that their own beliefs about nature are by no means self-evidently true (Horton and Finnegan, 1973; Elkana, 1977).

E. National Studies

Studies of non-Western science are but one variant form of the study of science in national cultures. Benjamin Martin long ago declared Newtonianism the "Philosophia Britannica." Despite such effusions, the spirit of the Enlightenment favored an emphasis upon the cosmopolitan character of contributions to the common good of the Republic of Letters. Only with the growth of nationalistic atavisms in the nineteenth century did commentators routinely fix their gaze on the nation state. For science, this development was foreshadowed in Charles Babbage's rambling work of social analysis, history, and distemper entitled *Reflections on the Decline of Science in England* (London, 1830). The spirit of national pride reached its first full flower in France on the eve of the Franco-Prussian war, with statements such as "Chemistry is a French science. It was founded by Lavoisier of immortal memory" (Wurtz, 1869, p. 1; see also Duhem, 1906, and Paul, 1972).

The desire of scientists to establish wide networks of collaboration and correspondence helped keep such crude feeling in check. On the other hand, the very growth in the size of national scientific communities made some attention to their individual histories all the more necessary. Marxism provided one possible basis for such studies (for Britain, see Crowther, 1935; for the U.S.A., Struik, 1948). National Socialism briefly offered another (Lenard, 1929; see Beyerchen, 1977). Since World War II, more eclectic work has slowly come to prominence. The enlargement of the scientific enterprise; the expansion of the historical research community; the changing roles of science in modern society—all have combined to favor the study of science in particular national contexts.

The preponderance of work in this as in other areas of the history of science lies with scholars in the United States. Not surprisingly, American science has been well studied in recent years. In the 1950s, the colonial era attracted most attention (see the essays and editor's remarks in Hindle, 1976). The very period in which the funding of scientific research by the federal government expanded by leaps and bounds also saw the widespread popularity of a historical thesis about "American indifference to basic research" (Shryock, 1948). That thesis has fallen into disfavor (Reingold, 1972a) as scholars concentrating on the nineteenth century have turned toward the documentary sources and produced a quickening flow of monographs on key individuals (e.g., Dupree, 1959, on Asa Gray; and Lurie, 1960, on Louis Agassiz); on science in the winning of the West (Goetzmann, 1966); on the funding of research (Miller, 1970); on the social uses of scientific ideas (Rosenberg, 1976); on particular scientific institutions and disciplines (Beardsley, 1964, on chemistry; Kohlstedt, 1976, on the A.A.A.S.; Sinclair, 1974, on the Franklin Institute); on agricultural chemistry (Rossiter, 1975); and in critical review of traditional interpretations (Daniels, 1972). Already in 1966 one close observer of the historical scene would telegraph: "The History of American Science: A Field Finds Itself" (Dupree, 1966). The claim is well borne out by subsequent developments and by the sophistication of such recent institutional studies as Oleson and Brown (1976) and Oleson (1978). The role of "interests," of communal traditions, and of ease of access to appropriate sources in creating this new complex of understandings is vividly highlighted by the fact that the last British scholar to write a book on the history of American science was Crowther (1937) and that no Frenchman has ever undertaken the task.

Possibly owing to their greater ethnic awareness and more diverse heritage, scholars in the United States have not hesitated to contribute to the history of science in other nations. Alexander Vucinich has produced a major survey of *Science in Russian Culture* (Vucinich, 1963–1970), and Loren Graham has elucidated the central theme of *Science and Philosophy in the Soviet Union* (Graham, 1972), while more selective studies range from an examination of *The Lysenko Affair* (Joravsky, 1970) to consideration of *Newton and Russia* (Boss, 1972) and of *Atomic Energy in the Soviet Union* (Kramish, 1959). French science has been well served by American scholars. A favorite approach to wider issues has been through biography, as in Guerlac (1961) on Lavoisier and chemistry; Hankins (1970) on d'Alembert and mechanics; Gillmor (1971) on Coulomb, physics, and engineering; and Baker (1975) on Condorcet and social science. Mesmerism has also come in for attention (Darnton, 1968). French and British scholars have contributed actively (e.g., Crosland, 1967, on the Society of Arcueil; Roger, 1963, on the life sciences; Taton, 1964, on the teaching of science; and Fox, 1973, on patronage). A recent review article (Crosland, 1973) offers useful access to the literature.

German science was little studied in the years following World War II, despite its obvious historical significance. Interest has recently revived. Ben-David (1971), Forman (1971), Gasman (1971), Gregory (1977), and Beyerchen (1977) offer the best entrée to nineteenth- and twentieth-century institutions and ideologies. For the eighteenth century, see Hufbauer (1971). Compared with its Continental neighbors, but not with the United States, Britain has always been well served. Scholars in the United States have found British science a congenial subject of study. Such early enquiries as Merton (1938) have been followed by studies of particular institutions (e.g., Schofield, 1963, on the Lunar Society); of individual biography (Williams, 1965, on Faraday; Manuel, 1968, on Newton; Wilson, 1972, on Lyell; Heilbron, 1974, on Moseley); of scientific ideas and controversy (Kargon, 1966, on seventeenth-century atomism; Burchfield, 1975, on the age of the earth); and of the wider cultural contexts of science (e.g., Thackray, 1974; Kargon, 1977; and Cannon, 1978). British contributions to the study of British science are rapidly increasing. Among works mentioned earlier, Cardwell (1957), Gowing (1964), Webster (1975), and Porter (1977) are especially important. Allen (1976) is a fine study of *The Naturalist in Britain*. For Britain as for Germany and the United States, there is a signal dearth of synthetic accounts utilizing recent advances in historical understanding.

The quickened sense of national identity in Canada, South Africa, and Australia has led to pioneering reviews of the development of the science of those countries in Levere and Jarrall (1974), Brown (1977), and Moyal (1975), while one aspect of science in a South American nation has finally received adequate treatment in Stepan (1976).

F. Discipline Histories

The histories of scientific disciplines are the form of history of science that has directly served the technical needs, intellectual interests, and psychological concerns of scientific practitioners. Such discipline histories are a correspondingly ancient genre. The esteem in which subjects like astronomy, chemistry, or biology have long been held is reflected in the extensive pedigree of their discipline histories. For example, for chemistry, after an uncertain beginning with Robert Vallensis' *De Veritate et Antiquitate Artis Chemicae* (Paris, 1561), the line runs from Oluf Borch's *Conspectus Scriptorum Chemicorum Illustriorum* (Copenhagen, 1696), through Gmelin (1797–1799), Trommsdorff (1806), Thomson (1830–1831), Kopp (1843–1847) and many later works, to such recent monumental compilations as Partington (1961–1970). There are even books devoted to the historiography of chemistry—for example, Strube (1974) and Weyer (1974). The maturation of new specialties and subspecialties in science has often been marked by appropriate historical exercises, as with Hoppe (1884) and Benjamin (1895) for electricity, Ostwald (1896) for electrochemistry, Stubbe (1963) for genetics, and Fruton (1972) for biochemistry.

Discipline history by scientists for scientists has usually been based on an individualistic epistemology, in keeping with the image of the scientist as one voyaging through strange seas of thought, alone. There are also individualistic property relations in science, giving importance to the adjudication of rival claims. One result has been an historical interest in questions of priority —of who first exposed "error" and established "right" answers, or who developed successful instruments and techniques. Characteristic of the motivating sentiments of such works is Partington's dry declaration that "digressions into general political and economic history, the history of learned societies, et cetera, topics usually called 'background material' and easily found in books and encyclopedias included in all public libraries, may be omitted" (Partington, 1961–1970, II, p. vi).

A focus on the establishment by individuals of correct facts and theories has its limitations, considered as history, however useful it may be to the large and ever growing cadres of scientific practitioners. This kind of discipline history has also been severely criticized for its philosophical naïveté about truth, facts, and scientific method (see, e.g., Agassi, 1963). Such criticism relates, *inter alia*, to a conflict of values between historians of science anxious to establish the autonomy of the field and scientists concerned to guard "their" history. Discipline history by scientists for scientists shows no sign of disappearing or of being unduly influenced by attention to its critics. Instead, scientists are concentrating on developments in their own lifetimes in specialties with which they are heavily identified (see, e.g., Keilin, 1966, on cell respiration; Ainsworth, 1976, on mycology; Tylecote, 1976, on metallurgy; also the brilliant personal history of the search for clues to the genetic code, in Watson, 1968). Recently there have been historical conferences of biochemists and computer specialists convened to review the origins of their fields. One outcome is *The Annals of the History of Computing*, a quarterly journal sponsored by the American Federation of Information Processing Societies.

Professional historians of science have turned to the radically different approach to discipline history implicit in Koyré's method of conceptual analysis. The ideas and theories used in astronomy, physics, and chemistry especially have been the object of extensive and intensive historical scrutiny in the last two decades. (See, e.g., the works cited in section I above; also McMullin, 1963 and 1977; Sabra, 1967; Klein, 1970; Fox, 1971; Elkana, 1974a; Brush, 1976; and the deeply learned works of Kuhn, 1978, and Heilbron, 1979.) So, to a lesser extent, have those of mathematics (Mahoney, 1973) and of biology and the sciences of man (e.g., Coleman, 1964; Mendelsohn, 1964; Provine, 1971; Holmes, 1974; Olby, 1974; and Winsor, 1976.) Important French contributions include Canguilhem (1955 and 1968) and Limoges (1970). This attention to conceptual history—to thought as reflected in texts, documents, experiments, and memoirs, whether "right" or "wrong"—has had a slow but profound and erosive effect upon previous philosophical assumptions. It is no longer possible to escape the implications of the extreme variability in "scientific" ideas over time, of the recurrent ambiguities in the connection between concepts and experimental evidence, and of the historical difficulties inherent in the ideas of scientific truth, scientific progress, and scientific method.

The dividing line between work in discipline history by scientists and by historians of science has not been clear cut; witness such conceptual studies as Rudwick (1972); Holton (1973 and 1978); Farley (1977); or Gould (1977). The division may become more pronounced as professional historians of science begin to attempt complex social histories of scientific disciplines in cultural context. In these histories, scientific ideas and research programs are being recognized not only as "given" by nature (or by God) but also as evolving responses to intellectual, institutional, financial, social, and career pressures and opportunities. Early examples of this important new genre include Edge and Mulkay (1976) on radio astronomy in Britain; Porter (1977) on British geology; Kevles (1978) on American physics; and the more sharply focused work of Geison (1978) on late Victorian British physiology. Lemaine et al. (1976) provides a cross-section of present research. Kohler (1975) gives an exemplary overview of work on the one discipline of biochemistry. Morrell (1972) opens up a comparative perspective by focusing on two schools of chemists, while Pyenson (1978) considers physics in two peripheral communities.

Discipline history offers an important meeting ground for traditional concerns with scientific ideas, new interests in scientists as social groups, the growing desire to treat modern periods and problems, and the fresh awareness of the importance of social and cultural factors. Forman (1971) provides one especially rewarding example of the intellectual insights to be gained by sophisticated work in this area. The history of the physical and natural science disciplines thus offers a continuing area of vitality and concentrated effort within the history of science: indeed there are flourishing specialist journals for the history of astronomy (*Journal of the History of Astronomy*); of biology (*Journal of the History of Biology; Studies in History of Biology*); chemistry (*Ambix*); mathematics (*Historia Mathematica*); natural history (*Bulletin of the British Museum: Natural History: Historical Series; Journal of the Society for the Bibliography of Natural History*) and the exact sciences, including physics (*Historical Studies in the Physical Sciences; Archive for History of Exact Sciences*).

The history of the social and behavioral science disciplines offers a striking contrast. Economic thought is the one area in which "discipline history by scientists for scientists" has been conducted on a high level of sophistication. The (U.K.) *History of Economic Thought Newsletter* serves to coordinate activity. Elsewhere, there are occasional works of subtlety and distinction, such as Rieff (1959) on Freud, Bramson (1961) on sociology, Haskell (1977) on the emergence of professional social science in late nineteenth century America, and Hughes (1958) on European social thought in the period 1890–1930. However, too much research depends on narrowly based interest groups of practitioners, concerned to fight contemporary battles of anthropology, psychology, or sociology with historical materials. An analysis of the strengths and weaknesses of the resulting work is provided by "Scholarship and the History of the Behavioural Sciences" (Young, 1966). The growing subtlety of articles in the *Journal of the History of the Behavioral Sciences* offers promise of better things, as does the creation of a *Journal of the History of Sociology*. Meanwhile Stocking (1968) provides an exemplary set of essays on the history of ideas in one science (anthropology); Young (1970) offers a pioneering conceptual analysis and Ross (1972) a major biography in a second (psychology); Clark (1973) gives important historical insights into the social context of a third (sociology) at a crucial stage of its development; and Decker (1977) and Sulloway (1979) begin the complex task of making a thorough historical evaluation of Freudian theory.

G. Science and Religion

The mutual interdependence, autonomy, or hostility of science and religion —depending on the viewpoint of the author—provides clear examples of the ways in which interests and values may shape and mold scientific historical writing. The assertion of the autonomy of science was central to the stance of the early Royal Society, even as theological issues and religious needs powerfully shaped the subsequent articulation of a Newtonian program for natural philosophy (Webster, 1975; Jacob, 1976). In early Victorian England, William Whewell and other Broad Church leaders sought to create histories, philosophies, and popular accounts in which science and theology were distinct but mutually reinforcing (Whewell, 1837; and Powell, 1834; see Cannon, 1978). In acute contrast, Auguste Comte and Karl Marx articulated moral visions in which science, untrammelled by religious dogma, would provide the key to man's secular salvation (Budd, 1977). Later in the nineteenth century, in America, John William Draper would advocate a "religion of science" (Draper, 1875) and Andrew Dickson White would produce a massive *History of the Warfare of Science with Theology in Christendom* (1896). One among many counterblasts was James Walsh's *The Popes and Science* (1908). Altogether in a different class was J. T. Merz's *Religion and Science: A Philosophical Essay* (1915).

That Victorian crisis of religious faith, which was so subtly interwoven with Darwinian science, cannot concern us here. What is rather of interest are the various early twentieth-century attempts to effect a rapprochement between changed scientific understandings (especially with regard to matter, energy, space, and time) and religious belief. Those attempts had been prefigured in such German-language authors as Haeckel, Mach, and Ostwald. Following World War I there was a fresh flow of works written in English. This literature, historical or philosophical in style, was concerned to reassert the importance of metaphysics as a common ground for science and religion (see, e.g., Whitehead, 1925; Burtt, 1925; Needham, 1925; Eddington, 1929; and Barnes, 1933). More recently there have been both scholarly and polemical attempts to demonstrate the importance of (Protestant) Christianity in the formation of modern science (Raven, 1942; Hooykaas, 1972) and sustained historical enquiries into particular conflicts between science and religion (Fleming, 1950; Westfall, 1958; Turner, 1974). There is within the historical sociology of science a tradition that connects Puritanism and science (e.g., Merton, 1938, and Webster, 1975). Russell (1973) provides a reader in current historical work.

H. Science, Medicine, and Technology

The history of medicine and the history of technology are two cognate subjects that have developed in tandem with, and to some extent out of, the history of science. Depending on definitions, much history of science lies within the history of medicine, and vice versa. Physicians and medical institutions played leading roles in nurturing natural philosophy in the Renaissance and the early modern era. Conversely, scientific ideas and scientific method have done much to shape medicine within the last hundred years. Furthermore, a succession of historians of medicine (Henry Sigerist, Richard Shryock, Walter Pagel, Owsei Temkin, and Erwin Ackerknecht) have had a powerful if often indirect influence on the discipline of the history of science. Through their writings, the German tradition of *Kulturgeschichte* has been made accessible. The symbiosis of science and medicine is such that the separate treatment of their histories reflects convenience and the present professional interests and values of the scientific and medical communities as much as it reflects historical patterns of development (see, e.g., the *Festschriften* for Pagel, edited by Debus, 1972, and for Temkin, edited by Stevenson and Multhauf, 1968; see also chapter 3 in this volume).

The divorce between science and technology was sharper in much of the historical past. The emergence of science-based technologies (dyestuffs; electric light and power) in the second half of the nineteenth century signalled a new state of affairs. Today the distinction between the two fields is difficult and often controversial (see Derek Price, 1965). Reflecting this state of affairs, there is a continuing historical debate about the extent to which early modern science was indebted to craft technique (see, e.g., Hall, 1959, and Webster, 1975) and about the degree to which the British Industrial Revolution was or was not dependent upon scientific knowledge (see, e.g., Clow and Clow, 1952; Bernal, 1953; Schofield, 1963; Musson and Robinson, 1969; Mathias, 1973). The question of the interactions of science and technology in the formation of American capitalism is also receiving attention (Noble, 1977). Whether such works of scholarship as Multhauf's *Neptune's Gift: A History of Common Salt* (1978) are classified as history of science or as history of technology is a matter of personal taste. What is indisputable is that the history of technology as an organized field of scholarly enquiry is very new, fast growing, highly variegated, and full of intellectual ferment. (For further discussion, see Kranzberg, 1962, and chapter 2 of this *Guide*, where the important subject of industrial research laboratories is also fully treated.)

I. Philosophy, Psychology, and Sociology of Science

Our knowledge of nature is an obvious subject for philosophical investigations. To the extent that those investigations lie within traditions of enquiry open to or based on historical argumentation and analysis, philosophy and history of science will be inextricably mixed. We have already noted how Comtean and neo-Kantian philosophical programs gave rise to works such as Mach ([1907] 1883) and Tannery (1912–1950). More recently, the idealistic currents of thought present in Meyerson and Koyré have powerfully influenced the professional practice of the history of science in the English-speaking world. Departments or programs in history and philosophy of science have been set up in Melbourne, Australia; Bloomington, Indiana (see Giere and Westfall, 1973, for a twentieth-anniversary celebration); Princeton, New Jersey; and Leeds and Cambridge, England, among other places. The use of history of ideas within philosophical argument has been convincingly displayed in a long series of works (e.g., Hanson, 1958; Hesse, 1966; Toulmin, 1972; Kuhn, 1977).

The programmatic arguments for history and philosophy of science as a composite subject were initially aimed at undermining the dominance, within Anglo-Saxon philosophy, of ahistoric, formal analyses. In this they were eminently successful (see, e.g., Stuewer, 1970, and Elkana, 1974b). Today spokesmen for "hps" find themselves fighting a kind of rear guard action against "irrational" and "hermetic" ideas, advanced at times from within their camp as in Feyerabend (1975), or from the movement toward social history (e.g., Hesse, 1973; Lakatos, 1974). The question of the proper relations between historians and philosophers of science continues to generate its own special literature, Burian (1977) being the most recent addition.

The relations between the history and sociology of science have been, to an extent, the mirror image of those between the history and philosophy of science. Interest in the social roots of science in the 1930s provided the context for the fruitful union of history and sociology in both British and American writings (see, e.g., Zilsel, 1942). The strength of the idealist program corresponded to a stagnation in the composite, "history and sociology of science"—a stagnation now ending as one consequence of the marriage between the new social history and a resurgent historical sociology. Works such as Barnes (1974), Edge and Mulkay (1976), Zuckerman (1977), and Mendelsohn et al. (1977) are among the first fruits of this union. (Chapter 7 in this volume reflects the earlier split, viewing Merton, for instance, more as a practitioner of the empirical sociology of science than as a historical sociologist.)

The psychology of science is a far less articulated field of enquiry. It does however possess significant historical roots and offers material important to historians within that tradition of enquiry into the psychosocial environment of great men begun by de Candolle (1873) and Galton (1874). Protection of the values of upper-class culture against the threats implicit in the existence of an urban proletariat was one powerful motivator to this early work. More recent psychological studies have been concerned with contemporary rather than historical investigation. Subjects of research include possible ways to encourage bright students into science (Roe, 1953), the understanding of personality differences between young scientists and nonscientists (Hudson, 1966) and the "moon scientists" at work (Mitroff, 1974). The value to the historian of such enquiries is not immediately obvious. More rewarding is the work of Erik Erikson. His psychological examination of creative genius has helped spawn the whole field of psychohistory. Within the history of science that field is well represented by a brilliant *Portrait of Isaac Newton* (Manuel, 1968).

J. "Great Man" Studies

The psychobiography of Newton may be seen as but the latest in a long line of "great man" studies in science. (No chauvinism is intended: it is simply fact that few great women scientists have been biographees.) Science has eponymized its heroes since the earliest days: Kepler's Laws, Hooke's modulus, the Linnaean system, the Ohm, Planck's constant—a plethora of examples springs to mind. The need to canonize certain figures has complex roots in the intellectual character and sociological formation of scientific theories. At the same time, the creation of heroes of science can be of great importance to national pride. Contemplation of the work of these masters serves useful pedagogic functions for students and neophytes. It affords intellectual challenge, inspiration, and recreation to mature practitioners. It also offers the opportunity of vicarious participation in high intellectual endeavor to the ordinary educated public. Where the great man in question had an unusual range of interests, was given to prolific publications, or left an especially rich *Nachlass* of manuscripts, the possibilities are virtually inexhaustible.

Bibliophilia, scientific learning, and meticulous scholarship have been the ingredients in several notable bibliographical studies of great men, as Fulton (1961) on Robert Boyle, and Duveen and Klickstein (1954 and 1965) on Antoine Lavoisier. Anniversaries of the birth, the death, or the main work of a great man may each serve to focus attention and rekindle interest. (See, e.g., Westman, 1976, for the Copernicus quincentenary.) A more immediately national pride may be seen at work in the monumental editions of the work of, for example, Galileo Galilei (Favaro, 1890–1909) or Christian Huygens (22 volumes, 1888–1950), or the current, more modest effort devoted to Joseph Henry (Reingold, 1972–   ). That same pride may be seen mixed with a great variety of other motives in the case of Isaac Newton. The importance to Victorian mythology of David Brewster's 1831 and 1855 biographies has been remarked upon. Newton has played a continuing symbolic role for his alma mater of Cambridge, attracting the attention of figures as varied as Joseph Edleston and John Maynard Keynes (see Edleston, 1850, and Keynes, 1946). His importance to physical scientists in other national traditions is apparent from, for example, Rosenberger (1895). Theologians and historians have seen Newton as their own (see, e.g., Mc-

Lachlan, 1950; Manuel, 1963), while much of the early professional work of Anglo-American historians of science in the 1950s and 1960s was focused on his work. (See, e.g., Cohen on *Franklin and Newton,* 1956; the essays on *Newton's Letters and Papers,* in Cohen, 1958; the multivolume edition of Newton's correspondence in Turnbull et al., 1959–1978; Hall and Hall on *Unpublished Newton,* 1962; the essays on *The Annus Mirabilis* tricentennial, in Palter, 1970; the deeply learned introduction to the *Principia* of Cohen [1971]; the magisterial multivolume edition of his *Mathematical Manuscripts* in Whiteside, 1967–1980; and the discussions of Newton's impact—on chemistry in Thackray, 1970a, on optics in Steffens, 1977, and the impact of his physics in Westfall, 1971, and his alchemy in Dobbs, 1975.) Newton studies have reached a stage where even his library has been minutely examined (Harrison, 1978), while review articles (e.g., Westfall, 1976) and a book-length bibliography (Wallis and Wallis, 1977) barely suffice to keep one abreast of recent scholarship.

The Newton industry represents the most developed but by no means only example of the fascination exerted by a great man of science. As the center of gravity of historical studies moves forward into the nineteenth and even the twentieth century, figures like Darwin, Freud, and Einstein are receiving steadily growing attention. Already a multivolume edition of Darwin's correspondence is under way. "Great man" studies necessitate an increased attention to documentary evidence and nourish a growing sophistication of analysis. The result is a revelation of many unsuspected aspects to the heroes of science. There is a fresh awareness of the subtle, elusive quality of their ideas and of the persistence and intricacy of their patterns of thought. That awareness has done much to challenge stereotypes of science as impersonal, value-free enquiry. Like all intellectual novelties, the challenge can be troubling to both students and settled practitioners in the sciences (see "Should the History of Science be Rated X?" by Brush, 1974).

III. THE HISTORIOGRAPHY OF SCIENCE REVISITED

As will by now be apparent, the history of science is a large subject with ramified roots, a long history, and proud traditions. Reviews, overviews, historical analyses, and historiographic defenses of one or another aspect of the subject are themselves diverse, multifaceted, and of considerable pedigree. This section of concluding remarks will be limited to recent statements.

One aspect of George Sarton's lifelong attempt to establish history of science as a fully professional discipline, at once autonomous yet central to the cultural history of mankind, appears in the repeated exposition of programmatic and historiographic arguments. Those arguments may be most conveniently approached through the essays collected together in Stimson (1962). Sarton's emphasis on cumulative, positive knowledge stands in sharp contrast to the vision of science as a part of the history of ideas, as articulated in Koyré (1955) and Butterfield (1959). The growing dominance of the history of ideas approach among professional historians of science may be seen in the shift of emphasis between Guerlac (1950) and Guerlac (1963), and in the vigorous polemics of Agassi (1963). The *Festschrift* for Alexandre Koyré (Koyré, 1964) reflects well the best work in this tradition, while Gillispie (1973) elegantly portrays Koyré's role as intellectual role model in the 1950s and 1960s. Again, the emergence of the "idealist paradigm" may be traced through the proceedings of three conferences that were seminal in shaping this style of history of science (see Shryock, 1955; Clagett, 1959b; Crombie, 1963). Also important was the decision of the National Science Foundation, following the first of these conferences, to downplay sociology and establish a research program uniting only history and philosophy of science.

One early sign of the decay of the idealist thrust may be found in Thomas S. Kuhn's 1968 historiographic article in the *International Encyclopedia of the Social Sciences*. There and in Kuhn (1971), the need to reunite "internal" and "external" history was cogently argued. Since that time, other authors have sketched out new approaches (Ravetz, 1974), vigorously criticized the dependence of history of science on philosophy of science, pointed out the dangers of settling for "historical science" (Gowing, 1975), hinted at the dangers of fashion and Federal funds (Hahn, 1975), resurveyed the field from a radical perspective (Young, 1973), and examined with erudition "The Many Faces of the History of Science" (Cohen, 1977).

That the field has many faces is a sign of its maturing as a part of history. The history of science may no longer aspire to the hegemony Sarton envisaged for it. However, in depth of scholarship, catholicity of interest, variety of approaches, and importance of results, it bears healthy witness to the place of science in modern civilization, and to the complexity of the continuing relation between man and nature.

Acknowledgements

In its final form this chapter owes much to John Clive, I. Bernard Cohen, Shmuel Eisenstadt, Yehuda Elkana, Gerald L. Geison, Margaret Gowing, A. Rupert Hall, Mary Hesse, Thomas S. Kuhn, David C. Lindberg, Robert K. Merton, Nathan Reingold, Dorothy Ross, Martin Rudwick, Steven Shapin, and Charles Webster. Colleagues in the Department of History and Sociology of Science at the University of Pennsylvania made many helpful suggestions. David Miller and Jeffrey L. Sturchio offered valuable aid.

BIBLIOGRAPHIC INTRODUCTION

A. Archives

Manuscripts germane to the history of science are too many, too variegated, and too scattered to be easily catalogued or even known about. Archival collections may be found in academies, colleges, and universities; in national, regional, and local libraries; and in scientific and specialist archives. The Bibliothèque Nationale, the British Museum, and the Library of Congress have rich resources. So too do the National Archives (Washington, D.C.) and the American Philosophical Society (Philadelphia). The Center for the History of Physics of the American Physical Society (New York) provides an exemplary resource. A "Conference on Science Manuscripts" is reported in *Isis* 53 (1962), no. 171. See also "Problems of Source Materials," symposium 7 in Forbes (1978). (References in this form in the bibliographic introduction are to the main bibliography for this chapter, below.)

B. Bibliographies

The history of science is well equipped with bibliographical aids. Since its foundation, *Isis* has carried regular critical bibliographies of the history of science. Those from 1913 to 1965 have been integrated with entries arranged under individuals, institutions and subjects, as *The Isis Cumulative Bibliography* (4 vols. London: Mansell, 1971–1980). This provides the starting point of bibliographic research and may be complemented by the bibliographies published annually in *Isis* since 1965. The *Bulletin Signalétique* (Paris) regularly publishes bibliographies of the field. Sarton's *Horus: A Guide to the History of Science* (1952) provides the most convenient, if dated, one-volume guide; it may be supplemented by Russo (1969) and Thornton and Tully (1972). Bibliographies in the history of technology and the history of medicine offer substantial coverage of the history of science; see chapters 2 and 3 in this volume.

C. Dictionaries and Encyclopedias

Authoritative accounts of the work of deceased scientists of all periods and all nations are available in the superb sixteen-volume *Dictionary of Scientific Biography* (Gillispie, 1970–1980). Compact studies of over 1,000 eminent scientists and engineers are available in *A Biographical Dictionary of Scientists* (Williams, 1969). The *World Who's Who in Science* (Debus, 1968) offers brief accounts of 30,000 scientists, the majority still living. Poggendorff (1863–present) provides biobibliographies of many first and second rank men of science in certain fields. Useful summary articles on particular subjects in the history of science may be found in the *Dictionary of the History of Ideas* (1973); the *Encyclopedia of Philosophy* (1967); and the *Encyclopedia of the Social Sciences* (1930–1934)—and to a lesser degree in the newer *International Encyclopedia of Social Sciences* (1968).

D. Journals

The largest circulation and widest coverage periodical in the field (and since 1924 the official journal of the History of Science Society) is Sarton's

Isis; Mieli's *Archivio* is now continued as the *Archives Internationales d'Histoire des Sciences* (see text, I. B.). *Annals of Science* (established 1936) and *History of Science* (1962) are other general-interest reviews. *Minerva* (established 1962) provides an excellent forum for organizational and policy related studies. Most European and several Asian countries possess their own national journals: the *British Journal for the History of Science* (established 1962); *Centaurus* (Denmark, 1950); *Gesnerus* (Switzerland, 1943); *Indian Journal of the History of Science* (1966); *Japanese Studies in the History of Science* (1962); *Lychnos* (Sweden, 1936); *Physis* (Italy, 1959); and the *Revue d'Histoire des Sciences* (France, 1947). In addition, there is an increasing array of periodicals devoted to special sciences, such as *Ambix* (chemistry, 1937) or the *Archive for History of Exact Sciences* (1961). Journals also exist to serve the history of given institutions (e.g., *Notes and Records of the Royal Society of London,* 1938), to cater to certain categories of author (e.g., *Synthesis,* 1972, for undergraduate and postgraduate students); and to pursue particular themes (e.g., *Zygon: Journal of Religion and science,* 1966). A guide to journals is available (see *Isis,* 1980, in bibliography).

E. Museums

See the discussion in chapter 2 of this volume.

F. Sourcebooks

The reprinting of the whole or part of classic scientific texts has long been an adjunct to the study of the history of science. The 256 volumes of Wilhelm Ostwald's *Klassiker der Exacten Wissenschaft* form one notable example. The *Alembic Club Reprints* reproduce important chemical papers, while the *Harvard Sourcebooks* offer brief excerpts from the sources in, e.g., *Ancient Science, Medieval Science, Chemistry,* and *Physics.* Westfall and Thoren (1968) provides a one-volume anthology; Knight (1972) surveys *Natural Science Books in English, 1600–1900.* The Readex Microprint Corporation has produced a collection of 3000 classic scientific texts. A microfiche collection is also offered by Microeditions Hachette of Paris. The Sources of Science Series (Johnson Reprint) includes significant histories of science, as well as scientific works. The Arno Press has published another series: Classics, Staples, and Precursors in the History, Philosophy and Sociology of Science.

Anthologies of *history of science* writings, include Wiener and Noland (1957; articles from the *Journal of the History of Ideas*); Barber and Hirsch (1962; though titled *The Sociology of Science,* the articles reprinted are mainly historical); Basalla (1968); Olson (1971); and Barnes (1972; once again, mainly historical writing under the title of *Sociology of Science*).

Also Hindle (1976); Reingold (1976); and Sivin (1977) all reprinting articles
from *Isis*.

G. Introductory Texts

A considerable number of one-volume texts on the history of science were
produced in the postwar era, as part of the social construction of a field
of university teaching and research. The more durable include Butterfield
(1949), Mason (1953), Hall (1954), Cohen (1960), Gillispie (1960), and Price
(1961). In recent years, attention has been directed toward monographs
that treat a limited period or a problem in some detail: see, e.g., Kuhn,
The Copernican Revolution (1957); Williams, *The Origins of Field Theory*
(1966); Westfall, *Mechanisms and Mechanics* (1971); Rudwick, *The Meaning of Fossils* (1972); and Allen, *The Life Sciences in the Twentieth
Century* (1975). As yet, few of these works move confidently beyond the
boundaries of scientific ideas. The spread of history of science as a field
of teaching is reflected in two recent conferences (Kauffman, 1971; Brush
and King, 1972) and in the report of a History of Science Society Committee
(Sharlin, 1975).

H. Classics

Certain works have achieved the status of classics in the field for the centrality of the problems they attack, the novelty and importance of their
insights, or the felicity and power of their expression. Any listing is necessarily perilous and invidious, but among the works of certain stature are
Burtt (1925), Lovejoy (1936), Clark (1937), Merton (1938a), and Koyré
(1939). In perusing them a student will learn much about the development
of the field. If a classic textbook were allowed, it would certainly be
Butterfield (1949).

I. Staple Reference Works

In addition to the works listed above, there are certain staples of enduring
value, for either their factual information, their encyclopedic coverage, or
their authority of interpretation. The line between works in this category
and those of more limited and specialist use is thin indeed, but most historians of science agree on the "indispensability" of Merz (1896); Sarton
(1927–1948); Thorndike (1923–1958); Wolf (1935 and 1939); Needham
(1954–   ); Daumas (1957); and Taton (1957–1964).

   Certain volumes of conference proceedings and the *Festschriften*
offered to leading scholars are also in this category. They not only include
important papers but give insights into the shifting texture of a living field
of enquiry. See Montagu (1944), the George Sarton *Festschrift;* Underwood
(1953), the Charles Singer *Festschrift;* Clagett (1959b), the "critical prob-

lems" conference; Crombie (1963), the "scientific change" conference; Debus (1972), *Festschrift* for Walter Pagel; Young (1973), *Festschrift* for Joseph Needham; and Forbes (1978), the Edinburgh International Congress.

BIBLIOGRAPHY

Agassi, Joseph. *Towards an Historiography of Science.* Monograph no 2, *History and Theory* (1963).

Ainsworth, Geoffrey G. *Introduction to the History of Mycology.* Cambridge: Cambridge University Press, 1976.

Allen, David E. *The Naturalist in Britain: A Social History.* London: Allen Lane, 1976.

Allen, Garland. *Life Science in the Twentieth Century.* New York: John Wiley, 1975.

Arago, François. *Biographies of Distinguished Scientific Men.* Tr. by Admiral W. H. Smyth et al. London, 1857.

Artz, Frederick B. *The Development of Technical Education in France, 1500–1850.* Cambridge, Mass.: M.I.T. Press, 1966.

Bailly, Jean S. *Histoire de l'astronomie ancienne depuis son origine.* Paris, 1775; 2d ed., 1781.

Baily, Francis. *An Account of the Life of the Rev. John Flamsteed: The First Astronomer Royal.* London, 1835; reprint ed., New York: Arno Press, 1975.

Baker, Keith M. *Condorcet: From Natural Philosophy to Social Mathematics.* Chicago: University of Chicago Press, 1975.

Barber, Bernard, and Hirsch, W., eds. *The Sociology of Science.* New York: Free Press, 1962.

Barnes, Barry. *Scientific Knowledge and Sociological Theory.* London: Routledge, 1974.

———, ed. *Sociology of Science: Selected Readings.* Baltimore: Penguin, 1972.

Barnes, Barry, and Shapin, S., eds. *Natural Order: Historical Studies of Scientific Culture.* London: Sage, 1979.

Barnes, Ernest W. *Scientific Theory and Religion.* Cambridge: Cambridge University Press, 1933.

Barnes, Harry E. "The Historian and the History of Science." *Scientific Monthly* 11 (1919): 112–126.

Basalla, George, ed. *The Rise of Modern Science: Internal or External Factors?* Lexington, Mass.: Heath, 1968.

Baxter, James P. *Scientists against Time.* Boston: Little, Brown, 1946.

Beardsley, Edward H. *The Rise of the American Chemistry Profession, 1850–1900.* Gainesville: Florida University Press, 1964.

Beckmann, Johann. *Geschichte der Erfindungen.* 4 vols. Göttingen, 1784–1805.

Bell, Whitfield J. *Early American Science: Needs and Opportunities for Study*. Chapel Hill: North Carolina University Press, 1955.

————. *John Morgan, Continental Doctor*. Philadelphia: University of Pennsylvania Press, 1965.

Ben-David, Joseph. *The Scientist's Role in Society: A Comparative Study*. Englewood Cliffs, N.J.: Prentice-Hall, 1971.

Benjamin, Park. *A History of Electricity (The Intellectual Rise in Electricity)*. New York, 1895; reprint ed., New York: Arno Press, 1975.

Berman, Morris. *Social Change and Scientific Organization: The Royal Institution, 1799–1844*. Ithaca, N.Y.: Cornell University Press, 1978.

Bernal, John D. *The Social Function of Science*. London: Routledge, 1939.

————. *Science and Industry in the Nineteenth Century*. London: Routledge, 1953.

————. *Science in History*. London: Watts, 1954.

Berthelot, Marcellin. *Collection des anciens alchimistes grecs*. 3 vols. Paris, 1888.

Beyerchen, Alan. *Scientists under Hitler: Politics and the Physics Community in the Third Reich*. New Haven: Yale University Press, 1977.

Bloor, David. *Knowledge and Social Imagery*. London: Routledge, 1976.

Boas, Marie. "The Establishment of the Mechanical Philosophy." *Osiris* 10 (1952): 412–541.

————. *Robert Boyle and Seventeenth-Century Chemistry*. Cambridge: Cambridge University Press, 1958.

Bonelli, M. L. Righini, and Shea, W. R., eds. *Reason, Experiment, and Mysticism in the Scientific Revolution*. New York: Science History Publications, 1975.

Boss, Valentin. *Newton and Russia: The Early Influence, 1698–1796*. Cambridge, Mass.: Harvard University Press, 1972.

Bramson, Leon. *The Political Context of Sociology*. Princeton: Princeton University Press, 1961.

Brasch, Frederick E. "The Teaching of the History of Science." *Science* 42 (1915): 746–760.

Brewster, Sir David. *The Life of Sir Isaac Newton*. London, 1831.

————. *Memoirs of the Life, Writings, and Discoveries of Sir Isaac Newton*. 2 vols. London, 1855.

Brown, A. C., ed. *A History of Scientific Endeavour in South Africa*. Rondebosch: Royal Society of South Africa, 1977.

Brunet, Pierre. *L'Introduction des théories de Newton en France au XVIIIe siècle*. Paris: Blanchard, 1931.

Brunschvicg, Leon. *Les Étapes de la philosophie mathématique*. Paris: Alcan, 1912; 3d ed., 1929.

Brush, Stephen G. "Should the History of Science Be Rated X?" *Science* 183 (1974): 1164–1172.

————. *The Kind of Motion We Call Heat*. 2 vols. New York: North-Holland, 1976.

Brush, Stephen G., and King, A. L. *History in the Teaching of Physics.* Hanover, N.H.: New England University Press, 1972.

Buckley, Arabella. *A Short History of Natural Science: For the Use of Schools and Young Persons.* London, 1875; 5th ed., 1894.

Budd, Susan. *Varieties of Unbelief: Atheists and Agnostics in English Society, 1850–1960.* London: Heinemann, 1977.

Burchfield, Joe D. *Lord Kelvin and the Age of the Earth.* New York: Science History Publications, 1975.

Burian, Richard M. "More than a Marriage of Convenience: On the Inextricability of History and Philosophy of Science." *Philosophy of Science* 44 (1977): 1–42.

Burns, Robert M. *A History of the Electrochemical Society, 1902–1976.* Princeton, N.J.: The Electrochemical Society, 1977.

Burtt, Edwin A. *The Metaphysical Foundations of Modern Physical Science: A Historical and Critical Essay.* New York: Harcourt Brace, 1925; variously reprinted.

Butterfield, Herbert. *The Origins of Modern Science, 1300–1800.* London: G. Bell and Sons, 1949; new ed., 1957; variously reprinted.

————. "The History of Science and the Study of History." *Harvard Library Bulletin* 13 (1959): 329–347.

Cajori, Florian. *A History of Physics, Including the Evolution of Physical Laboratories.* New York: Macmillan, 1899; reprint ed., 1924.

————. *A History of the Conceptions of Limits and Fluxions in Great Britain.* Chicago: Open Court, 1919.

Candolle, Alphonse de. *Histoire des sciences et des savants depuis deux siècles.* Geneva, 1873.

Canguilhem, Georges. *La Formation du concept de réflexe aux XVIIᵉ et XVIIIᵉ siècles.* Paris: Presses Universitaires de France, 1955.

————. *Etudes d'histoire et de philosophie des sciences.* Paris: J. Vrin, 1968.

Cannon, Susan F. *Science in Culture: The Early Victorian Period.* New York: Science History Publications, 1978.

Cantor, Moritz B. *Vorlesungen über Geschichte der Mathematik.* 4 vols. Leipzig, 1880–1908.

Cardwell, Donald S. L. *The Organization of Science in England.* London: Heinemann, 1957; rev. ed., 1972.

Carus, Julius V. *Geschichte des Zoologie.* Vol. 12 of *Geschichte der Wissenschaften in Deutschland.* Munich, 1872.

Cassirer, Ernst. *Die Philosophie des Aufklärung.* Tübingen: Mohr, 1932; English tr. Princeton, N.J.: Princeton University Press, 1951; variously reprinted.

Clagett, Marshall. *The Science of Mechanics in the Middle Ages.* Madison: University of Wisconsin Press, 1959a.

————. *Archimedes in the Middle Ages.* Vol. 1: *The Arabo-Latin Tradition.* Madison: University of Wisconsin Press, 1964.

————. *Archimedes in the Middle Ages.* Vol. 2: *The Translation from the*

Greek by William Moerbeke. Memoirs of the American Philosophical Society, no. 117. Philadelphia: American Philosophical Society, 1976.

————. *Archimedes in the Middle Ages,* vol. 3: *The Fate of the Medieval Archimedes, 1300–1565.* Memoirs of the American Philosophical Society, no. 125, 1978.

————, ed. *Critical Problems in the History of Science.* Madison: University of Wisconsin Press, 1959b.

Clark, George N. *Science and Social Welfare in the Age of Newton.* Oxford: Clarendon Press, 1937.

Clark, Ronald W. *Tizard.* Cambridge, Mass.: M.I.T. Press, 1965.

Clark, Terry N. *Prophets and Patrons: The French University and the Emergence of the Social Sciences.* Cambridge, Mass.: Harvard University Press, 1973.

Clerke, Agnes M. *A Popular History of Astronomy during the Nineteenth Century.* Edinburgh, 1885.

Clow, Archibald, and Clow, Nan. *The Chemical Revolution: A Contribution to Social Technology.* London: Batchworth Press, 1952.

Cohen, I. Bernard. *Franklin and Newton: An Inquiry into Speculative Newtonian Experimental Science.* Memoirs of the American Philosophical Society, no. 43. Philadelphia: American Philosophical Society, 1956.

————. *The Birth of a New Physics.* Garden City, N.Y.: Doubleday, 1960.

————. *Introduction to Newton's "Principia."* Cambridge: Cambridge University Press, 1971.

————. "The Eighteenth-Century Origins of the Concept of Scientific Revolution." *Journal of the History of Ideas* 27 (1976): 257–288.

————. "The Many Faces of the History of Science." In *The Future of History,* pp. 65–110. Edited by Charles F. Deltzell. Nashville, Tenn.: Vanderbilt University Press, 1977.

————, ed. *Isaac Newton's Papers and Letters on Natural Philosophy.* Cambridge: Cambridge University Press, 1958.

Cohen, Robert S.; Stachel, J. J.; and Wartofsky, M. W., eds. *For Dirk Struik: Scientific, Historical, and Political Essays in Honor of Dirk J. Struik.* Boston Studies in the Philosophy of Science, vol. 15. Dordrecht, Holland: Reidel, 1974.

Cole, Jonathan R., and Zuckerman, H. "The Emergence of a Scientific Specialty: The Self-Exemplifying Case of the Sociology of Science." In *The Idea of Social Structure: Papers in Honor of Robert K. Merton,* pp. 139–174. Edited by Lewis A. Coser. New York: Harcourt Brace Jovanovich, 1975.

Coleman, William. *Georges Cuvier, Zoologist: A Study in the History of Evolution Theory.* Cambridge, Mass.: Harvard University Press, 1964.

Crombie, Alistair C. *Robert Grosseteste and the Origins of Experimental Science.* Oxford: Clarendon Press, 1953.

————, ed. *Scientific Change*. London: Heinemann, 1963.

Crombie, Alistair C., and Hoskin, M. "A Note on History of Science as an Academic Discipline." In *Scientific Change,* pp. 757–764. Edited by Alistair C. Crombie. London: Heinemann, 1963.

Crosland, Maurice P. *Historical Studies in the Language of Chemistry.* London: Heinemann, 1962.

————. *The Society of Arcueil: A View of French Science at the Time of Napoleon I.* London: Heinemann, 1967.

————. "The History of French Science: Recent Publications and Perspectives." *French Historical Studies* 8 (1973): 157–171.

Crowther, James G. *British Scientists of the Nineteenth Century.* London: Paul, Trench, Trubner, 1935.

————. *Soviet Science.* New York: Dutton, 1936.

————. *Famous American Men of Science.* New York: Norton, 1937.

————. *Fifty Years with Science.* London: Barrie and Jenkins, 1970.

Daedalus. vol. 103, No. 3 (1974). "Science and Its Public: The Changing Relationship."

Daedalus. vol. 107, No. 2 (1978). "Limits of Scientific Inquiry."

Daniels, George H., ed. *Nineteenth Century American Science: A Reappraisal.* Evanston, Ill.: Northwestern University Press, 1972.

Darmstädter, Ludwig, and DuBois-Reymond, R. *4000 Jahre Pioneer-Arbeit in der exacten Wissenschaften.* Berlin, 1904; 2d ed., as *Handbuch zur Geschichte der Naturwissenschaften und der Technik.* Berlin, 1908.

Darnton, Robert. *Mesmerism and the End of the Enlightenment in France.* Cambridge, Mass.: Harvard University Press, 1968.

Daumas, Maurice. *Histoire de la Science.* Paris: Gallimard, 1957.

Debus, Allen G. "An Elizabethan History of Medical Chemistry." *Annals of Science* 18 (1962): 1–29.

————. *The English Paracelsians.* London: Oldbourne, 1965.

————. *The Chemical Philosophy.* 2 vols. New York: Science History Publications, 1977.

————, ed. *World Who's Who in Science.* Chicago: Marquis, 1968.

————, ed. *Science, Medicine, and Society in the Renaissance: Essays to Honor Walter Pagel,* 2 vols. London: Heinemann, 1972.

Decker, Hannah S. *Freud in Germany: Revolution and Reaction in Science, 1893–1907.* New York: International Universities Press, 1977.

Delambre, Jean B. J. *Histoire de l'astronomie ancienne.* 2 vols. Paris, 1817.

Delorme, Suzanne; Adam, A.; Couder, A.; Rosband, J.; and Robinet, A. *Fontenelle: Sa vie et son oeuvre.* Paris: Albin Michel, 1961.

Dijksterhuis, E. J. *The Mechanization of the World Picture.* Original Dutch ed., 1950; Oxford: Clarendon Press, 1961.

Dobbs, Betty J. T. *Foundations of Newton's Alchemy.* Cambridge: Cambridge University Press, 1975.

Douglas, Mary. *Natural Symbols.* London: Barrie and Jenkins, 1970.

———. *Implicit Meanings*. London: Routledge, 1975.

Draper, John W. *History of the Conflict between Religion and Science*. New York, 1875.

Duhem, Pierre. *La Théorie physique: Son objêt et sa structure*. Paris: Chevalier et Rivière, 1906; English tr. by P. P. Wiener, *The Aim and Structure of Physical Theory*. Princeton: Princeton University Press, 1954.

———. *Le Système du monde: Histoire des doctrines cosmologiques de Platon à Copernic*. 10 vols. Paris, 1913–1959.

Dupree, A. Hunter. *Science and the Federal Government: A History of Policies and Activities to 1940*. Cambridge, Mass.: Harvard University Press, 1957.

———. *Asa Gray, 1810–1888*. Cambridge, Mass.: Harvard University Press, 1959.

———. "The History of American Science: A Field Finds Itself." *American Historical Review* 71 (1966): 863–874.

Durkheim, Emile, and Mauss, M. "De Quelques formes primitives de classification: Contribution à l'étude des représentations collectives." *Année Sociologique* 6 (1901–2): 1–72. English tr. *Primitive Classification*. London: Cohen and West, 1963.

Duveen, Denis I., and Klickstein, H. S. *A Bibliography of the Works of Antoine Laurent Lavoisier (1743–1794)*. London: Dawson, 1954.

———. *Supplement to a Bibliography*. London: Dawson, 1965.

Eddington, Arthur S. *The Nature of the Physical World*. Cambridge: Cambridge University Press, 1929.

Edelstein, Ludwig. "Recent Trends in the Interpretation of Ancient Science." *Journal of the History of Ideas* 13 (1952): 573–604. Reprinted in *Roots of Scientific Thought*, pp. 90–121. Edited by P. Wiener and A. Noland. New York: Basic Books, 1957.

Edge, David O., and Mulkay, M. J. *Astronomy Transformed: The Emergence of Radio Astronomy in Britain*. New York: John Wiley, 1976.

Edleston, J. *Correspondence of Sir Isaac Newton and Professor Cotes Including Letters of Other Eminent Men*. London, 1850.

Elkana, Yehuda. *The Discovery of the Conservation of Energy*. London: Hutchinson, 1974a.

———. "The Distinctiveness and Universality of Science: Reflections on the Work of Professor Robin Horton." *Minerva* 15 (1977): 155–173.

———, ed. *The Interaction between Science and Philosophy*. Atlantic Highlands, N.J.: Humanities Press, 1974b.

Elkana, Yehuda; Lederberg, J.; Merton, R. K.; Thackray, A.; and Zuckerman, H., eds. *Toward a Metric of Science: The Advent of Science Indicators*. New York: John Wiley, 1978.

Engels, Friedrich. *Dialectics of Nature*. Berlin, 1927 (written 1872–1873; first published 1882); reprint ed., New York: International Publishers, 1940.

Farley, John. *The Spontaneous Generation Controversy from Descartes to Oparin*. Baltimore: John Hopkins University Press, 1977.

Farrington, Benjamin. *Greek Science: Its Meaning for Us*. Harmondsworth, Middlesex, England: Penguin Books, 1944; rev. ed., 1949; reprint ed., 1969.

Favaro, Antonio. *Le Opere de Galileo Galilei*. 20 vols. Florence, Italy, 1890–1909.

Feyerabend, Paul. *Against Method*. London: N.L.B., 1975.

Fischer, Johann K. *Geschichte der Physik*. 8 vols. Göttingen, 1801–1808.

Fleming, Donald. *John William Draper and the Religion of Science*. Original ed., 1950; reprint ed., New York: Octagon Books, 1972.

Forbes, Eric G. *Human Implications of Scientific Advance*. Proceedings of the XVth International Congress of the History of Science. Edinburgh: University of Edinburgh Press, 1978.

Forman, Paul. "Weimar Culture, Causality, and Quantum Theory." *Historical Studies in the Physical Sciences* 3 (1971): 1–116.

Forman, Paul; Heilbron, J. L.; and Weart, S. "Physics *circa* 1900: Personnel, Funding, and Productivity of the Academic Establishments." *Historical Studies in the Physical Sciences* 5 (1975).

Fox, Robert. *The Caloric Theory of Gases from Lavoisier to Regnault*. Oxford: Clarendon Press, 1971.

——. "Scientific Enterprise and the Patronage of Research in France, 1800–1870." *Minerva* 11 (1973): 442–473.

French, Richard, and Gross, M. "A Survey of North American Graduate Students in the History of Science, 1970–1971." *Science Studies* 3 (1973): 161–171.

Freund, Ida. *The Study of Chemical Composition: An Account of Its Method and Historical Development*. Cambridge: Cambridge University Press, 1904.

Fruton, Joseph S. *Molecules and Life: Historical Essays on the Interplay of Chemistry and Biology*. New York: Wiley-Interscience, 1972.

Fulton, John F. *A Bibliography of the Honourable Robert Boyle*. Original ed., 1932–1933; 2d ed., Oxford: Clarendon Press, 1961.

Galton, Francis. *English Men of Science: Their Nature and Nurture*. London: Macmillan, 1874.

Gasman, Daniel. *The Scientific Origins of National Socialism*. London: Macdonald, 1971.

Geertz, Clifford. *The Interpretation of Cultures*. New York: Basic Books, 1973.

Geison, Gerald L. *Michael Foster and the Cambridge School of Physiology: The Scientific Enterprise in Late Victorian Society*. Princeton: Princeton University Press, 1978.

Giere, Ronald, and Westfall, R. S., eds. *Foundations of Scientific Method: The Nineteenth Century*. Bloomington: Indiana University Press, 1973.

Gillispie, Charles C. *Genesis and Geology: A Study in the Relations of Scientific Thought, Natural Theology, and Social Opinion in Great Britain, 1790–1850*. Cambridge, Mass.: Harvard University Press, 1951.

————. "Science in the French Revolution." In *The Sociology of Science*, pp. 89–97. Edited by Bernard Barber and W. Hirsch. Glencoe, Ill.: Free Press, 1962; reprinted from *Behavioral Science* 4 (1959): 67–101.

————. *The Edge of Objectivity: An Essay in the History of Scientific Ideas*. Princeton: Princeton University Press, 1960.

————. *Lazare Carnot, Savant*. Princeton: Princeton University Press, 1971.

————. "Alexandre Koyré." In *Dictionary of Scientific Biography*, vol. 7, pp. 482–490. Edited by Charles C. Gillispie. New York: Charles Scribner's Sons, 1973.

————, ed. *Dictionary of Scientific Biography*, 16 vols. New York: Charles Scribner's Sons, 1970–1980.

Gillmor, C. Stewart. *Coulomb and the Evolution of Physics and Engineering in Eighteenth-Century France*. Princeton: Princeton University Press, 1971.

Gilpin, Robert. *France in the Age of the Scientific State*. Princeton: Princeton University Press, 1968.

Gmelin, Johann F. *Geschichte der Chemie*. 3 vols. Göttingen, 1797–1799.

Goetzmann, William H. *Exploration and Empire: The Explorer and the Scientist in the Winning of the American West*. New York: Alfred A. Knopf, 1966; Vintage Books ed., 1972.

Gould, Stephen J. *Ontogeny and Phylogeny*. Cambridge, Mass.: Harvard University Press, 1977.

Gowing, Margaret. *Britain and Atomic Energy, 1939–1945*. London: Macmillan, 1964.

————. *Independence and Deterrence: Britain and Atomic Energy, 1945–1952*. 2 vols. London: Macmillan, 1974.

————. "What's Science to History, or History to Science?" Inaugural Lecture. Oxford: Clarendon Press, 1975.

Graham, Loren R. *Science and Philosophy in the Soviet Union*. New York: Alfred A. Knopf, 1972.

Grant, Edward, ed. *Nicholas Oresme and the Kinematics of Circular Motion: Tractatus de Commensurabilitate vel Incommensurabilitate Motuum Celi*. Madison: University of Wisconsin Press, 1971.

————. *Sourcebook in Medieval Science*. Cambridge, Mass.: Harvard University Press, 1974.

Grant, Robert. *History of Physical Astronomy from the Earliest Ages to the Middle of the Nineteenth Century*. London, 1852; reprint ed., New York: Johnson, 1966.

Greene, John C. *The Death of Adam. Evolution and Its Impact on Western Thought*. Ames: Iowa State University Press, 1959.

Greenberg, Daniel. *The Politics of Pure Science*. New York: New American Library, 1967.

Gregory, Frederick. *Scientific Materialism in Nineteenth-Century Germany*. Dordrecht, Holland: Reidel, 1977.

Guerlac, Henry. "The History of Science." In *Rapports du IX congrès international des science historiques,* pp. 182–211. Paris, 1950.

———. *Lavoisier: The Crucial Year*. Ithaca, N.Y.: Cornell University Press, 1961.

———. "Some Historical Assumptions of the History of Science." In *Scientific Change,* pp. 797–812. Edited by Alistair C. Crombie. London: Heinemann, 1963; revised reprint in Guerlac, ed., *Essays and Papers in the History of Modern Science,* pp. 27–39 (1977, see next entry).

———, ed. *Essays and Papers in the History of Modern Science*. Baltimore: Johns Hopkins University Press, 1977.

Gunther, Siegmund, and Windelband, W. *Geschichte der Antiken Naturwissenschaft und Philosophie*. Nördlingen, 1888.

Habermas, Jürgen. *Knowledge and Human Interests*. Original German ed., 1968; Boston: Beacon Press, 1971.

Hahn, Roger. *The Anatomy of a Scientific Institution: The Paris Academy of Sciences, 1666–1803*. Berkeley: University of California Press, 1971.

———. "New Directions in the Social History of Science." *Physis* 17 (1975): 205–217.

Hall, A. Rupert. *Ballistics in the Seventeenth Century*. Cambridge: Cambridge University Press, 1952.

———. *The Scientific Revolution, 1500–1800*. London: Longmans Green, 1954.

———. "The Scholar and the Craftsman in the Scientific Revolution." In *Critical Problems in the History of Science,* pp. 3–23. Edited by Marshall Clagett. Madison: University of Wisconsin Press, 1959b.

———. "Merton Revisited." *History of Science* 2 (1963): 1–16.

Hall, A. Rupert, and Hall, M. B. *Unpublished Scientific Papers of Isaac Newton*. Cambridge: Cambridge University Press, 1962.

Hall, Cargill. *Lunar Impact: A History of Project Ranger*. Washington, D.C.: N.A.S.A., 1977.

Hankins, Thomas L. *Jean d'Alembert: Science and the Enlightenment*. Oxford: Clarendon Press, 1970.

Hannaway, Owen. *The Chemists and the Word: The Didactic Origins of Chemistry*. Baltimore: Johns Hopkins University Press, 1975.

Hannequin, Arthur. *Essai critique sur l'hypothèse des atomes dans la science contemporaine*. Paris, 1895.

Hanson, Norwood R. *Patterns of Discovery: An Inquiry into the Conceptual Foundations of Science*. Cambridge: Cambridge University Press, 1958; variously reprinted.

Harrison, J. *The Library of Isaac Newton*. Cambridge: Cambridge University Press, 1978.

Haskell, Thomas L. *The Emergence of Professional Social Science: The American Social Science Association and the Nineteenth-Century Crisis of Authority*. Urbana: University of Illinois Press, 1977.

Heilbron, John L. *H. G. J. Moseley: The Life and Letters of an English Physicist, 1887–1915*. Berkeley: University of California Press, 1974.

————. *Electricity in the 17th and 18th Centuries*. Berkeley: University of California Press, 1979.

Heller, Agost. *Geschichte der Physik*. 2 vols. Stuttgart, 1882–84.

Hesse, Mary B. *Forces and Fields: The Concept of Action at a Distance in the History of Physics*. London: Nelson, 1961.

————. *Models and Analogies in Science*. Notre Dame, Ind.: University of Notre Dame Press, 1966.

————. "Reason and Evaluation in the History of Science." In *Changing Perspectives in the History of Science: Essays in Honour of Joseph Needham*, pp. 127–147. Edited by Mikuláš Teich and Robert Young. London: Heinemann, 1973.

————. *The Structure of Scientific Inference*. London: Macmillan, 1974.

Hessen, Boris. "The Social and Economic Roots of Newton's *Principia*." In *Science at the Crossroads*, pp. 147–212. Reprint edited by P. G. Wersky. London: Cass, 1971. Original, 1931.

Hewlett, Richard G.; and Anderson, O. E. *The New World, 1939–1946*. Vol. 1 of *A History of the United States Atomic Energy Commission*. University Park: Pennsylvania State University Press, 1962.

Hewlett, Richard G., and Duncan, Francis. *Atomic Shield, 1947–1952*. Vol. 2 of *A History of U.S.A.E.C.* University Park: Pennsylvania State University Press, 1969.

Hiebert, Erwin N. *The Impact of Atomic Energy: A History of Responses of Governments, Scientists, and Religious Groups*. Newton, Kansas: Faith and Life Press, 1961.

Hindle, Brooke. *The Pursuit of Science in Revolutionary America, 1735–1789*. Chapel Hill: University of North Carolina Press, 1956.

————, ed. *Early American Science (Selections from Isis)*. New York: Science History Publications, 1976.

Hollinger, David A. "T. S. Kuhn's Theory of Science and Its Implications for History." *American Historical Review* 78 (1973): 370–393.

Holmes, Frederic. *Claude Bernard and Animal Chemistry*. Cambridge, Mass.: Harvard University Press, 1974.

Holton, Gerald. *Thematic Origins of Scientific Thought: Kepler to Einstein*. Cambridge, Mass.: Harvard University Press, 1973.

————. *The Scientific Imagination: Case Studies*. Cambridge: Cambridge University Press, 1978.

Hooykaas, R. *Religion and the Rise of Modern Science.* Edinburgh: Scottish Academic Press, 1972.

Hoppe, E. *Geschichte der Elektrizität.* Leipzig, 1884.

Hornberger, Theodore. "Halliwell-Phillips and the History of Science." *Huntington Library Quarterly* 12 (1949): 391–399.

Horton, Robin. "African Traditional Thought and Western Science." *Africa* 37 (1967): 50–71; 155–187.

————. "Lévy-Bruhl, Durkheim, and the Scientific Revolution." In Horton and Finnegan, eds. *Modes of Thought,* pp. 249–305 (1973; see next entry).

Horton, Robin, and Finnegan, R., eds. *Modes of Thought: Essays on Thinking In Western and Non-Western Societies.* London: Faber and Faber, 1973.

Hoskin, Michael. *William Herschel and the Construction of the Heavens.* London: Oldbourne, 1963.

Hudson, Liam. *Contrary Imaginations: A Psychological Study of the Young Student.* New York: Schocken Books, 1966.

Hufbauer, Karl. "Social Support for Chemistry in Germany in the Eighteenth Century: How and Why Did It Change?" *Historical Studies in the Physical Sciences* 3 (1971): 205–232.

Hughes, H. Stuart. *Consciousness and Society: The Reorientation of European Social Thought, 1890–1930.* New York: Alfred A. Knopf, 1958.

Isis. Directory of Members and Guide to Graduate Study. Philadelphia: History of Science Society, 1980; privately published.

Jacob, J. R. *Robert Boyle and the English Revolution: A Study in Social and Intellectual Change.* New York: Burt Franklin, 1977.

Jacob, Margaret C. *The Newtonians and the English Revolution, 1689–1720.* Ithaca, N.Y.: Cornell University Press, 1976.

Jones, Reginald. *The Wizard War: British Scientific Intelligence, 1939–1945.* New York: Coward, McCann and Geoghegan, 1978.

Joravsky, David. *The Lysenko Affair.* Cambridge, Mass.: Harvard University Press, 1970.

Kargon, Robert H. *Atomism in England from Hariot to Newton.* Oxford: Clarendon Press, 1966.

————. *Science in Victorian Manchester: Enterprise and Expertise.* Manchester: Manchester University Press, 1977.

Kästner, Abraham G. *Geschichte der Mathematik.* 4 vols. Göttingen, 1796–1800.

Kauffman, George B. *Teaching the History of Chemistry: A Symposium.* Budapest: Akadémiai Kiadó, 1971.

Keilin, David. *The History of Cell Respiration and Cytochromes.* Cambridge: Cambridge University Press, 1966.

Keller, Alex. "Zilsel, the Artisans, and the Idea of Progress in the Renaissance." *Journal of the History of Ideas* 11 (1950): 235–240. Reprinted in *Roots of Scientific Thought,* pp. 281–286. Edited by P. Wiener and A. Noland. New York: Basic Books, 1957.

Kevles, Daniel J. *The Physicists: The History of a Scientific Community in Modern America.* New York: Alfred A. Knopf, 1978.

Keynes, John M. "Newton, the Man." In *Newton Tercentenary Celebrations,* pp. 27–34. Cambridge: Cambridge University Press, 1946.

Kistiakowsky, George B. *A Scientist at the White House: The Private Diary of President Eisenhower's Special Assistant for Science and Technology.* Cambridge, Mass.: Harvard University Press, 1976.

Klein, Martin J. *Paul Ehrenfest.* Vol. 1: *The Making of a Theoretical Physicist.* New York: Elsevier, 1970.

Knight, David M. *Natural Science Books in English, 1600–1900.* London: Batsford, 1972.

Kobell, Franz R. von. *Geschichte der Mineralogie von 1650 bis 1860.* Vol. 2 of *Geschichte der Wissenschaften in Deutschland.* Munich, 1864.

Kohler, Robert. "The History of Biochemistry: A Survey." *Journal of the History of Biology* 8 (1975): 275–318.

Kohlstedt, Sally G. *The Formation of the American Scientific Community.* Urbana: University of Illinois Press, 1976.

Kopp, Hermann. *Geschichte der Chemie.* 4 vols. Braunschweig, 1843–1847.

Koyré, Alexandre. *Études galiléennes.* 3 parts. Paris: Hermann, 1939. English tr. Atlantic Highlands, N.J.: Humanities Press, 1978.

———. "Influence of Philosophic Trends on the Formulation of Scientific Theories." *Scientific Monthly* 80 (1955): 107–111.

———. *From the Closed World to the Infinite Universe.* Baltimore: Johns Hopkins University Press, 1957; variously reprinted.

———. *L'Aventure de l'esprit: Mélanges Alexandre Koyré.* 2 vols. Paris: Hermann, 1964.

———. *Metaphysics and Measurement: Essays in Scientific Revolution.* London: Chapman and Hall, 1968.

Kramish, Arnold. *Atomic Energy in the Soviet Union.* Stanford, Calif.: Stanford University Press, 1959.

Kranzberg, Melvin. "The Newest History: Science and Technology." *Science* 136 (1962): 463–469.

Kronick, David A. *A History of Scientific and Technical Periodicals, 1665–1790.* New York: Scarecrow Press, 1962; 2d ed., 1976.

Kuhn, Thomas S. *The Copernican Revolution: Planetary Astronomy in the Development of Western Thought.* Cambridge, Mass.: Harvard University Press, 1957.

———. *The Structure of Scientific Revolutions.* Chicago: University of Chicago Press, 1962; rev. ed., 1970.

———. "The History of Science." In *International Encyclopedia of the Social Sciences.* Edited by David L. Sills. New York: Macmillan, 1968; reprinted in Kuhn, *The Essential Tension,* pp. 105–126 (1977; see below).

———. "The Relations between History and the History of Science." *Daedalus* 100 (1971): 271–304.

———. *The Essential Tension: Selected Studies in Scientific Tradition and Change.* Chicago: University of Chicago Press, 1977.

———. *Black-Body Theory and the Quantum Discontinuity, 1894–1912.* Oxford: Clarendon Press, 1978.

Kuhn, Thomas S.; Forman, P.; Heilbron, J.; and Allen, L. *Sources for the History of Quantum Physics: An Inventory and Report.* Memoirs of the American Philosophical Society, no. 68. Philadelphia: American Philosophical Society, 1967.

Lakatos, Imré. "History of Science and Its Rational Reconstructions." In *The Interaction between Science and Philosophy,* pp. 195–241. Edited by Yehuda Elkana. Atlantic Highlands, N.J.: Humanities Press, 1974.

Lasswitz, K. *Geschichte der Atomistik von Mittelalter bis Newton.* 2 vols. Hamburg, 1890.

Lemaine, Gerard; MacLeod, R.; Mulkay, M.; and Weingart, P., eds. *Perspectives on the Emergence of Scientific Disciplines.* The Hague: Mouton, 1976.

Lenard, Philipp. *Grosse Naturforscher: Eine Geschichte der Naturforschung in Lebensbeschreifungen.* Munich: J. F. Lehmanns, 1929.

Lenz, Harold G. *Zoologie der Alten Griechen und Romer.* Gotha, 1856.

Levere, Trevor H., and Jarrall, R. A., eds. *A Curious Field-Book: Science and Society in Canadian History.* New York: Oxford University Press, 1974.

Lewes, G. H. *Aristotle: A Chapter from the History of Science.* London, 1864.

Libby, Walter. *An Introduction to the History of Science.* Boston: Houghton Mifflin, 1917.

Libes, Antoine. *Histoire philosophique des progrès de la physique.* 2 vols. Paris, 1810–1813.

Lilley, Samuel, ed. "Essays on the Social History of Science." *Centaurus* 3, nos. 1 and 2 (1953).

Limoges, Camille. *La Sélection naturelle: Étude sur la première constitution d'un concept (1837–1859).* Paris: P.U.F., 1970.

Lindberg, David. *Theories of Vision from al-Kindi to Kepler.* Chicago: University of Chicago Press, 1976.

———, ed. *Science in the Middle Ages.* Chicago: University of Chicago Press, 1978.

Lloyd, Geoffrey. *Early Greek Science: Thales to Aristotle.* London: Chatto and Windus, 1970.

———. *Greek Science after Aristotle.* London: Chatto and Windus, 1973.

Lovejoy, Arthur O. *The Great Chain of Being: A Study of the History of an Idea.* Cambridge, Mass.: Harvard University Press, 1936; variously reprinted.

Lukes, Steven. *Émile Durkheim, His Life and Work: A Historical and Critical Study.* New York: Harper & Row, 1972.

Lurie, Edward. *Louis Agassiz: A Life in Science.* Chicago: University of Chicago Press, 1960.

Lyons, Henry. *The Royal Society 1660–1940: A History of Its Administration under Its Charters.* Cambridge: Cambridge University Press, 1944.

Mach, Ernst. *The Science of Mechanics: A Critical and Historical Account of Its Development.* Original German ed., Leipzig, 1883; 3d American from 4th German ed., Chicago: Open Court, 1907.

McKie, Douglas, and Heathcote, N. J. de V. *The Discovery of Specific and Latent Heats.* London: Arnold, 1935; reprint ed., New York: Arno, 1975.

McLachlan, Herbert, ed. *Newton's Theological Manuscripts.* Liverpool: Liverpool University Press, 1950.

MacLeod, Roy M. "The Support of Victorian Science: The Endowment of Research Movement in Great Britain, 1868–1900." *Minerva* 4 (1971): 197–230.

———. "Changing Perspectives in the Social History of Science." In *Science, Technology, and Society: A Cross-Disciplinary Perspective,* pp. 149–195. Edited by Ina Spiegel-Rösing and Derek Price. London: Sage, 1977.

McMullin, Ernan, ed. *The Concept of Matter.* Notre Dame, Ind.: University of Notre Dame Press, 1963.

———. *The Concept of Matter in Modern Philosophy.* Notre Dame, Ind.: University of Notre Dame Press, 1977.

Maeyama, Y., and Saltzer, W. C., eds. *Prismata: Naturwissenschaftsgeschichtliche Studien, Festschrift für Willy Hartner.* Wiesbaden: Steiner, 1977.

Mahoney, Michael S. *The Mathematical Career of Pierre de Fermat (1601–1665).* Princeton: Princeton University Press, 1973.

Maier, Anneliese. *Studien zur Naturphilosophie der Spätscholastik.* 5 vols. Rome: Edizioni di storia e letteratura, 1949–1958.

Manuel, Frank. *Isaac Newton, Historian.* Cambridge, Mass.: Harvard University Press, 1963.

———. *A Portrait of Isaac Newton.* Cambridge, Mass.: Harvard University Press, 1968.

Marcuse, Herbert. *One-Dimensional Man: Studies in the Ideology of Advanced Industrial Society.* Boston: Beacon Press, 1964.

Marx, Karl. *The Poverty of Philosophy.* Written 1847; reprint ed., Moscow: Foreign Languages Publishing House, n.d.

Mason, Stephen F. *A History of the Sciences: Main Currents of Scientific Thought.* London: Routledge, 1953; some subsequent and American editions under different titles.

Mathias, Peter. "Who Unbound Prometheus?" In *Science and Society, 1600–1900*, pp. 54–80. Edited by Peter Mathias. Cambridge: Cambridge University Press, 1973.

Menard, Henry W. *Science: Growth and Change.* Cambridge, Mass.: Harvard University Press, 1971.

Mendelsohn, Everett. *Heat and Life: The Development of the Theory of Animal Heat.* Cambridge, Mass.: Harvard University Press, 1964.

Mendelsohn, Everett; Weingart, P.; and Whitley, R., eds. *The Social Production of Scientific Knowledge.* Dordrecht: Reidel, 1977.

Merton, Robert K. *Science, Technology, and Society in Seventeenth-Century England. Osiris* 4 (1938a): 360–632; reprinted with new introduction, New York: Howard Fertig, 1970; Humanities Press, 1978.

———. "Science and the Social Order." *Philosophy of Science* 5 (1938b): 321–337; reprinted in Merton, 1973, below.

———. "Science and Technology in a Democratic Order." *Journal of Legal and Political Sociology* 1 (1942): 115–126; reprinted in Merton, 1973, below.

———. "Paradigm for the Sociology of Knowledge." In *Twentieth Century Sociology,* pp. 336–405. Edited by George Gurvitch and Wilbert E. Moore. New York: Philosophical Library, 1945; reprinted in Merton, 1973, below.

———. *The Sociology of Science: Theoretical and Empirical Investigations.* Edited by Norman Storer. Chicago: University of Chicago Press, 1973.

———. "The Sociology of Science: An Episodic Memoir." In *The Sociology of Science in Europe,* pp. 3–141. Edited by R. Merton and J. Gaston. Carbondale: Southern Illinois University Press, 1977.

Merz, John T. *A History of European Thought in the Nineteenth Century.* Part I: *Scientific Thought.* 2 vols. London: Blackwood, 1896; reprint ed., New York: Dover Publications, 1965.

Metzger, Hélène. *Les Doctrines chimiques en France.* Paris: P.U.F., 1923.

———. *Attraction universelle et religion naturelle chez quelques commentateurs anglais de Newton.* 3 parts. Paris: Hermann, 1938.

Meyerson, Emile. *Identity and Reality.* Original French ed. 1908; London: Allen and Unwin, 1930; variously reprinted.

———. *Du Cheminement de la pensée.* 3 vols. Paris: Alcan, 1931.

Mieli, Aldo. *La Science arabe et son rôle dans l'évolution scientifique mondiale.* Leiden: Brill, 1938.

Miller, Howard S. *Dollars for Research: Science and Its Patrons in Nineteenth-Century America.* Seattle: University of Washington Press, 1970.

Miller, Samuel. *A Brief Retrospect of the Eighteenth Century: Containing a Sketch of the Revolutions and Improvements in Science.* 2 vols. New York: 1803.

Mitroff, Ian I. *The Subjective Side of Science.* New York: Elsevier, 1974.

Montagu, M. F. Ashley. *Studies and Essays in the History of Science and*

Learning, Offered in Homage to George Sarton. New York: Abelard-Schuman, 1944.

Montucla, Jean E. *Histoire des mathématiques: Jusqu'à nos jours.* 4 vols. Original ed. of vols. 1 and 2, Paris, 1758; Paris, 1799–1802.

Moore, Forris J. *A History of Chemistry.* New York: McGraw Hill, 1918; 3d ed., 1939.

Morell, Thomas. *Elements of the History of Philosophy and Science from the Earliest Authentic Records to the Commencement of the Eighteenth Century.* London, 1827.

Morrell, Jack B. "The Chemist Breeders: The Research Schools of Liebig and Thomas Thomson." *Ambix* 19 (1972): 1–46.

——. *A List of Theses in History of Science in British Universities in Progress or Recently Completed.* British Society for the History of Science, 1978; (privately published).

Mottelay, Paul F. *Bibliographical History of Electricity and Magnetism.* London: Griffin, 1922; reprint ed., New York: Arno Press, 1975.

Moyal, Ann Mozley, ed. *Scientists in Nineteenth-Century Australia: A Documentary History.* Melbourne: Cassell Australia, 1975.

Multhauf, Robert P. *Neptune's Gift: A History of Common Salt.* Baltimore: Johns Hopkins University Press, 1978.

Murdoch, John E., and Sylla, E. D., eds. *The Cultural Context of Medieval Learning.* Boston Studies in the Philosophy of Science, vol. 26. Dordrecht: Reidel, 1975.

Musson, A. E., and Robinson, E. *Science and Technology in the Industrial Revolution.* Manchester: Manchester University Press, 1969.

Nakayama, Shigeru; Swain, D. L.; and Eri, Y., eds. *Science and Society in Modern Japan: Selected Historical Sources.* Cambridge, Mass.: M.I.T. Press, 1974.

Nakayama, Shigeru, and Sivin, Nathan, eds. *Chinese Science: Explorations of an Ancient Tradition.* Cambridge, Mass.: M.I.T. Press, 1973.

Nasr, Seyyed H. *Science and Civilization in Islam.* Cambridge, Mass.: Harvard University Press, 1968.

Nauka. *Institute of the History of Science and Technology: U.S.S.R. Academy of Sciences.* Moscow: Central Department of Oriental Literature, Nauka Publishing House, 1977.

Needham, Joseph. *Science and Civilization in China.* 7 vols. Cambridge: Cambridge University Press, 1954–

——. *The Shorter Science and Civilization in China.* Abridgement by Colin A. Ronan. Cambridge: Cambridge University Press, 1978–

——, ed. *Science, Religion, and Reality.* London: Sheldon Press, 1925.

Needham, Joseph, and Pagel, W., eds. *Background to Modern Science.* Cambridge: Cambridge University Press, 1938.

Neugebauer, Otto. *The Exact Sciences in Antiquity.* Princeton: Princeton

University Press, 1952; 2d ed., augmented, Providence, R.I.: Brown University Press, 1957.

———. *A History of Ancient Mathematical Astronomy*. 3 vols. New York: Springer, 1975.

Noble, David F. *America by Design: Science, Technology, and the Rise of Corporate Capitalism*. New York: Alfred A. Knopf, 1977.

North, John D. *Richard of Wallingford*. 3 vols. Oxford: Clarendon Press, 1976.

Olby, Robert. *The Path to the Double Helix*. Seattle: University of Washington Press, 1974.

Oleson, Alexandra. *Knowledge in American Society, 1860–1920*. Baltimore: Johns Hopkins University Press, 1978.

Oleson, Alexandra, and Brown, Sanborn C., eds. *The Pursuit of Knowledge in the Early American Republic: American Scientific and Learned Societies*. Baltimore: Johns Hopkins University Press, 1976.

Olson, Richard. *Science As Metaphor*. Belmont, Calif.: Wadsworth, 1971.

Ornstein, Martha. *The Role of Scientific Societies in the Seventeenth Century*. Chicago: University of Chicago Press, 1913; 2d ed., 1928; reprint ed., New York: Arno Press, 1975.

Ostwald, F. Wilhelm. *Elektrochemie. Ihre Geschichte und Lehre*. Leipzig, 1896.

Pagel, Walter. *Paracelsus: An Introduction to Philosophical Medicine in the Era of the Renaissance*. Basel, Switzerland: Karger, 1958.

———. *William Harvey's Biological Ideas*. Basel, Switzerland: Karger, 1967.

Palter, Robert M. *The Annus Mirabilis of Sir Isaac Newton, 1666–1966*. Cambridge, Mass.: M.I.T. Press, 1970.

Partington, James R. *A History of Chemistry*. 4 vols. London: Macmillan, 1961–1970.

Paul, Harry W. *The Sorcerer's Apprentice: The French Scientist's Image of German Science, 1840–1919*. Gainesville: University of Florida Press, 1972.

———. "Scholarship versus Ideology: The Chair of the General History of Science at the Collège de France." *Isis* 67 (1976): 376–397.

Pedersen, Olaf, and Pihl, M. *Early Physics and Astronomy: A Historical Introduction*. New York: Elsevier, 1974.

Poggendorff, Johann C. *Biographisch-Literarisches Handwörterbuch zur Geschichte der Exacten Wissenschaften*. 2 vols. Leipzig, 1863, and later continuations; ongoing.

Poincaré, Henri. *La Science et l'hypothèse*. Paris: Flammarion, 1906.

Poincaré, Lucien. *La Physique moderne, son évolution*. Paris: Flammarion, 1920.

Porter, Roy. *The Making of Geology: Earth Science in Britain 1660–1815*. Cambridge: Cambridge University Press, 1977.

Powell, Baden. *An Historical View of the Progress of the Physical and Mathematical Sciences from the Earliest Ages to the Present Times.* London, 1834; new ed., Cabinet Cyclopedia, 1837.

Price, Derek J. de Solla. *Science since Babylon.* New Haven: Yale University Press, 1961; 2d ed., 1975.

———. *Little Science, Big Science.* New York: Columbia University Press, 1963.

———. "Is Technology Historically Independent of Science? A Study in Statistical Historiography." *Technology and Culture* 4 (1965): 553–568.

———. "A Guide to Graduate Study and Research in the History of Science and Medicine." *Isis* 58 (1967): 385–395.

———. "Who's Who in the History of Science: A Survey of a Profession." *Technology and Society* 5 (1969): 52–55.

Price, Don K. *The Scientific Estate.* Cambridge, Mass.: Harvard University Press, 1965.

Priestley, Joseph. *The History and Present State of Electricity with Original Experiments.* London, 1767.

———. *The History and Present State of Discoveries Relating to Vision, Light, and Colours.* London, 1772.

Provine, William B. *Origins of Theoretical Population Genetics.* Chicago: University of Chicago Press, 1971.

Pyenson, Lewis. "The Incomplete Transmission of a European Image: Physics at Greater Buenos Aires and Montreal, 1890–1920." *Proceedings of the American Philosophical Society* 122 (1978): 92–114.

Rattansi, Piyo. "Paracelsus and the Puritan Tradition." *Ambix* 11 (1963): 24–32.

———. "Some Evaluations of Reason in Sixteenth- and Seventeenth-Century Natural Philosophy." In *Changing Perspectives in the History of Science,* pp. 148–166. Edited by Mikuláš Teich and Robert Young. London: Heinemann, 1973.

Raven, Charles E. *John Ray, Naturalist.* Cambridge: Cambridge University Press, 1942.

Ravetz, Jerome R. *Scientific Knowledge and Its Social Problems.* Oxford: Clarendon Press, 1971.

———. "Science, History of." In *Encyclopaedia Britannica,* 15th ed., 1974.

Reingold, Nathan. "American Indifference to Basic Research: A Reappraisal." In *Nineteenth-Century American Science: A Reappraisal,* pp. 38–62. Edited by George M. Daniels. Evanston, Ill.: Northwestern University Press, 1972.

———. *Science in America Since 1820 (Selections from Isis).* New York: Science History Publications, 1976.

———, ed. *Science in Nineteenth-Century America: A Documentary History.* New York: Hill and Wang, 1964.

———, ed. *The Papers of Joseph Henry.* 15 vols. Washington, D.C.: Smithsonian Institution Press, 1972–

Rieff, Phillip. *Freud: The Mind of the Moralist.* New York: Viking, 1959.

Rigaud, Stephen P. *Historical Essay on the First Publication of Sir Isaac Newton's "Principia."* Oxford, 1838.

———. *Correspondence of Scientific Men of the Seventeenth Century.* 2 vols. Oxford, 1841.

Roe, Anne. *The Making of a Scientist.* New York: Dodd, Mead, 1953.

Roger, Jacques. *Les Sciences de la vie dans la pensée francaise du XVIIIe siècle.* Paris: Colin, 1963.

Rosenberg, Charles E. *No Other Gods: Science and American Social Thought.* Baltimore: Johns Hopkins University Press, 1976.

Rosenberger, Ferdinand. *Die Geschichte der Physik.* 3 vols. Braunschweig, 1882–1890.

———. *Isaac Newton und Seine Physicalischen Prinzipien.* Leipzig, 1895.

Ross, Dorothy. *G. Stanley Hall: The Psychologist as Prophet.* Chicago: University of Chicago Press, 1972.

Ross, Sydney. *"Scientist: The Story of a Word." Annals of Science* 18 (1962): 65–85.

Rossi, Paolo. *Francis Bacon: From Magic to Science.* Bari, Italy, 1957; English tr., Chicago: University of Chicago Press, 1967.

Rossiter, Margaret. *The Emergence of Agricultural Science: Justus Liebig and the Americans, 1840–1880.* New Haven: Yale University Press, 1975.

Rudwick, Martin J. S. *The Meaning of Fossils: Episodes in the History of Palaeontology.* London: Macdonald, 1972.

———. "The History of the Natural Sciences as Cultural History." Inaugural lecture. Amsterdam: The Free University, 1975.

Russell, Colin, ed. *Science and Religious Belief: A Selection of Recent Historical Studies.* London: Open University Press, 1973.

Russell, Colin; Coley, N. G.; and Roberts, G. K. *Chemists by Profession: The Origins and Rise of the Royal Institute of Chemistry.* London: Open University Press, 1977.

Russo, François. *Éléments de bibliographie de l'histoire des sciences et techniques.* Paris: Hermann, 1954; 2d ed., 1969.

Sabra, A. I. *Theories of Light from Descartes to Newton.* London: Oldbourne, 1967.

Sachs, Julius von. *Geschichte der Botanik von 16 Jahrhundert bis 1860.* Vol. 15 of *Geschichte der Wissenschaften in Deutschland.* Munich, 1875; English tr., Oxford: Clarendon Press, 1890; reprint ed., 1906.

Sanderson, Michael. *The Universities and British Industry, 1850–1970.* London: Routledge, 1972.

Sarton, George. *An Introduction to the History of Science.* 3 vols. in 5 parts. Baltimore: Williams and Wilkins, 1927–1948.

————. "Montucla (1725–1799): His Life and Works." *Osiris* 1 (1936): 519-567.

————. "Paul, Jules, and Marie Tannery." *Isis* 38 (1947): 33–50.

————. *Horus: A Guide to the History of Science.* Waltham, Mass.: Chronica Botanica, 1952.

Schmitt, Charles. *Critical Survey and Bibliography of Studies on Renaissance Aristotelianism, 1958–1969.* Padua: Antenore, 1971.

Schofield, Robert E. *The Lunar Society of Birmingham: A Social History of Provincial Science and Industry in Eighteenth-Century England.* Oxford: Clarendon Press, 1963.

Schorlemmer, Carl. *The Rise and Development of Organic Chemistry.* Manchester, 1879; French ed., 1885; German ed., 1889; 2d English ed., London, 1894.

Schrock, Robert R. *Geology at M.I.T., 1865–1965.* Vol. 1: *The Faculty and Supporting Staff.* Cambridge, Mass.: M.I.T. Press, 1977.

Schuster, Arthur. *The Progress of Physics during Thirty-Three Years.* Cambridge: Cambridge University Press, 1911.

Sedgwick, William T., and Tyler, H. *A Short History of Science.* New York: Macmillan, 1917; 2d ed., 1939.

Shapin, Steven. "Phrenological Knowledge and the Social Structure of Early Nineteenth-Century Edinburgh." *Annals of Science* 32 (1975): 219–243.

Shapin, Steven, and Thackray, A. "Prosopography as a Research Tool in the History of Science." *History of Science* 12 (1974): 1–28.

Sharlin, Harold I., ed. *Report on Undergraduate Education in the History of Science.* History of Science Society, 1975; privately published.

Shils, Edward. *The Intellectuals and the Powers and Other Essays.* Chicago: University of Chicago Press, 1972.

Shryock, Richard. *The Development of Modern Medicine: An Interpretation of the Social and Scientific Factors Involved.* Philadelphia: University of Pennsylvania Press, 1936.

————. "American Indifference to Basic Research in the Nineteenth Century." In *The Sociology of Science,* pp. 98–110. Edited by Bernard Barber and W. Hirsch. New York: Free Press, 1962; reprinted from *Archives Internationales d'Histoire des Sciences,* 28 (1948–1949): 3–18.

————. "Conference on the History, Philosophy, and Sociology of Science." *Proceedings of the American Philosophical Society* 99 (1955): 327–354.

Sinclair, Bruce. *Philadelphia's Philosopher Mechanics: A History of the Franklin Institute, 1824–1865.* Baltimore: Johns Hopkins University Press, 1974.

Sivin, Nathan. *Chinese Alchemy: Preliminary Studies.* Cambridge, Mass.: Harvard University Press, 1968.

Sivin, Nathan, ed. *Science and Technology in East Asia (Selections from Isis).* New York: Science History Publications, 1977.

Smith, Alice K. *A Peril and a Hope: The Scientists' Movement in America*

1945–47. Chicago: University of Chicago Press, 1965; rev. ed., Cambridge, Mass.: M.I.T. Press, 1970.

Smith, David E. *History of Modern Mathematics*. New York: John Wiley, 1906.

Smith, Edgar F. *Chemistry in America: Chapters from the History of the Sciences in the United States*. New York: Appleton, 1914.

Smith, Preserved. *A History of Modern Culture*. 2 vols. New York: Holt, 1930–1934.

Snow, A. J. *Matter and Gravity in Newton's Physical Philosophy*. London: Oxford University Press, 1926.

Southern, R. W. "Science and Technology in the Middle Ages: Commentary." In *Scientific Change*, pp. 301–306. Edited by Alistair C. Crombie. London: Heinemann, 1963.

Spiegel-Rösing, Ina, and Price, Derek de Solla, eds. *Science, Technology, and Society: A Cross-Disciplinary Perspective*. London: Sage, 1977.

Sprat, Thomas. *The History of the Royal Society of London for the Improving of Natural Knowledge*. Original ed., 1667; reprint ed., London: Routledge, 1958.

Stahl, William H. *Roman Science: Origins, Development, and Influence in the Later Middle Ages*. Madison: University of Wisconsin Press, 1962.

Steffens, Henry. *The Development of Newtonian Optics in England*. New York: Science History Publications, 1977.

Stepan, Nancy. *Beginnings of Brazilian Science: Oswaldo Cruz, Medical Research, and Policy, 1890–1920*. New York: Science History Publications, 1976.

Stern, Phillip M. *The Oppenheimer Case: Security on Trial*. New York: Harper & Row, 1969.

Stevenson, Lloyd G., and Multhauf, R. P., eds. *Medicine, Science, and Culture: Historical Essays in Honor of Owsei Temkin*. Baltimore: Johns Hopkins University Press, 1968.

Stewart, Dugald; Mackintosh, J.; Playfair, J.; and Leslie, J. *Dissertations on the History of Metaphysical and Ethical, and of Mathematical and Physical Science*. Edinburgh, 1835. These essays were originally published over the years 1815–1835; they were reprinted in 1842 as a volume of "preliminary discourses" that served as introduction to the 7th edition of the *Encyclopaedia Britannica*.

Stimson, Dorothy, ed. *Sarton on the History of Science*. Cambridge, Mass.: Harvard University Press, 1962.

Stimson, Dorothy; Guerlac, H.; Boas, M.; Cohen, I. B.; and Roller, D. H. Papers and commentaries on teaching the history of science. In *Critical Problems in the History of Science*, pp. 223–253. Edited by Marshall Clagett. Madison: University of Wisconsin Press, 1959.

Stocking, George. *Race, Culture, and Evolution: Essays in the History of Anthropology*. New York: Free Press, 1968.

Strube, Wilhelm. *Die Chemie und Ihre Geschichte: Forschungen zur Wirtschaftsgeschichte.* Vol. 5. Edited by Jurgen Kuczynski and Hans Mottek. Berlin: Akademie Verlag, 1974.

Struik, Dirk J. *Yankee Science in the Making.* Boston: Little Brown, 1948; new ed., New York: Collier, 1962.

Stubbe, Hans. *Kurze Geschichte der Genetik.* Jena: Fischer, 1963; rev. ed., 1965; English tr., Cambridge, Mass.: M.I.T. Press, 1972.

Stuewer, Roger H., ed. *Historical and Philosophical Perspectives of Science.* Minnesota Studies in the Philosophy of Science, vol. 5. Minneapolis: University of Minnesota Press, 1970.

Sulloway, Frank J. *Freud, Biologist of the Mind: Beyond the Psychoanalytic Legend.* New York: Basic Books, 1979.

Tannery, Paul. *Pour l'Histoire de la science héllène.* Paris, 1887.

————. *Mémoires scientifiques.* 17 vols. Paris: Gauthier-Villars, 1912–1950.

Taton, René, ed. *A General History of the Sciences.* 4 vols. Original French ed., Paris: P.U.F., 1957–1964; London: Thames and Hudson, 1963–1966.

————, ed. *Enseignement et diffusion des sciences en France au XVIIIe siècle.* Paris: Hermann, 1964.

Teich, Mikuláš, and Young, Robert, eds. *Changing Perspectives in the History of Science: Essays in Honour of Joseph Needham.* London: Heinemann, 1973.

Thackray, Arnold. *Atoms and Powers: An Essay on Newtonian Matter-Theory and the Development of Chemistry.* Cambridge, Mass.: Harvard University Press, 1970a.

————. "Science: Has Its Present Past a Future?" In *Historical and Philosophical Perspectives of Science,* pp. 112–127. Edited by Roger Stuewer. Minneapolis: University of Minnesota Press, 1970b.

————. *John Dalton: Critical Assessments of His Life and Science.* Cambridge, Mass.: Harvard University Press, 1972.

————. "Natural Knowledge in Cultural Context: The Manchester Model." *American Historical Review* 79 (1974): 672–709.

————. "Measurement in the Historiography of Science." In *Toward a Metric of Science,* pp. 11–30. Edited by Yehuda Elkana et al. New York: John Wiley, 1978.

Thackray, Arnold, and Merton, Robert K. "On Discipline Building: The Paradoxes of George Sarton." *Isis* 63 (1972): 473–495.

Thomson, Thomas. *The History of Chemistry.* 2 vols. London, 1830–1831.

Thorndike, Lynn. *A History of Magic and Experimental Science.* 8 vols. New York: Columbia University Press, 1923–1958.

Thornton, J. L., and Tully, R. I. J. *Scientific Books, Libraries, and Collectors.* London: Library Association, 1954; 3d rev. ed., 1972.

Toulmin, Stephen. "From Form to Function: Philosophy and History of Science in the 1950s and Now." *Daedalus* 106 (1977): 143–162.

_____. *Human Understanding.* Vol. 1: *General Introduction and Part 1.* Oxford: Clarendon Press, 1972.

Trommsdorff, Johann B. *Versuch einer allgemeinen Geschichte der Chemie.* 3 parts. Erfurt, 1806.

Turnbull, H. W.; Scott, J. F.; Hall, A. R.; and Tilling, L., eds. *The Correspondence of Isaac Newton.* 7 vols. Cambridge: Cambridge University Press, 1959–1978.

Turner, Frank M. *Between Science and Religion: The Reaction to Scientific Naturalism in Late Victorian England.* New Haven: Yale University Press, 1974.

Tylecote, Richard F. *A History of Metallurgy.* London: The Metals Society, 1976.

Underwood, E. Ashworth. *Science, Medicine and History: Essays in Honour of Charles Singer.* 2 vols. London: Oxford University Press, 1953; reprint ed., New York: Arno Press, 1975.

Veblen, Thorstein. "The Place of Science in Modern Civilization." *American Journal of Sociology* 11 (1906): 585–609; reprinted in *The Place of Science in Modern Civilization and Other Essays,* pp. 1–31. New York: Huebsch, 1919; variously reprinted.

Venable, F. P. *The Development of the Periodic Law.* Easton, Pa., 1896.

Vucinich, Alexander. *Science in Russian Culture.* 2 vols. Stanford, Calif.: Stanford University Press, 1963–1970.

Wade, Nicholas. "Thomas S. Kuhn: Revolutionary Theorist of Science." *Science* 179 (1977): 143–145.

Wallis, Peter, and Wallis, R. *Newton and Newtonia, 1672–1975: A Bibliography.* Folkestone: Dawson, 1977.

Watson, James D. *The Double Helix. A Personal Account of the Discovery of the Structure of DNA.* New York: Athenaeum, 1968; 2d ed., 1979.

Weber, Max. "Die Protestantische Ethik und der Geist des Kapitalismus." *Archiv für Sozialwissenschaft und Sozialpolitik,* 20 (1904–5): 1–54; 21 (1905): 1–110. Tr. by T. Parsons. New York: Scribners, 1930.

Webster, Charles. *The Great Instauration: Science, Medicine, and Reform, 1626–1660.* London: Duckworth, 1975.

Weiner, Charles. "Resource Centers and Programs for the History of Physics." In *History in the Teaching of Physics,* pp. 47–54. Edited by Stephen G. Brush and A. L. King. Hanover, N.H.: New England University Press, 1972.

Weld, Charles R. *A History of the Royal Society: Compiled from Authentic Documents.* 2 vols. London, 1848; reprint ed. New York: Arno Press, 1975.

Werskey, Paul G. *The Visible College: The Collective Biography of British Scientific Socialists of the 1930s.* New York: Holt, Rinehart, and Winston, 1979.

Westfall, Richard S. *Science and Religion in Seventeenth-Century England.*

New Haven: Yale University Press, 1958; reprint ed., Garden City, N.Y.: Doubleday Anchor, 1970.

————. *Force in Newton's Physics: The Science of Dynamics in the Seventeenth Century.* London: MacDonald, 1971.

————. *The Construction of Modern Science: Mechanisms and Mechanics.* New York: John Wiley, 1971.

————. "The Changing World of the Newtonian Industry." *Journal of the History of Ideas* 37 (1976): 175–184.

Westfall, Richard S., and Thoren, V. E. *Steps in the Scientific Tradition.* New York: John Wiley, 1968.

Westman, Robert S., ed. *The Copernican Achievement.* Berkeley: University of California Press, 1976.

Weyer, Jost. *Chemiegeschichtsschreibung von Wiegleb (1790) bis Partington (1970): Eine Untersuchung über ihre Methoden, Prinzipien und Ziele.* Arbor Scientiarum: Beiträge zur Wissenschaftsgeschichte, Reihe A: Abhandlungen, Bd, III. Hildesheim: Verlag Gerstenberg, 1974.

Whewell, William. *History of the Inductive Sciences: From the Earliest to the Present Time.* 3 vols. London, 1837; reprint ed., London: Cass, 1967.

White, Andrew Dickson. *A History of the Warfare of Science with Theology in Christendom.* 2 vols. New York, 1896; reprint ed., New York: Dover, 1960.

Whitehead, Alfred North. *Science and the Modern World.* New York: Macmillan, 1925; variously reprinted.

Whiteside, Derek T., ed. *The Mathematical Papers of Isaac Newton.* 7 vols. Cambridge: Cambridge University Press, 1967–1980.

Whittaker, Edmund T. *A History of the Theories of Aether and Electricity: From the Age of Descartes to the Close of the Nineteenth Century.* London: Nelson, 1910; rev. and enl. ed., New York: Harper and Brothers, 1951.

Wiener, Philip P., and Noland, Aaron, eds. *Roots of Scientific Thought: A Cultural Perspective.* New York: Basic Books, 1957.

Williams, L. Pearce. *Michael Faraday: A Biography.* New York: Basic Books, 1965.

————. *The Origins of Field Theory.* New York: Random House, 1966.

Williams, Trevor I. *A Biographical Dictionary of Scientists.* New York: Wiley-Interscience, 1969.

Wilson, Leonard G. *Charles Lyell, the Years to 1841: The Revolution in Geology.* New Haven: Yale University Press, 1972.

Winsor, Mary P. *Starfish, Jellyfish, and the Order of Life: Issues in Nineteenth-Century Science.* New Haven: Yale University Press, 1976.

Wohlwill, Emil. *Galilei und Sein Kampf für die Copernicanische Lehre.* 2 vols. Hamburg and Leipzig: Voss, 1909.

Wolf, Abraham. *A History of Science, Technology, and Philosophy in the Sixteenth and Seventeenth Centuries.* New York: Macmillan, 1935.

———. *A History of Science, Technology, and Philosophy in the Eighteenth Century.* New York: Macmillan, 1939.

Woolf, Harry. *The Transits of Venus: A Study of Eighteenth-Century Science.* Princeton: Princeton University Press, 1959.

Wurtz, Charles A. *Histoire des doctrines chimiques depuis Lavoisier jusqu'à nos jours.* Paris: Hachette, 1869; English tr., London: Macmillan, 1869.

Yates, Frances A. *Giordano Bruno and the Hermetic Tradition.* Chicago: University of Chicago Press, 1964.

———. *The Rosicrucian Enlightenment.* Boston: Routledge, 1972.

York, Herbert. *The Advisors: Oppenheimer, Teller, and the Superbomb.* San Francisco, Calif.: Freeman, 1975.

Young, Robert M. "Scholarship and the History of the Behavioral Sciences." *History of Science* 5 (1966): 1–51.

———. *Mind, Brain, and Adaptation in the Nineteenth Century.* Oxford: Clarendon Press, 1970.

———. "The Historiographic and Ideological Contexts." In *Changing Perspectives in the History of Science,* pp. 344–438. Edited by Mikuláš Teich and Robert Young. London: Heinemann, 1973.

Zeuthen, Hieronymus G. *Geschichte der Mathematik in Altertum und Mittelalter.* Danish ed., 1893; Copenhagen, 1896.

Zilsel, Edgar. "The Sociological Roots of Science." *American Journal of Sociology* 47 (1942): 544–562.

Zittel, Karl A. von. *Geschichte der Geologie und Palaeontologie bis Ende des 19. Jahrhunderts.* Munich, 1899.

Zuckerman, Harriet. *Scientific Elite: Nobel Laureates in the United States.* New York: Free Press, 1977.

Chapter 2
History of Technology

Carroll W. Pursell, Jr. UNIVERSITY OF CALIFORNIA
AT SANTA BARBARA

INTRODUCTION

A. History of Technology: Origin and Development

The history of technology, as it is practiced today, had its origins in and still encompasses several diverse types of study. Engineers, historians of science, economic historians, general social historians, and students of American studies have all found occasion to delve into the question of technology, each for their own purposes and from their own special perspectives. At their best, these studies remain valuable both to the communities of scholars who produced them and to the newer specialist, the historian of technology.

History is natural to the inventive enterprise. Modern patent law, as it was pioneered in the United States, places a predominant emphasis upon priority of discovery (analogous, of course, to the process in science). Since only the first discoverer could be awarded a patent for any device, the establishment of priority through what was essentially the amassing of an historical case study was critical to the hopeful inventor. Disputed priorities and the litigation that often resulted are historical arguments, and as such, with their attendant testimony, depositions, and other evidence, they still represent mines of data for the professional historian of technology. Added to this necessary exercise are the charm and utility of accounts of inventions as key points in the complex systems that populate the technological landscape. The drama of the significant invention and the heroic inventor, though lacking in analytic power, has provided a stirring and sometimes useful prehistory of the field. Related to these efforts is the pious attempt (again common to all fields) of practitioners to use history to legitimize their profession,

70

to celebrate a rapidly passing Golden Age, or to put down on paper the oral tradition of participants and survivors. Such piety on the part of engineers, for example, has served both to add to the body of protohistory of technology and to provide a mass of data for future studies of engineering ideology.

The historian of science, though sometimes springing from roots similar to those described above (see also chapter 1, I.A., in this volume), turns to technology not from piety or hope of gain but out of the necessity of treating an important, if tangential, branch of learning. The modern popular confusion between science and technology, together with the thesis that technology is merely applied science, justifies the inclusion of history of technology under the larger banner of history of science. Like alchemy and other "pseudosciences," technology has long been a special category in the outlines and bibliographies of the history of science. Such classic surveys as those of Wolf (1938) and Forbes and Dijksterhuis (1963) include "science and technology" as a part of their titles, and a standard early work, Struik (1948), though not mentioning technology in the title, nevertheless treats it freely in the text. Pioneering undergraduate courses often carried both words in their titles and syllabi. Such books form an important beginning of study in the field, and although technology remains secondary to their primary focus, and though "technology" often connotes only the part of technology that has been closely related to science, the basic work itself remains usually reliable and often brilliant (see chapter 1, II. H., in this volume).

No category of scholars has given more attention to the history of technology than the economic historian. Technology (or technological knowledge) is clearly a factor of production and can no more be ignored in a comprehensive study than can capital, labor, or raw materials. The massive interest in the British Industrial Revolution—which dates perhaps from Arnold Toynbee's influential *Lectures on the Industrial Revolution in England* (1884)—brought technology to the fore. Such works as Usher's *History of Mechanical Inventions* (1929) and Hunter's *Steamboats on the Western Rivers* (1949) are only signal examples of the realization that, although engineering may be more than mere science modified by economics, technology has had a profound influence on economics. It is by this same route that Marxist historians have been led to investigate the machine. Economics remains the "bottom line," but much of what is written above it is technology.

Another group that has made valuable contributions to the history of technology over the years is the old style (to differentiate them from the newer demographic and cliometric) social historians. Arthur Schlesinger, Sr.'s lengthy series of volumes on the history of American life explored the history of Americans when they were not engaged in politics, war, or diplomacy; along with dress and housing, food and entertainment, technology found an important and congenial forum. Some currently practicing historians of science trace their own intellectual heritage back to this style of social (and often anonymous) history. Sinclair's study of the Franklin Institute (1974) and Pursell's study of the stationary steam engine in America (1969) are examples of books drawing heavily on this tradition.

Finally, the field of American studies, though itself a relatively young discipline, has contributed significantly to the literature now defining the history of technology. By definition interdisciplinary (though often dominated by students of literature), American studies sought to integrate all facets of American life into a study of the national character. It was inevitable that so signal an aspect of American life as technology would demand inclusion in this synthesis. Such landmark books as Leo Marx (1964) on the pastoral image and technology, and the recent study by Kasson (1976) on technology and republican virtue, span an academic generation of contributions that have greatly enriched our understanding of both technology and its cultural milieu in America.

These are the major professional communities that have interested themselves in the writing of technological history. It remains to be pointed out that this, like all kinds of history, carries profound and usually implicit political and ideological responsibilities. The old battle between the Ancients and the Moderns hinged on an attempt to undermine the supposed superiority of classical civilizations and to demonstrate, instead, the progressive nature of modern times. Down to the present day, those who fight this battle have had recourse to the technological record. Against the ancient philosophers one can place modern inventors; against antique arts, a host of modern machines and devices; against the conspicuous wealth, leisure, and culture of a few, the widespread material possessions and satisfactions of the many in our own time. No example has so faithfully and effectively served the Idea of Progress as that of technology. The host of celebrants of the progressive nature of modern civilization have created a literature and an attitude that undergird the newer professional study of the history of technology.

Although occasional books that can be styled histories of technology date back to the Industrial Revolution itself, the field as such was undeveloped, and hardly even defined as such, until the twentieth century. Then, in 1909, Conrad Matschoss founded his annual *Beiträge zur Geschichte der Technik und Industrie* which, after its twenty-first volume became the quarterly *Technikgeschichte—Beiträge zur Geschichte der Technik und Industrie*. A similar effort, this time supported by a society, was launched in Great Britain a few years later.

In 1919 a group of British scholars, who included Rhys Jenkins and H. W. Dickinson—the Boulton and Watt scholar—and some of whom were associated with the Science Museum in London, met to form the Newcomen Society for the Study of the History of Engineering and Technology. The first volume of their famous *Transactions,* dated 1920–1921, defined the group's interests: the technology of industry and transportation in the British Isles. Led by Dickinson, who remained editor of the *Transactions* from 1920 until his death in 1952, this group made its range of interests a special branch of study, one might almost say a new academic discipline, and laid down what became the traditional way of presenting the subject. In his first presidential address to the Society, the engineer Arthur Titley recalled the momentum for the history of technology that had been built up by the 1919 Watt centenary meeting in Birmingham. He claimed that the Society hoped to attract to its ranks those who had already interested themselves in industrial history, and to stimulate similar interests in others. It was, according to Titley, a society whose aims were new, and which had no precedents for its guidance.

In the United States, the organized cultivation of the field dates to the Humanistic-Social Research Project (1953–1955) of the American Society for Engineering Education, which undertook to survey the humanistic and social science aspects of engineering curricula in the nation. Its report, issued in 1956, noted that the study of the history of technology was "of special interest and importance for engineering education" and recommended that this study be in some way encouraged. Melvin Kranzberg, serving as chair of the Humanistic-Social Division of the ASEE, then organized an advisory committee for technology and society that, at its second meeting in 1958, decided to establish a Society for the History of Technology. The new society and its quarterly journal, *Technology and Culture* (both of which were organized by Kranzberg), marked a significant change from the Newcomen Society model then a generation old. Drawing its membership from much the same sources (academics, practicing engineers, government employees, businessmen, and amateur enthusiasts), it had its primary base in academia and placed far greater emphasis on what has since been termed "technology and society"—that is, the social causes and effects of technology, including, but by no means limited to, industrial history. As with the Newcomen Society's *Transactions*, SHOT's *Technology and Culture* has been a faithful mirror of its supporters' special approach to the history of technology.

In the United States, it was not surprising that since the impetus for the organization of SHOT came out of a concern for curricular reform, courses in the history of technology were encouraged soon after. Starting as the home base of both SHOT and *Technology and Culture*, the Case Institute of Technology launched the first formal graduate program in the field about 1960. Scattered courses appeared elsewhere, especially in colleges of engineering and in institutes of technology. A report edited by Ezra Heitowit of Cornell University in 1977 disclosed that although nearly 400 different institutions offered over 2300 courses that could be classed as dealing with "science, technology, and society," only a very small number of these claimed to be histories of technology. During the early 1970s, private foundations and the National Endowment for the Humanities gave considerable financial support to "technology studies" programs at various schools, but again the history of technology played only a minor role in this effort. By the mid-1970s, one could sense increasing interest in teaching the subject on the part of scholars trained in closely related areas (general history, American studies, engineering, history of science, etc.), but no statistical evidence existed that could be brought to bear upon the subject.

Although societies and journals first appeared outside the United States, the teaching of the history of technology to undergraduates appears to have lagged. In Great Britain, Manchester and Bath both have active centers, and the Open University, especially, has exhibited an interest in the field. In Scandinavia, regular courses are given only in Sweden.

Private scholarly efforts have been supplemented by official government programs in several countries. The Soviet Union's Institute of History of Science and Technology, established within the framework of the Academy of Sciences, was perhaps the first. In both Great Britain and the United States the identification, and sometimes preservation, of historical monuments has been an activity recently extended to cover engineering and technological sites. The British Ministry of Public Buildings and Works began an Industrial Monuments Survey in 1963, giving impetus to the foundation, a year later, of the *Journal of Industrial Archeology*. In the United States the Historic American Engineering Record, a branch of the Department of the Interior, has led the way in making inventories and surveys of technological sites and structures. This example, coupled with a new insistence that historical as well as environmental impact reports precede major construction or destruction projects, has provided a new opportunity to assess and preserve the best of the nation's engineering record. The process, however, is still in its infancy.

The major international organization concerned with the history of technology is ICOHTEC—the International Cooperation in History of Technology Committee of the International Union of the History and Philosophy of Science, organized in 1965. ICOHTEC sponsors, among other activities, symposia and conferences on various topics in the member nations and meets with the triennial International Congress of the History and Philosophy of Science meetings. On a more continuing basis, *Technology and Culture,* which styles itself an "international quarterly," makes an effort, through its appointment of advisory editors, its broad coverage of subjects, and its frequent publication of work by authors from outside the United States, to give responsible coverage to the international scholarly community.

B.   Principal Traditions in Writing

The history of technology has been written in such fundamentally different ways that the very boundaries of the field are sometimes blurred by their diversity. Principal among these have been: the internal approach, which concentrates on the careful development of the hardware of technology; the business or economic approach, which concentrates on the support for and strategies of technological activities; the social approach, which deals with the several social systems that interact with technologies; and the artifactual approach, which concentrates on particular tools, machines, sites, and so on. In one way or another, most history of technology as written is a variation of one, or a combination of several, of these approaches.

The internal history of technology is the most easily distinguished. The French historian Maurice Daumas has quoted with approval Lucien Febvre's dictum that the primary goal of the history of technology is to "establish a technical history of techniques" (Daumas, 1976, p. 90). Such studies, in his mind, were often tainted with a lack of technical expertise on the part of the author, as well as a certain national pride in identifying firsts for one's own country. Nevertheless, solid histories of such devices as the steam engine or windmill, computer or automobile, are clearly desirable and too often lacking. As often as not, the student of particular devices is thrown back upon surveys, such as Singer's (1955–1958) and Klemm's (1964), that not only concentrate on particular devices and techniques, but even borrow their organization from such machines and processes. Even clearly economic history like Usher's (1929) uses as chapter headings such subjects as waterwheels and windmills, water clocks and mechanical clocks, machine tools and quantity production, and attempts to provide an authentic account of "a technical history of techniques."

At their best, such studies rise far above a mere chronology of machine development. Mayr (1970), in his history of the concept of feedback in machines, provides nothing less than an intellectual history of a profound concept in addition to a clear description of its incorporation in succeeding generations of particular devices.

Perhaps the most common type of history of technology is that which stresses business strategies and economic changes. Within this category may be placed most of the studies that concentrate on the transformation of invention to innovation—that is, the embodiment of new ideas in readily available products. Classics of this sort include Strassmann (1959), with his concentration on the business community's calculation of risk in innovation, and Passer (1953), with his study of entrepreneurship in the electrical industry and its relationship to competition and economic growth. The relation between technological change and market conditions is detailed with great ingenuity by Schmookler (1962), who was convinced that invention followed market incentives.

Since, in the United States at least, new technologies appear through the auspices of private business (or private business in cooperation with government), it is not surprising that studies of particular business people and enterprises can cast much light on the origin and fate of technological change. Jenkins' (1975) history of the development of photography nicely links each major technical change in the process with the rise and dominance of new firms in the industry, with consequent changes in the market structure. His concentration on the work of George Eastman in this case is both inevitable and enlightening. As is always the case, the best books of any type transcend the form. Hughes's (1971) biography of the engineer-inventor-entrepreneur Elmer Sperry investigates the innovations of one entrepreneur who not only worked in a business context, but successfully manipulated the volatile mixture of politics and economics that marks the intersection of government, military, and business activities.

Arthur Schlesinger's American life series of a generation ago systematically included science and technology with the other aspects of social history discussed in its several volumes, and an entire American tradition (indeed, perhaps *the* American tradition) of the history of technology has followed this path. The underlying assumption of most of these studies is that technology is a significant aspect of society, both acting upon and being acted upon by the larger context.

One type of social history of technology may be termed "institutional." Sinclair's (1974) history of the Franklin Institute is an outstanding example of such a study, demonstrating the urban dynamics that create and shape a technical institution. Birr's (1957) history of General Electric's research laboratory is in the same tradition. Baxter's (1946) story of the World War II Office of Scientific Research and Development won a Pulitzer prize for its rich narrative of "scientists against time." Mechanics' institutes, research laboratories, and government agencies, not less than churches, philanthropic foundations, and fraternal orders have helped shape our society and are available for study on much the same terms.

A second type—perhaps less appropriately called social history (contrast the usages in chapters 1 and 3 of this volume)— may be termed "intellectual." "Technocracy," "rational control," and "feedback" are similar to "the great chain of being," "democracy," and "liberalism" in being ideas amenable to the technique of intellectual history. Histories of ideas of this sort may be traced against a broad social background. Layton (1971a, 1976) has masterfully outlined the tenets of engineering ideology in the United States, and has focused especially on the concept, within the profession, of social responsibility. Carey and Quirk (1970), in discussing the "mythos of the Electronic Revolution," dissect the ideology (in both its meanings) of the "post-industrial" hypothesis and compare it to the political uses of previous suggestions that technology had created historical discontinuities. Leo Marx (1964) has created a classic model of the intellectual history of technology with his discussion of the pastoral ideal in nineteenth-century American literature. And, finally, Mumford (1934, 1967, 1970) has come close to making technology itself an idea that can be traced through history through the plotting of its material and social manifestations.

Along with American social history in general, the social history of technology has produced a newer type of study dealing less with institutions and ideas than with social structure. Despite promising beginnings, the field has not yet produced a literature with the rich statistical and demographic density that characterizes the "new social history" in general. The growing interest in structures, however, promises to broaden our understanding of the subject significantly. Calvert's (1967) treatment of the rise of the mechanical engineering profession in terms of a clash between what he called "shop" and "school" cultures made an exciting beginning in this direction. More recently, Merritt Roe Smith's prize-winning (1977) study of the Harper's Ferry Federal Arsenal provides a careful analysis of technical change (and, more important, resistance to technical change) in terms of the positive and negative reinforcements of various social groups within a closely knit community. Noble's study (1977) of science and technology in the rise of the modern industrial corporation in the United States paints a similar picture on a broader canvas: engineers, captains of industry, educators, and foundation administrators all molding a people-machine nexus in which patents, research, schooling, and the work environment are all rationalized to accomplish a common end.

Finally, it should be kept in mind that biography continues to play an important role in the history of technology. Ever since Smiles's (1862) Victorian sketches of heroic technologists did so much to publicize the Whiggish notion of inventor as culture hero, scholars have continued to investigate the lives of great women and men with often rich results. Hunt's (1912) life of Ellen Swallow Richards, the first woman to graduate from the Massachusetts Institute of Technology, is still a precious source of insight not only into the life of this remarkable person but also into the general difficulty of women in a field dominated by men. Hamlin's (1955) monumental life of Benjamin Henry Latrobe, the British-born American engineer, remains a monument of scholarship; its dimensions will become clear with the publication of the Latrobe papers announced by the Yale University Press. Recently, Bruce (1973) has published what will no doubt remain the definitive life of Alexander Graham Bell, and has made it clear, for example, that as an inventor Bell proceeded quite differently from Sperry (see Hughes, 1971) Eastman (see Jenkins, 1975), and, most certainly, Edison (see Josephson, 1959). It is astonishing how few prominent technologists have attracted modern scholarly biographers. In part, perhaps, it is a natural reaction against the hagiography of amateur engineering history that will dissipate as the history of technology becomes more established as a professional field.

A last category of the social history of technology may be called artifactual—that is, it includes studies that deal with particular artifacts: tools, machines, or significant sites. The rise during the 1960s of the field of industrial archeology has added a new dimension to the older literature, which

tended to be heavily drawn from amateur enthusiasts or the curatorial concerns of museum personnel. Such studies as that of Comp (1975) on the
Toole Copper and Lead Smelter draw heavily on the kind of site interpretations pursued by the Historic American Engineering Record. Using physical remains as a document, Condit (1968) was able to demonstrate how the
Ingalls Building in Cincinnati incorporated both problems and solutions
that advanced the art of building construction in the area of reinforced concrete skyscrapers. Studies like these demonstrate conclusively that general
knowledge can be extracted from particular cases, and that the material
embodiment of technology provides an additional and fruitful dimension
to the study of the history of technology.

C.   Dividing the History of Technology

As in so many other aspects of the field, there is no generally agreed-upon
system for dividing a survey of the history of technology. Aside from the
necessity of imposing some chronological pattern upon the subject, the matter is still open to debate. Even in terms of chronology, however, it is not
clear that the common divisions appropriate to political or economic history
are equally suitable to technology. For example, are the traditional rubrics
of ancient, medieval, Renaissance, early modern, and modern appropriate as
a set, or are some appropriate and others not? Multhauf (1974) has gone so
far as to decry the Industrial Revolution as an organizing device in the field,
arguing that it obscures and constricts as much as it illuminates and expands
our understanding of technical change during these and subsequent years.
Mumford's (1934) classic division of the history of technology (and indeed of
the world) into eotechnic, paleotechnic, and neotechnic, is a clear refusal to
force technology into political, economic, or artistic molds.

On the other hand, in his collection of readings in the history of
American technology, Pursell (1969) deliberately chose such familiar categories as the colonial period, the Civil War, the winning of the West, and
so forth, to underline the idea that technology is so tied in with other aspects
of American life and culture that it is no less appropriately studied in these
categories than are other aspects. Sinclair (1974), in his similar volume covering Canadian technology, also arranges his documents in categories familiar
to students of the more traditional aspects of Canadian history: discovery,
settlement, expansion, war, depression, and so on. Lynn White has based
his full career on a study of medieval technology, and other technologies are
frequently labeled Renaissance, Victorian, ancient, and so forth.

The Pursell and Sinclair anthologies suggest another possible division,
into national categories. It has been charged that the general surveys of the
subject have been excessively nationalistic—that Singer (1955–1958) pays
undue attention to British practice and novelty, while Kranzberg and Pursell (1967) dwell at too great length on American developments. It is a

matter of some debate whether technologies reflect national styles, but it can hardly be denied that the fate of certain technologies differs in different nations. There is a degree, of course, to which technology, like science, is international or at least the property of an international community. It is also true, however, that, like any other form of human behavior, technology often shows marked variation from country to country. Thermodynamics may be the same everywhere, but steam engines are not found everywhere, and where they are found they are sometimes very different in design and use. It is even possible that smaller than national divisions may display similar technological discontinuities. The growing number of state inventories by the Historic American Engineering Record, for example, may eventually provide evidence to show that American technology is not homogeneous from coast to coast. Though nationalistic passions are all too easily aroused over such questions as priority of invention, it seems clear that national histories of technology are a reasonable enterprise.

A third possible system of categorization is by function. If technology is the way people get things done, it appears reasonable to organize its discussion in terms of the tasks to be undertaken. Singer's volume on the Industrial Revolution (1958), for example, is divided into six sections labeled "Primary Production," "Forms of Energy," "Manufacture," "Static Engineering," "Communications," and "Scientific Basis of Technology." These sections (except for the last) are in turn subdivided: that on primary production into "Agriculture: Farm Implements," "Agriculture: Techniques of Farming," "Fish Preservation," "Metal and Coal Mining," "Extraction and Production of Metals: Iron and Steel," and "Extraction and Production of Metals: Nonferrous Metals." This scheme, of course, shares with any such topical approach to history a certain violation of basic chronology that somewhat dims the student's view of any given historical epoch.

Another system for dividing the field is by device. Such work as Bryant's (1976) on the Diesel engine or Ellis' (1975) on the machine gun depend upon the selection of particular devices for special consideration. The same logic applied to surveys can produce such works as the classic *History of Mechanical Inventions* by Usher (1929), in which there are separate chapters on water wheels and windmills, water and mechanical clocks, and machine tools. Usher is perhaps wise, however, in not depending entirely upon such divisions; in lesser hands this approach sometimes degenerates into a mere catalog of appliances. Instead, alongside the chapters on devices (the clock), there are others on functions (printing and textile manufacture), on an era (preChristian antiquity and its mechanical equipment), or biography (Leonardo da Vinci, engineer and inventor), and on broad topics, such as "The Place of Technology in Economic History" and "The Emergence of Novelty in Thought and Action." As Usher's pioneer effort clearly illustrates, there are several ways to divide the history of technology, each with its own advantages and drawbacks.

I. SOME PROBLEM AREAS OF SPECIAL INTEREST

A. Relation of Science to Technology

There is no more common or difficult—and perhaps more politically explosive—issue in the entire historiography of technology than its relationship to science. One popular model has it that science and technology, after centuries of separation because of different class bases, first came together in the seventeenth century, at the beginning of the Scientific Revolution. The union then became fruitful in the nineteenth century, when accumulating scientific data, the increasing control of those data by professional scientists, and the increasing intransigence of technological problems all combined to make science a fertile and perhaps necessary source of new technological applications. In the twentieth century, or at least certainly since World War II, science has become the major source of new technologies—so much so that they are usually regarded, in fact, as applied science. However well this serves the interests of scientists, who invoke usefulness to attract patronage, and of technologists, who want to wrap themselves in the mantle of immunity granted by society to science, this model has proved increasingly unsatisfactory to historians of technology.

One area for investigation that seems to hold promise for sorting out the relationship between science and technology is that of the industrial research process. Kendall Birr (1957) pioneered the scholarly study of the subject with his investigation of the General Electric Laboratory, one of the earliest and most successful industrial research groups in the United States. Although Birr identifies three preconditions to "the marriage of science and technology" (the maturation of science, its acceptance by business, and the proper institutional innovations to provide for it in industry), he is interested mainly in institutional history rather than in the science-technology relationship as such. The same is true of his useful survey of the rise of industrial research as an activity in the United States (1966). Fagen's massive study of engineering and science in the Bell Telephone system (1975) is even more anecdotal and institutional, although it inevitably contains suggestive episodes and relationships. More pointed is Hugh Aitken's *Syntony and Spark* (1976), which focuses on the transfer of knowledge between the three corporate subcultures of science, technology, and economics in the early radio industry. Reich (1977) has recently studied the role of industrial research and patenting in closing out, rather than advancing, technical programs.

It is rarely claimed, however, that there was no relationship between science and technology before the building of the first industrial research laboratory, and several scholars have made useful case studies in which one might expect the two to influence one another. Hughes (1976) on high-voltage power transmission; Multhauf on geology, chemistry, and the production of common salt (1976); and Bryant (1973) on thermodynamics and the evolution of the heat engine, are excellent cases in point. Bryant, for example, shows with great precision how, although the steam engine preceded a modern understanding of thermodynamics, significant improvements in heat engines followed it. In the wonderfully unpredictable way of history, however, Otto made no use of such knowledge in developing his gasoline engine, while Diesel quite deliberately did.

At least three important efforts have been made to rescue technology from its supposed subordination to science without denying a close and interactive relationship between the two. Buchanan has chosen to invoke what he calls (1976) a "Promethean Revolution" in which modern science and technology are both manifestations of a unifying Western faith in material progress. During the late middle ages there was a fusion between speculative thought and practical action that unbound Prometheus and produced "the thinking hand," a type that persisted until the nineteenth century, when the rise of professionalism created the distinction between science and technology with which contemporary scholars and bureaucrats struggle. Cardwell, in his study of *Turning Points in Western Technology* (1972), claims for technology its own autonomy and sees it as the camp follower of neither science nor society. Technology, for him, marches to its own internal logic, as does science; and the history of each is essentially a history of ideas.

Perhaps the most widely noted version of the classic relationship is that presented by Layton in his concept of science and technology as "mirror-image twins." In this (1971) scheme, both become professionalized during the nineteenth century, sharing a common reliance on the scientific method, but develop separate and distinct cultures, each with appropriate language, literature, style, concerns, and personnel; each is the mirror-image twin of the other. Technology, therefore, is not subordinate to or dependent upon science; both are related to each other, though they are separated by cultural barriers that are crossed only infrequently and with great difficulty.

It is instructive to compare two handy benchmarks in the historiography of the relationship between science and technology: the fall 1961 issue of *Technology and Culture,* devoted to "Science and Engineering"; and the October 1976 issue of that same journal, devoted to "The Interaction of Science and Technology in the Industrial Age." One obvious difference between the two issues, published a decade and a half apart, is that engineering has been replaced by a concept of technology large enough to include the ecology of grassland management. More important: although both sets of papers attempt to free technology from the popular notion of applied science, the more recent scholars go almost so far as to suggest that the very formulation of the question is a hindrance to progress in the field.

B. Professionalization of Engineering

Until fairly recently, most of the literature on the structure of professionalism was produced by occupational sociologists. Three American engineering specialties now have their own histories, however, and these histories have begun to provide a basis for discovering changes through time. Calhoun (1960), looking at civil engineers in the United States in the first half of the nineteenth century, emphasized the bureaucratic nature of the field and emphasized the conflict felt by engineers who threaded their way through the thicket of professional responsibility, public policy, and private profit. Calvert's study (1967) of mechanical engineers also emphasizes conflict, this time between the proponents of shop culture and those of school culture, both of which sought to dominate the emerging profession. Spence (1970) has provided us with a fine account of mining engineers in the American West.

We have yet to see a major engineering society treated by a trained historian, but Pursell (1976) traces the rise and fall of a regional society (in San Francisco) at the turn of the century. Rae (1975) has begun a statistical effort to portray the profession, and the same author (1954) has suggested its ambivalent relationship to business. The most suggestive book on the subject is that of Layton (1971), which not only delves into the engineering view of social responsibility, but probes the engineers' sometimes uneasy commitment to both science and business and the ideology that helps bridge that gap. Merritt (1969) contains many useful insights and suggestions for the period 1850 to 1875 in America, including an especially good section on the rise of urban engineers. Noble (1977) provides a rich and detailed description of the role of engineers in their service to the rising industrial corporation during the first decades of the twentieth century, emphasizing the application of scientific technology to the task of industrial research, engineering education, scientific management, and the new field of management generally. Prescott's (1954) account of the history of the Massachusetts Institute of Technology is a reminder that we still know too little of the history of specific technical schools.

C. Medieval Technology

No field has undergone a more dramatic change in the past generation than the study of medieval technology. Reminiscing about his long career in the field, Lynn White (1975) remarked that our understanding of medieval technology lagged a generation behind that of medieval science, and that previous to 1933 he had run across no discussion of any aspect of the topic. It was the reading of the work of the anthropologist Alfred Kroeber that made him determine to "apply the methods of cultural anthropology to the middle ages"; this began his career as the premier scholar in the field.

The long identification of the middle ages with technological (as well as cultural and intellectual) stagnation was first dramatically broken by Lewis Mumford with his study of *Technics and Civilization* (1934). In that great work he identified "the clock, not the steam engine" as "the key machine of the modern industrial age" and the medieval monastery as the seat of a new desire for "order and power" that formed a basis of the new technology (pp. 13, 14).

No aspect of medieval history has more than a few active scholars, and its technology is no exception. L. R. Shelby's work on medieval masons' tools (1961, 1965) is characteristic of the careful and useful studies recently under way. There is no doubt, however, that it is White who has dominated the field and redefined the problem. His basic work on *Medieval Technology and Social Change* (1962) concentrates on such fundamental devices as the stirrup, the plow, the three-field system of cultivation, and the mill, and describes in sweeping terms the social changes (shock combat, the feudal system, increased food supply and therefore more urbanization, etc.) that they brought about. Such examples are further suggested in his strictures against a too sanguine belief in the powers of technology assessment (1974): Who could have guessed, for example, that our cult of the child would arise from the invention of knitting, or class hostility from the invention of the chimney? His widely influential essay "The Historical Roots of Our Ecological Crisis" (1967) traces the idea of dominion over nature to the very beginnings of the Judeo-Christian tradition, but concentrates on the medieval linking of work and piety and its fantasies of power. It is largely through work like this, meticulous in research but sweeping in interpretation, that the middle ages have come to be seen as the seedbed of modern technology.

D. The American System of Manufactures

No element of American technology is more famous than the American system of manufactures: the "principles and practice of quantity manufacture of standardized products characterized by interchangeable parts and the use of a growing array of machine tools and specialized jigs and fixtures, along with power, to substitute simplified and, as far as possible, mechanized operations for the craftsman's art" (Sawyer, 1954, p. 369). The standard account traces this development back to the original musket contract of Eli Whitney (1798), then forward through the spread of the system, first through the small arms industry, then outward to clocks, sewing machines, bicycles, and other goods made up of standardized small metal parts. The traditional case for the rise of the American system and Whitney's part in it may be found in Green (1956). The Whitney story has been brilliantly attacked by Woodbury (1960), who emphasizes his long record of failure to manufacture successfully even so simple a device as the cotton gin; his failure to deliver muskets within the terms of his government contract; the certain fact that surviving Whitney muskets are not interchangeable; and the complete lack of evidence concerning the actual technological machines and practices within the Whitney factory.

The social milieu within which the "American system" was developed, and the society that proved so accepting of uniformity, is investigated in Sawyer's "The Social Basis of the American System of Manufacturing"

(1954). Following the lead of contemporary European observers, Sawyer emphasizes "the absence of rigidities and restraints of class and craft; the freedom from hereditary definitions of the tasks or hardened ways of going about them; the high focus on personal advancement and drives to higher material welfare; and the mobility, flexibility, adaptability of Americans and their boundless belief in progress" (p. 376). As Fisher (1967) has shown, these attributes were not merely reactions to the American experience but in part preoccupations imposed upon it by Europeans eager to find in America the pastoral ideal that Leo Marx (1964) has examined in a literary context. More recently, Kasson (1976) has investigated the tension between republican values and the machine and its resolution in favor of technology as social control. The most useful British observations of the mature American system, the 1855 report of Whitworth and Wallis, are reproduced with an excellent commentary and notes by Rosenberg (1969). One important result, in part of these reports, was the introduction of aspects of the system into British industry. The key case of small arms in Great Britain has been carefully investigated by Fries (1975), who concludes that although the manufacture of private arms changed very little, the manufacture of military arms "changed greatly almost overnight" (p. 377).

If the American system did not spring exclusively from the brow of Eli Whitney (a claim even the traditional accounts were careful not to emphasize), what then were its sources? Uselding (1974) has added greatly to our knowledge of the contributions of Elisha K. Root, factory foreman and general superintendent at the Colt Armory, especially his previous work on the die-forging of axes. More fundamentally, Smith (1977) has challenged at once several notions: that American workmen welcomed the American system, that it lowered costs, and that it necessarily involved interchangeability. His study of the Harper's Ferry Armory shows that not only did its craftspeople resist the new system, but that their products were competitive in economic terms. Interchangeability was so expensive that only arms made on military contracts were done up to that standard. At the same time, Smith adds to our knowledge of the contributions made by Thomas Blanchard and John Hall, and he reminds us of the measure of social control over technological pace and direction exercised by such institutions as church, school, state, and business organizations.

The actual machines that were developed and deployed as a part of the American system are covered in the standard survey by Roe (1916). More recently, careful studies by Woodbury of the gear-cutting machine (1958), the grinding machine (1959), the milling machine (1960), and the lathe (1961) have added immeasurably to our knowledge.

E.   Technology and the Rise of the Modern Corporate State

No aspect of modern life in America is more pervasive or less understood than the industrial corporation. The economists' view—that blind market forces dictate the rational behavior of business enterprises—and an intellectual aversion to "conspiracies," have combined to obscure the origins and workings of this most powerful influence in modern life. More particularly, economic and business historians have often assumed that technology, like the market, was a given that explained behavior but that itself did not have to be explained except in the most simplistic and mechanistic terms. Alfred Chandler (1977) has gone far toward demonstrating how deliberate management (his "visible hand") has replaced blind market forces in directing our modern economy. However, in Chandler's scheme, technology remains something of an independent variable.

Yet the connection between technology and the rising corporation in the United States has attracted increasing attention. Samuel Hays (1959) fundamentally revised the notion of the movement to conserve natural resources by drawing attention to the cooperation between large resource corporations, the scientists and engineers who served them, and government regulators. The reformer and writer Jim Hightower (1972) traced the intertwining of academic science, corporate strategies, and government subsidy in the development of modern agribusiness. Ruth Cowan (1976) has begun a study of the role of corporate planning, advertising, invention, and ideology in defining the nature of the American home and women's place in it. Leonard Reich (1977) has investigated the efforts of AT&T to use research and patents to try to dominate the early radio industry, in part to protect its vested interest in telephony.

The most comprehensive study to date is that of David Noble, *America by Design* (1977). Looking particularly at the role of engineers, he traces the successful attempt of large corporations to rationalize both machines and people. Industrial research, scientific management, patent control, and the subsidy of engineering education all combine to create a business environment in which corporate managers can plan ahead for growth and increasing dominance. The tensions that this approach created within the engineering community—dedicated at once to science and profits—has been explained by Edwin Layton (1971). As he shows, the primacy of service to profits has been built not only into the very structure of professional association and education, but into the ideology of engineering as well. The result of this activity, as these studies clearly show, has been a significant shift from what Lewis Mumford (1964) called "democratic" to "authoritarian" technics; that is, one largely under the control of a small and politically irresponsible group, characterized by large, simple, interconnected, powerful but unstable features.

The influence of this corporate connection in Europe has been investigated by Kuisel (1967)—in terms of the life of the French technocrat Ernest Mercier—and by Charles Maier (1970) in his study of the impact of scientific management and technocracy on evolving European ideologies and politics (particularly those of the right) during the 1920s. Both studies show an alarming tendency toward fascism and the usefulness of modern scientific technology to those who own the means of production.

F.   Technology and Ideology

The word "ideology" means both the study of ideas (or science of ideas), and also the assertions put forward as the "official" ideas of a group or institution. An excellent beginning of this study with regard to technology is put forth by Layton (1976) for American science and technology in general, and (1971) for engineers as a group in particular. Haber (1964) is an intellectual history of the efficiency craze of the early twentieth century, a craze given its direction and content by engineers. The uses of this particular ideology in Europe between the wars is ably described by Maier (1970), and Ludwig (1974) has traced the compatibility of German engineers and technology with the Third Reich. Bailes (1974) has shown that in the Soviet Union Stalin saw engineering ideology as a mask for counterrevolutionary forces and suppressed it during the 1930s. The subservience of American engineers to corporate liberalism is chronicled in detail by Noble (1977), who shows the ideological links between the growing engineering profession and the industrial corporation. Carey and Quirk (1970), describing the ideological message behind the twentieth-century idea of an electronic revolution, have linked it suggestively to the "mechanical sublime" which proved abortive in the nineteenth century. Although he deals primarily with people who identified themselves as scientists, Tobey (1971) has studied the attempts to place science at the center of American culture; this study has proved useful for historians of technology as well.

G.   The Development Phase of Technological Change

In recent years, some of the attention once directed toward "invention" has been shifted to the development phase of technological change. Introducing a special issue of *Technology and Culture* (July 1976) on this theme, Hughes points out that historians of technology are increasingly dissatisfied with the simple linear progression from invention to development to innovation carried out by the inventor or scientist, engineer, and entrepreneur. (See chapter 5, II. A. 2., in this volume.) Defining development tentatively as "a move away from the elegant but abstract concepts associated with invention to the construction and testing of models" so that design changes can be made to "respond to the demands of the environment"—that is, its uses in the real world—Hughes points out that this process is one with which engineers are comfortable. Hughes's own full-scale study of Elmer Sperry (1971) follows this process in detail. Jewkes et al. (1961), although in the older tradition of a study primarily of invention, also looks at the development phase of the fifty cases studied.

Starting from a concern over international economic (and therefore technological) development, especially with regard to the Third World, Layton

(1977) moves backward from economic growth to technological change to the development of technology. This progression he defines as "not . . . a simple linear sequence, but . . . a flow which includes both social and technical events, and an important amount of interaction between the various categories of events" (p. 205). Although he asserts that "the interdisciplinary research laboratory has been the most significant institution for modern technological development" (p. 211), Layton also warns that "a major portion of modern industry is quite unrelated to the science-technology complex" (p. 215) and that this last sector has been more successful in developing process innovations (as opposed to product innovations) than has the research laboratory. Despite their strictures against the simple linear model of technological change, recent scholars have more often modified than rejected it outright. The true value of their criticisms has been to focus attention on the rich and complex process by which ideas become realities in the world of technology.

H.   Technology Transfer

The idea of technology transfer—that is, the movement of a technology from one country to another or from one industry or application to another —has been a matter of concern to government policy makers for many years. Policy studies on the effectiveness of efforts to stimulate economic growth by shifting Western technology to developing nations, or new technologies from one industry to another (such as spinoffs from aerospace to other industries), have cast a new light on what have been traditional concerns of the historian of technology. Merritt Roe Smith's study (1977) of Harper's Ferry Armory is a story of change and resistance to transfer within a single industry. Danhof (1969) has made a careful study of the way agricultural innovations were diffused through rural communities in the mid-nineteenth century United States. Rosenberg's study of innovation in the machine tool industry (1963) directs attention to a key source of technological change, which radiated out to influence many different industries.

International transfer is no new thing, of course. Curti and Birr (1954) studied American efforts to export technology to developing nations in the nineteenth and early twentieth centuries. Pursell (1969) has provided a study of the transfer of steam engine technology to America in the late eighteenth and early nineteenth centuries, and his description (1964) of the efforts of Thomas Digges to send textile mechanics to America in the 1790s gives particularity to a migration the prevention of which was an important part of British national policy (see also Jeremy, 1977). Rosenberg (1972) uses the rise of America as an exporter rather than merely a borrower of technology as an organizing principle of his economic study of American technology, and Fries (1975) gives a case study in the transfer of the American System to the British small arms industry.

For more recent times, Sutton has devoted three volumes (1968, 1971,

1973) to a detailed account of the transfer of Western technology to the Soviet Union between 1917 and 1965. Such works demonstrate conclusively that while there may be national styles of technology, there is also an international community of technology, and that technology does and always has flowed across continental, national, industrial, and corporate boundaries. The introduction of the stirrup into medieval Europe from the East (White, 1962), the American building of railroads in Czarist Russia in the nineteenth century, and Project Paperclip, which brought German rocket technicians to America after World War II (Lasby, 1971), are all examples of this restless flow.

I. Technology Assessment

The idea of technology assessment has been gaining increasing attention since the introduction of legislation in 1967 to establish some such capability within the Congress. (An Office of Technology Assessment was finally set up in 1972.) Although the idea is usually cast in terms of looking forward to identify future technologies and assess their likely social and economic impact, some effort has been made by the government to stimulate so-called "retrospective" technology assessments: that is, to try to determine what contemporary information was available about past technologies when they were first introduced, and how well that information was used to foresee the results that later became apparent. It is too soon to evaluate this effort with any degree of certainty, but the concept of technology assessment is useful in directing the attention of the historian to past efforts to anticipate the effects of new technologies.

The best beginning is with Kranzberg (1970), who focuses on the history of efforts to anticipate consequences of new technologies. Inouye and Susskind (1977) have made a study of one signal effort, the 1937 report of the Federal government on *Technological Trends*. The background of the 1972 legislation setting up the OTA has been summarized by Pursell (1974a). In Pursell (1974b), the debate over a suggested moratorium on research (1927–1937) is viewed as a political struggle between humanists and scientists-technologists over social control of new technologies. A necessary cautionary note is struck by White (1974) in his presidential address to the American Historical Association: historians cannot accurately assess new technologies, but they can show just how complicated and difficult that process will be. A fascinating study that seems to indicate that "retrospective" technology assessment is perhaps not much different from good history is Tarr and McMichaels' account (1977) of evolving urban wastewater technologies in this country.

J. Women and Technology

Ruth Cowan has suggested (1977) that women interact with technology differently from men in four ways: (1) as females (that is, biologically); (2) as workers in the marketplace; (3) as workers in the home; and (4) ideologically, as people excluded from full participation in modern scientific technology. The first aspect has been little studied: Kennedy's (1970) book on the history of contraceptive devices is one start. Wertz and Wertz (1977) have traced the history of childbirth in the United States, and though this is obviously a "natural" act it is also (Arms, 1975) one that has been increasingly mechanized, with results only now being studied. For women as workers in the market economy, Baker (1964) remains the classic work. The study of housework was suggested by Ravetz (1965). This was ably carried forward by Cowan (1976), in a study of the results of household electrification in the 1920s that will certainly become a classic.

Where Cowan concentrates on the industrialization of the home during the 1920s, Andrews and Andrews (1974) cover the nineteenth century. Sklar (1973) has written an excellent biography of Catherine E. Beecher, an American pioneer in the movement to upgrade home technology, and Baldwin (1949) is a handy though thin and official chronology of the development of the home economics movement. Most of these studies concentrate on the middle-class housewife, but Kleinberg (1976) describes the political (and ultimately economic) policies in late nineteenth century Pittsburgh that determined the extent to which working-class wives would benefit from new urban technologies, such as sewers and paved streets. In this, as in so many other areas, Giedion (1948) retains his usefulness as a source of ideas and striking examples.

Female participation in the making of the technological world has certainly been limited: in the United States, for example, less than 1 percent of engineers are women. Hunt (1912) records the life of Ellen H. Richards, who, as the first woman to graduate from the Massachusetts Institute of Technology, is a notable exception. Gilbreth (1970) gives an informal account of the most famous female engineer in America, Lillian Gilbreth. Anderson (1931), in a curious and impressionistic mixture of reportage and social speculation, suggests that women not only are excluded from, but are immune to, the effects of the machine, and that, being outside technological society, they will be able to save men from the consequences of their own emasculation. This type of reaction, and whatever reality may be behind it, badly needs study.

K. Technology and Literature

The classic beginning point for a study of technology and American literature is Leo Marx (1964). Covering writers from Shakespeare to F. Scott Fitzgerald, he traces what he calls the "machine in the garden"—that is, the pastoral ideal in the face of increasing industrialization. For British literature Sussman (1968) discusses authors from Thomas Carlyle to Rudyard Kipling under the rubric, "Victorians and the Machine." A less successful study of American authors in the twentieth century is West (1967). A suggestive study of Mark Twain's *Connecticut Yankee in King Arthur's Court* is provided in Smith (1964). Hank Morgan, the representative of nineteenth-century American success in the areas of representative government and industrial technology, proves capable of destroying an entire culture—along with thousands of people—but is unable to replace it with anything better. As Smith shows, the experience of writing this novel so disillusioned Twain that he never again wrote a long work of the first rank.

Kasson (1976), as a part of his study of technology and republican virtues in nineteenth-century America, provides an important discussion of the most prominent utopias and dystopias from American authors of the last century. Like Marx and Smith, working in the tradition of American studies, Kasson establishes the key role played by technology in the thinking and plans of the American leaders of the early republic. Ultimately, as he shows, they failed to shape a technological society "consonant with republican ideals." Trachtenberg (1965) skillfully blends literary criticism with art, business practices, and technology in his study of the unique place the Brooklyn Bridge has had in American thought. The science fiction subculture is usefully surveyed by Philmus (1970), who covers writers from Francis Godwin to H. G. Wells, and by Kingsley Amis (1960). Berger (1972) provides a study of John W. Campbell, long-time editor of *Astounding Science Fiction,* one of the most successful and influential of the science fiction magazines.

L.   Technology and the Environment

Nothing could be more obvious than the fact that technology has an impact upon the natural environment: indeed it is almost the expectation of such an impact that defines the idea of technology. Only recently, however, has this become a major and separate area of study, especially with reference to untoward impacts. White's influential suggestion (1967) that the roots of our present ecological crisis go back into the Judaeo-Christian tradition gives the subject a compelling thesis, big enough to argue against. Rosenberg (1971) provides an economic context for the problem in recent times, concerning himself largely with the problems of external costs. Several suggestive studies have been made of urban pollution: Te Brake (1975) traces the interaction of growing population, a fuel crisis, and air pollution in London from 1250 to 1650; Cain (1974) deals with the efforts of Chicago to divert its sewage from its water supply; and Tarr and McMichael (1977) trace the fascinating interaction of new technologies and subsequent new problems in sewage disposal in American cities during the last and present centuries.

Tobey (1976) relates the rise of grasslands ecology in America in the 1880s to the needs of Midwestern farmers. The greatest attention of historians generally has been focused on the conservation crusade that crested during the Progressive era in America. Hays (1959), in revising the traditional view of that reform movement, roots it firmly in the broad technological effort to rationalize and make efficient the various factors of industrial production. For the more recent period, Commoner (1971) blames the environmental crisis largely on technological changes since World War II, rather than on the simple fact of overpopulation. Graham (1970) chronicles the reaction of the self-interested technical community to the antiDDT message of Rachel Carson's *Silent Spring*.

II. RELATION OF THE HISTORY OF TECHNOLOGY TO OTHER AREAS

It would be neither possible nor desirable to try to isolate the history of technology from related disciplines. There have been and will continue to be topics that cross disciplinary lines, as well as investigators who cross back and forth between logically distinct fields of inquiry; and even for those who do not, the study of one will suggest insights and methodologies that can, by analogy, enrich others. Distinctions between fields covered by this volume arise primarily out of the phenomena to be investigated, and these differences in turn lead to the creation of different supporting systems of peers, colleagues, journals, patronage, academic organization, audience, and so forth. Although such social ramparts are often difficult to breach, the intellectual concerns of the various fields need show no such territoriality.

A. History of Science

The history of science is not only one of the points of origin for the history of technology; it is a continuing source of ideas, techniques, colleagues, and general inspiration (see chapter 1, especially II, H., in this volume). A topic such as the history of industrial research laboratories will interest historians of both science and technology, and one will expect to find the two working shoulder to shoulder to understand this important institution. In another area, an organization like the National Academy of Engineering will eventually attract the efforts of a historian of technology, and the quality of that study will be enhanced by an awareness of work being done on the history of the National Academy of Sciences. If similar effects have similar causes, the "mirror-image twins" of science and technology, as Layton has called them, will provide endless instances of the need for historians of science and technology to have an intimate knowledge of each others' work. At the same time, it would be a mistake to minimize the fact that definitions (and similar limiting devices) have severe political implications. To the extent that historians of science and technology consciously or unconsciously internalize and profess the ideologies of, for example, the scientific and engineering communities, they will be tempted to engage in territorial disputes no less bitter than those that animate the subjects of their study. Whether science or engineering won the war, reached the moon, or can claim responsibility for current luxury (or danger) are matters of no small import. Standing at one remove from such disputes, historians of both fields are better equipped to transcend them and may even have a role in tempering them with scholarship.

At the same time it must be emphasized that science and technology are separate enterprises, and despite the fact that the history of science was one of the roots of the history of technology, the latter has over the past few years developed interests and concepts appropriate to its own enterprise. Despite some overlap, the bodies of literature of the two fields are quite different, and it is no longer realistic to expect the master of one to be automatically familiar with the other. Although some scholars still attempt to keep abreast of both fields, increasingly separate evolutions are making this more and more difficult.

B.   Sociology of Science

What has been said about the relationship between the historians of technology and of science is equally true of the former and the sociologist of science (see chapter 7 in this volume). Here the added division of quite different methodologies introduces both a potential barrier and a benefit. Historians operate as amateur sociologists quite as often as the latter try their hand at history, and the results in both cases are improved when the attempt is conscious and informed by mutual respect and awareness. Often the interaction between science and technology is in precisely those areas where the sociologist is most active: in the applications of knowledge, in institution building, and in group interaction. Methodologies for which historians have neither the taste nor the training, and vocabularies that are novel, are barriers as often imagined as met with, and are usually worth surmounting in the interest of the enrichment of historical study.

C.   History of Medicine

Medicine, in both its historical and sociological aspects, is a special problem for historians of technology. In the past their paths have seldom crossed those of the historian of medicine, except when both have been led through the history of science. Represented by different traditions and institutions, audiences and enthusiasms, those who study medicine and those who study technology tend to go their separate ways (see chapter 3 in this volume). Yet a case can be made for the thesis that technology and medicine are more closely related in real life than are technology and science (see chapter 8 in this volume). Neither the physician nor the technologist has ever been afforded the luxury of mere speculation; both are measured (as they expect to be) by the practical results of their activities. Both use tools to effect a desired result in nature. Both draw somewhat upon science (more now than in the past), and yet both practice more than just applied science. The growing elaboration of medical hardware, no less than the increasing possibility of genetic engineering, dramatize the natural conjunction between the history of technology and the history and sociology of medicine. One may confidently predict that in the future, despite the formidable barriers of institutional separation, these two specialties will grow increasingly interactive.

D. Popular Social Criticism of Technology

Whereas in the study of both science and medicine the disciplines of history and sociology are cultivated distinguishably from each other, the same is not true of technology. There is no body of scholars who style themselves sociologists of technology, and the trails blazed by such past giants as Ogburn and Gilfillan have seen little traffic in recent times. The excellence of their work, however, and the obvious utility of covering that territory, make inevitable the resurrection of that specialty (see chapter 7). In its place we have seen a continuation of what may be called the social criticism of technology, or a kind of popular sociology of technology. Such widely read investigations of machine civilization as those of Stuart Chase and Charles A. Beard from the late 1920s have been followed in our own time by the work of Ivan Illich and Theodore Roszak. Such studies have done a great deal to alert the public to questions of technology and society, to suggest fruitful hypotheses for academic sociologists, and to provide grist for the mill of those historians of technology who are interested in cultural styles. (This is also the closest point of contact with the philosophy of technology; see chapter 5.)

E. Science Policy Studies

Finally, there is an expanding common ground between historians of technology and the new bands of scholars specializing in science policy studies. As Layton (1977) has pointed out, the science policies pursued and debated by nations are often concerned, in fact, with technology: nuclear power, energy independence, technology transfer, mutual defense, stimulation of technological innovation, the numbers and distribution of technical specialists, and so forth. The very idea that science (or technology) is something about which nations have policies is a vivifying concept that can provide fresh perspective on British attempts in the eighteenth century to prevent the expatriation of technicians (a kind of brain drain), or on official American encouragement of engineering development in emerging nations during the nineteenth century (a kind of technology transfer). Every good historian will, of course, resist the temptation to read present concerns into the past. But every good historian of technology will also gain fresh insight from the science-technology policy studies that are beginning to appear in this country as well as in several others (see chapter 9).

F. General History

The place of the history of technology in relation to the general field of history is not, at the present, satisfactory. In theory, it now stands alongside such other specialized fields as the history of religion, the history of ideas, diplomatic history, and agricultural history. In fact, it probably has had less impact on the writing and teaching of broad synthetic history than any of these. History, like any other academic discipline, is faddish, with first one field then another enjoying a vogue, only to slip over into a kind of genteel decline leading in some cases to virtual disappearance as a living tradition. In part, this must certainly be due to vogues and levels of awareness in the society at large. In a period of intense international commitment and competition, for example, one will expect a broad interest in foreign affairs to lead to new popularity (and new books and new teaching positions) in diplomatic history. There is some evidence that the current interest in technology in the United States has turned the minds of students and academic administrators toward an interest in the history of technology. There is some evidence also that articles in this area are now more welcome in general historical journals than they once were, and a book in the history of technology (Smith, 1977) won the 1976 Frederick Jackson Turner award of the Organization of American Historians.

At the same time, it must be admitted that the history of technology is not yet integrated into the body of general history. Textbooks and surveys may devote a superficial and pious note to a few heroic inventors or the American System of Manufactures before moving on to the "more important" history of politics and diplomacy. Yet even here there is some reason for optimism. In his 1974 presidential address to the Organization of American Historians, John Higham suggested that by the midnineteenth century the defining paradigm of American community had passed from primordial associations, to ideological, to technological. Whether this hypothesis will meet with general acceptance is less important than the hope it gives that technology may move out of the footnote and segregated chapter of history and, for America at least, become a necessary part of understanding our entire national experience.

The integration of the history of technology into general histories of Europe could be no less profound. White's work on the medieval period (1962), Cipolla's on the age of exploration (1965), any number of traditional accounts of the Industrial Revolution, and more recent works such as Maier (1970) on technocracy and European ideologies during the 1920s and 1930s all suggest the central role of technological developments in charting the course of European history. One may anticipate with some confidence that, sooner or later, the profound influence of technological development on human society will be integrated into the histories of all nations and cultures.

III. DEVELOPMENT OF SCHOLARSHIP OUTSIDE THE UNITED STATES

Although such works as Smiles's biographies of engineers and industrialists helped develop a kind of prehistory of technology in Great Britain during the Victorian age, an organized field began to develop only with the creation in 1919 of the Newcomen Society for the History of Engineering and Technology. The special blend of technical histories and the history of manufactures represented by the Newcomen Society has remained a dominant trait of the field in Britain to the present. Recently this approach was underscored by the creation in 1963 of the British Ministry of Public Buildings and Works and the founding of the *Journal of Industrial Archeology* a year later. At the present time, there can be little doubt that the United Kingdom leads in the pursuit of industrial archeology and the precise, specific kind of history of technology that can be written from such sources of information.

Recently, also, several British authors have taken the lead in attempting to carve out for the history of technology an intellectual territory separate from but analogous to that long occupied by the history of science. Cardwell, in his *Turning Points in Western Technology* (1972), and Pacey, in his book *The Maze of Ingenuity* (1975), both insist on viewing the history of technology as a sweeping flow acting as handmaiden to no superior force, but driven forward by the internal dynamism of its own ideas.

An even more severe commitment to the autonomy of technology is inherent in the writing of A. Rupert Hall, whose annual *History of Technology* (the first volume of which appeared in 1976) gives voice to the historians of both ideas and techniques. The specter of both the Industrial Revolution and the classic history of science hangs low over the field. Perhaps the most significant recent development has been the spread of centers of study, particularly in the red brick universities. Bath University in Birmingham, the University of Manchester Institute of Science and Technology, and now the University of London all have developed a greater or lesser competence in the field, and the Open University has made technology something of a special interest.

The formal study of the history of technology in Germany can be dated from 1909. In that year, the first teaching post devoted to the subject (created for Carl Matschoss at the Technische Hochschule in Berlin) was established, and two major journals were begun: the *Beiträge zur Geschichte der Technik und Industrie* in Berlin and the *Archiv für die Geschichte der Naturwissenschaften und der Technik* in Leipzig. The latter soon concentrated on the history of science; the former, however, being sponsored by the Verein Deutscher Ingenieure (VDI) marked the emergence of that group as a primary force behind German leadership in the field (see also the philosophy of technology efforts of the VDI reported in chapter 5).

A second step forward was taken with the opening of the Deutsches Museum in Munich in 1925. Planned since 1903, this great museum immediately became a focus of scholarly interest and, in 1963, was made the home of the new Research Institute for the History of Science and Technology. Groups of various complexion and tenure have been important for many years. A prime example was the Fachgruppe für die Geschichte der Technik, which was organized in 1931 by Matschoss within the Verein Deutscher Ingenieure. In 1962 German historians, meeting in their twenty-fifth annual conference, agreed to establish a section on "Technology and History."

Government recognition of the field has come more slowly. In 1957 a study group, Geschichte der Produktivkräfte, was established in the German Democratic Republic as a part of the Academy of Sciences, but it has proved to be something of a false start. In the Federal Republic of Germany, despite the heroic efforts of such scholars as Friedrich Klemm, government support has gone little beyond a 1962 memorandum from the Standing Committee of State Ministries of Education, urging schools to place a greater stress on the history of technology.

Recent writing on the subject (as well as the best source of the history of the subject) is discussed by Rurup (1974), in an excellent bibliographic overview of continuing German achievement in this field.

The history of technology has always been an important study in the Soviet Union, in large part because the Marxist-Leninist philosophy of history places great emphasis upon systems of social production. In 1921 a Commission on the History of Science, Philosophy, and Technology was established by the Soviet government. It was soon metamorphosed into the Institute of the History of Science and Technology of the Academy of Sciences at Leningrad. In 1929 the problem of educating "new, proletarian specialists" led the Central Committee of the Communist Party to urge a course in "the Marxist history of technology" as a part of the required training of this group (quoted in Joravsky, 1961, p. 5). After an eclipse during the late 1930s and the war years, a new Institute was made a part of the Academy of Sciences in 1945. Since that time academic work in the field has been continuous and productive (Joravsky, 1961).

The Soviet view as expressed by a leading practitioner is that "natural science only indicates possible solutions for technical problems, but by itself it can neither guide technology nor determine the scope or pace of its development. It is exclusively the economic laws of a given social system which in the long run determines the behavior of people and which guide the direction and pace of the progress of technology," a progress that can be defined as "the triumph of human thought in its fight against the forces of nature" (Zvorikine, 1961, pp. 2–4). The fullest statement of the Soviet work is contained in the survey by Zvorikine (1960), and its working out in the recent period may be followed in Mogilev (1973).

France established an early tradition of interest in the history of technology in the eighteenth century, which saw both the publication of Diderot's *Encyclopedie* and the foundation, in 1794, of the Conservatoire National des Arts et Métiers in Paris. The field as a subject of serious scholarship, however, was slow to develop, and it was not until the 1950s and 1960s that Maurice Daumas, along with Bertrand Gille, began the organization of the field in that country. A special issue of the *Annales d'histoire économique et sociale* (No. 36, November 30, 1935) on the subject of "Technology, History and Life" gave expression to its editors' (Lucien Febvre and Marc Bloch) conviction that the history of technology must be, first and foremost, a history of techniques. This priority has informed the critical work of Daumas (1976) and is clearly expressed in his multivolume *Histoire générale des techniques* (two volumes translated into English, 1962).

The Canadian experience is in many ways similar to that discovered in other countries. In 1974 Sinclair, writing the introduction to his collection of articles on the history of technology in that country, lamented the fact that "curiously enough, the subject has been little studied" (1974, p. 3). Nevertheless, for some years the Institute for the History and Philosophy of Science and Technology, founded by John W. Abrams, has been teaching classes and training graduate students at the University of Toronto, and a growing number of these are concerned with technology rather than science.

Most nations, both West and East, have recently shown a new interest in the history of technology as a field of scholarly inquiry. University courses, study groups, official institutes, sections of historical or engineering organizations, new journals or new access to old ones, and, most important, the persistent efforts of individual scholars in Japan, Italy, Poland, Czechoslovakia, India, and a number of other nations are helping establish the field as one of real importance and interest.

A new journal, *Technologia Bruxellensis*, appeared in March 1978 under the auspices of the Brussels Association of Engineers. Articles in both Dutch and French treat industrial archeology as well as the history of technology. In Syria an Institute for the History of Arabic Science has been established at the University of Aleppo, and in May 1977 the first issue of a *Journal for the History of Arabic Science* appeared, containing material on the history of technology as well. Japan played host to the XIVth International Congress of the History and Philosophy of Science; and recently a Japanese translation of the second volume of Kranzberg and Pursell (1967) was published and widely reviewed.

V. VALUES AND THE HISTORY OF TECHNOLOGY

Technology is a form of human behavior, and questions of ethical and value implications, therefore, strike to its core. This view runs counter, of course, to the common perception that technology is somehow "neutral" and therefore beyond ethics. The question is more than one of simple academic interest. The frequently heard slogan "guns don't kill people, people kill people" reminds us that public policy can be intimately tied to one or another view of technology as a force in our society.

The argument for neutrality rests on the obvious belief that ethics and values are relevant only to action and that cold machines cannot act. Willful action is restricted to human beings, and it is they, therefore, who are responsible for the results of that action. As the mere agent of that action, technology cannot be held responsible. This belief is reinforced by the common observation that a gun, for example, can be used either for aggression (bad) or for defense (good), carelessly (bad) or to provide food (good).

The argument against neutrality stems from a basically Aristotelian belief that form, material, and purpose (function) are necessarily interrelated in the world made by human beings. Thus a chair and a strategic bomber have certain purposes, and these are built in, as it were, to their form and influence the material from which they will be built. Indeed, if we know the form and material of a tool, we can make a shrewd guess about its purpose, and if we know its purpose we can say something about its likely form and fabric. According to this argument, therefore, the purposes (ethics and values) of our society are built into the very form and fabric of our technology, and the latter does not exist in some neutral sphere divorced from that purpose.

These arguments are endlessly elaborated and intertwined and are mixed with questions of direction and control. They have grave implications, also, for the way in which the history of technology is studied and taught. If technology is neutral or, at the other extreme, necessarily progressive and benign, one may well be satisfied with an internal description of the elaboration of machines and their components and uses, setting this story in a vague framework of self-congratulation and optimism about the future. If, on the other hand, technology embodies the values of the society that creates it, or, at the other extreme, has its own internal and inherent values, then the history of technology should encompass not only the elaboration of devices but also those values and their impact on the society in which they exist.

One advantage of the former approach is that it gives the history of technology a field of its own to till; it resists the argument that the historian interested in the largest and most important questions will pass over technology as merely symptomatic to deal directly with human beings themselves. An advantage of the latter approach is that it demonstrates that, although human beings are the proper study of history, they can best be understood through their behavior, and one fundamental expression of their behavior is technology.

On this count, as on others, the history of technology is still intimately bound up with the history of science. The late Paul Goodman once suggested that technology is not simply an inferior kind of science but an important branch of moral philosophy. It was his suggestion that as a form of behavior, technology had ethically to be prudent, and that it was only an historic confusion with science that allowed the technician to claim (by association) a kind of immunity that society was beginning to grant to scientists. Technology, properly viewed, was to be seen as one of the humanities—just as Lynn White (1974) has reminded us that quantity is only one of the qualities.

Historians alone are not likely to solve the ethical dilemmas of the modern world, but the historian of technology can and should, at the very least, make it clear that not every technology that can be is, that no two machines are ever entirely interchangeable, and that the history of technology is therefore a record of choices made—choices that, quite obviously, grow out of values, and that are entirely meaningless outside that value context.

BIBLIOGRAPHIC INTRODUCTION

Note: Items listed in the bibliographic introduction are not repeated in the main bibliography.

A.   Physical Sites

There is no way to list the important physical sites in the history of technology. In the United States, such places as the Old Slater Mill in Pawtucket, R.I., the site of the first cotton spinning mill with Arkwright machinery in America; Hopewell Village, a reconstructed nineteenth-century iron furnace maintained by the National Park Service in east central Pennsylvania; the Old Bale Mill (a gristmill dating from the 1840s) in Sonoma County, California; and the seventeenth-century ironworks rebuilt at Saugus, Massachusetts, are outstanding examples of outdoor museums that must be experienced to be appreciated. In England the Ironbridge Gorge Museum Trust, Ironbridge, Telford, Salop, has done an outstanding job of preserving and restoring the single most important British site associated with the changing role of iron in the Industrial Revolution.

Perhaps the best guide to such sites is through the growing literature on industrial archeology. Neil Cossons, *The BP Book of Industrial Archaeology* (London: David & Charles, 1975) contains a seven-page guide to "Museums of Industry" (pp. 451–457) that gives useful information on British sites. The following works may also be consulted:

Buchanan, R. A. *Industrial Archeology in Britain.* Harmondsworth, Middlesex: Penguin, 1972.

Hudson, Kenneth. *A Guide to the Industrial Archeology of Europe.* Cranbury, N.J.: Fairleigh Dickinson University Press, 1971.

Rees, D. Morgan. *Mines, Mills, and Furnaces: Industrial Archaeology in Wales.* London: Her Majesty's Stationers Office, 1969.

Sande, Theodore Anton. *Industrial Archeology—A New Look at the American Heritage*. Brattleboro: Stephen Greene Press, 1976.

It should be noted that the Historic American Engineering Record, a branch of the National Park Service (Washington, D.C. 20240) is engaged in a massive effort to inventory and record all significant engineering sites and structures in the United States. Thus far, inventories have been published for the Lower Michigan Penninsula, the Lower Merrimack River Valley, Long Island, North Carolina, Delaware, New England, and a number of other locations.

B. Museums

Museums of science, technology, and industry are invaluable aids in the study of the history of technology. More than is the case with some other branches of historical study, the understanding of material things is significantly advanced by a close study of objects as well as written documents. Outstanding museums for the history of technology include the Science Museum in London, the Deutsches Museum in Munich, the Conservatoire des Arts et Métiers in Paris, the Instituto e Museo di Storia della Scienze in Florence, and the Tekniska Museet in Stockholm. In the United States, the Smithsonian Institution's Museum of History and Technology has the highest attendance of any such museum in the world; it is followed by the Museum of Science and Industry in Chicago. Other excellent American museums include the Franklin Institute in Philadelphia, the Henry Ford Museum and Greenfield Village in Dearborn, Michigan, the Corning Glass Museum in Corning, New York, the Merrimack Valley Textile Museum in North Andover, Massachusetts, and the Hagley Museum, near Wilmington, Delaware.

The best introduction to technical museums is the special theme issue of *Technology and Culture* 6 (Winter 1965), especially the following:

Bedini, Silvio A. "The Evolution of Science Museums," pp. 1–29 (which includes a table, "Early Science Collections and Museums through the Nineteenth Century");

Brown, John J. "Museum Census: A Survey of Technology in Canadian Museums," pp. 83–98;

Ferguson, Eugene S. "Technical Museums and International Exhibitions," pp. 30–46;

Finn, Bernard S. "The Science Museum Today," pp. 74–82 (which includes a table, "Some Modern Science Museums").

C. Archives and Special Collections

Manuscript resources for the history of technology are everywhere and nowhere: nowhere because no institution has yet declared itself to be the special repository of such material and undertaken an aggressive policy toward acquisition; everywhere, because, like every other human activity, technology leaves its written trail, which is commingled with that of other human activities. Manuscript resources, therefore, are found in many, perhaps most, of the archives, large and small, that historians usually visit. Two special locations demand particular attention. The Historic American Engineering Record, besides its published inventories, has amassed a sizable collection of drawings and photographs of structures built more than fifty years ago. One way into this material is through the *Historic American Engineering Record Catalog, 1976,* compiled by Donald E. Sackheim (Washington, D.C.: National Park Service, 1976). A second resource is the biographical file of American engineers maintained by the Smithsonian Institution and housed at the Museum of History and Technology. Hundreds of engineers who flourished in the nineteenth century are documented in this collection.

Since the federal government has always had a particular interest in technological matters, the National Archives in Washington, D.C., are a prime source of manuscript materials. Two key agencies, which bracket much of this interest, are described in Nathan Reingold, "U.S. Patent Office Records as Sources for the History of Invention and Technological Property," *Technology and Culture* 1 (Spring 1960): 156–167; and Meyer Fishbein, "Archival Remains of Research and Development during the Second World War," in *World War II: An Account of Its Documents,* pp. 163–179, edited by James E. O'Neill and Robert W. Krauskopf (Washington, D.C.: Howard University Press, 1976), which deals with the Office of Scientific Research and Development, among other agencies. Material on the history of technology may be found in such diverse record groups as those for the Department of State (dispatches from foreign service personnel abroad) and the Department of Agriculture.

The Library of Congress contains many collections of interest, including those of Vannevar Bush, head of the wartime Office of Scientific Research and Development, and Gen. Goethals, who directed the American Panama Canal effort. Schools such as the Massachusetts Institute of Technology have historical archives (in this case, the Historical Collection), as do many industrial corporations such as the General Electric Company. The Franklin Institute in Philadelphia has a fine archive that is of particular importance because of its leadership in the Mechanics' Institute movement during the first half of the nineteenth century. The Eleutherian Mills Historical Library near the Hagley Museum in Delaware has a first-rate archive of industrial, technological, and business papers as well as a research library and impressive photograph collection. Finally, special collections of old trade catalogs are maintained at several locations, most importantly the Smithsonian Institution on the east coast and the University of California, Santa Barbara, on the west.

D. Bibliographies, Directories, Indexes

American Society of Civil Engineers. *A Biographical Dictionary of American Civil Engineers.* New York: American Society of Civil Engineers, 1972.
  Includes 170 entries for those born before the Civil War, and list of those to be included in future volumes.

Bell, Samuel Peter, comp. *A Biographical Index of British Engineers in the 19th Century.* New York: Garland Publishers, 1975.

Ezell, Edward C. "Science and Technology in the Nineteenth Century." In *A Guide to the Sources of United States Military History*, pp. 185–215. Edited by Robin Higham. Hamden, Conn.: Archon Books, 1975.

Ferguson, Eugene S. *Bibliography of the History of Technology.* Cambridge, Mass.: Society for the History of Technology and M.I.T. Press, 1968.

———, comp. *Cassiers's Magazine, Engineering Monthly, 1891–1913.* Iowa State University Bulletin 63, no. 10. Ames: College of Engineering, Iowa State University, 1964.

Goodwin, Jack. "Current Bibliography in the History of Technology." *Technology and Culture.*
  Appears in every April issue.

Hacker, Barton C., ed. *An Annotated Index to Volumes 1 through 10 of "Technology and Culture," 1959–1969.* Chicago: University of Chicago Press, 1976.

Higgins, Thomas James. "A Biographical Bibliography of Electrical Engineers and Electrophysicists," Parts 1 and 2. *Technology and Culture* 2 (Winter 1961): 28–32; 2 (Spring 1961): 146–156.

Hounshell, David A. "A Guide to Manuscripts in Electrical History." *Technology and Culture* 15 (October 1974): 626–627.

Hindle, Brooke. *Technology in Early America: Needs and Opportunities for Study.* Chapel Hill: University of North Carolina Press, 1966.

History of Science Society. "Critical Bibliography of the History of Science and Its Cultural Influences." *Isis.*
  Appears yearly; 100th appeared in Vol. 66, Part 5, 1975. Contains subsection on history of technology.

Miles, Wyndham D., ed. *American Chemists and Chemical Engineers.* Washington, D.C.: American Chemical Society, 1976.
  Biographies of 517 deceased individuals.

Pursell, Carroll W., Jr. "Science and Technology in the Twentieth Century." In *A Guide to the Sources of United States Military History*, pp. 269–291. Edited by Robin Higham. Hamden, Conn.: Archon Books, 1975.

E. Journals

Articles on the history of technology can be found in many different general and specialized historical journals. *Agricultural History,* for example, occasionally carries articles on agricultural technology; *American Quarterly*, the journal of the American Studies Association, has had a longstanding interest in the history of technology; and state and local historical journals often carry pieces on local inventors or innovations. The journals listed below, however, are most consistent in providing such articles.

Dějiny věd a techniky (1968–   ) [Czech].

History of Technology (1976–   ) [British].

IA: The Journal of the Society for Industrial Archeology (1975–   ) [U.S.].

Isis (1913–   ) [U.S.].

Journal of Industrial Archeology (1964–   ) [British].

Le Machine: Bollettino dell' Instituto Italiano per la Storia della Technica (1967–   ) [Italian].

Technikgeschichte (1909–   ) [German].

Newcomen Society for the Study of the History of Engineering and Technology, *Transactions* (1920–   ) [British].

Technology and Culture (1959–   ) [U.S.].

F. Historiographic Discussions

Buchanan, Angus. "Technology and History." *Social Studies of Science* 5 (November 1975): 489–499.

Burke, John G. "The Complex Nature of Explanations in the Historiography of Technology." *Technology and Culture* 11 (January 1970): 22–26.

Daniels, George H. "The Big Questions in the History of American Technology." *Technology and Culture* 11 (January 1970): 1–21.

Daumas, Maurice. "L'Histoire Generale des Techniques." *Technology and Culture* 1 (Fall 1960): 415–418.

———. "The History of Technology: Its Aims, Its Limits, Its Methods." In *History of Technology*, 1st annual vol., 1976, pp. 85–112. London: Mansell, 1976.

Dupree, A. Hunter. "The Role of Technology in Society and the Need for Historical Perspective." *Technology and Culture* 10 (October 1969): 528–534.

Ferguson, Eugene S. "Toward a Discipline of the History of Technology." *Technology and Culture* 15 (January 1974): 13–30.

Heilbroner, Robert L. "Do Machines Make History?" *Technology and Culture* 8 (July 1967): 335–345.

Jones, Howard Mumford. "Ideas, History, Technology." *Technology and Culture* 1 (Winter 1959): 20–27.

Joravsky, David. "The History of Technology in Soviet Russia and Marxist Doctrine." *Technology and Culture* 2 (Winter 1961): 5–10.

Layton, Edwin T., Jr. "The Interaction of Technology and Society." *Technology and Culture* 11 (January 1970): 27–31.

———. "Technology as Knowledge." *Technology and Culture* 15 (January 1974): 31–41.

————. "Conditions of Technological Development." In *Science, Technology, and Society: A Cross-Disciplinary Perspective,* pp. 197–222. Edited by Ina Spiegel-Rösing and Derek de Solla Price. Beverly Hills: Sage Publications, 1977.

Lenzi, Giulio. "Storia della Technica dal Medioevo al Rinascimiento, by Umberto Forti." *Technology and Culture* 1 (Fall 1960): 419–420.

Mayr, Otto. "The Science-Technology Relationship as a Historiographic Problem." *Technology and Culture* 17 (October 1976): 663–672.

Multhauf, Robert P. "Some Observations on the State of the History of Technology." *Technology and Culture* 15 (January 1974): 1–12.

Price, Derek de Solla. "On the Historiographic Revolution in the History of Technology." *Technology and Culture* 15 (January 1974): 42–48.

Rurup, Reinhard. "Reflections on the Development and Current Problems of the History of Technology." *Technology and Culture* 15 (April 1974): 161–193.

Schlesinger, Arthur M., Sr. "An American Historian Looks at Science and Technology." *Isis* 36 (October 1946): 162–166.

"The Society for the History of Technology." Organizational Notes, in *Technology and Culture* 1 (Winter 1959): 106–108.
        Written by Melvin Kranzberg.

Woodbury, Robert S. "The Scholarly Future of the History of Technology." *Technology and Culture* 1 (Fall 1960): 345–348.

Zvorikine, A. "The Soviet History of Technology." *Technology and Culture* 1 (Fall 1960): 421–425.

————. "The History of Technology as a Science and a Branch of Learning: A Soviet View." *Technology and Culture* 2 (Winter 1961): 1–4.

G. Surveys

Burlingame, Roger. *Engines of Democracy: Inventions and Society in Mature America.* New York: Charles Scribner's Sons, 1940.

————. *March of the Iron Men: A Social History of Union through Invention.* New York: Charles Scribner's Sons, 1943.

————. *Backgrounds of Power: The Human Story of Mass Production.* New York: Charles Scribner's Sons, 1949.

Daumas, Maurice, ed. *A History of Technology and Invention: Progress Through the Ages.* 2 vols. New York: Crown Publishers, 1962.

Derry, T. K., and Williams, Trevor I. *A Short History of Technology from the Earliest Times to A.D. 1900.* New York: Oxford University Press, 1961.

Finch, James Kip. *The Story of Engineering.* Garden City, N.Y.: Doubleday, 1960.

Forbes, R. J. *Man the Maker: A History of Technology and Engineering.* New York: Henry Schuman, 1950.

————. *Studies in Ancient Technology.* 9 vols. Leiden: E. J. Brill, 1955–1964; New York: William S. Heinman, 1964–1972.

Forbes, R. J., and Dijksterhuis, E. J. *A History of Science and Technology.* 2 vols. Baltimore: Pelican Books, 1963.

Giedion, Siegfried. *Mechanization Takes Command: A Contribution to Anonymous History.* New York: Oxford University Press, 1948.

Klemm, Friedrich. *A History of Western Technology.* Cambridge, Mass.: M.I.T. Press, 1964.

Kranzberg, Melvin, and Pursell, Carroll W., Jr. *Technology in Western Civilization.* 2 vols. New York: Oxford University Press, 1967.

Landes, David S. *The Unbound Prometheus: Technological Change and Industrial Development in Western Europe from 1750 to the Present.* New York: Cambridge University Press, 1969.

Mumford, Lewis. *Technics and Civilization.* New York: Harcourt, Brace & World, 1934; reprint ed., New York: Harcourt Brace Jovanovich, 1963.

——. *The Myth of the Machine: Technics and Human Development.* New York: Harcourt, Brace & World, 1967.

——. *The Myth of the Machine: The Pentagon of Power.* New York: Harcourt Brace Jovanovich, 1970.

Needham, Joseph. *Science and Civilisation in China.* New York: Cambridge University Press, 1954–.

   A multi-volume work in progress.

Singer, Charles; Holmyard, E. J.; Hall, A. R.; and Williams, Trevor I., eds. *A History of Technology.* 5 vols. Oxford: Oxford University Press, 1955–1958.

Struik, Dirk J. *Yankee Science in the Making.* Boston: Little, Brown, 1948.

Toynbee, Arnold. *Lectures on the Industrial Revolution in England.* London, 1884.

Usher, Abbott Payson. *A History of Mechanical Inventions.* Cambridge, Mass.: Harvard University Press, 1929.

Wolf, A. *A History of Science, Technology, and Philosophy in the 18th Century.* New York: Macmillan, 1938; reprint ed., Gloucester, Mass.: Peter Smith, n.d.

——. *A History of Science, Technology, and Philosophy in the 16th and 17th Centuries.* New York: Macmillan, 1935; 2nd ed., 1950; reprint ed., Gloucester, Mass.: Peter Smith, n.d.

H. Anthologies

Hughes, Thomas Parke. *Changing Attitudes Toward American Technology.* New York: Harper & Row, 1975.

Kranzberg, Melvin, and Davenport, William H. *Technology and Culture: An Anthology.* New York: Shocken Books, 1972.

Layton, Edwin T., Jr. *Technology and Social Change in America.* New York: Harper & Row, 1973.

Pursell, Carroll W., Jr. *Readings in Technology and American Life.* New York: Oxford University Press, 1969.

Saul, S. B. *Technological Change: The United States and Britain in the 19th Century.* London: Methuen, 1970.

Sinclair, Bruce; Ball, Norman R.; and Peterson, James O., eds. *Let Us Be Honest and Modest: Technology and Society in Canadian History.* Toronto: Oxford University Press, 1974.

Smith, Cyril Stanley. *Sources for the History of the Science of Steel, 1532–1786.* Cambridge, Mass.: Society for the History of Technology and M.I.T. Press, 1968.

BIBLIOGRAPHY

Aitken, Hugh G. J. *Syntony and Spark: The Origins of Radio.* New York: John Wiley, 1976.

Alford, L. P. *Henry Laurence Gantt: Leader in Industry.* New York: Harper & Bros., 1934.

Amis, Kingsley. *New Maps of Hell: A Survey of Science Fiction.* New York: Harcourt, Brace, 1960.

Andrews, William D., and Andrews, Deborah C. "Technology and the Housewife in Nineteenth-Century America." *Women Studies Abstracts* 2 (Summer 1974): 309–328.

Arms, Suzanne. *Immaculate Deception: A New Look at Women and Childbirth in America.* Boston: Houghton Mifflin, 1975.

Art, Robert J. *The TFX Decision: McNamara and the Military.* Boston: Little, Brown, 1968.

Artz, Frederick B. *The Development of Technical Education in France, 1500–1850.* Cambridge, Mass.: Society for the History of Technology and M.I.T. Press, 1966.

Bailes, Kendall E. "The Politics of Technology: Stalin and Technocratic Thinking among Soviet Engineers." *American Historical Review* 79 (April 1974): 445–469.

———. *Technology and Society under Lenin and Stalin: Origins of the Soviet Technical Intelligentsia, 1917–1941.* Princeton: Princeton University Press, 1978.

Baker, Elizabeth Faulkner. *Technology and Woman's Work.* New York: Columbia University Press, 1964.

Baldwin, Keturah E. *The AHEA Saga: A Brief History of the Origin and Development of the American Home Economics Association and a Glimpse at the Grass Roots from Which It Grew.* Washington, D.C.: American Home Economics Association, 1949.

Bateman, Fred. "Improvement in American Dairy Farming, 1850–1910: A Quantitative Analysis." *Journal of Economic History* 28 (June 1968): 255–273.

Baxter, James Phinney III. *Scientists Against Time.* Boston: Little, Brown, 1946; reprint ed., Cambridge, Mass.: M.I.T. Press, 1968.

Beard, Charles A., ed. *Whither Mankind: A Panorama of Modern Civilization*. New York: Longmans, Green, 1928.

Bedini, Silvio A. *Thinkers and Tinkers: Early American Men of Science*. New York: Charles Scribner's Sons, 1975.

Beer, John Joseph. *The Emergence of the German Dye Industry*. Illinois Studies in the Social Sciences, vol. 44. Urbana: University of Illinois Press, 1959.

Bell, Daniel. *Work and Its Discontents: The Cult of Efficiency in America*. New York: League for Industrial Democracy, 1970.

Berger, Albert I. "The Magic That Works: John W. Campbell and the American Response to Technology." *Journal of Popular Culture* 5 (Spring 1972): 867–943.

Birr, Kendall. *Pioneering in Industrial Research: The Story of the General Electric Research Laboratory*. Washington, D.C.: Public Affairs Press, 1957.

———. "Science in American Industry." In *Science and Society in the United States*, pp. 35–80. Edited by David D. Van Tassel and Michael G. Hall. Homewood, Ill.: Dorsey Press, 1966.

Blake, Nelson Manfred. *Water for the Cities: A History of the Urban Water Supply Problem in the United States*. Syracuse: Syracuse University Press, 1956.

Boorstin, Daniel J. *The Americans: The Democratic Experience*. New York: Random House, 1973.

Bridenbaugh, Carl. *The Colonial Craftsman*. New York: New York University Press, 1950.

Brilliant, Ashleigh E. "Some Aspects of Mass Motorization in Southern California, 1919–1929." *Historical Society of Southern California Quarterly* 47 (June 1965): 191–208.

Brownell, Blaine A. "A Symbol of Modernity: Attitudes Toward the Automobile in Southern Cities in the 1920s." *American Quarterly* 24 (March 1972): 20–44.

Bruce, Robert V. *Lincoln and the Tools of War*. Indianapolis: Bobbs-Merrill, 1956.

———. *Bell: Alexander Graham Bell and the Conquest of Solitude*. Boston: Little, Brown, 1973.

Bryant, Lynwood. "Rudolf Diesel and His Rational Engine." *Scientific American* 221 (August 1969): 108–118.

———. "The Role of Thermodynamics in the Evolution of Heat Engines." *Technology and Culture* 14 (April 1973): 152–165.

———. "The Development of the Diesel Engine." *Technology and Culture* 17 (July 1976): 432–446.

Buchanan, R. A. *The Industrial Archeology of Bath*. Bath: Bath University Press, 1969.

———. "The Promethean Revolution: Science, Technology, and History."

In *History of Technology,* 1st annual vol., pp. 73–83. Edited by A. Rupert Hall and Norman Smith. London: Mansell, 1976.

Buchanan, R. A., and Cossons, Neil. *The Industrial Archaeology of the Bristol Region.* New York: Augustus M. Kelley, 1969.

Burke, John G. "Bursting Boilers and the Federal Power." *Technology and Culture* 7 (Winter 1966): 1–23.

Burns, Alfred. "Ancient Greek Water Supply and City Planning: A Study of Syracuse and Acragas." *Technology and Culture* 15 (July 1974): 389–412.

Butti, Ken, and Perlin, John. "Solar Water Heaters in California, 1891–1930." *The CoEvolution Quarterly* 15 (Fall 1977): 1–13.

Cain, Louis P. "Unfouling the Public's Nest: Chicago's Sanitary Diversion of Lake Michigan Water." *Technology and Culture* 15 (October 1974): 594–613.

Calhoun, Daniel Hovey. *The American Civil Engineer: Origins and Conflict.* Cambridge, Mass.: The Technology Press, M.I.T., 1960.

Calvert, Monte A. *The Mechanical Engineer in America, 1830–1910.* Baltimore: Johns Hopkins University Press, 1967.

Cardwell, Donald S. L. *From Watt to Clausius: The Rise of Thermodynamics in the Early Industrial Age.* Ithaca: Cornell University Press, 1971.

——. *Turning Points in Western Technology: A Study of Technology, Science, and History.* New York: Science History Publications, 1972.

Carey, James W., and Quirk, John J. "The Mythos of the Electronic Revolution." *American Scholar* 39 (Spring 1970): 219–241, and 39 (Summer 1970): 395–424.

Carlson, Robert E. "British Railroads and Engineers and the Beginnings of American Railroad Development." *Business History Review* 34 (Summer 1960): 137–149.

Cavert, William L. "The Technological Revolution in Agriculture, 1910–1955." *Agricultural History* 30 (January 1956): 18–27.

Chandler, Alfred D., Jr. "Anthracite Coal and the Beginnings of the Industrial Revolution in the United States." *Business History Review* 46 (Summer 1972): 141–181.

——. *The Visible Hand: The Managerial Revolution in American Business.* Cambridge, Mass.: Harvard University Press, 1977.

Chase, Stuart. *Men and Machines.* New York: Macmillan, 1929.

Christie, Jean. "The Mississippi Valley Committee: Conservation and Planning in the Early New Deal." *The Historian* 32 (May 1970): 449–469.

Cipolla, Carlo M. *Guns, Sails and Empires: Technological Innovation and the Early Phases of European Expansion, 1400–1700.* New York: Pantheon, 1965.

Clow, Archibald, and Clow, Nan. *The Chemical Revolution: A Contribution to Social Technology.* London: Batchworth Press, 1952.

Colman, Gould P. "Innovation and Diffusion in Agriculture." *Agricultural History* 42 (July 1968): 173–187.

Commoner, Barry. *The Closing Circle: Nature, Man & Technology.* New York: Alfred A. Knopf, 1971.

Comp, T. Allan. "The Toole Copper and Lead Smelter." *IA: The Journal of the Society for Industrial Archeology* 1 (Summer 1975): 29–46.

Condit, Carl W. *American Building Art: The Nineteenth Century.* New York: Oxford University Press, 1960.

——. *American Building Art: The Twentieth Century.* New York: Oxford University Press, 1961.

——. "The First Reinforced-Concrete Skyscraper: The Ingalls Building in Cincinnati and Its Place in Structural History." *Technology and Culture* 9 (January 1963): 1–33.

Cooper, Grace Rogers. *The Invention of the Sewing Machine.* Museum of History and Technology Bulletin 245. Washington, D.C.: Smithsonian Institution, 1968.

Copley, Frank Barkley. *Frederick W. Taylor: Father of Scientific Management.* 2 vols. New York: Harper & Bros., 1923; reprint ed., Clifton N.J.: Augusta Kelley, n.d.

Cowan, Ruth Schwartz. "The 'Industrial Revolution' in the Home: Household Technology and Social Change in the 20th Century." *Technology and Culture* 17 (January 1976): 1–23.

Curti, Merle, and Birr, Kendall. *Prelude to Point Four: American Technical Missions Overseas, 1838–1938.* Madison: University of Wisconsin Press, 1954.

Danhof, Clarence H. "Gathering the Grass." *Agricultural History* 30 (October 1956): 169–173.

——. *Change in Agriculture: The Northern United States, 1820–1870.* Cambridge, Mass.: Harvard University Press, 1969.

Davis, Robert B. " 'Peacefully Working to Conquer the World': The Singer Manufacturing Company in Foreign Markets, 1854–1889." *Business History Review* 43 (Autumn 1969): 299–325.

Dickinson, H. W. *James Watt: Craftsman & Engineer.* Cambridge: Cambridge University Press, 1935; reprint ed., North Pomfret, Vt.: David and Charles, n.d.

——. *Matthew Boulton.* Cambridge: Cambridge University Press, 1936.

Durand, William Frederick. *Robert Henry Thurston: A Biography. The Record of a Life of Achievement as Engineer, Educator, and Author.* New York: American Society of Mechanical Engineers, 1929.

Ellis, John. *The Social History of the Machine Gun.* New York: Pantheon, 1975.

Elsner, Henry, Jr. *The Technocrats: Prophets of Automation.* Syracuse: Syracuse University Press, 1967.

Enos, John Lawrence. *Petroleum Progress and Profits: A History of Process Innovation.* Cambridge, Mass.: M.I.T. Press, 1962.

Esper, Thomas. "The Replacement of the Longbow by Firearms in the English Army." *Technology and Culture* 6 (Spring 1965): 382–393.

Etzioni, Amitai. *The Moon-Doggle: Domestic and International Implications of the Space Race.* Garden City, N.Y.: Doubleday, 1964.

Fagen, M. D., ed. *A History of Engineering and Science in the Bell System: The Early Years (1875–1925).* Murray Hill, N.J.: Bell Telephone Laboratories, 1975.

Feller, Irwin. "The Diffusion and Location of Technological Change in the American Cotton-Textile Industry, 1890–1970." *Technology and Culture* 15 (October 1974): 569–593.

Ferguson, Eugene S., ed. *Early Engineering Reminiscences (1815–1840) of George Escol Sellers.* Museum of History and Technology Bulletin, no. 238. Washington, D.C.: Smithsonian Institution, 1965.

———. "The Mind's Eye: Nonverbal Thought in Technology." *Science* 197 (26 August 1977): 827–836.

Fisher, Bernice M. "Public Education and 'Special Interest': An Example from the History of Mechanical Engineering." *History of Education Quarterly* 6 (Spring 1966): 31–40.

Fisher, Marvin. *Workshops in the Wilderness: The European Response to American Industrialization, 1830–1860.* New York: Oxford University Press, 1967.

Fitch, James Marston. *American Building: The Historical Forces That Shaped It.* 2d ed. Boston: Houghton Mifflin, 1966.

Flink, James J. *America Adopts the Automobile, 1895–1910.* Cambridge, Mass.: M.I.T. Press, 1970.

———. "Three Stages of Automobile Consciousness." *American Quarterly* 24 (October 1972): 451–473.

———. *The Car Culture.* Cambridge, Mass.: M.I.T. Press, 1975.

Fox, Frank W. "The Genesis of American Technology, 1790–1860: An Essay in Long-Range Perspective." *American Studies* 15 (Fall 1976): 29–48.

Fries, Russell I. "British Response to the American System: The Case of the Small Arms Industry after 1850." *Technology and Culture* 16 (July 1975): 377–403.

Fuller, Wayne E. "The Ohio Road Experiment, 1913–1916." *Ohio History* 74 (Winter 1965): 13–28.

Gilbreth, Frank B., Jr. *Time Out for Happiness.* New York: Thomas Y. Crowell, 1970.

Gilfillan, S. C. "An Attempt to Measure the Rise of American Inventing and the Decline of Patenting." *Technology and Culture* 1 (Summer 1960): 201–214.

Gorman, Mel. "Charles F. Brush and the First Public Electric Street Lighting System in America." *Ohio Historical Quarterly* 70 (April 1961): 128–144.

Graham, Frank, Jr. *Since Silent Spring*. Boston: Houghton Mifflin, 1970.

Green, Constance McL. *Eli Whitney and the Birth of American Technology*. Boston: Little, Brown, 1956.

Habakkuk, H. J. *American and British Technology in the Nineteenth Century: The Search for Labor-Saving Inventions*. Cambridge: Cambridge University Press, 1962.

Haber, Samuel. *Efficiency and Uplift: Scientific Management in the Progressive Era, 1890–1920*. Chicago: University of Chicago Press, 1964.

Hacker, Barton C. "Greek Catapults and Catapult Technology: Science, Technology, and War in the Ancient World." *Technology and Culture* 9 (January 1968): 34–50.

Hamlin, Talbot. *Benjamin Henry Latrobe*. New York: Oxford University Press, 1955.

Hays, Samuel P. *Conservation and the Gospel of Efficiency: The Progressive Conservation Movement, 1890–1920*. Cambridge, Mass.: Harvard University Press, 1959.

Hewlett, Richard G. "Beginnings of the Development in Nuclear Technology." *Technology and Culture* 17 (July 1976): 465–478.

Hewlett, Richard G., and Anderson, Oscar E., Jr. *The New World, 1939–1946*. Vol. 1 of *A History of the United States Atomic Energy Commission*. University Park: Pennsylvania State University Press, 1962.

Hewlett, Richard G., and Duncan, Francis. *Nuclear Navy, 1946–1962*. Chicago: University of Chicago Press, 1974.

Hightower, Jim. "Hard Tomatoes, Hard Times: Failure of the Land Grant College Complex." *Society* 10 (November-December 1972): 10–11, 14, 16–22.

Hill, Forest G. *Roads, Rails, and Waterways: The Army Engineers and Early Transportation*. Norman: University of Oklahoma Press, 1957.

Hills, R. L., and Pacey, A. J. "The Measurement of Power in Early Steam-Driven Textile Mills." *Technology and Culture* 13 (January 1972): 25–43.

Hindle, Brooke, ed. *America's Wooden Age: Aspects of Its Early Technology*. Tarrytown, N.Y.: Sleepy Hollow Restorations, 1975.

Hughes, Thomas Parke. *Elmer Sperry: Inventor and Engineer*. Baltimore: Johns Hopkins University Press, 1971.

———. "The Science-Technology Interaction: The Case of High-Voltage Power Transmission Systems." *Technology and Culture* 17 (October 1976): 646–662.

Hunsaker, J. C. "Forty Years of Aeronautical Research." In *Annual Report of the Board of Regents of the Smithsonian Institution. Publication 4232. Showing the Operations, Expenditures, and Condition of the Institution for the Year Ended June 30, 1955*, pp. 241–271. Washington, D.C.: Government Printing Office, 1956.

Hunt, Caroline L. *The Life of Ellen H. Richards*. Boston: n.p., 1912.

Hunter, Louis C. *Steamboats on the Western Rivers: An Economic and Technological History*. Cambridge, Mass.: Harvard University Press, 1949.

Illich, Ivan. *Tools for Conviviality*. New York: Harper & Row, 1973.

Inouye, Arlene, and Susskind, Charles. " 'Technological Trends and National Policy,' 1937: The First Modern Technology Assessment." *Technology and Culture* 18 (October 1977): 593–621.

Jardin, Anne. *The First Henry Ford: A Study in Personality and Business Leadership*. Cambridge, Mass.: M.I.T. Press, 1970.

Jenkins, Reese V. *Images and Enterprise: Technology and the American Photographic Industry, 1839 to 1925*. Baltimore: Johns Hopkins University Press, 1975.

――――. "Technology and the Market: George Eastman and the Origins of Mass Amateur Photography." *Technology and Culture* 16 (January 1975): 1–19.

Jeremy, David J. "Innovation in American Textile Technology during the Early 19th Century." *Technology and Culture* 14 (January 1973): 40–76.

――――. "British Textile Technology Transmission to the United States: The Philadelphia Region Experience, 1770–1820." *Business History Review* 47 (Spring 1973): 24–52.

――――. "Damming the Flood: British Government Efforts to Check the Outflow of Technicians and Machinery, 1780–1843." *Business History Review* 51 (Spring 1977): 1–34.

Jervis, John B. *The Reminiscences of John B. Jervis: Engineer of the Old Croton*. Edited by Neal FitzSimons. Syracuse: Syracuse University Press, 1971.

Jewkes, John; Sawers, David; and Stillerman, Richard. *The Sources of Invention*. London: Macmillan, 1961.

Josephson, Matthew. *Edison: A Biography*. New York: McGraw-Hill, 1959.

Kakar, Sudhir. *Frederick Taylor: A Study in Personality*. Cambridge, Mass.: M.I.T. Press, 1970.

Kasson, John F. *Civilizing the Machine: Technology and Republican Values in America, 1776–1900*. New York: Grossman, 1976.

Kelley, Robert L. "Forgotten Giant: The Hydraulic Gold Mining Industry in California." *Pacific Historical Review* 23 (November 1954): 343–356.

Kendall, Edward C. "John Deere's Steel Plow." *U.S. National Museum Bulletin* 218 (1959): 15–25. Washington, D.C.: Smithsonian Institution, 1959.

Kennedy, David M. *Birth Control in America: The Career of Margaret Sanger*. New Haven: Yale University Press, 1970.

Kennett, Lee, and Anderson, James La Verne. *The Gun in America: The Origins of a National Dilemma*. Westport, Conn.: Greenwood Press, 1975.

Kevles, Daniel J. "Federal Legislation for Engineering Experiment Stations: The Episode of World War I." *Technology and Culture* 12 (April 1971): 182–189.

Kleinberg, Susan J. "Technology and Women's Work: The Lives of Working Class Women in Pittsburgh, 1870–1900." *Labor History* 17 (Winter 1976): 58–72.

Kouwenhoven, John A. *Made in America.* New York: Doubleday, 1948; reprint ed., in paperback, new title, *The Arts in Modern American Civilization,* New York: W. W. Norton, 1967.

Kranzberg, Melvin. "Historical Aspects of Technology Assessment." In *Technology Assessment,* pp. 380–388. Hearings before the Subcommittee on Science, Research, and Development of the Committee on Science and Astronautics, 91st Cong., 1st sess., November 18–December 12, 1969, House of Representatives. Washington, D.C.: Government Printing Office, 1970.

Kuisel, Richard F. *Ernest Mercier: French Technocrat.* Berkeley: University of California Press, 1967.

―――. "Technocrats and Public Economic Policy: From the Third to the Fourth Republic." *Journal of European Economic History* 2 (Spring 1973): 53–99.

Kujovich, Mary Yeager. "The Refrigerator Car and the Growth of the American Dressed Beef Industry." *Business History Review* 44 (Winter 1970): 460–482.

Lasby, Clarence G. *Project Paperclip: German Scientists and the Cold War.* New York: Atheneum, 1971.

Layton, Edwin T., Jr. *The Revolt of the Engineers: Social Responsibility and the American Engineering Profession.* Cleveland: Press of Case Western Reserve University, 1971.

―――. "Mirror-Image Twins: The Communities of Science and Technology." *Technology and Culture* 12 (October 1971): 562–580.

―――. "The Diffusion of Scientific Management and Mass Production from the U.S. in the Twentieth Century." *Proceedings no. 4, 14th International Congress of the History of Science,* pp. 377–386. Tokyo: n.p., 1974.

―――. "American Ideologies of Science and Engineering." *Technology and Culture* 17 (October 1976): 688–701.

Ludwig, Karl-Heinz. *Technik und Ingenieure im Dritten Reich.* Düsseldorf: Droste Verlag, 1974.

Lytle, Richard H. "The Introduction of Diesel Power in the United States, 1897–1912." *Business History Review* 42 (Summer 1968): 115–148.

McCullough, David. *The Great Bridge.* New York: Simon & Schuster, 1972.

Machon, John K. "Anglo-American Methods of Indian Warfare, 1676–1794." *Mississippi Valley Historical Review* 45 (September 1958): 254–275.

Maclaren, Malcolm. *The Rise of the Electrical Industry During the Nine-teenth Century.* Princeton: Princeton University Press, 1943; reprint ed., Clifton, N.J.: Augusta Kelley, n.d.

Maier, Charles S. "Between Taylorism and Technocracy: European Ideologies and the Vision of Industrial Productivity in the 1920's." *Contemporary History* 5 (1970): 27–61.

Malone, Patrick M. "Changing Military Technology among the Indians of Southern New England, 1600–1677." *American Quarterly* 25 (March 1973): 48–63.

Man, Science and Technology: A Marxist Analysis of the Scientific-Technological Revolution. Moscow and Prague: Academia Prague, 1973.

Marx, Leo. *The Machine in the Garden: Technology and the Pastoral Ideal in America.* New York: Oxford University Press, 1964.

Mayr, Otto. *The Origins of Feedback Control.* Cambridge, Mass.: M.I.T. Press, 1970.

———. "Adam Smith and the Concept of the Feedback System: Economic Thought and Technology in 18th-Century Britain." *Technology and Culture* 12 (January 1971): 1–22.

———. "Yankee Practice and Engineering Theory: Charles T. Porter and the Dynamics of the High-Speed Steam Engine." *Technology and Culture* 16 (October 1975): 570–602.

Mazlish, Bruce, ed. *The Railroad and the Space Program: An Exploration in Historical Analogy.* Cambridge, Mass.: M.I.T. Press, 1965.

Merritt, Raymond H. *Engineering in American Society, 1850–1875.* Lexington: University Press of Kentucky, 1969.

Morison, Elting E. *Men, Machines, and Modern Times.* Cambridge, Mass.: M.I.T. Press, 1966.

———. *From Know-How to Nowhere: The Development of American Technology.* New York: Basic Books, 1974.

Multhauf, Robert P. "Geology, Chemistry, and the Production of Common Salt." *Technology and Culture* 17 (October 1976): 634–645.

Mumford, Lewis. "Tools and the Man." *Technology and Culture* 1 (Fall 1960): 320–334.

———. "Authoritarian and Democratic Technics." *Technology and Culture* 5 (Winter 1964): 1–8.

Musson, A. E. "The 'Manchester School' and the Exportation of Machinery." *Business History* 14 (January 1972): 17–50.

Nelson, Daniel. "Scientific Management, Systematic Management, and Labor, 1880–1915." *Business History Review* 48 (Winter 1974): 479–500.

Noble, David F. *America by Design: Science, Technology, and the Rise of Corporate Capitalism.* New York: Alfred A. Knopf, 1977.

Pacey, Arnold. *The Maze of Ingenuity: Ideas and Idealism in the Development of Technology.* New York: Holmes & Meier, 1975.

Parsons, William Barclay. *Engineers and Engineering of the Renaissance.* Introduction by Robert S. Woodbury. Reprint ed., Cambridge, Mass.: M.I.T. Press, 1968.

Passer, Harold C. *The Electrical Manufacturers, 1875–1900: A Study in Competition, Entrepreneurship, Technical Change, and Economic Growth.* Cambridge, Mass.: Harvard University Press, 1953.

Paul, Rodman Wilson. "Colorado as a Pioneer of Science in the Mining West." *Mississippi Valley Historical Review* 47 (June 1960): 34–50.

Philmus, Robert M. *Into the Unknown: The Evolution of Science Fiction from Francis Godwin to H. G. Wells.* Berkeley: University of California Press, 1970.

Pickett, Calder M. "Technology and the New York Press in the 19th Century." *Journalism Quarterly* 37 (Summer 1960): 398–407.

Plummer, Kathleen Church. "The Streamlined Moderne." *Art in America* 62 (January–February 1974): 46–54.

Post, Robert C. *Physics, Patents, and Politics: A Biography of Charles Grafton Page.* New York: Science History Publications, 1976.

Prescott, Samuel C. *When M.I.T. Was "Boston Tech": 1861–1916.* Cambridge: M.I.T. Press, 1954.

Price, Derek J. de Solla. "Automata and the Origins of Mechanism and Mechanistic Philosophy." *Technology and Culture* 5 (Winter 1964): 9–23.

———. "Is Technology Historically Independent of Science? A Study in Statistical Historiography." *Technology and Culture* 6 (Fall 1965): 553–568.

———. "The Relations between Science and Technology and Their Implications for Policy Formation." In *Science and Technology Policies,* pp. 149–172. Edited by G. Strasser and E. Simons. Cambridge, Mass.: Ballinger, 1973.

Pursell, Carroll W., Jr. "Thomas Digges and William Pearce: An Example of the Transit of Technology." *William and Mary Quarterly* 21 (October 1964): 551–560.

———. *Early Stationary Steam Engines in America: A Study in the Migration of a Technology.* Washington, D.C.: Smithsonian Institution Press, 1969.

———. "Belling the Cat: A Critique of Technology Assessment." *Lex et Scientia* 10 (October–December 1974a): 130–145.

———. " 'A Savage Struck by Lightning': The Idea of a Research Moratorium, 1927–37." *Lex et Scientia* 10 (October–December 1974b): 146–161.

———. "The Technical Society of the Pacific Coast, 1884–1914." *Technology and Culture* 17 (October 1976): 702–717.

Rae, John B. "The Engineer as Business Man in American Industry." *Explorations in Entrepreneurial History* 7 (December 1954): 94–104.

———. "The 'Know-How' Tradition in American History." *Technology and Culture* 1 (Spring 1960): 139–150.

———. *The American Automobile: A Brief History.* Chicago: University of Chicago Press, 1965.

———. *Climb to Greatness: The American Aircraft Industry, 1920–1960.* Cambridge, Mass.: M.I.T. Press, 1968.

———. *The Road and the Car in American Life.* Cambridge, Mass.: M.I.T. Press, 1971.

———. "Engineers Are People." *Technology and Culture* 16 (July 1975): 404–418.

Raistrick, Arthur. *Industrial Archeology: An Historical Survey.* London: Eyre Methuen, Eyre & Spottiswoode, 1972.

Rasmussen, Wayne D. "Advances in American Agriculture: The Mechanical Tomato Harvester as a Case Study." *Technology and Culture* 9 (October 1968): 531–543.

Ravetz, Alison. "Modern Technology and an Ancient Occupation: Housework in Present-Day Society." *Technology and Culture* 6 (Spring 1965): 256–260.

Reich, Leonard S. "Research, Patents, and the Struggle to Control Radio: A Study of Big Business and the Uses of Industrial Research." *Business History Review* 51 (Summer 1977): 208–235.

Reingold, Nathan. "Alexander Dallas Bache: Science and Technology in the American Idiom." *Technology and Culture* 11 (April 1970): 163–177.

Reti, Ladislao. "Leonardo and Ramelli." *Technology and Culture* 13 (October 1972): 577–605.

Robinson, Eric. "James Watt and the Law of Patents." *Technology and Culture* 13 (April 1972): 115–139.

Robinson, Eric, and Musson, A. E. *James Watt and the Steam Revolution: A Documentary History.* New York: Augustus M. Kelley, 1969.

Roe, Joseph Wickham. *English and American Tool Builders.* New Haven: Yale University Press, 1916.

———. *James Hartness: A Representative of the Machine Age at Its Best.* New York: American Society of Mechanical Engineers, 1937.

Rogin, Leo. *The Introduction of Farm Machinery in Its Relation to the Productivity of Labor in the Agriculture of the United States During the Nineteenth Century.* University of California Publications in Economics, no. 9. Berkeley: University of California Press, 1931.

Rolt, L. T. C. *Isambard Kingdom Brunel: A Biography.* New York: St. Martin's Press, 1959.

———. *The Railway Revolution: George and Robert Stephenson.* New York: St. Martin's Press, 1962.

———. *A Short History of Machine Tools.* Cambridge, Mass.: M.I.T. Press, 1965.

Rosenberg, Nathan. "Technological Change in the Machine Tool Industry,

1840–1910." *Journal of Economic History* 23 (December 1963): 414–443.

————. "Economic Development and the Transfer of Technology: Some Historical Perspectives." *Technology and Culture* 11 (October 1970): 550–575.

————. "Technology and the Environment: An Economic Exploration." *Technology and Culture* 12 (October 1971): 543–561.

————. *Technology and American Economic Growth*. New York: Harper & Row, 1972.

————. *Perspectives on Technology*. Cambridge: Cambridge University Press, 1976.

————, ed. *The American System of Manufactures*. Edinburgh: University of Edinburgh Press, 1969.

Rosenbloom, Richard S. "Men and Machines: Some 19th-Century Analyses of Mechanization." *Technology and Culture* 5 (Fall 1964): 489–511.

Ross, Earle D. "Retardation in Farm Technology Before the Power Age." *Agricultural History* 30 (January 1956): 11–18.

Rossiter, Margaret W. *The Emergence of Agricultural Science: Justus Liebig and the Americans, 1840–1880*. New Haven: Yale University Press, 1975.

Roszak, Theodore. *Where the Wasteland Ends: Politics and Transcendence in Postindustrial Society*. Garden City, N.Y.: Doubleday, 1972.

Rothschild, Emma. *Paradise Lost: The Decline of the Auto-Industrial Age*. New York: Random House, 1973.

Sawyer, John E. "The Social Basis of the American System of Manufacturing." *Journal of Economic History* 14 (Fall 1954): 361–379.

Scherer, F. M. "Invention and Innovation in the Watt-Boulton Steam-Engine Venture." *Technology and Culture* 6 (Spring 1965): 165–187.

Schiff, Eric. *Industrialization Without National Patents: The Netherlands, 1869–1912, Switzerland 1850–1907*. Princeton: Princeton University Press, 1971.

Schmitz, Andrew, and Seckler, David. "Mechanized Agriculture and Social Welfare: The Case of the Tomato Harvester." *American Journal of Agricultural Economics* 52 (November 1970): 569–577.

Schmookler, Jacob. "Economic Sources of Inventive Activity." *Journal of Economic History* 22 (March 1962): 1–20.

————. *Invention and Economic Growth*. Cambridge, Mass.: Harvard University Press, 1966.

Scott, Lloyd N. *Naval Consulting Board of the United States*. Washington, D.C.: Government Printing Office, 1920.

Sharlin, Harold I. *The Making of the Electrical Age: From the Telegraph to Automation*. London: Abelard-Schuman, 1963.

Shelby, L. R. "Medieval Masons' Tools: The Level and Plumb Rule." *Technology and Culture* 2 (Spring 1961): 127–130.

———. "Medieval Masons' Tools: Compass and Square." *Technology and Culture* 6 (Spring 1965): 236–248.

———. *John Rogers, Tudor Military Engineer*. New York: Oxford University Press, 1967.

———. "Mariano Taccola and His Books on Engines and Machines." *Technology and Culture* 16 (July 1975): 466–475.

Sinclair, Bruce. "At the Turn of a Screw: William Sellers, the Franklin Institute, and a Standard American Thread." *Technology and Culture* 10 (January 1969): 20–34.

———. "The Promise of the Future: Technical Education." In *Nineteenth-Century American Science: A Reappraisal*, pp. 249–272. Edited by George H. Daniels. Evanston, Ill.: Northwestern University Press, 1972.

———. *Philadelphia's Philosopher Mechanics: A History of the Franklin Institute, 1824–1865*. Baltimore: Johns Hopkins University Press, 1974.

Sklar, Kathryn Kish. *Catherine Beecher: A Study in American Domesticity*. New Haven: Yale University Press, 1973.

Sloane, Eric. *A Reverence for Wood*. New York: Funk & Wagnalls and Thomas Y. Crowell, 1965; reprint ed., Westminster, Md.: Ballantine Books, 1973.

Smiles, Samuel. *Selections from Lives of the Engineers*. Edited by Thomas P. Hughes. Cambridge, Mass.: M.I.T. Press, 1966.

Smith, Cyril Stanley. "Art, Technology, and Science: Notes on Their Historical Interaction." *Technology and Culture* 11 (October 1970): 493–549.

Smith, Henry Nash. *Mark Twain's Fable of Progress: Political and Economic Ideas in "A Connecticut Yankee."* New Brunswick, N.J.: Rutgers University Press, 1964.

Smith, Merritt Roe. "John H. Hall, Simeon North, and the Milling Machine: The Nature of Innovation among Antebellum Arms Makers." *Technology and Culture* 14 (October 1973): 573–591.

———. *Harper's Ferry Armory and the New Technology: The Challenge of Change*. Ithaca, N.Y.: Cornell University Press, 1977.

Smith, Thomas M. "Project Whirlwind: An Unorthodox Development Project." *Technology and Culture* 17 (July 1976): 447–464.

Spence, Clark C. *God Speed the Plow: The Coming of Steam Cultivation to Great Britain*. Urbana: University of Illinois Press, 1960.

———. *Mining Engineers and the American West: The Lace-Boot Brigade, 1849–1933*. New Haven: Yale University Press, 1970.

Strassmann, W. Paul. *Risk and Technological Innovation: American Manufacturing Methods during the Nineteenth Century*. Ithaca, N.Y.: Cornell University Press, 1959.

Sussman, Herbert L. *Victorians and the Machine: The Literary Response to Technology*. Cambridge, Mass.: Harvard University Press, 1968.

Sutton, Anthony C. *Western Technology and Soviet Economic Development, 1917 to 1930*. Stanford, Calif.: Hoover Institute Press, 1968.

———. *Western Technology and Soviet Economic Development, 1930 to 1945*. Stanford, Calif.: Hoover Institute Press, 1971.

———. *Western Technology and Soviet Economic Development, 1945 to 1965*. Stanford, Calif.: Hoover Institute Press, 1973.

Tarr, Joel A., and McMichael, Francis Clay. "Decisions about Wastewater Technology: 1850–1932." *Journal of the Water-Resources Planning and Management Division of the American Society of Civil Engineers* 103 (May 1977): 46–61.

Te Brake, William H. "Air Pollution and Fuel Crisis in Pre-Industrial London, 1250–1650." *Technology and Culture* 16 (July 1975): 337–359.

Temin, Peter. *Iron and Steel in Nineteenth-Century America: An Economic Inquiry*. Cambridge, Mass.: M.I.T. Press, 1964.

Terborgh, George. *The Automation Hysteria*. New York: W. W. Norton, 1966.

Thomas, Daniel H. "Pre-Whitney Cotton Gins in French Louisiana." *Journal of Southern History* 31 (May 1965): 135–148.

Thomis, Malcolm I. *The Luddites: Machine-Breaking in Regency England*. Hamden, Conn.: Shoe String Press, 1970.

Thompson, Robert Luther. *Wiring a Continent: The History of the·Telegraph Industry in the United States, 1832–1866*. Princeton: Princeton University Press, 1947.

Tobey, Ronald C. *The American Ideology of National Science, 1919–1930*. Pittsburgh, Pa.: University of Pittsburgh Press, 1971.

———. "Theoretical Science and Technology in American Ecology." *Technology and Culture* 17 (October 1976): 718–728.

Trachtenberg, Alan. *Brooklyn Bridge: Fact and Symbol*. New York: Oxford University Press, 1965.

Trombley, Kenneth E. *The Life and Times of a Happy Liberal: A Biography of Morris Llewellyn Cooke*. New York: Harper & Bros., 1954.

Turnbull, Archibald Douglas. *John Stevens: An American Record*. New York: Century, 1928; reprint ed., Plainview, N.Y.: Books for Libraries, 1972.

Uselding, Paul. "Elisha K. Root, Forging, and the 'American System.'" *Technology and Culture* 15 (October 1974): 543–568.

Ward, John W. "The Meaning of Lindbergh's Flight." *American Quarterly* 10 (Spring 1958): 3–16.

Watkins, George. *The Stationary Steam Engine*. Devon, England: Newton Abbot; and North Pomfret, Vt.: David & Charles, 1968.

Weber, Gustavus A. *The Patent Office: Its History, Activities, and Organization*. Baltimore: Johns Hopkins University Press, 1924; reprint ed., New York: AMS Press, n.d.

Wertz, Richard W., and Dorothy C. Wertz. *Lying-In: A History of Child-birth in America.* New York: Free Press, 1977.

West, Thomas Reed. *Flesh of Steel: Literature and the Machine in American Culture.* Nashville, Tenn.: Vanderbilt University Press, 1967.

White, John H., Jr. *American Locomotives: An Engineering History, 1830–1880.* Baltimore: Johns Hopkins University Press, 1968.

White, Lynn, Jr. *Medieval Technology and Social Change.* Oxford: Oxford University Press, 1962.

———. "The Historical Roots of Our Ecological Crisis." *Science* 155 (10 March 1967): 1203–1207.

———. *Machina ex Deo.* Cambridge, Mass.: M.I.T. Press, 1968.

———. "Technology Assessment from the Stance of a Medieval Historian." *American Historical Review* 79 (February 1974): 1–13.

———. "Medieval Engineering and the Sociology of Knowledge." *Pacific Historical Review* 44 (February 1975): 1–21.

———. "The Study of Medieval Technology, 1924–1974: Personal Reflections." *Technology and Culture* 16 (October 1975): 519–530.

Wik, Reynold M. "Steam Power on the American Farm, 1830–1880." *Agricultural History* 25 (October 1951): 181–186.

———. *Steam Power on the American Farm.* Philadelphia: University of Pennsylvania Press, 1953.

———. "Henry Ford's Tractors and American Agriculture." *Agricultural History* 38 (April 1964): 79–86.

———. *Henry Ford and Grass-Roots America.* Ann Arbor: University of Michigan Press, 1972.

Woodbury, Robert S. *History of the Gear-Cutting Machine: A Historical Study in Geometry and Machines.* Cambridge, Mass.: The Technology Press, M.I.T., 1958.

———. *History of the Grinding Machine: A Historical Study in Tools and Precision Production.* Cambridge, Mass.: The Technology Press, M.I.T., 1959.

———. *History of the Milling Machine: A Study in Technical Development.* Cambridge, Mass.: The Technology Press, M.I.T., 1960.

———. "The Legend of Eli Whitney and Interchangeable Parts." *Technology and Culture* 1 (Summer 1960): 235–253.

———. *History of the Lathe to 1850: A Study in the Growth of a Technical Element of an Industrial Economy.* Cleveland: Society for the History of Technology, 1961.

Chapter 3
History of Medicine

Gert H. Brieger, UNIVERSITY OF CALIFORNIA AT SAN FRANCISCO

INTRODUCTION

Self-assessment is an important activity for any group of scholars. It seems particularly appropriate for historians of medicine because of their disparate backgrounds and interests. By the same token it becomes a more difficult task. Perhaps this is why, until very recently, the status of the field has been assessed primarily in terms of utility and teaching, but not often in terms of historiography. (See, for instance, Galdston, 1957, and Rosen, 1948).

The objectives of this chapter are to delineate the field of history of medicine, describe its great breadth, and characterize its relationships to other fields. Starting from a fairly narrow base that was often restricted to the discussion of great men and their important ideas, the history of medicine has, in recent times, shifted more of its interests to the study of health and disease in their social, economic, and cultural settings. Techniques in use by other historians for some time are now increasingly adopted by researchers in the history of medicine. The relationship of the history of medicine to its sister field, the history of science, is as close as it is important. Many of the interests of specialized fields of history, such as family and women's history, overlap with those of the historian of medicine. The sociologist, the anthropologist, the geographer, and the artist provide methods and conceptual frameworks important for the history of medicine. Finally, the chapter briefly covers teaching and training efforts in the field and raises the question of the qualifications of the historian of medicine.

The bibliography is aimed especially at those who wish to undertake research problems utilizing medical-historical literature or who need a basic reading and teaching guide. It is important to keep in mind that our

goal is to guide, to point in the right directions. The chapter is not encyclopedic, nor is it intended to be a review of the entire literature. Since the
readership of the *Guide* will presumably be found mainly in North America,
it is the American scene that is most heavily emphasized. If favorite books
remain unmentioned it should not be taken as a denigration of their value.
The main concentration has been on seminal writings or those that contain
useful further references.

Justice cannot be done to a particular finding or argument, or to the
assessment of a particular work, in a mere summary. The major purpose
is to call attention to the variety of aspects of the history of medicine, as well
as to indicate some of its prospects. Details are left to those who wish to
follow the leads provided.

Some of these leads will direct scholars to the increasing number of
major manuscript collections that are so important to the history of medicine. The Library of Congress, the National Archives, the National Library
of Medicine, the New York Academy of Medicine, the American Philosophical Society, the Welch Library of the Johns Hopkins University School
of Medicine, the Countway Library of Harvard Medical School, and the
Yale Medical Library are the most important of many. The Rockefeller
Archives, recently opened to scholars, as well as the libraries of state and
local history societies, also contain rich resources for the history of medicine,
as do the libraries of local medical societies and hospitals. It has been those
scholars most familiar with archival and manuscript sources who have
carried the discipline of the history of medicine (as is true in the history
of science) beyond the old biobibliographical traditions of the field.

To encompass the history of medicine, one must first define medicine
itself. In recent decades we have returned increasingly to the social definitions of medicine that were prominent before and during the first half
of the nineteenth century. Holistic medicine and community medicine are
by no means new concepts. From the time of Hippocrates to that of the
physicians involved in the European revolutions of 1848, medicine was
viewed as a social science. Geography, economics, politics, cultural patterns—all were perceived as integral to the problems and definitions of
health and disease.

Medical concerns have traditionally revolved around the promotion
of health, prevention and cure of disease, and rehabilitation from illness
or injury. Anatomy, physiology, therapeutics, and hygiene have always
been the focus of historical studies. A few subjects have continued to be
mainly the province of the historians of medicine. These include all those
subjects dealing with the physician's more technical role, such as diagnosis,
therapy, and education. Much of the so-called internal history of medicine—
medical theory, understanding of disease, evolution of technical apparatus

and instruments, history of all the surgical approaches to therapy—would fall under this heading.

In recent decades, especially with the growing sophistication of the history of biology (see chapter 1, II. E. and F., in this volume), there has been a great deal of overlap between the history of medicine and the history of science. Harvey scholars, for instance, may and do come from either field. The social implications of health and disease, so well illustrated in the writings of Charles Rosenberg and George Rosen, to mention just two authors discussed below, provide an obvious overlap between social history and the history of medicine.

Henry Sigerist, doyen of the history of medicine in our time, stressed that the history of medicine is infinitely more than the history of the great doctors and their books. His theme is nowhere better stated than in the introduction to the first volume of his *A History of Medicine* (1951):

> In consulting the past of medicine, we are interested not only in the history of health and disease, of the physician's actions and thoughts, but also in the social history of the patient, of the physician, and of their relationship. What was the position of the sick man in the various societies? What did disease mean to the individual, how did it affect his life? . . . Where were patients treated? In the streets, in the homes, in temples, or in hospitals? How easily available were medical services in the cities, in the country? What was the cost of medical care? Was it such that only the rich could afford a doctor? If this was the case, what medical care, if any, was available to the poor? (pp. 15–16).

Sigerist's protégé, Owsei Temkin, long-time distinguished professor of the history of medicine at Johns Hopkins, has epitomized this position in an aphorism: One should consider all that is historical in medicine as well as all that is medical in history (Temkin, 1968).

State-of-the-field articles have been rare in the history of medicine; some of these are worthy of examination. Probably the most wide ranging and useful one—besides the collection in Edwin Clarke's *Modern Methods in the History of Medicine* (1971)—was originally given as George Rosen's presidential address to the American Association of the History of Medicine in 1966 ("People, Disease, and Emotion: Some Newer Problems for Research in Medical History," 1967). Here Rosen took the social character of medicine as his point of departure. As had Sigerist before him, Rosen stressed the necessity for a shift in our angle of vision to enable us to view the history of medicine as the history of human societies and their efforts to deal with problems of health and disease. The patient then would be more prominently the focus of study than would the physician.

I. HISTORY OF THE HISTORY OF MEDICINE

As Sigerist has pointed out, the history of medicine is relatively young among historical disciplines. In 1804 René Laennec, inventor of that great symbol of modern clinical medicine, the stethoscope, wrote his doctoral dissertation on the usefulness of the Hippocratic doctrines of fever. In 1839 Emile Littré announced in the preface to his ten-volume translation into French of the Hippocratic works that he wished to make them available as medical books, to be studied by students for their useful medical contents. Twelve years later, in 1851, another French physician-historian, Charles Daremberg, instructed readers that the translations and collections in his *Corpus Medicorum Graecorum* would principally serve historians and philologists. Thus the change: Until the middle of the nineteenth century the history of medicine was primarily medicine; thereafter, medical materials began to be studied for their historical value.

In the earlier periods, classical medical works were read by scholars or physicians interested not only in the history of the profession but also in the use of the old books in teaching medicine. Medical teachers and their students assumed that the historical approach was one valid way to gain medical knowledge needed by the practicing physician.

With the increasing emphasis on science in the training of physicians in the latter half of the nineteenth century, the history of medicine tended to be viewed more and more as a dead and useless past, a repository of mistakes, outworn traditions, and superstitions. As medicine became more scientific, so did history. The history of medicine, the history of art, and the history of science became important subspecialties of history, requiring specialized technical knowledge and training. Practitioners of the history of medicine included physicians and philologists, as well as specialized historians.

This is not to say that prior to the nineteenth century no history of medicine was written. Since ancient times physicians had included historical portions in their works; by the latter part of the seventeenth century, they were writing histories of medicine that were intended to explain as well as describe developments in medical thought and practice.

A.   Early Works

By the beginning of the eighteenth century, under a reigning ideology that was distinctly antisuperstitious, medical histories dominated by ideas of progress and pragmatism began to appear. The French physician Daniel LeClerc, whose 1696 *Histoire de la médecine* was the first of these modern histories, incorporated the idea of progress in the subtitle of his work, ". . . *in which are seen the origin and the progress of this art, from century to century. . . .*" LeClerc attempted to go beyond the mere chronicle; he wished as well to interpret the context and the importance of medical thought and practice. Unfortunately he covered history only through antiquity.

The British physician John Freind, while briefly imprisoned in the Tower of London on charges of treason, planned a history of medicine that would continue where LeClerc had stopped. In 1725 and 1726 Freind published his *History of Physick from the Time of Galen to the Beginning of the Sixteenth Century,* the first such major survey in English.

By the end of the eighteenth century, the pragmatic ideal of using the history of medicine for the education of the physician reached its greatest height with the publication of Kurt Sprengel's *Versuch einer pragmatische Geschichte der Arzneykunde* (1792–1803). Sprengel was a botanist and physician as well as an accomplished medical historian; his work laid the foundation for the many who followed him in the nineteenth century.

In the nineteenth century the history of medicine continued to be pursued by such German physician-scholars as Heinrich Haeser, J. F. K. Hecker, and August Hirsch—all of whom wrote on the history of epidemic diseases. In England Michael Foster, D'arcy Power, and others wrote biographies for the Masters in Medicine series that appeared around 1900.

The history of the history of medicine has been described by a number of authors. Sigerist sketched it briefly in volume 1 of his *History of Medicine* (1951). Edith Heischkel carefully reviewed it, in German, in Walter Artelt's *Einführung in die Medizinhistorik* (1949). A convenient summary in English, with a large annotated bibliography, may be found as "An Exhibit on the History of Medical Historiography" in the *Bulletin of the History of Medicine* 26 (May–June 1952): 277–287. Noel Poynter's Garrison Lecture, "Medicine and the Historian" (1956), also gives a summary.

Formal teaching and research institutes for the history of medicine had their beginning in Germany. Julius Pagel joined the faculty of the University of Berlin as an historian of medicine in 1891. His 1898 *Einführung in die Geschichte der Medizin* (a subsequent edition was brought out by Karl Sudhoff) became the premier textbook for the early decades of the twentieth century. (For Julius Pagel's views on the history of medicine, see Walter Pagel, 1951.) Karl Sudhoff gave up his medical practice to head a newly endowed Institute for the History of Medicine at the University of Leipzig in 1905. An institute in Vienna, with Max Neuburger as head, soon followed. By the middle decades of this century, every West German medical school had a department of medical history, some with two or more scholars. Several universities in Switzerland and France also served as centers for medicohistorical research and teaching. Such progress was not to be seen in the English-speaking world for some time.

In the United States, beginning in colonial times, physicians began to write about the history of their profession. Some, including James Thacher, Stephen Williams, and Samuel D. Gross, compiled useful biographical dictionaries of medical men. Benjamin Rush, Joseph Gallup, Daniel Drake, and Elisha Bartlett wrote accounts of epidemic diseases. In the years since the Civil War, the United States Army has been an important source of books about medical history, as well as of scholars who have contributed much to the understanding of our historical evolution. Joseph J. Woodward, John Shaw Billings, and Fielding H. Garrison are three representative examples. Billings taught an early course in medical history at the Johns Hopkins University in the late 1870s.

In the years following the Civil War, Joseph M. Toner, a Washington, D.C., physician, collected valuable material in writing about many aspects of American medical history (see Bell, 1973). To celebrate the American centennial in 1876, the editor of the Philadelphia-based *American Journal of the Medical Sciences* ran a year-long, four-part series entitled "A Century of American Medicine"; it was prepared by some of the country's leading physicians. The essays on surgery, internal medicine, literature, and obstetrics-gynecology included names of many American physicians who had taught or written about medicine. The celebratory tone of these and similar works of the late nineteenth century came to typify medical history in America. (According to some critics, we have yet to outgrow this tradition.)

B. The Johns Hopkins Institute of the History of Medicine

Henry E. Sigerist, Swiss-born, a student of Karl Sudhoff at the history of medicine institute in Leipzig, succeeded Sudhoff in 1925. Not until Sigerist came to take charge of the recently founded Institute of the History of Medicine at Johns Hopkins in 1932 did medical history assume any academic standing in the United States. Much has been written about Sigerist by his students. (See Temkin, 1957, 1958; Rosen, 1958; Falk, 1958.) Sigerist's own extensive writings, including parts of an autobiography (see Sigerist, 1958), provide many details about his intellectual and social life. It was not only as a scholar that Sigerist helped shape a new era in the history of medicine. He guided a series of scholars who have been influential to this day. He also had remarkable organizational capacities.

Prior to Sigerist's arrival in America, William H. Welch toured Europe to find historical books for the Hopkins medical library; he also occupied a chair in the history of medicine. From the beginning, Welch, William Osler, Howard Kelly, and their Hopkins colleagues had taught about historical aspects of medicine; their history, however, was usually integrated into their medical teaching. They also published some excellent historical articles, many of which appeared in the early pages of the *Bulletin of the Johns Hopkins Hospital*. In 1890, the year after the hospital opened its doors to patients, the newly constituted faculty founded a history of medicine club, which occasioned many of their papers.

When Sigerist and his student Owsei Temkin arrived in Baltimore, they inherited this rich tradition as well as an institute in the medical school that was sufficiently endowed to form a base for attracting scholars. Within a year after their arrival, Sigerist and Temkin, with a few others, began the *Bulletin of the Institute of the History of Medicine;* it was issued initially as a supplement to the *Bulletin of the Johns Hopkins Hospital.* The first two volumes, in 1933 and 1934, continued to use the large parent journal, but in 1935 the *Bulletin* declared its independence. In 1938 the American Association for the History of Medicine decided to publish its papers and transactions in the *Bulletin,* and in 1939 the journal became simply the *Bulletin of the History of Medicine,* the official organ of both the Institute and the Association. The editor of the *Bulletin* has usually been the William H. Welch Professor of the History of Medicine at Johns Hopkins. He has also had the option of first refusal or acceptance for the publication of any paper presented at the annual meetings of the Association.

Thus Sigerist's organizational ability was initially put to use running the only full-time department in the country devoted to the history of medicine, as well as editing the major American journal in the field. He later devoted his energy to reforming the American Association for the History of Medicine. Founded by a group of physician-scholars in Philadelphia in 1925, the association met once or twice a year, almost always in conjunction with the Association of American Physicians at their Atlantic City meeting. A half day of historical papers and a dinner session comprised the usual schedule. Under Sigerist's influence the meetings of the Association were extended and made to stand on their own. Beginning in the late 1940s the group began to meet yearly in various cities throughout the country, usually in April or May.

Henry Sigerist was a superb teacher. Many students who attended the Johns Hopkins Medical School between 1932 and 1947 have fond memories of his courses. For effective teachers, teaching is not confined to the classroom. Sigerist's broad interests and extensive knowledge of medicine's past, as well as his enthusiasm for teaching and for people, were communicated widely. He exerted a broad influence upon his colleagues and junior associates, including Owsei Temkin, the classicist Ludwig Edelstein, and Erwin Ackerknecht (who had studied medical history with Sigerist in Leipzig but was also trained in medicine and in anthropology). During his Hopkins years Sigerist had few graduate students. Only one, Ilza Veith, actually completed her Ph.D. degree under his direction. Another, Genevieve Miller, was his research assistant but went to Cornell for her Ph.D. However, a number of young physicians with interests in social medicine came under his tutelage and influence. The outstanding example is George Rosen; Milton Roemer and Leslie Falk are two others. All took up the cause of social medicine with a distinctly historical flavor. (See G. Miller, 1958 and 1979.)

It would be impossible to review here the totality of Sigerist's work, but his major book, *A History of Medicine* (vol. 1, 1951), deserves special mention. Sigerist looked upon the history of medicine in social and economic as well as in scientific and technical terms. It was his hope to retire from active teaching and administration to write a broadly based overall history of medicine in eight volumes. He lived to finish only the first volume and the better part of a second. When volume 1, *Primitive and Archaic Medicine,* was published, George Rosen hailed it as "the new history of medicine." The older histories of medicine had been primarily biobibliographical or philosophical. A few historians at the turn of the twentieth century, especially J. H. Baas, Julius Pagel, and Max Neuburger, paid some attention to social and cultural factors, but these played little role in their interpretation of medicine's past. Sigerist's book devoted much space to cultural as well as social factors—to the way people lived, dressed, ate, and worked; this was a true social history of medicine. Unfortunately Sigerist did not live to reach his goal. His first volume, which is thoroughly integrated into the history of civilization, provides a model form for the history of medicine. It also serves as a handy reference source for the older medical history literature.

Sigerist's earlier books may also be read with profit. Many still believe, for example, that the best one-volume overall history of medicine that one can recommend to those interested in the field is *The Great Doctors.* Published originally in German in 1931, it quickly became available in an English translation, albeit a rather stiff and precious one. It has since also appeared in various paperback editions. Sigerist was at his best in popularizing the history of medicine in *The Great Doctors,* which is composed of a series of short biographical sketches of the work and thought of great physicians from Imhotep to William Osler. In essence, however, it is far more than a collection of short biographies. For each of the physicians, Sigerist weaves in cultural history as well as material on the state of the art of medicine. In fact, some readers coming to *The Great Doctors* without any historical background do not initially appreciate the depth of Sigerist's analysis.

Sigerist returned to Switzerland in 1947 to work on his projected eight volumes but had not finished the work when he died in 1957.

In 1949 Richard H. Shryock, an American historian who had taught at Duke University and the University of Pennsylvania, became the second William H. Welch Professor at the Institute of the History of Medicine at Johns Hopkins. Early in his career, Shryock had turned to the social history of medicine. His extraordinary productivity, beginning in 1923 and continuing for fifty years, is demonstrated in the bibliography of his work compiled for a special Shryock number of the *Journal of the History of Medicine* (January 1968). He is best known for his studies of American medicine, and his *Development of Modern Medicine* (1936, 1947) remains a useful single volume dealing with the international picture from the seventeenth to the twentieth century.

Shryock's work on early American medicine (1960), on medical research (1947), on medical licensing (1967), and his collection of essays that discuss not only people and events in American medicine, but some of the important historiographic problems as well (1966), are all standard sources. It has happened to many who have followed in Shryock's footsteps, that when they think they have a good, new idea, it will often have already been clearly stated in one of his books or papers.

In 1958 Owsei Temkin became the third William H. Welch Professor at Johns Hopkins, to be followed a decade later by his student Lloyd G. Stevenson.

II. COMMON HISTORICAL APPROACHES

Scholarship of all kinds has been fundamentally transformed in this century. Although the changes in the history of science have probably been more far-reaching, the history of medicine too has witnessed a maturation and a growth compared to the pre-World War II decades. The strengths and weaknesses of the common historical approaches will become apparent as we discuss their characteristics in the following sections.

For the history of medicine there is nothing available resembling John Higham's *History* (1965), which treats the historical profession, theories of history, and the important works. Edwin Clarke's *Modern Methods in the History of Medicine* (1971) deals almost exclusively with varying methods, as its title indicates.

A.   Biography and the History of Medicine

The biobibliographic tradition, strong in the history of medicine since the Renaissance, has been replaced in recent decades by more traditional historical analysis. Despite this development, one must pay particular attention to existing medical biographies, memoirs, and autobiographies. Although it is true that some historians look upon biography and autobiography as literature rather than history, such works have in the past provided the history of medicine with a wealth of material. (See Stevenson, 1955b.)

Distinguished and informative biographies have been written about physicians since the time of the ancients. Ironically, perhaps the best-known medical name of all, Hippocrates, must continue to live out his fame with only the sparsest of hard data available about him. We know he was small of stature, taught medicine for a fee, and was well known in his own time. We do not know his precise life dates, so that when one sees 470–390 B.C., or some such figures, one knows they are really only guesswork. We do know he lived and flourished around 400 B.C. and that he was a contemporary of Socrates and Plato. We do not know which, if any, of the seventy or so books that make up the *Corpus Hippocraticum* were actually written by him (Sigerist, 1934).

Among the medical biographies and biographical dictionaries listed in the bibliography below, there is only space enough here to mention a few representative examples. These may prove useful as a starting point for reading in the history of medicine. In addition to biographical medical dictionaries, there are also literally thousands of full-length biographies. Although any number are exercises in hagiography and advertisement, others give extraordinary insight into the development of the medical profession, medical education, and medical thought. Many were written by physicians or surgeons.

One outstanding physician-biographer was the noted neurological surgeon Harvey Cushing. Cushing not only wrote memoirs of his experiences as a surgeon during World War I, but he produced an extraordinary biobibliography of Andreas Vesalius, the noted Renaissance anatomist. He traced in detail the development of Vesalius' career as an anatomist and physician against the background of the numerous editions of Vesalius' works. Cushing capped his historical work with a two-volume study, *The Life of Sir William Osler* (1925). Cushing's biography is a study not only of Osler's career as a clinician but also of the development of medical education in Canada, the United States, and England from the end of the nineteenth century. While Cushing's work is unique, it is not singular. Many physicians have created biographies of equal import and value. Simon Flexner's *William Henry Welch and the Heroic Age of American Medicine* (1941), John F. Fulton's *Harvey Cushing* (1946), G. A. Lindeboom's biography of Herman Boerhaave (1968), and Geoffrey Keynes's *William Harvey* (1964), are other examples of the physician as historical craftsman. And physicians are not the only ones who have tried their hands at medical biography: Charles O'Malley's *Andreas Vesalius* (1964) is a magnificent book that goes beyond biography to historical cultural synthesis—thus belying the view of some historians who maintain that biography is literature and not history.

B. The History of Ideas

The history of ideas is a major component of medical history as it is of the history of science (see chapter 1, I. E., in this volume). The conceptual approach to the study of disease, of abnormal and normal bodily functions, and of therapeutics has always been important. (See, e.g., W. Riese, *The Conception of Disease*, 1953.)

Owsei Temkin, in a recent essay on the historiography of the history of medical ideas, points out that as more nonphysicians have become actively interested in the history of medicine, medical historians have moved away from an "iatrocentric" (physician-related) approach to an increasing interest in the context in which such ideas flowered.

Temkin himself wrote "An Historical Analysis of the Concept of Infection." It first appeared in a volume honoring a pioneer in the history of ideas, Arthur O. Lovejoy of Johns Hopkins University (Temkin, 1953, reprinted in *The Double Face of Janus,* 1977). Here Temkin discusses lay as well as medical ideas about the cause and the nature of infection. Two other examples of the approach may be found in Temkin's collected essays, *The Double Face of Janus.* For the *Dictionary of the History of Ideas* (1973), Temkin had written a masterful essay, "Health and Disease," that is reprinted there and covers the whole history of medicine. "The Scientific Approach to Disease: Specific Entity and Individual Sickness," another of these essays, surveys the long debate between those who believed in an ontological approach to the study of disease (that is, that diseases exist as independent entities) and the nonontological belief that disease is to be viewed as disordered physiology. Late nineteenth-century pathologists were much concerned with this question. The overly simple phrase, "There are no diseases, only sick people," is typical of this long debate. (See also Temkin, *The Falling Sickness,* 1945, 1971.)

Another notable practitioner of this approach to the historical understanding of medicine is Lester King, a Chicago pathologist who teaches history of medicine at the University of Chicago and who was for many years on the editorial staff of the *Journal of the A.M.A.* His book *The Medical World of the Eighteenth Century* (1958) clearly describes the problems faced by physicians of that day. In *Growth of Medical Thought* (1963) he surveys the development of medical science and changing patterns of medical doctrines from the time of Hippocrates through nineteenth-century pathology. (See also King, 1970 and 1978.)

In the history of biology, Thomas Hall's two-volume *Ideas of Life and Matter* (1969), reprinted in paperback as *A History of General Physiology* (1975), provides equally rich sources for the historian of medicine. Hall's work discusses in great detail differing points of view on the properties of living matter, the long and intricate debate over the doctrine of vitalism, and the evolution of the concept of irritability. Although Hall has a facility for bringing complex philosophical as well as biological matters into clear and simple focus, his work is typical of the genre; at times the reader is no longer firmly anchored in time or space.

In a somewhat different tradition, the surgeon Leo Zimmerman and the medical historian Ilza Veith, in their *Great Ideas in the History of Surgery* (1961, 1967), combine introductory and interpretive historical and cultural material with excerpts from surgical writers beginning with the *Edwin Smith Papyrus* to Ferdinand Sauerbruch and the beginnings of thoracic surgery early in the twentieth century. Actually, the Zimmerman and Veith volume can serve as a good introductory text to the history of medicine as a whole, not just the history of surgery.

The concept of inflammation, and ideas about the process of wound healing in ancient times East and West, have recently been extensively explored by a pathologist, Guido Majno, in *The Healing Hand* (1975).

An older book in this tradition, unfortunately generally neglected, is by the philosopher Scott Buchanan, *The Doctrine of Signatures* (1938). His work is important in the history of therapeutics, but Buchanan also effectively demonstrates how one can use the doctrine of signatures—the idea that certain plants and animals have characteristic marks or colors that identify them for use as remedies—to discuss a broader philosophical approach to medicine.

Particular medical specialties have suggested a variety of subjects for those interested in examining the development of medical ideas. For example, physiology—especially that aspect of it concerned with the functions of digestion, respiration, growth, and metabolism, or such concepts as homeostasis and irritability—has been widely investigated. Pathology and therapeutics have also offered the historian of ideas such interesting concepts for research as the cell theory, inflammation and wound healing, the healing power of nature (for the latter, see especially Neuburger, 1926), and the genesis of cancer (Rather, 1978).

A few of the great names in the history of medicine have been studied by history-of-ideas methods. This is true especially for William Harvey on the circulation of the blood (see especially Pagel, 1965); Louis Pasteur and others who unraveled the concept of infection (Dubos, 1950, 1976); and Claude Bernard on the theory of homeostasis (see especially Olmsted and Olmsted, 1952; Virtanen, 1960; and Holmes, 1974).

The year 1978 marked the 400th anniversary of Harvey's birth and the 350th of the publication of the *De motu cordis*. Articles about his place in the history of medicine have appeared in large numbers. Harvey's work not only demonstrated the circulation of the blood; it may also be viewed as a landmark in the history of the scientific method as it applies to medicine.

A new translation of Harvey's *De motu cordis* has recently been completed by Gweneth Whitteridge. She and her British colleagues Walter Pagel and Sir Geoffrey Keynes are the foremost Harvey scholars of our time. A younger American historian of medicine, Jerome Bylebyl of Johns Hopkins, wrote the extensive article on Harvey for the *Dictionary of Scientific Biography* and is another outstanding Harvey scholar. In his essay Bylebyl not only summarizes recent scholarship but describes Harvey's intellectual origins in admirable fashion. (See also the review by Stevenson, 1976.) In a series of papers that appeared in the British historical journal *Past and Present,* several scholars debated whether or not Harvey's work was influenced by the fact that he was a Royalist and not a Parliamentarian in mid-seventeenth-century British upheavals. These papers, and others about Puritanism and science, have been put together in a convenient collection by Charles Webster (1974). (See also Webster's monumental study, *The Great Instauration: Science, Medicine, and Reform 1626–1660* [1975], which no serious scholar working in this period can afford to ignore.)

One final reference under this heading: John Theodore Merz's four-volume *A History of European Thought in the Nineteenth Century*. First published between 1904 and 1912 and reprinted in 1965, Merz's work covers science extensively in the first two volumes, especially concepts such as vitalism and cell theory, as well as other physiological concepts. For historians of medicine who deal with the evolution of key concepts in the history of physiology and pathology, Merz may still serve as a convenient starting place.

The history of ideas has long been an active and exciting aspect of the history of medicine. As Temkin succinctly put it, "Asking why others thought as they did challenges us to ask why we think as we do" (in Clarke, 1971, p. 15).

C.   Social History (the Professions; Hospitals; Psychiatry; Public Health; Quackery and Sectarianism)

A definition of social history has been given by J. Jean Hecht in the *Encyclopedia of the Social Sciences* (1968): It "is the study of the structure and process of human action and interaction as they have occurred in socio-cultural contexts in the recorded past." This is a rather ponderous definition, but behind it lies a rich tradition of historical scholarship. The sources generally used by social historians are virtually unlimited; the same is true for social historians of medicine. However, as mentioned earlier, it is only in recent decades that medical historians have followed the lead of the few early workers in the field, such as Erwin Ackerknecht, in his study of malaria, and Richard Shryock, in many books on aspects of medical history in America.

Today there is great interest in the social history of medicine. A recent issue of the *Journal of Social History* (July 1977) was devoted entirely to papers in the history of medicine. Topics included (among others): medical practice in France around 1890; the hospital and the patient in nineteenth-century America; the eighteenth-century hospital in Denmark and Norway; women and health reform; the Russian cholera epidemic of 1892–1893; and popular healers in nineteenth-century France. These subjects are typical of the concerns of younger historians of medicine, along with such problems as the impact of disease on society, the structure of the medical profession, the evolution of various medical organizations, and the legitimization of medical care.

Gerald Grob, in a survey of the prospects of the social history of medicine in the United States (1977), stresses the importance of describing more precisely the nature of the social context in which medical events occurred. Grob's concerns are not new. Charles Rosenberg of the University of Pennsylvania has voiced similar concerns. In a recent volume of his essays, *No Other Gods* (1976), Rosenberg examines the interface between science, social thought, and the institutional structure of science. He explores the ways in which medical images and ideas are formed, the ideological function of communicating conventional understanding of social realities, and the way science and medicine become a source of authority. Medical sociologists, especially Eliot Freidson in *Professional Dominance* (1970), have explored similar themes from a different perspective.

In Britain, the recently founded Society for the Social History of Medicine has become the focal point for much of this sort of work. In a long introductory essay to a recently published symposium, John Woodward and David Richards (1977) survey the state of the field and point out the many possibilities for further research. In his inaugural lecture to that society in 1970, Thomas McKeown, professor of social medicine in Birmingham (and a leading historical demographer whose work is described below) spoke on "A Sociological Approach to the History of Medicine." McKeown defined his own view of the history of medicine in terms similar to those used by Sigerist some forty years earlier. While Sigerist stressed that the medical historian should be a physician interested in grappling with the problems of medicine and society, McKeown defined the problems in terms of present issues. Sigerist, for all his sociological interests, was a historian at heart. McKeown, for all his historical interests, is a professor of social medicine.

While McKeown ought to be praised for injecting the historical approach into his work, he invited criticism by writing:

> The social history of medicine is much more than a blend of social history and medical history, more than medical developments seen in the social context of their period; it is essentially an operational approach which takes its terms of reference from difficulties confronting medicine in the present day. It is lack of such insight, derived from contemporary experience, which makes a good deal of medical history so sterile for the uninitiated (p. 342).

At least one British historian, J. F. Hutchinson, immediately charged that McKeown's basic purpose was not to understand the past, "but to provide necessary information for reforming present evils" (Hutchinson, 1973; see also Stocking, 1965).

What McKeown, who is singled out here only as a convenient example, has not appreciated is that it is precisely this rather unsophisticated approach to history that the uninitiated will accept. It is the professional such as Hutchinson who will say that the purpose of such an endeavor is ahistorical. Furthermore, McKeown in his lecture described the traditional tasks of the physician as diagnosis, understanding of disease, prevention, cure, prognosis, and palliation, maintaining that only the last was generally achieved before the end of the nineteenth century. McKeown's view is unfortunately all too typical of the present-day physician's attitude about the effectiveness of medicine. Such people measure medicine only in terms of therapeutic potency. Beyond this appraisal, the doctor-patient relationship—which is so vital to the medical encounter—may be said to be timeless in a social and psychological sense, while differing according to time and place.

Some important work in the social history of medicine is quantitative; there is a brief discussion below of related advances, stemming from the work of the *Annales* school in France, on population studies. However, there is much work that is not quantitative. For example, two English scholars, the historian R. S. Roberts and the sociologist S. W. F. Holloway, have made important contributions.

Roberts, reviewing medicine and its practitioners in Tudor and Stuart England (1962, 1964), gives a careful analysis of the social context of practice in the provinces and in London. To do this he had to go to sources that were not exclusively medical in content, such as county and school registries, in order to make sense of the education and distribution patterns of the varied types of existing medical practitioners. Such materials were, of course, intimately related to the educational preparation as well as the later practice of these physicians.

Holloway, in his work on the Apothecaries Act of 1815 (1966), also stresses the importance of understanding the nature of medical education in order to chart the varied contexts of medical practice and thought. (See also Holloway, 1964.)

Similar studies by the British sociologists Ivan Waddington (1977) and N. D. Jewson (1976) focus on the changing character of the medical profession and the effects upon the place of the patient. Jewson particularly, in describing "the disappearance of the sick man from medical cosmology, 1770–1870," emphasizes the interrelationship between the mode of production of medical thought or knowledge and the social structure of medical services. He concludes that the eclipse of bedside medicine, first by hospital and then by laboratory medicine, represents a shift from a person-oriented toward an object- or disease-oriented cosmology. While provocative, this analysis says little about the much larger group of physicians, the everyday practitioners; rather, it focuses on those who *taught* medicine.

An area of special interest to the social historian of medicine is that of professions and the process of professionalization. Carlo Cipolla, in a provocative article, "The Professions: The Long View" (1973), has shown that professionals were the most influential element in the emerging middle class of medieval and Renaissance European cities. However, to get perspective on the medical profession one must look at the other professions—the so-called service sector of an economy, such as notaries. Cipolla ascribes the growth of this service sector in the late Middle Ages to three main factors: improved standards of living, movement to towns, and the increasing secularization of society. While these are neither new nor startling conclusions, what is of importance is that these are not the facets of history with which the medical historian generally works. It is of parallel interest that (nonmedical) social historians, dealing with the rise of the middle class in late nineteenth- and early twentieth-century America, have now begun to analyze the role of the medical profession within the development of a larger profession-based class structure. (See for instance, Bledstein, 1976, and Wiebe, 1967.)

As the problem of providing sufficient medical care to the American people at prices that will not bankrupt them has increasingly become the subject of public discussion and debate, social historians of medicine have begun to analyze the evolution of health care institutions and their effects on patients. In one sense this is a new development; in the past such institutions were frequently examined only for their role in the training and education of the physician. Although there are a number of good histories of individual hospitals, there is as yet no overall analytic social history of American hospitals. Europeans have been better served by their historians. (See B. Abel-Smith, 1964, for the British, and D. Jetter, 1966–1972, for the Continent. Abel-Smith's primary focus is on the administrative aspects of the British hospitals until the advent of the National Health Service. Jetter, on the other hand, has concentrated more on architectural and space matters.)

There are stirrings among American medical historians, particularly in recent articles by Morris Vogel (1976) and Charles Rosenberg (1977). Rosenberg indicates that factors affecting the increasing prominence of hospitals and hospital care in the twentieth century may be found among social and ecological changes and among changes involving complicated laboratory and radiological equipment. Technology, as others besides Rosenberg have stressed, has also affected the patterns of disease in such developments as refrigeration, canning, and cleaning compounds. Thus the challenges and complexities for the social historian of medicine are many, and the field is slowly becoming more sophisticated. As it does, its appeal should be correspondingly wider, for social responses to illness have always been inherently fascinating.

None of this is to say that such studies are to be found only in the recent literature. As long ago as 1945, Erwin Ackerknecht showed the richness and complexity of the factors affecting malaria in the upper Mississippi Valley. His monograph has long served as a model, but unfortunately his model has not yet been frequently enough copied. George Rosen has also published much on the social history of medicine since the mid-1930s. For a review and appreciation of his work see the Rosen memorial issue of the *Journal of the History of Medicine* 33 (July 1978), and Rosenberg (1979).

Social historians have also long been interested in the history of psychiatry and psychiatric care. General histories of psychiatry have been appearing periodically since early in the twentieth century. Emil Kraepelin's *One Hundred Years of Psychiatry* (1962) is available in an English translation; it is of interest because he was one of the early workers in the field. An American psychiatrist, Gregory Zilboorg, provided a much fuller summary, describing especially the period from the sixteenth to the eighteenth century when psychological medicine was emancipating itself from demonology. The later periods of Zilboorg's *A History of Medical Psychology* (1941) have a distinctly Freudian flavor.

Zilboorg did not neglect the ancient period of Greece and Rome; neither did George Rosen, later, in *Madness and Society* (1968). A very important source for this period, one really more in the tradition of the history of ideas than of social history, was a series of Sather Lectures by the British classical scholar E. R. Dodds, published as *The Greeks and the Irrational* (1951).

Erwin Ackerknecht's *Short History of Psychiatry* (1959, 1968) has long been the standard source for a quick and reliable overview of the entire subject. George Mora (1965) and Mora and Jeanne Brand (1970) have described the historiography of psychiatry. They include in their analysis much of the older non-American work.

Some of the most interesting work currently under way in this country deals with the care of the mentally ill. What were the diagnostic categories used for admission? What was the fate of patients? How were they treated? And what were the underlying theories of mental function, of the causation of mental illness? In an attempt to answer some of these questions, Barbara Rosenkrantz of Harvard University is currently working on a large computer-assisted study using the case records of several New England state and private asylums in the nineteenth century.

Concepts of insanity prior to the Civil War have been thoroughly discussed in a book by Norman Dain (1964). While he describes changing views of the etiology of mental illness, Dain's main emphasis is on the rise of moral therapy. Eric Carlson and his coworkers have also published papers dealing with this subject. One of the most successful examples of medical history written by a practitioner may be found in J. S. Bockoven's small book, *Moral Treatment in American Psychiatry* (1963). With perceptions enriched by his own experience as a state hospital physician, Bockoven traces the rise and fall of moral therapy in nineteenth-century American psychiatry. He carefully examines the work of Pliny Earle that led to the downfall of what Albert Deutsch, in *Care of the Mentally Ill* (1949), called the cult of curability. Bockoven's description of conditions in the mid-twentieth century are invaluable depictions of the complexities of the mental hospital culture and environment. The all-important role and status of the attendant, for instance, must be understood if one is to clarify the picture of large asylums that were built with increasing frequency in the middle decades of the last century.

Historians have written about a number of psychiatric hospitals. None of these histories is better than the one on Worcester State Hospital by Gerald Grob (1966). Grob went on to study mental institutions in general in the United States. A first volume of this research was published as *Mental Institutions in America: Social Policy to 1875* (1973; see also Rothman, 1971).

These few references by no means exhaust the list. The Freud industry, for example, is large and flourishing. Most of Freud's own work is now available in convenient paperback editions. A major work of value to all students of the history of psychiatry is Henri Ellenberger's *The Discovery of the Unconscious: The History and Evolution of Dynamic Psychiatry* (1970). Describing theories of the mind and various therapeutic techniques from the primitive healers to the twentieth century, Ellenberger traces the work of the major psychologists prior to the advent of psychoanalysis, as well as the work of Jung and others who came after Freud, in a volume of more than nine hundred pages. Ilza Veith's *Hysteria: History of a Disease* (1965) covers the whole field of psychiatry as well as a great deal of the history of medicine. Finally, two books that describe the status of psychiatry in the eighteenth and nineteenth century deserve special mention. In *George the Third and the Mad Business* (1969), Ida Macalpine and Richard Hunter attempt to prove that our late lamented king was suffering from a metabolic form of insanity caused by porphyria. Whether one accepts or rejects the arguments for this diagnosis is not the important point; the latter half of the book clearly outlines the "mad business" in England in the late eighteenth and early nineteenth century. The second book, *The Trial of the Assassin Guiteau: Psychiatry and Law in the Gilded Age* (1968) by Charles Rosenberg, describes the complexities of arguments over the etiology of mental illness. It is a clear statement of psychiatric thought and practice in 1881, at the time of President Garfield's assassination.

The history of public health has been another area of interest to scholars in social medicine for over a hundred years. Because the British were the first to experience the results of the Industrial Revolution, they were also among the earliest of the Western nations to find a need for an organized sanitary system. A leader of the sanitary reform movement, Sir John Simon, wrote a history of the field, *English Sanitary Institutions* (1890). (Royston Lambert's *Sir John Simon,* 1963, is one of the best medical biographies in recent times.)

A classic in the history of public health, Edwin Chadwick's *Report on the Condition of the Labouring Population* first appeared in 1842. It served as the model for Friedrich Engels' exposé of the living conditions of workers two years later, and also for sanitary surveys in Massachusetts and New York. The Chadwick report was reprinted in 1965 with a perceptive and useful introduction by the British historian M. W. Flinn. (See also earlier biographies of Chadwick by Lewis, 1952, and Finer, 1952.)

Lemuel Shattuck's Report of the *Sanitary Commission of Massachusetts* (1850) described the need for a health department in that state. It is better known to historians in our day than it was to Shattuck's contemporaries. Shattuck's reasons why his suggestions would meet the resistance of legislators, physicians, and citizens of his state have a modern ring. (For Massachusetts health conditions see Blake, 1959, for the colonial period, and Rosenkrantz, 1972, for the period 1842–1936.)

An 1865 "Report of the Citizens' Association" about the sanitary conditions of New York was successful—as had been its predecessor in London twenty-odd years earlier—in stimulating the creation of a health department. Cholera, or the threat of a renewed outbreak of the dreaded disease, played its part as well. The Metropolitan Health Bill, passed in 1866, established a department that would prove to be a model of its kind. (See Brieger, 1966; also John Duffy's two-volume history of the department, 1968, 1974. The Arno Press has republished forty-four titles on public health in America under the editorship of Barbara Rosenkrantz.)

Published collections of papers and biographies have been useful to historians of public health. Fielding Garrison's memoir of John Shaw Billings (1915), C. E. A. Winslow's biography of Herman Biggs of New York (1929), and James Cassedy's more recent study of Charles V. Chapin of Rhode Island (1962) are three examples. The papers of John B. Grant (edited by Conrad Seipp, 1963) and Joseph Mountain (1956), along with the pellagra studies of Joseph Goldberger (edited by Milton Terris, 1964), are three more examples of this kind of useful historical material.

Erwin Ackerknecht's long paper on "Hygiene in France" (1948), Alfons Fischer's two splendid volumes on German public health developments (1933), and Erna Lesky's 1959 monograph on Austria, as well as her recent translation of a part of the monumental work of Johann Peter Frank, are good studies representing work outside the English and American scene.

George Rosen's *A History of Public Health* (1958) has not yet been superseded as the most useful single-volume general history of this field. See also his more recent *Preventive Medicine in the United States 1900–1975* (1975).

For the history of epidemiology, another important field, the literature consists mainly of the history of epidemic diseases. Major Greenwood, a British epidemiologist, summarizes some of the British work on vital statistics in his *Medical Statistics from Graunt to Farr* (1948) and *Some British Pioneers of Social Medicine* (1948). For the American scene, a series of essays edited by Franklin H. Top (1952) describes the state of epidemiology as well as the leading causes of death from colonial times to the middle of the present century. For reference purposes, the most useful source of information about changing patterns of disease and life expectancy in the years 1900 to 1960 is Monroe Lerner and Odin Anderson's *Health Progress in the United States* (1963).

Charles Rosenberg's *The Cholera Years* (1962) deals with the social response to cholera epidemics in the United States in 1832, 1849, and 1866. His discussion of the movement toward health reform in New York City, for instance, is extensive. Other social historians, while not setting out exclusively to chart society's response to diseases, have nevertheless not neglected health, diet, and medical care. In fact, public health, in all its aspects, has in recent decades been incorporated into the work of general historians. A number of examples of the works of general historians who have paid attention to medical themes may be found in the bibliography.

Other themes have been pursued by social historians of medicine. For the American scene, one that deserves discussion but can only be alluded to here is the history of medical sectarianism and quackery. A number of writers on medical topics of the nineteenth century have incorporated sections on homeopathy, popular health and hygiene, Thomsonianism, and other aspects on the fringes of so-called "regular" medicine. (See particularly books by Kett, 1968, and Rothstein, 1972.) Martin Kaufman (1971) and Harris Coulter (1973) have written book-length studies of homeopathy in America. John Davies (1955) has described phrenology both as a fad and as a serious attempt to understand human behavior in the nineteenth century. Alex Berman's articles remain the standard source on the botanical sect known as Thomsonians (1951, 1956). Ronald Deutsch has written about food faddism and food quackery in his *Nuts Among the Berries* (1961 and 1977). The most extensive work on quackery in America may be found in two books by James Harvey Young of Emory University. *The Toadstool Millionaires* (1961) covers the story up to the food and drug law of 1906. Subsequent events, including pressure in 1938 and 1962 for amendments to strengthen the law, are discussed in *Medical Messiahs* (1967).

II. NEW TECHNIQUES IN THE HISTORY OF MEDICINE

Although the approaches discussed in the preceding section also have their newer aspects, the title "new techniques" is reserved here for a number of approaches to the history of medicine that are relatively new or that have experienced a significant resurgence in recent years. These include the use of oral history to provide data for the historian, population studies using the methods of the demographer, paleopathology, contemporary history, and so-called psychohistory. All but the last receive separate discussion. Medical historians have always been interested in the psychological dimension, but they have not generally been drawn into the many discussions regarding the validity of psychoanalytic techniques applied to historical inquiries. There is now a journal devoted to that subject, the *Journal of Psychohistory*, and there are noteworthy studies, by Erik Erikson and others.

The interested reader may wish to consult the introduction to the collection edited by Bruce Mazlish (1971) as a good starting point.

A. Oral History

Oral history is one of the oldest methods of passing on historical lore from generation to generation; recently it has been appropriated as a tool of the medical historian. In *The Gateway to History* (1938), Allan Nevins of Columbia University urged fellow historians to preserve not only the rich treasures of papers and written materials, but oral memoirs as well. After World War II, when portable tape recorders became readily available, an Oral History Office was created at Columbia. Today, many libraries, archives, and history departments conduct similar programs. (See the article by Benison in Clarke, 1971.)

Despite burgeoning oral history enterprises, there is much misunderstanding about the technique. Actually, the oral historian works in much the same manner as the traditional historian. First he collects primary and secondary materials relevant to a given individual's life. Then he uses those data in such a way as to stir the memory of the individual who is interviewed. After the interview, the taped material is transcribed and edited. Thus, it says pretty much what the interviewer wants it to say, so that actually the oral history memoir is just a different kind of historical document. It is a product of both the memory of the person interviewed and the historian attempting to understand the events in which the interviewee has participated. The historian, by formulating relevant historical questions on the basis of his prior research, by serving as an audience to the person being interviewed, and finally by editing the transcript, helps to shape the ultimate document that emerges. An oral history document is an interpretation of a mass of primary and secondary data filtered through memory at a particular point in time. It may contain new information that has never previously been recorded, but it is essentially an interpretive document and must be read as such. Oral history has great virtue when the interviewer has done extensive research, developed relevant questions, and, not least, furnished relevant documents to help the interviewee refurbish his or her memory. When the interviewer has not prepared such background context, the document that emerges is all too often merely anecdotal, or worse, self-serving hagiography.

One of the most interesting recent oral history memoirs in the history of medicine is Saul Benison's *Tom Rivers: Reflections on a Life in Medicine and Science* (1967). This oral history charts the development of a pioneer virologist's career. It contains material not only on Rivers' research in virology, but on his association with the Johns Hopkins Medical School, the Rockefeller Institute for Medical Research, and the National Foundation. Rivers' memories of the National Foundation not only serve as a

contribution to the internal history of one of the important medical foundations in the United States, but also provide an informative account of the
development of the Salk and Sabin polio vaccines. Benison's memoir is
frequently cited. The American Association for the History of Medicine
awarded him its most prestigious honor, the William H. Welch medal, thus
indicating some of the recognition given this useful new technique in the
history of medicine.

B. Demography

Of the newer emphases in historical research, none has provoked more
heated discussion than the resort to numbers. Some have charged that quan-
tification too often substitutes for real understanding. Charges and counter-
charges may be put aside here, because one of the quantifier's preeminent
interests—historical population studies—overlaps with the history of medicine in such an important way that it cannot be ignored. Demographers are
interested in birth rates, death rates, morbidity and mortality rates in
various societies, in fertility, marriage patterns, and food supplies. All these
subjects are central to the history of medicine as well.

A decade or so ago, the so-called "demographic transition" was much
discussed. This transition, from high to low mortality and lower fertility,
occurred mainly in Western Europe, North America, and Japan. Its causes
have been the subject of discussion in economic history and demography,
and more recently, occasionally in the medical-historical literature. David
Landes, an economic historian at Harvard, wrote in *Daedalus* more than a
decade ago (1968) that the demographic transition was a central phenomenon of modern history, a core aspect of the process of modernization itself.
But historical demography is far more complex than the transition phenomenon alone. Early demographers, such as John Graunt and William
Petty in late seventeenth-century England, reported deaths by causes. Much
more data, especially census returns, will be required before demography
can aid the medical historian to any significant degree in analyzing rates
and causes of mortality.

Thomas McKeown, mentioned earlier in terms of his sociological approach to the history of medicine, has devoted himself for two decades to
problems involved in the decline of death rates and the growth of population. His work is now conveniently summarized in book form in *The
Modern Rise of Population* (1976), dealing mainly with the British part of
the story. McKeown, with the assistance of R. G. Brown, published a provocative essay, "Medical Evidence Related to English Population Changes
in the Eighteenth Century" (1955), in which it was argued that medical
measures contributed little to declining mortality rates. After analyzing
the role of physicians, the use of hospitals and dispensaries, and various

remedies utilized in combating disease, McKeown and Brown came to the conclusion that, contrary to the estimates of earlier demographers, the decline of mortality from diseases should more properly be ascribed to improved living conditions and better nutrition. Although they admit that a decline in the virulence of some infections may have played a part in lowering mortality rates, they maintain that specific medical measures did not.

McKeown continued his analysis with a paper in 1962 on the decline of mortality in England and Wales during the nineteenth century, again arguing that medical measures had little direct impact. A decade later, with his coworkers R. G. Brown and R. G. Record, McKeown summarized his work in a long article in which the conclusion was that population growth was not influenced by sanitary measures before about 1870. A comparison of the British data with figures from Sweden, France, and Ireland led to similar conclusions: Improved diet and better food production were the key factors in Europe's population growth.

P. E. Razzell (1974) has criticized McKeown's argument, maintaining that better food supplies are not sufficient to account for the population increase, but a reduction of smallpox and the common gastrointestinal infections could. (See also individual articles by D. V. Glass and E. A. Wrigley in *Daedalus* Spring 1968; J. Barzun's critique of quantifiers, *Clio and the Doctors,* 1975; and Langer, 1963.)

Since the issue of modern population shifts has been brought into focus by economic historians and demographers, McKeown claims that medical historians can no longer afford to ignore it. At least one medical historian who has not is James Cassedy of the National Library of Medicine. In 1969 he published the first portion of a multivolume study of demography in America, covering the colonial and postrevolutionary periods. Cassedy deals more with problems of statistics than with the population trends investigated by McKeown. Nonetheless, for the American story, this is an important first step.

Arthur Imhof and Oivind Larsen (1977), two European historians of medicine using Scandinavian materials, have recently written:

> The discussion and the subsequent harmonization of views which had to take place between the historian, trained in economic and demographic history, and the medical researcher, interested in medical history, as well as preventive and social medicine, proved very fruitful in relation to geographical and historical epidemiology.

Gerald Grob (1978) has written a biography of Edward Jarvis, a mid-nineteenth-century Massachusetts physician who was involved in problems of the federal census as well as with care of the mentally ill. William Farr, the most important of the nineteenth-century British physicians involved in reforming methods of collecting and using vital statistics, has recently been studied by John Eyler of the University of Minnesota (1976), including a full-length biography of Farr (1979). (See also Greenwood, 1948.)

The gold that awaits historians of medicine who mine the demographic lode has been richly exemplified in two recent books by the economic historian and demographer Carlo M. Cipolla. Author of the widely used *Economic History of World Population* (1962 and subsequent editions), Cipolla has mined rich Italian archives for data about health matters in the seventeenth century. First in *Cristofano and the Plague* (1973), and more recently in *Public Health and the Medical Profession in the Renaissance* (1976), Cipolla has demonstrated the potential that lies waiting in archival materials with respect to community responses to everyday diseases as well as epidemic outbreaks. In *Cristofano,* Cipolla concentrated on one community, Prato in Tuscany, and the work of the public health officer there in attempting to contain an outbreak of plague in 1630. In the later book, Cipolla describes the establishment and functions of health boards in numerous northern Italian communities; on this basis he outlines the character and distribution of early seventeenth-century Tuscan medical practitioners. He used Milanese bills of mortality, begun in 1492 and issued continuously after 1503. The famous London bills of the seventeenth century were derived from these earlier Italian models. The Italian bills, almost unknown until now, did not have a John Graunt to study them and to popularize their usefulness. (See Graunt, 1662, 1939.)

C. Paleopathology

Strictly speaking, paleopathology is the study of prehistoric diseases; it is also, however, a meeting ground for a variety of disciplines, including archeology, pathology, immunology, chemistry, radiology, and history. (See Kerley and Bass, 1967.) In the nineteenth century, the German pathologist and anthropologist Rudolf Virchow was an early worker in this field. In the twentieth century a number of Near Eastern and American excavations led to extensive work by Warren Dawson, Marc Ruffer, Arles Hrdlicka, and Roy Moodie. (See Sigerist's *A History of Medicine,* 1951; Jarcho, 1966; an article by Kerley in Clarke, 1971; and Wells, 1964.) Recently there has been a trend toward broadening the studies from a description of individual mummies or Indian finds to a consideration of the disease picture of entire populations.

Using the techniques of the scientific historical detective, including complex immunochemical assays, microradiographical analyses, electrophoresis of proteins, and careful microscopic study, paleopathologists have begun to supply interesting information to the historian. Knowledge of the distribution of diseases among prehistoric peoples can, for instance, aid in the understanding of hereditary relationships, population movements, and environmental conditions.

One of the topics most often debated in the earlier medico-historical literature concerned the origin of syphilis. Was it a New World disease brought back to Europe by the sailors of Columbus (and possibly others)? Or did it in fact exist in another form in the Old World long before 1492? There are complex historical and biological questions involved here. Prehistoric New World remains with syphilitic bone lesions have been found; as yet none have been found in the Old World. The fact that syphilis is only one of a family of treponemal diseases further complicates the story and the research. An old review of the literature on the subject is that of Williams (1932). Newer books on paleopathology, especially Brothwell and Sandison (1967), also deal with it. Crosby, in the *Columbian Exchange* (1972), provides an extensive discussion—a fine starting place for anyone interested in the origins of syphilis. (See also Rosebury, *Microbes and Morals,* 1971; Sigerist, *Civilization and Disease,* 1943, 1962; and Guerra, 1978.)

Disease is not the only interest of the paleopathologist. Surgical treatment, especially trephining of the skull, has also been the subject of extensive study. While much trephination (or trepanation, as it is also called) had its origins in ritual, it was practiced as a medical measure as well. Skulls with well-healed trephine sites indicate that some ancient patients survived their operations at least long enough for the skull bones to show signs of healing. Trephining, a procedure still used frequently in modern neurosurgery, was carried out for a variety of religious and medical reasons and by a variety of techniques. These are all explained and illustrated in one of the issues of the first volume of the *Ciba Symposia* (1939); these symposia are in general a rich source of medico-historical materials. The study of this early surgical procedure has wide implications for our understanding of related problems such as techniques of surgery, instruments, wound healing, and patient selection. (See especially G. Majno, *The Healing Hand,* 1975.)

The paleopathologist's and the archeologist's findings have forced us on occasion to reassess the role and the effectiveness of the earliest physicians. Finding the remains of prehistoric men with well-healed fractures of the limbs led to the assumption that primitive surgeons or bonesetters knew their business well. Adolph Schultz, however, in a study of 118 adult wild gibbons (1939), has shown that there were 42, or 36%, with well-healed fractures. Many of these had excellent union as well as maintenance of length. Nature, Schultz and others have concluded, has its own effective healing powers.

Since the broad history of diseases has always been central to the interests of the medical historian, it is obvious that the study of prehistoric disease should be very much a part of the history of medicine. The techniques may be those of the paleontologist, archeologist, or pathologist, but the results ought to be of interest to historians as well.

D. Contemporary History

"Every true history is contemporary history," wrote the early twentieth-century Italian philosopher of history Benedetto Croce, and the aphorism has rich implications for today's rapidly changing, highly technical medicine. Much contemporary history turns out to be either propaganda or a simple review of the scientific literature, but this should not deter us from the enterprise. If contemporary history is to be a fruitful approach for the historian of medicine, discoveries or developments must be placed in a historical context, and even events that did not lead to fruitful results, or that were not on the path of progress (as interpreted by the reviewer), must be included and evaluated.

British historian Geoffrey Barraclough notes that contemporary history has different meanings for different people. What is history to the young person, born after the introduction of antibiotics, was part of the "contemporary" life of an older generation. Questions that have arisen about medicine in the 1970s—particularly the shift from a caring and helping profession to a highly scientific all-too-often-impersonal endeavor—require for their understanding a careful delineation of the complex social and technical aspects of the society in which we live. If we accept Barraclough's definition, that "contemporary history begins when the problems which are actual in the world today first take visible shape," is it possible for us to describe the historical context in such a way as to provide better understanding? If the contemporary-history approach is to help, it should at the very least aid us in asking better questions.

What distinguishes history of medical science from a review of the scientific literature is an awareness, along with an analysis, of the context in which the scientific work has been carried out. Wrong turns and blind alleys of medical thought and practice enrich the history of medicine as much as still useful scientific theories and the results of successful experiments. The historian must examine the education of the physician, the institutions in which research has been carried out, and the personality of the researcher, as well as relations with other scientists, scientific organizations, and governments.

Until recently, monographs on the history of American medicine virtually all focused on the first hundred years after independence. In recent years there have been attempts to correct this deficiency, in dealing with the complex developments of the last fifty to seventy years—the period in which American scientific medicine came of age. (See, e.g., Duffy, 1976, and Bordley and Harvey, 1976.)

IV. THE HISTORY OF MEDICINE AND ITS
RELATIONSHIP TO OTHER FIELDS

No intellectual endeavor exists entirely independent of other fields; this is as true of the history of medicine as of any field.

A.   The History of Science

An obvious and close relationship exists between the history of medicine and the history of science (see chapter 1 in this volume). The two disciplines once followed similar paths, which have since crossed, then diverged. The history of science in the United States is now far more advanced, academically, than the history of medicine.

Problems that once were solely in the domain of the history of medicine have now become the province of historians of science as well. For instance, work on Renaissance anatomy or seventeenth-century physiology today may be preferred by representatives of either group. Nonetheless, there are some areas that still seem to be primarily the preserve of historians of medicine: the history of therapeutics, of medical institutions and medical practice, of twentieth century medicine in general. (See the article by M. Boas in Clarke, 1971; also G. Clark, 1966.)

During the 1930s there were fierce debates between the leaders of the two disciplines. "It is a remarkable fact," wrote George Sarton in 1935, "that the history of medicine has been studied more systematically and by a larger number of scholars than the history of any branch of science." However, as Sarton continued, it did not follow that the history of medicine would suffice to explain the general development of science, as some doctors seemed to assume. Nor, Sarton's objection continued, was the history of medicine the best part of the history of science. In his opinion, quantity of work did not automatically assure quality—a view shared by historians of all sorts in later years. Much of the history of medicine, Sarton warned, was "not of a high standard." Further, he complained that the history of science was not being supported as well as the history of medicine. Although medicine might justifiably be called the mother of all sciences, the large majority of scientific discoveries had no medical applications and were worthy of study in their own right. Thus the historian of medicine who imagined himself *ipso facto* a historian of science was laboring under a gross delusion. It was, Sarton ended his querulous editorial, like the play of Hamlet with Hamlet left out. (See Sarton, 1935; also chapter 1, I. C., in this volume.)

This attack from the leading historian of science of the day did not go unanswered. Early in 1936 Henry Sigerist, editor of the *Bulletin of the History of Medicine*—then beginning its fourth volume, in contrast to the twenty-three of *Isis* edited by Sarton—wrote an open letter to Sarton in the name of himself and his colleagues in the history of medicine. Sigerist agreed with some of Sarton's charges. He acknowledged, for instance, that a good deal of work in the history of medicine was done by physician-amateurs doing historical studies in their spare time. Although admittedly much of this work was amateurish, there were any number of physicians who produced histories that measured up to a very high standard. For the rest, Sigerist pointed out that the amateur's role was important; most new subjects started in an amateur way, with standards being raised gradually. (On the place of the amateur in the history of medicine, see McDaniel, 1939, and Wilson, 1978.)

Sigerist further disagreed with Sarton's claim that the history of medicine was both popular and well endowed. Its popularity, he pointed out, was as a hobby of retired physicians, not as a rigorous discipline, historical or medical. "Very little money is required to make a medicohistorical institute a valuable medical department." Sigerist continued, "But you have no idea how difficult it is to find that little money" (p. 4).

Medical history, in Sigerist's view, was a proper part of medicine. He felt that it should be written by historians who, in addition, were physicians living in close touch with the problems of modern medicine. Here Sigerist was a follower of Benedetto Croce—recall his dictum that "every true history is contemporary history"—and the increasing interest given to medical history could be explained by developments in medicine itself. (Sigerist's view of medicine was very broad indeed: he could also say [1948], "Medical history is to a large extent economic history.")

There is much in the history of science that is germane to the history of medicine. Ironically, Sarton was a scholar who contributed to both fields. His multivolume *Introduction to the History of Science* (1927–1948) contains much biobibliographic material on physicians, from ancient times to the fourteenth century. His later two-volume work, *A History of Science* (1952–1959), contains a great deal of material on the Hippocratic corpus and on Hellenistic medicine generally.

Similarly, the eight volumes of Lynn Thorndike's *A History of Magic and Experimental Science* (1923–1958) discuss medieval physicians, their work and thought. This is a particularly rich source for the history of alchemy and astrology in their relationship to medicine.

More recently, the work of Allen G. Debus—now conveniently collected in his two-volume *The Chemical Philosophy* (1977)—has clearly shown the close relationship between the history of chemistry and the history of medicine. The work of Paracelsus, Van Helmont, and the "iatrochemists" of the seventeenth century, as Debus shows, cannot be claimed as the exclusive province of either group of scholars. (See also Hannaway, 1975.)

There is also much that is pertinent for the historian of medicine in the history of technology (see chapter 2, II. C., in this volume). Maurice

Daumas, in his history of scientific instruments in the seventeenth and eighteenth centuries (1972), includes thermometers and microscopes. Bradbury, in *The Evolution of the Microscope* (1967), traces the history of that instrument to its pre-seventeenth-century origins; then, in great detail, he covers the story from developments in the early nineteenth century to the electron microscopes of the twentieth. Excellent illustrations and optical diagrams make this a handy reference volume.

The British physician-historian Charles Singer wrote extensively on the history of science and technology as well as the history of medicine. Singer's history of anatomy and physiology from the Greeks to Harvey (1925; 1957) was long a standard source; his *A Short History of Medicine* (second edition, with E. A. Underwood, 1962) is probably the most satisfactory one-volume text available. A two-volume *Festschrift* for Singer, *Science, Medicine, and History* (see Underwood, 1953), claims unity for all three fields.

B.   General History

General historians have become more interested in the history of diseases, especially in their social and biological consequences. In addition, historians working on scientific institutions, professionalism, social reform, and poverty, have all contributed to our understanding of health and welfare in former times.

Rarely is the coverage of medical history in general histories more than minimal. Although certain subjects, such as the scientific revolution of the seventeenth century, receive attention, others equally important are overlooked. For example, the *Cambridge Modern History* (1903–1912), as planned by Lord Acton around the turn of the century, devotes a chapter to science, including human anatomy and physiology, in volume V, "The Age of Louis XIV"; and William Harvey's name, as discoverer of the circulation of the blood, appears in the index. But one searches these volumes in vain for an extended discussion of such topics as disease, medical thought and practice, or health. This is also generally true of the *New Cambridge Modern History* (1957–1970).

In the large general encyclopedias, the history of medicine fares better. Often the history of surgery receives a separate article. The eleventh—the "literary"—edition of the *Encyclopaedia Britannica* (1910), for example, has an extensive article tracing the evolution of medicine; so does the most recent, much-revised edition (1975). The *Dictionary of American History* (1940; 1976) has numerous articles that are of interest to the historian of medical thought and practice, including some general surveys of the history of American medicine. The article by Owsei Temkin on "Health and Disease" in the *Dictionary of the History of Ideas* (1973, 1977) is an impressive overall survey of the history of medicine.

In the last two decades, some general historians have begun to stake claims in the history of medicine. In so doing, they have helped direct some medical historians away from antiquarian discussions of great physicians and the various institutions they were associated with, such as medical schools, hospitals, and professional societies, toward broader social concerns. Too often the focus of traditional medical histories had been narrow, overlooking the interplay between some particular medical subject and societal needs and norms.

Today numerous historical journals carry articles of interest to practitioners in our field. For instance, as already mentioned, in the summer of 1977 an entire issue of the *Journal of Social History* was devoted to medical subjects. The *Journal of Interdisciplinary History* and *Comparative Studies in Society and History* frequently carry articles of medical interest. The *Journal of American History,* which carries a bibliography of recent articles, includes as one category of the bibliography "Science, Medicine, and Technology."

French historians of the *Annales* school have long been practitioners of the interdisciplinary approach to the study of history. Founded in 1929 by Lucien Febvre and Marc Bloch, the *Annales d'Histoire Economique et Sociale* was devoted to bringing the methods of the social sciences to bear on historical research. Later, many of the contributors turned to demographic studies, and the editors have always been interested in biological and medical questions. Some examples: Fernand Braudel, in *Capitalism and Material Life, 1400–1800* (1973), deals extensively with housing, clothing, food, and population pressures; contents of a recent collection of eight translations from the *Annales, Biology of Man in History* (R. Forster and O. Ranum, 1975), include Evelyne Patlagen's "Birth Control in the Early Byzantine Empire"; E. LeRoy Ladurie's "Famine Amenorrhea"; and Michele Bordeaux's "Blazing a Trail to a History of Customary Law by Means of Geographic Hematology."

Nineteenth-century scholars like J. F. K. Hecker, Heinrich Haeser, and August Hirsch, elucidating the history and geography of disease, produced studies that were almost encyclopedic in nature. In the twentieth century—especially after the conception of "disease entities" gained new relevance with discoveries of bacterial agents—medical historians tended to restrict their work mostly to the medical aspects of disease. There are limits to such an approach, as George Rosen pointed out long ago. One must also look upon disease in its social setting; indeed, it is through this perspective that disease becomes more than a medical phenomenon.

For historians who have paid attention to disease in history, plague and cholera epidemics have been especially fruitful subjects. F. P. Wilson's *The Plague in Shakespeare's London* (1927), for instance, examined the disrupting effects of repeated epidemics upon the stage and London life.

A topic of recurrent fascination for general historians as well as historians of medicine is the so-called Black Death of 1347–1348. New historical analytic tools and hitherto unused archives are being brought to bear on the history of plague, and medievalists and Renaissance historians have joined medical historians in making it the object of important new studies. Nowhere may this be better seen than in *The Black Death, A Turning Point in History?* (1971). Here William M. Bowsky has provided for college history courses a collection of interpretations as well as a bibliography that will be of use to all historians of medicine. The reader will see how historians who agree upon data or long-term trends may arrive at quite different conclusions. In all the popular college studies series, this is the only one devoted to a subject in the history of medicine.

Two recent additions to our specialized knowledge of the effects of plague are Jean-Noël Biraben, *Les Hommes et la peste en France et dans les pays européens et mediterranées* (1975); and Michael Dols, *The Black Death in the Middle East* (1977). John Norris (1977) has traced the geographic origin and path of the plague that brought the Black Death to Europe. Philip Ziegler's *The Black Death* (1969) is one of the better popular studies of plague; it also stresses the accompanying social pathology.

Cholera has also been a source of concern to historians, as for instance in Charles Rosenberg's *The Cholera Years* (1962). Rosenberg combines the best of medical history with the best of social history in analyzing the effects of outbreaks of the disease on American society in 1832, 1849, and 1866. Medical, public health, and governmental responses to the spread of the disease are all analyzed. The social response was partly religious, partly rational in nature; Rosenberg provides a good picture of American urban society in the middle decades of the nineteenth century. The book is readily available in a paperback reprint and is widely used in courses.

Other diseases have been studied as well. C. E. A. Winslow's *The Conquest of Epidemic Disease* (1943), an example of work by a practicing epidemiologist, traces the course of malaria, yellow fever, smallpox, and other diseases. A more recent book—this one by a general historian—is William H. McNeill's *Plagues and Peoples* (1975); this is a major work that attempts to examine the effects of disease on the course of history. Although McNeill's work invites historians to pay more attention to biological factors in history, it is not entirely successful. Many more detailed monographic studies on separate diseases need to be carried out. One monograph is Alfred Crosby's *Columbian Exchange* (1972). It reviews the biological consequences of the New World explorations, dealing extensively with the origins of syphilis—now believed to involve, as well, Yaws and Pinta, the other treponemal diseases.

In American history, one could cite a number of examples in addition to Rosenberg: John Duffy's *Epidemics in Colonial America* (1953) and *The Sword of Pestilence* (1966), for example. Elizabeth Etheridge, in *The Butterfly Caste* (1972), carefully examines the role of pellagra in the poor rural

southern United States in the early decades of this century. Joseph Goldberger's difficulties in establishing the dietary-deficiency theory of pellagra,
and professional as well as social and economic resistance to his ideas, are all
well described. One final example: The American colonial historian Peter
Wood of Duke University has carefully described the complex subject of
sickle-cell trait and the resistance to malaria in a chapter of *The Black
Majority* (1974), his book on slavery in colonial South Carolina. A number
of books on slavery deal explicitly with health problems and medical care.
(See, e.g., K. Stampp, *The Peculiar Institution,* 1956; Willie Lee Rose,
Rehearsal For Reconstruction, 1964; R. Fogel and S. Engerman, *Time on
the Cross,* 1974; and Todd Savitt, *Medicine and Slavery,* 1978.)

C. Specialized Fields (Family History; Childhood and Child
Rearing; Women's History; History of Education
and Medical Education)

A number of specialized fields in history deal with problems of health and
disease. For example, a recent upsurge of interest in the history of the family
and childrearing practices has produced studies that are relevant to the
historian of medicine. Three recent collections of essays (Rabb and Rotberg, 1973; Rosenberg, 1975; and *Daedalus,* Spring 1977) deal with such
topics as youth, child rearing, adolescence, the age of sexual maturity.
Lawrence Stone's monumental work, *The Family, Sex, and Marriage in
England 1500–1800* (1977) is also germane. Earlier, Philippe Ariès's path-
breaking book, *Centuries of Childhood* (1962), had discussed the development of the concept of childhood from the Middle Ages on, dealing with
dress, schooling, discipline, and role in the family.

Social historians—including Ivy Pinchbeck, who with Margaret Hewitt
published a two-volume study of *Children in English Society* (1969–1973)—
often deal with infant mortality, foundling hospitals, and diseases of children. The Pinchbeck and Hewitt book is a study of the evolution of social
concern, of the way the state and the family provided for the care and
welfare of children in English society from the Tudor Period through the
nineteenth century. The book also addresses the problems caused by urbanization and industrialization. (For a history of children in the United
States, see Robert Bremner's five-volume work, 1970–1974). Elsewhere,
Pinchbeck has assessed the impact of the industrial revolution on the economic and social position of women. Her *Women Workers in the Industrial
Revolution, 1750–1850* (1969) is a portrait of English working-class family
life during this period.

The new *Journal of Psychohistory* (earlier, *The History of Childhood
Quarterly*) publishes many articles on children and the family. Its founding
editor, Lloyd deMause, has collected ten useful articles on comparative
aspects of childrearing at different times and in varying national cultures;
see *The History of Childhood* (1974).

In the last decade, much of the literature on medical care and the diseases of women has sprung from a renewed interest in women's studies generally. Martha H. Verbrugge (1976) has reviewed this literature up to 1975. She cites, in addition to the recent literature, such older work as Harvey Graham on the history of obstetrics and gynecology (1951); Norman Himes on contraception (1936); and Richard Shryock (1950) and John Blake (1965) on women in American medicine. (See also A. Davis, 1974.)

While the newer work tends to have a broader perspective, some of it has suffered from a feminist ideology based on a conspiracy theory of history. This view is typified by Ann Douglas Wood (1973) and G. Barker-Benfield (1976); according to them, women's health care has always mirrored a male-dominated world. Victorian physicians are perceived as maintaining a *status quo* by keeping women subordinate in domestic and maternal roles. Regina Morantz (1974) answers the article by Wood (and others like it); both pieces appear in Hartman and Banner eds., *Clio's Consciousness Raised* (1974)—a good place to begin the study of women's medical care.

Gail Parsons (1977) objects to the focus on women and their sexuality to the exclusion of male disorders. Men, too, complained of neglect even in the nineteenth century. Parsons also objects to the idea of a medical conspiracy of male physicians against their female patients; it is difficult, she points out, to explain an entire profession's behavior on such flimsy grounds.

Making extensive use of her papers, David M. Kennedy (1970) has written about Margaret Sanger's career and the birth control movement in America. In a more sweeping study of the social history of contraception, James Reed (1978) deals with the development of birth control technology as well as with the lengthy efforts by Margaret Sanger, Robert Dickinson, and Clarence Gamble to gain acceptance of birth control reforms. (For a review article, see John C. Burnham, "American Historians and the Subject of Sex," 1972.)

The role of women as professionals, doctors, nurses, and midwives has come under increasing scrutiny. Articles by Shryock (1950) and Blake (1965) have already been mentioned; in addition, a number of books have explored the theme of male professional dominance and the difficulties women have experienced in attempting to achieve a fair share of positions as well as due respect for their work. JoAnn Ashley in *Hospitals, Paternalism, and the Role of the Nurse* (1976); Jean Donnison's *Midwives and Medical Men* (1977); and Mary Roth Walsh in *Doctors Wanted: No Women Need Apply* (1977), all explore this theme. (For nursing, see M. A. Nutting and Lavinia Dock, *A History of Nursing*, 4 vols., 1907–1912; C. Woodham-Smith, *Florence Nightingale, 1820–1910*, 1951; and R. Kalisch and B. J. Kalisch, *The Advance of American Nursing*, 1978.)

On the history of medical education, older works by physicians (such as Theodor Puschmann, 1891) tended to be general histories of medical theories as they were taught to physicians. A better focus is found in works such as Charles Newman's on Britain in the nineteenth century (1957); this has recently been broadened by Jeanne Peterson's consideration of the problems of professionalization in Victorian London (1978). Other important work includes Erwin Ackerknecht's summary of early nineteenth-century Paris hospital medicine (1967) and Thomas N. Bonner's *American Doctors and German Universities* (1963). Several edited volumes are rich sources on medical education in particular periods or national settings; see, e.g., Poynter, 1966, for Britain; and O'Malley, 1970, which gives a broader survey.

Historians of American medical education have long looked upon W. F. Norwood's book on the story up to the Civil War (1944) as the standard source. Shafer (1936), Shryock (1960), Whitfield Bell (1975), Joseph Kett (1968), and William Rothstein (1972) have all described various aspects of medical education in the course of their work on the medical profession in America. Individual schools and hospitals have also frequently been the subject of study; among the best of these are Chesney's three volumes on the Johns Hopkins Hospital and Medical School (1943–1963); Turner (1974); and George Corner's *Two Centuries of Medicine* (1965), which describes the first school to be founded in the American colonies, at the University of Pennsylvania. (See also Kaufman, 1976.)

Medical historians have generally not followed the lead of general historians in the field, so opportunities for further research are numerous. No adequate assessment, for instance, has been made of medical education in the context of higher education in America. Although revisionist work on the Flexner Report is well under way (Hudson, 1972; Berliner, 1975, 1977), many questions remain to be answered. The interesting two-day discourse delivered by John Morgan at the opening of the medical school in Philadelphia in 1765, now widely cited and quoted—and certainly important for an understanding of Morgan and his time—may be one of those cases where historians have contributed to the production of historical artifacts by reprinting and overemphasizing documents that were not nearly so influential in their own day.

A list of research questions about medical education that remain to be answered might include: What were the characteristics of medical students and what were the necessary qualifications for the study of medicine? What were the qualifications of teachers? What was taught? What was the organization of learning and teaching? How did it vary from time to time and place to place? What was the relationship of medical schools to communities, to medical societies, and to universities? What were the legal, social, and ethical conditions for learning, teaching, practicing, and licensing? How was medical education financed? Who paid for it, and how much did it cost?

Bernard Bailyn, in his prevocative little book *Education in the Forming of American Society* (1960), raised many problems yet to be solved by historians of education. His list could apply equally to the history of medical education.

D. Sociology

Little has been written about the importance of the sociological point of view for the history of medicine (see, for example, chapter 8 in this volume). In the excellent guide edited by Edwin Clarke, *Modern Methods in the History of Medicine* (1971), there is no separate chapter devoted to the social sciences and the history of medicine. However, in his chapter on "The Medical Profession, Medical Practice, and the History of Medicine," Charles Rosenberg points out the complex relationship between the physician's social functions and his medical thought. That is where he maintains "the best medical history is to some degree always a historical sociology of medical knowledge" (Clarke, p. 29).

Some years ago, in a volume entitled *Sociology and History: Methods* (1968), the American historian Richard Hofstadter joined with sociologist Seymour Martin Lipset to discuss the fruitfulness of sociological methods for historians. After surveying the two fields and discussing methodological problems, they devote the bulk of the book to papers demonstrating the uses of a combined approach in the study of philanthropy, census returns, religion and class structure, political polling, and the economics of investments.

Almost a decade before, the sociologist C. Wright Mills had devoted a chapter to the uses of history in the *Sociological Imagination* (1959). Mills firmly believed that the social sciences are historical disciplines: "All sociology worthy of the name is historical sociology" (p. 146). "To eliminate such materials—the record of all that man has done and become—from our studies would be like pretending to study the process of birth but ignoring motherhood" (p. 147).

Although the sociology of medicine and the history of medicine thus overlap, they do not coincide, as Barber notes (1968).

Some sociologists have used the historical approach extensively. One of the best examples of this is the work of the Israeli sociologist Joseph Ben-David. In *The Scientist's Role in Society* (1971), a culmination of essays published during the previous dozen years, he studies the development of higher education, scientific productivity, and professional roles in Europe and America (see chapter 7 in this volume); for the medical historian, his focus on the physician's role in society is of especial interest.

A number of medical historians have brought the sociological approach into their historical research. George Rosen and Henry Sigerist are the best known, but others, especially those with a concern for the development of social medicine, have also addressed themselves to historical elements in sociological trends. The Belgian physician René Sand, in *The Advance to Social Medicine* (1952), used both history and sociology to illustrate the rise of preventive medicine, social hygiene, industrial medicine, and public assistance, as well as the advent of social medicine in the various countries of the world.

Shortly after Henry Sigerist's death in 1957, Milton Roemer collected Sigerist's papers on the sociology of medicine (1960). They included studies of socialized medicine, an outline for a history of the development of hospitals, and various essays on the organization and financing of medical care. One particularly important paper is Sigerist's discussion of German health insurance legislation during the 1880s, under Bismarck's chancellorship.

The pioneer medical sociologist Bernhard Stern devoted himself to the study of social factors that have tended to inhibit medical innovation. Opposition to such discoveries as Harvey's theory of the circulation of the blood; resistance to techniques such as Auenbrugger's percussion, and vaccination to prevent smallpox are typical of his concerns. (See Stern's collection of essays, *Historical Sociology*, 1959; also Barber, 1961.) George Rosen wrote his Ph.D. dissertation in sociology, examining the technological and institutional factors leading to medical specialization, with ophthalmology as his particular example (1944).

E. Anthropology

Anthropology's ties to medicine go back to the nineteenth century when physician-researchers Rudolf Virchow, in Germany, and Paul Broca, in France, were also well-known physical anthropologists. Physical anthropologists are often still to be found teaching anatomy.

Ethnographers have often been interested in medical subjects—such as belief systems about health and disease, folk cures, and so on. The recently published *Asian Medical Systems* (1976), edited by Charles Leslie, immediately conveys the breadth of the field as well as the possibilities for the historian of medicine. Wayland Hand's *American Medical Folklore* (1976) brings together a number of representative articles.

Primitive medicine, covered concisely in Sigerist's *A History of Medicine,* volume 1 (1957), has been more extensively investigated by Erwin Ackerknecht. Sigerist's chapter relies on Ackerknecht's work of the 1940s, which had been published in a series of articles, mostly in the *Bulletin of the History of Medicine.* Many of these articles were later collected in one volume, *Medicine and Ethnology* (1971), with a new introduction by Ackerknecht himself.

Works on shamanism interest the historian of medicine as much as they do the anthropologist. (See Mircea Eliade's classic, *Shamanism,* 1964; and I. M. Lewis's *Ecstatic Religion,* 1971; also, for the historian interested in a broader understanding of the meaning of uncleanliness, see Mary Douglas' *Purity and Danger: An Analysis of Concepts of Pollution and Taboo,* 1966.)

The work of historians who have made extensive use of anthropological methods and literature in their historical work includes, as an early example, Marc Bloch's *The Royal Touch: Sacred Monarchy and Scrofula in England*

and France (1973). Originally published in French fifty years earlier, the book discusses the royal healing power and in particular the medieval ritual of kings touching patients with scrofula, a tuberculous infection of lymph nodes of the neck. In studying this widespread affliction, Bloch assessed the general basis of French and English royal power as well as the medical belief systems of the medieval and Renaissance periods. In a more recent work, British historian Keith Thomas has made an extensive study of the belief systems of sixteenth- and seventeenth-century England in his *Religion and the Decline of Magic* (1971). Witchcraft, magical healing, and astrological beliefs all have a bearing on health practices as well as on beliefs about diseases and their cures. (See also MacFarlane, 1970.)

A relatively new approach to an old problem is called biohistory. The complexity of the relationship between the ecosystem and cultural and historical factors has often been unappreciated. In a recent book, *Keepers of the Game* (1978), a young historian, Calvin Martin of Rutgers University, shows the devastating effects of disease upon the traditional religious beliefs of eastern North American Indian tribes. The Indians, Martin argues, attributed the advent of epidemic diseases to evil spirits within the game. This erroneous assumption—that it was the animals and not the white man who brought the scourge of epidemic disease upon them—caused the Indians to change their sensible attitude about hunting to one that condoned the reckless slaughter of game. Participation in the white man's economic activities of fur trading for profits, in the hope of destroying the evil spirits that supposedly brought the new diseases, simultaneously annihilated both the game and the former customs of religious belief of these Indians.

F. Public Policy

Prior to World War II, historians showed little interest in the relationship between government and science. After 1945, with the growing perception of the role of government in the development of scientific research, studies began to appear. Richard Shryock's *American Medical Research Past and Present* (1947) described research trends in medicine as well as the changing patterns of financial support, private and governmental. Howard Miller casts a wider net in his *Dollars for Research: Science and Its Patrons in Nineteenth-Century America* (1970). Miller integrates the questions of research support and support of higher education. Hunter Dupree's *Science in the Federal Government* (1957) has long been a standard source for studying the relationship between science and government in the United States. This comprehensive survey, with a good chapter on medicine and public health, generally takes the story only to 1940.

Rosemary Stevens' *American Medicine and the Public Interest* (1971) is a valuable resource on the medical profession and its changing patterns of organization and administration. Stevens concentrates on the growth of medical specialization, tracing its implications for medical education and medical care, and discussing the effects of specialization on the funding of health services.

Also appearing since World War II, there is a long shelf of medical histories of the U. S. Army Medical Corps. The many books in this series provide rich detail on the organization of services as well as on technical developments in all the medical specialties.

Shyrock and some of his students had long debated the question of American indifference to basic research. This question has finally been put to rest by Nathan Reingold, another Shryock student, in a well-argued reappraisal (1972). Reingold's extensive review of the literature is a rich source on American science policy. He believes that the widely discussed thesis of American indifference is sterile, historiographically, because there is not sufficient evidence either to refute or to confirm it. (See also Comroe and Dripps, 1976. George Corner's thorough history of the Rockefeller Institute, 1964, is an account of the growth of medical science and research in America in the first half of this century. For recent government support of medical science see Shannon, 1967; Turner, 1967; and Swain, 1962.)

A number of historians have provided detailed histories of the evolution of British social insurance (see especially B. Gilbert, 1966). On the origins of government medical services, see Ruth Hodgkinson's major work on the *Origins of the National Service* (1967), which examines the growth of state medical services in the nineteenth century. Hodgkinson discusses the New Poor Law (1831–1871) and the relationship of the Poor Law medical officer to the various institutions providing for the care of the sick, from the workhouse to voluntary hospitals and public dispensaries.

Jeanne Brand, in her *Doctors and the State* (1965), examines the role of the medical officers of the local government boards in England between 1870 and 1912 in passing sanitary legislation and extending the national government's responsibility for the health of its citizens. Her discussion of the National Insurance Bill of 1911 is important: the bill was the prelude to the British National Health Service, which came immediately after World War II.

National health insurance, now standard in many countries, has been an issue of intense debate in the United States since the early decades of this century. The American story of health insurance may be found in Odin Anderson's *The Uneasy Equilibrium* (1968) and Ronald Numbers' *Almost Persuaded* (1978); the social security story is treated in books such as Roy Lubove, *The Struggle for Social Security, 1900–1955* (1968), and Edwin E. Witte, *The Development of the Social Security Act* (1963).

Three books about the political process involved in health legislation also deserve mention. The best documented of the three is Stephen Strickland's *Politics, Science, and Dread Disease: A Short History of United States Medical Research Policy* (1972); it is primarily an analysis of cancer research and the growth of the National Cancer Institute, but it also discusses the workings of the health lobbyists in Washington.

The other two books are less scholarly but not less important. The first, by Richard Harris on the Medicare legislation of the early 1960s, appeared initially as a series of articles in *The New Yorker* magazine. An expanded version appeared in 1969 as *A Sacred Trust*. Beginning with a concise history of attempts to pass national health insurance legislation, Harris concentrates on the politicking needed to pass a variety of bills in the early 1960s, culminating in the passage of health-care-for-the-elderly Medicare legislation in 1965. Harris' account is not well documented, but the discussion of the political tradeoffs involved in the legislative process is unfailingly fascinating.

Eric Redman's *The Dance of Legislation* (1973) is another volume in this genre. Redman, as a young Harvard graduate, spent time as staff assistant to Senator Warren Magnuson in 1969–1970. He chronicles the intricacies of writing, introducing, and seeing through the legislative process the bill that created the National Health Service Corps.

Some historians have criticized the Harris and Redman books as mere journalism. Whether these *are* journalism (at its best) or contemporary history is beside the point; the books are excellent introductions to issues that are important for an understanding of the relationship between government and contemporary medical thought and practice.

G.   Health and Human Values (including Bioethics)

An increasingly complex technology involved in the care of sick patients has tended to make encounters between doctors and patients less and less personal; this raises problems of medical ethics and the humane behavior of physicians. The rekindled interest in questions of medical ethics and of the humaneness of medical care, however defined, must be viewed in the context of history. We cannot successfully deal with these problems except against a background of: Nazi medical atrocities committed in the name of clinical investigation; the conservation-ecology movement; the Civil Rights and antiwar movements of the late 1960s; and the growing consumer rights movement.

With few exceptions, those who have written in the broad field of bioethics have not generally approached their subjects historically. The work of Ludwig Edelstein on Greek medicine is a prime example of an exception to this sweeping generalization. In 1943 he examined the Hippocratic oath historically, relating its various strictures and precepts to the beliefs of one group, the Pythagorean sect. He then elaborated on his discussion of the oath in a broader paper on the "Professional Ethics of the Greek Physician" (1956). (The oath, the paper on ethics, and other papers of Edelstein have been collected in *Ancient Medicine*, 1967.)

Nineteenth-century medical ethics—often really medical etiquette, as Chauncey Leake stressed (1927)—has been discussed by a number of historians, including Chester Burns (1977) and Donald Konold (1962), as well as Leake. The professional, legal, and sociological implications of ethical codes have been described for the British and American scene by Jeffrey Berlant in *Profession and Monopoly* (1975). The *Journal of Medicine and Philosophy* has included some historical articles, though they have been in the distinct minority. The multivolume *Encyclopedia of Bioethics* (1978) contains some historical articles on the development of ethical codes and on the history of human experimentation. For these and other aspects of the history of biomedical ethics, see chapter 6 of this volume.

Although interest in the topic dates back at least a hundred years, no full-scale historical study of human experimentation has yet been carried out. A number of source books on human experimentation contain important historical material. (See particularly Katz, 1972; Freund, 1969; and Ladimer and Newman, 1963.) Richard French (1975) has clearly shown the importance of the British antivivisection movement in the latter decades of the nineteenth century and its effects upon the development of experimental medicine, particularly physiological research. (See also Stevenson, 1955.)

Similarly, while there have been sociological, legal, and philosophical discussions of informed consent, there has been little attempt to approach the subject historically. For instance, just to raise one question: How or why did preoperative permission granted by patient to surgeon evolve? There is also a long and varied history of student use of patients for learning purposes that has not yet been carefully elucidated. The Roman writer Martial commented on the problem: "I lay sick, but you came to me, Symmachus, with a train of a hundred apprentices. A hundred hands frosted by the north wind touched me. I had no fever before Symmachus; now I have." (Quoted by Drabkin, 1944, p. 336.)

Changing patterns of medical education—from apprenticeship to didactic lectures in large amphitheaters to bedside work in the modern medical center—have obviously introduced changes into the ethical dimensions of clinical learning. These too need further historical work.

Perhaps with anthologies that include large sections of the older literature and early codes—such as the ones recently published by Reiser, Dyck, and Curran (1977), and by Burns (1977)—more historians will pay attention to the opportunities that await them in this currently fashionable field.

H. Geography

The Hippocratic work *Airs, Waters, and Places* is usually deemed to be the earliest medical writing to deal with the effects of climate and terrain upon diseases. Clarence Glacken, in his *Traces on the Rhodian Shore* (1967), examines the history of ideas of nature in Western thought from the time of Hippocrates to the end of the eighteenth century.

The geography of disease was studied widely by nineteenth-century medical historians. At least two of them, August Hirsch (1883) and Charles Creighton (1891–1894) produced classic, multivolume studies. Erwin Ackerknecht, in a short monograph (1965), has concisely summed up this tradition.

Medical historians still find the geographical literature helpful. Kenneth Thompson, professor of geography at the University of California at Davis, for instance, has published numerous studies: "Trees as a Theme in Medical Geography and Public Health" (1978); "Wilderness and Health in the Nineteenth Century" (1976); "Irrigation as a Menace to Health in California: A Nineteenth-Century View" (1969); and "Early California and the Causes of Insanity" (1976). Thompson's interests are not unique.

In the years following World War II, the American Geographical Society established a medical geography section under the direction of Jacques May, and a number of volumes have appeared on the ecology of human disease (1958, 1961) and on the ecology of malnutrition. These represent an effort to link together geography, social science, and medicine in order to explain the ecology of diseases such as amebiasis, leprosy, tuberculosis, and cholera.

I. The History of Art

Henry Sigerist explored the relationship between the history of medicine and the history of art in a chapter of his *Civilization and Disease* (1943). Fourteen years earlier, in a provocative article on "William Harvey's Position in the History of European Thought," Sigerist had postulated that in the period just prior to Harvey, in the Baroque art of the sixteenth century, the new ways of viewing nature had begun to emerge. One of these new perspectives included motion, and Harvey the physician, in Sigerist's view, was simply extending the concept of motion into his study of the human heart.

Medicine and art are related in a variety of ways. Drawing and painting, the most obvious way, are used to illustrate medical books. With the advent of moveable type, not only did books become more readily available, but anatomical drawings could be more accurately reproduced. Robert Herrlinger's *History of Medical Illustration from Antiquity to A.D. 1600* (1970) is a magnificent collection. Earlier, Fielding Garrison had discussed the *Principles of Anatomic Illustration before Vesalius* (1926), describing the importance of the medieval five-picture series; this topic has been discussed more recently by Ynez O'Neill (1969, 1977). In the early 1950s the medical historian Charles D. O'Malley and the anatomist-historian John B. deC. M. Saunders collaborated in bringing out accessible reprints of the important illustrations in the works of Vesalius (1950) and Leonardo

(1952), together with translations from their works and concise biographical sketches.

Art has also been used to illustrate the history of medicine. Otto Bettman's *A Pictorial History of Medicine* (1956), drawn from the rich Bettman Archives in New York, is a well-known example. Earlier, there were similar works by René Dumesnil (1935) and André Hahn and Paul Dumaître (1962).

Art historians have also addressed medical themes. Eugen Hollander, in the early years of this century, published four large books whose titles tell their contents: *Die Medizin in der klassichen Mallerei* (1903; second edition, 1913: depiction by major Renaissance artists of the physician, medical institutions, and diseases); *Die Karikature und Satire in der Medizin* (1905, 1921: caricature as applied to medicine from antiquity to the twentieth century); *Plastik und Medizin* (1912: medicine and sculpture); and *Wunder Wundergeburt und Wundergestalt aus deutschen Flügblattern des fünfzehnten bis achtzehnten Jahrhunderts* (1921: wonders such as half-man, half-animal figures, various birth defects, and magical beliefs relating to disease and cure).

Some notable recent works: Millard Meiss, *Painting in Florence and Siena after the Black Death* (1951, 1973); Henri Mondor's reprint collection, *Doctors and Medicine in the Works of Daumier* (1960); William Heckscher's *Rembrandt's Anatomy of Dr. Nicolaas Tulp* (1958); and the elaborate study by Raymond Klibansky, Erwin Panofsky, and Fritz Saxl of Dürer's *Melancholy* (1964). Readers of the latter two examples will be rewarded by excellent discussions of Renaissance anatomy by Heckscher and of sixteenth-century medical theory, especially that of the humors, by Klibansky, Panofsky, and Saxl.

There are a number of collections of medical art; of these, the Philadelphia Museum's collection of prints and paintings has been capably described by Carl Zigrosser (1955).

V. RESEARCH PROBLEMS

By now it must be obvious that problems for research are legion. A partial list would include the following nine:

1. Health practices—including nutrition, clothing, housing, and sanitation—of ordinary urban dwellers at any given time and place. Health and the city ought to be a major area for continuing study. Rural people should be studied as well.

2. Reasons for the acceptance of and resistance to medical ideas and practices.

3. The impact of technology upon medicine, and the development of medical instrumentation.

4. Health values and the meaning of disease in cultural, social, economic, and philosophical terms at various times and places.

5. The relationship of health to economics, especially the influence of poverty upon health and health care.

6. Changing patterns of medical care in terms of financing, organization, and specialization.

7. The relationship of medicine to biology, especially in the twentieth century.

8. Changing patterns of medical recruitment and medical education.

9. Historical perspectives on the relations between physicians and other health practitioners.

A final caveat about the approach that medical historians take to their work: In 1949 George Rosen wrote "Levels of Integration in Medical Historiography," an essay-review of several general histories of medicine then newly published. Rosen stressed that all historians, no matter what their theoretical standpoints might be, employ some sort of principles in the arrangement and integration of their materials. Historians of medicine, he shrewdly observed, especially if they happen also to be physicians, write from a decidedly *iatrocentric* viewpoint:

> Medical theory, literature, and practice occupy focal points in the presentation; less attention is paid to the history of the medical profession and still less to the history of community health. While the significance of social factors and conditions are recognized, this aspect is relegated to the very periphery of the picture (p. 465).

The patient, Rosen warned, clearly deserves a more prominent place in the history of medicine.

A similar warning was raised by Erwin Ackerknecht in a brief paper in 1967, "A Plea for a Behaviorist Approach in Writing the History of Medicine." Medical history, Ackerknecht said, is too often based on what elite physicians thought and did; it was they who wrote articles and books and gave speeches to students. But what the average physician, treating the average patient, thought and did is of equal importance. "We must admit to ourselves," Ackerknecht wrote, "that we often do not know the most elementary facts of either medical practices or of the social aspects of medical practice even for periods not very far removed at all" (p. 212). Ackerknecht called this model a "behaviorist" approach, urging historians of medicine to consider what doctors actually did, not merely what they wrote about. This program has elements of what some general historians refer to as history "from the bottom up," from the point of view of what actually happened to the people of a given time or place. It may be harder for the medical historian to characterize such people, but this does not excuse us from making the attempt.

VI. TEACHING AND TRAINING IN THE HISTORY OF MEDICINE

The earliest survey of American efforts in the teaching of the history of medicine was made in 1904 by a Baltimore physician, Eugene Cordell. When Henry Sigerist resurveyed the field in 1939, he found less activity. The most recent study of teaching and library facilities is that of Genevieve Miller, covering 1967 and 1968 (Miller, 1969). Miller's study reveals that, of the ninety-five medical schools surveyed, only fourteen (15%) reported obligatory courses in medical history. These varied from just a few hours to fifty at Johns Hopkins—a figure that has subsequently been reduced. Three schools in Canada reported that they have required instruction in their medical curriculum.

This is a relatively small percentage of North American medical students exposed to required courses—although an additional twenty-six schools reported elective courses in the history of medicine. The figures show a steady decline, from the 1904 survey to the 1939 survey (thirty-five schools) to a 1952 survey (twenty-six) to 1968 (fourteen schools). Considering that the number of medical schools rose from seventy-seven in 1939 to over one hundred in 1968, the decrease in the number of American medical students exposed to formal education in the history of medicine is serious.

Although precise figures are not available, a few more schools have instituted courses since Miller's report. In addition, many American universities and colleges now offer courses that discuss medicine and society, and a number of these are taught by historians. Some of these historians —for example, Saul Benison at the University of Cincinnati, Charles Rosenberg at the University of Pennsylvania, John Duffy at the University of Maryland, Gerald Grob at Rutgers, Barbara Rosenkrantz at Harvard, John Burnham at Ohio State, and Harvey Young at Emory—are among the most distinguished American medical historians, even though their primary academic department is not in a school of medicine. Their influence, and that of a score of others teaching in similar circumstances, could be important for the future.

There are also a number of medical faculty members who have, for years, offered course work, or at least reading and discussion sessions, but who do not call themselves historians of medicine.

According to Miller's survey, nine North American schools offer graduate degrees in the history of medicine; at the time she wrote, another ten had full-time professors. In the decade since, a few schools have added to the list of full-time chairs.

Estimating that about one third of the medical schools on the North American continent spend some time dealing with historical aspects of the profession with their students, one cannot claim title to an enviable record with this statistic, especially when compared to those of European schools.

What students experience in all this is difficult to judge, but certainly some exposure to the history of medicine is preferable to none at all.

An increasing number of undergraduates profess an interest in the health sciences; the numbers climb above fifty percent in some entering freshman classes. The history of medicine, in this situation, could assume increasing importance. In any case, much could be accomplished by using history as an introduction to the broad field of medicine.

Miller found ambivalence toward medical history among medical educators. Relatively few students in medical schools, she believes, are imbued with a positive attitude toward the history of medicine. Although antiquarianism and mediocre teaching have always retarded the progress of medical history, there may be some reason for hope. As the rest of this chapter demonstrates, an increasingly sophisticated and more competent historiography in medicine has come into existence in recent years. It is clearly reflected by publications in the field. Has there been a correlative improvement in teaching? This is harder to gauge. Judging from course syllabi, the level of courses today is high indeed. (For discussions of the place of medical history in medical education, see Blake, 1966; Galdston, 1957; Rosen, 1948; and King, Risse, Burns, and Hudson, 1975.)

A question that has vexed many over the course of the last forty years has to do with the qualifications of the historian of medicine. The issue turns on whether or not one needs to be a physician, or at least to have firsthand acquaintance with medicine or medical research, and if the latter only, to what degree?

There has been much misunderstanding about this issue. If it helps, one should recall that one of the most highly esteemed scholars in the history of medicine—Ludwig Edelstein (brought to Johns Hopkins by Henry Sigerist)—was a classicist, not a physician. (See his *Ancient Medicine*, 1967.)

Again, much of American medical history has been well covered by nonphysicians like Shryock, Henry Shafer, John Duffy, and Charles Rosenberg. On the other side, that so much of the latter work stops around 1870 may not have been chance. As the medical sciences have become more complicated and sophisticated, the need for specialized knowledge has become correspondingly more obvious.

There is no need to set down rules as to proper qualifications. Neither is it necessary to protect "turf"; the field is so broad and there is so much that needs to be done that the more diversity there is, the better off all will be. There are, of course, some aspects of the history of medicine, such as those involving patient care, that might benefit from direct experience with medical work. But how much medical experience is necessary is a difficult question. Perhaps the matter was best summed up by Owsei Temkin in 1968: "For what counts in the end is the man's ability to live up to his task. This cannot be judged in advance, and we lack sufficient experience to point to statistical results" (p. 56). From the surveys in this chapter, as well as the bibliography, it should be apparent that historians of medicine who have come to its study from history have, in the last two decades, contributed as much as their colleagues who have come to history from medicine. In the list of the American Association for the History of Medicine's Garrison lecturers and William H. Welch medalists since 1955, non-medically trained historians outnumber those with M.D. degrees. (See the *Bulletin of the History of Medicine* 50 [1976]: 118–119.)

A final point: Many members of the American Association for the History of Medicine seem to feel quite strongly that nonprofessional historians of medicine—that is, those who do not earn their living by history alone—are a necessary and vital part of the organization. Physicians who, in the time they can find, read and write about their profession's past should be encouraged in every way possible. In the history of science, the amateur has largely been displaced. Although some historians of medicine would like to see a similar trend in our field, we would probably then no longer speak as well as we do now to the needs of physicians and medical students. The field would, accordingly, be weaker and narrower as a result.

BIBLIOGRAPHIC INTRODUCTION

One frequently noted characteristic of recent science is "information over-load," the proliferation of scientific communications to a point where it is difficult for the individual scientist to keep up. The medical-historical literature has not yet grown to such an unmanageable scale. The medical sciences—which have grown unmanageable—are aided by the National Library of Medicine's MEDLINE computer-based bibliography. This has now been joined by HISTLINE, which provides rapid, large-scale literature searches for the history of medicine. That will become *the* place for the researcher to begin.

The balance of this bibliographic introduction is divided into five sections listing journals, major bibliographies, bibliographic collections and dictionaries, collections of readings, and certain representative works in general history. After that will come the main bibliography, an alphabetical listing of all the works cited or discussed in the foregoing chapter, except for some older classics that have appeared in *so* many editions they are not repeated here. Also included are representative biographies and general histories of medicine and allied subjects.

A.   Major Journals in the History of Medicine and Related Fields

Annals of Medical History. New York: 1917–1942.
Archiv für Geschichte der Medizin. Leipzig: 1907–1929. After 1929 known as *Sudhoff's Archiv.*

Bulletin of the History of Dentistry. Batavia, New York: 1958–
Bulletin of the History of Medicine. Baltimore: 1933– .
Bulletin of the Society for the Social History of Medicine. Nottingham, England: 1970– .
Centaurus. Copenhagen: 1950– .
Ciba Symposia: Summit, N.J.: 1939–1951.
Clio Medica. Oxford: 1965– .
Gesnerus. Aaran, Switzerland: 1943– .
History of Science. New York: 1962– .
Isis. New York: 1913– .
Janus. Amsterdam: 1896– .
Journal of Medicine and Philosophy. Chicago: 1976– .
Journal of the History of Behavioral Sciences. Brandon, Vermont: 1965– .
Journal of the History of Biology. Cambridge, Mass.: 1968– .
Journal of the History of Medicine and Allied Sciences. New Haven: 1946– .
Kyklos. Leipzig: 1928–1932.
Medical History. London: 1957– .
Medical Life. New York: 1920–1938.
Medizinhistorisches Journal. Hildesheim: 1966– .
Mitteilungen zur Geschichte der Medizin und der Naturwissenschaften. Hamburg and Leipzig: 1902–1942.
Pharmacy in History. Madison, Wis.: American Institute of the History of Pharmacy, 1959– .
Studies in History of Biology. Baltimore: 1977– .

B.   Major Bibliographies

Austin, Robert B. *Early American Medical Imprints: A Guide to Works Printed in the United States 1668–1820.* Washington: Department of H.E.W., 1961.
Bloomfield, Arthur L. *A Bibliography of Internal Medicine: Communicable Disease.* Chicago: University of Chicago Press, 1958.
Bloomfield, Arthur L. *A Bibliography of Internal Medicine: Selected Diseases.* Chicago: University of Chicago Press, 1960.
Blake, John B. and Roos, Charles. *Medical Reference Works 1679–1966: A Selected Bibliography.* Chicago: Medical Library Association, 1967.
Callisen, Adolph C. P. *Medizinisches Schriftsteller-Lexicon der jetzt leben-den Ärzte, Wundärzte, Geburtshelfer, Apotheker, und Naturforscher aller gebildeten Völker.* 33 vols. Copenhagen, 1830–1845.
Choulant, Ludwig. *History and Bibliography of Anatomic Illustration.* Tr. by M. Frank. Rev. ed., New York: Henry Schuman, 1945.
Current Work in the History of Medicine: An International Bibliography. London: The Wellcome Institute for the History of Medicine, 1954– .

Ebert, Myrl. "The Rise and Development of the American Medical Periodical, 1797–1850." *Bulletin of the Medical Library Association* 40 (July 1952): 243–276.

Garrison, Fielding H. "The Medical and Scientific Periodicals of the 17th and 18th Centuries." *Bulletin of the History of Medicine.* 2 (July 1934): 285–343. Addenda by D. A. Kronick 32 (September-October 1958): 456–474.

Garrison, Fielding H. "Revised Students' Check-List of Texts Illustrating the History of Medicine with References for Collateral Reading." *Bulletin of the History of Medicine.* 1 (November 1933): 333–434.

Gilbert, Judson B. *Diseases and Destiny: A Bibliography of Medical References to the Famous.* London: Dawson, 1962.

Guerra, Francisco, comp. *American Medical Bibliography 1639–1783.* New York: Lathrop C. Harper, 1962.

Haller, Albrecht von. *Bibliotheca medicinae practicae.* 4 vols. Basel: J. Schweighauser, 1776–1788.

Index Catalog of the Library of the Surgeon General's Office. 4 series. Washington: U.S. Government Printing Office, 1880–1961.

Index Medicus. 21 vols. New York, 1879–1899.

Isis. Cumulative Bibliography. 3 vols. London: Mansell, 1971–1976.

Kelly, Emerson C. *Encyclopedia of Medical Sources.* Baltimore: Williams and Wilkins, 1948.

Miller, Genevieve, ed. *Bibliography of the History of Medicine of the United States and Canada, 1939–1960.* Baltimore: Johns Hopkins University Press, 1964.

Morton, Leslie T. *A Medical Bibliography* (Garrison and Morton). 3d ed. London: André Deutsch, 1970.

National Library of Medicine. Bibliography of the History of Medicine. Washington: 1964–

Osler, Sir William. *Bibliotheca Osleriana.* Oxford: Clarendon Press, 1929.

Smit, Peter. *History of the Life Sciences: An Annotated Bibliography.* New York: Hafner, 1974.

Thornton, John L. *Medical Books, Libraries, and Collectors: A Study of Bibliography and the Book Trade in Relation to the Medical Sciences.* 2d ed. London: André Deutsch, 1966.

Trautmann, Joanne, and Pollard, Carol, comp. *Literature and Medicine: Topics, Titles, and Notes.* Philadelphia: Society for Health and Human Values, 1975.

C.   Biographic Collections and Dictionaries

Bayle, A. L. J., and Thillaye, A. J. *Biographie Médicale.* 2 vols. Paris: A. Delahaye, 1855.

Debus, Allen G., ed. *World Who's Who in Science: A Biographical Dic-*

tionary of Notable Scientists from Antiquity to the Present. Chicago: Marquis Who's Who, 1968.

Fischer, Isidor. *Biographisches Lexikon der hervorragenden Ärzte der letzten fünfzig Jahre.* 2 vols. Berlin: Urban and Schwarzenberg, 1932–1933.

Gillispie, Charles, ed. *Dictionary of Scientific Biography.* 14 vols. New York: Charles Scribner's Sons, 1970–1976.

Gross, Samuel D. *Lives of Eminent Physicians and Surgeons of the Nineteenth Century.* Philadelphia: Lindsay and Blakiston, 1861.

Hirsch, August. *Biographisches Lexikon der hervorragenden Ärzte aller Zeiten und Völker.* 6 vols. Vienna: Urban and Schwarzenberg, 1884–1888.

Hutchinson, Benjamin. *Biographica Medica; or, Historical and Critical Memoirs of the Lives and Writings of the Most Eminent Medical Characters that Have Existed from the Earliest Account of Time to the Present Period; with a Catalog of their Literary Productions.* 2 vols. London: J. Johnson, 1799.

Kelly, Howard A., and Burrage, Walter L., eds. *Dictionary of American Medical Biography.* New York: Appleton, 1928.

MacMichael, William. *The Gold-Headed Cane.* London: J. Murray, 1827; new ed., Springfield, Ill.: Charles C. Thomas, 1953.

Munk, William. *Roll of the Royal College of Physicians of London.* 5 vols. London: The College, 1878–1968.

Pagel, Julius. *Biographisches Lexikon herrovorragenden Ärzte des neunzehnten Jahrhunderts.* 5 vols. Berlin: Urban and Schwarzenberg, 1901.

Plarr, Victor G. *Plarr's Lives of the Fellows of the Royal College of Surgeons of England.* Revised by Sir D'Arcy Power. 2 vols. Bristol: John Wright and Sons, 1930.

Sourkes, Theodore L., and Stevenson, Lloyd G. *Nobel Prize Winners in Medicine and Psychology 1901–1965.* Rev. ed., New York: Abelard-Schuman, 1966.

Talbott, John H. *A Biographical History of Medicine.* New York: Grune and Stratton, 1970.

Thacher, James. *American Medical Biography.* 2 vols. Boston: Richardson, 1828.

Wickersheimer, Ernest. *Dictionnaire biographique des médecins en France au moyen âge.* 2 vols. Paris: E. Droz, 1936.

D. Collections of Readings

Brieger, Gert H., ed. *Medical America in the Nineteenth Century: Readings from the Literature.* Baltimore: Johns Hopkins University Press, 1972.

——. *Theory and Practice in American Medicine.* New York: Science History Publications, 1976.

Brock, A. J. *Greek Medicine: Being Extracts Illustrative of Medical Writing from Hippocrates to Galen.* Translated and annotated. New York: E. P. Dutton, 1929.

Burns, Chester R., ed. *Legacies in Ethics and Medicine.* New York: Science History Publications, 1977.

———. *Legacies in Law and Medicine.* New York: Science History Publications, 1977.

Camac, Charles, N. B. *Epoch-Making Contributions to Medicine, Surgery, and the Allied Sciences.* Philadelphia: W. B. Saunders, 1909.

Clendening, Logan, comp. *Source Book of Medical History.* New York: Dover, 1960.

Doyle, Paul A., ed. *Readings in Pharmacy.* New York: John Wiley, 1962.

Earle, A. Scott, ed. *Surgery in America: From the Colonial Era to the Twentieth Century: Selected Writings.* Philadelphia: W. B. Saunders, 1965.

Fulton, John F., and Wilson, Leonard G., eds. *Selected Readings in the History of Philosophy.* 2d ed. Springfield, Ill.: Charles C. Thomas, 1966.

Haymaker, Webb, and Schiller, Francis. *The Founders of Neurology.* 2d ed. Springfield, Ill.: Charles C. Thomas, 1970.

Holmstedt, B., and Liljestrand, G., eds. *Readings in Pharmacology.* New York: Macmillan, 1963.

Hunter, Richard, and Macalpine, Ida, eds. *Three Hundred Years of Psychiatry, 1535–1860.* London: Oxford University Press, 1963.

Hurwitz, Alfred, and Degenshein, George A., eds. *Milestones in Modern Surgery.* New York: Hoeber-Harper, 1958.

Kelly, Emerson C., comp. *Medical Classics.* Baltimore: Williams and Wilkins, 1936–1941.

King, Lester S., ed. *A History of Medicine: Selected Readings.* Baltimore: Penguin Books, 1971.

Leavitt, Judith W., and Numbers, Ronald L., eds. *Sickness and Health in America: Readings in the History of Medicine and Public Health.* Madison: University of Wisconsin Press, 1978.

Lechevalier, Hubert A., and Soltorovsky, Morris, eds. *Three Centuries of Microbiology.* New York: McGraw-Hill, 1965.

Long, Esmond R., ed. *Selected Readings in Pathology.* 2d ed. Springfield, Ill.: Charles C. Thomas, 1961.

Major, Ralph H., ed. *Classic Descriptions of Disease.* 3d ed. Springfield, Ill.: Charles C. Thomas, 1945.

Willius, Frederick, A. and Keys, Thomas E., eds. *Classics of Cardiology.* 2 vols. Rev. ed., New York: Dover, 1961.

Zimmerman, Leo M., and Veith, Ilza. *Great Ideas in the History of Surgery.* 2d ed. New York: Dover, 1967.

E. **Representative Works in General History with Sections or
Chapters on Aspects of the History of Medicine or
Public Health**

Blumenthal, Henry. *American and French Culture, 1800–1900: Interchanges in Art, Science, Literature, and Society.* Baton Rouge: Louisiana State University Press, 1975.

Boorstin, Daniel J. *The Americans.* 3 vols. New York: Random House,
1958–1973.

Bridenbaugh, Carl. *Cities in the Wilderness: The First Century of Urban
Life in America, 1625–1742.* New York: Ronald Press, 1938.

———. *Cities in Revolt: Urban Life in America, 1743–1776.* New York:
Alfred A. Knopf, 1955.

Braudel, Fernand. *Capitalism and Material Life 1400–1800.* Translated by
M. Kochan. New York: Harper & Row, 1973.

Buer, M. C. *Health, Wealth, and Population in the Early Days of the
Industrial Revolution.* London: George Routledge and Sons, 1926.

Davis, Natalie Zemon. *Society and Culture in Early Modern France.* Stanford, Calif: Stanford University Press, 1975.

Dick, Everett N. *The Sodhouse Frontier, 1854–1890: A Social History of
the Northern Plains.* Lincoln: Johnson, 1954.

Eggleston, Edward. *The Transit of Civilization From England to America
in the Seventeenth Century.* New York: Appleton, 1900; reprint ed.
Boston: Beacon Press, 1959.

Fogel, Robert William and Engerman, Stanley L. *Time on the Cross: The
Economics of American Slavery.* 2 vols. Boston: Little, Brown, 1974.

Gay, Peter. *The Enlightenment: An Interpretation.* 2 vols. New York: Random House, 1966–1969.

Hale, J. R. *Renaissance Europe: Individual and Society, 1480–1520.* London: William Collins, 1971.

Laslett, Peter. *The World We Have Lost: England Before the Industrial
Age.* 2d ed. New York: Charles Scribner's Sons, 1971.

Lubove, Roy. *The Progressives and the Slums: Tenement House Reform
in New York City, 1890–1917.* Pittsburgh: University of Pittsburgh
Press, 1962.

Miller, Perry. *The New England Mind: From Colony to Province.* Cambridge, Mass.: Harvard University Press, 1953.

Rose, Willie Lee. *Rehearsal for Reconstruction: The Port Royal Experiment.* Indianapolis: Bobbs-Merrill, 1964.

Stampp, Kenneth M. *The Peculiar Institution: Slavery in the Ante-Bellum
South.* New York: Alfred A. Knopf, 1965.

Thomas, Keith. *Religion and the Decline of Magic.* New York: Charles
Scribner's Sons, 1971.

Thompson, E. P. *The Making of the English Working Class*. New York: Random House, 1963.

Warner, Sam Bass, Jr. *The Urban Wilderness: A History of the American City*. New York: Harper & Row, 1972.

Wiebe, Robert H. *The Search for Order 1877–1920*. New York: Hill and Wang, 1967.

Wood, Peter H. *Black Majority: Negroes in Colonial South Carolina from 1670 through the Stono Rebellion*. New York: Alfred A. Knopf, 1974.

Zeldin, Theodore. *France, 1848–1945*. 2 vols. Oxford: Clarendon Press, 1973–1977.

BIBLIOGRAPHY

Abel-Smith, Brian. *The Hospitals, 1800–1948*. London: Heinemann, 1964.

Ackerknecht, Erwin H. *Malaria in the Upper Mississippi River Valley 1760–1900. Bulletin of the History of Medicine*, supplement no. 4. Baltimore: Johns Hopkins University Press, 1945.

———. "Hygiene in France." *Bulletin of the History of Medicine* 22 (March-April 1948): 117–155.

———. *Rudolf Virchow: Doctor, Stateman, Anthropologist*. Madison: University of Wisconsin Press, 1953.

———. "Paleopathology." In *Anthropology Today: An Encyclopedic Inventory*, pp. 120–126. Edited by A. L. Kroeber. Chicago: University of Chicago Press, 1953.

———. "Recollections of a Former Leipzig Student." *Journal of the History of Medicine* 13 (April 1958): 147–150.

———. "A Plea for a Behaviorist Approach in Writing the History of Medicine." *Journal of the History of Medicine* 22 (July 1967): 211–214.

———. *Medicine at the Paris Hospital, 1794–1848*. Baltimore: Johns Hopkins University Press, 1967.

———. *A Short History of Psychiatry*. 2d ed. Tr. by Sula Wolff. New York: Hafner, 1968.

———. *Medicine and Ethnology*. Edited by H. H. Walser and H. M. Koebling. Baltimore: Johns Hopkins University Press, 1971.

———. *Therapeutics From the Primitives to the 20th Century*. New York: Hafner, 1973.

Adelman, Howard B. *Marcello Malpighi and the Evolution of Embryology*. 5 vols. Ithaca: Cornell University Press, 1966.

Allen, Garland E. *Life Science in the Twentieth Century*. New York: John Wiley, 1975.

Anderson, Odin W. *The Uneasy Equilibrium: Private and Public Financing of Health Services in the United States, 1875–1965*. New Haven: College and University Press, 1968.

Ariès, Philippe. *Centuries of Childhood: A Social History of Family Life.* New York: Alfred A. Knopf, 1962.

Artelt, Walter. *Einführung in die Medizin-historik.* Stuttgart: Ferdinand Enke Verlag, 1949.

Ashley, JoAnn. *Hospitals, Paternalism, and the Role of the Nurse.* New York: Teachers College Press, 1976.

Baas, Johann H. *Outlines of the History of Medicine and the Medical Profession.* Translated by H. E. Handerson. New York: J. H. Vaul, 1889.

Bailyn, Bernard. *Education in the Forming of American Society: Needs and Opportunities for Study.* Chapel Hill: University of North Carolina Press, 1960.

Barber, Bernard. "Resistance by Scientists to Scientific Discovery." *Science* 134 (September 1, 1961): 596–602; also in *The Sociology of Science,* pp. 539–556. Edited by Bernard Barber and Walter Hirsch. Glencoe, Ill.: Free Press, 1962.

————. "Science: The Sociology of Science." In *International Encyclopedia of the Social Sciences,* vol. 14, pp. 92–100. New York: Free Press, 1968.

Barker-Benfield, G. J. *The Horrors of the Half-Known Life: Male Attitudes Toward Women and Sexuality in Nineteenth-Century America.* New York: Harper & Row, 1976.

Barraclough, Geoffrey. *An Introduction to Contemporary History.* New York: Basic Books, 1964.

Barzun, Jacques. *Clio and the Doctors: Psycho-History, Quanto-History, and History.* Chicago: University of Chicago Press, 1974.

Bell, Whitfield J. Jr. *John Morgan, Continental Doctor.* Philadelphia: University of Pennsylvania Press, 1965.

————. "Joseph M. Toner (1825–1896) as a Medical Historian." *Bulletin of the History of Medicine* 47 (January-February 1973): 1–24.

————. *The Colonial Physician and Other Essays.* New York: Science History Publications, 1975.

Ben-David, Joseph. *The Scientist's Role in Society: A Comparative Study.* Englewood Cliffs, N.J.: Prentice-Hall, 1971.

Benison, Saul. "Oral History: A Personal View." In *Modern Methods in the History of Medicine,* pp. 286–305. Edited by Edwin Clarke. London: Athlone Press, 1971.

————, ed. *Tom Rivers: Reflection on a Life in Medicine and Science: An Oral History Memoir.* Cambridge, Mass.: M.I.T. Press, 1967.

Berlant, Jeffrey L. *Profession and Monopoly: A Study of Medicine in the United States and Great Britain.* Berkeley: University of California Press, 1975.

Berliner, Howard S. "A Larger Perspective on the Flexner Report." *International Journal of Health Services* 15 (Fall 1975): 573–592.

————. "New Light on the Flexner Report: Notes on the A.M.A.-Carnegie Foundation Background." *Bulletin of the History of Medicine* 51 (Winter 1977): 601–609.

Berman, Alex. "The Thomsonian Movement and Its Relation to American Pharmacy and Medicine." *Bulletin of the History of Medicine* 25 (September-October 1951): 405–428; (November-December 1951): 503–518.

————. "Neo-Thomsonianism in the United States." *Journal of the History of Medicine* 11 (April 1956): 133–155. Also in *Theory and Practice in American Medicine,* pp. 149–72. Edited by G. H. Brieger. New York: Science History Publications, 1976.

Bettmann, Otto L. *A Pictorial History of Medicine.* Springfield, Ill.: Charles C. Thomas, 1956.

Biraben, Jean-Noel. *Les Hommes et la peste en France et dans les pays européens et mediterranées.* Paris: Mouton, 1975.

Blake, John B. *Public Health in the Town of Boston, 1630–1822.* Cambridge, Mass.: Harvard University Press, 1959.

————. "Women and Medicine in Ante-Bellum America." *Bulletin of the History of Medicine* 39 (March-April 1965): 99–123.

————, ed. *Education in the History of Medicine: Report of a Macy Conference.* New York: Hafner, 1968.

Bledstein, Burton J. *The Culture of Professionalism: The Middle Class and the Development of Higher Education in America.* New York: W. W. Norton, 1976.

Bloch, Marc. *The Royal Touch: Sacred Monarchy and Scrofula in England and France.* Tr. by J. E. Anderson. Montreal: McGill University Press, 1973.

Boas, Marie. *The Scientific Renaissance 1450–1630.* New York: Harper & Row, 1962.

Bockoven, J. Sanbourne. *Moral Treatment in American Pychiatry.* New York: Springer, 1963.

Bonner, Thomas N. *American Doctors and German Universities: A Chapter in International Intellectual Relations 1870–1914.* Lincoln: University of Nebraska Press, 1963.

Bordley, James and Harvey, A. McGehee. *Two Centuries of American Medicine 1776–1976.* Philadelphia: W. B. Saunders, 1976.

Bowsky, William M., ed. *The Black Death: A Turning Point in History?* New York: Holt, Rinehart, and Winston, 1971.

Brand, Jeanne L. *Doctors and the State: The British Medical Profession and Government Action in Public Health, 1870–1912.* Baltimore: Johns Hopkins University Press, 1965.

Braudel, Fernand, *Capitalism and Material Life 1400–1800.* Translated by Miriam Kochan. New York: Harper & Row, 1973.

Bremner, Robert H., ed. *Children and Youth in America: A Documentary*

History. 5 vols. Cambridge, Mass.: Harvard University Press, 1970–1974.

Brieger, Gert H. "Sanitary Reform in New York City: Stephen Smith and the Passage of the Metropolitan Health Bill." *Bulletin of the History of Medicine* 40 (September-October 1966): 407–429.

Brooks, Chandler McC. and Cranefield, Paul F., eds. *The Historical Development of Physiological Thought.* New York: Hafner, 1959.

Brothwell, Don, and Sandison, A. T., eds. *Diseases in Antiquity: A Survey of the Diseases, Injuries, and Surgery of Early Populations.* Springfield, Ill.: Charles C. Thomas, 1967.

Browne, E. G. *Arabian Medicine.* Cambridge: Cambridge University Press, 1921.

Buchanan, Scott. *The Doctrine of Signatures: A Defence of Theory in Medicine.* London: Kegan Paul, Trench, and Trubner, 1938.

Bulloch, William. *The History of Bacteriology.* London: Oxford University Press, 1938.

Bullough, Vern L. *The Development of Medicine as a Profession: The Contribution of the Medieval University to Modern Medicine.* New York: S. Karger, 1966.

Burnham, John C. *Psychoanalysis and American Medicine, 1894–1918: Medicine, Science, and Culture.* Monograph no. 20, *Psychological Issues* (1967).

———. "American Historians and the Subject of Sex." *Societas* 2 (Autumn 1972): 307–316.

Burns, Chester R., ed. *Legacies in Ethics and Medicine.* New York: Science History Publications, 1977.

Burrow, James G. *Organized Medicine in the Progressive Era: The Move Toward Monopoly.* Baltimore: Johns Hopkins University Press, 1977.

Butterfield, Herbert. *The Origins of Modern Science 1300–1800.* London: G. Bell and Sons, 1957.

Caldwell, Charles. *Autobiography of Charles Caldwell, M.D.* Philadelphia, 1855; reprint ed., New York: DeCapo Press, 1968.

Cartwright, Frederick F. *The Development of Modern Surgery.* New York: Thomas Y. Crowell, 1967.

Cassedy, James H. *Charles V. Chapin and the Public Health Movement.* Cambridge, Mass.: Harvard University Press, 1962.

———. *Demography in Early America: Beginnings of the Statistical Mind, 1600–1800.* Cambridge, Mass.: Harvard University Press, 1969.

Castiglioni, Arturo. *A History of Medicine.* 2d ed. Tr. by E. B. Krumbhaar. New York: Alfred A. Knopf, 1947.

Chesney, Alan M. *The Johns Hopkins Hospital and the Johns Hopkins University School of Medicine: A Chronicle.* 3 vols. Baltimore: Johns Hopkins University Press, 1943–1963.

Churchill, Edward D., ed. *To Work in the Vineyard of Surgery: The Remi-*

niscences of J. Collins Warren (1842–1927). Cambridge, Mass.: Harvard University Press, 1958.

Cipolla, Carlo M. *The Economic History of World Population.* Baltimore: Penguin Books, 1962.

———. "The Professions: The Long View." *Journal of European Economic History* 2 (Spring 1973): 37–52.

———. *Cristofano and the Plague: A Study in the History of Public Health in the Age of Galileo.* Berkeley: University of California Press, 1973.

———. *Public Health and the Medical Profession in the Renaissance.* Cambridge: Cambridge University Press, 1976.

Clark, Sir George. "The History of the Medical Profession: Aims and Methods." *Medical History* 10 (July 1966): 213–220.

Clarke, Edwin. *Modern Methods in the History of Medicine.* London: Athlone Press, University of London, 1971.

Coleman, William. *Biology in the Nineteenth Century: Problems of Form, Function, and Transformation.* New York: John Wiley, 1971.

Coulter, Harris. *Divided Legacy. A History of the Schism in Medical Thought.* 3 vols. Washington, D.C.: Wehawken, 1973–1977.

Corner, George W. *A History of the Rockefeller Institute, 1901–1953: Origins and Growth.* New York: Rockefeller University Press, 1964.

———. *Two Centuries of Medicine: A History of the School of Medicine, University of Pennsylvania.* Philadelphia: Lippincott, 1965.

Comroe, Julius H., Jr., and Dripps, Robert D. "Scientific Basis for the Support of Biomedical Science." *Science* 192 (April 9, 1976): 105–111.

Cordell, Eugene F. "The Importance of the Study of the History of Medicine." *Medical Library and Historical Journal* 2 (October 1904): 268–282.

Creighton, Charles. *A History of Epidemics in Britain.* 2 vols. Cambridge: The University Press, 1891–94. Reprinted with additional material. New York: Barnes and Noble, 1965.

Crosby, Alfred W., Jr. *The Columbian Exchange: Biological and Cultural Consequences of 1492.* Westport, Conn.: Greenwood Press, 1972.

Cushing, Harvey. *The Life of Sir William Osler.* 2 vols. New York: Oxford University Press, 1925.

Daedalus. Vol. 106, No. 2 (1977). "The Family."

Dain, Norman. *Concepts of Insanity in the United States, 1789–1865.* New Brunswick: Rutgers University Press, 1964.

Davies, John D. *Phrenology: Fad and Science; A 19th Century American Crusade.* New Haven: Yale University Press, 1955.

Debus, Allen G. *The Chemical Philosophy: Paracelsian Science and Medicine in the Sixteenth and Seventeenth Centuries.* 2 vols. New York: Science History Publications, 1977.

———, ed. *Medicine in Seventeenth-Century England.* Berkeley: University of California Press, 1974.

De Mause, Lloyd, ed. *The History of Childhood*. New York: Harper Torchbooks, 1975.

Deutsch, Albert. *The Mentally Ill in America: A History of their Care and Treatment from Colonial Times*. 2d ed. New York: Columbia University Press, 1949.

Deutsch, Ronald M. *The Nuts Among the Berries*. New York: Ballantine Books, 1961; 2d ed. *The New Nuts Among the Berries*. Palo Alto: Bull, 1977.

Dewhurst, Kenneth. *Dr. Thomas Sydenham (1624–1689): His Life and Original Writings*. Berkeley: University of California Press, 1966.

Diepgen, Paul. *Geschichte der Medizin: Die historische Entwicklung der Heilkunde und des ärztlichen Lebens*. 2 vols. Berlin: W. de Gruyter, 1949–1955.

Dodds, E. R. *The Greeks and the Irrational*. Berkeley: University of California Press, 1951.

Dols, Michael. *The Black Death in the Middle East*. Princeton: Princeton University Press, 1977.

Donnison, Jean. *Midwives and Medical Men: A History of Inter-Professional Rivalries and Women's Rights*. New York: Schocken Books, 1977.

Douglas, Mary. *Purity and Danger: An Analysis of Concepts of Pollution and Taboo*. London: Routledge and Kegan Paul, 1966.

Drabkin, I. E. "On Medical Education in Greece and Rome." *Bulletin of the History of Medicine* 15 (April 1944): 333–351.

Dubos, René. *Louis Pasteur: Free Lance of Science*. Boston: Little, Brown, 1950; 2d ed., 1976.

Duffy, John. *Epidemics in Colonial America*. Baton Rouge: Louisiana State University Press, 1953.

———. *Sword of Pestilence: The New Orleans Yellow Fever Epidemic of 1853*. Baton Rouge: Louisiana State Press, 1966.

———. *A History of Public Health in New York City*. 2 vols. New York: Russell Sage Foundation, 1968–1974.

———. *The Healers: The Rise of the Medical Establishment*. New York: McGraw-Hill, 1976; Urbana: University of Illinois Press, 1979.

Dumesnil, René. *Histoire illustrée de la médecine*. Paris: Libraire Plon, 1935.

Dupree, A. Hunter. *Science in the Federal Government: A History of Policies and Activities to 1940*. Cambridge, Mass.: Harvard University Press, 1957.

Edelstein, Ludwig. *The Hippocratic Oath: Text, Translation, and Interpretation. Bulletin of the History of Medicine*, supplement no. 1. Baltimore: Johns Hopkins University Press, 1943.

———. "The Professional Ethics of the Greek Physician." *Bulletin of the History of Medicine* 30 (September-October 1956): 391–419.

———. *Ancient Medicine*. Edited by Owsei Temkin and C. Lilian Temkin. Baltimore: Johns Hopkins University Press, 1967.

Eliade, Mircea. *Shamanism: Archaic Techniques of Ecstasy*. Princeton: Princeton University Press, 1964.

Ellenberger, Henri F. *The Discovery of the Unconscious: The History and Evolution of Dynamic Psychiatry*. New York: Basic Books, 1970.

Erikson, Erik. *Young Man Luther: A Study in Psychoanalysis and History*. New York: W. W. Norton, 1958.

Etheridge, Elizabeth W. *The Butterfly Caste: A Social History of Pellagra in the South*. Westport, Conn.: Greenwood Press, 1972.

Eyler, John M. "Mortality Statistics and Victorian Health Policy: Program and Criticism." *Bulletin of the History of Medicine* 50 (Fall 1976): 335–355.

————. *Victorian Social Medicine: The Ideas and Methods of William Farr*. Baltimore: Johns Hopkins University Press, 1979.

Faber, Knud. *Nosography in Modern Internal Medicine*. New York: Paul Hoeber, 1923.

Falk, Leslie A. "Medical Sociology: The Contributions of Dr. Henry E. Sigerist." *Journal of the History of Medicine* 13 (April 1958): 214–228.

Figlio, Karl. "The Historiography of Scientific Medicine: An Invitation to the Human Sciences." *Comparative Studies in Society and History* 19 (July 1977): 262–286.

Finer, Samuel E. *The Life and Times of Sir Edwin Chadwick*. London: Methuen, 1952.

Fischer, Alfons. *Geschichte des deutschen Gesundheitswesens*. 2 vols. Berlin: F. A. Herbig, 1933.

Flexner, Simon, and Flexner, James T. *William Henry Welch and the Heroic Age of American Medicine*. New York: Viking, 1941.

Flinn, Michael W., ed. *Report on the Sanitary Condition of the Labouring Population of Great Britain by Edwin Chadwick*. Edinburgh, Edinburgh University Press, 1965.

Fogel, Robert W., and Engerman, Stanley L. *Time on the Cross: The Economics of American Negro Slavery*. 2 vols. Boston: Little, Brown, 1974.

Forssmann, Werner. *Experiments on Myself: Memoirs of a Surgeon in Germany*. Tr. by Hilary Davis. New York: St. Martin's Press, 1974.

Forster, Robert, and Ranum, Orest, eds. *Biology of Man in History: Selections from the Annales Economies, Sociétés, Civilisations*. Baltimore: Johns Hopkins University Press, 1975.

Foster, Sir Michael. *Lectures on the History of Physiology during the Sixteenth, Seventeenth, and Eighteenth Centuries*. Cambridge: Cambridge University Press, 1906; reprint ed., New York: Dover, 1970.

Foster, W. D. *A Short History of Clinical Pathology*. Edinburgh: E. and S. Livingstone, 1961.

Foucault, Michel. *Madness and Civilization: A History of Insanity in the Age of Reason*. Tr. by R. Howard. New York: Random House, 1965.

Frank, Johann Peter. *A System of Complete Medical Police*. Edited by Erna

Lesky. Tr. by E. Vilim. Baltimore: Johns Hopkins University Press, 1976.

Freidson, Eliot. *Professional Dominance: The Social Structure of Medical Care.* New York: Atherton Press, 1970.

French, Richard D. *Antivivisection and Medical Science in Victorian Society.* Princeton: Princeton University Press, 1975.

Freund, Paul, ed. *Experimentation with Human Subjects.* New York: George Braziller, 1970.

Fulton, John F. *Harvey Cushing: A Biography.* Springfield, Ill.: Charles C. Thomas, 1946.

Galdston, Iago, ed. *On the Utility of Medicine.* New York: International Universities Press, 1957.

Garrison, Fielding H. *John Shaw Billings: A Memoir.* New York: G. P. Putnam's Sons, 1915.

————. *The Principles of Anatomic Illustration before Vesalius.* New York: Paul Hoeber, 1926.

————. *An Introduction to the History of Medicine.* 4th ed. Philadelphia: W. B. Saunders, 1929.

Geison, Gerald L. *Michael Foster and the Cambridge School of Physiology: The Scientific Enterprise in Late Victorian Society.* Princeton: Princeton University Press, 1978.

Gilbert, Bently B. *The Evolution of National Insurance in Great Britain: The Origins of the Welfare State.* London: Joseph, 1966.

Glacken, Clarence J. *Traces on the Rhodian Shore: Nature and Culture in Western Thought from Ancient Times to the End of the Eighteenth Century.* Berkeley: University of California Press, 1967.

Glass, D. V. "Notes on the Demography of London at the End of the Seventeenth Century." *Daedalus* 97 (Spring 1968): 581–592.

Gnudi, Martha T., and Webster, Jerome P. *The Life and Times of Gaspare Tagliacozzi, Surgeon of Bologna 1545–1599.* New York: Herbert Reichner, 1950.

Godlee, Sir Rickman J. *Lord Lister.* London: Macmillan, 1917.

Goodfield, G. June. *The Growth of Scientific Physiology.* London: Hutchinson, 1960.

Graham, Harvey [Flack, Isaac H.]. *Eternal Eve: The History of Gynecology and Obstetrics.* Garden City, N.Y.: Doubleday, 1951.

Graunt, John. *Natural and Political Observations Mentioned in a following Index, and made upon the Bills of Mortality.* London: Thomas Roycroft, 1662; reprint ed., Baltimore: Johns Hopkins University Press, 1939.

Greenwood, Major. *Some British Pioneers of Social Medicine.* London: Oxford University Press, 1948.

————. *Medical Statistics from Graunt to Farr.* Cambridge: Cambridge University Press, 1948.

Grob, Gerald N. *The State and the Mentally Ill: A History of Worcester State Hospital in Massachusetts, 1830–1920.* Chapel Hill: University of North Carolina Press, 1966.

——. *Mental Institutions in America: Social Policy to 1875.* New York: Free Press, 1973.

——. "The Social History of Medicine and Disease in America: Problems and Possibilities." *Journal of Social History* 10 (Summer 1977): 391–409.

——. *Edward Jarvis and the Medical World of Nineteenth-Century America.* Knoxville: University of Tennessee Press, 1978.

Gross, Samuel D. *Autobiography: With Sketches of His Contemporaries.* 2 vols. Philadelphia: George Barrie, 1887.

Guerra, Francisco. "The Dispute over Syphilis: Europe versus America." *Clio Medica* 13 (June 1978): 39–61.

Haeser, Heinrich. *Lehrbuch der Geschichte der Medizin und der epidemischen Krankheiten.* 3d ed. 3 vols. Jena: F. Manke, 1875–1882.

Hahn, André, and Dumaître, Paul. *Histoire de la médecine et du livre médical à la lumiére des collections de la Bibliotheque de la Faculté de Médecine de Paris.* Paris: Olivier Perrin, 1962.

Hall, A. Rupert. *The Scientific Revolution 1500–1800: The Formation of the Modern Scientific Attitude.* London: Longmans, Green, 1954.

Hall, Marie Boas. "History of Science and History of Medicine." In *Modern Methods in the History of Medicine,* pp. 157–72. Edited by Edwin Clarke. London: Athlone Press, 1971.

Hall, Thomas S. *Ideas of Life and Matter: Studies in the History of General Physiology 600 B.C. to 1900 A.D.* 2 vols. Chicago: University of Chicago Press, 1969.

Hand, Wayland D., ed. *American Folk Medicine: A Symposium.* Berkeley: University of California Press, 1976.

Hannaway, Owen. *The Chemists and the Word: The Didactic Origins of Chemistry.* Baltimore: Johns Hopkins University Press, 1975.

Harris, Richard. *A Sacred Trust.* Baltimore: Penguin Books, 1969.

Hartman, Mary, and Banner, Lois W., eds. *Clio's Consciousness Raised: New Perspectives on the History of Women.* New York: Harper Torchbooks, 1974.

Hecht, J. Jean. "History: Social History." In *International Encyclopedia of the Social Sciences,* vol. 6, pp. 455–62. New York: Free Press, 1968.

Hecker, Justus F. K. *The Epidemics of the Middle Ages.* London: The Sydenham Society, 1844.

Heckscher, William S. *Rembrandt's Anatomy of Dr. Nicolaas Tulp. An Iconological Study.* New York: New York University Press, 1958.

Heischkel, Edith. "Die Geschichte der Medizingeschichtschreibung." In *Einführung in die Medizinhistorik,* pp. 203–37. Edited by Walter Artelt. Stuttgart: Ferdinand Enke Verlag, 1949.

Herrlinger, Robert. *History of Medical Illustration from Antiquity to A.D. 1600.* Tr. by G. Fulton-Smith. London: Pitman Medical, 1970.

Higham, John; Krieger, Leonard; and Gilbert, Felix. *History.* Englewood Cliffs, N.J.: Prentice-Hall, 1965.

Himes, Norman E. *Medical History of Contraception.* Baltimore: Williams and Wilkins, 1936.

Hirsch, August. *Handbook of Geographical and Historical Pathology.* 3 vols. Tr. by Charles Creighton. London: The New Sydenham Society, 1883–1886.

Hodgkinson, Ruth G. *The Origins of the National Health Service: The Medical Services of the New Poor Law, 1834–1871.* Berkeley: University of California Press, 1967.

Hollander, Eugen. *Plastik und Medizin.* Stuttgart: F. Enke, 1912.

———. *Die Karikature und Satire in der Medizin.* 2d ed. Stuttgart: F. Enke, 1921.

———. *Wunder, Wundergeburt, und Wundergestalt: In Einblattdrucken des fünfzehnten bis achtzehnten Jahrhunderts.* Stuttgart: F. Enke, 1921.

———. *Die Medizin in der klassichen Mallerei.* Stuttgart: F. Enke, 1923.

Holloway, S. W. F. "Medical Education in England, 1830–1858: A Sociological Analysis." *History* 49 (October 1964): 299–324.

———. "The Apothecaries Act, 1815: A Reinterpretation." *Medical History* 10 (April 1966): 107–129; (July): 221–236.

Holmes, Frederic L. *Claude Bernard and Animal Chemistry: The Emergence of a Scientist.* Cambridge, Mass.: Harvard University Press, 1974.

Hudson, Robert P. "Abraham Flexner in Perspective: American Medical Education 1865–1910." *Bulletin of the History of Medicine* 46 (November-December 1972): 545–561.

Hughes, Arthur. *A History of Cytology.* New York: Abelard-Schuman, 1959.

Hughes, Sally Smith. *The Virus: A History of the Concept.* New York: Science History Publications, 1977.

Hurd-Mead, Kate Campbell. *A History of Women in Medicine from the Earliest Times to the Beginning of the Nineteenth Century.* Haddam, Conn.: Haddam Press, 1938.

Hutchinson, J. F. "Historical Method and the Social History of Medicine." *Medical History* 17 (October 1973): 423–428.

Imhof, Arthur E., and Larson, Oivind. "Social and Medical History: Methodological Problems in Interdisciplinary Quantitative Research." *Journal of Interdisciplinary History* 7 (Winter 1977): 493–498.

Jarcho, Saul, ed. *Human Palaeopathology.* New Haven: Yale University Press, 1966.

Jetter, Dieter. *Geschichte des Hospitals.* 3 vols. Wiesbaden: Franz Steiner Verlag, 1966–1972.

Jewson, N. D. "The Disappearance of the Sick Man from Medical Cosmology, 1770–1870." *Sociology* 10 (May 1976): 225–244.

Jones, Ernest. *The Life and Work of Sigmund Freud.* 3 vols. New York: Basic Books, 1953.

Kalisch, Phillip A., and Kalisch, Beatrice J. *The Advance of American Nursing.* Boston: Little, Brown, 1978.

Katz, Jay M., comp. *Experimentation with Human Beings.* New York: Russell Sage, 1972.

Kaufman, Martin. *Homeopathy in America. The Rise and Fall of a Medical Heresy.* Baltimore: Johns Hopkins University Press, 1971.

———. *American Medical Education: The Formative Years, 1765–1910.* Westport, Conn.: Greenwood Press, 1976.

Kearney, Hugh. *Science and Change 1500–1700.* New York: McGraw-Hill, 1971.

Keele, Kenneth D. *The Evolution of Clinical Methods in Medicine.* Springfield, Ill.: Charles C. Thomas, 1963.

Kennedy, David M. *Birth Control in America: The Career of Margaret Sanger.* New Haven: Yale University Press, 1970.

Kerley, Ellis R. "Recent Advances in Paleopathology." In *Modern Methods in the History of Medicine,* pp. 135–156. Edited by Edwin Clarke. London: Athlone Press, 1971.

Kerley, Ellis R., and Bass, William M. "Paleopathology: Meeting Ground for Many Disciplines." *Science* 157 (August 11, 1967): 638–644.

Kett, Joseph F. *The Formation of the American Medical Profession: The Role of Institutions, 1780–1860.* New Haven: Yale University Press, 1968.

Keynes, Sir Geoffrey. *The Life of William Harvey.* Oxford: Clarendon Press, 1966.

King, Lester S. *The Medical World of the Eighteenth Century.* Chicago: University of Chicago Press, 1958.

———. *The Growth of Medical Thought.* Chicago: University of Chicago Press, 1963.

———. *The Road to Medical Enlightenment 1650–1695.* New York: American Elsevier, 1970.

———. *The Philosophy of Medicine: The Early Eighteenth Century.* Cambridge, Mass.: Harvard University Press, 1978.

King, Lester S.; Risse, Guenter B.; Burns, Chester R.; and Hudson, Robert P. "Viewpoints in the Teaching of Medical History." *Clio Medica* 10 (June 1975): 129–160.

Klibansky, Raymond; Panofsky, Erwin; and Saxl, Fritz. *Saturn and Melancholy: Studies in the History of Natural Philosophy, Religion, and Art.* New York: Basic Books, 1964.

Kobler, John. *The Reluctant Surgeon: A Biography of John Hunter.* Garden City, N.Y.: Doubleday, 1960.

Konold, Donald E. *A History of American Medical Ethics, 1847–1912.* Madison: The State Historical Society of Wisconsin, 1962.

Kraepelin, Emil. *One Hundred Years of Psychiatry*. New York: Citadel Press, 1962.

Kremers, Edward and Urdang, George. *History of Pharmacy*. 4th ed. Revised by Glenn Sonnedecker. Philadelphia: Lippincott, 1976.

Kuhn, Thomas S. "Science: The History of Science." In *International Encyclopedia of the Social Sciences*, vol. 14, pp. 74–83. New York: Free Press, 1968.

Ladimer, Irving, and Newman, Roger W., eds. *Clinical Investigation in Medicine: Legal, Ethical, and Moral Aspects*. Boston: Law-Medicine Research Institute, Boston University, 1963.

Lambert, Royston. *Sir John Simon, 1816–1904, and English Social Administration*. London: Macgibbon and Kee, 1963.

Landes, David. "The Treatment of Population in History Textbooks." *Daedalus* 97 (Spring 1968): 363–384.

Langer, William L. "Europe's Initial Population Explosion." *American Historical Review* 69 (October 1963): 1–17.

Leake, Chauncey D., ed. *Percival's Medical Ethics*. Baltimore: Williams and Wilkins, 1927.

Lerner, Monroe, and Anderson, Odin W. *Health Progress in the United States 1900–1960*. Chicago: University of Chicago Press, 1963.

Lesky, Erna. *Österreichisches Gesundheitswesen im Zeitalter des aufgeklärten Absolutismus*. Vienna: Rudolf M. Rohrer, 1959.

———. *The Vienna Medical School of the 19th Century*. Tr. by L. Williams and I. S. Levij. Baltimore: Johns Hopkins University Press, 1976.

Leslie, Charles, ed. *Asian Medical Systems: A Comparative Study*. Berkeley: University of California Press, 1976.

Lewis, I. M. *Ecstatic Religion: An Anthropological Study of Spirit Possession and Shamanism*. Baltimore: Penguin Books, 1971.

Lindeboom, G. A. *Herman Boerhaave: The Man and His Work*. London: Methuen, 1968.

Lewis, Richard A. *Edwin Chadwick and the Public Health Movement 1832–1854*. London: Longmans, Green, 1952.

Lipset, Seymour M., and Hofstadter, Richard, eds. *Sociology and History: Methods*. New York: Basic Books, 1968.

Long, Esmond R. *A History of Pathology*. 2d ed. New York: Dover, 1965.

Lubove, Roy. *The Struggle for Social Security 1900–1935*. Cambridge, Mass.: Harvard University Press, 1968.

Macalpine, Ida, and Hunter, Richard. *George III and the Mad-Business*. New York: Random House, 1969.

McCollum, Elmer V. *A History of Nutrition*. Boston: Houghton Mifflin, 1957.

McDaniel, W. B. "The Place of the Amateur in the Writing of Medical History." *Bulletin of the History of Medicine* 7 (July 1939): 687–695.

Macfarlane, A. *Witchcraft in Tudor and Stuart England.* New York: Harper
     & Row, 1970.
McKeown, Thomas. "A Sociological Approach to the History of Medicine."
     Medical History 14 (October 1970): 342–351.
──── . *The Modern Rise of Population.* New York: Academic Press, 1976.
McKeown, Thomas, and Brown, R. G. "Medical Evidence Related to
     English Population Changes in the Eighteenth Century." *Population
     Studies* 9 (July 1955): 119–141.
McKeown, Thomas, and Record, R. G. "Reasons for the Decline of Mor-
     tality in England and Wales during the Nineteenth Century." *Popu-
     lation Studies* 16 (March 1962): 94–122.
McKeown, Thomas; Brown, R. G.; and Record, R. G. "An Interpretation of
     the Modern Rise of Population in Europe." *Population Studies* 26
     (November 1972): 345–382.
MacKinney, Loren C. *Early Medieval Medicine: With Special Reference
     to France and Chartres.* Baltimore: Johns Hopkins University Press,
     1937.
McNeill, William H. *Plagues and Peoples.* Garden City, N.Y.: Doubleday
     Anchor, 1976.
Majno, Guido. *The Healing Hand: Man and Wound in the Ancient World.*
     Cambridge, Mass.: Harvard University Press, 1975.
Major, Ralph H. *A History of Medicine.* 2 vols. Springfield: Charles C.
     Thomas, 1954.
Malgaigne, J. F. *Surgery and Ambroise Paré.* Translated and edited by W. B.
     Hamby. Norman: University of Oklahoma Press, 1965.
Marshall, Helen E. *Mary Adelaide Nutting: Pioneer of Modern Nursing.*
     Baltimore: Johns Hopkins University Press, 1972.
Martin, Calvin. *Keepers of the Game: Indian-Animal Relationships and the
     Fur Trade.* Berkeley: University of California Press, 1978.
May, Jacques M. *The Ecology of Human Disease.* New York: M.D. Publica-
     tions, 1958.
──── . *Studies in Disease Ecology.* New York: Hafner, 1961.
Mazlish, Bruce, ed. *Psychoanalysis and History.* Englewood Cliffs: Prentice-
     Hall, 1963; 2d ed., 1971.
Meade, Richard H. *An Introduction to the History of General Surgery.*
     Philadelphia: W. B. Saunders, 1968.
Meiss, Millard. *Painting in Florence and Siena after the Black Death: The
     Arts, Religion, and Society in the Mid-Fourteenth Century.* Princeton:
     Princeton University Press, 1951.
Merz, John Theodore. *A History of European Thought in the Nineteenth
     Century.* 4 vols. London: Blackwood 1904–1912; reprint ed., New York:
     Dover, 1965.
Mettler, Cecilia C. *History of Medicine: A Correlative Text Arranged*

According to Subjects. Edited by F. A. Mettler. Philadelphia: Blakiston, 1947.

Miller, Genevieve. "Backgrounds of Current Activities in the History of Science and Medicine." *Journal of the History of Medicine* 13 (April 1958): 160–178.

———. "The Teaching of Medical History in the United States and Canada." *Bulletin of the History of Medicine* 43 (1969): 259–267; 344–375; 444–472; 553–586.

———. "The Teaching of Medical History in the United States and Canada: Historical Resources in Medical School Libraries." *Bulletin of the History of Medicine* 44 (May-June 1970): 251–278.

———. "The Teaching of Medical History in American Medical Schools." In *The Education of American Physicians.* Edited by Ronald Numbers. Berkeley: University of California Press, 1980.

Miller, Howard S. *Dollars for Research: Science and Its Patrons in Nineteenth-Century America.* Seattle: University of Washington Press, 1970.

Mills, C. Wright. *The Sociological Imagination.* New York: Oxford University Press, 1959.

Mondor, Henri. *Doctors and Medicine in the Works of Daumier.* Tr. by C. de Chabanne. Boston: Boston Book and Art Shop, 1960.

Mora, George. "The Historiography of Psychiatry and Its Development: A Re-Evaluation." *Journal of the History of the Behavioral Sciences* 1 (January 1965): 43–52.

Mora, George, and Brand, Jeanne L., eds. *Psychiatry and Its History: Methodological Problems in Research.* Springfield, Ill.: Charles C. Thomas, 1970.

Morantz, Regina. "The Lady and Her Physician." In *Clio's Consciousness Raised: New Perspectives on the History of Women,* pp. 38–53. Edited by Mary Hartman and Lois W. Banner. New York: Harper Torchbooks, 1974.

Mountain, Joseph W. *Selected Papers of Joseph W. Mountain, M.D.* Joseph W. Mountain Memorial Committee, n.p., 1956.

Neuburger, Max. *History of Medicine.* 2 vols. Tr. by Ernest Playfair. London: H. Frowde, 1910–1925.

———. *Die Lehre von der Heilkraft der Natur im Wandel der Zeiten.* Stuttgart: F. Sula, 1926. Tr. by Linn S. Boyd as *The Doctrine of the Healing Power of Nature Throughout the Course of Time,* New York, 1933.

Nevins, Allan. *The Gateway to History.* Rev. ed., Garden City, N.Y.: Doubleday Anchor, 1962.

Newman, Charles. *The Evolution of Medical Education in the Nineteenth Century.* London: Oxford University Press, 1957.

Nordenskiöld, Eric. *The History of Biology*. New York: Alfred A. Knopf, 1928.

Norwood, William F. *Medical Education in the United States before the Civil War*. Philadelphia: University of Pennsylvania Press, 1944. Reprint ed., New York: Arno Press, 1971.

Numbers, Ronald L. *Almost Persuaded: American Physicians and Compulsory Health Insurance, 1912–1920*. Baltimore: Johns Hopkins University Press, 1978.

Nutting, M. A., and Dock, Lavinia. *A History of Nursing*. 4 vols. New York: G. P. Putnam's Sons, 1907–1912.

Olmsted, J. M. D. *Francois Magendie: Pioneer in Experimental Physiology and Scientific Medicine in XIX Century France*. New York: Henry Schuman, 1944.

Olmsted, James M. D., and Olmsted, E. Harris. *Claude Bernard and the Experimental Method in Medicine*. New York: Henry Schuman, 1952.

O'Malley, Charles D. *Andreas Vesalius of Brussels 1514–1564*. Berkeley: University of California Press, 1965.

————, ed. *The History of Medical Education*. Berkeley: University of California Press, 1970.

O'Malley, Charles D., and Saunders, J. B. deC. M. eds. *Leonardo da Vinci on the Human Body*. New York: Henry Schuman, 1952.

O'Neill, Ynez V. "The Fünfbilderserie Reconsidered." *Bulletin of the History of Medicine* 43 (May-June 1969): 236–245.

————. "The Fünfbilderserie: A Bridge to the Unknown." *Bulletin of the History of Medicine* 51 (Winter 1977): 538–549.

Pagel, Julius. *Einführung in die Geschichte der Medizin*. Berlin: S. Karger, 1898.

Pagel, Walter. "Julius Pagel and the Significance of Medical History for Medicine." *Bulletin of the History of Medicine* 25 (May-June 1951): 207–225.

————. *Paracelsus: An Introduction to Philosophical Medicine in the Era of the Renaissance*. Basel: S. Karger, 1958.

————. *William Harvey's Biological Ideas: Selected Aspects and Historical Background*. New York: S. Karger, 1967.

Parsons, Gail Pat. "Equal Treatment for All: American Medical Remedies for Male Sexual Problems: 1850–1900." *Journal of the History of Medicine* 32 (January 1977): 55–71.

Penfield, Wilder. *The Difficult Art of Giving: The Epic of Alan Gregg*. Boston: Little, Brown, 1967.

Peterson, M. Jeanne. *The Medical Profession in MidVictorian London*. Berkeley: University of California Press, 1978.

Pinchbeck, Ivy. *Women Workers and the Industrial Revolution, 1750–*

1850. London: Routledge & Sons, 1930; reprint ed., New York: A. M. Kelley, 1969.

Pinchbeck, Ivy, and Hewitt, Margaret. *Children in English Society*. 2 vols. London: Routledge and Kegan, Paul, 1969–1973.

Poynter, F. N. L. "Medicine and the Historian." *Bulletin of the History of Medicine* 30 (September-October 1956): 420–435.

———, ed. *The Evolution of Medical Education in Britain*. Baltimore: Williams and Wilkins, 1966.

———. *Medicine and Science in the 1860s*. London: Wellcome Institute, 1963.

Puschmann, Theodor. *A History of Medical Education from the Most Remote to the Most Recent Times*. Tr. by E. H. Hare. London: H. K. Lewis, 1896; reprint ed., New York: Hafner, 1966.

Puschmann, Theodor; Neuburger, Max; and Pagel, Julius. *Handbuch der Geschichte der Medizin*. 3 vols. Jena: Gustav Fischer, 1902–1905.

Rabb, Theodore K., and Rotberg, Robert I., eds. *The Family in History: Interdisciplinary Essays*. New York: Harper & Row, 1973.

Rather, L. J. *The Genesis of Cancer: A Study in the History of Ideas*. Baltimore: Johns Hopkins University Press, 1978.

Razzell, P. E. "An Interpretation of the Modern Rise of Population in Europe: A Critique." *Population Studies* 28 (March 1974): 5–18.

Redman, Eric. *The Dance of Legislation*. New York: Simon and Schuster, 1973.

Reed, James. *From Private Vice to Public Virtue: The Birth Control Movement and American Society since 1830*. New York: Basic Books, 1978.

Reingold, Nathan. "American Indifference to Basic Research: A Reappraisal." In *Nineteenth-Century American Science: A Reappraisal*, pp. 38–62. Edited by George Daniels. Evanston: Northwestern University Press, 1972.

Reiser, Stanley J. *Medicine and the Reign of Technology*. Cambridge: Cambridge University Press, 1978.

Reiser, Stanley J.; Dyck, Arthur J.; and Curran, William J., eds. *Ethics in Medicine: Historical Perspectives and Contemporary Concerns*. Cambridge, Mass.: MIT Press, 1977.

Richardson, Robert G. *Surgery: Old and New Frontiers*. New York: Charles Scribner's Sons, 1968.

Riese, Walther. *The Conception of Disease: Its History, Its Versions and Its Nature*. New York: Philosophical Library, 1953.

Roberts, R. S. "The Personnel and Practice of Medicine in Tudor and Stuart England. Part I. The Provinces." *Medical History* 6 (October 1962): 363–382; "Part II. London." 8 (July 1964): 217–234.

Rose, Willie Lee. *Rehearsal for Reconstruction: The Port Royal Experiment*. Indianapolis: Bobbs-Merrill, 1964.

Rosebury, Theodore. *Microbes and Morals: The Strange Story of Venereal Disease.* New York: Viking, 1971.

Rosen, George. "A Theory of Medical Historiography." *Bulletin of the History of Medicine* 8 (May 1940): 655–665.

———. "The Place of History in Medical Education." *Bulletin of the History of Medicine* 22 (September-October 1948): 594–627. Also in Rosen, *From Medical Police to Social Medicine,* pp. 3–36. New York: Science History Publications, 1974.

———. "Levels of Integration in Medical Historiography: A Review." *Journal of the History of Medicine* 4 (Autumn 1949): 460–467.

———. "The New History of Medicine: A Review." *Journal of the History of Medicine* 6 (Autumn 1951): 516–522.

———. "Toward a Historical Sociology of Medicine: The Endeavor of Henry E. Sigerist." *Bulletin of the History of Medicine* 32 (November-December 1958): 500–516.

———. *A History of Public Health.* New York: M.D. Publications, 1958.

———. "People, Disease, and Emotions: Some Newer Problems for Research in Medical History." *Bulletin of the History of Medicine* 41 (January-February 1967): 5–23.

———. *Madness in Society: Chapters in the Historical Sociology of Mental Illness.* Chicago: University of Chicago Press, 1968.

———. *Preventive Medicine in the United States 1900–1975: Trends and Interpretations.* New York: Science History Publications, 1975.

———. "Social Science and Health in the United States in the Twentieth Century." *Clio Medica* 11 (December 1976): 245–268.

Rosenberg, Charles E. *The Cholera Years: The United States in 1832, 1849, and 1866.* Chicago: University of Chicago Press, 1962.

———. *The Trial of the Assassin Guiteau: Psychiatry and Law in the Gilded Age.* Chicago: University of Chicago Press, 1968.

———. "The Medical Profession, Medical Practice, and the History of Medicine." In *Modern Methods in the History of Medicine,* pp. 22–35. Edited by Edwin Clarke. London: Athlone Press, 1971.

———. *No Other Gods: On Science and American Social Thought.* Baltimore: Johns Hopkins University Press, 1976.

———. "And Heal the Sick: The Hospital and the Patient in the 19th Century America." *Journal of Social History* 10 (Summer 1977): 428–447.

———. "The Therapeutic Revolution: Medicine, Meaning, and Social Change in Nineteenth-Century America." *Perspectives in Biology and Medicine* 20 (Summer 1977): 485–506.

———, ed. *The Family in History.* Philadelphia: University of Pennsylvania Press, 1975.

———. *Healing and History: Essays for George Rosen.* New York: Science History Publications, 1979.

Rosenkrantz, Barbara G. *Public Health and the State: Changing Views in*

Massachusetts, 1842–1936. Cambridge, Mass.: Harvard University Press, 1972.

Rothman, David J. *The Discovery of the Asylum: Social Order and Disorder in the New Republic.* Boston: Little, Brown, 1971.

Rothstein, William G. *American Physicians in the Nineteenth Century: From Sects to Science.* Baltimore: Johns Hopkins University Press, 1972.

Rothschuh, Karl E. *History of Physiology.* Original German ed., 1953; edited and tr. by G. B. Risse. Huntington, N.Y.: Robert E. Krieger, 1973.

Sand, René. *The Advance to Social Medicine.* London: Staples Press, 1952.

Sarton, George. *Introduction to the History of Science.* 3 vols. in 5 parts. Baltimore: Williams and Wilkins, 1927–1948.

———. "The History of Science Versus the History of Medicine." *Isis* 23 (September 1935): 313–320.

———. *A History of Science.* 2 vols. Cambridge, Mass.: Harvard University Press, 1952–1959.

Saunders, J. B. deC. M. *The Transitions from Ancient Egyptian to Greek Medicine.* Lawrence: University of Kansas Press, 1963.

Saunders, J. B. deC. M., and O'Malley, Charles D., eds. *The Illustrations from the Works of Andreas Vesalius of Brussells.* Cleveland: World, 1950.

Savitt, Todd L. *Medicine and Slavery: The Diseases and Health Care of Blacks in Antebellum Virginia.* Urbana: University of Illinois Press, 1978.

Schouten, J. *The Rod and Serpent of Asklepios: Symbol of Medicine.* New York: Elsevier, 1967.

Schullian, Dorothy M., and Schoen, Max, eds. *Music and Medicine.* New York: Henry Schuman, 1948.

Schultz, Adolph H. "Notes on Diseases and Healed Fractures of Wild Apes and Their Bearing on the Antiquity of Pathological Conditions in Man." *Bulletin of the History of Medicine* 7 (June 1939): 571–582.

Shafer, Henry B. *The American Medical Profession, 1783–1850.* New York: Columbia University Press, 1936.

Shannon, James A. "The Advancement of Medical Research: Twenty-Year View of the Role of the National Institutes of Health." *Journal of Medical Education* 42 (February 1967): 97–108.

Shattuck, Lemuel. *Report of a General Plan for the Promotion of Public and Personal Health.* Boston: Dutton and Wentworth, 1850.

Shryock, Richard H. *The Development of Modern Medicine.* 2d ed. New York: Alfred A. Knopf, 1947.

———. *American Medical Research Past and Present.* New York: The Commonwealth Fund, 1947.

———. "Women in American Medicine." *Journal of the American Medical*

Women's Association 5 (September 1950): 371–379. Also in Shryock, *Medicine in America: Historical Essays,* pp. 177–99. Baltimore: Johns Hopkins University Press, 1966.

——. *Medicine and Society in America 1660–1860.* New York: New York University Press, 1960.

——. *Medicine in America: Historical Essays.* Baltimore: Johns Hopkins University Press, 1966.

Sigerist, Henry E. *The Great Doctors: A Biographical History of Medicine.* Tr. by E. & C. Paul. New York: W. W. Norton, 1933.

——. "On Hippocrates." *Bulletin of the History of Medicine* 2 (May 1934): 190–214. Also in *Henry E. Sigerist on the History of Medicine,* pp. 97–119. Edited by F. Marti-Ibanez. New York: M.D. Publications, 1960.

——. "The History of Medicine And the History of Science." *Bulletin of the History of Medicine* 4 (January 1936): 1–13.

——. *Civilization and Disease.* Ithaca: Cornell University Press, 1943; reprint ed., University of Chicago Press, Phoenix, 1962.

——. "Medical History in the United States: Past-Present-Future." *Bulletin of the History of Medicine* 22 (January-February 1948): 47–64. Also in *Henry E. Sigerist on the History of Medicine,* pp. 233–250. Edited by F. Marti-Ibanez. New York: M.D. Publications, 1960.

——. *A History of Medicine.* 2 vols. New York: Oxford University Press, 1951–1961.

——. "William Harvey's Position in the History of European Thought." In *Henry E. Sigerist on the History of Medicine,* pp. 184–92. Edited by F. Marti-Ibanez. New York: M.D. Publications, 1960.

——. *Henry E. Sigerist on the Sociology of Medicine.* Edited by Milton I. Roemer. New York: M.D. Publications, 1960.

——. *Henry E. Sigerist: Autobiographical Writings.* Comp. by Nora Sigerist Beeson. Montreal: McGill University Press, 1966.

Simon, Sir John. *English Sanitary Institutions: Reviewed in Their Course of Development and in Some of their Political and Social Relations.* London: Cassell, 1890.

Sims, J. Marion. *The Story of My Life.* New York: Appleton, 1884; reprint ed., New York: DeCapo Press, 1968.

Singer, Charles, and Underwood, E. Ashworth. *A Short History of Medicine.* 2d ed. New York: Oxford University Press, 1962.

Stampp, Kenneth M. *The Peculiar Institution: Slavery in the Ante-Bellum South.* New York: Alfred A. Knopf, 1956.

Stern, Bernhard J. *Historical Sociology.* New York: Citadel Press, 1959.

Stevens, Rosemary. *Medical Practice in Modern England: The Impact of Specialization and State Medicine.* New Haven: Yale University Press, 1966.

——. *American Medicine and the Public Interest.* New Haven: Yale University Press, 1971.

Stevenson, Lloyd G. "Science down the Drain." *Bulletin of the History of Medicine* 29 (January-February, 1955a): 1–26.

———. "Biography versus History: With Special Reference to the History of Medicine." *College of Physicians of Philadelphia Transactions* 23 (August 1955b): 83–93.

———. "William Harvey and the Facts of the Case." *Journal of the History of Medicine* 31 (January 1976): 90–97.

Stocking, George W. "On the Limits of 'Presentism' and 'Historicism' in the Historiography of the Behavioral Sciences." *Journal of the History of the Behavioral Sciences* 1 (July 1965): 211–217.

Stone, Lawrence. *The Family, Sex, and Marriage in England 1500–1800.* New York: Harper & Row, 1977.

Strauss, Maurice B., ed. *Familiar Medical Quotations.* Boston: Little, Brown, 1968.

Strickland, Stephen P. *Politics, Science, and Dread Disease: A Short History of United States Medical Research Policy.* Cambridge, Mass.: Harvard University Press, 1972.

Swain, Donald C. "The Rise of a Research Empire: NIH, 1930–1950." *Science* 138 (14 December 1962): 1233–1237.

Talbot, Charles H. *Medicine in Medieval England.* New York: American Elsevier, 1967.

Temkin, Owsei. "An Historical Analysis of the Concept of Infection." In *Studies in Intellectual History*, pp. 123–147. Baltimore: Johns Hopkins University History of Ideas Club, 1953. Also in *The Double Face of Janus*, pp. 456–471. Baltimore: Johns Hopkins University Press, 1977.

———. "In Memory of Henry E. Sigerist." *Bulletin of the History of Medicine* 31 (July-August 1957): 295–299.

———. "Henry E. Sigerist and Aspects of Medical Historiography." *Bulletin of the History of Medicine* 32 (November-December 1958): 485–499.

———. "Who Should Teach the History of Medicine." In *Education in the History of Medicine*, pp. 53–60. Edited by John B. Blake. New York: Hafner, 1968.

———. "The Historiography of Ideas in Medicine." In *Modern Methods in the History of Medicine*, pp. 1–21. Edited by Edwin Clarke. London: Athlone Press, 1971.

———. *Galenism: Rise and Decline of a Medical Philosophy.* Ithaca: Cornell University Press, 1973.

———. "Health and Disease." *Dictionary of the History of Ideas*, vol. 2, pp. 395–407. New York: Charles Scribner's Sons, 1973. Also in *The Double Face of Janus*, pp. 419–40 (1977; see next entry).

———. *The Double Face of Janus and Other Essays in the History of Medicine.* Baltimore: Johns Hopkins University Press, 1977.

Terris, Milton, ed. *Goldberger on Pellagra.* Baton Rouge: Louisiana State University Press, 1964.

Thomas, Keith. *Religion and the Decline of Magic.* New York: Charles Scribner's Sons, 1971.

Thompson, Kenneth. "Irrigation as a Menace to Health in California, A Nineteenth Century View." *The Geographical Review* 59 (1969): 195–214.

———. "Early California and the Causes of Insanity." *Southern California Quarterly* 58 (Spring 1976): 45–62.

———. "Wilderness and Health in the Nineteenth Century." *Journal of Historical Geography* 2 (1976): 145–161.

———. "Trees as a Theme in Medical Geography and Public Health." *Bulletin of the New York Academy of Medicine* 54 (May 1978): 517–531.

Thorndike, Lynn. *A History of Magic and Experimental Science.* 8 vols. New York: Macmillan, 1923–1958.

Top, Franklin H., ed. *The History of American Epidemiology.* St. Louis: C. V. Mosby, 1952.

Turner, Thomas B. "The Medical Schools Twenty Years Afterwards: Impact of the Extramural Research Support Programs of the National Institutes of Health." *Journal of Medical Education* 42 (February 1967): 109–118.

———. *Heritage of Excellence: The Johns Hopkins Medical Institutions, 1914–1947.* Baltimore: Johns Hopkins University Press, 1974.

Underwood, E. Ashworth, ed. *Science, Medicine, and History: Essays on the Evolution of Scientific Thought and Medical Practice Written in Honor of Charles Singer.* 2 vols. London: Oxford University Press, 1953.

Vallery-Radot, René. *The Life of Pasteur.* Tr. by R. L. Devonshire. 2 vols. New York: Doubleday, 1901; reprint ed., New York: Dover, 1960.

Veith, Ilza. *Hysteria: The History of a Disease.* Chicago: University of Chicago Press, 1965.

Verbrugge, Martha H. "Women and Medicine in Nineteenth-Century America." *Signs: Journal of Women in Culture and Society* 1 (Summer 1976): 957–972.

Virtanen, Reino. *Claude Bernard and His Place in the History of Ideas.* Lincoln: University of Nebraska Press, 1960.

Vogel, Morris J. "Patrons, Practitioners, and Patients: The Voluntary Hospital in MidVictorian Boston." In *Victorian America,* pp. 121–138. Edited by D. W. Howe. Philadelphia: University of Pennsylvania Press, 1976.

Waddington, Ivan. "General Practitioners and Consultants in Early Nineteenth-Century England: The Sociology of Intra-Professional Conflict." In *Health Care and Popular Medicine in Nineteenth-Century England,* pp. 164–88. Edited by John Woodward and David Richards. New York: Holmes and Meier, 1977.

Walsh, Mary Roth. *Doctors Wanted: No Women Need Apply; Sexual*

Barriers in the Medical Profession, 1835–1975. New Haven: Yale University Press, 1977.

Wangensteen, Owen H., and Wangensteen, Sarah D. *The Rise of Surgery: From Empiric Craft to Scientific Discipline*. Minneapolis: University of Minnesota Press, 1978.

Webster, Charles, ed. *The Intellectual Revolution of the Seventeenth Century*. London: Routledge and Kegan Paul, 1974.

————. *The Great Instauration: Science, Medicine, and Reform 1626–1660*. London: Duckworth, 1975.

Wells, Calvin. *Bones, Bodies, and Disease: Evidence of Disease and Abnormality in Early Man*. New York: Praeger, 1964.

Whitteridge, Gweneth. *William Harvey and the Circulation of the Blood*. New York: American Elsevier, 1971.

Wiebe, Robert H. *The Search for Order 1877–1920*. New York: Hill and Wang, 1967.

Williams, H. U. "The Origin and Antiquity of Syphilis: The Evidence from Diseased Bone: A Review with Some New Material from America." *Archives of Pathology* 13 (May 1932): 779–814; (June 1932): 931–83.

Wilson, F. P. *The Plague in Shakespeare's London*. Oxford University Press, 1927.

Wilson, Leonard G. "Editorial: History Versus the Historians." *Journal of the History of Medicine* 33 (April 1978): 127–128.

Winslow, Charles E. A. *The Life of Hermann M. Biggs, M.D., D.Sc., LL.D. Physician and Statesman of Public Health*. Philadelphia: Lea and Febiger, 1929.

————. *The Conquest of Epidemic Disease: A Chapter in the History of Ideas*. Princeton: Princeton University Press, 1943.

Withington, Edward T. *Medical History from the Earliest Times*. London: Scientific Press, 1894.

Witte, Edwin E. *The Development of the Social Security Act*. Madison: University of Wisconsin Press, 1963.

Wood, Ann Douglas. "The Fashionable Diseases: Women's Complaints and their Treatment in Nineteenth-Century America." *Journal of Interdisciplinary History* 4 (Summer 1973): 25–52. Also in *Clio's Consciousness Raised: New Perspectives on the History of Women*, pp. 1–22. Edited by Mary Hartman and Lois W. Banner. New York: Harper Torchbooks, 1974.

Wood, Peter H. *Black Majority: Negroes in Colonial South Carolina From 1670 through the Stono Rebellion*. New York: Alfred A. Knopf, 1974.

Woodham-Smith, Cecil. *Florence Nightingale, 1820–1910*. New York: McGraw-Hill, 1951.

Woodward, John, and Richards, David, eds. *Health Care and Popular Medicine in Nineteenth-Century England: Essays in the Social History of Medicine*. New York: Holmes and Meier, 1977.

Wrigley, E. A. "Mortality in Pre-Industrial England: The Example of Colyton, Devon, over Three Centuries." *Daedalus* 97 (Spring 1968): 546–580.

Wrigley, Edward A.; Eversley, D. E. C.; and Laslett, Peter. *An Introduction to English Historical Demography from the Sixteenth to the Nineteenth Century.* New York: Basic Books, 1966.

Young, James Harvey. *The Toadstool Millionaires: A Social History of Patent Medicine in America before Federal Regulation.* Princeton: Princeton University Press, 1961.

———. *The Medical Messiahs: A Social History of Health Quackery in Twentieth-Century America.* Princeton: Princeton University Press, 1967.

Ziegler, Philip. *The Black Death.* London: Collins, 1969.

Zigrosser, Carl, comp. *Ars Medica: A Collection of Medical Prints Presented to the Philadelphia Museum of Art by Smith Kline and French Laboratories.* Philadelphia: Philadelphia Museum of Art, 1955.

Zilboorg, Gregory. *A History of Medical Psychology.* New York: W. W. Norton, 1941.

Zimmer, Henry R. *Hindu Medicine.* Baltimore: Johns Hopkins University Press, 1948.

Zimmerman, Leo M., and Veith, Ilza. *Great Ideas in the History of Surgery.* 2d ed. New York: Dover, 1967.

Part II
Philosophy

Chapter 4
Philosophy of Science
Historical, Social, and Value Aspects

Alex C. Michalos, UNIVERSITY OF GUELPH, ONTARIO

INTRODUCTION: PHILOSOPHY AS CRITICAL THINKING

In this chapter a number of basic issues in contemporary philosophy of science are reviewed, along with a variety of approaches to the subject. Our aim is to discover their possible relevance for ethical and evaluative issues in science. Some conceptual corners have been cut, dogmatic assertions substituted for carefully prepared arguments, and well-known researchers neglected. Although some of these deficiencies are corrected by the bibliography, others are not.

Briefly, the chapter follows this outline. After some introductory remarks here, on philosophy as critical thinking, Section I contains three overviews of our subject. A brief history of the philosophy of science is presented in section I.A. Following that, section I.B. gives a country-by-country review of work in the philosophy of science in the twentieth century. In section I.C., several different logically possible ways to organize our subject are considered.

The basic question before us in this essay is sharpened in Section II, on values in science. Section III, dealing with logic, provides us with the overall orientation to the basic tool for critical analysis. Section IV then takes up several major issues in the Anglo-American tradition of analytical philosophy of science, looking at them in terms of value issues. Section V concludes the chapter with a discussion of science and social responsibility.

Plato (d. 347 B.C.) characterized a philosopher as one who is disposed to look for the one in the many and the many in the one. In the hands of Aristotle (384–322 B.C.), this trait became a disposition to look for similarities where there are apparently only differences and differences where there are apparently only similarities. Broadly speaking, Plato and Aristotle seem to have been interested in characterizing philosophy as an activity, a process rather than a product, and one would not be too far off the mark in identifying this activity with critical thinking. At any rate, that rough characterization will do for our purposes.

Virtually everything about which a philosopher is likely to think critically may be subsumed under one or more of three general rubrics—namely, metaphysics, epistemology, and axiology. Each of these terms designates a field of inquiry that has evolved as a result of the pursuit of answers to three distinct questions: What exists? How do you know? What good is it?

Like philosophy generally, the philosophy of science is a mixture of the metaphysics, epistemology, and axiology of science. Although some philosophers occasionally announce their intention to provide one or another of these views of science and to eschew the others, it is doubtful that there are any purebred philosophies of science. The move from an epistemological to an ontological question may be virtually imperceptible. For example, the epistemological question "How does a scientist know?" may be easily transformed into the ontological question "What is it a scientist knows?" Furthermore, all questions and answers may be more or less colored with evaluations.

If one thinks of the philosophy of science as critical thinking about metaphysical, epistemological, and axiological issues in science, then there are still several ways to proceed. In the next section, a wide variety of issues and points of view will be reviewed. In section I.A. our aim will be to trace the historical antecedents of the philosophy of science. Since until the end of the seventeenth century much of science was indistinguishable from philosophy, any history of our subject is in large measure a history of philosophy. Even in the twentieth century, it is difficult to decide whether this or that particular work is philosophy of science or something else. Nevertheless, in section I.B. an attempt is made to review recent work in the subject in several different countries. Finally, section I.C. offers several ways of organizing or thinking about these various approaches.

OVERVIEWS OF THE SUBJECT

A. Historical Overview of the Philosophy of Science

This brief review of the history of the philosophy of science generally follows that of Laurens Laudan in "Theories of Scientific Method from Plato to Mach" (1968). Another useful study is John Losee's *A Historical Introduction to the Philosophy of Science* (1972). William A. Wallace's *Causality and Scientific Explanation* (vol. 1, 1972; vol. 2, 1974) has excellent analyses of some of the views of virtually all of the philosophers that will be mentioned below. Joseph J. Kockelmans' *Philosophy of Science: The Historical Back-*

ground (1968) has excellent selections from the work of most of the writers after Kant who are mentioned here and several more.

Plato's *Republic, Sophist, Timaeus,* and *Theatetus* contained rather more than the germs of fundamental problems in the philosophy of science. His speculations on the relative reliability and intelligibility of sense perception, formal certainty, being and becoming, the ultimate matter of the world, space and time, the importance of mathematics, are all mainstream issues. Karl Popper (1963), for example, has found Plato highly relevant; see also F. M. Cornford (1957).

Aristotle's contribution to the subject is recognized by everyone. He considered all of the issues that Plato had treated. In his *Topics* he provided careful analyses of principles of classification and definition and explained the direct relation between precision of expression and the possibility of refutation. The *Posterior Analytics* contains discussions of scientific explanation, four kinds of causation, axiomatics, and induction. His view of the diversity of sciences and their different fundamental principles (e.g., physics and astronomy) represented a nonreductionist, antiunity-of-science point of view. Since his philosophy of science was based on his general theory of knowledge, he held the latter to be normative for science. On this aspect of Aristotle's philosophy of science, see Ernan McMullin (1969). His biological discussions in *De Anima (On the Soul)* still merit attention; for example, see Marjorie Grene (1972). (Other relevant features of Aristotle's views are mentioned in chapters 1, 5, and 6 of this volume; see also F. Solmsen, 1960; M. G. Evans, 1958; and R. P. McKeon, 1947.)

Claudius Galen (ca. 129–199) should be remembered for his insistence on observation and experiment, especially in physiology. His contemporary, Claudius Ptolemy (d. 140), is usually regarded as the first great instrumentalist; his geocentric astronomy dominated the field for over a thousand years. His instrumentalism, the idea that astronomic science only had to be concerned with saving appearances, was challenged by men like Averroes (ca. 1126–1198). Averroes' mysticism and neoplatonism contrasted sharply with Aristotle's quasiempirical approach; where Aristotle saw diversity, Averroes saw a unified Being hierarchically structured into causally connected levels. The unity of science would have been conceivable to Averroes. More on Galen's views may be found in W. L. H. Duckworth's translation of Galen's *On Anatomical Procedures* (1962). For more on Averroes and Ptolemy, see Marshall Clagett (1955); and Thomas S. Kuhn (1957).

Moses Maimonides (ca. 1135–1204) held the instrumentalist view of astronomy, as did Thomas Aquinas (ca. 1224–1274). For the former, theoretical reason was supposed to be superior to practical reason, and both were consistent with revelation. Aquinas' reflections on scientific matters ranged as widely as Aristotle's, and they often came to similar conclusions. Aquinas' faculty psychology was more developed than Aristotle's, and his epistemology was more complicated because of the theological demands put upon it. Aquinas had to accommodate faith, revelation, and knowledge of such supersensibles as angels. Reviews of recent Thomistic philosophies of science may be found in William Wallace (1967).

Roger Bacon's (ca. 1214–1292) *Opus Majus* contains methodological discussions stressing the resistance of authority, the need for experimentation, and the possibility of certainty with the help of divine illumination. He had views on mathematics, the tides, heat, refraction and the principles of convex lenses. His contemporary Albert the Great (d. 1280) wrote commentaries on Aristotle's works, recommended close observation, and generally appealed to experience as a criterion of truth. Albert is reported to have shown through actual tests that ostriches would not eat iron, but would not turn down moderately sized stones and bones. He was a geocentrist and an alchemist. More on Bacon and Albert may be found in A. C. Crombie (1959); see also S. C. Easton (1952).

William of Ockham (1285–1349) has probably received more attention for his work in formal logic than in scientific method. He was an empiricist and a nominalist, and he regarded logic as the study of signs. Science was supposed to be concerned with universally quantified propositions, as proposed by Aristotle. Scientific arguments were to have demonstrable conclusions following from premises that were at least hypothetically necessary, self-evident, and indemonstrable. Causal relations could be apprehended only through experience and observation. The induction of such relations required a principle of the uniformity of nature similar to those that were popularized in the nineteenth century. Good analyses may be found in P. Boehner (1958); E. A. Moody (1965); and S. Shapiro (1957).

William of Ockham may be regarded as an early Renaissance figure. *The* Renaissance figure is of course Leonardo da Vinci (1452–1519). His notebooks contain discussions of hydraulics, mechanics, anatomy, geology, architecture, and military engineering. His contemporary Nicholas Copernicus (1473–1543) gave us the heliocentric system, which he regarded as a more or less realistic description of the heavens. In this departure from the Ptolemaic instrumentalist view, he was joined by Johannes Kepler (1571–1630) and Galileo Galilei (1564–1642). Kepler was an interesting mixture of empiricist astronomer and mystic astrologer. Like Galileo, he insisted upon observation and experiment, although for all authors in and around the period, it is not clear how much actual experimentation was undertaken. Occasionally, incoherent experiments requiring nonexistent apparatus were recommended. Moreover, the experiments of some authors required communication with God. Galileo regarded the heliocentric system as not only true, but demonstrably true, which is what led to his confrontation with Cardinal Robert Bellarmine. Kuhn's *The Copernican Revolution* (1957) is a classic study of the astronomical views of all these authors except Leonardo.

More on the latter may be found in Crombie (1959); more on the nature of experiments in this period may be found in Kuhn (1977).

Perhaps the best known methodologist of this period is Francis Bacon (1561–1626). His *Novum Organum* (1620) seems to have served as an inspiration to most inductivist textbook writers on scientific method through the nineteenth century. He proposed, but could not complete, a Great Instauration involving the classification of all sciences, development of a theory of inductive logic, collection of data and experimentation, a set of examples of sound methodology, a list of all verified generalizations, and a fully developed new Science of Nature. He claimed that sense experience was fallible and traced the main sources of its fallibility in his famous Four Idols. His three rules of presence, absence, and differing degrees were later popularized by John Stuart Mill as the inductive methods of agreement, disagreement, and concomitant variation. These rules underlie some of the most sophisticated statistical techniques available today, including multivariate regression and correlation. (See F. H. Anderson, 1948; P. Rossi, 1968; and C. J. Ducasse, 1960.)

William Harvey's classic, *An Anatomical Disquisition on the Motion of the Heart and Blood in Animals,* was published in 1628. It provided not only a sound defense of his theory of the circulation of the blood, but a veritable mine of examples of experimental and theoretical research at its best. Galen's theory of the ebbing and flowing of the blood had stood for over a thousand years. Harvey had to drive it out with the most meticulous account of the superior explanatory power, simplicity, and external and internal consistency of his new theory. More on Harvey may be found in Crombie (1959); A. C. Michalos (1970); and John Passmore (1958). (Also see chapter 3 in this volume.)

The great rationalist philosopher of this period, René Descartes (1596–1650), is remembered for several methodological and scientific investigations. His *Rules for the Direction of the Mind* (ca. 1628) and *Discourse on the Method of Rightly Conducting the Reason and Seeking for Truth in the Sciences* (1637) have been regarded as models of rationalism, although Descartes himself was hardly oblivious to the benefits of experience. Aspects of the hypothetico-deductive method have been found in his work. This method was also advanced by Descartes' more youthful contemporary, Blaise Pascal (1623–1662). As Descartes laid the foundations of analytic geometry, Pascal laid the foundations of the theory of probability. For Descartes' philosophy of science, see W. Doney (1967); L. J. Beck (1952); and G. Buchdahl (1963). See I. Todhunter (1949) for more on Pascal.

The rationalism and logico-mathematical brilliance of Gottfried Wilhelm Leibniz (1646–1716) cannot be ignored, although, as in Descartes' case, one suspects there was more empiricism in the man's work than textbook writers care to recall. For a purer case of rationalism one should

probably turn to his contemporary, Benedict Spinoza (1623–1677). Reviews
of Leibniz' work may be found in Nicholas Rescher (1967) and George
Gale (1970).

Six years after the death of Francis Bacon, the next great English
empiricist, John Locke (1632–1704), was born. Locke was by no means the
methodologist or scientist that Bacon or Leibniz was. However, his view of
the human mind as a blank tablet awaiting the imprints of experience set
the stage for Leibniz's exhaustive and in many ways more subtle discussion
of innate ideas. One need not swallow the problematic metaphysics or
psychology of either philosopher to appreciate the fundamental problem
they were addressing. More on Locke's psychology and methodology may
be found in John Yolton (1956 and 1970); Laudan (1967); and R. M.
Yost (1951).

One cannot avoid mentioning Isaac Newton (1642–1727) in this period,
although, as Laudan (1968) suggests, it is difficult to know what to make of
his methodological *obiter dicta*. He preached rigorous inductivism, but
apparently practiced some hypothetico-deductivism. One thing is certain:
he could not have used Bacon's or anyone else's inductive methods to induce
the Law of Gravitation or the Laws of Motion. That is equally true of his
empiricist contemporaries—e.g., Robert Hooke (1635–1703) and Robert
Boyle (1627–1691)—and their laws. More on these people may be found in
R. E. Butts and J. W. David (1970); E. J. Dijksterhuis (1961); and R. M.
Blake (1960). (See also the discussion of the British physician and epidemiologist, Thomas Sydenham, in chapter 6 of this volume.)

David Hume (1711–1776) gave us, in addition to his famous problem
of justifying inductive inferences (to be discussed at length below), associa-
tionist psychology and a clear statement of the regularity view of causation.
Some of his empiricism was shared by his idealist contemporary, George
Berkeley (1685–1753). The latter held that to be was to be perceived; he
rejected all nonexperiential physical and mathematical claims—e.g., the idea
of infinitely divisible lines. In instrumentalist fashion, he held that science
was concerned with correlations rather than causal relations. Work on
Hume is more than a minor industry; see, for example, V. C. Chappell
(1966); A. Flew (1961); J. Yolton (1963); and D. C. Stove (1973). On
Berkeley, see Martin and Armstrong (1968); Popper (1953); and G. J. Whitrow, (1953).

Immanuel Kant's (1724–1804) critical rationalism made the contribution of the human mind to the realm of the phenomenal world unavoidable
—indeed, constitutive of that world. Beyond the phenomenal world there
was posited a noumenal world which was in some sense responsible for the
former. His *Critique of Pure Reason* (1781) represented an attempt to give
the Newtonian view of the universe in the phenomenal world an *a priori*
justification. For analyses of Kant's work, see L. W. Beck (1965); G. Bird
(1962); S. Körner (1960); P. Strawson (1966); and G. Buchdahl (1965).

One of the most influential methodological treatises to come out of the nineteenth century was John F. W. Herschel's *Preliminary Discourse on the Study of Natural Philosophy* (1830). Herschel was a Baconian inductivist who sharpened his predecessor's rules and also addressed problems of causality and hypothesis formation. He endorsed the hypothetico-deductive method of presenting theories, as one might expect from the leading Newtonian astronomer of his time. He distinguished the context of discovery from the context of justification, separating questions of the origin of hypotheses from questions of their acceptability. Laws of nature were regarded as generalizations with specifiable boundary conditions, and theories were regarded as systems of connected laws. (See Ducasse, 1960; and W. F. Cannon, 1961.)

The other great inductivist manifesto of the nineteenth century, John Stuart Mill's *A System of Logic, Ratiocinative and Inductive* (1843) is heavily indebted to Herschel. Mill's inductivism extended all the way to mathematical statements. His explanation of the use of the canons of induction for cases of multiple causation was more sophisticated than anything that had come before. He accepted the hypothetico-deductive method, but urged the use of inductive procedures to obtain plausible hypotheses to begin with. Unlike Bacon, he believed that some empirical generalizations could be completely verified. More on Mill may be found in R. P. Anschutz (1953); Ducasse (1960); and A. Ryan (1970).

The major assault on naive inductivism in the nineteenth century came from William Whewell, especially in his *Philosophy of the Inductive Sciences* (1840). With Kant and Hume, he saw that the universality and necessity of laws and theories could not be apprehended by experience. One discovered laws and theories by collecting facts together in more comprehensive notions, not in a straightforward inductive way. Through a process he called the "consilience of inductions," more abstract and inclusive generalizations could be built from, and be indirectly supported by, less abstract generalizations. He believed that the best science was that which explained phenomena in different areas—and that he felt therefore had a vera causa at its heart. Although he was guided to this position by his rationalist metaphysics, he argued convincingly that only a philosophy like his could show the virtues of the then conquering wave theory of light over Newton's corpuscular theory. He developed a theory of theory change that is remarkably similar to Thomas Kuhn's widely discussed twentieth-century theory of paradigm shifts. For more, see R. E. Butts (1965 and 1968); Ducasse (1951); A. W. Heathcote (1953); E. W. Strong (1965); and H. T. Walsh (1962). (See also chapter 1 in this volume.)

In *The Principles of Science* (1874), William Jevons develops a point of view similar to that of Whewell. Augustus De Morgan (1806–1871) should be remembered for his *Essay on Probability* (1838), along with John Venn for his *The Logic of Chance* (1866).

Auguste Comte (1798–1857) is generally regarded as the inventor of positivism, although much of the hard-headed empiricism that is characteristic of his philosophy had been around for several years. As we will see below, philosophy of science in the first half of the twentieth century has been dominated by positivism. (See chapter 1 in this volume; also Alvin W. Gouldner, 1970; and Richard von Mises, 1951.)

Ernst Mach (1838–1916) came on the scene about forty years after Comte preaching roughly the same sort of empiricist, inductivist, antimetaphysical positivism. Mach was a clear instrumentalist with respect to laws and theories. He wanted to reconstruct Newtonian mechanics on a phenomenological basis; indeed, on a phenomenologically *given* basis. In effect there would be a fundamental division between empirical and logicotheoretical statements in the system. More on Mach may be found in R. S. Cohen and R. J. Seeger (1970); P. Frank (1961); and von Mises (1951).

The most influential thinker of the nineteenth century was Karl Marx (1818–1883). Marx held that true theories are reflections of things that have existence independent of observers. All things are finally related and constantly changing. So complete knowledge of anything involves everything else. The scientific enterprise consists of approximations to the truth, but the approximations are bound to be biased by class interests. Although Marx advocated materialism, it was not the sort that implied the reducibility of mental to physical phenomena. It was more a matter of rejecting otherworldly goods in favor of the goods and services of this world. Practical scientific knowledge, he thought, was a form of human wealth because such knowledge was directly related to the forces of production. Indeed, to have knowledge at all, one must have something that can alter the world in a fairly straightforward sense. Thus, Marx's view of knowledge has strong pragmatic overtones. A good review of Marx's views may be found in John McMurtry (1978). (See also chapters 1 and 5 in this volume.)

This is a good place to end our brief overview. There are obviously many well-known names that have been omitted—e.g., Charles Darwin, Ludwig Boltzmann, Justus Liebig, Charles Peirce, and so on. That is inevitable in the space available. Some less familiar names with significant national rather than international historical importance will be introduced in the country-by-country overview that follows.

B.   Twentieth-Century Work outside the Anglo-American School: A Country-by-Country Survey (Austria, Belgium, Denmark, Finland, France, Germany, Italy, Japan, Norway, Poland, Rumania, Sweden and Switzerland)

Since its first volume in 1970, *Zeitschrift für allgemeine Wissenschaftstheorie* has included brief essays summarizing recent work in the philosophy of science. This section is based on those essays, supplemented by sundry other sources cited in the text. Because most of the analyses in the rest of this chapter are based on the work of writers in Canada, the United Kingdom, and the United States, reviews of those countries are not included in this section. The interested reader may want to consult Robert Butts, "Philosophy of Science in Canada" (1974); Theodore Kisiel and Galen Johnson, "New Philosophies of Science in the U.S.A.: A Selective Survey" (1974); and Colin Howson and John Worrall, "The Contemporary State of Philosophy in Britain" (1974). To simplify the discussion, all titles are given in English unless the non-English titles would be easily decipherable by English-speaking readers.

Austria. Victor Kraft was already writing on the foundations of mathematics and science before the initiation of the Vienna Circle (1929), but he continues to work in the discipline. He has published, among other works, *Mathematik, Logik und Erfahrung* (1947), *Der Wiener Kreis* (1952), and *Erkenntnislehre* (1960). His colleague at the University of Vienna, Béla Juhos, is a former member of the Vienna Circle and the most productive Austrian philosopher of science. His research is devoted to applied logic, probability and induction, and basic concepts of physics. For examples, see Juhos, "Methodologie der Naturwissenschaften" (1968) and his bibliography at the end of that article. Other philosophers of science at Vienna include Curt Christian (formal sciences, model theory, etc.) and Heinz Zemanek, at the Institute of Technology (computer theory and the foundations of cybernetics).

An International Research Center for Foundations of the Sciences is located at the University of Salzburg. Paul Weingartner's research there has been in the foundations of logic and mathematics, modal logic, and the relations between metaphysics and science. Several international colloquia have been held at the Center since 1963.

At the University of Graz, Rudolf Haller specializes in linguistics and metaphysics. Rudolf Freundlich works on the logical and semantic structure of natural languages. Ernst Tropitsch is a sociologist and philosopher of culture who has written on the concept of law in social science and legal theory. Johann Mokre, also a sociologist, has done research on the theory of law.

At the University of Innsbruck, Gerhard Frey's publications cover philosophical logic, philosophy of language, information theory, and foundations of mathematics (Frey, 1958 and 1965). His colleagues, Rudolf Wohlgenannt and Reinhard Kamitz, work in formal logic; Bernulf Kantscheider, on the applicability of geometrical structures to the reality of experience; and Otto Muck and Vladimir Richter have applied methods of the theory of science to metaphysics and theology. (See Zecha, 1979.)

Belgium. Philosophy of science in Belgium began after World War II with Ferdinand Renoirte's *Eléments de critique des sciences et de cosmologie*

(1947). Renoirte's position was positivistic, with laws and theories being regarded as statements of measurable relationships. Metaphysics and science were isolated from one another just as physics and astronomy had been isolated in antiquity. Moreover, the role of the philosopher of science was not to be that of a normative critic, but of a faithful scribe; his aim was to describe and follow scientific procedures as carefully as possible.

S. Dockx believes that Renoirte overemphasizes mathematics and measurable relationships, and underemphasizes the need for conceptual knowledge in science. Biology, for example, involves a different sort of knowledge than physics, not an inferior sort.

Jean Ladrière succeeded Renoirte at the Catholic University of Louvain and brought his own brand of positivism. Like Renoirte, he holds that scientific explanation requires deduction, and laws and theories are statements of algebraic relationships. Purely formal sciences are supposed to have a sort of internal intelligibility that Plato might have appreciated. By stressing the formal and operational aspect of science, Ladrière is able to accommodate metaphysics and theology. (See Ladrière, *Les Limitations internes des formalismes,* 1957.)

G. Hirsch is a mathematician who works on philosophical problems of mathematics in the positivist tradition. He believes that formalization increases understanding and that derivations from theories are explored not primarily to test theories but to discover their full meaning and scope. L. Apostle also came to the philosophy of science through positivism, having worked with Rudolf Carnap at the University of Chicago. His work involves the foundations of logic, information theory, and cybernetics. J. Ruytinx (1962) has provided a logical and historical study of problems related to the positivist program of a unified science. Finally, Ch. Perelman and L. Olbrecht-Tyteca's *La Nouvelle rhétorique: Traité de l'argumentation* (1958) is a classic study of the theory of argumentation.

More on Belgium may be found in Paul Gochet (1975).

Denmark.   Niels Bohr (1885–1962) and Jørgen Jørgensen (1894–1969) dominated the Danish scene in the first half of the twentieth century. According to Bohr's famous Copenhagen interpretation of quantum theory, there is inescapable system-instrument interaction in all experiments with microobjects, so the autonomous existence of the latter is at least questionable. Bohr used the term "complementarity" to designate this relationship (see Aage Petersen, 1968). Jørgensen's *A Treatise of Formal Logic: Its Evolution and Main Branches with Its Relation to Mathematics and Philosophy* (1931), is a classic. Earlier, he was committed to the idea of a unified science based on the grammar and principles of logic, mathematics, and physics; later in his career he regarded psychology as the fundamental unifying discipline. Jørgensen was a realist, accepting the autonomous existence of an external world as a verified hypothesis. He also accepted principles of induction, continuity, legality, real causes, and successive approximations to truth.

Peter Zinkernagel has investigated the foundations of language and attempted to show that the language of classical physics was inescapable if one wanted to correctly describe physical reality. Johs Witt-Hansen has examined the relation of this problem to some of the views of Marx and Lenin. (See Johs Witt-Hansen, "Philosophy of Science [Wissenschaftstheorie] in Denmark," 1970.)

Finland.   The two most influential philosophers of science in Finland in the first half of the twentieth century were Eino Kaila (1890–1958) and G. H. von Wright. The former was a participant in the Vienna Circle, while the philosophical roots of the latter were grounded in Cambridge, where he worked with C. D. Broad and J. M. Keynes. Von Wright's output in the fields of probability, induction, practical reason, and action theory is prodigious; some examples: *The Logical Problem of Induction* (1941); *A Treatise on Induction and Probability* (1951); *The Logic of Preference* (1963); *Norm and Action* (1963); *The Varieties of Goodness* (1963); and *Causality and Determinism* (1974).

Jaakko Hintikka spends part of his academic life at Stanford University but continues to be a commanding presence in his home country. His formal work on induction and probability, although in the Cambridge tradition, is much more sophisticated and fruitful than any other work in that tradition. He has developed a two-dimensional continuum of inductive methods, of which Carnap's single continuum may be seen as a special case. The important methodological difference between the aims of Carnap and Hintikka is this: while Carnap sought an optimum value *a priori* for his index of caution, Hintikka approaches the selection of his index-values pragmatically, on the basis of given problems. Aspects of Hintikka's system have been developed by Risto Hilpinen (1968); by Raimo Tuomela (1966); and in several papers in Hintikka and Patrick Suppes (1970).

Tuomela and Juhani Pietarinen have mixed some of Hintikka's system with some information-theoretic formalism to obtain subtle measures of the empirical adequacy of behavioral theories. Tuomela (1973) represents further developments and possibilities. He has also made a substantial contribution to action theory (Tuomela, 1977).

See Hintikka, "Philosophy of Science [Wissenschaftstheorie] in Finland" (1970).

France.   The positivist-empiricist tradition in French philosophy of science has been strong. Although Claude Bernard (1813–1878) rejected philosophic systems, he belonged to this positivist tradition (see Bernard, 1865). He insisted that progress in science required theories and hypotheses, and that data collecting in the absence of guides was worthless. He seemed to hold that mental and physical phenomena were finally reducible to physicochemical processes.

Jules Henri Poincaré (1854–1912) belonged to the same tradition. He believed that there is an orderly universe existing independently of human perception, but that science could not penetrate to that reality. Scientific hypotheses, laws, and theories are instruments that allow us to connect and relate phenomena. Discovery generally proceeds along the lines of inductive generalization, and the aim is to find simple, unifying theories. On the foundations of mathematics, Poincaré was an intuitionist, rejecting attempts like Bertrand Russell's to reduce mathematics to logic, and accepting the indefinability of the integers. The axioms of geometry were supposed to be conventions or definitions (Poincaré, 1902).

Henri Bergson (1859–1941) tended to philosophize on or with the aid of science, rather than to explore and develop its methodology. Through intuition, he held, one could apprehend reality and discover metaphysical truth, while the scientific enterprise proceeded through the intellect and could obtain knowledge only of appearances. He accepted evolution, but rejected the mutation-with-natural-selection mechanism advocated by Darwin. In its place he posited a vital impetus that permeates and purposively drives the universe, apparently working out God's plan. (See Bergson, 1907, 1919, and 1932; also chapter 5, I.B., in this volume.) The more mystical evolutionary views of Pierre Teilhard de Chardin (1955) are in the Bergsonian tradition.

Pierre Duhem's *La Théorie physique: Son objet, sa structure* (1906) is a classic instrumentalist philosophy of science. Only metaphysical views are explanatory; those of science deal merely with appearances. One can penetrate reality with pure reason, he held, but not with the methods of science. Scientific theories provide more or less economic (simple) systems of laws that allow us to anticipate the results of experiments. Theories must be posited because they cannot be induced. Moreover, there cannot be any crucial tests of theories because of the multitude of connections among theories and auxiliary hypotheses. (As we will see later, a great deal of contemporary discussion has been devoted to this Duhemian thesis.)

Gaston Bachelard (1884–1962) was a prolific writer who advocated a "dialectical" rationalism that accommodated reason, experience, and imagination. He rejected the idea of experience as given and insisted upon the creative contribution of human beings. In general he saw the scientific enterprise as a cumulative one in which new theories saved what was worthwhile in older theories and extended our understanding (G. Bachelard, 1932, 1934, 1938, 1949).

Contemporary philosophers of science in France tend to have an appreciation for the history of the subject. For example, Robert Blanché (1966 and 1967) deals with the theory of argumentation by examining classical and modern views; and Jules Vuillemin (1955) traces the scientific sources of Kantian criticism. (See also Vuillemin, 1962, 1968, and 1971.)

Suzanne Bachelard approaches the history of mathematical physics from a phenomenological perspective derived from Maurice Merleau-Ponty (see S. Bachelard, 1958; and M. Merleau-Ponty, 1942 and 1945). O. Costa de Beauregard (1957, 1963a, and 1963b) has written broadly on issues in contemporary physics. Also, Jacques Merleau-Ponty has made a historical and epistemological study of theories of cosmology (1965, and, with B. Morando, 1971).

Strongly influenced by G. Bachelard, Georges Canguilhem (1955, 1965, 1966, 1968) has shown the inseparability of the history and philosophy of science. Much of his work is in the history and philosophy of medicine. Michel Foucault's famous *Naissance de la clinique* (1963) and *Histoire de la folie* (1961) are in the same tradition. (See chapters 1 and 6 in this volume.)

Contributions to the philosophy of biology have been made by François Meyer (1954); Raymond Ruyer (1954); François Jacob (1970) and Andre Lwoff (1969). Gilles-G. Granger (1955, 1960) has examined the role of models in social science, especially in economics.

Reviews may be found in Denis Zaslawsky, "La Philosophie des sciences (Wissenschaftstheorie) en France (1950–1971)" (1971); and in G. Granger and J. Vuillemin, "Tendances de la philosophie des sciences en France depuis 1950" (1968).

Germany. To talk about contemporary philosophy in Germany, one must sooner or later talk about phenomenology. Moreover, if one can believe the central figure of that philosophy, Edmund Husserl, there might not have been a phenomenological movement without Franz Brentano (1838–1917). Brentano practiced what he called "descriptive psychology" or, at first, "descriptive phenomenology." Mental phenomena could be distinguished from all others in virtue of their inherent intentional reference to objects beyond themselves. Such reference might be of three distinct sorts—namely, mere attention, judgment (affirmation or denial), and affection (attraction or repulsion, love or hate). Brentano held a nonpropositional theory of judgment. According to this theory, when one judges, for example, that a dog is wet, the object of one's judgment is not a proposition or even an idea of some sort, but the particular wet dog. The immediate consequence of this theory was that it allowed Brentano to reject an important class of abstract entities (propositions)—i.e., it allowed him to be a nominalist. However, he accepted the idea of immortal souls, God, and creation *ex nihilo*. (See Wolfgang Stegmüller, *Main Currents in Contemporary German, British, and American Philosophy*, 1969.)

Edmund Husserl (1859–1938) distinguished the world of scientific knowledge from the world known to everyone, the *Lebenswelt*. The latter constituted an infallible foundation on which to build the total body of human knowledge. One gained access to this foundation through an intuitive process called "transcendental-phenomenological reduction." The knowledge thus obtained would not be historically relative, but would be eternal and absolutely certain. It would be, in the first place, knowledge of a transcendental ego or pure consciousness. In his later years, he came to regard the world as a phenomenon available not to a single ego but to an intersubjective community of individuals. A unified science could finally be anchored in this communal "data base." More on Husserl may be found in Theodore J. Kisiel (1970a, 1970b); Joseph J. Kockelmans (1970); Herbert Spiegelberg (1960); and Stegmüller (1969).

Max Scheler (1874–1928) applied a type of eidetic reduction, but came to somewhat different conclusions from Husserl. Scheler classified knowledge into three types according to its function. What he regarded as scientific knowledge was knowledge obtained in the interest of controlling nature. Knowledge of essences and of the "categories of Being" was a second sort, and ultimately the sort on which metaphysics and genuine science could be built. The motive for obtaining this sort of knowledge was love of understanding. One who pursued it was very close in spirit to the etymological roots of philosophy—i.e., was a lover of wisdom. By some kind of synthesis of the first two sorts of knowledge, one was able to obtain a third, namely, knowledge of "Being itself" and salvation. By introducing this third sort of knowledge, Scheler paved the way for Martin Heidegger. Scheler was sympathetic to the scientific enterprise but insisted that human beings had a spiritual aspect that could not be apprehended by scientific methods. (See Spiegelberg, 1960; Manfred S. Frings, 1965; and Stegmüller, 1969.)

Martin Heidegger (1889–1976) is given an extended analysis in this volume (chapter 5, II. A. 4). However, a few comments are in order here. Heidegger was above all a metaphysician interested in analyzing the nature of Being, and he regarded an analysis of the nature of human Being as a necessary condition of his primary aim. His application of the methodology of Husserl was set in what William James might have regarded as the mood of a "twice-born" Christian before the second birth. His writings are full of pronouncements of dread, anxiety, guilt, the need for salvation, and so on. Unlike Husserl, he believed that one's historicity is inescapable. On this score he followed Wilhelm Dilthey (1833–1911). One is a child of one's times, as the saying goes, and one's actions can be understood only relative to those times. Heidegger did not distinguish pure and applied science, and he objected to science generally on the grounds that it requires specialization, rigorous classification, and a fixation on tidy, manageable research problems. For more, see Stegmüller (1969); M. Grene, *Heidegger* (1957); and Kockelmans (1965).

Several distinguished German mathematicians and physicists have made fundamental contributions to the study of the foundations of science. For example, Max Planck (1858–1947) believed in the existence of a world independent of human thoughts and regarded science as our best hope of obtaining knowledge of that world. For his discovery of the "quantum of action" he received the Nobel prize in 1918. Hermann Weyl (1885–1955) was a mathematician and physicist who addressed philosophical problems in *Philosophie der Mathematik und Naturwissenschaft* (1927). He thought that quantum theory made the idea of objective measurement, without human interference in whatever is measured, impossible; that is, he followed the Copenhagen view. Another Nobel prize physicist (1932) is Werner Heisenberg, whose Gifford lectures were published as *Physics and Philosophy* (1959). David Hilbert (1862–1943) and Gottlob Frege (1848–1925) are both famous for their mathematical brilliance. For more on Frege, see G. E. M. Anscombe and Peter Geach (1961), and Jeremy Walker (1965). On Hilbert, see *Die Naturwissenschaften* I (1922; the entire issue is devoted to him).

Jürgen Habermas' *Erkenntnis und Interesse* (1968) is an interesting development of many strands of philosophy. By reflecting on the methodology of the sciences, he believes that one is able to see that human reason has an interest in being emancipated from its particular environment. Particular interests are in some way constitutive of knowledge. The interest of technological control is supposed to be constitutive of "empirical-analytic" science, and the interest of improved human interaction is supposed to be constitutive of "historical-hermeneutic" sciences. Reason itself is supposed to follow "an emancipatory cognitive interest" that, roughly speaking, seems to mean that people find pleasure in personal freedom and general independence. Moreover, these goals may be obtained through self-reflection. (For more on Habermas and his Frankfurt school colleagues, see chapter 5, II. B. 8 in this volume.)

Wolfgang Stegmüller's *Main Currents* (1969) has been used here several times as a source. He has also written on the philosophy of science, mainly from a positivistic-analytic point of view. His papers collected and published in two volumes in 1977 will give English readers greater access to his work.

There are several prolific German writers in the area of formal decision theory. For example, Günter Menges (1974) has written on various kinds of inductive acceptance rules, probability, statistics, and semantic information; H. J. Skala has recently expounded his *Nonarchimedian Utility Theory* (1975); and Helmut Jungermann and Gerard De Zeeuw have edited an excellent collection of papers, *Decision Making and Change in Human Affairs* (1977).

For more on Germany, see Kockelmans and Kisiel (1970); Stegmüller (1969); and Hans Kohn, *The Mind of Germany* (1960). (See also chapter 5, I. B., in this volume for German philosophers' outstanding work on technology.)

Italy.   It is curious that Giambattista Vico (1668–1744) seems to have reached conclusions that became popular only after they were independently discovered by other people in later periods and in other lands. In his *Scienza Nuova* (1725) he claimed, for example, that the truths of mathematics might be known with demonstrable certainty because mathematical statements are essentially conventions. Propositions of physical sciences involved nonconventional features and therefore could not be known to be true with the same certainty. Like Aristotle, he believed that different areas of inquiry required different methods of investigation. He also believed that historical investigation could yield results that had greater certainty than results yielded by any physical sciences, because the former dealt with human experience while the latter did not. In a sense, then, he anticipated Dilthey and the *Verstehen* theorists of the early twentieth century. When he insisted that historians must try to recapture the modes of thought of the historical figures they studied, he was anticipating Dilthey and, later, Kuhn. According to the latter, a historian must recapture "out-of-date ways of reading out-of-date texts" (Kuhn, 1977, p. xiii). Although it is surely sensible to try to understand an author in his own terms, one must avoid slipping from that to the unwarranted view that all or even most authors are right (i.e., saying things that are true) in their own terms. The former is a constructive methodological principle, but the latter is a dubious empirical claim at best or a destructive methodological principle of epistemological relativism at worst.

Giuseppe Peano (1858–1932) worked in a field that could hardly be expected to make one popular, but his impact was more immediate than that of Vico. Peano made original contributions in the foundations of arithmetic, projective geometry, logic, and set theory. His famous five postulates served as the axiomatic basis for arithmetic. Some of his ideas and formal notation are still with us through their influence on Bertrand Russell and Alfred North Whitehead's *Principia Mathematica* (1910–1913).

For much of the first half of the twentieth century, philosophy in Italy was dominated by two idealists, Giovanni Gentile (1875–1944) and Benedetto Croce (1866–1952). Both of them were extremely prolific and not especially sympathetic to the scientific enterprise. However, in 1947 Ludovico Geymonat founded the Centro di Studi Metodologici at Turin, which represented the official introduction of nonidealist (mainly neopositivist) philosophy of science to Italy. Five years later, the Società Italiana di Logica e Filosofia della Scienze was founded in Rome. Geymonat was a student of Peano and had an early interest in the foundations of mathematics. Geymonat sees the rational procedures of science as a good model for the rest of society. In all cases, we can only approximate complete knowledge of the real world existing independently of us (Geymonat, 1945, 1953, 1960, 1970–1972).

Giulio Preti (1952, 1957) has attempted to develop a scientific philosophy along the lines recommended by, for example, Hans Reichenbach. Paolo Filiasi Carcano's interests in science are similarly broad. His aim has been to try to identify the attitudes or motivating forces behind science, and to locate these in a more or less complete account of all human activities (Carcano, 1941, 1957). The approach is analogous to that of Jürgen Habermas.

Research in logic and metamathematics has been undertaken by A. Pasquinelli (1964); E. Casari (1964); and Evandro Agazzi (1961, 1969). Agazzi's views are conventionalistic, and he believes that axiomatization helps establish the meaning of theoretic terms. Scientific theories are interpreted in realist terms as approximately true descriptions of the real world. Bruno De Finetti's well-known contributions to the foundations of probability theory are reviewed briefly in De Finetti "Probability: The Subjectivistic Approach" (1968).

Vittorio Somenzi has applied an operationalist and physicalist philosophy to problems of cybernetics and human action. Specifically, he hopes to be able to define consciousness operationally, using only concepts and methods of the physical sciences (Somenzi, 1965, 1969).

For surveys, see Evandro Agazzi, "Recent Developments of the Philosophy of Science in Italy" (1972); Piero Caldirola, "Physics and Philosophy in Italy" (1968); and Ettore Casari, "La Logique en Italie" (1968).

Japan.   Japanese philosophy of science is dominated by logicians and analytic philosophers. Some philosophers, like Nobushige Sawada (1964, 1969), seem to be more interested in developing scientific philosophy than in philosophizing about science.

S. Ohmori advocates a phenomenological approach to the philosophy of physics, arguing for the primacy of perceptual or common-sense descriptions over sophisticated scientific theory-laden descriptions. Shigeo Nagai and Hiroshi Kurosaki (1967) explore such basic issues as the semantic notion of truth, causality, laws, theories, models, and determinism. Hidekichi Nakamura (1969) mixes analytic philosophy and Marxian dialectic. Similarly, Mitsuo Taketani (1946–1952) has adapted Marx's theory of dialectical development to the development of science.

Hiroshi Nagai's profilic writings have dealt mainly with the foundations of mathematics and the unity of science (Nagai, 1955, 1962, 1963). Other philosophers of mathematics include Shoji Maehara and Setsuya Seki. The former tends to hold a Platonist view of mathematical objects, while the latter is a formalist. Chikio Hayashi is a statistician who works on the unification of statistical methods.

Philosophy of biology and medicine is pursued along Bergsonian lines by Hisayuki Omodaka (1955 and 1964). Ryuichi Yasugi is well known for his *History and Methodology of the Theory of Evolution* (1965).

There are several philosophers of physics and physicists working on philosophic problems. For example, Hideki Yukawa is a physicist who has tried to relate his ideas about the fundamental material of the universe to those of the Chinese philosopher Chuang Tsu. Takahiko Yamanouchi (1965, 1970), another physicist, is an optimist about applying scientific methods to all activities.

For surveys, see Hiroshi Nagai, "Recent Trends in Japanese Research on the Philosophy of Science" (1971); T. Yamanouchi, "Physics and Philosophy in Japan" (1968) and S. Maehara, "Logic in Japan" (1968).

Norway. Philosophy of science in Norway began with the work of Arne Naess. He was given the chair in philosophy at the University of Oslo in 1939 and was the only professor of philosophy in the country for twenty years. Naess distrusted formal logic and in his early years hoped to be able to develop a purely "behavioristic science of science." The task of a philosopher of science was to describe and explain accurately the behavior of scientists, not to evaluate it. Through the years his radical behaviorism has mellowed, and the intentions of scientists have also become the focus of his attention. In his most recent work (1972), he advocates a plurality of competing research programs and a humanistic approach to science. That is, science is no longer viewed in neutral positivist terms, but as a human artifact with social significance that should be guided by broader human purposes.

For other work by Norwegians, see H. Skjervheim (1959); Johan Galtung (1967); and Yngvar Løchen (1970); as well as Tore Nordenstam and Hans Skjervheim (1973).

Poland. In the first half of the twentieth century, Poland contributed more than its share of first-rate logicians, metamathematicians, and scientific methodologists. Many of them, as has often been the case for Europeans, have emigrated to other countries. Of those who remained in Poland, one might mention first A. Mostowski, who demonstrated the undecidability of arithmetic and the incompleteness of set theory. He also proved that the axiom of choice is independent of the other axioms of set theory. J. Łos developed a theory of logical matrices and simplified many proofs of theorems in model theory. He was one of the first to work with chronological logic. Jan Łukasiewicz did pioneering work in modal logic, intuitional logic, and many-valued logic. J. Słupecki constructed a three-valued system of sentential logic. J. Kalinowski applied three-valued logic to normative sentences. T. Kubinski (1968) developed erotetic logic—that is, the logic of questions. W. Marciszewski tried to develop a logic of persuasion.

Adam Schaff (1951, 1960, 1964) believed that the analytic philosophers' approach to philosophic problems reduced the latter to mere linguistic problems, and that the approach was connected to Western ideology.

Tadeusz Kotarbiński's *Praxiology: An Introduction to the Sciences of Efficient Action* (1965; see also 1968) is well known (see chapter 5, II. A. 2). The study of efficient action is supposed to be independent of morality, and

Kotarbiński tries to articulate rules of action that increase efficiency. Aspects of Kotarbiński's views are challenged in Michalos (1972).

The important positivist writings of Kasimierz Ajdukiewicz have been edited by Jerzy Giedymin (1977). Ajdukiewicz had a theory of meaning that entailed the existence of nonintertranslatable languages. Thus, he envisioned a plurality of world perspectives and rejected the idea of a single unified science. He developed the foundations of a logic of questions and a classification scheme for axiomatic systems. His approach to the acceptance of scientific theories was pragmatic, rather than merely syntactic or semantic. (See also M. Kokoszyńska, 1968.)

Under the influence of Marxism, several writers have tried to discover laws of the development of scientific theories. Some "philosophers of humanities" have argued that there is no essential difference between the methods of the natural sciences and the humanities. A textbook by Stefan Nowak (1977) has recently been translated and published.

Recent reviews of Polish work may be found in Stanislaw Kaminski, "The Development of Logic and the Philosophy of Science in Poland after the Second World War" (1977); Vladimir Zeman, "The Philosophy of Science in Eastern Europe: A Concise Survey" (1970); and Slupecki, "Logic in Poland" (1968).

Romania.   The emphasis in Romania is on logic and metamathematics. Anton Dumitriu (1966, 1968) has published two foundational studies; his view is that mathematics is not just a set of symbolic formulae but is somehow grounded in empirical sciences. To understand mathematics, therefore, one must examine its history and its connections with other sciences. He has also investigated logic as a theory, a science, and a system (1969, 1973); he concludes that logic cannot be regarded as a science or a system, but that it is a theory of knowledge apprehended by intuition. A. Joja (1904–1972) presented a theory of dialectical logic (1960); his view of logic was also historicist. G. C. Moisil (1965) has contributed to many-valued logics and the use of mathematical logic for automatic mechanisms.

Petre Botezatu's logical investigations are referred to as studies in "natural logic" and are similar to what others have called the theory of argumentation or informal logic. The aim is to provide an accurate description of logical operations as they occur in natural languages and actual thinking. Deduction is regarded as a process of analysis and synthesis. Formally the conclusions of valid deductive arguments are given in the premises, but materially there is supposed to be the possibility of novelty. Botezatu presents five "antinomies of axiomatization," including, for example, the "antinomy of simplification," which says roughly that if the axiomatic basis of a system is simplified, then the derivation of theorems will become more complicated. Aristotle's old principle of the inverse relation between the extension and intention of terms is another one of Botezatu's antinomies. (See Botezatu, 1969.)

Several works in the philosophy of physics have also been published by Romanians: Ion Tudosescu (1971); T. Toró (1973); Gheorghe Bârsan (1973). On the behavioral sciences, Vasile Pavelcu's *The Drama of Psychology* (1964) is an attempt to persuade psychologists not to abandon the particularities of their subject in the interests of trying to imitate the physical sciences' generalizations. Achim Mihu tries to get sociologists to abandon any pretentions of eliminating "subjective elements" from knowledge concerning society. For a survey, see Dima, "The Philosophy of Science in Romania" (1975).

Sweden. The most active group of philosophers of science in Sweden is affiliated with the Institute for the Theory of Science at the University of Göteborg. The members of the group come from a wide variety of backgrounds, including government and business, but their common interest is in "research science"—i.e., the science whose object of investigation is scientific research. Their approach is interdisciplinary. Historical, sociological, economic, philosophic, and purely formal techniques are used to systematically describe and evaluate research processes and products. The director of the institute is Håken Törnebohm. His papers and those of others who have contributed to the regular colloquia at the institute have been published in a series beginning in 1968.

In his internationally known *Continental Schools of Metascience* (1968), Gerard Radnitzky gives an overview of recent work in the theory of science. Radnitzky's own position is essentially that of Habermas, and he calls it "hermeneutic-dialectics."

Switzerland. Since the founding of the International Society for the Cultivation of Logic and Philosophy of Science in 1946, research in these subjects has been carried on at a high level in Switzerland. The official journal of the Society is *Dialectica*. Karl Dürr (1888–1970), one of its founders, did important work on logic, especially the history of logic (Dürr, 1949). Paul Bernays, another founder of the society, is well-known for his studies in the foundations of logic and mathematics. With David Hilbert he wrote *Grundlagen der Mathematik* (vol. 1, 1934, vol. 2, 1939). He has written many technical papers on axiomatic set theory and model theory.

Ferdinand Gonseth is a mathematician and philosopher with a wide range of methodological concerns. Broadly speaking, he regards his position as similar to the critical realism of Karl Popper, who was a cofounder of the society. However, Gonseth would characterize his own efforts as directed toward developing an "ethic of research" or, perhaps, a "phenomenology of the searching mind." His "open philosophy" is based on dialectics and a radical empiricism. He rejects both conventionalism and the idea of absolute truth (see Gonseth, 1945–1956 and 1964).

For more on philosophy of science in Switzerland, see Henri Lauener, "Wissenschaftstheorie in der Schweiz" (1971).

C. Frameworks for Philosophy of Science

In this section our aim is to examine alternative approaches to the philosophy of science from a logical point of view. Given the historical and sociological welter of views discussed, we want to be able to think about this or that particular question or approach as the realization of a logical possibility. Alternatively, one might say that our aim is to examine classificatory schemes that might be used to organize or systematize one's thoughts about the philosophy of science.

Short introductions to the philosophy of science include Peter Caws (1965) and Stephen Toulmin (1953). More substantial introductions may be found in Abraham Kaplan (1964) and Marx Wartofsky (1968).

Some works by now have become classics; for instance Karl Popper, *The Logic of Scientific Discovery* (1959); R. B. Braithwaite, *Scientific Explanation* (1953); Ernest Nagel, *The Structure of Science* (1961); and Thomas Kuhn, *The Structure of Scientific Revolutions* (1962).

Finally, two good introductory anthologies are available: Alex Michalos, *Philosophical Problems of Science and Technology* (1974); and Baruch Brody, *Readings in the Philosophy of Science* (1970). Reviews of recent work include Raymond Klibansky, *Contemporary Philosophy* (1968); and Peter Asquith and Henry Kyburg, eds., *Current Research in Philosophy of Science* (1979). A large number of the survey articles in the bibliography are from these two volumes.

Seven sets of approaches will be discussed here.

1. Activities vs. Results. In the first place, one may approach our subject assuming that it is science as a human activity or set of activities that should be the focus of one's attention. The philosopher should be interested in discovering the sorts of things that scientists typically do. So one would raise questions like, How are scientific discoveries made?, and, How are theories established or discredited? An activities view is given in Peter Achinstein (1968).

Alternatively, one might focus on science as a set of results that are generated by human activities. Then the center stage would be occupied by questions like, According to contemporary science, what is the nature of space and time? The objects of philosophers' attention in this case are not space and time themselves, which are the objects of scientists' attention, but scientists' views about space and time. For a results view, see Rudolf Carnap, *Philosophical Foundations of Physics* (1966). What Roman Catholic philosophers have traditionally called the "philosophy of nature" would be compatible with the examination of science as a set of results. Recall the work of Bergson mentioned above, for example, or David Hawkins (1964), or

Errol Harris (1965). Insofar as one is committed to philosophical reflection about the world, scientific procedures may be less interesting than scientific results. One may speculate about the great questions of theology and philosophy as philosophers and theologians have always speculated, but one must try to accommodate the results of science in one's speculations. Of course, if one abandons all attempts at accommodation, then one's philosophy of nature or science will be indistinguishable from purely speculative metaphysics. A taxonomy of philosophies of nature and science is given by Ernan McMullin (1968, 1970) and by William Wallace (1967).

2. **Functional and Formal.**   Second, one may proceed from a functional or a formal point of view. In the former case, one would raise questions such as, How do scientific laws function? or What role do laws play in the scientific enterprise? In the latter case one might ask, What is the logical structure of scientific laws and theories? A systematic structural analysis of the key concepts in science would yield what some authors have referred to as a "theory of science." It would provide not only unifying threads among diverse disciplines, but clear structural or formal lines of demarcation. A formalist view is espoused by Henry Kyburg (1968); for a "theory of science" view see Israel Scheffler (1963). A functionalist view is given in N. R. Hanson (1961). Kisiel and Johnson (1974) have interesting things to say on the functional point of view, and include a brief literature review. Patrick Suppes (1978) and Kyburg (1978) review recent work in the formal tradition.

3. **Broad Synthetic vs. Narrow Analytic.**   Third, one may approach science from a narrow analytic or a broad synthetic point of view. That is, one could see one's task as primarily dissecting various features of science to see how they look or work, or as primarily locating science within a broader framework of human activities and artifacts. Questions about, for example, the form and function of scientific explanations would be relevant to the former (analytic) approach, while questions about the relevance of morality to science would merit special attention in the latter (synthetic) approach. ("Analytic" and "synthetic" as used here should not be confused with the technical terms as sometimes used by philosophers.) Several authors have been mentioned in our historical and country-by-country surveys whose works are largely in the synthetic tradition: e.g., Scheler, Habermas, Geymonat, Preti, Sawada, Yamanouchi, Løchen, Radnitzky, and Harris. Similarly, several analysts have already been cited here: e.g., Nagel, Hempel, Carnap, and Scheffler.

Perhaps one of the reasons for the lack of communication between Anglo-American and European philosophers of science is that most of the former have been analytic and most of the latter have been synthetic in their approaches. Where the analysts have been interested in investigating science as a splendid truth-finding enterprise, the synthesists have been interested in giving a critique of science as one cultural artifact among many others, with peculiar advantages and disadvantages for other cultural artifacts and for human development. Where the analysts have sought to understand the object of their admiration, the synthesists have sought to put that object to work for a wide variety of interests of human beings. In a sense, then, the synthesists have been more pragmatic in their aims than the analysts. Although, because their immediate goals have been so different, the two schools have found little to talk about, recent efforts by Anglo-Americans to investigate the evaluative and ethical implications of science suggest that there may be more opportunities for dialogue in the future. See Kuhn (1977, p. xv); additional reviews of recent work in Europe may be found in Gary Gutting (1978) and Patrick Heelan (1978).

4. **Descriptive and Prescriptive.** Fourth, one may think critically about science as it is (or has been) on the one hand, or as it ought to be on the other. In the former case, one's task is essentially that of accurately reporting the nature of the enterprise, while in the latter one is prescribing the nature of the enterprise as it should be in order to satisfy certain normative requirements. Kuhn's *The Structure of Scientific Revolutions* (1962) views science in historical perspective. A view of science through contemporary practice is presented in Carl Hempel's *Aspects of Scientific Explanation* (1965).

As one might expect, it is often difficult to determine whether a particular philosophical account of some feature of science is an accurate report of what is the case or a persuasive characterization of what ought to be the case. Logical positivists or empiricists often talk about providing "rational reconstructions" of various aspects of science. It remains an open question, however, whether any particular reconstruction does what it is supposed to do or merely provides a reasonable but largely fictional account of some event in the history of science. For example, although it is possible to describe the process of theory appraisal as subjecting theories to severe tests in the interests of destroying bad theories as soon as possible, it is questionable whether most scientists are conscious of such interests. An explication of the concept of theory corroboration in terms of severe testing with particular aims in mind may rationalize but fictionalize scientific practice. Discussions of these issues are available in Finocchiaro (1973); Imre Lakatos (1968); and Michalos (1971).

Since other chapters in this volume are devoted entirely to historical studies, there is no need to dwell on historical methods here. Nevertheless it is worthwhile to mention that because it is difficult to distinguish historical from other kinds of case studies, there is a tendency to inflate the role of historical studies in the philosophy of science. For reviews of recent work, see Laudan (1978), and McMullin (1978). (See also the discussion in chapter 1 of this volume.)

5. Syntactic, Semantic, and Pragmatic. Fifth, in the task of analyzing or explicating key concepts in science, three distinct approaches have
been devised, with technical names: "syntactic," "semantic," and "pragmatic." For example, suppose one wanted to explicate the concept of probability. A syntactical explication would be purely formal and would consist
of a set of principles indicating the mathematical structure of probability. It
would, in fact, be virtually identical with the mathematician's calculus of
probability.

A semantic explication of probability would begin with an analysis of
the meaning of the term in an informal sense. Assuming that there is general agreement about the formal properties of probability, semanticists focus
their attention on what it is that probability statements are supposed to be
about. As we will see below, some people hold that when we talk about
probability we are talking about a logical relation; others insist that the
term designates physical features of the world; and still others are committed
to the view that it designates diverse psychological states. Semanticists put
flesh of one sort or another on syntactical skeletons.

A pragmatic explication of probability would attempt to particularize
or historicize various semantic explications. It is unfortunate that the term
"pragmatic" has been used in this context, because it carries some unfortunate historical connotations. In the present context the term is used to
designate the view that words do not have meanings in the abstract. Words
are used by people in particular sociohistorical situations, and they mean
whatever those people take them to mean. The pragmatist accepts the need
for the robust explications of the semanticists, but insists that they must be
bound by spatiotemporal and cultural constraints. For example, instead of
talking about *the* meaning of probability, pragmatists would prefer to talk
about what this or that group of people mean by the term probability, or
how they use the term. For a purely syntactic analysis see Karl Popper
(1961); for a semantic analysis, see Rudolf Carnap, *Logical Foundations of
Probability* (1950); and for a pragmatic analysis, see Stephen Toulmin
(1956).

**6. Approaches Emphasizing the Philosophical or Scientific Litera-
ture.** Sixth, one may proceed to think critically about science primarily
from a consideration of relevant philosophical literature and practice, or
from relevant scientific literature and practice. Since the 1940s probably,
and the 1950s certainly, it has been possible for students to specialize in the
philosophy of science within traditional philosophy departments. Moreover,
philosophers of science tend to read the work of other philosophers and to
write for them as well. However, some philosophers of science would prefer
to work with scientists rather than with philosophers. They try to identify
and solve problems in science that they, and perhaps scientists, regard as
philosophical. Thus, for example, philosophically oriented philosophers of

science have shed a lot of ink on the so-called "paradox of the ravens" and, what is worse, the "grue-bleen" problem; while scientifically oriented philosophers have emptied their pens on the proper interpretation of quantum theory. It should come as no surprise, then, to find that people who are inclined to one or the other camp are occasionally inclined to look upon what the other side is doing as irrelevant if not downright wasteful. Scientifically oriented philosophers complain that what the other side is doing has nothing to do with science, while philosophically oriented philosophers complain that what the other side is doing has nothing to do with philosophy. However, together, these two orientations have contributed a richer stock of literature to the philosophy of science than either could have contributed alone. For an example of philosophy of science proceeding from scientific literature, see Mario Bunge (1968); Nelson Goodman (1954) proceeds from philosophical literature.

7. The Philosophy of Particular Sciences, Methods, or Theories. Seventh and finally, one may approach the philosophy of science from a single scientific discipline or from several. Some people specialize in the philosophy of mathematics, physics, biology, psychology, history, economics, or sociology. Others move up a notch conceptually to work in the philosophy of the physical or of the social sciences generally.

Examples of philosophies of specific sciences include: Max Black, *The Nature of Mathematics* (1933); Mario Bunge, *Philosophy of Physics* (1973); Jerry Fodor, *The Language of Thought* (1975); David Harvey, *Explanation in Geography* (1969); Stephan Nowak, *Methodology of Sociological Research* (1977); Alexander Rosenberg, *Microeconomic Laws* (1976); Michael Ruse, *The Philosophy of Biology* (1973); and William Todd, *History as Applied Science* (1972). A nonempiricist who has offered penetrating analyses of particular sciences from a neo-Kantian perspective is Ernst Cassirer, for instance in *Determinism and Indeterminism in Modern Physics* (1956) and in *Substance and Function and Einstein's Theory of Relativity* (1923).

Examples of philosophies of social science include: Abraham Kaplan, *The Conduct of Inquiry* (1964); Richard Rudner, *Philosophy of Social Science* (1966); and Peter Winch, *The Idea of a Social Science* (1958). Examples of philosophies of physical science are Carl Hempel, *Philosophy of Natural Science* (1966), and Peter Achinstein, *Concepts of Science* (1968).

Some important issues arise in certain sciences and have virtually no discernible implications outside those sciences. Thus, an exhaustive analysis of issues that are common to several disciplines will necessarily omit any discussion of interesting but idiosyncratic topics. Since we have an enormous variety of more or less common ideas to draw upon (e.g., discovery, theory, model, measurement, etc.), there are actually fewer significant unique topics than one might imagine. For example, the concept of unconscious mental acts may be treated as a unique problem of psychology or psychoanalytic theory, or as a species of the more general problem of unobservable entities. The Heisenberg indeterminacy relation may be regarded as a unique problem of quantum theory or as a particular kind of measurement problem, and so on.

The most notable examples of issues that have received considerable attention by philosophers of science and are not easily subsumed under more general types of problems concern the concepts of space and time, and the foundations of mathematics. As we saw in our country-by-country overview, there is an enormous amount of work going on around the world related to these subjects. Alternative analyses of the notions of space and time have been offered since the time of Plato and Aristotle, but they apparently have few implications outside of physics, metaphysics, and the philosophy of physics. Most of the work has been devoted to the question of whether they are absolute, relative, or relational. These questions are mainly ontological rather than epistemological. Adolf Grünbaum (1973) gives a good analysis of space and time, as well as of many issues in the philosophy of physical science. Alternatives include Bas Van Fraassen (1970) and Lawrence Sklar (1974). Many nonempiricist or general philosophical interpretations of space and time have also been written, e.g., Milic Čapek (1961).

Recent work in the foundations of mathematics has tended to cluster about four distinct views. So-called "logicists" claim that mathematics is derivable from logic, provided that set theory is regarded as a branch of logic. "Platonists" take the ancient view that mathematics is finally grounded in some special realm of objects which is apprehended directly by the mind. "Formalists" take an alternative ancient view, nominalism; for them, mathematics must be grounded in more or less visual, finite, and discriminable entities. "Intutionists" hold that mathematics is a creation of the mind, and that there are mathematical assertions that cannot be proven true or false and must be regarded as neither. Paul Benacerraf and Hilary Putnam (1964) have provided a good introductory anthology; Stephen Barker's work (1964) is an introductory text; Max Black's *The Nature of Mathematics* (1933) is more advanced. Hilary Putnam (1978) contains a good review of recent work in this area.

II. VALUES IN SCIENCE

The fundamental problem of this volume is not ontological or epistemological but axiological; philosophic discussions of various aspects of science are to be related to value and ethical questions. In recent philosophic work on science, what, if anything, is at stake from the point of view of values and ethics? We have approached this question a number of times in our historical, sociological, and logical overviews. In the following sections the analyses will be more direct and detailed. The extra detail will provide a clearer view than our brief overviews could of the nature of contemporary philosophical analyses of science. Hopefully, the extra detail will help to show what it is to do philosophy of science.

III. LOGIC

If one sets out to think critically about science, one is likely to prefer reasonable to unreasonable criticism. As any witness to a debate in the Canadian House of Commons might attest, there is little virtue in undisciplined criticism. At a minimum, critical thinking ought to be guided by generally accepted principles of logic. In this section we will take a brief look at some of these principles, just enough to indicate a few points at which values or judgments of value enter into logic. One may have good reason to claim indubitability, universal acceptance, et cetera, for some principles of logic; but one would have to pick one's terms and principles carefully. Even such apparently innocuous principles as the Principle of Identity (everything is identical to itself) and the Principle of Excluded Middle (every meaningful declarative sentence is either true or false) have given rise to serious disagreements.

Since most readers of this volume are likely to be familiar with the terms "induction" and "deduction," it will be instructive to consider them in some detail. There are two plausible and popular ways to distinguish inductive from deductive arguments, neither of which is entirely satisfactory. On what one may call the *formalist* account, an argument is said to be deductive, by definition, if its premises imply its conclusion. Such arguments are also said to be truth-preserving, since whatever truth is in their premises is passed on to their conclusions. On the contrary, the premises of inductive arguments do not imply their conclusions, but merely make them more or less probable, well-supported, plausible, et cetera. In other words, inductive arguments are not truth-preserving; the truth of their premises is not necessarily passed on to their conclusions.

On what may be called a *suppositional* account, an argument is said to be deductive if it is supposed to imply its conclusion and inductive if it is supposed to make its conclusion more or less probable, acceptable, et cetera. Thus, when an argument is characterized as deductive or inductive on this view, it is not being evaluated in any way. To say that an argument is deductive or inductive is merely to say what it is supposed to do, whether or not it does. Wesley Salmon (1966) gives a formalist account of this distinction, Stephen Barker (1965) offers a suppositional account.

Since the two approaches are not equivalent, a choice must be made. Moreover, the choice will almost certainly be made for reasons that decision makers will regard as good reasons. One or the other approach will, for some reason or other, be regarded as preferable, all things considered. There will be an evaluation, a judgment of value, at the core of a deliberate decision to draw the inductive-deductive distinction in one way rather than another.

Furthermore, it should be remembered that a sound argument or proof is an argument that is both materially and formally correct. Hence, since formal correctness is dependent upon what one regards as an argument satisfying deductive or inductive canons, how one cuts the inductive-deductive pie determines to some extent how good or bad any given argument is going to appear. In short, there can be no serious discussion of good argumentation that does not presuppose or rely upon a value judgment concerning the inductive-deductive distinction. Howson and Worrall (1974) discuss crusaders against one or another form of logic; and Botezatu's efforts (1969) in developing a "natural logic" have already been cited.

IV. TOPICS IN ANGLO-AMERICAN ANALYTICAL PHILOSOPHY OF SCIENCE

The original list of topics to be taken up here was quite long: scientific significance, theories, observation, explanation, laws, reduction, models, measurement, induction, probability, and utility. Some topics were combined, and others were reluctantly dropped because of space limitations.

Part of the thinking behind including some topics and excluding others had to do with historical developments in the analytical tradition. Challenges to the traditional logical empiricist view came first from Thomas Kuhn, Karl Popper, and their followers, with their views centering around issues of discovery and growth in science; topics listed as "current" are really perennial—issues that have persisted through all the recent changes in analytical philosophy of science and that continue to be hotly debated today.

A second reason for the choices had to do with the focus of the *Guide* on value issues in the sciences. It seemed that more value issues could be demonstrated in terms of the issues retained. Nonetheless, some loss was occasioned by the particular choices made, and the nature of this loss warrants illustration.

In an earlier version of the chapter, a section on scientific models was included. It covered general theories of models (e.g., Mary Hesse's *Models and Analogies in Science,* 1966), as well as such specific types as representational, ideal, semantic, theoretical, and behavioral models. Treating behavioral models offered the chance to note how the adoption of or insistence on the use of one or another type of model can have profound evaluative and ethical implications for science.

Behavioristic psychologists had to struggle fiercely for their existence within the established community of psychologists. However, once behavioral models caught on, they tended to sweep everything before them; association offices, editorial boards of journals, granting agencies, and graduate departments became dominated by behaviorists. A similar phenomenon took place in political science. At stake was not merely the future of a couple of academic disciplines and some prestigious positions in scientific communities. The issue was, and still is in both fields, the right way to conceptualize the essence of human nature. The real difference between behaviorist and nonbehaviorist models of human nature is that the former choose to ignore precisely what the latter refuse to ignore—namely, that humans are thinking, feeling creatures.

A. Central Issues in the Logical Empiricist Tradition

1. **Scientific Significance.** It was a basic tenet of positivist or empiricist philosophers of science in the 1930s that all cognitively meaningful propositions were either logically true or false, or else in principle experimentally testable. Logically true propositions are often referred to as "analytic" and are characterized as having self-contradictory denials, as being true in all possible worlds merely in virtue of the meanings of the terms employed in them.

Nonanalytic propositions are often referred to as "synthetic" or "empirical," and a lot of philosophical ink has been shed trying to find a precise criterion of meaningfulness for them. One plausible suggestion was the requirement of complete verifiability: A sentence is empirically meaningful if and only if it is not analytic and is implied by a finite logically consistent set of observation sentences. Observation sentences are sentences in which observable characteristics are attributed to objects: e.g., "This chair is green" or "John is taller than Frank." A. J. Ayer, in *Language, Truth, and Logic* (1936), applied this criterion to a variety of philosophic issues.

The trouble with the verifiability criterion is that it makes some scientific laws empirically meaningless. Some laws of nature apply to more objects than anyone could ever observe. For example, there will never be a set of observation sentences that will logically imply the unrestricted generalization "All men are mortal." This means that some scientific laws (unrestricted generalizations that are commonly believed to be not only empirically meaningful but true) cannot be regarded as empirically meaningful. Since these laws are quite secure, the criterion has to be abandoned.

Another candidate to be the criterion of empirical meaningfulness is the requirement of *complete falsifiability:* A sentence is empirically meaningful if and only if its denial is not analytic and is implied by a finite logically consistent set of observation sentences. Unfortunately, this criterion suffers the same fate as the other. Although it allows some unrestricted generalizations to be empirically meaningful, it makes their denials meaningless. This is certainly anomalous because if a given claim is meaningful and therefore true or false, then anyone who denies it must be making a claim that is equally meaningful only false or true, depending on the status of the former. Still, that is just the sort of sour pickle this criterion breeds. So it too has been abandoned. Karl Popper (1963) discusses some confusions about this criterion.

Weaker criteria of confirmability and testability were suggested, but they also turned out to be objectionable. The criteria always excluded or included too much. Apparently meaningful sentences were ruled out and apparently meaningless sentences were ruled in. Thus, it seemed to some philosophers by the late 1940s that the only way to solve this problem was to design an artificial language whose vocabulary and grammar would prohibit all the unwanted and permit all the wanted sentences. Once again in the history of human affairs, what began as a philosophic sanitation problem was transformed into a philosophic capital works project. Instead of a swift clean-up job, a long-drawn-out development project was proposed. Good reviews of the problem of empirical significance criteria may be found in Carl Hempel (1950) and Israel Scheffler (1963).

The analytic-synthetic distinction, explained above, allows us to classify all propositions as follows:

	A priori	*A posteriori*
Analytic	propositions in formal sciences, such as logic, mathematics	none
Synthetic	metaphysical propositions	propositions in empirical sciences, such as physics, sociology, etc.

In this scheme, metaphysical claims are regarded as claims about the world whose truth status may be investigated without experience or observation. Formal claims are not claims about the world; their truth status may also be investigated without experience. Empirical scientific claims are claims about the world whose truth status requires experiential or observational investigation. For a review of issues related to the analytic-synthetic distinction, see John Woods and L. W. Sumner (1969).

Most philosophers of science who have labored over the problem of finding an empirical meaningfulness criterion have been trying to find decisive identifying characteristics for empirical scientific claims, to specify the necessary and sufficient conditions for classifying any proposition as worthy of consideration by techniques and personnel they already regarded as scientific (see Mulkay, 1977, p. 100). Some fields of study struck these philosophers as intellectual and practical dead ends—e.g., theology, esthetics, ethics and, worst of all, metaphysics. Some philosophers even claimed that these fields were downright pernicious, that they gave people an unwarranted sense of security and stifled any inclinations toward intellectual progress or social reform. How sweet it would be, they thought, to have an ironclad empirical meaningfulness criterion to bash the brains of the merchants of soporific slush, Even today a scientist may be heard to throw down the gauntlet to the uninitiated: "That's not scientific!" he may exclaim, as if a knockdown criterion had been found. Alas, it has not been found.

It may be possible not only to waste one's time but to employ it in destructive ways in the interests of worthless fields of study. However, it is doubtful that a principle distinguishing empirical meaningfulness from meaninglessness, empirical science from nonscience, would also serve to separate out worthwhile from worthless fields of study. Insofar as the exclusion of worthless or dangerous investigations was dear to the hearts of those who labored for an empirical meaningfulness criterion, the labor has been in vain. No one has been able to produce such a criterion; and even if anyone had, it would not have been sufficient for the evaluative task. (Alternative approaches to the problem of specifying the defining characteristics of science may be found in chapters 1, 5, and 7 of this volume.)

2. Theory and Observation. According to the "received view" of scientific theories, they may be characterized as axiomatic systems consisting of a set of

1. Formation rules for constructing well-formed formulas in the system
2. Transformation rules (i.e., rules of inference)
3. Three distinct vocabularies for logical, observational, and theoretic axioms
4. Correspondence rules connecting observation and theoretic terms

There are very few scientific theories that are cast in this form, and there are plenty of other views of the nature of theories. Outside of logic and mathematics, the classic illustrations are Joseph Woodger's (1950) formalization of aspects of biology (cytology, embryology, genetics, and taxonomy) and C. L. Hull's (1952) axiomatized learning theory. Nevertheless, this view occupied the center of the stage for at least thirty years in the philosophy of science. It is clearly in the positivist formalistic tradition; i.e., the idea is to analyze the structure of theories in syntactic and semantic terms. For reviews of recent work on theories, see Achinstein (1968); Robert L. Causey (1977 and 1978); C. A. Hooker (1975a and 1975b); Anatol Rapoport (1958); Wolfgang Stegmüller (1976); and Frederick Suppe (1974 and 1978). Suppe provides the best summary of the "received view" of theories.

As a simple illustration of the received view, one may consider a physical theory of probability. In particular, consider it as a theory about the relative frequency of the occurrence of events in repetitive sequences, e.g., flips of a coin, rolls of a die, and so on. Then the formation and transformation rules would include the rules of algebra and set theory, supplemented by some special rules related to the use of a probability operator. The usual battery of logical and mathematical terms would be included in the logical vocabulary. The observation vocabulary would include terms like "coin," "die," and "toss." The basic theoretic term required is "probability" itself, and that would be defined and measured by the ratio of the number of favorable events to the total number of events in a selected sequence of repetitive events. Insofar as this definition functions as a correspondence rule, the *definiendum* (defined part) should have only theoretic terms and the *definiens* (defining part) should have only observational terms. Unfortunately, here as elsewhere, it is difficult to determine exactly which terms are theoretic in every context. The axioms of the theory might be the standard three, concerning the addition and multiplication of probabilities and the constraints put on numerical values. Armed with this theory, then, one may crank out new postulates and test them in the real world of actual coin tosses and crap shooting. The classic axiomatization of this theory is in Andrei Kolmogorov (1950). For other examples of theories cast in axiomatic form, see Mario Bunge (1973); Merton S. Krause (1972); Henry Kyburg (1968); Michael Ruse (1975); Herbert A. Simon (1970); and Mary Williams (1970).

There is some difficulty in maintaining a distinction between observation terms and theoretical terms. That problem has driven some people to replace it with a distinction between old and new terms, where what is regarded as old or new depends on the particular problem being investigated. See, for example, Suppe (1974); and Raimo Tuomela (1973).

Some people have rejected the axiomatic view of theories altogether. Some hold, for instance, that although a theory must organize or systematize a body of knowledge, this can be accomplished without rigorous axiomatization. Many good textbooks provide systematic presentations of scientific material without bundling it up in an axiomatic framework. Scientific theories may be thought of as organizing or systematizing material in this looser sense. For example, see Anatol Rapoport (1958).

Alternatively, some writers seem to regard theories as nothing more than unrestricted universal generalizations to which people are deeply committed. For these people, there would seem to be no difference between theories and laws from a structural point of view. Perhaps, for them, only a difference in attitude separates laws from theories. For an example, see Popper (1959).

Theories have also been regarded as essentially answers to questions, again with structure playing no significant role. Close to this view is the idea that a theory is a collection of sentences that provides intuitive understanding of something for someone. For the "question-answer" view, see Sylvain Bromberger (1963); for the "understanding" view, see Fred R. Dallmayr and Thomas A. McCarthy (1977).

An instrumentalist view of theories has also been adopted by some. On this view, theories are typically regarded as sets of inference rules. See Alexander (1958) and Pierre Duhem ([1906] 1954).

According to a semantic view of theories, the latter are regarded as sets of statements that have the logical status of definitions without empirical significance. That is, they define a particular sort of natural system without asserting that any given system is of that sort. Theoretical claims, on the other hand, assert that a given system is of this or that theoretical sort. Thus, theoretical claims can be true or false. For example, the axioms of preference theory may be said to define a coherent set of preferences, while it is a matter of fact whether this or that individual's set of preferences satisfy the axioms. This view of theories may be found in Ronald Giere (1979), Joseph Sneed (1971), and Stegmüller (1977).

In spite of this fairly impressive array of alternative views of the nature of theories, much more has been written about particular theories in diverse sciences. Consider, for instance, the vast critical literature devoted to the theories of Karl Marx, Charles Darwin, and Sigmund Freud. It would not be an exaggeration to say that the philosophy of any particular science is always dominated by the philosophical problems of the dominant theories of that science. For example, contemporary philosophy of physics is largely devoted to problems related to quantum theory and relativity theory. After winning their wings in defense of the scientific status of the synthetic theory of evolution, philosophers of biology have flown fearlessly into current battles over sociobiological theories. Philosophers of statistics have joined the battle between Bayesians and Neyman-Pearson theorists. Chomsky's theory of innate linguistic capacities, Skinner's behavioral theory, and practically all theories of mental illness have been the subjects of severe philosophical criticisms. Theories of rational decision-making abound in the literature. Clearly, a review of the philosophy of science in the twentieth century could be overwhelmed by a review of philosophical discussions of particular features of particular theories of particular sciences.

An example of the overwhelming character of the situation can be seen in philosophical discussions of particular physical theories. See, for instance, Jeffrey Bub (1974); Robert Cohen and Marx Wartofsky (1973); Lindley Darden and Nancy Maull (1977); C. A. Hooker (1973); Peter Mittelstaedt (1976); and Kristin Shrader-Frechette (1977).

Here are a few examples of work on theories in other areas of science: for biological theories, see David Hull (1973) and Michael Ruse (1969, 1971a, 1971b, 1973, and 1977); on theories in social science and statistics, see Allan Birnbaum (1977), Alex Michalos (1978a and 1978b), and Mark Richelle (1976).

Advocates of the received view of theories hold that observation terms and theoretical terms should be clearly distinguished—indeed, that the latter should be explicitly defined by the former. General reviews of issues related to the observation term-theoretical term distinction may be found in Causey (1978); Hempel (1958); Ilkka Niiniluoto and Raimo Tuomela (1973); Tuomela (1973); and Suppe (1974).

The impossibility of constructing explicit definitions of all theoretical terms by observation terms is easy to demonstrate. The proof turns on the nature of disposition-designating terms like "magnetic," "intelligent," and "soluble." It is characteristic of such terms that they refer to properties that are beyond any referred to by any set of observation terms. As long as one insists that observation terms must designate properties that are in principle observable, and that disposition-designating terms must designate properties that are in principle one step beyond any given observation, one has logically blocked the possibility of exhaustively capturing all the fish of the latter in a net designed by the former. The most one can hope for is a new net. (For a full discussion of this sort of skepticism, see Frederick Will, 1974.)

As it happens, a new net was invented in the late 1920s—namely, operational definitions. Instead of trying to capture the full meaning of disposition-designating terms by a set of terms designating observable properties, one appeals to a set of operations. For example, an operational definition of "mentally deficient" might run as follows: One is mentally deficient *if and only if,* when one is given a Stanford-Binet IQ test, one scores below 70. The trouble with operational definitions in this form is that, given standard interpretations of the logical connectives involved, everything turns out to have the defined property as long as no tests are ever performed.

We have to chuck either conventional logic or this view of operational definitions. The easiest thing to do is to rearrange the various parts of the definition so the test comes first. Then we get what is called a *reduction sentence:* If one is given a Stanford-Binet IQ test, then one is mentally deficient, by definition, if one scores below 70. Now whenever anyone fails to be tested, nothing follows with respect to her or his mental deficiency. We know *a priori* that one is mentally deficient if one has a score below 70, but this is useless information in the absence of any tests. In fact, its usefulness is strictly limited to testable cases. That is, the meaning of disposition-designating terms like "mentally deficient" is specified only partially, for testable cases, by reduction sentences. As other tests are constructed, the meaning of such terms can be extended. But it can never be exhausted by such a procedure because, as intimated above, disposition-designating terms designate properties that are in principle one step beyond anything right here, right now. The classic study of operational definitions and the development of reduction sentences is Carnap's "Testability and Meaning" (1936, 1937). Hempel's "Fundamentals of Concept Formation in Empirical Science" (1952) and "Problems and Changes in the Empiricist Criterion of Meaning" (1950) are also instructive.

Insofar as the distinction between observational and theoretical terms disguised a hankering for certainty concerning empirical matters, it was bound to fail. Corrigibility is logically built into empirical claims, including those involving the simplest of observables. Two thousand years of epistemology should have been sufficient to alert philosophers of science to the total darkness at the end of that tunnel.

On the other hand, some people have tried to maintain the distinction without having any illusions about the certainty of observation reports. These people just wanted to have some neutral facts available now and then, especially when theories had to be tested. Insofar as observation reports or terms were burdened with theoretical overtones, they reasoned, such reports and terms would be biased. After all, a theory-laden report or term is literally a report or term whose very meaning is at least partially determined by some theory; thus it cannot be regarded as a neutral datum to be used in the appraisal of alternative theories.

In the worst of all possible worlds, every term, and therefore every report, would be theory-laden. So it would be impossible to administer an unbiased test of any theory. What is worse, people committed to diverse theories would find it difficult, and occasionally even impossible, to communicate with one another. Their official views of the world might be so different that they could be correctly described as living in quite different worlds. A good review of the literature on theory-laden terms and relativism is found in Carl R. Kordig (1971).

Given the possibility of this undesirable scenario, it is easy to appreciate the motivation of those who insist upon an observation term–theoretical term distinction. They would like to have some assurance that we are not in fact living in the worst of all possible worlds. Fortunately, the place is crawling with such assurances. When psychologists test alternative theories about the behavior of rats in various kinds of experimental setups, there is typically no question about the observable facts. When political scientists test alternative theories accounting for voter turnout, the same statistical facts are usually employed. The whole point of the exercise is to try to provide a satisfactory account of the facts. If the facts changed to suit every theory, the exercise would effectively lose its point. Similarly, demographers, epidemiologists, criminologists, and geographers typically have to dip into the very same pool of statistical time series—i.e., facts—in order to test, confirm, or disconfirm their theories. Moreover, becoming a demographer, et cetera, implies learning to interpret such common pools of facts in the proper ways.

All the preceding problems with the observable-theoretical distinction are central to philosophy of science, but they represent the tip of an iceberg of problems in the philosophy of psychology or philosophical psychology. The problems of innate capacities, mental illness, behaviorism, rational action, goal-directed behavior, theories of the unconscious, and deterministic laws of human behavior have already been mentioned. One might also mention the question of the role of inference in perception; the nature of mental representation; what it is to have a concept; whether mental images exist, and, if so, how; the nature of emotion, feelings, and cognition, and their relationships; the problem of split brains, or, more generally, of personal identity; intentionality; mind-body relationships; consciousness; privacy and the so-called problem of "other minds." Most of these problems are classics of epistemology and metaphysics, and it would be difficult to distinguish contemporary approaches to them from enlightened approaches taken by our predecessors, although there is, of course, a vast difference in the availability of a wide variety of experimental results and theories.

For recent work on perception, see Michael P. Bradie (1976); Hector-Neri Castañeda (1977); Michael Goldman (1977); D. W. Hamlyn (1969); N. R. Hanson (1969); R. J. Nelson (1976); George Pitcher (1971); William Powers (1973); and J. R. Royce and William Rozeboom (1972).

For mind-body problems, see D. M. Armstrong (1973); Herbert G. Bohnert (1974); Herbert Feigl (1958); Norman Geschwind (1974); Gordon G. Globus (1972); Keith Gunderson (1969); Roland Puccetti (1973); and Michael Radner and Stephen Winokur (1970).

On sensations and images, see Joseph Agassi (1975); Peter Alexander (1963); D. M. Armstrong (1964); Rudolf Arnheim (1971); Alistair Hannay (1971); Robert Howell (1976); and Elliott Sober (1976).

The significance of some of these issues for the question of values in science has already been suggested. Without a clear distinction between observation claims and theoretical claims, one road to an incorrigible foundation for science is closed. Moreover, if one thinks of the total body of scientific knowledge and beliefs along the lines of the received view of theories, the whole edifice may appear to be jeopardized by the elimination of suitably stable (i.e., incorrigible) axioms. Some people would find that prospect terrifying. Nevertheless, the loss of certainty is not the loss of science.

To some extent the fear of flying with theory-laden terms and reports is also rooted in the quest for certainty of procedures. Without theory-free facts, it is felt, one cannot be sure that one's testing procedures are unbiased. No one imagines that the possession of theory-free facts is sufficient for impartiality, but, as illustrated above, it is not obviously necessary either. It is plainly logically possible for statements of fact to be free enough of theoretical overtones to permit an appraisal of theories that is impartial enough for the scientific enterprise to flourish. Furthermore, the practice and growth of science for at least the past couple of hundred years must be accepted as evidence that it is not only logically possible, but a matter of fact. Research procedures, like observation reports, do not have to be infallible in order to be good or even reliable and valid. Like democracy, they may occasionally even be bad, but they are better than any alternatives and adequate for the job to be done.

B.   Discovery and Growth

Perhaps nothing about science captures the imagination of everyone, scientist and lay person, as much as the idea of a scientific discovery. Ordinary people have opinions, views, or hunches about the way the world works, but scientists have discoveries. It would seem that any reasonable initiation into any scientific community would have to pay some attention to the logical structure of discoveries. After all, without discoveries there would be no growth; and without growth, the scientific enterprise would grind to a sterile halt.

To the casual observer it may seem as if scientific discoveries are nothing more than generalizations from particular observations. Indeed, as we saw repeatedly in our historical and sociological overviews, this view has been held by observers who have been considerably more than casual. The theory is called *induction by enumeration*. One notices, for example, that a particular thief was raised in a one-parent household in which economic, social, and personal security were in short supply. As time passes, one encounters, reads, or hears about other thieves coming from similar backgrounds. Then the generalization emerges: All thieves are raised in socio-economically deprived households. By examining the backgrounds of many thieves, a scientist—sociologist, criminologist, or psychologist, perhaps—discovers a regular relationship between deprivation and theft. He discovers something that looks as if it may be a law of nature.

It is easy to see why one might think that induction by enumeration is virtually sufficient to account for the logic of scientific discoveries. Just as one noticed a relation between deprivation and thieves, one might notice a similar relation between deprivation and drug pushers, murderers, kidnappers, and rapists. Thus, one moves up a level, conceptually, still working within the pattern of enumerative induction. That is:

   − All thieves are raised in deprived households.
   − All kidnappers are raised in deprived households.
   − All counterfeiters are raised in deprived households.
   − Thieves, kidnappers, counterfeiters, and the like, are criminals.
   − So, all criminals are raised in deprived households.

Notice, however, that the higher-level generalization was logically mediated by the conceptual link between thieves and the like, and criminals. If one did not know that thieves, et cetera, were all species of the genus criminal, then one could not logically have arrived at the higher-level generalization. Moreover, one does not discover that, say, thieves are criminals the way one discovers that thieves come from deprived households. The conceptual link between thieves and criminals is forged while one is learning to speak English. The deprivational link is, on the inductivist view, built up from observations. For good accounts of induction by enumeration, see Stephen Barker (1957) and William Kneale (1949).

Just as one might design a conceptual link between thieves and criminals, one might design hypotheses that link theft not to some kind of deprivation but to rational calculation or illness. Such an approach might be patterned after a method known as *induction by elimination*.

For example, one might reason that theft may be explained by deprivation, illness, or rational calculation, and rule out the first two possibilities for some affluent white-collar criminals. Wherever the alternative hypotheses originate, on the eliminative inductivist view truth is discovered by the elimination of falsehood. Furthermore, it is usually the case that such inductivists do not know and do not care about the origin of scientific hypotheses. Their concern is the discovery of a true hypothesis from a set of options—whereas the enumerative inductivist was concerned with the discovery of lawlike generalizations from lower-level generalizations or from particular claims. A good account of induction by elimination may also be seen in Barker (1957). An interesting mixture of enumerative and eliminative induction is the Hintikka-Hilpinen system, in Risto Hilpinen (1968).

A third analysis of the logical structure of discovery proceeds from *analogy*. Freud argued, for example, that since people often daydream about events they would like to see happen, and night dreams are similar to daydreams, night dreams probably have a similar source—namely wish fulfillment. Interestingly enough, an analogical argument based on physiological (rather than psychological) considerations led to a quite different hypothesis concerning the origin of dreams. It has been argued that since a certain amount of muscle tension remains after strenuous exercise, and mental exercise is similar to physical exercise, (night) dreams are probably caused by a delayed shutdown of conscious processes. Instead of having one's mental faculties stop altogether when one falls asleep, some of the machinery continues to operate in a disjointed way, producing more or less coherent dreams. For our purposes it does not matter whether the psychological or physiological hypothesis is nearer to the truth. The point is that occasionally scientific hypotheses are discovered by means of analogical arguments. (See N. R. Hanson, 1961; Mary Hesse, 1966; and Michael Ruse, 1973. Analogy is one of a larger set of "plausibility" patterns of discovery in George Polya, 1945, 1954, and Paul Durbin, 1968; see also N. Rescher, 1976.)

It has also been argued that some discoveries consist in seeing familiar things in unfamiliar or new ways. For example, although people were familiar with shadows and with the phenomenon of straight sticks appearing bent when submerged in water, the idea of explaining such things by thinking of light travelling in straight lines was quite novel. In order to conceive of the Principle of the Rectilinear Propagation of Light, one had to begin to think of shadows and optical illusions as the effects of something. Once this discovery was made, additional questions concerning the source, direction, and speed of light were bound to emerge. This is the view of Stephen Toulmin (1953).

A similar view has been advocated by Thomas Kuhn (1962, 1977). However, Kuhn stresses the social aspect of scientific investigation and discovery. In his view discoveries do not usually arise at particular points of time as a result of a single scientist's efforts. Discoveries or innovations are essentially social phenomena with vague spatial and temporal identities. Moreover, and more importantly, they are not usually the result of bold new flights of fantasy. On the contrary, they are supposed to result from extremely constrained, tradition-bound puzzle solving that is concentrated on generally accepted theories and procedures. The paradox of discovery is that a community of severely indoctrinated scientists whose basic aim is to extend their pet theories as far as possible (rather than to seek new ones) finds itself dragged willy nilly to new and incompatible theories. (For more on the sociology of this view, see chapter 7 in this volume.)

Just as eliminative inductivists tend to bypass the question of discovering hypotheses and focus on the question of discovering a true hypothesis in a set of plausible alternatives, some philosophers have focused on the question of discovering growth-ensuring hypotheses. They have been quite opposed to the idea of a logic of discovery along the lines of enumerative induction and quite content to leave the question of the ultimate source of discoveries to psychologists. Their problem has been to explain the growth of science in the absence of enumerative induction. Their solution is essentially an appeal to trial and error. Scientists make conjectures and submit them to tests. If a conjecture fails to pass its tests, it is rejected. If it passes, it is provisionally accepted.

As indicated in our historical and sociological overviews, this trial-and-error procedure, often referred to as the hypothetico-deductive method, has been recommended by several philosophers of science. For a review of the origins of the hypothetico-deductive method in the seventeenth century, see Laudan (1968). In this century Karl Popper has been its main champion; see *The Logic of Scientific Discovery* (1959) and subsequent works.

Schematically, the procedure runs like this:

- Given hypothesis H and initial conditions C, phenomenon P should occur.
- If P occurs, the conjuction "H and C" is provisionally acceptable.
- If P does not occur, "H and C" is falsified.
- If "H and C" is false, then either H or C is false or both are false.

Four points should be emphasized about this procedure. First, it is basically patterned after a deductive argument form known as *modus tollens*. Second, it is not asserted that the hypothesis is true when it passes its test. It is merely asserted that because it passed its test, we have no good reason to reject it. Third, strictly speaking, the hypothesis and its initial test conditions stand or fall together following a test. Following a passed test, no one is interested in the fact that the initial conditions were as they were supposed to be. However, following a failed test, proponents of the hypothesis may have serious doubts about the conditions. Since a faulty setup may have been the cause of the unsuccessful prediction, the setup must be inspected. If it is clean, then the hypothesis has to bear the brunt of the failure. For the time being at least, it is provisionally rejected or at least regarded as doubtful.

Due to the influence of Pierre Duhem, this third point about the hypothetico-deductive method has received an extraordinary amount of attention in the past seventy years. Duhem's thesis (sometimes called the "D-thesis") was that since isolated hypotheses cannot be tested and a single test cannot identify which statement (H or C) in an experiment is false, there can be no crucial experiments of any hypothesis. That is, all hypothesis testing must lead to inconclusive results.

As has already been suggested, if one is virtually certain that one's experimental setup is clean then one's hypothesis can be seriously undermined. The fact that the sympathetic theory of gravity, the phlogiston theory of combustion, the pangenetic theory of inheritance, and many others have been thoroughly discredited and abandoned leaves no doubt about that point. Although empirical hypotheses are, from a logical point of view, always subject to reappraisal, it is not always rational or wise to do so. Just as one may have very good reasons to believe that a particular prediction has been successful or not, one may have good reasons to believe that a particular initial condition or hypothesis has been faulty. Thus, the D-thesis is unwarranted. For a review of recent work on the D-thesis, see Sandra Harding (1976) and R. M. Yoshida (1975).

The fourth point to be noticed about the hypothetico-deductive method is that there is nothing in the procedure described so far that ensures the growth of science. What has to be insisted upon in this scenario is the inverse relationship between empirical *content* and logical *im*probability. If we want bold new scientific hypotheses with an abundance of explanatory power, then our conjectures must be highly improbable. Security can be bought only at the price of timidity. Highly probable hypotheses are highly probable only because they depart minimally from already accepted evidence and beliefs. The hypothetico-deductive method, supplemented by the requirement of testing only logically improbable hypotheses, ensures the continued growth of science. For a nineteenth-century view of the significance of improbable hypotheses, see W. S. Jevons, *The Principles of Science* (1874). In this century, the issues have been pressed most forcefully by Popper in *The Logic of Scientific Discovery* (1959) and *Conjectures and Refutations* (1962). For a criticism, see Michalos, *The Popper-Carnap Controversy* (1971).

Another contemporary view of discovery and growth in science has roots that are nearly 150 years old. According to this view, in the process of becoming a scientist one is exposed to certain models or exemplars of good practice. One recreates standard experiments in order to develop an appreciation of the scientific enterprise as it is practiced by physicists, anthropologists, or whatever. Little by little one is brought to the frontiers of one's discipline and one develops an appreciation of one's ignorance. In short, one is socialized into a scientific community that may be roughly characterized as a set of individuals committed to what is called a "paradigm." A scientific paradigm consists of a set of accepted models of good practice, hypotheses, laws, theories, and rules of behavior or regulative principles. The roots of this view may be found in William Whewell; see R. E. Butts (1968). The chief contemporary proponent is Thomas Kuhn (1962, 1977); Gernot Böhme (1977) is also helpful.

Scientific discoveries and growth, in this view, may be looked at in two ways. Normal or evolutionary discoveries are essentially mopping-up exercises undertaken without any threatened assault on a paradigm. They have the effect of enriching a paradigm by adding to its stock of models, laws, and so on. Revolutionary discoveries are paradigm-testing and, hence, threatening. Such discoveries are supposed to be brought about by a steady accumulation of anomalies within an accepted paradigm. So new ones are sought. For example, the abandonment of the Ptolemaic geocentric view of the universe in favor of the less cumbersome Copernican heliocentric view was a revolutionary change, a change in scientific paradigms. On the other hand, within the geocentric paradigm, the addition or subtraction of epicycles to the constructed orbits of planets would be regarded as evolutionary changes. An excellent discussion of the difference between evolutionary and revolutionary views in Kuhn and Popper's views may be found in Imre Lakatos and Alan Musgrave (1970).

Finally, we must ask how these views of scientific discoveries and growth are relevant to evaluative issues in science. At least one of them, induction by enumeration, has enjoyed enormous popularity as a rough (but very serious) account of what many textbook authors have described as *the* scientific method. A scientist is supposed to observe, form a generalization, test, observe again, and so on. Whether or not such sequences ever occurred in science as it is actually practiced, a lot of students have been taught (more, one hopes, in the last century than in this one) that they ought to occur. Thus, this particular view of the path to scientific discoveries has served as a model or exemplar for budding scientists. It has provided the uninitiated with an alleged behavior pattern that is proper, reasonable, and accepted by the scientific community. It has given visible form to a scientist's way of pursuing the truth. In a world where appearance, form, and style are often as important as reality, the possession of codified institutions cannot be underestimated. Scientists do not proceed by consulting a holy book, by going to court, by a public election, by breaking bread, rolling dice, dealing cards, or reading the stars. They have their own method (so the story has gone)—the scientific method.

If this approach sounds plausible, then it ought to be easy to appreciate the resistance one encounters from scientists when someone tries to introduce new methods into science. For more than a hundred years, those who have recommended the hypothetico-deductive method or defended induction by enumeration have been engaged in a struggle over scientific credentials. Although the players and procedures have changed today, the principle is the same. To a significant extent, science is a set of accepted procedures; to be a good scientist is, in the first place, to master these procedures. Moreover, since a good scientist is a person with high status in the social hierarchy of the scientific community, and since high status is generally preferable to low status, from a purely personal or ego-enhancing point of view, most scientists would like to be regarded as good scientists. Hence, scientists have personal as well as professional reasons for trying to articulate proper procedures. (Further analyses along these lines may be found in Mulkay, 1977, as well as in chapters 5 and 7 of this volume.)

C. Current Issues

1. **Explanation.** Just as some procedures leading to scientific discoveries and growth are supposed to have a privileged status within the scientific community, some patterns of explanation are supposed to be similarly regarded. The pattern that has occupied the center of the academic stage for at least thirty years is called the *covering law model*. The best general discussion is given by one of the originators of the model, Hempel, in *Aspects of Scientific Explanation* (1965). There is also a good review in Joseph Hanna (1978).

Every explanation may be divided into two parts, an explaining part (*explanans*) and a part to be explained (*explanandum*). In the *deductive nomological* form of the covering law model, the explanans consists of a set of premises that imply the explanandum. Ideally these premises would be true and would contain some general scientific laws. In other words, the model or exemplar for this form of the covering law model is a sound argument or proof (above). Explaining something scientifically—logically or structurally—in this view means providing a sound argument whose conclusion is whatever needed explaining.

To illustrate this type of explanation, we can consider a scientific explanation of the bursting of a water pipe in the winter:

- C_1 Last winter the temperature fell below 32°F.
- C_2 An outdoor water pipe contained water.
- C_3 If water expands sufficiently in a pipe then it bursts.
- L_1 Water freezes at temperatures below 32°F.

    – L_2 Water expands when it freezes.
    – E Hence, the pipe burst.

All the conditions for a deductive nomological explanation are satisfied. The argument form is deductively valid. The premises are true. Three of the premises state initial conditions ($C_1 - C_3$) and two state general laws (L_1 and L_2). Both laws, as is typically the case in explanations of this sort, are causal—i.e., in L_1 freezing water is supposed to be the effect of the temperature's falling below $32°F$, and in L_2 expanding water is supposed to be the effect of its freezing. Hence, the whole account may be regarded as a causal explanation of the bursting of a water pipe. See Hempel (1965) and Thomas Nickles (1971).

When the model was first proposed, it was claimed that the only difference between an explanation and a prediction is where one begins: If one has only the premises, they may be used to predict the bursting of the pipe. If one has only the concluding statement of fact, it may be explained by producing the premises. Logically or structurally, explanations and predictions were supposed to be symmetric.

The trouble is that it is easy to think of premises that would be excellent for predictive but worthless for explanatory purposes. For example, if we had known nothing about water pipes except that one bursts every winter, then we would have been able to predict the bursting of the pipe last winter. However, once the pipe burst, it would have explained nothing to point out that it bursts every winter. The fact that it bursts every winter is merely a summary of precisely the sort of fact that requires explanation, namely, that it burst last winter.

For reasons of this sort, the symmetry thesis was abandoned early, but that was only the beginning. Regardless of anyone's commitment to scientific laws, it remains logically possible that as empirical propositions any of them will be false. Besides, some laws are accepted not because they are believed to be true, but just because they have a lot of support or, perhaps, merely more support than any alternative. So there was some sympathy for the idea of requiring premises of acceptable scientific explanations to be merely well supported instead of true. This was the thin end of the wedge. (See Israel Scheffler, 1957.)

Once one begins to think of the role of inductive logic in relation to explanation, the attraction of an inductive analogue of deductive nomological explanation becomes irresistible. This is especially true when one considers the fact that so-called statistical laws are inescapable in science. Thus, in the *inductive-statistical* form of the covering law model, statistical laws are substituted for universal generalizations, inductive validity is substituted for deductive validity, and inductive support of some sort is substituted for the truth of premises. An example of this form of explanation might run as follows.

- L_1 All plants have a pair of factors controlling their height.
- L_2 The offspring of sexual parents receive one factor from each parent in a random fashion.
- C Jones planted seeds that were the offspring of plants having an equal share of both factors.
- E Hence, Jones probably had a crop of plants with $3/4$ displaying the effects of one of these factors and $1/4$ displaying the effects of the other.

Here we have an explanation with two laws (L_1 and L_2) and one statement of initial conditions (C). The first law (L_1) is an unrestricted generalization. The second (L_2) is a version of Mendel's First Law. Since it refers to a random distribution of factors in plants, it is essentially statistical. Hence, this feature of the explanans is transmitted to the explanandum. Thus, if one knew that Jones had produced a crop described as in the conclusion (E), then one could use the premises of this argument as a probabilistic explanation of that crop. On the other hand, if one had only the premises of this argument, then one could use them to make a probabilistic prediction of the crop distribution described in the conclusion. Thorough discussions are found in Hempel (1965).

The second form of the covering law model seems perfectly plausible to most philosophers of science, but some people with special interests in social sciences still think it is too rigid. Since there is still no general agreement about the criteria that must be satisfied for a proposition to count as a law and there is a tendency for contemporary scientists to avoid the term altogether, little would be lost by abandoning the requirement that one of the premises should be a law. Then arguments like the following could be regarded as acceptable scientific explanations.

- Political views are normally distributed (i.e., in bell-shaped distribution curves) in any large population.
- Politicians try to maximize votes.
- Most people are favorably inclined toward people they perceive to be like themselves.
- Most people will vote for people toward whom they are favorably inclined.
- Hence, most politicians will try to appear as middle-of-the-roaders whether or not they are.

This is a fairly standard explanation of the usual bland or ambiguous type of speech that flows out of politicians. All the premises are well supported by empirical evidence, but none of them has been honored by the title of "law." Moreover, this example from political science is typical of explanatory accounts in the social sciences. Thus, again one either has to weaken one's criteria to accommodate widespread practice or to justify one's criteria on *a priori* grounds. No one doubts that such explanations are logically weaker than the original deductive nomological type, but some people regard them as acceptable models of scientific explanations while others regard them as mere "sketches" or truncated versions of acceptable models. Reviews of these issues may be found in Robert Borger and Frank Cioffi (1970) and in Patrick Gardiner (1959).

Explanations in the biological sciences have also raised special problems. These sciences are supposed to employ *teleological explanations* that in one way or another are supposed to account for phenomena in terms of purposes or functions. Teleological explanations may be conveniently divided into two types—namely, goal-directed and functional explanations. In *goal-directed explanations* the explanandum is some sort of goal-directed behavior; e.g., the pecking of woodpeckers is explained by reference to the aim of the activity, namely, finding larvae of insects. In a "rational action" species of this sort of explanation, human action is explained by reference to the aim of maximizing estimated utility. In *functional explanations* the explanandum is some feature of a physical system that is explained in terms of its function in the system; e.g., the presence of gills in fish is explained in terms of their respiratory function. For reviews of this literature, see Peter Achinstein (1977); Marjory Grene and Everett Mendelsohn (1976); Ernest Nagel (1977); William C. Wimsatt (1972); and Larry Wright (1976).

Alternative accounts of both types of teleological explanations may be found in the literature, but there seems to be general agreement that goal-directed explanations are easily interpreted as special sorts of causal explanations. Functional explanations, however, seem to be a distinct type. So a fuller analysis of these explanations is warranted.

Michael Ruse has suggested the following example of a functional explanation.

- Cows are well adapted (i.e., they have a good chance of surviving and reproducing).
- Cows are well adapted only if they have the means to feed their very young.
- Cows can feed their very young only if they have udders.
- Therefore, cows have udders.

In this example, the presence of udders on cows is explained in terms of their function of supplying food for the very young. Although the presence of udders is in a sense the cause of cows being able to feed their young, it is the presence of udders that is explained in this explanation. Typically, however, it is some *effect* that is explained in a causal explanation. For example, the bursting of a water pipe was the effect of the water freezing and the other phenomena mentioned in the explanans of our first illustration. Thus, insofar as the roles of cause and effect are structurally reversed in functional explanations, they are logically distinguishable from causal explanations (see Ruse, 1973).

For all the types of explanations reviewed so far it is assumed that an explanation is essentially an argument. This assumption is explicitly rejected in the *statistical relevance model*. According to this view, a statistical explanation of an event is a probability assessment or statement concerning that event, rather than an argument with a statement of the event as its conclusion. Perhaps the easiest way to appreciate the idea behind this model is to compare it with that of the inductive-statistical form of the covering law model. According to the latter, we were able to *argue* to the conclusion (explain the explanandum) with some probability that Jones would have a crop displaying a particular distribution of effects. According to the statistical relevance model, we would be able to *assert* that the explanandum was probable to a certain degree relative to the explanans. The formal methods used in both models could be identical. The difference is in the way one describes what one is doing when one explains an event, namely, arguing or making probability judgments. The statistical relevance model effectively eliminates the possibility of misunderstanding, although the rejection of the idea of an inductive explanation as a kind of argument seems to be a high price to pay for such instruction. This model has been developed by Wesley Salmon (1971; see also his 1975).

Some philosophers think that the attempt to specify syntactic and semantic conditions of adequacy for scientific explanations is fundamentally misguided. The crucial feature of any explanation, they claim, is its capacity to convey understanding. Indeed, as explained in our overviews, some philosophers have held that the fundamental difference between social and physical sciences is that the former aim at an empathic understanding (*Verstehen*) of human action while the latter aim at causal explanations. For *Verstehen* theorists, the syntactic and semantic structures of the sentences that serve as the vehicle for empathic understanding are beside the point. According to these people, an explanation that does not enlighten anyone is as worthless as an argument that does not convince anyone; and social science without *Verstehen* is as worthless as physical science without causal explanations. A good review of the *Verstehen* literature may be found in Fred R. Dallmayr and Thomas A. McCarthy (1977). A view of explanations as "good reasons" may be found in William Dray (1957). On pragmatic aspects of explanation, see Bas Van Fraassen (1977) and Kuhn (1977, pp. 16–18).

What has been presented in the preceding paragraph is an extreme softline view of explanation in stark form. The summary makes it easier to appreciate the relevance of these issues to value questions in science. It is clear that rigorous codification of the structure of scientific explanations would serve the same social purposes as rigorous codification of the logic of scientific discovery. The more the community is able to articulate its accepted models or exemplars, the easier it is to distinguish good science and scientists from everything else, and to distribute rewards and punishments accordingly. Insofar as the view of the extreme softliners can be sustained, the line between the scientific approach toward human understanding and any other approach tends to turn extremely soft. (See Peter Winch, 1958; Mark Roberts, 1976; Charles Taylor, 1971; and Kuhn, 1977; also chapters 7, 8, and 9 in this volume.)

Discussions of the structure of explanation have been regarded as having fundamental significance for the ethical neutrality of science. Insofar as human behavior may be explained in accordance with the deductive nomological form of the covering law model, it would appear that such behavior is completely determined. That is, insofar as the model is satisfiable in its strong form, human freedom of choice is supposed to be eliminated. Furthermore, since such freedom is a necessary condition of the ability to hold anyone responsible for his actions, it is a necessary condition of morality. Thus, the satisfiability of the strong form of the covering law model with respect to the explanation of human behavior has been regarded as having devastating consequences for morality. Those who assume, or merely hypothesize and try to prove, that human behavior can be explained in a deductive nomological form are engaged in the devil's own work. If they succeed, the argument runs, the moral fabric of all societies will be irreparably torn. Thus, at the very least, no one who adopts the hardline position should assume that his research is morally neutral. (See, for example, Paul Ricoeur, 1973; and Charles Taylor, 1964.)

It is perhaps also worth mentioning here that one has a moral obligation to pursue the truth and avoid falsehood. Without it, the obligation to speak the truth is at best silly and at worst virtually impossible to satisfy. Honesty requires a reasonable attempt to find out just what sort of world one lives in. If one happens to be engaged in research as a more or less primary activity, then the obligation is even clearer. Thus, no matter what position one adopts with respect to the structure of explanations, one's research is not going to be morally neutral. For a more thorough discussion of this view, see Michalos (1976).

Finally, some philosophers have argued that the irreducibility of teleological explanations to causal explanations tends to make biological organisms unique sorts of physical entities. However, as indicated above, there now seems to be general agreement that goal-directed explanations—including rational explanations as an important special case—can be analyzed as causal explanations. Even if it is granted that functional explanations do not admit of such analysis, it is not clear what sort of uniqueness that confers on living things. More precisely, it is not clear what evaluative or ethical implications would follow from the fact that, say, some features of planaria, pigs, people, and potted plants all require functional explanations (see Taylor, 1964).

2. Induction, Probability, and Utility

i. *Induction.*   The concepts of induction and deduction were introduced earlier. After problems of explanation, problems of induction are the most popular subjects for university courses in the philosophy of science. Introductory reviews of the literature may be found in Marguerite Foster and M. L. Martin (1966); Henry Kyburg (1957 and 1966); and Brian Skyrms (1966).

The fundamental problem of induction is known as "Hume's Problem," after David Hume, mentioned earlier in our historical survey. It arises as follows. The premises of valid deductive arguments imply their conclusion; that is, one is guilty of self-contradiction to accept the premises but not the conclusions. On the other hand, valid inductive arguments do not provide such a warrant; there is no self-contradiction in accepting the premises while rejecting the conclusion. But, Hume argued, we do accept the conclusions of inductive arguments. What is our warrant for doing so? This is Hume's problem.

There are several ways to attack the problem. First, it may be said that the warrants for accepting the conclusions of valid inductive and deductive arguments are exactly the same. There are rules of inference that warrant conclusions *either* certainly *or* with a certain degree of probability, confirmation, etc. To use these rules and then turn around and reject either the certain conclusions or conclusions with some degree of probability is equally inappropriate. Warranted inferences are inferences made in accordance with accepted rules. Whether the rules permit inductive or deductive inferences, the warranting game is the same. This is basically the view of P. F. Strawson (1952).

Some philosophers have objected to this view on the ground that there are really two ways to warrant an inference—namely, by showing that it is permitted by an accepted rule (as above), and by showing that it is at least as good as any other in an appropriate sense of "good." Moreover, formally correct inductive inferences may be shown to be at least as good as any others. This is a good warrant for inductive procedures generally and, although it is significantly different from the preceding solution to Hume's problem, it is certainly compatible with that solution. Neither precludes the other logically. This pragmatic approach has been recommended by several authors and reviewed by Salmon (1957, 1963).

There are three other proposed solutions that merit some attention. Some people have thought that the way to solve Hume's problem is to add an appropriate universal premise to all inductive arguments. Examples might be "Nature is everywhere uniform," "The unobserved is like the observed," or "The future is like the past." But the addition of such premises to inductive arguments transforms them into deductive arguments, so this solution is more drastic than it may seem. It really amounts to the elimination of inductive in favor of deductive arguments. Moreover, it is not really a solution to the problem of finding a satisfactory warrant for inductive inferences. It is like getting rid of the baby with the bath water. The problem was to find a *warrant* for a particular kind of inference, not just an easy escape. Reviews of this approach may be found in C. D. Broad, (1968); Salmon (1953); G. H. von Wright (1941); and J. O. Wisdom (1952).

Another way to tackle Hume's problem by rejecting inductive inference altogether is to employ the method of "conjectures and refutations." Instead of trying to justify claims that go beyond one's evidence, one should merely try to falsify them. If one seriously tests one's claims, whether they are generalizations or not, one need not worry about providing a warrant for accepting them. Until they are falsified, one has no good reason to reject them. However, and this is the crucial point, one need not accept them as anything but testable hypotheses. One certainly does not have to judge the degree to which they are supported by any accumulated evidence. On the contrary, one merely accepts a commitment to continue to apply the most severe tests one can find to the boldest hypotheses one can find. So the body of scientific knowledge and beliefs will grow on the corpses of discarded hypotheses. As reported above, this is the view of Popper (1959, 1962). Reviews of the approach may be found in Imre Lakatos (1968a and 1968b) and in Lakatos and Alan Musgrave (1970).

Finally, it has been suggested that the way to solve Hume's problem is to codify current inductive practices. Those who hold this view believe quite simply that acceptable rules of inductive inference are just those that are consistent with the kinds of inductive inferences that are presently routinely accepted. For example, the fact that people are prepared to accept arguments that are patterned after the statistical syllogism as inductively valid is, in this view, enough to regard them as warranted. Warranted inferences are inferences that are generally allowed. In fact, the more they are allowed, the more secure their warrants become. In other words, inductive procedures and patterns of inductive inference feed upon themselves. The more they are used, the more they are regarded as useable and, therefore, the more they are used, and so on. This is another view that has been around awhile, at least since Nelson Goodman's *Fact, Fiction, and Forecast* (1954).

Just as Hume found himself embroiled in the problem of induction while attempting to put morality on a sound evidential basis, philosophers of science have suffered the same fate trying to perform the same task for science. Some philosophers have regarded the failure to solve this problem as "the scandal of philosophy," while others have found it equally scandalous that so many mountains have been made out of such a molehill. If one assumes that in one way or another the scientific enterprise is based on inductive procedures, then the whole edifice would be jeopardized by the failure to provide a rational warrant for such procedures. That is precisely the assumption many people have made, with fear for the life of science being an immediate consequence. As indicated above, however, such fears seem unnecessary.

Some other valuable sources on induction include: R. J. Bogdan (1976); Rudolf Carnap and Richard Jeffrey (1971); L. J. Cohen (1970); Jaakko Hintikka and Patrick Suppes (1966); and Grover Maxwell and R. M. Anderson (1975).

ii. *Probability.*   Most of the philosophers of science who have tried to codify patterns of warranted inductive inferences have appealed to the theory of probability. Although rigorous axiomatizations of probability theory were not available until the early part of the twentieth century, the basic principles were well known for over a hundred years before that. Most philosophical work related to probability theory did not center on the principles. That was left to mathematicians and, later, statisticians. Philosophers focused on three fundamental questions. (1) What is the meaning of the term "probability"? (2) How should one obtain numerical probability values? (3) How should the various interpretations, principles, and values be applied to the problem of warranting inductive inferences?

Although much has been written on the paradoxes of the ravens, of lotteries, and of "grue" and "bleen," these technicalities will not be reviewed here. The paradox of the ravens is mainly relevant to a particular view of the inductive probability of theories; lotteries, to a particular sort of rule of probability maximization; and "grue bleen," to predictions involving objects with a particular sort of property.

For general reviews of the literature on probability, see Rudolf Carnap's classic *Logical Foundations of Probability* (1950); also Marguerite Foster and M. L. Martin (1966); Henry Kyburg (1961 and 1970); Alex Michalos (1969); and Nicholas Rescher (1973). On the three paradoxes, see Risto Hilpinen (1968); Israel Scheffler (1963); and Henry Smokler (1966).

What is the meaning of "probability"? Alternatively, one may ask: How should probability statements be interpreted? Examining the possible sources, grounds, or origins of such statements, a threefold genetic classification of possible answers to these questions may be obtained. Probability statements may arise out of and be based upon physical, psychological, or logical considerations. If the statements are generated from an analysis of certain physical properties of the world, probability is given a *physical interpretation*. If the statements are generated from an analysis of certain psychological properties of someone (e.g., a decision maker), probability has a *psychological interpretation*. If the statements are generated from an analysis of certain logical properties of a set of statements in some language, probability has a *logical interpretation*.

Several classifications of interpretations of probability and probability statements have been used in the literature, usually with no criterion of demarcation being systematically employed. The most popular division is the dyadic one consisting of so-called objective and subjective probability. The interpretation referred to here as physical is generally classified as objective. The logical interpretation is sometimes regarded as objective and sometimes as subjective. People who adopt the logical interpretation often use the term "confirmation" instead of "probability."

For those who believe that probability statements are about logical relations holding among statements in some language, two measurement theories merit attention. According to the *classical theory,* one measures the initial probability of an event by determining the ratio of the number of cases that are favorable (for some purpose) to the total number of equally possible events of a certain kind. Since the entire analysis is carried out *a priori,* it is clear that the sort of possibility referred to in the phrase "equally possible" must be logical, and that strictly speaking what the classical theorist is basing his initial probability appraisals on is not the relations among certain events but the relations among statements describing those events. The classical theory is usually attributed to Pierre Laplace, *Théorie analytique des probabilités* (1812).

According to the *logical range theory,* on the other hand, one measures the initial probability of a statement by determining its logical range and adding the initial probabilities of all the state descriptions in that range— the terms "logical range" and "state description" being part of an enormous stock of technical devices that do not lend themselves readily to any but technical interpretation. This is the theory of Rudolf Carnap in *Logical Foundations of Probability* (1950) and *The Continuum of Inductive Methods* (1952).

For those who believe that probability statements are about the physical attributes of certain features of the world, three measurement theories merit attention. According to the *finite frequency theory,* one measures the initial probability of an event regarded as favorable by determining the ratio of the number of such cases to the total number of events in an *observed* sequence of repetitive events of a certain kind. There are three fundamental differences between this theory and those considered so far. In the first place, the analysis is carried out *a posteriori* on the basis of observations. Second, the kinds of events involved are necessarily repetitive. An event is repetitive if it is repeatable an unlimited number of times, e.g., drawing a card from a deck with replacement, rolling a six with a fair die, and kicking a football. In such cases, strictly speaking, it is the kind, sort, or type that is repeated, not any particular event; e.g., one may enjoy as many kicks as one pleases, but one cannot repeat, say, the first kick. In contrast to kicking a football as a kind of repetitive event, the first kick is unique. Unique events, by definition, are not repeatable. Finally, this theory differs from the others considered because it requires sequences of repetitive events. This is the view of Andrei Kolmogorov (1950). Alternative frequency views based on the limit of relative frequencies in random sequences may be found in Hans Reichenbach, *The Theory of Probability* (1949) and Richard von Mises, *Probability, Statistics, and Truth* (1939). Some defects of limiting frequency views are reviewed in Bas Van Fraassen (1977).

According to the *propensity theory*, one measures the initial probability of an event regarded as favorable by determining, on the basis of one's experimental setup and scientific theories, the expected relative frequency of such cases in a sequence of events of a certain kind. The basic difference between this theory and the frequency theory just cited is that here probability statements are not reports about observed sequences, but reports about the tendencies of experimental setups to generate certain kinds of sequences. While the frequentist focuses merely on the outputs of experimental setups, the propensity theorist focuses on the setups themselves and on whatever scientific theories are available to account for their outputs. This is the view of Popper (1959). It is reviewed in Tom Settle (1974 and 1975).

A physical analogue of the logical range theory has also been constructed. Although the *physical range theory* is much more primitive and has never received the attention of its logical counterpart, regarding it as an analogue of the latter is an accurate and easy way to suggest its basic nature. Instead of talking about logical possibilities and ranges, a physical range theorist talks about physical possibilities and ranges. Instead of carrying out his analysis *a priori*, he proceeds *a posteriori*. (See William Kneale, 1949.)

Finally, for those who believe that probability statements are about the psychological properties—attitudes, beliefs, and the like—of someone, two measurement theories merit attention. According to what may be called a *simple judgment theory*, one measures the initial probability of an event by simply assigning it a numerical value on the interval between zero and one, inclusive, in accordance with one's intuitive judgment concerning its occurrence or nonoccurrence.

Alternatively, one may employ a *personal odds theory* according to which, if one is willing to give odds of m to n in favor of some event, then the probability of that event is $m/m + n$. For personal odds views, see Bruno De Finetti (1972); I. J. Good (1975); Henry Kyburg and Henry Smokler (1964); and Leonard J. Savage (1954).

Until around the middle of this century, philosophers haggled about *the* proper interpretation of probability. Since then, one is more likely to find people simply explaining their preference for a particular view, rather than explaining the alleged inadmissibility of alternatives. Although contemporary writers may be inclined to put constraints on probability functions that earlier writers were not aware of, most people seem willing to admit that it may be impossible to obtain initial values in one way and especially useful to obtain them in some other way. The lesson here is familiar. The richer one's stock of alternative theories, the greater one's chances for success. Insofar as philosophical analyses of probability have contributed to the enrichment of our stock of warranted inductive procedures, such analyses have increased our chances of articulating and, hence, critically evaluating inductive procedures in science and daily affairs. Concerning new constraints, see Isaac Levi (1978); Soshichi Uchii (1973); and Van Frassen (1977).

iii. *Utility.*   At least five other authors in the *Guide* have addressed the basic problem of this section in their own terms. A brief review of these other discussions should help set the stage for the philosophical discussion to follow.

Gaston in chapter 7 considers questions of the influence of social phenomena on science, social versus technical norms of scientists' behavior, the value of various risks attached to the acceptance of hypotheses, and internalist-externalist analyses of the growth of science. Engelhardt and Erde in chapter 6 consider the ethical aspects of definitions of diseases, the adaptive value of truth, and the role of values in the choice of diagnostic procedures. Aiken and Freeman in chapter 8 address the problem of the function of knowledge-building versus pragmatic values. Crane in chapter 9 reviews the work of authors who are divided on the questions of research project selection and funding on the basis of narrow discipline oriented versus economically oriented values. Finally, Thackray in chapter 1 discusses intellectual versus social factors in the development of both science and the history of science. From a logical point of view, the central problem underlying all these discussions is the conflict between cognitive and pragmatic (or social) utilities or values—i.e., the subject of this section.

Anyone who has an ordered set of preferences that may be exhaustively measured on an interval scale is said to have a utility function. For some limited areas, provided that they do not contain more than half a dozen items, one may be expected to have such a utility function. However, given the wide variety of things that people value, it would be a rare person indeed who could neatly order her or his total set of preferences. Most people do not have, and probably do not miss, utility functions for all their preferences. Recent thorough discussions of and guides to the literature on utility measurement may be found in W. Edwards and Amos Tversky (1967); P. C. Fishburn (1964); Richard Jeffrey (1965); Robert D. Luce and H. Raiffa (1964); Alex Michalos (1978); and H. J. Skala (1975).

Since preferences are, by anyone's reckoning, closely related to values, it is often assumed that insofar as one has a utility function, one's values are measured on an interval scale. Moreover, by combining utility and probability values, it is possible to increase substantially the variety of one's inductive procedures. The method of combination is straightforward—involving a Maximization of Expected Utility rule—but not necessary to explain here.

It has been suggested that the idea of utility considered here is too general to serve the specific interests of science. After all, the argument runs, the values that are of particular concern to scientists represent only a subset of all the values that people hold. Moral, political, aesthetic, religious, economic, and social values, for example, are supposed to be irrelevant to the scientific enterprise. Hence, if one is going to use the Maximization of Expected Utility Rule to determine the acceptability of scientific hypotheses, one is going to have to put some constraints on one's utility function. More precisely, one must distinguish epistemic from pragmatic utility, and employ only the former in science. Pragmatic utility may be identified with the broader concept with which this discussion began. Epistemic utility requires a bit more explanation.

The epistemic utility or value of a hypothesis is its utility or value from the point of view of the aims of pure or basic science. Without getting bogged down in a debate about the difference between pure and applied (or "mission-oriented") science, one may safely assume that truth is near the top of the list of aims of pure science. Besides truth, defenders of this position claim, there are other epistemic values—e.g., explanatory power, coherence, and precision. So far as the expected utility of a scientific hypothesis is concerned, then, these are the only kinds of values that should be taken into account. Basic literature on epistemic and pragmatic utility or value includes: Hempel (1965); Hilpinen (1968); Isaac Levi (1967); and Michalos (1970).

As one might expect, there is some dispute about the matter. Some people believe that pragmatic values must be considered in the determination of the acceptance of scientific hypotheses. According to these people, the decision to accept or reject a hypothesis is always based, for instance and among other things, on the seriousness of making a mistake. One must take into account the expected utility of accepting a hypothesis that may turn out to be false, and the utility must be as pragmatic as the actions one is likely to perform under the influence of a false belief. That is, because one's scientific beliefs influence one's actions beyond the realm of science, one's assessment of the consequences of holding those beliefs must include an appraisal of the consequences beyond this realm. Hence, the evaluation of the expected utility of scientific hypotheses must be based on pragmatic as well as epistemic utility. Reviews of the "seriousness of mistakes" literature may be found in Baruch Brody (1970) and Philipp Frank (1954).

See also John C. Harsanyi (1976); H. Jungermann and G. de Zeeuw (1977); J. Leach (1968); Leach, Butts, and Pearse (1973); Isaac Levi (1977); Michael Martin (1973); Alex Michalos (1973); Ernest Nagel (1961); George J. Stack (1976); Patrick Suppes (1969); and D. Wendt and C. Vlik (1975).

V. SOCIAL RESPONSIBILITY

The preceding section has taken us slightly beyond the threshold of a discussion of the social responsibilities of scientists *as* scientists. Although we have discussed half a dozen evaluative issues in science to which fundamental problems in the philosophy of science are relevant, the social responsibilities question has not been faced head-on. In one way or another, the question has been addressed in nearly all the chapters of this volume.

As scientists, what, if any, special social responsibilities do scientists have? Since no one has been able to provide precise necessary and sufficient conditions for distinguishing the scientific enterprise from everything else, one should not expect a logically tight answer to this question. Still, several worthwhile points may be made.

In the first place, it must be appreciated that scientists are not immune to the buck-passing syndrome. Most of them will almost certainly be inclined to narrow the range of activities for which they are prepared to accept responsibility and, at the same time, widen the range of activities for which they are prepared to accept authority. Notwithstanding the psychological theory of cognitive dissonance, most human beings seem to manage this particular pair of incompatible inclinations.

Although people in business seem to be the only group blessed with the analytic aphorism, "The business of business is business," others certainly try to have their way in the same fashion, namely, by fiat. In the case of science, the inclination is to come down very hard on the *as scientist* part of our question, thereby paving the way for the narrowest possible purview. Scientists, after all, are not moralists, politicians, social workers. So they need not have the concerns of moralists, politicians, and so on. So the answer to our question is a flat no; scientists *as* scientists have *scientific* responsibilities and that is that.

The trouble with this argument is that it assumes that all concerns or problems can be uniquely sorted into mutually exclusive pigeonholes. On the contrary, most concerns or problems can be regarded as species of several genera. For example, unemployment is an economic, moral, political, and scientific, as well as a social, problem. The task of "correctly" measuring the number of unemployed people in a country or region continues to haunt official and unofficial researchers around the world. In fact, about this problem there remains a considerable disparity of views from one country to the next. Officially unemployed people may be eligible for compensation. Unofficially unemployed people—e.g., housewives—will not usually be eligible. Hidden unemployed people are surely unemployed but not officially unemployed and not eligible for compensation. To be counted as a member of the hidden unemployed is to be counted as a person without hope at best and as a slacker at worst. In either case, because they are no longer trying to find work, they are not officially regarded as unemployed. Their official status thus depends on their desires and the activities in which they engage in the interest of satisfying those desires. Or rather, it depends on some interviewer's perception of those desires and activities. Needless to say, the self-images of the hidden unemployed and unemployed housewives are affected by their employment classification. Indeed, it is unlikely that the self-image of anyone in a work-oriented society is unaffected by her or his employment status. Clearly, then, the question, "Who ought to be regarded as unemployed?" is as much moral, political, economic, and social as it is scientific. Hence, anyone who sets out to measure unemployment scientifically must be aware of, and must make decisions concerning, the propriety and consequences of a number of alternatives. Anyone attempting to measure unemployment without regard for the presumably nonscientific facts of unemployment would be a poor scientist. A good scientist *as* a scientist would address the problem in all its richness. He or she may not be able to manage the problem in that form and may have to introduce arbitrary restrictions in order to manage it at all. But that is not the same as refusing to grapple with its richness on the grounds of its unscientific character, whatever that may be.

Apart from the fact that more or less scientific and unscientific aspects of problems are inextricably woven together, there are independent arguments for attributing special social responsibilities or obligations to scientists *as* scientists.

First, since the results of scientific investigation may be used intentionally to influence or control human action, investigators should at least be required to share some of the responsibility for aberrant uses. Although one may balk at the suggestion that Pavlov should be condemned for all the immoral uses to which operant conditioning has been put, one should not be oblivious to the unseemly side of the social impact of his discovery. Undesirable consequences unleashed by scientific discoveries may be as real as desirable consequences.

Second, if allegedly scientific claims are used to legitimize socioeconomic policies, then the scientists making those claims in behalf of those policies should be held partly responsible for the consequences of the policies if they are put into effect. For example, those who recommended separate tracks in school for black and white or bright and dull students on the basis of their research should be held responsible for the costs as well as the benefits that followed the development of programs consistent with those policies. Whenever social programs are initiated on the strength of the recommendations of scientists, whose recommendations would not be heeded at all if they were not made *as* scientific, the scientists must share the responsibility for the consequences of the programs. If scientists are not held accountable for the consequences of their scientific pronouncements then they will be encouraged to be irresponsible, and they will enjoy an unwarranted social privilege that most people cannot and should not enjoy. These two arguments are used in the document, *Scientific Freedom and Responsibility* (1975), produced by the A.A.A.S. Committee on Scientific Freedom and Responsibility.

Introductory and review literature on science and social responsibility may be found in A. Cornelius Benjamin (1965); Joseph Haberer (' ᷓ9);

A. R. Hall (1956); I. L. Horowitz (1971); W. Truitt and T. Solomons (1974); Harry Woolf (1964); and F. E. West (1974).

Finally, because scientists draw from the same limited resource pool from which the rest of the human race draws, they have an obligation to try to make their demands reasonable from the point of view of the public interest. The assumption behind this argument is that there is no invisible hand operating to allocate the world's resources equitably or even efficiently. Moreover, it is demonstrably certain that if everyone attends only to what he or she perceives as his or her own interests, a socially self-destructive result may occur. That is the clear message of so-called "prisoner's dilemma" studies. It is also the message of two children in a playpen who finally tear the toys apart rather than share them. (See André Cournand, 1977; Alex Michalos, 1978; and Ian Mitroff, 1974.)

Scientists *as* scientists must look beyond their own interests in order to preserve those interests. They must try to assess the total demands on the resource pool that they are tapping in order to avoid what one author has called "the tragedy of the commons." Here, as on our roadways, one must drive defensively. To assume that the "other guy," an elected representative, civil servant, or kind-hearted citizen, is going to be wise enough or morally good enough to balance all interests equitably and efficiently is to reject the lessons of history. The public good is the business of everybody—scientist and nonscientist alike.

See also Paul Feyerabend (1975); Jürgen Habermas (1971); Garrett Hardin (1968); Gerald Holton and William Blanpied (1976); K. D. Knorr, H. Strasser, and H. G. Zilian (1975); Gerard Radnitzky (1968); and L. B. Young and W. J. Trainor (1971).

Acknowledgments

I would like to thank Michael Ruse and Peter Asquith for their efforts with this project; and Paul Durbin, without whose patience, persistence, and extremely hard work this chapter would not be here.

BIBLIOGRAPHIC INTRODUCTION

Although the literature in philosophy of science is vast—too vast to comprehend in a single bibliography—very little of the standard literature bears directly on values in science or in the philosophy of science. A decision was made to let the chapter deal with the value issues and the bibliography present straight philosophy of science. Even so, "straight" or "standard" ought not here to be taken in too literal a sense; an attempt has been made to include in the bibliography all the non-Anglo-American references in section I. B. of this chapter.

The dimensions of the bibliography thus engendered quickly grew beyond tolerable limits for this volume, and another decision had to be made.

It was arbitrarily decided to exclude most references to works not cited in the text, as well as all references prior to 1940.

A.   Journals, Centers, and Societies

A list of philosophy of science journals, centers of study, and societies may be found in Peter Asquith and Henry Kyburg, eds., *Current Research in Philosophy of Science* (1979). That volume, sponsored by the Philosophy of Science Association, is an excellent supplementary source for all aspects of work in philosophy of science in North America.

B.   A List of Basic Works

Here is a list of eleven basic works that ought to count as classics or near-classics on almost anybody's list:

Braithwaite, R. B. *Scientific Explanation*. Cambridge: Cambridge University Press, 1953.

Bunge, M. *Causality*. Cleveland: World, 1963.

Carnap, R. *Logical Foundations of Probability*. Chicago: University of Chicago Press, 1950.

Duhem, P. *The Aim and Structure of Physical Theory*. Princeton: Princeton University Press, 1954. French original 1906.

Habermas, P. *Knowledge and Human Interests*. Boston: Beacon Press, 1971. German original 1968.

Hempel, C. G. *Aspects of Scientific Explanation and Other Essays in the Philosophy of Science*. New York: Free Press, 1965.

Kolmogorov, A. N. *Foundations of the Theory of Probability*. New York: Chelsea, 1950.

Kuhn, T. S. *The Structure of Scientific Revolutions*. Chicago: University of Chicago Press, 1962.

Nagel, E. *The Structure of Science*. New York: Harcourt Brace and World, 1961.

Popper, K. R. *The Logic of Scientific Discovery*. New York: Basic Books, 1959. German original 1935.

Tarski, A. *Logic, Semantics, and Metamathematics*. Oxford: Clarendon Press, 1956.

Note: These works are not cited in the main bibliography.

BIBLIOGRAPHY

A.A.A.S. Committee on Scientific Freedom and Responsibility. *Scientific Freedom and Responsibility*. Washington, D.C.: American Association for the Advancement of Science, 1975.

Abel, T. *The Foundation of Sociological Theory*. New York: Random House, 1970.

Achinstein, P. *Concepts of Science: A Philosophical Analysis*. Baltimore: Johns Hopkins University Press, 1968.

———. *Law and Explanation: An Essay in the Philosophy of Science*. Oxford: Oxford University Press, 1971.

———. "Function Statements." *Philosophy of Science* 44 (1977): 341–367.

Achinstein, P., and Barker, S., eds. *The Legacy of Logical Positivism*. Baltimore: Johns Hopkins University Press, 1969.

Agassi, J. *Science in Flux*. Boston Studies in the Philosophy of Science, vol. 28. Dordrecht: D. Reidel, 1975.

Agazzi, E. *Introduzione ai problemi dell' assiomatica*. Milan, 1961.

———. *Temi e problemi di filosofia della fisica*. Milan, 1969.

———. "Recent Developments of the Philosophy of Science in Italy." *Zeitschrift für allgemeine Wissenschaftstheorie* 3 (1972): 359–371.

Alexander, H. G. "General Statements as Rules of Inference." In *Concepts, Theories, and the Mind-Body Problem*, pp. 309–329. Edited by H. Feigl, M. Scriven, and G. Maxwell. Minneapolis: University of Minnesota Press, 1958.

Alexander, P. *Sensationalism and Scientific Explanation*. London: Routledge and Kegan Paul, 1963.

Anderson, F. H. *The Philosophy of Francis Bacon*. Chicago: University of Chicago Press, 1948.

Anschutz, R. P. *The Philosophy of J. S. Mill*. Oxford: Clarendon Press, 1953.

Anscombe, G. E. M., and Geach, P. *Three Philosophers*. Oxford: Oxford University Press, 1961.

Armstrong, D. M. *Bodily Sensations*. London: Routledge and Kegan Paul, 1964.

———. "Epistemological Foundations for a Materialist Theory of the Mind." *Philosophy of Science* 40 (1973): 178–193.

Arnheim, R. *Visual Thinking*. Berkeley: University of California Press, 1971.

Asquith, P., and Kyburg, H., eds. *Current Research in Philosophy of Science*. East Lansing, Mich.: Philosophy of Science Association, 1979.

Ayer, A. J. *The Problem of Knowledge*. Middlesex: Penguin Books, 1956.

———. "The Conception of Probability as a Logical Relation." In *Observation and Interpretation*, pp. 12–17. Edited by S. Körner. New York: Dover, 1957.

Bachelard, G. *Le Rationalisme appliqué*. Paris, 1949.

Bachelard, S. *La Conscience de rationalité*. Paris: P.U.F., 1958.

Băncilă, O. *Causality in Philosophy and Science*. Bucharest: Editura ştiinţifică, 1969.

Barker, S. F. *Induction and Hypothesis*. Ithaca: Cornell University Press, 1957.

———. *Philosophy of Mathematics*. Englewood Cliffs: Prentice-Hall, 1964.

————. *The Elements of Logic.* New York: McGraw-Hill, 1965.

Bârsan, G. *Time in Science and Philosophy.* Bucharest: Editura ştiinţifică, 1973.

Beck, L. J. *The Method of Descartes: A Study of the Regulae.* Oxford: Clarendon Press, 1952.

Beck, L. W. *Studies in the Philosophy of Kant.* Indianapolis: Bobbs-Merrill, 1965.

Beckner, M. *The Biological Way of Thought.* New York: Columbia University Press, 1959.

Benacerraf, P., and Putnam, H., eds. *Philosophy of Mathematics.* Englewood Cliffs: Prentice-Hall, 1964.

Benjamin, A. C. *Science, Technology, and Human Values.* Columbia: University of Missouri Press, 1965.

Bird, G. *Kant's Theory of Knowledge.* New York: Humanities Press, 1962.

Birnbaum, A. "The Neyman-Pearson Theory as Decision Theory and as Inference Theory: With a Criticism of the Lindley-Savage Argument for Bayesian Theory." *Synthese* 36 (1977): 19–50.

Black, D. *The Theory of Committees and Elections.* Cambridge: Cambridge University Press, 1963.

Blackwell, R. J. *Discovery in the Physical Sciences.* Notre Dame: University of Notre Dame Press, 1969.

Blake, R. M. "Isaac Newton and the Hypothetico-Deductive Method." In *Theories of Scientific Method: The Renaissance Through the Nineteenth Century*, pp. 119–143. Edited by R. M. Blake, C. J. Ducasse and E. H. Madden. Seattle: University of Washington Press, 1960.

Blanché, R. *Recherches sur quelques concepts et méthodes de l'algèbre.* Paris: P.U.F., 1962.

————. *Structures intellectuelles.* Paris: Vrin, 1966.

————. *Raison et discours.* Paris: Vrin, 1967.

————. *Leçons sur la première philosophie de Russel.* Paris: Flammarion, 1967.

————. *La Logique et le monde sensible.* Paris: Flammarion, 1971.

Block, N. "Fictionalism, Functionalism, and Factor Analysis." In *PSA 1974*, pp. 127–142. Edited by R. S. Cohen et al. Dordrecht: D. Reidel, 1976.

————. "Philosophy of Psychology." In *Current Research in Philosophy of Science*, pp. 450–463. Edited by P. Asquith and H. Kyburg. East Lansing, Mich.: Philosophy of Science Association, 1979.

Block, N., and Dworkin, G., "I.Q., Heritability, and Inequality." 2 parts. *Philosophy and Public Affairs* 3 (1974): 331–409; 4 (1974): 40–99.

Boehner, P. *Collected Articles on Ockham.* Edited by E. M. Buytaert. St. Bonaventure, N.Y.: Franciscan Institute Publications, 1958.

Bogdan, R. J., ed. *Local Induction.* Dordrecht: D. Reidel, 1976.

Böhme, G. "Models for the Development of Science." In *Science, Technology and Society: A Cross-Disciplinary Perspective*, pp. 319–351. Edited by I. Spiegel-Rösing and D. de Solla Price. London: Sage, 1977.

Bohnert, H. G. "The Logico-Linguistic Mind-Brain Problem and a Proposed Step Towards Its Solution." *Philosophy of Science* 41 (1974): 1–14.

Borger, R., and Cioffi, F., eds. *Explanation in the Behavioural Sciences.* Cambridge: Cambridge University Press, 1970.

Botezatu, P. *Outline of a Natural Logic.* Bucharest: Editura ştiinţifică, 1969.

Bradie, M. P. "The Causal Theory of Perception." *Synthese* 33 (1976): 41–74.

Brand, M. *The Nature of Causation.* Urbana: University of Illinois Press, 1976.

———, ed. *The Nature of Human Action.* Glenview: Scott, Foresman, 1970.

Broad, C. D. *Induction, Probability, and Causation.* Dordrecht: D. Reidel, 1968.

Brodbeck, M., ed. *Readings in the Philosophy of the Social Sciences.* New York: Macmillan, 1968.

Brody, B. A., ed. *Readings in the Philosophy of Science.* Englewood Cliffs: Prentice-Hall, 1970.

Bromberger, S. "A Theory about the Theory of Theory and about the Theory of Theories." In *Philosophy of Science,* pp. 79–105. The Delaware Seminar, vol. 2, 1962–1963. Edited by B. Baumrin. New York: John Wiley, 1963.

———. "Why-Questions." In *Mind and Cosmos,* pp. 86–111. Edited by R. G. Colodny. Pittsburgh: University of Pittsburgh Press, 1966.

Bub, J. *The Interpretation of Quantum Mechanics.* Dordrecht: D. Reidel, 1974.

Buchdahl, G. "The Relevance of Descartes's Philosophy for Modern Philosophy of Science." *British Journal for the History of Science* 1 (1963): 227–249.

———. "Causality, Causal Laws, and Scientific Theory in the Philosophy of Kant." *British Journal for the Philosophy of Science* 16 (1965): 187–208.

———. *Metaphysics and the Philosophy of Science.* Oxford: Basil Blackwell, 1969.

Buck, R. C., and Cohen, R. S., eds. *PSA 1970: In Memory of Rudolf Carnap.* Boston Studies in the Philosophy of Science, vol. 8. Dordrecht: D. Reidel, 1971.

Bunge, M. *Metascientific Queries.* Springfield, Ill.: Charles C. Thomas, 1959.

———. "The Weight of Simplicity in the Construction and Assaying of Scientific Theories." *Philosophy of Science* 28 (1961): 120–149.

———. *The Myth of Simplicity.* Englewood Cliffs: Prentice-Hall, 1963.

———. *Scientific Research.* 2 vols. New York: Springer Verlag, 1968.

———. *Method, Model, and Matter.* Dordrecht: D. Reidel, 1973a.

———. *Philosophy of Physics*. Dordrecht: D. Reidel, 1973b.

Burks, A. *Chance, Cause, Reason*. Chicago: University of Chicago Press, 1978.

Butts, R. E. "Necessary Truth in Whewell's Philosophy of Science." *American Philosophical Quarterly* 2 (1965): 161–181.

———. "Philosophy of Science in Canada." *Zeitschrift für allgemeine Wissenschaftstheorie* 2 (1974): 341–358.

———, ed. *William Whewell's Theory of Scientific Method*. Pittsburgh: University of Pittsburgh Press, 1968.

Butts, R. E., and Davis, J. W., eds. *The Methodological Heritage of Newton*. Toronto: University of Toronto Press, 1970.

Caldirola, P. "Physics and Philosophy in Italy." In *Contemporary Philosophy: A Survey*, vol. 2: *Philosophy of Science*, pp. 223–231. Edited by R. Klibansky. Florence: La Nuovà Italia Editrice, 1968.

Cannon, W. F. "John Herschel and the Idea of Science." *Journal of the History of Ideas* 22 (1961): 215–239.

Canguilhem, G. *La Formation du concept de réflexe aux XVIIe et XVIIIe siècles*. Paris: P.U.F., 1955.

———. *La Connaissance de la vie*. Paris: Vrin, 1965.

———. *Essai sur quelques problèmes concernant le normal et le pathologique*. Paris: P.U.F., 1966.

———. *Etudes d'histoire et de philosophie des sciences*. Paris: Vrin, 1968.

Čapek, M. *The Philosophic Impact of Contemporary Physics*. Princeton: Princeton University Press, 1961.

———. *Bergson and Modern Physics*. Boston Studies in the Philosophy of Science, vol. 7. Dordrecht: D. Reidel, 1971.

Carcano, P. F. *Antimetafisica e sperimentalismo*. Rome, 1941.

———. *La methodologica nel rinnovarsi del pensiero contemporaneo*. Naples, 1957.

Carnap, R. *The Continuum of Inductive Methods*. Chicago: University of Chicago Press, 1952.

———. *Philosophical Foundations of Physics*. New York: Basic Books, 1966.

Carnap, R., and Jeffrey, R. C., eds. *Studies in Inductive Logic and Probability*. Berkeley: University of California Press, 1971.

Cartwright, N. "Philosophy of Physics." In *Current Research in Philosophy of Science*, pp. 381–385. Edited by P. Asquith and H. Kyburg. East Lansing, Mich.: Philosophy of Science Association, 1978.

Casari, E. *Questioni di filosofia della matematica*. Milan, 1964.

———. "La Logique en Italie." In *Contemporary Philosophy: A Survey*, vol. 1: *Logic and Foundations of Mathematics*, pp. 224–227. Edited by R. Klibansky. Florence: La Nuova Italia Editrice, 1968.

Cassirer, E. *Determinism and Indeterminism in Modern Physics*. New Haven: Yale University Press, 1956.

Castañeda, H. "Perception, Belief, and the Structure of Physical Objects of Consciousness." *Synthese* 35 (1977): 235–252.

Causey, R. L. *Unity of Science*. Dordrecht: D. Reidel, 1977.

——. "Theory and Observation." In *Current Research in Philosophy of Science*, pp. 187–206. Edited by P. Asquith and H. Kyburg. East Lansing, Mich.: Philosophy of Science Association, 1978.

Caws, P. *The Philosophy of Science*. Princeton: Van Nostrand, 1965.

Chappell, V. C., ed. *Hume*. Garden City, N.Y.: Doubleday, 1966.

Claggett, M. *Greek Science in Antiquity*. New York: Collier Books, 1955.

Cohen, L. J. *The Implications of Induction*. London: Methuen, 1970.

Cohen, R. S.; Hooker, C.; Michalos, A. C.; and Van Evra, J., eds. *PSA 1974*. Boston Studies in the Philosophy of Science, vol. 32. Dordrecht: D. Reidel, 1976.

Cohen, R. S., and Seeger, R. J., eds. *Ernst Mach: Physicist and Philosopher*. Boston Studies in the Philosophy of Science, vol. 6. Dordrecht: D. Reidel, 1970.

Cohen, R. S., and Wartofsky, M. W., eds. *Logical and Epistemological Studies in Contemporary Physics*. Boston Studies in the Philosophy of Science, vol. 13. Dordrecht: D. Reidel, 1973.

——. *Methodological and Historical Essays in the Natural and Social Sciences*. Boston Studies in the Philosophy of Science, vol. 14. Dordrecht: D. Reidel, 1974.

Colodny, R. G., ed. *Frontiers of Science and Philosophy*. University of Pittsburgh Series in the Philosophy of Science, vol. 1. Pittsburgh: University of Pittsburgh Press, 1962.

——. *Mind and Cosmos*. University of Pittsburgh Series in the Philosophy of Science, vol. 3. Pittsburgh: University of Pittsburgh Press, 1966.

——. *The Nature and Function of Scientific Theories*. Pittsburgh: University of Pittsburgh Press, 1970.

——. *Paradigms and Paradoxes: The Philosophical Challenge of the Quantum Domain*. Pittsburgh: University of Pittsburgh Press, 1972.

Cornford, F. M. *Plato's Cosmology*. New York: Liberal Arts Press, 1957.

Costa de Beauregard, O. *Théorie synthétique de la relativité restreinte et des quanta*. Paris: Gautier-Villars, 1957.

——. *La Notion de temps, équivalence avec l'espace*. Paris: Hermann, 1963.

——. *Le Second principe de la science du temps*. Paris: Seuil, 1963.

Cournand, A. "The Code of the Scientist and Its Relationship to Ethics." *Science* 198 (1977): 699–705.

Crombie, A. C. *Medieval and Early Modern Science*, vol. 1: *Science in the Middle Ages: V–XIII Centuries;* vol. 2: *Science in the Later Middle Ages and Early Modern Times: XIII–XVII Centuries*. Garden City, N.Y.: Doubleday Anchor, 1959.

Cummins, R. "Functional Analysis." *The Journal of Philosophy* 72 (1975): 741–765.

Cunningham, F. *Objectivity in Social Science*. Toronto: University of Toronto Press, 1973.

Dagognet, F. *La Raison et les remèdes: Essai sur l'imaginaire et le réel dans la therapeutique contemporaine.* Paris: P.U.F., 1964.

———. *Méthode et doctrine dans l'oeuvre de Pasteur.* Paris: P.U.F., 1967.

Dallmayr, F. R. and McCarthy, T. A., eds. *Understanding and Social Inquiry.* Notre Dame: University of Notre Dame Press, 1977.

Danto, A. C. *Analytical Philosophy of History.* Cambridge: Cambridge University Press, 1965.

Danto, A. C., and Morgenbesser, S., eds. *Philosophy of Science.* New York: Meridian Books, 1960.

Darden, L., and Maull, N. "Interfield Theories." *Philosophy of Science* 44 (1977): 43–64.

De Finetti, Bruno. "Probability: The Subjectivistic Approach." In *Contemporary Philosophy: A Survey,* vol. 2: *Philosophy of Science,* pp. 45–53. Edited by R. Klibansky. Florence: La Nuova Italia Editrice, 1968.

———. *Probability, Induction, and Statistics.* New York: John Wiley, 1972.

Diesing, P. *Patterns of Discovery in the Social Sciences.* Chicago: Aldine-Atherton, 1971.

Dijksterhuis, E. J. *The Mechanization of the World Picture.* Oxford: Clarendon Press, 1961.

Dima, T. "The Philosophy of Science in Romania." *Zeitschrift für allgemeine Wissenschaftstheorie* 6 (1975): 355–368.

———. *Inductive Methods.* Bucharest: Editura ştiinţifică. 1975.

Doney, W., ed. *Descartes: A Collection of Critical Essays.* Garden City, N.Y.: Doubleday, 1967.

Dray, W. H. *Laws and Explanation in History.* London: Oxford University Press, 1957.

Dretske, F. I. "Laws of Nature." *Philosophy of Science* 44 (1977): 248–268.

Ducasse, C. J. "Whewell's Philosophy of Scientific Discovery." *Philosophical Review* 60 (1951): 56–69.

———. "John Stuart Mill's System of Logic." In *Theories of Scientific Method: The Renaissance Through the Nineteenth Century,* pp. 218–232. Edited by R. M. Blake, C. J. Ducasse, and E. H. Madden. Seattle: The University of Washington Press, 1960.

———. "Francis Bacon's Philosophy of Science." In *Theories of Scientific Method,* pp. 50–74. (See preceding entry.)

———. "John F. W. Herschel's Methods of Experimental Inquiry." In *Theories of Scientific Method,* pp. 153–182. (See preceding entry.)

Dumitriu, A. *The Solution of Logico-Mathematical Paradoxes.* Bucharest: Editura Academiei, 1966.

Durbin, P. *Logic and Scientific Inquiry.* Milwaukee: Bruce, 1968a.

———. *The Logical Mechanism of Mathematics.* Bucharest: Editura Academiei, 1968.

———. *History of Logic.* Bucharest: Editura didactică şi pedagogică, 1969.

———. *Theory of Logic.* Bucharest: Editura Academei R.S.R., 1973.

————, ed. *Philosophy of Science: An Introduction*. New York: McGraw-Hill, 1968b.

Dürr, K. *Leibniz' Forschungen im Gebiet der Syllogistik*. Berlin: Walter de Gruyter, 1949.

Easton, S. C. *Roger Bacon and His Search for a Universal Science*. New York: Columbia University Press, 1952.

Eberle, R. A. *Nominalistic Systems*. Dordrecht: D. Reidel, 1970.

Edwards, P., ed. *The Encyclopedia of Philosophy*. New York: Macmillan, 1967.

Edwards, W., and Tversky, A., eds. *Decision Making*. Middlesex: Penguin, 1967.

Enescu, G. *Logic and Truth*. Bucharest: Editura politică, 1967.

————. *Philosophy and Logic*. Bucharest: Editura ştiinţifică, 1973.

Evans, M. G. "Causality and Explanation in the Logic of Aristotle." *Philosophy and Phenomenological Research* 19 (1958): 466–485.

Feigl, H. *The "Mental" and the "Physical."* Minneapolis: University of Minnesota Press, 1958.

Feigl, H., and Brodbeck, M., eds. *Readings in the Philosophy of Science*. New York: Appleton-Century-Crofts, 1953.

Feigl, H., and Maxwell, G., eds. *Scientific Explanation, Space, and Time*. Minnesota Studies in the Philosophy of Science, vol. 3. Minneapolis: University of Minnesota Press, 1962.

Feigl, H., and Scriven, M., eds. *The Foundations of Science and the Concepts of Psychology and Psychoanalysis*. Minnesota Studies in the Philosophy of Science, vol. 1. Minneapolis: University of Minnesota Press, 1956.

Feigl, H.; Scriven, M.; and Maxwell, G., eds. *Concepts, Theories, and the Mind-Body Problem*. Minnesota Studies in the Philosophy of Science, vol. 2. Minneapolis: University of Minnesota Press, 1958.

Fetzer, J. A. "Reichenbach, Reference Classes and Single-Case Probabilities." *Synthese* 34 (1977): 185–217.

Feyerabend, P. K. "Explanation, Reduction, and Empiricism." In *Scientific Explanation, Space, and Time*, pp. 28–97. Edited by H. Feigl and G. Maxwell. Minneapolis: University of Minnesota Press, 1962.

————. "How to be a Good Empiricist: A Plea for Tolerance in Matters Epistemological." In *Philosophy of Science*, pp. 3–40. The Delaware Seminar, vol. 2, 1962–1963. Edited by B. Baumrin. New York: John Wiley, 1963.

————. "On the 'Meaning' of Scientific Terms." *The Journal of Philosophy* 62 (1965): 266–274.

————. " 'Science': The Myth and Its Role in Society." *Inquiry* 18 (1975): 167–181.

Finocchiaro, M. A. *History of Science as Explanation*. Detroit: Wayne State University Press, 1973.

Fishburn, P. C. *Decision and Value Theory*. New York: John Wiley, 1964.

Flew, A. *Hume's Philosophy of Belief*. New York: Humanities Press, 1961.

Fodor, J. A. *Psychological Explanation*. New York: Random House, 1968.

———. *The Language of Thought*. New York: Crowell, 1975.

Foster, M. H., and Martin, M. L., eds. *Probability, Confirmation, and Simplicity*. New York: Odyssey Press, 1966.

Foucault, M. *Histoire de la folie*. Paris: Plon, 1961.

———. *Naissance de la clinique*. Paris: P.U.F., 1963.

Frank, P. G., ed. *The Validation of Scientific Theories*. New York: Beacon Press, 1954.

Frey, G. *Gesetz und Entwicklung in der Natur*. Hamburg, 1958.

———. *Erkenntnis der Wirklichkeit; Philosophische Folgerungen der Modernen Naturwissenschaften*. Stuttgart, 1965.

Fried, Y., and Agassi, J. *Paranoia: A Study in Diagnosis*. Dordrecht: D. Reidel, 1976.

Frings, M. S. *Max Scheler: A Concise Introduction into the World of a Great Thinker*. Pittsburgh: University of Pittsburgh Press, 1965.

Gaa, J. "The Replacement of Scientific Theories: Reduction and Explication." *Philosophy of Science* 42 (1975): 349–373.

Gale, G. "The Physical Theory of Leibniz." *Studia Leibnitiana II* 2 (1970): 114–127.

Galtung, J. *Theory and Methods of Social Research*. Oslo: Universitetsforlaget, 1967.

Gardiner, P., ed. *Theories of History*. New York: Free Press, 1959.

Giedymin, J., ed. *Kazimierz Adjukiewicz: The Scientific World-Perspective and Other Essays 1931–1963*. Dordrecht: D. Reidel, 1977.

Giere, R. N. "The Epistemological Roots of Scientific Knowledge." In *Induction, Probability, and Confirmation*, pp. 212–261. Edited by G. Maxwell and R. M. Anderson. Minneapolis: University of Minnesota Press, 1975.

———. "Testing Versus Information Models of Statistical Inference." In *Logic, Laws, and Life*, pp. 19–70. Edited by R. G. Colodny. Pittsburgh: University of Pittsburgh Press, 1977.

———. *Understanding Scientific Reasoning*. New York: Holt, Rinehart and Winston, 1979.

Geschwind, N. *Selected Papers on Language and the Brain*. Boston Studies in the Philosophy of Science, vol. 16. Dordrecht: D. Reidel, 1974.

Geymonat, L. *Studi per un nuovo razionalismo*. Turin, 1945.

———. *Saggi di filosofia neo razionalistica*. Turin, 1953.

———. *Filosofia e filosofia della scienza*. Milan, 1960.

Globus, G. G. "Biological Foundations of the Psychoneural Identity Hypothesis." *Philosophy of Science* 39 (1972): 291–301.

Gochet, P. "Recent Trends in Philosophy of Science in Belgium." *Zeitschrift für allgemeine Wissenschaftstheorie* 6 (1975): 145–163.

Goldman, M. "Perceptual Objects." *Synthese* 35 (1977): 257–284.

Gonseth, F. *La Géometrie et le problème de l'espace*. Neuchâtel, 1945–1956.

———. *Le Problème du temp*. Neuchâtel, 1964.

Good, I. J. "Explicativity, Corroboration, and the Relative Odds of Hypotheses." *Synthese* 30 (1975): 39–74.

Goodman, N. "The Logical Simplicity of Predicates." *The Journal of Symbolic Logic* 14 (1949): 32–41.

———. *Fact, Fiction, and Forecast*. London: Athlone Press, 1954.

Goudge, T. *The Ascent of Life*. Toronto: University of Toronto Press, 1961.

Gouldner, A. W. *The Coming Crisis of Western Sociology*. New York: Basic Books, 1970.

Granger, G. *Méthodologie économique*. Paris: P.U.F., 1955.

———. *Pensée formelle et sciences de l'homme*. Paris: Aubier, 1960.

Granger, G., and Vuillemin, J. "Tendances de la philosophie des sciences en France depuis 1950." In *Contemporary Philosophy: A Survey*, vol. 2: *Philosophy of Science*, pp. 161–163. Edited by R. Klibansky. Florence: La Nuova Italia Editrice, 1968.

Grene, M. *Heidegger*. New York, 1957.

———. "Aristotle and Modern Biology." *Journal of the History of Ideas* 30 (1972): 395–424.

———. *The Understanding of Nature*. Boston Studies in the Philosophy of Science, vol. 23. Dordrecht: D. Reidel, 1974.

Grene, M., and Mendelsohn, E., eds. *Topics in the Philosophy of Biology*. Boston Studies in the Philosophy of Science, vol. 27. Dordrecht: D. Reidel, 1976.

Grünbaum, A. *Geometry and Chronometry in Philosophical Perspective*. Minneapolis: University of Minnesota Press, 1968.

———. "Space, Time, and Falsifiability, Part 1." *Philosophy of Science* 37 (1970): 469–588.

———. *Philosophical Problems of Space and Time*. Boston Studies in the Philosophy of Science, vol. 12. Dordrecht: D. Reidel, 1973.

Gunderson, K. "Cybernetics and Mind-Body Problems." *Inquiry* 12 (1969): 406–419.

Gutting, G. "Continental Philosophy of Science." In *Current Research in Philosophy of Science*, pp. 94–117. Edited by P. Asquith and H. Kyburg. East Lansing, Mich.: Philosophy of Science Association, 1979.

Haberer, J. *Politics and the Community of Science*. New York: Van Nostrand Reinhold, 1969.

Hacking, I. *Logic of Statistical Inference*. Cambridge: Cambridge University Press, 1965.

Hall, E. W. *Modern Science and Human Values*. Princeton: Van Nostrand, 1956.

Hamlyn, D. W. *The Psychology of Perception*. London: Routledge and Kegan Paul, 1969.

Hanna, J. F. "An Interpretive Survey of Recent Research on Scientific Explanation." In *Current Research in Philosophy of Science,* pp. 291–316. Edited by P. Asquith and H. Kyburg. East Lansing: Philosophy of Science Association, 1978.

Hannay, A. *Mental Images: A Defense.* New York: Humanities Press, 1971.

Hanson, N. R. "Is There a Logic of Discovery?" In *Current Issues in the Philosophy of Science,* pp. 20–34. Edited by H. Feigl and G. Maxwell. New York: Holt, Rinehart and Winston, 1961.

———. *Patterns of Discovery.* Cambridge: Cambridge University Press, 1961.

———. *Perception and Discovery.* San Francisco: Freeman, Cooper, 1969.

Hardin, G. "The Tragedy of the Commons." *Science* 162 (1968): 1243–1248.

Harding, S. G., ed. *Can Theories Be Refuted?* Dordrecht: D. Reidel, 1976.

Harris, E. J. *The Foundations of Metaphysics in Science.* New York: Humanities Press, 1965.

Harsanyi, J. *Essays on Ethics, Social Behavior, and Scientific Explanation.* Dordrecht: D. Reidel, 1976.

Harvey, D. *Explanation in Geography.* London: Edward Arnold, 1969.

Hawkins, D. *The Language of Nature.* San Francisco: W. H. Freeman, 1964.

Heathcote, A. W. "William Whewell's Philosophy of Science." *British Journal for the Philosophy of Science* 4 (1953): 302–314.

Heelan, P. A. "Continental Philosophy and Philosophy of Science." In *Current Research in Philosophy of Science,* pp. 84–93. Edited by P. Asquith and H. Kyburg. East Lansing, Mich.: Philosophy of Science Association, 1979.

Heisenberg, W. *Physics and Philosophy.* New York: Harper Torchbooks, 1959.

Hempel, C. G. "Problems and Changes in the Empiricist Criterion of Meaning." *Revue Internationale de Philosophie* 4 (1950): 41–63.

———. "Fundamentals of Concept Formation in Empirical Science." In *International Encyclopedia of Unified Science,* pp. 1–93. Edited by O. Neurath, R. Carnap, and C. Morris. Chicago: University of Chicago Press, 1952.

———. "The Theoretician's Dilemma." In *Concepts, Theories, and the Mind-Body Problem,* pp. 37–98. Edited by H. Feigl, M. Scriven, and G. Maxwell. Minneapolis: University of Minnesota Press, 1958.

———. *Philosophy of Natural Science.* Englewood Cliffs: Prentice-Hall, 1966.

Hesse, M. B. *Models and Analogies in Science.* Notre Dame: University of Notre Dame Press, 1966.

———. *The Structure of Scientific Inference.* London: Macmillan, 1974.

Hilpinen, R. *Rules of Acceptance and Inductive Logic.* Acta Philosophica Fennica, vol. 22. Amsterdam: North-Holland, 1968.

Hintikka, J. "Philosophy of Science (Wissenschaftstheorie) in Finland." *Zeitschrift für allgemeine Wissenschaftstheorie* 1 (1970) 119–132.

Hintikka, J., and Suppes, P., eds. *Aspects of Inductive Logic.* Amsterdam: North-Holland, 1966.

———. *Information and Inference.* Dordrecht: D. Reidel, 1970.

Holton, G. *Thematic Origins of Scientific Thought: Kepler to Einstein.* Cambridge, Mass.: Harvard University Press, 1973.

———. "On Being Caught Between Dionysians and Apollonians." In *Science and Its Public: The Changing Relationship,* pp. 65–28. (1976; See next entry.)

Holton, G., and Blanpied, W. A., eds. *Science and Its Public: The Changing Relationship.* Boston Studies in the Philosophy of Science, vol. 33. Dordrecht: D. Reidel, 1976.

Hooker, C. A., ed. *Contemporary Research in the Foundations and Philosophy of Quantum Theory.* University of Western Ontario Series in the Philosophy of Science, vol. 2. Dordrecht: D. Reidel, 1973.

———. "Philosophy and Meta-Philosophy of Science: Empiricism, Popperianism, and Realism." *Synthese* 32 (1975a): 177–232.

———. "On Global Theories." *Philosophy of Science* 42 (1975b): 152–179.

Howell, R. "Ordinary Pictures, Mental Representations, and Logical Forms." *Synthese* 33 (1976): 149–174.

Howson, C., and Worrall, J. "The Contemporary State of Philosophy in Britain." *Zeitschrift für allgemeine Wissenschaftstheorie* 2 (1974): 363–374.

Hull, C. L. *A Behavior System.* New Haven: Yale University Press, 1952.

Hull, D. L. "Certainty and Circularity in Evolutionary Taxonomy." *Evolution* 21 (1967): 147–189.

———. "The Operational Imperative: Sense and Nonsense in Operationism." *Systematic Zoology* 17 (1968): 438–457.

———. "Reduction in Genetics: Biology or Philosophy." *Philosophy of Science* 39 (1972): 491–499.

———. *Darwin and His Critics: The Reception of Darwin's Theory of Evolution by the Scientific Community.* Cambridge, Mass.: Harvard University Press, 1973.

———. *Philosophy of Biological Science.* Englewood Cliffs: Prentice-Hall, 1974.

———. "Are Species Really Individuals?" *Systematic Zoology* 25 (1976): 174–191.

———. "Informal Aspects of Theory Reduction," In *PSA 1974,* pp. 653–670. Edited by R. S. Cohen et al. Dordrecht: D. Reidel, 1976.

———. "Philosophy of Biology." In *Current Research in Philosophy of Science,* pp. 421–435. Edited by P. Asquith and H. Kyburg. East Lansing, Mich.: Philosophy of Science Association, 1978.

Jacob, F. *La Logique du vivant.* Paris: Gallimard, 1970.

Jeffrey, R. C. *The Logic of Decision.* New York: McGraw-Hill, 1965.

Jobe, E. K. "Some Recent Work on the Problem of Law." *Philosophy of Science* 34 (1967): 363–381.

Johnstone, H. W. "Theory of Argumentation." *In Contemporary Philosophy: A Survey,* vol. 1: *Logic and Foundations of Mathematics,* pp. 177–184. Edited by R. Klibansky. Florence: La Nuova Italia Editrice, 1968.

Joja, A. *Studii de logică.* Bucharest: Editura Academiei, 1960.

Jørgensen, J. *A Treatise of Formal Logic, Its Evolution and Main Branches, with Its Relation to Mathematics and Philosophy.* New York: Russell and Russell, 1962.

Juhos, B. "Methodologie der Naturwissenschaften." In *Contemporary Philosophy: A Survey,* vol. 2: *Philosophy of Science,* pp. 108–120. Edited by R. Klibansky. Florence: La Nuova Italia Editrice, 1968.

Jungermann, H., and de Zeeuw, G., eds. *Decision Making and Change in Human Affairs.* Dordrecht: D. Reidel, 1977.

Kaminski, S. "The Development of Logic and the Philosophy of Science in Poland after the Second World War." *Zeitschrift für allgemeine Wissenschaftstheorie* 8 (1977): 163–171.

Kaplan, A. *The Conduct of Inquiry.* San Francisco: Chandler, 1964.

Kisiel, T. J. "Husserl on the History of Science." In *Phenomenology and the Natural Sciences,* pp. 68–92. Edited by J. J. Kockelmans and T. J. Kisiel. Evanston: Northwestern University Press, 1970a.

———. "Phenomenology as the Science of Science." In *Phenomenology and the Natural Sciences,* pp. 5–44. (1970b; see preceding entry.)

Kisiel, T., and Johnson, J. "New Philosophies of Science in the U.S.A.: A Selective Survey," *Zeitschrift für allgemeine Wissenschaftstheorie* 1 (1974): 138–191.

Klibansky, R., ed. *Contemporary Philosophy: A Survey,* vol. 1: *Logic and Foundations of Mathematics;* vol. 2: *Philosophy of Science.* Florence: La Nuova Italia Editrice, 1968.

Kneale, W. *Probability and Induction.* Oxford: Clarendon Press, 1949.

———. *The Development of Logic.* Oxford: Clarendon Press, 1962.

Knorr, K. D.; Strasser, H.; and Zilian, H. G. eds. *Determinants and Controls of Scientific Development.* Dordrecht: D. Reidel, 1975.

Kockelmans, J. J. *Martin Heidegger: A First Introduction to His Philosophy.* Pittsburgh: Duquesne University Press, 1965.

———. *Phenomenology and Physical Science.* Pittsburgh: Duquesne University Press, 1966.

———. "The Mathematization of Nature in Husserl's Last Publication, *Krisis.*" In *Phenomenology and the Natural Sciences,* pp. 45–67. Edited by J. J. Kockelmans and T. J. Kisiel. Evanston: Northwestern University University Press, 1970.

———, ed. *Philosophy of Science: The Historical Background.* New York: Free Press, 1968.

Kockelmans, J. J., and Kisiel, T. J., eds. *Phenomenology and the Natural Sciences*. Evanston: Northwestern University Press, 1970.

Kohn, H. *The Mind of Germany*. New York: Charles Scribner's Sons, 1960.

Kokoszyńska, M. "Kasimierz Ajdukiewicz." In *Contemporary Philosophy: A Survey*, vol. 2: *Philosophy of Science*, pp. 202–209. Edited by R. Klibansky. Florence: La Nuova Italia Editrice, 1968.

Kordig, C. R. "The Comparability of Scientific Theories." *Philosophy of Science* 38 (1971a): 467–485.

———. *The Justification of Scientific Change*. Dordrecht: D. Reidel, 1971b.

Körner, S. *Kant*. Middlesex: Penguin, 1960.

Kotarbiński, T. *Praxiology: An Introduction to the Sciences of Efficient Action*. Oxford: Oxford University Press, 1965.

———. "L'Évolution de la praxéologie en Pologne." In *Contemporary Philosophy: A Survey*, vol. 2: *Philosophy of Science*, pp. 438–451. Edited by R. Klibansky. Florence: La Nuova Italia Editrice, 1968.

Kraft, V. *Mathematik, Logik und Erfahrung*. Wien, 1947.

———. *Der Wiener Kreis*. Wien, 1952.

———. *Erkenntnislehre*. Wien, 1960.

Krause, M. S. "An Analysis of Festinger's Cognitive Dissonance Theory." *Philosophy of Science* 39 (1972): 32–50.

Kubiński, T. "The Logic of Questions." In *Contemporary Philosophy: A Survey*, vol. 1: *Logic and Foundations of Mathematics*, pp. 185–190. Edited by R. Klibansky. Florence: La Nuova Italia Editrice, 1968.

Kuhn, T. S. *The Copernican Revolution*. Cambridge, Mass.: Harvard University Press, 1957.

———. "Logic of Discovery or Psychology of Research." In *Criticism and the Growth of Knowledge*, pp. 1–22. Edited by I. Lakatos and A. Musgrave. Cambridge: Cambridge University Press, 1970.

———. *The Essential Tension*. Chicago: University of Chicago Press, 1977.

Kyburg, H. E. *Probability and the Logic of Rational Belief*. Middletown, Conn.: Wesleyan University Press, 1961.

———. *Philosophy of Science: A Formal Approach*. New York: Macmillan, 1968.

———. *Probability and Inductive Logic*. New York: Macmillan, 1970.

———. "The Application of Formal Methods in Philosophy of Science." In *Current Research in Philosophy of Science*, pp. 28–39. Edited by P. Asquith and H. Kyburg. East Lansing, Mich.: Philosophy of Science Association, 1978.

Kyburg, H. E., and Smokler, H. E., eds. *Studies in Subjective Probability*. New York: John Wiley, 1964.

Ladrière, J. *Les Limitations internes des formalismes*. Louvain: Nauwelaerts, 1957.

Lakatos, I. "Changes in the Problem of Inductive Logic." In *The Problem of Induction,* pp. 315–417. Edited by I. Lakatos. Amsterdam: North-Holland, 1968a.

——. "Falsification and the Methodology of Research Programmes." In *Criticism and the Growth of Knowledge,* pp. 91–195. Edited by I. Lakatos and A. Musgrave. Cambridge: Cambridge University Press, 1970.

——, ed. *The Problem of Inductive Logic.* Amsterdam: North-Holland, 1968b.

Lakatos, I., and Musgrave, A., eds. *Problems in the Philosophy of Science.* Amsterdam: North-Holland, 1968.

——. *Criticism and the Growth of Knowledge.* Cambridge: Cambridge University Press, 1970.

Laudan, L. "The Nature and Sources of Locke's Views on Hypotheses." *Journal of the History of Ideas* 28 (1967): 211–223.

——. "Theories of Scientific Method from Plato to Mach." *History of Science* 6 (1968): 1–63.

——. "Two Dogmas of Methodology." *Philosophy of Science* 43 (1976): 585–597.

——. *Progress and Its Problems: Toward A Theory of Scientific Growth.* Berkeley: University of California Press, 1977.

——. "Historical Methodologies: An Overview and Manifesto." In *Current Research in Philosophy of Science,* pp. 40–54. Edited by P. Asquith and H. Kyburg. East Lansing, Mich.: Philosophy of Science Association, 1978.

Lauener, H. "Wissenschaftstheorie in der Schweiz." *Zeitschrift für allgemeine Wissenschaftstheorie* 2 (1971): 291–317.

Leach, J. "Explanation and Value Neutrality." *British Journal for the Philosophy of Science* 19 (1968): 93–108.

Leach, J.; Butts, R.; and Pearce, G., eds. *Science, Decision, and Value.* University of Western Ontario Series in Philosophy of Science. Dordrecht: D. Reidel, 1973.

Leblanc, H. *Statistical and Inductive Probabilities.* Englewood Cliffs: Prentice-Hall, 1962.

Lehman, H. "Functional Explanation in Biology." *Philosophy of Science* 32 (1965): 1–20.

——. *Introduction to the Philosophy of Mathematics.* Oxford: Basil Blackwell, 1978.

Levi, I. *Gambling With Truth.* New York: Alfred A. Knopf, 1967.

——. "Epistemic Utility and the Evaluation of Experiments." *Philosophy of Science* 44 (1977): 368–386.

——. "Inductive Appraisal." In *Current Research in Philosophy of Science,* pp. 339–352. Edited by P. Asquith and H. Kyburg. East Lansing, Mich.: Philosophy of Science Association, 1978.

Løchen, Y. *The Sociologist's Dilemma.* Oslo: Gyldendal Norsk Forlag, 1970.

Losee, J. *A Historical Introduction to the Philosophy of Science*. London: Oxford University Press, 1972.

Luce, H. D., and Raiffa, H. *Games and Decisions: Introduction and Critical Survey*. New York: John Wiley, 1964.

Lwoff, A. *L'Ordre biologique*. Paris: Laffont, 1969.

MacIntyre, A. *The Unconscious*. London: Routledge and Kegan Paul, 1962.

Mackie, J. L. *The Cement of the Universe: A Study of Causation*. Oxford: Oxford University Press, 1974.

Maehara, S. "Logic in Japan." In *Contemporary Philosophy: A Survey*, vol. 1: *Logic and Foundations of Mathematics*, pp. 228–231. Edited by R. Klibansky. Florence: La Nuova Italia Editrice, 1968.

Mare, C. *Determinism and Modern Physics*. Bucharest: Editura politică, 1966.

Martin, C. B., and Armstrong, D. M., eds. *Locke and Berkeley*. Garden City, N.Y.: Doubleday, 1968.

Martin, M. "Value Judgments and the Acceptance of Hypotheses in Science and Science Education." *Philosophic Exchange* 1 (1973): 83–102.

Maxwell, G. "The Ontological Status of Theoretical Entities." In *Scientific Explanation, Space, and Time*, pp. 3–27. Edited by H. Feigl and G. Maxwell. Minneapolis: University of Minnesota Press, 1962.

Maxwell, G., and Anderson, R. M., eds. *Induction, Probability, and Confirmation*. Minnesota Studies in the Philosophy of Science, vol. 6. Minneapolis: University of Minnesota Press, 1975.

Maxwell, N. "The Rationality of Scientific Discovery. Part 1: The Traditional Rationality Problem." *Philosophy of Science* 41 (1974): 123–153.

——. "The Rationality of Scientific Discovery, Part 2: An Aim Oriented Theory of Scientific Discovery." *Philosophy of Science* 41 (1974): 247–295.

Mayr, E. "Teleological and Teleonomic: A New Analysis." In *Methodological and Historical Essays in the Natural and Social Sciences*, pp. 91–118. Boston Studies in the Philosophy of Science, vol. 14. Edited by R. S. Cohen and M. W. Wartofsky. Dordrecht: D. Reidel, 1974.

McKeon, R. P. "Aristotle's Conception of the Development and the Nature of Scientific Method." *Journal of the History of Ideas* 8 (1947): 3–44.

McLaughlin, A. "Rationality and Total Evidence." *Philosophy of Science* 37 (1970): 271–278.

McMullin, E. "The Nature of Scientific Knowledge: What Makes It Science." In *Philosophy in a Technological Culture*, pp. 28–54. Edited by G. McLean. Washington, D.C.: Catholic University of America Press, 1964.

——. "Recent Work in the Philosophy of Science." *New Scholasticism* 40 (1966): 478–518.

——. "Philosophies of Nature." *New Scholasticism* 43 (1968): 29–74.

——. "The History and Philosophy of Science: A Taxonomy." In *Histori-*

cal and Philosophical Perspectives of Science, pp. 12–67. Edited by R. H. Stuewer. Minneapolis: University of Minnesota Press, 1970.

——. "Historical Methodologies." In *Current Research in Philosophy of Science,* pp. 55–83. Edited by P. Asquith and H. Kyburg. East Lansing, Mich.: Philosophy of Science Association, 1978.

McMurtry, J. *The Structure of Marx's World-View.* Princeton: Princeton University Press, 1978.

Mehlberg, H. *The Reach of Science.* Toronto: University of Toronto Press, 1958.

Menges, G., ed. *Information, Inference, and Decision.* Dordrecht: D. Reidel, 1974.

Merleau-Ponty, J. *Cosmologie du XXe siècle.* Paris: Gallimard, 1965.

Merleau-Ponty, J., and Morando, B. *Le Trois étapes de la cosmologie.* Paris: Laffont, 1971.

Merleau-Ponty, M. *La Structure du comportement.* Paris, 1942.

——. *Phénoménologie de la perception.* Paris, 1945.

Meyer, F. *Problématique de l'évolution.* Paris: P.U. F., 1954.

Michalos, A. C. "Positivism versus the Hermeneutic-Dialectic School." *Theoria* 35 (1969): 267–278.

——. *Principles of Logic.* Englewood Cliffs: Prentice-Hall, 1969.

——. *Improving Your Reasoning.* Englewood Cliffs: Prentice-Hall, 1970.

——. "Cost-Benefit Versus Expected Utility Acceptance Rules." *Theory and Decision* 1 (1970): 61–88.

——. "The Impossibility of an Ordinal Measure of Acceptability." *The Philosophical Forum* 2 (1970): 103–106.

——. *The Popper-Carnap Controversy.* The Hague: Martinus Nijhoff, 1971.

——. "Efficiency and Morality." *The Journal of Value Inquiry* 2 (1972): 137–143.

——. "Values in Science and Science Education." *Philosophic Exchange* 1 (1973): 103–108.

——. "Measuring the Quality of Life." In *Values and the Quality of Life,* pp. 24–37. Edited by J. King-Farlow and W. R. Shea. New York: Science History Publications, 1976a.

——. "The Morality of Cognitive Decision-Making." In *Action Theory,* pp. 325–340. Edited by M. Brand and D. Walton. Dordrecht: D. Reidel, 1976b.

——. *Foundations of Decision-Making.* Ottawa: Canadian Association for Publishing in Philosophy, 1978.

——. "Philosophy of Social Science." In *Current Research in Philosophy of Science,* pp. 463–502. Edited by P. Asquith and H. Kyburg. East Lansing, Mich.: Philosophy of Science Association, 1979.

——, ed. *Philosophical Problems of Science and Technology.* Boston: Allyn and Bacon, 1974.

Mihu, A. *The A.B.C. of Sociological Investigation.* Cluj-Napoca: Editura Dacia, 1971.

Mitroff, I. I. "Solipsism: An Essay in Psychological Philosophy." *Philosophy of Science* 38 (1971): 376–394.

——. *The Subjective Side of Science.* New York: American Elsevier, 1974.

Mittelstaedt, P. *Philosophical Problems of Modern Physics.* Boston Studies in the Philosophy of Science, vol. 18. Dordrecht: D. Reidel, 1976.

Moisil, G. C. *Old and New Essays of Non-Classical Logic.* Bucharest: Editura ştiinţifică, 1965.

——. "La Logic mathématique dans le République Socialiste de Roumanie." In *Contemporary Philosophy: A Survey,* vol. 1: *Logic and Foundations of Mathematics,* pp. 220–223. Edited by R. Klibansky. Florence: La Nuova Italia Editrice, 1968.

Moody, E. A. *The Logic of William of Ockham.* New York: Russell and Russell, 1965.

Mulkay, M. J. "Sociology of the Scientific Research Community." In *Science, Technology and Society: A Cross-Disciplinary Perspective,* pp. 93–148. Edited by I. Spiegel-Rösing and D. de Solla Price. London: Sage, 1977.

Munson, R. "Biological Adaptation." *Philosophy of Science* 38 (1971): 200–215.

——, ed. *Man and Nature: Philosophical Issues in Biology.* New York: Delta, 1971.

Naess, A. *The Pluralist and Possibilist Aspect of the Scientific Enterprise.* London: Allen and Unwin, 1972.

Nagai, H. *Formation of the Modern Philosophy of Science.* Tokyo, 1955.

——. *The Ontological Foundation of Mathematics.* Sobunsha, 1962.

——. *Studies in Contemporary Philosophy of Nature.* Sobunsha, 1963.

——. "Recent Trends in Japanese Research on the Philosophy of Science." *Zeitschrift für allgemeine Wissenschaftstheorie* 2 (1971): 101–114.

Nagai, S., and Kurosaki, H. *Fundamentals of Philosophy of Science.* Tokyo, 1967.

Nagel, E. "Teleology Revisited." *The Journal of Philosophy* 74 (1977): 261–302.

Nagel, E.; Suppes, P.; and Tarski, A., eds. *Logic, Methodology, and Philosophy of Science: Proceedings of the 1960 International Congress.* Stanford, Calif.: Stanford University Press, 1962.

Nakamura, H. *Basis of the Philosophy of Science.* Tokyo, 1969.

Nelkin, D. *Science Textbook Controversies and the Politics of Equal Time.* Cambridge, Mass.: M.I.T. Press, 1977.

Nelson, R. J. "On Mechanical Recognition." *Philosophy of Science* 43 (1976): 24–52.

Neurath, O.; Carnap, R.; and Morris, C., eds. *Foundations of the Unity of Science.* 2 vols. Chicago: University of Chicago Press, 1970.

Nickles, T. "Covering Law Explanation." *Philosophy of Science* 38 (1971): 542–561.

Niiniluoto, I., and Tuomela, R. *Theoretical Concepts and Hypothetico-Inductive Inference.* Dordrecht: D. Reidel, 1973.

Nordenstam, T. and H. Skjervheim. "Philosophy of Science in Norway." *Zeitschrift für allgemeine Wissenschaftstheorie* 4 (1973): 147–164.

Nowak, S. *Methodology of Sociological Research.* Dordrecht: D. Reidel, 1977.

Omodaka, H. *An Introduction to Science.* Tokyo, 1955.

———. *The Philosophy of Medicine.* Tokyo, 1964.

Partington, J. R. *A Short History of Chemistry.* London: Macmillan, 1937.

Pasquinelli, A. *Nuovi principi di epistemologia.* Milan, 1964.

Passmore, J. A. "William Harvey and the Philosophy of Science." *Australasian Journal of Philosophy* 36 (1958): 85–94.

Pavelcu, V. *The Drama of Psychology.* Bucharest: Editura didactică şi pèdagogică, 1964.

Perelman, Ch., and Olbrecht-Tyteca, L. *La Nouvelle rhétorique: Traité de l'argumentation.* Paris, 1958.

Peterson, A. "Niels Bohr and the Philosophy of Science." In *Contemporary Philosophy: A Survey,* vol. 2: *Philosophy of Science,* pp. 277–285. Edited by R. Klibansky. Florence: La Nuova Italia Editrice, 1968.

Piaget, J. *The Child's Conception of Number.* London: Routledge and Kegan Paul, 1952.

———. *The Child's Conception of Space.* London: Routledge and Kegan Paul, 1956.

———. *The Growth of Logical Thinking.* London: Routledge and Kegan Paul, 1958.

———. *The Child's Conception of Geometry.* London: Routledge and Kegan Paul, 1960.

Pitcher, G. *A Theory of Perception.* Princeton: Princeton University Press, 1971.

Polya, George. *How to Solve It.* Princeton: Princeton University Press, 1945; 2d rev. ed., Garden City, N.Y.: Doubleday, 1957.

———. *Mathematics and Plausible Reasoning.* 2 vols. Princeton: Princeton University Press, 1954.

Popa, C. *The Theory of Definition.* Bucharest: Editura ştiinţifică 1972.

Popper, K. R. "A Note on Berkeley as Precursor of Mach." *British Journal for the Philosophy of Science* 4 (1953): 26–36.

———. *The Poverty of Historicism.* London: Routledge and Kegan Paul, 1957.

———. "The Propensity Interpretation of Probability." *The British Journal for the Philosophy of Science* 10 (1959): 25–42.

———. *Conjectures and Refutations.* New York: Basic Books, 1962.

———. "The Demarcation Between Science and Metaphysics," In *The Phi-*

losophy of Rudolf Carnap, pp. 183–226. The Library of Living Philosophers. Edited by P. A. Schilpp. LaSalle: Open Court, 1963.

———. "Remarks on the Problems of Demarcation and of Rationality." In *Problems in the Philosophy of Science,* pp. 88–102. Edited by I. Lakatos and A. Musgrave. Amsterdam: North-Holland, 1968.

———. *Objective Knowledge.* Oxford: Oxford University Press, 1972.

Powers, W. T. *Behavior: The Control of Perception.* Chicago: Aldine-Atherton, 1973.

Preti, G. *Idealismo e positivismo.* Milan, 1952.

———. *Praxis ed empirismo.* Turin, 1957.

Puccetti, R. "Brain Bisection and Personal Identity." *The British Journal for the Philosophy of Science* 24 (1973): 339–355.

Putnam, H. *Philosophy of Logic.* New York: Harper & Row, 1972.

———. *Mind, Language, and Reality.* Cambridge: Cambridge University Press, 1975.

———. *Mathematics, Matter, and Method.* Cambridge: Cambridge University Press, 1976.

———. "Philosophy of Mathematics: A Report." In *Current Research in Philosophy of Science,* pp. 386–398. Edited by P. Asquith and H. Kyburg. East Lansing, Mich.: Philosophy of Science Association, 1978.

Quine, W. V. O. *Ontological Relativity.* New York: Columbia University Press, 1969.

———. *Philosophy of Logic.* Englewood Cliffs: Prentice-Hall, 1970.

Radne, M. and Winokur, S., ed. *Analyses of Theories and Methods of Physics and Psychology.* Minnesota Studies in the Philosophy of Science, vol. 4. Minneapolis: University of Minnesota Press, 1970.

Radnitzky, G. *Continental Schools of Metascience.* Göteborg: Akademiforlaget, 1968.

Radnitzky, G.; Törnebohm, H.; and Wallén, G. "Wissenschaftstheorie als Forschungswissenschaft." *Zeitschrift für allgemeine Wissenschaftstheorie* 2 (1971): 115–119.

Rapoport, A. "Various Meanings of Theory." *American Political Science Review* 52 (1958): 972–988.

Reeves, C. *The Psychology of Rollo May.* San Francisco: Jossey-Bass, 1977.

Reichenbach, H. *The Theory of Probability.* Berkeley: University of California Press, 1949.

Renoirte, F. *Eléments de critique des sciences et de cosmologie.* Louvain, 1947.

Rescher, N. *Introduction to Logic.* New York: St. Martin's Press, 1964.

———. *The Philosophy of Leibniz.* Englewood Cliffs: Prentice-Hall, 1967.

———. *Introduction to Value Theory.* Englewood Cliffs: Prentice-Hall, 1969.

———. *The Coherence Theory of Truth.* Oxford: Oxford University Press, 1973.

———. *Plausible Reasoning.* Atlantic Highlands, N.J.: Humanities, 1976.

Richelle, M. "Formal Analysis and Functional Analysis of Verbal Behavior: Notes on the Debate Between Chomsky and Skinner." *Behaviorism* 4 (1976): 209–221.

Ricoeur, P. "A Critique of B. F. Skinner's *Beyond Freedom and Dignity*." *Philosophy Today* 17 (1973): 166–175.

Roberts, M. J. "On the Nature and Condition of Social Science." In *Science and Its Public: The Changing Relationship,* pp. 47–64. Edited by G. Holton and W. A. Blanpied. Dordrecht: D. Reidel, 1976.

Rosen, D. A. "An Argument for the Logical Notion of a Memory Trace." *Philosophy of Science* 42 (1975): 1–10.

Rosenberg, A. *Microeconomic Laws: A Philosophical Analysis.* Pittsburgh: University of Pittsburgh Press, 1976.

Rossi, P. *Francis Bacon: From Magic to Science.* London: Routledge and Kegan Paul, 1968.

Royce, J. R., and Rozeboom, W. W., eds. *The Philosophy of Knowing.* New York: Gordon and Breach, 1972.

Rozeboom, W. W. "Dispositions Revisited." *Philosophy of Science* 40 (1973): 59–74.

Rudner, R. S. *Philosophy of Social Science.* Englewood Cliffs: Prentice-Hall, 1966.

Runciman, W. G. *A Critique of Max Weber's Philosophy of Social Science.* Cambridge: Cambridge University Press, 1972.

Ruse, M. E. "Confirmation and Falsification of Theories of Evolution." *Scientia* 104 (1969): 1–29.

——. "Are There Laws in Biology?" *Australasian Journal of Philosophy* 48 (1970): 234–246.

——. "The Revolution in Biology." *Theoria* 36 (1970): 1–22.

——. "Is the Theory of Evolution Different?" *Scientia* 106 (1971): 1–42.

——. "Natural Selection in *The Origin of Species*." *Studies in History and Philosophy of Science* 4 (1971): 311–351.

——. "Two Biological Revolutions." *Dialectica* 25 (1971): 17–38.

——. "Reduction, Replacement, and Molecular Biology." *Dialectica* 25 (1971): 39–72.

——. "The Nature of Scientific Models: Formal vs. Material Analogy." *Philosophy of the Social Sciences* 3 (1973): 63–80.

——. *The Philosophy of Biology.* London: Hutchinson, 1973.

——. "Woodger on Genetics." *Acta Biotheoretica* 24 (1975): 1–13.

——. "Reduction in Genetics." In *PSA 1974,* pp. 633–651. Edited by R. S. Cohen et al. Dordrecht: D. Reidel, 1976.

——. "Sociobiology: Sound Science or Muddled Metaphysics?" In *PSA 1976,* vol. 2, pp. 48–76. Edited by F. Suppes and P. D. Asquith. East Lansing, Mich.: Philosophy of Science Association, 1977.

Russell, B. *Human Knowledge.* New York: Simon and Schuster, 1948.

Ruyer, R. *Neo-finalisme.* Paris: P.U.F., 1954.

Ruytinx, J. *La Problématique de l'unité de la science.* Paris, 1962.

Ryan, A. *The Philosophy of John Stuart Mill.* London: Macmillan, 1970.

Rychlak, J. F. *A Philosophy of Science for Personality Theory.* Boston: Houghton Mifflin, 1968.

Salmon, W. C. "The Uniformity of Nature." *Philosophy and Phenomenological Research* 14 (1953): 39–48.

———. "Should We Attempt to Justify Induction?" *Philosophical Studies* 8 (1957): 33–48.

———. "On Vindicating Induction." *Philosophy of Science* 30 (1963): 252–261.

———. *The Foundations of Scientific Inference.* Pittsburgh: University of Pittsburgh Press, 1966.

———. "Theoretical Explanation." In *Explanation,* pp. 118–145. Edited by S. Körner. Oxford: Basil Blackwell, 1975.

———, ed. *Statistical Explanation and Statistical Relevance.* Pittsburgh: University of Pittsburgh Press, 1971.

Savage, L. J. *The Foundations of Statistics.* New York: John Wiley, 1954.

Sawada, N. *Philosophy and Logic in Our Time: Logical Analysis and Philosophical Synthesis.* Tokyo, 1964.

———. *The Structure of Knowledge: Conquest of Dogma and Scientific Thought.* Tokyo, 1969.

Schaff, A. *Some Problems of the Marxist Theory of Truth.* Warsaw: KiW, 1951.

———. *An Introduction to Semantics.* Warsaw: P.W.N., 1960.

———. *Language and Cognition.* Warsaw: P.W.N., 1964.

Schaffner, K. "Approaches to Reduction." *Philosophy of Science* 34 (1967): 137–147.

———. "The Watson-Crick Model and Reductionism." *British Journal for the Philosophy of Science* 20 (1969); 325–348.

———. "Reductionism in Biology: Prospects and Problems." In *PSA 1974,* pp. 613–632. Edited by R. S. Cohen et al. Dordrecht: D. Reidel, 1976.

Schaffner, K., and Cohen, R. S., eds. *PSA 1972.* Boston Studies in the Philosophy of Science, vol. 20. Dordrecht: D. Reidel, 1974.

Scheffler, I. "Explanation, Prediction, and Abstraction." *The British Journal for the Philosophy of Science* 7 (1957): 293–309.

———. *The Anatomy of Inquiry.* New York: Alfred A. Knopf, 1963.

———. *Science and Subjectivity.* New York: Bobbs-Merrill, 1967.

Scherrer, W. *Exakte Begriffe: Eine Kurzgefasste Erkenntnistheorie der Exakten Wissenschaften.* Berne, 1958.

Schilpp, P. A. *The Philosophy of Ernst Cassirer.* New York: Tudor, 1949.

Schilpp, P. A., ed. *The Philosophy of Rudolf Carnap.* The Library of Living Philosophers. LaSalle, Ill.: Open Court, 1963.

Scriven, M. "The Key Property of Physical Laws Is Inaccuracy." In *Current*

Issues in the Philosophy of Science, pp. 91–101. Edited by H. Feigl and G. Maxwell. New York: Holt, Rinehart and Winston, 1961.

——. "Explanations, Predictions, and Laws." In *Scientific Explanation, Space, and Time.* pp. 170–230. Edited by H. Feigl and G. Maxwell. Minneapolis: University of Minnesota Press, 1962.

——. "The Frontiers of Psychology: Psychoanalysis and Parapsychology." In *Frontiers of Science and Philosophy,* pp. 81–106. Edited by R. Colodny. Pittsburgh: University of Pittsburgh Press, 1962.

Seeger, R. J., and Cohen, R. S., eds. *Philosophical Foundations of Science.* Boston Studies in the Philosophy of Science, vol. 11. Dordrecht: D. Reidel, 1974.

Settle, T. W. "Induction and Probability Unfused." In *The Philosophy of Karl R. Popper,* pp. 697–749. Edited by P. A. Schilpp. LaSalle, Ill.: Open Court, 1974.

——. "Presuppositions of Propensity Theories of Probability." In *Induction, Probability, and Confirmation,* pp. 388–415. Edited by G. Maxwell and R. M. Anderson. Minneapolis: University of Minnesota Press, 1975.

——. *In Search of a Third Way.* Toronto: McClelland and Stewart, 1976.

Shapere, D. "Meaning and Scientific Change." In *Mind and Cosmos,* pp. 41–85. Edited by R. Colodny. Pittsburgh: University of Pittsburgh Press, 1966.

Shapiro, H. *Motion, Time, and Place According to William Ockham.* St. Bonaventure, N.Y.: Franciscan Institute Publications, 1957.

Shrader-Frechette, K. "Atomism in Crisis: An Analysis of the Current High Energy Paradigm." *Philosophy of Science* 44 (1977): 409–440.

Simon, H. A. *Models of Man.* New York: John Wiley, 1957.

——. "The Axiomatization of Physical Theories." *Philosophy of Science* 37 (1970): 16–26.

——. "Does Scientific Discovery Have a Logic?" *Philosophy of Science* 40 (1973): 471–480.

Skala, H. J. *Non-Archimedean Utility Theory.* Dordrecht: D. Reidel, 1975.

Skjervheim, H. *Objectivism and the Study of Man.* Oslo: Universitetsforlaget, mimeo, 1959.

Sklar, L. "Types of Inter-Theoretic Reduction." *The British Journal for the Philosophy of Science* 18 (1967): 109–124.

——. *Space, Time, and Spacetime.* Berkeley: University of California Press, 1974.

Skyrms, B. *Choice and Chance.* Belmont, Calif.: Dickenson, 1966.

Słupecki, J. "Logic in Poland." In *Contemporary Philosophy: A Survey,* vol. 1: *Logic and Foundations of Mathematics,* pp. 190–201. Edited by R. Klibansky. Florence: La Nuova Italia Editrice, 1968.

Smart, J. J. C. *Between Science and Philosophy.* New York: Random House, 1968.

Smokler, H. "Goodman's Paradox and the Problem of Rules of Acceptance." *American Philosophical Quarterly* 3 (1966): 71–76.

Sneed, J. D. *The Logical Structure of Mathematical Physics.* Dordrecht: D. Reidel, 1971.

Sober, E. "Mental Representations." *Synthese* 33 (1976): 101–148.

Solmsen, F. *Aristotle's System of the Physical World.* Ithaca: Cornell University Press, 1960.

Somenzi, V. *La filosifia degli automi.* Turin, 1965.

———. *La fisica della mente.* Turin, 1969.

Spector, M. "Models and Theories." *The British Journal for the Philosophy of Science* 16 (1965): 121–142.

———. *Methodological Foundations of Relativistic Mechanics.* Notre Dame: University of Notre Dame Press, 1972.

———. "Russell's Maxim and Reduction as Replacement." *Synthese* 32 (1975): 135–176.

Spiegelberg, H. *The Phenomenological Movement.* The Hague: Mouton, 1960.

Stack, G. J. "Value and Sociological Inquiry." *The Journal of Value Inquiry* 10 (1976): 220–233.

Stegmüller, W. *Main Currents in Contemporary German, British, and American Philosophy.* Dordrecht: D. Reidel, 1969.

———. *The Structure and Dynamics of Theories.* New York: Springer-Verlag, 1976.

———. *Collected Papers on Epistemology, Philosophy of Science, and History of Philosophy.* Dordrecht: D. Reidel, 1977.

Stove, D. C. *Probability and Hume's Inductive Scepticism.* Oxford: Clarendon Press, 1973.

Strawson, P. F. *Introduction to Logical Theory.* New York: John Wiley, 1952.

———. *The Bounds of Sense: An Essay on Kant's "Critique of Pure Reason."* London: Methuen, 1966.

Strong, E. W. "William Whewell and John Stuart Mill: Their Controversy about Scientific Knowledge." *Journal of the History of Ideas* 16 (1955): 209–231.

Stuewer, R. H., ed. *Historical and Philosophical Perspectives of Science.* Minnesota Studies in the Philosophy of Science, vol. 5. Minneapolis: University of Minnesota Press, 1970.

Suppe, F. R. "What's Wrong with the Received View on the Structure of Scientific Theories." *Philosophy of Science* 39 (1972): 1–20.

———. "Theory Structure." In *Current Research in Philosophy of Science,* pp. 317–338. Edited by P. Asquith and H. Kyburg. East Lansing, Mich.: Philosophy of Science Association, 1978.

———, ed. *The Structure of Scientific Theories.* Urbana: University of Illinois Press, 1974.

Suppe, F. R., and Asquith, P. D., eds. *PSA 1976*. 2 vols. East Lansing, Mich.: Philosophy of Science Association, 1976, 1977.

Suppes, P. *Studies in the Methodology and Foundations of Science*. Dordrecht: D. Reidel, 1969.

———. *A Probabilistic Theory of Causality*. Acta Philosophica Fennica, vol. 24. Amsterdam: North-Holland, 1970.

———. "Is Visual Space Euclidean?" *Synthese* 35 (1977): 397–422.

———. "Formal Methodologies." In *Current Research in Philosophy of Science*, pp. 16–27. Edited by P. Asquith and H. Kyburg. East Lansing, Mich.: Philosophy of Science Association, 1979.

Swoyer, C. and T. C. Monson. "Theory Confirmation in Psychology." *Philosophy of Science* 42 (1975): 487–502.

Taketani, M. *Problems on Dialectics*. Tokyo, 1946–1952.

Taylor, C. *The Explanation of Behaviour*. London: Routledge and Kegan Paul, 1964.

———. "Interpretation and the Sciences of Man." *Review of Metaphysics* 25 (1971): 3–51.

Teilhard de Chardin, P. *Le Phénomène humaine*. Paris, 1955.

Todd, W. "Intentions and Programs." *Philosophy of Science* 38 (1971): 530–541.

———. *History as Applied Science: A Philosophical Study*. Detroit: Wayne State University Press, 1972.

Todhunter, I. *A History of the Mathematical Theory of Probability from the Time of Pascal to that of Laplace*. New York: Chelsea, 1949.

Törnebohm, H. *Concepts and Principles in the Space-Time Theory with Einstein's Special Theory of Relativity*. Stockholm: Almqvist and Wicksell, 1964.

———. "A Foundational Study of Einstein's Space-Time Theory." *Scientia* 104 (1969): 1–13.

Törnebohm, H., and Radnitzky, G. "Research as an Innovative System." *Zeitschrift für allgemeine Wissenschaftstheorie* 2 (1971): 273–290.

Toró, T. *Modern Physics and Philosophy*. Timişoara: Editura Faela, 1973.

Toulmin, S. *The Philosophy of Science*. London: Hutchinson, 1953.

———. "Probability." In *Essays in Conceptual Analysis*, pp. 157–192. Edited by A. Flew. London: Macmillan, 1956.

———. *Human Understanding*, vol. 1: *General Introduction and Part 1*. Princeton: Princeton University Press, 1972.

Truitt, W. H., and Solomons, T. W. G. eds. *Science, Technology, and Freedom*. Boston: Houghton Mifflin, 1974.

Tudosescu, I. *Determinism and Science*. Bucharest: Editura Stiinţifică, 1971.

Tuomela, R. "Inductive Generalization in an Ordered Universe." In *Aspects of Inductive Logic*, pp. 155–174. Edited by J. Hintikka and P. Suppes. Amsterdam: North-Holland, 1966.

———. *Theoretical Concepts*. Vienna: Springer-Verlag, 1973.

———. "Dispositions, Realism, and Explanation." *Synthese* 34 (1977): 457–478.

———. *Human Action and Its Explanation.* Dordrecht: D. Reidel, 1977.

Uchii, S. "Higher Order Probabilities and Coherence." *Philosophy of Science* 40 (1973): 373–381.

Van Evra, J. W. "Traditional Philosophy of Science: A Defense." In *Basic Issues in the Philosophy of Science,* pp. 58–73. Edited by W. R. Shea. New York: Science History Publications, 1976.

Van Fraassen, B. C. *An Introduction to the Philosophy of Time and Space.* New York: Random House, 1970.

———. "Relative Frequencies." *Synthese* 34 (1977a): 133–166.

———. "The Pragmatics of Explanation." *American Philosophical Quarterly* 14 (1977b): 143–150.

Van Rootselaar, B., and Staal, J. F., eds. *Logic, Methodology, and Philosophy of Science III: Proceedings of the Third International Congress for Logic, Methodology, and Philosophy of Science, Amsterdam 1967.* Amsterdam: North-Holland, 1968.

von Mises, R. *Positivism: A Study in Human Understanding.* Cambridge, Mass.: Harvard University Press, 1951.

Vuillemin, J. *Physique et métaphysique Kantiennes.* Paris: P.U.F., 1955.

———. *Mathématique et métaphysique chez Descartes.* Paris: P.U.F., 1960.

———. *Recherches sur quelques concepts et méthodes de l'algèbre.* Paris: P.U.F., 1962.

———. *Leçons sur la première philosophie de Russel.* Paris: Colin, 1968.

———. *La Logique et le monde sensible.* Paris: Flammarion, 1971.

Walker, J. *A Study of Frege.* Oxford: Oxford University Press, 1965.

Wallace, W. A. "Philosophy of Science." In *New Catholic Encyclopedia,* pp. 1215–1219. Edited by W. J. McDonald. New York: McGraw-Hill, 1967.

———. *Causality and Scientific Explanation,* vol. 1: *Medieval and Early Classical Science;* vol. 2: *Classical and Contemporary Science.* Ann Arbor: University of Michigan Press, 1974.

Walsh, H. T. "Whewell and Mill on Induction." *Philosophy of Science* 29 (1962): 279–284.

Wartofsky, M. W., ed. *Proceedings of the Boston Colloquium for the Philosophy of Science, 1961–1962.* Boston Studies in the Philosophy of Science, vol. 1. Dordrecht: D. Reidel, 1963.

Weber, M. *On the Methodology of the Social Sciences.* Glencoe, Ill.: Free Press, 1949.

Weingartner, R. H. "The Quarrel about Historical Explanation." *Journal of Philosophy* 58 (1961): 29–45.

Wendt, D., and Vlek, C., eds. *Utility, Probability, and Human Decision-Making.* Dordrecht: D. Reidel, 1975.

West, F. E. *Science for Society: A Bibliography.* Washington, D.C.: American Association for the Advancement of Science, 1974.

Whewell, W. "Of the Transformation of Hypotheses in the History of Science." In *William Whewell's Theory of Scientific Method,* pp. 251–264. Edited by R. E. Butts. Pittsburgh: University of Pittsburgh Press, 1968.

Whitrow, G. J. "Berkeley's Philosophy of Motion." *British Journal for the Philosophy of Science* 4 (1953): 37–45.

Will, F. L. *Induction and Justification: An Investigation of Cartesian Procedure in the Philosophy of Knowledge.* Ithaca: Cornell University Press, 1974.

Williams, M. B. "Deducing the Consequences of Evolution: A Mathematical Model." *Journal of Theoretical Biology* 29 (1970): 343–385.

Wimsatt, W. C. "Teleology and the Logical Structure of Function Statements." *Studies in History and Philosophy of Science* 3 (1972): 1–80.

——. "Reductive Explanation: A Functional Account." In *PSA 1974,* pp. 671–710. Edited by R. S. Cohen et al. Dordrecht: D. Reidel, 1976.

Winch, P. *The Idea of a Social Science.* New York: Humanities Press, 1958.

Wisdom, J. O. *Foundations of Inference in Natural Science.* London: Methuen, 1952.

——. "General Explanation in History." *History and Theory* 15 (1976): 257–266.

Witt-Hansen, J. "Philosophy of Science (Wissenschaftstheorie) in Denmark." *Zeitschrift für allgemeine Wissenschaftstheorie* 1 (1970): 264–283.

Woodfield, A. *Teleology.* New York: Cambridge University Press, 1976.

Woodger, J. H. *The Technique of Theory Construction.* International Encyclopedia of Unified Science, vol. 2, no. 5. Chicago: University of Chicago Press, 1950.

Woods, J., and Sumner, L. W., eds. *Necessary Truth: A Book of Readings.* New York: Random House, 1969.

Woods, J., and Walton, D. "Petitio Principii." *Synthese* 31 (1975): 107–128.

Woolf, H. *Science as a Cultural Force.* Baltimore: Johns Hopkins University Press, 1964.

Wright, L. "Explanation and Teleology." *Philosophy of Science* 39 (1972): 204–218.

——. "Mechanisms and Purposive Behaviour III." *Philosophy of Science* 41 (1974): 345–360.

——. *Teleological Explanations.* Berkeley: University of California Press, 1976.

Wright, G. F. von. *The Logical Problem of Induction.* Oxford: Basil Blackwell, 1941.

——. *A Treatise on Induction and Probability.* London: Routledge and Kegan Paul, 1951.

——. *The Logic of Preference.* Edinburgh: University of Edinburgh Press, 1963a.

——. *Norm and Action.* London: Humanities Press, 1963b.

———. *The Varieties of Goodness*. London: Routledge and Kegan Paul, 1963c.

———. *Causality and Determinism*. New York: Columbia University Press, 1974.

Yamanouchi, T. *Man and Machine*. Tokyo, 1965.

———. "Physics and Philosophy in Japan." In *Contemporary Philosophy: A Survey*, vol. 2: *Philosophy of Science*, pp. 255–259. Edited by R. Klibansky. Florence: La Nuova Italia Editrice, 1968.

———. *On Understanding of Modern Physics: The World a Physicist Looks At*. Tokyo, 1970.

Yasugi, R. *History and Methodology of the Theory of Evolution*. Tokyo, 1965.

Yolton, J. W. *John Locke and the Way of Ideas*. Oxford: Clarendon Press, 1956.

———. "The Concept of Experience in Locke and Hume." *Journal of the History of Philosophy* 1 (1963): 53–72.

———. *Locke and the Compass of Human Understanding*. Cambridge: Cambridge University Press, 1970.

Yoshida, R. M. "Five Duhemian Theses." *Philosophy of Science* 42 (1975): 29–45.

———. *Reduction in the Physical Sciences*. Philosophy in Canada: Monograph Series, vol. 4. Halifax: Dalhousie University Press, 1977.

Yost, R. M. "Locke's Rejection of Hypotheses about Sub-microscopic Events." *Journal of the History of Ideas* 12 (1951): 111–130.

Young, L. B., and Trainor, W. J., eds. *Science and Public Policy*. Dobbs Ferry, N.Y.: Oceana, 1971.

Zaslawsky, D. "La Philosophie des sciences (Wissenschaftstheorie) en France (1950–1971)." *Zeitschrift für allgemeine Wissenschaftstheorie* 2 (1971): 318–325.

Zecha, G. "Die gegenwärtige Situation der Wissenschaftstheorie in Österreich." *Zeitschrift für allgemeine Wissenschaftstheorie* 1 (1970): 284–291.

Zeman, V. "The Philosophy of Science in Eastern Europe: A Concise Survey." *Zeitschrift für allgemeine Wissenschaftstheorie* 1 (1970): 133–141.

Zinov'ev, A. A. *Foundations of the Logical Theory of Scientific Knowledge*. Boston Studies in the Philosophy of Science, vol. 9. Dordrecht: D. Reidel, 1972.

Chapter 5
Philosophy of Technology
Carl Mitcham, ST. CATHARINE COLLEGE,
ST. CATHARINE, KENTUCKY

INTRODUCTION

If technology is taken to be the making and using of artifacts in the most general sense, then the philosophy of technology is an attempt to give a reasoned account of the nature and meaning of this basic human activity. As such, philosophy of technology appears initially to be a kind of practical philosophy related to ethics, or the rational account of human doing and its ends. Such issues as the social effects of industrialization, the dangers of nuclear weapons, environmental pollution, and the moral dilemmas of biomedical engineering reinforce this impression.

However, because modern technology includes the scientific making of objects, philosophy of technology also has a strong theoretical component. A comprehensive attempt to answer fundamental questions about the nature and meaning of technology will involve some analysis of the cognitive structure of engineering science and technique, as well as an account of the kind of being or reality artifacts have as distinct from natural objects. Along with an ethical evaluation of making and its practical consequences, philosophy of technology includes an epistemology of technical knowledge and an onotology of artifacts. Philosophical discussion of technology spans both practice and theory.

To appreciate the philosophy of technology in this full sense it is helpful to begin with its historical development.

I. HISTORY OF THE PHILOSOPHY OF TECHNOLOGY

A. Background

Philosophy of technology can readily be traced back to reflections on the human activity of making or fabricating put forth at the beginning of the modern period in Western history. Prior to the Renaissance, making did not occupy a large enough place in human consciousness to become a theme for systematic reflection; for the ancients, making, even in the form of art, was often mistrusted as inimical to virtue or the pursuit of the highest good because it focused attention on material reality. The only situations that justified technological innovation were situations of necessity, particularly military necessity. Making was not considered important as a contribution to the understanding either of the ends of human life or of the first principles of being.

Aristotle, for instance, had distinguished making (*poiesis*) from doing (*praxis*), in order to focus attention on doing. We *make* ships, houses, statues or money; we *do* sports, politics, or philosophy. The end of making is an object different from its act; the end of doing is the act itself, well performed. Life is a kind of doing, different forms of life being distinguishable on the basis of what they can do only or best. The fundamental question for man concerns what kind of doing or action is most properly human. As Aristotle outlines in his *Ethics,* the answer is that sports and politics probably, and the lives of sensual pleasure and material production certainly, fall short of what is most fitting to human capabilities. That which is most proper to man is the doing of philosophy, the free or detached contemplation of nature.

As the inheritor of this Greek perspective, Christian antiquity continued to be distrustful of technology. The Latin Middle Ages, nevertheless, came to see in the arts—even when their purposes were "superfluous, perilous and pernicious"—the exercise, as St. Augustine suggested, of "an acuteness of intelligence of so' high an order that it reveals how richly endowed our human nature is." The "progress and perfection which human skill has reached" is also a sign of divine benevolence (*City of God* XXII, 24). The Benedictine monastic tradition discovered in manual labor a ready spiritual ascesis for the mortification of human pride; it called men to a disciplined life of *ora et labora.* Later, the contemplative affirmation of the virtuous character of technology was extended to a recognition of the contributions that technology could make toward practicing corporal works of mercy. No doubt both insights prepared the way for a thoroughgoing reevaluation of human making. But from the perspective of the medieval dedication to the pursuit of an interior personal transformation at the expense of an external worldly one, making remained clearly subordinate to doing. The core difference between the heavenly and the earthly cities, according to St. Augustine, is that "the former make use of the world in order to enjoy God, whereas the latter would like to make use of God in order to enjoy the world" (*City of God* XV, 7).

With the radical critique of classical thought that emerges in the work of Nicolo Machiavelli, Francis Bacon, and René Descartes, this attitude is called directly into question. Machiavelli rejects the traditional

Christian virtues of "happiness in humility, lowliness, and contempt for worldly objects" in favor of "grandeur of soul, strength of body, and all such other qualities as render men formidable." He reinterprets Christianity as a religion that will reinforce these ideals and encourage men to assert their rightful liberty (*Discourses* II, 2). Bacon maintains that the inventions of printing, gunpowder, and the compass have done more to benefit humanity than all political conquests or philosophical disputes (*New Organon* I, 129). Descartes' methodological doubt aims not only at the construction of an indubitable system of thought, but also at making men "the masters and possessors of nature" (*Discourse* VI).

For the moderns, life is construed no longer primarily as doing, but as making. Thus there develops an epistemology that ties knowing to making, and a politics that reassesses the value of making as a means to human happiness. Previous knowledge is criticized as "barren of works." To repair this defect, Bacon proposes a reconstruction of science to produce "a line and race of inventions that may in some degree subdue and overcome the necessities and miseries of humanity." No longer should the mind approach nature as a passive observer willing to let "her work her own way." Mind must utilize art and hand until nature "is forced out of her natural state, and squeezed and moulded," because "the nature of things betrays itself more readily under the vexations of art than in its natural freedom" (*The Great Instauration,* proem and plan). Kant, having inscribed his *Critique of Pure Reason* (1787) with an epigram from Bacon, argues that the issue is not so much one of practical efficacy as one of self-consciousness. The ideal of the contemplative observer is an illusion. In fact, "reason has insight only into that which it produces after a plan of its own," and thus "it must not allow itself to be kept, as it were, in nature's leading-strings." This new attitude achieved an initial theoretical flowering in the Enlightenment program for the unification of science and the arts. Its practical fruition was the industrial revolution.

In the questioning of this typically modern attitude philosophy of technology proper begins to take form. Historically, this questioning was initiated by the Romantic movement. Jean-Jacques Rousseau, in his *Discourse on the Arts and Sciences* (1750), criticizes the Enlightenment idea that scientific and technological progress automatically contribute to the advancement of society by bringing about a union of wealth and virtue. Enlightenment optimism had been predicated on a reinterpretation of virtue as power. Rousseau, in his critique, appeals to a different understanding of virtue as sincerity or innocence, not to say freedom—which is presented as an ancient ideal obscured by the artificialities and conventions of civilization. Subsequent Romantic authors, such as the poet William Blake, flesh out the argument with vivid protests against the evils of the industrial revolution. They also extend the critique by taking issue with the adequacy of scientific and technical knowledge. Imagination, argues Samuel Taylor

Coleridge, is a vital faculty of mind that has access to deeper truths about the world than does the rational intellect.

Rousseau's criticism of civilization as insincere or devoid of true freedom—that is, the freedom of individuals to act authentically in accord with their own inner reality—can be extended to technology. In the nineteenth century it became the foundation for the Marxist critique of capitalism. In this critique, modern scientific technology is questioned not because of the ends it pursues, but because of its failure to realize these ends—at least in its present social manifestation. Just as Rousseau had argued that civilization corrupts the true nature of man, Karl Marx argues that a particular form of civilization, namely, capitalism, corrupts modern technology. Technology in its capitalist form is constrained, unable to achieve its true end; capitalist technology is inauthentic, unfree. To become what it should be, the liberation of man, technology must itself be set free from its social bonds.

Such criticisms of technology have functioned as important elements in modern social theory and played significant roles in various literary and political movements. But it is not until the twentieth century that technology becomes an explicit focus of philosophical reflection. The transition came about primarily through the influence of existentialism and efforts of engineers to analyze the nature of their work. Secondary influences include attempts by historians to understand the technological component in history as well as social science studies of industrial society. In the course of this transition from implicit to explicit reflection, three schools or traditions—West European, Anglo-American, and Soviet-East European—have made major contributions to philosophy of technology as currently practiced.

B.   The West European School

The European (meaning primarily German and French) tradition of philosophy of technology is the oldest and most variegated. It includes existential, sociological, engineering, and theological reflections on the nature and meaning of technology with a variety and depth not to be found in other traditions. Its weaknesses include a lack of internal synthesis in comparison with the East European school, and a failure to make good use of historical knowledge and empirical social science research as found in the Anglo-American school.

Some early twentieth-century existentialists argue that man, as a being who makes himself, is preeminently a technological creature. Man's basic nature is productive and historical as contrasted with the passive and ahistorical being-in-the-world of natural objects. This analysis of the "human life-world" undermines the popular Romantic distinction between natural and technological man. The Spanish philosopher José Ortega y Gasset, in "Thoughts on Technology" (1933 [1939, 1972]), presents technology as an essential part of human nature. Technology is "the system of activities through which man endeavors to realize the extranatural program that is himself," a material activity in the service of some human ideal ("Thoughts on Technology," 1933 [1972, p. 301]). The ideal in question is not necessarily personal. The human self-conception at the basis of any culture requires an appropriate technology for its worldly realization. Ortega's work is particularly seminal in this attempt to ground material invention and production in a prior spiritual invention, in the self-creation or willing of some human ideal, as well as in his anthropological account of the structure of this relationship.

One corollary of Ortega's argument is the idea that modern technology is in the process of engendering a world-historical transformation of the human condition which goes beyond the specific intentions of its users. Because it so effectively realizes the goal of making or work, scientific technology tends to undermine the value of making anything in particular, to equalize all making. Modern man recognizes that before he can have a particular technology he must have technology in general; "technologies are nothing but concrete realizations of the general technical function of man" (p. 311). By taking general technology as the object of conscious investigation and systematic development, modern man has unwittingly undercut his own strength. Modern man has no goal other than making, although making cannot be an end in itself. As Ortega succinctly puts it: "To be an engineer and nothing but an engineer means to be potentially everything and actually nothing. . . . That is why our time, being the most intensely technical, is also the emptiest in all human history" (p. 310).

On this interpretation, modern technology exhibits a transpersonal, not to say autonomous character. Insofar as its effects transcend the intentions of its users, technology cannot be characterized as a neutral means. The problem with technology is not the uses to which it might be put but the world-historical effects it engenders, independently of any particular use or user. Similar suggestions can be found in the historico-philosophical speculations of other existentialist thinkers, particularly Karl Jaspers (1949). The existentialist emphasis on social reality helped stimulate another development, that of a European sociology of technology. It is in this social theorizing that the question of the autonomy of technology plays its most central role. The sociological approach of taking technology as a dominant feature of the contemporary situation, then proceeding to examine its effects on man and the sociopolitical order, naturally raises this issue. The industrial sociology of Georges Friedmann and the Marxist revisionism of the Frankfurt School of Social Research are two cases in point. Friedmann has produced an abundance of empirical studies to provide scientific confirmation for the insights and criticisms voiced by Romantic writers about the effects on the worker of the division of labor. Representing the Frankfurt school, Herbert Marcuse stresses the controlling influence of modern technology in a host of other areas—from politics to art, literature, and even philosophy. At the same time, these theorists try to see technology in its particular historical form and not talk about technology *per se*. Technology is autonomous only within a certain definite historical context.

A different interpretation of technological society is offered by the positivist tradition of European sociology. According to August Comte, Henri de Saint Simon, and their followers, evils that others attribute to the autonomy of technology arise because technology is not in fact autonomous enough. Society needs to be made more rational in its organization; politicians need to be replaced by technicians. Since the engineer is the only person who is really on top of technology, raising him to a position of authority will bring its effects under control. The relation between this argument and the ideas of Thorstein Veblen and the technocracy movement in the United States should be obvious.

Langdon Winner (1977) has done an extensive study of the idea of autonomous technology, particularly as it has influenced political thinking. As he notes correctly, it is Jacques Ellul who has presented the question of the autonomy of technology in its most uncompromising and influential form. For Ellul, on the one hand, the positivist theory misunderstands the engineer. The engineer is not on top of technology; he is the instrument of technological rationality. On the other hand, Marxist theory fails to appreciate the extent to which technology has similar effects in socialist and capitalist countries. In fact, because of its ideological commitment to freedom through technology, socialism has become a utopian illusion.

Although they are related, the approaches of existentialists and European social theorists need to be distinguished. This can be done by contrasting the positions of Ellul and Martin Heidegger. Ellul, the best of the social theorists, views technology as *"l'enjeu du siècle,"* the bet of the age. Technology is a wager man has made with a force he does not fully understand—a force so powerful it can overwhelm him. Heidegger, the greatest of the existentialists, speaks in a similar vein of "the global movement of modern technology" as "a force whose scope in determining history can scarcely be overestimated" (1976, p. 276). As a consequence, he sees "the task of thought to consist in helping man in general, within the limits allotted to thought, to achieve an adequate relationship to the essence of technology" (p. 280). Heidegger seeks an understanding of the way technology is intimately bound up with the nature of man—a procession, as it were, of unconscious forces. Ellul, by contrast, is concerned not so much with the essence of technology or its anthropological foundations as with its social relationships and the need to become conscious of objective principles operating in these relationships. Social theory stresses technology as an external power unconsciously determining much of man's life; the existential approach stresses technology as itself an unconscious expression of some aspect of human nature.

The attempts by existentialists and social theorists to bring technology into consciousness by means of disciplined analysis and insight can both reasonably be called philosophy of technology. Yet neither approach ever uses the phrase "philosophy of technology." This usage arose among manufacturers, engineers, and economists attempting to reflect on the nature of the technological undertaking in which they were engaged.

The phrase "philosophy of technology" has some antecedents in the Scotch chemical technologist Andrew Ure's idea of a "philosophy of manufactures." In a book of that title (1835), Ure describes his purpose as "an exposition of the general principles on which productive industry should be conducted with self-acting machines." The exposition generates a number of conceptual issues that still concern philosophers of technology: distinctions between making and manufacturing, the classification of machines, and the possibility of rules for invention. But because Ure's technical discussion is coupled with an enthusiastic apology for the factory system, his more analytic expositions are often missed. Subsequent technical studies of the nature of machinery and industrial organization lead in two directions. One terminates in cybernetics and systems theory, which is an implicit philosophy of technology. The other gives birth to "philosophy of technology" in the strict sense as a defense of modern industrial engineering.

The first work to bear the title "philosophy of technology" is Ernst Kapp's *Grundlinien einer Philosophie der Technik* (1877). Kapp was a left-wing Hegelian with considerable practical experience of tools and machinery. In a way that foreshadows later discussions, he brings together detailed analyses of technological instruments with speculations about their human or cultural significance. His particular conclusion, that tools function as organ projections, is less important than two of his general principles: that machines deserve detailed philosophical attention, and that technology needs more sophisticated criticism than the external judgments offered by social or literary critics. Based on Kapp's usage, the phrase "philosophy of technology" has come to refer especially to engineering attempts to defend the profession against the hostility of the Romantic tradition, but also to efforts to come to terms with the full social implications of engineering practice by means of a careful consideration of engineering experience.

This early engineering-related discussion of the inherent structure of technical practice and the social significance of technology provided the initial formulation for a large number of issues that have since become central themes in philosophy of technology. The engineer Eberhard Zschimmer, in the second book to bear the title "philosophy of technology" (1913), proposes a neo-Hegelian interpretation of technology as "material freedom." Later Friedrich Dessauer, in the third book entitled a "philosophy of technology," argues for invention as the most important aspect of technology. In the contrast between Dessauer's stress on invention and Zschimmer's on the practical freedom derived from machines and technical products, there is disagreement about whether making or using is the fundamental category of technological practice. This distinction foreshadows the contemporary situation in which critics of technology focus on its social consequences, whereas defenders of technology appeal to the act of invention as a creative experience similar to that of artistic composition.

The period between the world wars witnessed an enlarging of the engineering discussion in various directions. World War I forced engineers to pay more serious attention to social criticisms of technology, such as those articulated by the Italian sociologist Gina Lombroso (1930). Because of its importance to industry, economists began to analyze the nature of engineering, which in turn stimulated engineering analyses of economic factors. It was during this period, too, that various experiments were made to bridge the gap between technology and the arts. The Bauhaus school of industrial design, in which engineers and artists sought to discover an aesthetics of technology, is the best known. As the engineer Albert Speer makes clear, National Socialism was also seen by some as a solution to the social and aesthetic problems created by technology. In a study focusing on this period, Donald Thomas (1978) sums up many of these trends by describing the tension between technological optimism and Romantic criticism in the social ideas of Rudolph Diesel, inventor of the diesel engine, and his son Eugene. Although originally quite hopeful, the elder Diesel began increasingly to despair of technology, and ended up committing suicide. His son, in consequence, turned from technology to literature in order to express his father's intuitions.

Following World War II engineering-related philosophy of technology went through another period of growth. In Germany the first formal organizational developments took place within the society of German engineers, the Verein Deutscher Ingenieure (VDI), which conducted a series of conferences on themes related to the philosophy of technology. In 1956 the VDI established a special "Mensch und Technik" study group, which was broken down into working committees on education, religion, language, sociology, and philosophy—all in relation to technology.

In France, the engineering analysis of industrial production initiated earlier by Jacques Lafitte's *Réflexions sur la science des machines* (1932) was extended by another engineer, Gilbert Simondon, in *Du Mode d'existence des objects techniques* (1958). Both books display considerable rigidity in attempting to be true to engineering experience while formulating an abstract interpretation of technological phenomena. Simondon, for instance, distinguishes between parts, devices, and systems as kinds of technological objects, and proposes a theory of technological evolution, on the basis of detailed references to examples like that of the internal combustion engine. In the Netherlands, the engineer Hendrik van Riessen began a second career in philosophy with *Filosofie en Techniek* (1949); the book provided the most comprehensive historico-philosophical survey of the field up to that time. Such developments point toward the Europeanization of what was previously a German discussion.

The single most important figure in engineering-philosophy discussions, both before and immediately after World War II, was Friedrich Dessauer (1881–1963). As a research engineer who pioneered in the development of X-ray therapy, and as a Christian social democrat who openly opposed Nazism, Dessauer made basic contributions to philosophy of technology at the same time that he sought to open up dialogue with existentialists, social theorists, and theologians. His major works run from *Technische Kultur?* (1907) and *Philosophie der Technik* (1927) to *Seele im Bannkreis der Technik* (1945) and *Streit um die Technik* (1956). It is the work of Dessauer that is most often cited in those instances where philosophers of science mention philosophy of technology—see, for example, Bernard Bavink (1932) or Alwin Diemer (1964).

A convenient way to summarize Dessauer's philosophy of technology is to contrast it with standard philosophies of science. The latter either analyze the structure and validity of scientific knowledge in general or consider the implications of specific theories. For Dessauer, both approaches fail to recognize the power of scientific-technical knowledge, which has become, through modern engineering, a totally new form of making. Dessauer attempts to explain in Kantian terms the transcendental preconditions of this power, as well as to reflect on the ethical implications of its application.

To the three Kantian critiques of scientific knowing, moral doing, and aesthetic feeling Dessauer adds a fourth critique—of technological making. Kant, in the *Critique of Pure Reason,* had argued that scientific knowledge is necessarily limited to the world of appearances (the "phenomenal" world); it can never make contact with "things-in-themselves" ("noumena"). Critical metaphysics is, however, able to delineate *a priori* forms of appearances, and to postulate behind phenomena the existence of some "noumenal" reality. The *Critique of Practical Reason* (of moral doing), and the *Critique of Judgment* (concerned with aesthetic evaluation), go farther; they affirm the existence of a "transcendent" reality beyond appearances as a precondition of the exercise of moral duty and the sense of beauty. Practical and aesthetic experience, nevertheless, do not make positive contact with this transcendent reality; neither can the critiques of these realms of experience articulate noumenal structures.

Dessauer argues that making, particularly in the form of invention, does establish contact with things-in-themselves. This contact, for Dessauer, is confirmed by two facts: that the invention, the object made, was not previously found in existence; and that, when it makes its phenomenal appearance, it works. An invention is not just something dreamed up, with no power; it is the result of cognitive contact with a realm of preestablished solutions to technical problems. Technological invention involves "real being from ideas" (1956, p. 234)—that is, the engendering of "existence out of essence," the material embodying of a transcendent reality.

Although other Kantians have found weaknesses in Dessauer's argument, and his conclusions have not been widely accepted, it is well not to overlook a sense in which Dessauer does extend the Kantian point of view. For Kant, all reasoning is oriented toward the practical; the more practical it is, the closer experience comes to a decisive transcending of its own "phenomenal" limitations. Dessauer locates the penetration of appearances precisely in a kind of practical experience that Kant had failed to recognize as worthy of separate analysis—partly because he was writing before the advent of modern technological making. This insight suggests that there may be something philosophically unique about modern technology.

On the basis of his metaphysical analysis of technology, Dessauer also constructs an ethical and a political theory. Typically, modern making (as well as doing) is grounded in a rejection of the knowledge of transcendent reality. "Experiments and historical observations we may have, from which we may draw advantages of ease and health, and thereby increase our stock of conveniences for this life," John Locke had written, "but beyond this I fear our talents reach not, nor are our faculties . . . able to advance" (*An Essay Concerning Human Understanding* IV, xii, 10). Marxists, criticizing technology in the particular historical form it takes under capitalism, question whether modern practice leads to human welfare as immediately as Locke assumes. Some liberal reformers join the Marxists on this matter, calling for a radical extension of practice, from the making of material objects to the making of social structures as well. Such extensions leave the modern commitment to practice grounded in purely practical reasons. Dessauer goes farther. For him, the autonomous, world-transforming power of modern technology is witness to its transcendent foundations. Man creates technology, but its power—which resembles that of "a mountain range, a river, an ice age, or a planet"—goes beyond what man expected; it brings into play more than this-worldly forces. Because of its transcendent foundations, modern practice is no longer to be conceived of simply as "the relief of man's estate" (Bacon); it becomes, instead, a "participating in creation, . . . *the greatest earthly experience of mortals*" (1927, p. 68).

Donald Brinkmann has argued (in *Mensch und Technik*, 1945) that technology is a this-worldly religiosity; that in it man seeks to establish his own salvation in a Promethean or Faustian way. One need not adhere either to Dessauer's unusually optimistic Roman Catholicism, or to Brinkmann's religious-fundamentalist pessimism, to argue that some kind of "mysticism of technology" is present at the core of the modern commitment to practice.

Since Dessauer's death, philosophy of technology in Germany has become less metaphysical. The most formative influence has been the work of the philosopher Simon Moser. In an important essay first published in

1958, Moser criticizes the metaphysical ideas of Dessauer as philosophically naive, at the same time that he argues against the metaphysical interpretations of Brinkmann, Heidegger, and others as lacking in technological sophistication. What is needed, he maintains, is closer cooperation between engineers and philosophers, in which engineers respect philosophical rigor and philosophers pay more attention to real engineering practice.

Since the late 1950s, philosophy of technology in Germany has developed primarily in conjunction with engineering universities and the VDI (see Huning, 1979). Within this group, discussion has focused on such conceptual issues as definitions of science and technology, the methodology of engineering design, or on education and the distinctive characteristics of technological activity in advanced industrial societies. The result has been the creation of what Friedrich Rapp (1978) calls "analytic philosophy of technology."

The major achievement of engineering philosophy of technology has been to focus attention on technological making as a philosophically interesting subject. Yet there is another aspect; technical analysis of the structure of this activity has, in the cases of Dessauer and van Riessen, also been intimately associated with theological interpretation. The fact that Dessauer's ideas were picked up by other engineers may indicate something about religious tendencies latent in engineering consciousness.

A fourth significant strand in European reflection on technology is that undertaken explicitly from the point of view of theology and morality. The most positive religious interpretation is, of course, Dessauer's technological mysticism, his idea of invention as a continuation of divine creation. This position has been echoed by the French Catholic philosopher Emmanuel Mounier (1948) in his defense of technology as man's "demiurgic function." Something similar is implicit in Pierre Teilhard de Chardin's well known theory of the "hominization" of the world. It can even be found operating to some extent in the theology of as eminent a figure as Karl Rahner and his defense of genetic engineering (1972) as well as in the so-called "theology of hope" and "theology of liberation" movements. Radically conservative Catholic critiques have been put forward by Eric Gill (1940) in England and by Georges Bernanos (1947) in France. For Gill, machines and commercialism are antithetical to the craftsman's pursuit of quality and the Christian's call to poverty. For Bernanos, modern technology is a manifestation of the sin of pride. However, despite papal pronouncements against artificial contraception and nuclear weapons, few Catholic thinkers have been less than sanguine in their assessment of the theological dimensions of modern technology.

It is ironic in light of Max Weber's thesis about the dependence of industrial civilization on the Protestant ethic that Protestant theologians have voiced some of the strongest objections to modern technology. The Lutheran bishop Hans Lilje (1928) and the Swiss Reformed theologian Heinrich Emil Brunner (1949) are examples. For Brunner, "Modern technics is, to put it crudely, the expression of the world-voracity of modern man, and . . . his inward unrest, the disquiet of the man who is destined for God's eternity, but has himself rejected this destiny" (p. 5). Ellul (e.g., 1975) has written theological works extending his sociological studies which share these sentiments and reflect the Calvinism of the French Reformed tradition. In opposition, the Neo-Calvinist assessments of van Riessen and his student Egbert Schuurman (1972) are much more positive; Schuurman argues that technology is a "reverent service to God" that answers the call to cultural creation.

In France, moral and religious discussions of technology have taken on a character independent of particular religious traditions partly because of the work of Henri Bergson. In the last chapter of *The Two Sources of Morality and Religion* (1932) Bergson argues that the problems of mechanistic civilization can be overcome by a revival of mysticism, which is at once ascetic (against luxuries) and charitable (eliminating inequalities). This idea of the need for a "moral supplement" to meet the problems of technology was reinforced in Pierre-Maxime Schuhl's *Machinisme et philosophie* (1938) and repeated by Mounier and others. Daniel Cérézuelle maintains, in "Fear and Insight in French Philosophy of Technology" (1979), that this idea has become the standard response to technology—one that could even appeal to Friedmann (1970) when he became disenchanted with Marxism. Another interpretation, that technology is an attempt to escape the limits of the human condition, has been advanced by the French philosopher Jean Brun (1961). In general, if Germany can be described as the European center for the analysis of technical processes, France can be called the center for the discussion of moral issues.

C.   The Anglo-American School

American philosophy of technology differs from the West European version in both historical depth and topical breadth. Despite the indigenous development of American pragmatism, sometimes interpreted as a technological philosophy, philosophy of technology in America has not been as closely associated with engineering as it has been in Germany. Until recently, the only explicit pragmatist attempt at a philosophy of technology was a 1955 essay by Joseph Cohen. In England and America, philosophy of technology has grown out of sociological and historical approaches to the understanding of technology.

The American school can be traced back at least to the sociological studies of William F. Ogburn. In *Social Change with Respect to Nature and Culture* (1922), Ogburn formulated an idea that is now virtually an intellectual cliché: the concept of a "cultural lag" between technological and social development. Technology develops more rapidly than social institutions, causing disharmonies. Assumptions of the inevitability or desirability of technological change bring about the need for corresponding institutional transformations. In the following decades a series of sociologically and historically inspired analyses of technology were published, arguing one or more aspects of this idea.

However, it was the publication in 1934 of Lewis Mumford's *Technics and Civilization* that produced the most lasting impact. This was due not only to Mumford's engaging style and subsequent prolific production, but to his historical imagination and philosophical aspirations as well. *Technics and Civilization* distinguishes various technical objects and practices with a view toward a clearer understanding of the periods of technological development and the full impact of this development on culture. Yet this descriptive aspect is subordinate to the articulation of a philosophical theory of the nature of man that proposes to establish the rightful place of technology in human life.

Mumford's theory of man is part of the American tradition of worldly idealism that extends from Emerson to Paul Goodman. The tradition is worldly in its concern with the ecology of the American environment—the harmonies of urban life, the preservation of wilderness, and sensitivity to organic realities. It is idealist in insisting that material nature is not the basis of organic activity, at least in its human form. The basis of human action is mind and man's struggle for creative self-realization. As Mumford makes the argument later, in *The Myth of the Machine* (2 vols., 1967 and 1970), although man is rightly engaged in worldly activities, he is not properly understood as *Homo faber* (tool maker) but as *Homo sapiens* (mind maker). Against what Mumford considers a technological-materialist image of man, he maintains that technics in the narrow sense of tool making and using has not been the main agent in human development. All of man's technical achievements are "less for the purpose of increasing food supply or controlling nature than for utilizing his own immense organic resources . . . to fulfill more adequately his superorganic demands and aspirations." The elaboration of symbolic culture through language, for instance, "was incomparably more important to further human development than the chipping of a mountain of hand-axes." For Mumford "man is preeminently a mind-making, self-mastering, and self-designing animal" (1967, pp. 8–9).

Using this anthropology, Mumford constructs a distinction between two kinds of technology: polytechnics and monotechnics. Poly- or biotechnics is the primordial form of making; at the beginning (logically if not historically), technics was "broadly life-oriented, not work-centered or power-centered" (1967, p. 9). This is the kind of technology that is in harmony with the polymorphous needs of life, and it functions in a democratic manner to realize a diversity of human potentials. In contrast, mono- or authoritarian technics is "based upon scientific intelligence and quantified production, directed mainly toward economic expansion, material repletion, and military superiority"—in short, toward power (1970, p. 155).

Although modern technology is a primary example of monotechnics, this authoritarian form did not originate in the industrial revolution of the eighteenth century, which was based on machines. Its origins go back five thousand years to the discovery of what Mumford calls the "megamachine"— that is, rigid social organization. The standard example of the megamachine is the army or some organized work force such as the one that created the pyramids or the Great Wall of China. The megamachine often brings with it striking material benefits, but at the expense of a limiting of human activities and aspirations that is dehumanizing. The consequence is the "myth of the machine," or the notion that megatechnics is both irresistible and ultimately beneficent. This is a myth and not reality because the mega-machine *can* be resisted and is not ultimately beneficial. Mumford's work as a whole is an attempt to demythologize megatechnics and thus bring about a radical reorientation of mental attitudes in order to transform mono-technical civilization.

One of the remarkable features of Mumford's work is that his negative critique is complemented by a positive study of urban life which culminated in the widely acclaimed *The City in History* (1961). Mumford is clearly not arguing for a simple-minded rejection of all technology. He seeks to make a reasoned distinction between two kinds of technology, one of which, because of its social relations, is in accord with human nature, the other of which is not. It is unfortunate that despite its considerable literary influence, Mumford's argument has not been accorded much philosophical attention. (The only exception is an important but limited study by James Carey and John Quick, 1970.) His theory both reflects a widely adopted intellectual position and raises critical philosophical issues. Not the least of these issues is the difficulty of basing a critique of technology on the metaphysical idea of man as a self-creative being.

In conjunction with the invention and development of computers there has arisen another, more specialized approach to the philosophical analysis of one kind of technology—which is nevertheless properly part of philosophy of technology in general. Here the stress is less on social or historical issues and more on conceptual and analytic ones, in harmony with the dominant character of Anglo-American philosophy itself. This incidentally has been the occasion for a limited interaction between philosophy and engineering of the sort called for in Germany by Moser.

The groundwork for this interaction was the mathematization of logic initiated at the turn of the century by figures like A. N. Whitehead and Bertrand Russell, together with the influence of logical positivism in Anglo-American philosophy. Because computer engineering utilizes developments in modern logic, many Anglo-American philosophers remained conversant with this aspect of technology. Especially since the 1940s, analyses of feedback control in cybernetics (see especially Norbert Wiener, 1948), information theory, game and decision theory, operations research, computer simulation of human behavior, and other developments in computer technology have raised questions about the relations between human and artificial intelligence that are relevant to the philosophy of mind. In 1950 the British logician A. M. Turing, in a now classic article, argued that if a machine could be constructed capable of certain outputs that, under specified conditions, were not distinguishable from the outputs of the human mind, then we would have to concede that the machine could think. In the 1950s and 1960s this and related topics were widely discussed, as can be gauged from such collections as Feigenbaum and Feldman (1963), Anderson (1964), Crosson and Sayre (1967), and Crosson (1970).

In the early 1970s, the work of Keith Gunderson and Hubert Dreyfus can be taken as representative. Gunderson (1971) gives a historicophilosophical analysis of the man-machine relation, arguing that one cannot in principle deny the possibility of machines exhibiting human intelligence. Dreyfus (1972), arguing from a phenomenological perspective, maintains just the opposite. Later, the computer scientist Joseph Weizenbaum (1976) would argue for moral limitations on computer development; Aaron Sloman (1978) replies that computers are a means of human self-expression. Kenneth Sayre (1976) uses information theory to give a general interpretation of biological phenomena from the material to the mental. Because such discussions have been subsumed under the categories of philosophy of science or philosophy of mind, these developments did not lead to a clearly defined area of Anglo-American scholarship called "philosophy of technology."

One other contribution to American philosophy of technology is theological interest in problems of technology. In the 1960s two conferences were held at the Catholic University of America. The first, *Technology and Christian Culture* (proceedings edited by R. P. Mohan, 1960), was a reflection of European Catholic views; the second, *Philosophy in a Technological Culture* (edited by George F. McLean, 1964) is an uneven treatment of a broader range of subjects, from technology in philosophy of science to ethical issues generated by technology. The proceedings of an ecumenical consultation held at Louvain in 1961, with papers reflecting both American and European thinking, has been edited by Hugh White (1964). This volume constitutes the single best introduction to theological assessments of technology during this period. Finally, Harold E. Hatt's monograph *Cybernetics and the Image of Man* (1968) draws on the work of Brunner in a Protestant attempt to assess the philosophical dimensions of artificial intelligence.

All this is background. As a self-conscious pursuit, philosophy of technology in England and America can conveniently be dated from two events. In March 1962, the Center for the Study of Democratic Institutions in Santa Barbara, California, hosted the Encyclopaedia Britannica Conference on the Technological Order. The conference devoted a day each to four topics: (1) the idea of technology, examining major contemporary positions; (2) the technical act, focusing on the historical and social conditions surrounding technical activity; (3) relations between nature, science, and technology; and (4) technology and the emerging nations. Participants included a diversity of scholars: Ellul, the Russian historian Alexandr Zvorikine, two American Jesuit scholars, W. Norris Clarke and Walter Ong; British political scientist Ritchie Calder; the publisher of *Scientific American*, Gerard Piel; as well as historians Lynn White, Jr., A. Rupert Hall, and Melvin Kranzberg.

A second seminal event was the summer 1966 issue of *Technology and Culture*, which included a set of papers under the heading "Toward a Philosophy of Technology." This symposium included a paper by Mumford, along with contributions by a group of philosophers who have since come to be recognized as among the leading English-language philosophers of technology: Joseph Agassi, Mario Bunge, and Henryk Skolimowski. This was the first use of the phrase "philosophy of technology" as a title in English. (The title had originally been that of the contribution by Bunge, an Argentinian philosopher with intimate knowledge of West European discussions of technology.) As *Technology and Culture* editor Melvin Kranzberg described the symposium in a prefatory note: "There is the questioning of technology in terms of human values; there is the attempt to define technology by distinguishing it from or by identifying it with other related fields; there is the epistemological analysis of technology; and there is the investigation of the rationale for technological developments" (p. 301).

Since these two events, scholarly activity in this area has increased rapidly. In 1969 the journal *Daedalus* devoted an issue to the theme "Ethical Aspects of Experimentation with Human Subjects." In 1971 *Philosophy Today* published a second symposium under the title "Toward a Philosophy of Technology" containing translations of work by two important European authors, Brinkmann and Moser, as well as an important historicophilosophical study by Hans Jonas. In 1973 an international symposium on the history and philosophy of technology was held at the University of Illinois at Chicago Circle (see G. Bugliarello and D. Doner, 1979); 1973 was also when another major sourcebook was published—Carl Mitcham and Robert Mackey's *Bibliography of the Philosophy of Technology*, a companion to their collection of readings *Philosophy and Technology* (1972). Finally, in 1975 Paul Durbin organized a conference at the University of Delaware to coalesce the philosophy of technology movement, one result of which was the initiation of *Research in Philosophy & Technology* (vol. 1, 1978; vol. 2, 1979), an annual with an international editorial board. The American school thus exhibits a strong international thrust and a broad, rather than narrow, focus.

Parallel with the rise of an explicit American discipline called "philosophy of technology" there has developed an intense interest in bioethics or the study of moral problems associated with biomedical technology.

Bioethics as a separate subject is covered in chapter 6. One of the prominent figures in that field, Hans Jonas, has argued that bioethics is ultimately part of a general ethics of technology. Since the publication of "The Practical Uses of Theory" (1959), Jonas has been among the most active English-language philosophers pursuing a comprehensive understanding of technology.

At some point bioethics begins to shade into another less well-defined area, the multifaceted environmental issue. Philosophical questions clustering around problems of pollution, ecology, energy, the social responsibility of scientists, technology assessment, and alternative technology all relate to philosophy of technology. In fact, they are likely to become primary foci.

D.   The Soviet-East European School

This is the most internally unified of the three schools under review, and the only one that can be said to have a doctrine. The doctrine is grounded in the thought of Karl Marx (1818–1883), with his analysis of the process of production as the primary human activity and the foundation of society and history. Currently, discussion centers around the concept of the "scientific-technological revolution" (STR), or the unification of science and technology to bring about what in the West has sometimes been called the second industrial revolution.

Marx's analysis of the production process is based on the modern understanding of man; indeed, it gives expression to that understanding in one of its most straightforward forms. For Marx, human life is essentially *"sensuous activity, practice"* ("Theses on Feuerbach," number 1). This activity "appropriates particular nature-given materials to particular human wants (*Capital,* 1867 [1967 p. 42]). In fact, "Where Nature is too lavish, she 'keeps [man] in hand, like a child in leading-strings' " (p. 513). Marx's argument is made initially as a correction to Hegel, in an attempt to give the Hegelian dialectic of consciousness real-world foundations. Hegel had grasped *"labor* as the *essence* of man," but "the only labor which Hegel knows and recognizes is *abstractly mental* labor" ("Critique of the Hegelian Dialectic and Philosophy as a Whole"). The substitution of real for abstract labor Marx worked out in the early *Economic and Philosophic Manuscripts of 1844* and in the *Grundrisse* manuscripts of 1857–1858. Marx's mature thought, however, takes the form of a comprehensive critique of political economy that includes a detailed examination of the production process in its most advanced form under capitalism.

When Marx subtitles *Capital* a "critique of political economy," he draws attention to his uniting of the "critical" tradition of Kant and subsequent German philosophy with the practical world of politics and economics. Political economy, the theory of government that flourished in England in the eighteenth and nineteenth centuries, takes its bearings from the modern reevaluation of technology associated with the industrial revolution. Classical political thought was concerned with how to minimize wealth and restrict its pernicious influence in the just state. Political economy inquires into the nature and causes of wealth with an eye toward formulating governmental policies designed to maximize social production. Marx's critique of political economy claims to expose the fundamental preconditions of this modern political theory and to correct its mistakes.

According to Marx, political economy rests on two mistaken ideas. Although it recognizes the primacy of making over doing, it fails to appreciate that making is always social; and it takes commodities or products to be things independent of the making process. The elementary factors of production are labor, the material worked on, and the instruments of labor. But the material worked on is—except with extractive industries such as mining—always the product of some previous production process; the same goes for the instruments of labor. Furthermore, both are simply means within a current form of the production process. Thus what is crucial is "not the articles made, but how they are made, and by what instruments" (*Capital*, p. 180). Materials and instruments are always related to some production process; the process itself is primary. Political economy, because of its uncritical acceptance of the individuality of production and the primacy of commodities, has not been a true inquiry into the production of wealth. Instead, it has analyzed production only from the point of view of that particular class, the bourgeois, who are individualists and property owners. Political economy has been tied to limited class interests.

Marx's liberation of political economy from bourgeois class interests entails a new analysis of the production process. It examines "how the instruments of labor are converted from tools into machines" (p. 371) and the way machines themselves tend to become organized into a system in which "the subject of labor goes through a connected series of detailed processes" (p. 379). The basis of this latter transformation is "the new science of technology" (p. 486), which analyzes the production process into its constituent functions. Marx is at his best in describing the way this "new technology" undermines the traditional skills and satisfactions of craft production and places over against the worker the specter of an autonomous factory in which labor functions have become equal and interchangeable.

Marx argues, however, that technology also reveals the "fitness of the laborer for varied work, consequently the greatest development of his varied aptitudes" (p. 488). If all jobs are equal, then a worker should be able to do anything he wants. The problem is that in a capitalist economy, where individuals own the means of production, workers are wage slaves in the production process. In a communist society, where the means of production are no longer privately owned, a worker will be free to "become accomplished in any branch he wishes"; it will be possible for him to "do one thing today and another tomorrow" (*German Ideology* I, 1, a). Carried through to its perfection and liberated from the capitalist mode of production, modern technology makes possible true human freedom and the exercise of traditional craft activities on a scale greater than ever before in history.

The STR idea extends and develops Marx's insights with respect to the transformation of the production process by modern science and technology. The term "scientific-technological revolution" was originally suggested in the early 1950s by the Western Marxists J. D. Bernal and Victor Perlo; Marxists in the socialist countries took up the discussion late in the same decade. Communist party ideologues at first criticized the concept (as they did cybernetics, with which it is closely associated), maintaining that, because it was derived from "bourgeois philosophy," it was a deviation from pure Marxism-Leninism. Although some social theorists considered the concept an important one—pointing as it did toward a significant change in the nature of the production process—it was only in the early 1960s, with a change in the official position of the Russian Communist party, that the STR concept became widely accepted in the social sciences or in politics. As Julian Cooper (1977) has argued in one of the best available studies on the STR concept in the Soviet Union, the idea that science and technology play a major role in the revolutionary transformation of society has been a central theme in Marxist-Leninist thought since 1917. The theory that the STR is a crucial factor in the construction of socialism and the transition to communism is simply its most recent manifestation.

The early defense and development of the STR concept occurred in East European criticisms of West European philosophy of technology. In the late 1950s and early 1960s East German philosophers characterized Dessauer's ideas specifically as bourgeois ideology in the service of reactionary imperialist forces. West European philosophy of technology in general was rejected as too conceptual in emphasis, as failing to recognize technology as a productive force influenced by social and economic conditions. It was condemned as a pessimistic concentration on the negative aspects of technological progress. And, finally, it was blamed for assuming that technological development would of itself bring about the gradual convergence of communism and capitalism. Hermann Ley in *Dämon Technik* (1961) argued that West European thinking constituted a "new witch's trial of technology" and stigmatized the idea of a "second industrial revolution" as an opiate.

In 1965 a conference was held in East Berlin on "Marxist-Leninist Philosophy and the Technological Revolution"; it undertook to formulate a more constructive Marxist position. Bringing together a number of East European philosophers, the conference focused on six questions associated with the idea of technological revolution: its essence and history, the role of science, the socialist image of man, planning in the technological revolution, methodological problems of modern science, and philosophical problems of industrial science. That same year Erwin Herlitzius' bibliography on "Technik and Philosophie" was published. Both led to renewed philosophical discussions of technology in the German Democratic Republic.

Two years later the Czech social philosopher Radovan Richta edited *Civilization at the Crossroads* (1967), a cooperative work in which sixty sociologists, economists, psychologists, historians, engineers, scientists, politicians, and philosophers dealt specifically with the nature of the STR. In contrast to the industrial revolution, which was based on mechanical power and factory organization, the STR rests on principles of automation and cybernetic management. The two revolutions thus have different internal structures and social consequences. In the first, science and technology remained relatively independent; in the second, technology becomes a scientific enterprise and science is revealed as having immediate technological implications. Whereas the industrial revolution increased the demand for manual labor, the STR diminishes it. What the STR demands is highly educated workers, so that the development of man and his creative capacities becomes the most effective way to increase production. While recognizing many of the bad side effects of technological developments, Richta and his coauthors do not criticize technology but call for its improvement. What they condemn is a one-sided technological progress under social conditions that fail to provide for man's general self-realization. This position includes an explicit critique of capitalist social organization and an implicit rejection of Stalinist centralization. (The analysis and program of *Civilization at the Crossroads* had a significant influence on the reform plans of the Dubček government during the "Prague Spring" of 1968.)

Between 1950 and 1965, Soviet discussions of technology stressed questions related to automation and cybernetics. The key concept in cybernetics is information. Wiener's description of information as something *sui generis,* neither matter nor energy, initially led some Marxists to suspect cybernetics of being a new form of idealism. In 1955 this misunderstanding was corrected in two important articles in *Voprosy filosofii:* "Basic Features of Cybernetics," by S. L. Sobolev, A. I. Kitov, and A. A. Liapunov; and "What is Cybernetics?" by Ernst Kol'man. The resulting interest in cybernetics and its transformation of technology contributed to the Russian understanding of the STR.

The developing Soviet analysis of the STR has involved the publication of a considerable number of works. One of the most influential was a book by A. A. Kuzin and S. V. Shukhardin (1967), which grew out of a paper delivered at a 1964 conference. The latter conference, in turn, resulted from the organization in 1962 of an STR section under the direction of Shukhardin in the Academy of Sciences Institute of the History of Natural Science and Technology in Moscow. Kuzin and Shukhardin distinguish between a technical revolution in tools and machines and a production revolution in social organization. Technical revolutions do not, of themselves, lead to production revolutions if there is no corresponding social revolution. In England in the eighteenth century, for instance, new textile machinery constituted a technical revolution; moreover, in conjunction with changes in the social class structure, it led to a revolution in textile production. The STR thus can be described, in a restricted sense, as a technical revolution in the instruments of production; and, in a broad sense, as a basic change in the organization of the production process once a social revolution has taken place that makes possible the full utilization of the new means.

The most sustained articulation of communist thinking on the STR is contained in the interdisciplinary volume *Man, Science, Technology: A Marxist Analysis of the Scientific and Technological Revolution* (1973). This work is a cooperative effort of the Soviet and the Czechoslovakian Academies of Science; in many ways it echoes the earlier Czech analysis of *Civilization at the Crossroads*. (Richta is an anonymous contributor to *Man, Science, Technology.*)

Beginning with analyses of science and technology, their internal relations and influence on production, *Man, Science, Technology* seeks to clarify the meaning of the term "revolution" as applied to science and technology. Revolution is defined as involving "radical qualitative changes in social structures in the course of the progressive development of society" (p. 19). In this sense scientific revolutions are the result either of "the discovery of fundamentally new phenomena or laws," or of "the utilization of new methods and technical means" (p. 20). In technology the "substitution of new technical media for old ones, by employing quite different principles, signifies revolution" (p. 21). What has now happened is that these two kinds of revolution have merged: technology is a new cognitive method for science; science offers new technical media for technology. The STR in a "narrow sense"—that is, as applied to science and technology in isolation from their social and economic conditions—refers to "a convergence of revolutionary changes in science and revolutionary changes in technology into a united process, science being the leading factor with respect to technology and production and paving the way for their further development" (p. 24). In conclusion, the STR becomes the contemporary unification of science-technology as a productive force; the STR "places between man and nature, not tools or machines, but self-regulating, self-adjusting processes of production" (p. 369).

These conceptual analyses do not remain at the level of abstractions, but are fleshed out with numerous references to developments in physics, chemistry, electronics, engineering, cybernetics, space technology, and biology. There is also a description of the socioeconomic consequences of the STR, which lead to a more general conception of the STR as involving changes in the labor force, in business organization, and in the social order.

Finally, there is what the authors term a philosophical discussion of influences on culture, religion, scientific thinking, and the future world development.

Interspersed throughout *Man, Science, Technology* is the argument that

> only under socialism [can] all these social and personal consequences of the scientific and technological revolution [be] given the chance to develop progressively, in the interests of man himself, while under capitalism they take on ugly shapes due to their tendency to a particularly one-sided development which results in their becoming their very opposites, acting to the detriment of man and society (p. 369).

Since technology, according to the STR analysis, is a kind of production process, it produces goods and services in accord with the interests of the social class that controls it. However, of itself it has a tendency to bring about economic and social effects in harmony with the interests and aspirations of only one class, the proletariat. Thus the STR inevitably contributes to the building up of socialism and the final transition from socialism to communism.

The doctrine enunciated in *Man, Science, Technology* is not, however, the only analysis of technology to be found in the Soviet tradition. Following World War I, in *Vestnik inzhenerov,* the journal of the All Russian Association of Engineers (VAI), and especially in the writings of the Russian engineer, P. K. Engelmeier, a philosophy of technology is articulated that has much in common with German engineering conceptions of this same period. Engelmeier, like many Russian intellectuals prior to the 1917 revolution, had been in close contact with West European thought. In the 1890s he published two articles in German periodicals emphasizing the need for a philosophical exploration and development of the engineering attitude toward the world. In 1911 he contributed a paper on this theme to the IVth International Congress of Philosophy held at Genoa, Italy.

With the founding of the VAI in 1917, Engelmeier became an influential member. His first contribution to *Vestnik inzhenerov* appeared in 1922. In 1927 he was made chairman of a "Circle on General Questions of Technology," formed "to work out a whole new world view, fully adapted to contemporary technical culture." In 1929 *Vestnik inzhenerov* published a paper by Engelmeier, "Is a Philosophy of Technology Necessary?," in which he outlined a full program for the philosophy of technology. The program would, he argued, involve investigation of "the concept 'technology,' the principles of contemporary technology, technology as a biological phenomenon, technology as an anthropological phenomenon, the role of technology in the history of culture, technology and the economy, technology and art, technology and ethics, and other social factors."

Engelmeier's comprehensive vision of a technocratic world view devoid of an overt recognition of Marxist philosophy constituted an implicit challenge to Communist party ideology. In fact, the same issue of *Vestnik inzhenerov* carried an accusatory article by Vladimir Markov (a Stalinist writer who would soon attack Bukharin for revisionist tendencies) to this effect. Kendall Bailes (1974), in an examination of the arguments between engineers and politicians in this period, points out that Engelmeier's philosophy of technology became involved in a general conflict between "expert" and "red." The conflict culminated in the "industrial party affair of 1930" in which large numbers of the Russian technical intelligentsia were tried and executed for counterrevolutionary tendencies. Engelmeier himself does not seem to have escaped this fate.

Still another philosophical analysis of technology, of some prominence in the Soviet-East European school, and one that has not shared the fate of Engelmeier's approach, is the one that has grown out of the "praxiology" of the Polish analytic philosopher Tadeusz Kotarbiński. Praxiology is defined as "the general theory of efficient action"; as such it is markedly sympathetic to various aspects of engineering. At the same time, it extends and develops certain notions first found in Marx. By analyzing technology as a species of efficient action, Kotarbiński is able to show the proper subordination of technology to social organization.

E.   Tensions and Limitations

From this historical survey it becomes apparent that philosophy of technology can have two different meanings. When "of technology" is understood in a subjective sense, as the subject or agent, philosophy of technology is an attempt by technologists or engineers to elaborate a technological philosophy. When "of technology" is understood as object, philosophy of technology points toward attempts by philosophers to take technology as a theme for systematic reflection.

Each approach has its inherent difficulties. The basic problem of technological philosophy is reductionism. Engineers (e.g., Dessauer or Engelmeier) have a strong tendency to view everything in technological terms and to fail to appreciate the nontechnological character of other aspects of reality. The basic problem with philosophical reflection on technology and the attempt to bring the technological phenomenon to consciousness is one of interpretation or "hermeneutics." The philosopher must strive to remain true to the object with which he is dealing, to think it through in a way that neither underinterprets nor overinterprets its inherent significance. Many engineers claim that philosophical analyses of technology do not display a full awareness of what technology means to the technologist himself, that in developing their interpretations philosophers have gone beyond the bounds of what is warranted by technology itself.

The foregoing historical review of schools in the philosophy of technology is limited in more than one respect. It fails to take into account recent contributions to the field by Japanese philosophers. Neither does it give adequate recognition to reflections on the nature and meaning of technology in technologically developing countries such as Argentina, Venezuela, Brazil, and India. There are even African contributions to this new field, especially on so-called alternative technologies.

Concentrating on independent philosophical traditions in Western Europe, the United States, and Eastern Europe fails to do justice to the international character of philosophy of technology. This limitation can be seen most readily through an examination of philosophy of technology in succeeding international congresses of philosophy. Engelmeier contributed the first explicit paper at the fourth congress in 1911. From that time on, scattered papers appeared relevant to the topic. By 1968, at Vienna, there was a whole colloquium devoted explicitly to the subject "Cybernetics and the Philosophy of Technical Science." In 1973, in large measure due to a Soviet and East European impetus, the main theme of the fifteenth congress in Varna, Bulgaria, was "Science, Technology, and Man." Finally, at the sixteenth congress, in Düsseldorf (1978), philosophy of technology had become internationally recognized as a new but important philosophic discipline.

II. PROBLEMS IN THE PHILOSOPHY OF TECHNOLOGY

Philosophical problems of technology are of two kinds. One kind deals with theoretical questions concerning the nature of technology, its relation to science, the structure of technological action, the essence of machines, the difference between machines and men—all of which may conveniently be termed "epistemological" or "metaphysical" issues. The other is more practical, concentrating on the ethical analysis of such problems as industrial alienation, nuclear weapons, pollution, and professional engineering practice.

The two kinds of problems are not necessarily separate. Practical problems engender theory; theory reflects on and stimulates practice. In philosophy there is a persistent need to integrate theory and practice. In philosophy of technology, where a primary question is whether this integration will take on a theoretical or a practical character, this need is felt in a singularly critical form. On the one hand, how can philosophy hope to remain theoretical in the face of such overwhelming practical problems as are raised by modern technology? And should it? On the other hand, how can practical problems be dealt with adequately in the absence of a theory of technology or at least a theory as to how technology engenders problems?

A. Metaphysical Issues

Metaphysics begins with conceptual clarification and logical analysis. The basic conceptual problem in philosophy of technology centers on the term "technology," which has two basic types of meaning. It can *denote* a broad range of things, from tools, assembly lines, and consumer products, to engineering sciences, bureaucratic organizations, and human aspirations. It can also *connote* certain attributes or relations that belong to these different kinds of things in different ways. The variegated denotations make for ready disagreements. One author, thinking of technology primarily as a kind of knowledge, may affirm its neutrality; another, construing technology as will to domination, may assert its nonneutrality. In such cases the difference turns more on definition than on any issue of fact or argument. Stipulative definition solves the problem only by ignoring it. Metaphysical investigation of technology thus places particular emphasis on the development of an adequate definition, one which acknowledges the various denotations ("extensional" usages) while identifying the central connotation (or "intension") that accounts for their being linked together in word or deed. One attempt to recognize the broad extensional range of the term "technology" has distinguished four basic types: (1) technology as object, (2) technology as process, (3) technology as knowledge, and (4) technology as volition (see Mitcham, 1978). Although this attempt stops short of arguing for some specific meaning present in all the related usages, the categories provide a convenient framework for a survey of metaphysical issues.

1. **Technology as Object.** To identify technology with particular artifacts, such as tools, machines, electronic devices, consumer products, and the like, is the commonsense view. Developed philosophically, it involves a classification of technological objects into various types and ultimately the articulation of an ontology of artifacts.

The classification of technological objects began in antiquity with the attempt to distinguish nature from artifice and to describe the principles of simple machines. Natural objects, according to Aristotle, are a substantial union of form and matter such that they have within themselves the principle of motion (change) and rest (resistance to or cessation from change). With artifacts, the relation between form and matter exists at a more superficial level. If a bed were to sprout, says Aristotle, "it would not be another bed that would come up, but wood" (*Physics* II, 2). The science of mechanics, in the form of an analysis of the principles of the five simple "machines"—lever, wedge, wheel and axle, pully, and screw—also began in antiquity, with the pseudo-Aristotelian *Mechanical Problems* and Archimedes' *On Balances or Levers.*

In the Middle Ages, the concept of a machine was extended from an ingenious, hand-operated instrument to manual instruments that, because of their need for power, required more than one man to operate them. With the substitution of animal or natural sources of mechanical energy, as in the case of the horse-drawn plow or windmill, the concept of a machine underwent still another transformation.

The modern concept of a machine—as an instrument in some sense independent of human power or direction—contrasts with the earlier idea of a tool that remains under man's power and control. Franz Reuleaux (1875) and Jacques Lafitte (1932) offer two attempts at classifying machines in this modern sense. Reuleaux, like Aristotle, first distinguishes machines from natural systems of mechanical motion. Contrasting circular motion in the solar system with that in a crank, he argues that, in the "kosmical system," motion takes on definite form because of opposing external forces, whereas in a crank the motion is constrained by internal forces embodied in the rigid parts out of which the machine is constructed. A machine is thus a "closed kinematic chain" or "a combination of resistant bodies so arranged that by their means the mechanical forces of nature can be compelled to do work accompanied by certain determinate motions" (1875, pp. 502–503). In the resulting engineering analysis of machines, machines are distinguished according to the kinds of determinate motion involved. Lafitte maintains that Reuleaux has correctly identified only what he calls "active machines." There are also "passive machines" that constrain motion and transmit force—e.g., posts, walls, and most architectural structures. And there are "reflexive" or self-regulating machines that, since Lafitte's time, have come to be called cybernetic devices.

Lafitte's passive machines include not only architectural structures, but what Mumford (1934) calls utensils, apparatus, and utilities. The proper understanding of premodern technology, for Mumford, depends on a recognition of utensils (baskets, tables, and chairs), apparatus (dye vats and brick kilns), and utilities (reservoirs, aqueducts, and roads) as distinct technological objects; they were the primary manifestations of technology before the coming of the machine. In a later work, Mumford (1952) classifies objects of art as symbolic artifacts; they are "special means of perpetuating, of recalling, of sharing with others [man's] own essential experience of life." Later still, Mumford (1967) argues that large social organizations, such as armies or the Egyptian work force that constructed the pyramids, constitute technological objects that he calls "megamachines."

These conceptual distinctions among various types of technological objects call for some ontology, for an account of the different kinds of reality that each might possess. Kapp (1877), as already mentioned, develops such a theory on the basis of an anthropological analysis. After extensive comparisons between human anatomy and technological inventions, Kapp concludes that weapons and tools are essentially projections of the human body by way of what he terms "systematic restriction"—that is, the isolating or perfecting of one special material or process. Clothes and houses, for example, are extensions of skin and body hair, the catapult an extension of the throwing arm. This notion of technological objects as extensions of the human body is the most widely adopted theory about the ontological status of artifacts. Marx, even before Kapp, appeals to this idea in his interpretation of machines. Mumford, as noted, views objects of art as a special kind of extension, a symbolic extension of interior experience. And Marshall McLuhan (1964) suggests that just as mechanical technology extends the human body, so electronic media extend the nervous system. Debates about computer simulations of cognitive processes, or about whether computers can think, are attempts to determine in what sense, if any, artificial intelligence is an extension of the human mind.

   2. Technology as Process. It is possible that what is fundamental about technology is not so much objects, things made and used, as processes, making and using. To take process or activity as the fundamental category of technology is characteristic of two different professional groups, engineers and social scientists. Engineers, in focusing on process, place stress on the *making* aspect, social scientists on *using*. For engineers what is fundamental is invention and design—making in the originative sense. For social scientists it is production and utilization, the public use of technology, that is most significant. In either case, approaching technology as a process eventually engages the philosophy of technology with theories of the nature of human action and its institutionalizations.

   Human action has traditionally been divided into two types, making and doing. Until the eighteenth century, making and using were commonly denoted by the term "art." The arts, in turn, were divided into servile and liberal, depending on whether the activity involved was primarily manual or mental. This distinction is not the same as that between the useful and fine arts, which is grounded in a distinction between utilitarian and aesthetic purposes. Aristotle suggests another possible difference, dividing the arts into those of "cultivation" and "construction." Cultivation is making or using, which helps nature to produce either more quickly or more perfectly what it could produce by itself; examples are the arts of medicine, teaching, and farming. Construction constrains nature to produce things it would not otherwise produce, the primary example being architecture.

   In Aristotle's cultivation-construction distinction, design would appear to be foreign to the arts of cultivation. In the arts of construction, by contrast, design is central. Today engineering is commonly defined, by engineers themselves, as systematic knowledge of how to design and develop artifacts. The modern engineering curriculum incorporates some mathematics and pure science, the so-called "engineering sciences" (strength of materials, thermodynamics, et cetera), and is actualized by social or economic needs. But it is design, operating under the technical ideal of efficiency—as contrasted with beauty, the ideal of aesthetic design—that orders or integrates the other elements. Engineering design mediates between scientific knowledge and social exigencies. (It should also be noted that design is distinct from the actual manufacturing of products, even though engineers sometimes refer to that as technology in a restricted sense.)

   One process that needs to be distinguished from design is invention.

Invention is often opposed to discovery. As such it can be defined as the creative core of Aristotle's arts of construction, where discovery is the creative core of the arts of cultivation. One discovers the cure for a disease; one invents a new artifact or production process. As the creative core of engineering, invention involves conceptual insight and operational testing. It was not Sir George Cayle, the founder of the science of aerodynamics, who invented the airplane by describing how it might be constructed. The airplane was invented by the Wright brothers, who designed, built, and flew one at Kitty Hawk in 1903. The movement from conceptualization to operational testing can proceed in either an unconscious, intuitive manner or in a rational, systematic way. The latter is what Alfred North Whitehead called "the greatest invention of the nineteenth century, . . . the invention of a method of invention." Thomas Edison's "invention factory" at Menlo Park, New Jersey, set up in the 1870s, is often taken as the prototype.

Design as a human activity has been studied by engineers, psychologists, business management theorists, researchers in artificial intelligence, historians—and to some extent by philosophers. These studies tend to emphasize either the systematic or the intuitive and imaginative aspects of design processes. Herbert Simon (1969) and some of the authors whose essays are in the Rapp collection (1974) attempt to elaborate a methodology of engineering design. Edwin Layton (1974) and Eugene Ferguson (1977) point out the ways in which engineering design involves imaginative and preconceptual processes. "Engineers display a plastic, geometrical, and to some extent nonverbal mode of thought that has more in common with that of artists than that of philosophers," according to Layton (p. 36). In either case, design differs from strictly conceptual thinking by being a kind of miniature construction. The question is, to what extent does this miniature construction take place by conceptual modeling or by more intuitive forms of representation?

The social science approach to technology as process stresses not the miniature construction of design but the macroconstruction of production processes and the corresponding institutions for the social utilization of products. The works of Marx and Friedmann are good examples, but Jacques Ellul's *The Technological Society* (1954 [1964]) is the most comprehensive, contemporary example.

Ellul begins by specifically rejecting any identification of technology with technological objects such as machines. Instead, modern technology is identified with "technique," which is defined as a kind of action whose fundamental characteristic is the rational pursuit of efficiency. Ellul's definition has roots in Max Weber's observation (in *Wirtschaft und Gesellschaft*, 1922) that there are techniques of every conceivable human activity—from prayer and thought to education, politics, artistic production, and performance. Engineering design, in this sense, is just one among many techniques.

The problem for the social scientist is to explain what Harold Lasswell terms "technicalization" (and others call "technicization"). Traditional techniques were "hedged about by sanctions" and therefore "included in the social order"—that is, in "the pattern of mores and countermores of institutions." In traditional society, for example, animals could be eaten only if butchered in a ritually prescribed manner. In modern society techniques have ceased "to be treated or symbolized as subject to sanctions"; at best they are subject to calculations of overall efficiency, as for instance in contemporary programs of technology assessment. Technicization is thus the transformation "from involvement with mores and countermores to expediency alone" (see Lasswell and Kaplan, *Power and Society,* 1950, pp. 50–51).

Ellul subjects this rudimentary observation to thoroughgoing analysis. His analysis rests on a distinction between techniques and technology—or what he calls "technical operations" and the "technical phenomenon." Technical operations are any human activities "carried out in accordance with a certain method in order to attain a particular end" (*Technological Society,* p. 19). They correspond, at least in part, to what psychologists variously refer to as habits, action strategies, behavioral schemes, or modules. To emphasize the connection between technical operations and the processes of doing and making, they are also sometimes referred to as skills. The development of skills is generally an unconscious result of practical experience, but it is readily subject to conscious direction. When this occurs, the result is the birth of technology—the "technical phenomenon." Technology takes what "was previously tentative, unconscious, and spontaneous and brings it into the realm of clear, rational, and reasoned concepts" (p. 20).

Although different from the traditional techniques of doing and making, the technical phenomenon is not unique to the modern period. According to Ellul it can be found in limited forms in premodern civilization—he mentions the Roman judicial system and the scholastic method of disputation. Prior to the modern period the technical phenomenon was restricted to specific subjects and was geographically circumscribed. The concept of the "one best way in the world" had not yet been formulated; instead, "it was a question of the one 'best way' in a given locality" (p. 70). Because of such limitations human freedom was preserved. Societies of quite different types were able to coexist, and the individual could repudiate the technical phenomenon and survive. In its distinctly modern form, the technical phenomenon is determined historically by quantitative proliferation and a resultant set of so-called "external relations."

For Ellul what has become crucial is not the inner structure of modern technology; what is decisive is "the characteristics of the relations between the technical phenomenon and society" (p. 63). This new relationship is described as giving to modern technology seven key features: (1) rationality, (2) artificiality, (3) automatism, (4) self-augmentation, (5) monism, (6) universalism, and (7) autonomy. (Ellul apparently does not mean this list to be either definitive or exhaustive; see his 1963, 1975, and 1977.) The bulk of

Ellul's argument consists of an examination of these seven features, especially as they are manifested in economic activity, social organizations, and what he calls the "human techniques," e.g., of medicine or education—the traditional arts of cultivation. One way of restating Ellul's conclusion is to say that techniques appropriate to the arts of construction have been extended, by way of an enlargement of the concept of efficiency, to the arts of cultivation. In the technological society, doing is planned and designed in the same manner as making.

Ellul's work has been both widely praised and roundly criticized (usually for his conception of the autonomy of modern technology), but has been subjected to little serious philosophical analysis. Mitcham and Mackey (1971) have made an initial attempt to sort out some of the basic conceptual issues. They argue that Ellul's technique-technology distinction is not wholly consistent; that his differentiation between the "technical phenomenon" in ancient and modern times assumes the priority of external over internal relations; and that Ellul does not acknowledge the way in which modern technology, more than a process, is a total attitude toward the world. David Menninger (1975) maintains that Ellul's apparent pessimism is the result of a dialectical approach, which is meant to be a call to liberty and action. David Lovekin (1977) contends that Ellul does not view technology primarily as a process but as a kind of consciousness. Ellul's sociological descriptions are meant merely to illustrate the workings of this consciousness. The technical phenomenon is a projection of the "logic of technology."

Ellul has recently published *Le Système technicien* (1977), the first part of a comprehensive revision of *The Technological Society*. Where *The Technological Society* begins from a number of factual observations and ends with the description of a system, *Le Système technicien* begins with the concept of a system and tries to see what more can be revealed from this perspective. It thus places more emphasis on conceptual issues; it also adds considerable factual updating (there were not many computers around in the early 1950s) and extends discussions of later writers (e.g., Habermas, Illich, Richta). The central issue remains the ideal of efficiency as this is realized in social terms through the all-encompassing technological system.

The engineering and social science analyses of technology as process unite in pointing toward the central importance of the ideal of efficiency. Modern engineering design is miniature construction undertaken in pursuit of efficiency. Roman engineers designed aqueducts by trial-and-error methods that resulted in many failures and much massive overbuilding. Modern design methods allow the engineer first to construct on paper, then to test out by means of mathematical or graphic representations, before proceeding to the construction of progressively more full-scale models. All this is done in order that neither more nor less material and energy will be put into a particular project or process than is necessary. In fact, the design process itself must be evaluated in terms of its efficiency in the context of some project. For Ellul, all using, insofar as it becomes incorporated in the "technical phenomenon," partakes of a similar conscious pursuit of efficiency. Business management, operations research, and much recent political science analysis are examples of this technicization of using and doing.

Attempts to give conceptual clarity to the ideal of efficiency have been undertaken from a number of different perspectives. As early as 1930 Bernard Bavink located the essence of technology as process in the technical ideal of "fitness for a purpose." This ideal functional fitness or efficiency is distinct from economic efficiency. When he designs a bridge, a civil engineer, as engineer, is not concerned that it cost as little as possible, but that concrete and steel not be wasted. But if labor is more expensive than concrete, it might be cheaper to build fewer forms and waste some concrete. Henryk Skolimowski (1966), following the lead of Kotarbiński's general science of efficient action, has tried to specify more completely the forms of efficiency found in different branches of engineering. Some problems related to the meaning of efficiency can also be found in pragmatist and instrumentalist discussions of truth.

The description of technicization as comprehensive efficiency sometimes raises questions about the relation between efficient action and fully human action. When arguments about technicism and technocracy go beyond debates about the adequacy of specific sociological descriptions, the most philosophically interesting question concerns the humanization or dehumanization of man through technology. A critique of technological efficiency as dehumanizing hangs on efficiency's obscuring of a more important aspect of human nature and human action. Using Kantian terms, a Romantic could argue that the problem with the technical phenomenon is that it is *no more* than phenomenon or appearance; it is not reality. Reality, Kant's "noumenon," or thing-in-itself, lies outside the realm of efficiency; indeed, it is covered over by its proliferation. What is more real, more truly human than efficiency, is freedom, which is disclosed to man neither in his thinking nor in his making but only in his (moral) doing. In a world in which all human action takes on the character of efficient making, man is in danger of losing his freedom.

The counterargument for humanization by technology generally stresses the material benefits of technology and the creative experience of the design activity. A recent vigorous statement of this position is given by Samuel Florman in *The Existential Pleasures of Engineering* (1974). For Florman efficiency and artifice are in themselves worthy objects of emotional response and aesthetic appreciation. Application of the procedures of design to the general planning of human action, and the inevitable formulation of issues in technical terms, is not something to be rejected. It is a means for human fulfillment and the humanization of the world. "Analysis, rationality, materialism, and practical creativity do not preclude emotional fulfillment; they are pathways to such fulfillment. They do not 'reduce' experience . . .;

they expand it" (p. 101). The Romantic rejection of technology as dehumanizing, according to Florman, hinges on a soft-headed fear of efficiency. Two related arguments are those of the anthropologist André Leroi-Gourhan (1943, 1945) and the philosopher James K. Feibleman (1977).

3. Technology as Knowledge. A third approach to technology looks upon it as a kind of knowledge. This is the view that has so far received the most scrutiny from "hard-nosed" philosophers. Mario Bunge (1967) and Stanley Carpenter (1974) have argued for a recognition of the following types of technology as knowledge:

First, there is unconscious *sensorimotor skill* in making or using artifacts. Since sensorimotor awareness is unconscious, it does not qualify as knowledge in the strict sense. A further result is that it is teachable only by means of the kind of intuitive training through practice and example that comes from apprenticeship to a master. In a related argument, Andrew Harrison (1978) maintains that intelligence of a specific sort, which he designates as "attention," plays a primary role in the exercise of making skills.

Second, technical *maxims* (Carpenter) or *rules of thumb* of prescientific work (Bunge) articulate generalizations about successful making or using operations. Most cookbook recipes, and other directions for making anything from clothes to model airplanes, are composed of such rules. The use of the term "rule" in this context is significant. Technological rules differ from scientific laws. Laws, in the scientific sense, are descriptive of reality; rules are prescriptive of action. This distinction needs to be kept in mind because even when the term "law" is used to describe a kind of technological knowledge, there is usually some relation to rules.

Third, descriptive *laws* (Carpenter) or "nomopragmatic statements" (Bunge) take the form, "If A then B," with concrete reference to experience. In Carpenter's words, descriptive laws "are like scientific laws in being explicitly descriptive and only implicitly prescriptive of action, but they are not yet scientific in that the theoretical framework which could explain the law is not yet explicit" (p. 165). They are empirical laws generalized on the basis of experience, from which we can infer technological rules of the form, "To get B, do A." Coulomb's empirical laws for constructing retaining walls —formulated on the basis not of engineering geology and physics but of observations about the sizes and shapes that hold up under certain conditions—are an example. There are also many descriptive laws of use, such as those developed by Frederick W. Taylor, based on his time-motion studies at the Watertown arsenal.

Fourth, there are technological *theories,* which either systematize descriptive laws or provide a conceptual framework to explain them. Technological theories, according to Bunge, are of two types, substantive and operative. "Substantive technological theories are essentially applications, to nearly real situations, of scientific theories." Aerodynamics or the theory of flight, for example, is an application of fluid dynamics. Substantive theories constitute the so-called "engineering sciences"; they are applied science in the strict sense. "Operative" technological theories "are from the start concerned with the operations of men and men-machine complexes in nearly real situations" (pp. 62–63). Examples include decision theory and operations research. Herbert Simon's *The Sciences of the Artificial* (1969) is a classic investigation of the structure of such theories. Substantive theory employs both the content and the method of science; operative theory, using only the method of science, applies it to problems of action in order to develop "scientific theories of action." Substantive technological theories are usually tied to making; operative, to using.

To view technology as a kind of knowledge not only invites epistemological analysis; it transforms technology from an extension of man into an inherent constituent of human nature. This makes the reflections of psychology, ethnology, and anthropology on the scope and structure of human thinking relevant to philosophy of technology.

The cognitive psychologist Jean Piaget, for instance, distinguishes four different stages in the development of intelligence: sensorimotor cognition or the development of the psychomotor complex (birth to age 2), preoperational or imaginative thinking (ages 2–7), concrete operational thinking or the interiorization of objective functional relationships (ages 7–11), and formal operational or abstract thinking (from puberty on). The stages of this sequence do not exhibit a simple one-to-one correspondence with the four types of technology as knowledge. Nevertheless, skills are characteristic of the earliest stage, technological theories of the latest— implying that the latter are representative of mature, adult cognition.

Studies of mythology and symbolic thinking, however, raise questions about any simple identification of modern technology as knowledge with fully developed human intelligence. The ethnologist and philosopher Lucien Levy-Bruhl, for instance, distinguishes between prelogical and logical thinking. The former is "corporate" thinking and exhibits an interest in mystical participation with the life of the natural world that is quite unlike the analytic habits of scientific or logical thinking. It is a mistake, Levy-Bruhl and others argue, to see primitive thinking as failed scientific thinking. Prelogical intelligence has its own integrity, its own structures and achievements, which are different from and not necessarily inferior to or simply preparatory for those of logical thinking. Although Levy-Bruhl was eventually persuaded by the presence of rational and empirical elements in primitive thinking to abandon the hard and fast distinction of his earlier work, his basic contrast remains influential—in the work of Mumford, for instance. It poses, in dramatic terms, the question: In what sense is technological thinking constitutive of *Homo sapiens*?

Cybernetics—and the related fields of information theory, systems theory, automata theory, et cetera—need to be considered in this context. Alternatively defined as the science of "control and communication in the animal and in the machine" (Norbert Wiener), or as the science of all possible machines (Eric Ashby), cybernetics is properly an engineering science, a kind of technological knowledge. During the period of its initial formulation in the 1940s, cybernetics was closely associated with neurophysiology on the hypothesis that negative feedback mechanisms are basic to the workings of the central nervous system. As a general theory of artifacts (from thermostats and self-tracking radar to prosthetic limbs and computers) and operations (from corrective neurosurgery to business management), cybernetics proposes to give a unified explanation of material, social, and mental phenomena. In its most comprehensive form, cybernetics gives an extremely general account of both objects and processes, natural and artificial.

Cybernetics, as Wiener's definition indicates, implies a fundamental identity between animal and machine. In traditional theory the difference between living and nonliving objects was that living things are self-moving, nonliving things are not. One aspect of the self-moving character of living things is that they are alleged to possess a source of motion within themselves or are otherwise able to draw energy out of the larger universe on their own initiative. Machines, despite their sometimes apparently self-moving character, cannot provide or acquire energy for themselves. Early modern technology is a power technology, focusing on ways of producing and transmitting energy. In cybernetics, the emphasis shifts from sources of power to determinate operations; the availability of energy is taken for granted. Cybernetics is the science of "all forms of behavior insofar as they are regular, or determinate, or reproducible" (Ashby, 1956). Since both men and machines exhibit this regularity of behavior, cybernetics tends to obscure traditional differences between man and machine, between the living and the nonliving.

Whether, in the light of this reduction of animal and machine to patterns of determinate behavior, one views machines as extensions of animals (including man), or animals as complex machines, seems to be a question of interpretation. Keith Gunderson (1971) has pointed out the ways in which the issues raised here can be traced back at least to La Mettrie's L'Homme machine (1747), which argued for a mechanistic interpretation of human behavior. Nineteenth-century debates between mechanists and vitalists in biology reflected similar issues in the man-machine question. So do current arguments about the implications of Kurt Gödel's incompleteness theorem, the limitations of artificial intelligence, and the validity of computer simulations of human cognitive processes. The vitalist thesis (or one of its variants) is calculated to preserve the presumed superiority of man, whereas the mechanist approach claims to serve as a better stimulus for scientific and technological research.

Cybernetics is an instance of what was referred to above as "techno-logical philosophy" and is subject to a corresponding reductionist tendency. The most general statement of cybernetic reductionism concerns, not the distinction between men and machines, but that between objects and processes. Cybernetics claims to reduce objects to processes; what is important is not what a thing is but how it behaves. The foundation of regular or ordered behavior is the technical concept of "information." Information may be described informally as a determination of the possibilities of behavior. In classical mechanics a machine is a mechanical linkage arranged so that any energy input into the system results in certain determinate motions with as little loss of energy through resistance as possible. A cybernetic device, by extension, is a communication linkage arranged so that any information input results in certain determinate behavior with as little information loss through "noise" as possible. A machine is n longer a "closed kinematic chain" (Reuleaux); it is a "closed information linkage."

Cybernetics, as the theory of the way information states interact with one another to produce certain behaviors, explains the nature of techno-logical objects in terms of information processing and becomes able to guide or direct this processing. "Cybernetics," in fact, means knowledge of control; Wiener derived the term from the Greek word for steersman.

4. Technology as Volition. The control of a process is only partially dependent on exact knowledge of the systematic functioning of that process; it also depends on the aims, intentions, desires, and choices of the "steers-man." Cybernetics may provide the cognitive means of control; it does not provide much help in determining the uses or purposes of control. As Wiener acknowledges, cybernetics provides "know-how" but not "know-what" (1950 and 1964).

The most commonly heard cliché about technology is that technology in itself is neither good nor bad. It is neutral, with its value wholly dependent on the uses men make of it. To speak of using is, in traditional terms, to speak of willing or volition—that is, of some affective or practical response to what is cognitively present to consciousness. The most unformulated, the most difficult to formulate, and yet the most commonly assumed viewpoint on technology is that it is grounded in some human act of the will.

The chief difficulty in formulating a theory of technology as volition is that the concept of will is one about which the history of philosophy provides precious little consensus. The term "will" does not occur as such in Greek thought; it enters Western intellectual history by way of the Christian philosophical tradition. In modern times it has been subject to both excessive emphasis (Friedrich Nietzsche) and extreme criticism (Gilbert Ryle). Nevertheless, as Paul Ricoeur has argued at length, and as Hannah Arendt has recently reaffirmed, the issue of volition remains central to any philosophy of action—and thus to any comprehensive understanding of technology. Moreover, it is a concept implicit in many philosophies of technology that do not make it an explicit issue.

Technology is almost always distinguished from science on the basis of ends or intentions: science is said to aim at knowing the world, technology at controlling or manipulating it. Skolimowski (1968), for instance, asserts that "technology is a form of human knowledge," then distinguishes it from scientific knowledge by saying that "science concerns itself with what *is;* technology with what *is to be*" because "our technological pursuits consist in providing means for constructing objects according to our desires and dreams" (p. 554). Others have distinguished between technology and moral action on the basis of the objects of volition. In technological making there is a will to transform the world; in moral doing the will is directed toward transformation of self.

Again, Ortega's existentialist analysis and Mumford's historical approach make implicit reference to the notion of technology as volition. For Ortega, technological objects, knowledge, and processes are grounded in a willed self-realization. Creation, at the level of material invention, is preceded by a self-creative affirmation. To adapt Sartre's well-known formulation: existence (as a willing subject) precedes essence (the willing of anything specific). For Mumford the problem with "mono-technics" is that it exemplifies a will to power at odds with a will broadly oriented toward life; a single purpose dominates and excludes all others.

What is lacking in these formulations is a comprehensive articulation of the concept of volition; there is, consequently, no explicit application of such a theory of willing to problems of technology. Paul Ricoeur, in his phenomenological description of willing, identifies three levels: "I will" can mean "I desire," "I move my body," or "I consent." Technology can be analyzed in at least these three volitional senses, as desire, as motivation or movement, and as consent. Such an analysis could begin to elucidate the dialectic of "technological eros" (see Jakob Hommes, 1955). As is documented by the sociological literature, technological desires engender and are reinforced by technological movements, which through the creation of objects, processes, and knowledge, reflect back onto and are supported by a consent to the technological presence.

One analysis of technology that points in this direction, although it makes no use of Ricoeur's categories, is that of Martin Heidegger. Heidegger's analysis of human existence is pregnant with implications for understanding technology as volition. As one of the most influential philosophies of the twentieth century, it deserves to be considered at some length.

In his early work Heidegger makes few explicit references to either technology or volition as such. Yet it is against the background of his early phenomenological description of human existence that Heidegger's later explicit arguments are developed. Heidegger begins with a phenome-

nological description of the existential features of *Dasein* ("to be here"),
a kind of Cartesian consciousness nonetheless characterized by worldly
involvement. In *Being and Time* (1927) Heidegger presents human existence or *Dasein* as inherently a being-in-the-world, and then proceeds to
investigate different aspects of this being-in-the-world in order to disclose
the distinctive character of human existence.

The world of *Dasein* is characterized primarily by a practical concern
for or care about *(Besorgen)* manipulating things and putting them to use.
Things encountered in the process Heidegger calls "equipment" or "gear."
"In our dealings we come across equipment for writing, sewing, working,
transportation, measurement." Breaking down the writing-equipment totality into its constituents discloses "inkstand, pen, ink blotting pad, table,
lamp, furniture, windows, doors, room" *(Being and Time,* p. 68). The distinguishing character of all equipment is that it is fundamentally context
dependent. It is not individual utensils that are first grasped and then
assembled into equipment totalities. Equipment is present initially as part
of a larger framework or system; then within this framework, individual
items can be recognized as pieces of equipment.

The world within which human existence as *Dasein* operates is a
system of relationships that Heidegger calls readiness-to-hand. Within this
system, materials are defined by their usability, tools by their serviceability.
Common sense tells us that things are first simply present-at-hand as things
in their own right; but according to Heidegger this is not the case. The
description of a hammer as "for hammering" is more primordial than any
conceptual description of the hammer as being of some particular size,
shape, weight, and color. To conceptualize a utensil as a neutral object that
is present-at-hand is to abstract from its given state. Generalized, this insight
means that science is an abstraction from technology, knowledge for its own
sake an abstraction from practical knowledge. The difficulties encountered
by a philosopher such as Descartes—who begins with pure consciousness and
tries to derive the practical world from it—confirm this insight. Pure consciousness derives from practical consciousness and not vice versa.

In the human realm, *Dasein's* being-in-the-world turns into a being-
with-others. Practical involvement takes on the new form of solicitude or
care for *(Fürsorge)* members of the social community.

Finally, in its relation of being-in, *Dasein* is characterized by two fundamental structures or "existentialia" as Heidegger calls them: mood and
understanding. Heidegger's special use of the term "understanding"—not
something theoretical but something practical, related to mood—under-
scores again the centrality of technological activity in his analysis of human
existence. Heidegger is at pains to present making and using skills as true
forms of knowledge, with conscious theoretical knowledge being derived
from this preconscious nontheoretical base. The reason is that he is trying
to analyze knowledge as constitutive of the world in the sense required by

the distinctly modern conception of man as maker. This view necessitates his interpretation of what in traditional terms has been called practical knowledge—the kind of knowledge that is intimately tied up with volition (and therefore temporality)—as the fundamental form of knowledge. Only this kind of knowledge forms both the world and men in it. More deeply than any previous philosopher, Heidegger reveals the essence of *Homo faber*.

At the end of the first part of *Being and Time*, Heidegger concludes his existential analytic by arguing that care *(Sorge)* is the essence of *Dasein*. Being-in-the-world necessarily entails a multifaceted practical involvement with the entities of the world. Although "care" cannot be translated simply as volitional activity, Heidegger explicitly states that it is care that makes willing possible. Although Heidegger does not use the terms "volition" and "will" frequently, *Being and Time* presents technology as object, process, and knowledge as fundamentally related to volition.

In Heidegger's celebrated turn away from the phenomenological method of *Being and Time*, his thinking does not retreat on this point. It becomes even more explicit. In an essay on "European Nihilism" (1940, but not published until 1961), Heidegger argues that modern technology could only arise in a world which has become nihilistic or forgetful of Being. The nihilism of Nietzsche's "will to power," Heidegger claims, is the culmination of Western subjectivism and leads to the pure "will to will" of the technological age.

Heidegger's most important later discussion of the subject is "The Question Concerning Technology" (1954 [1977]). In this essay, which grew out of lectures given in 1949 and 1950, Heidegger rejects the common ideas of technology as pure means and human activity. Instead, he argues that technology is a kind of truth, a kind of revealing or disclosing of what is. This view is still very much in harmony with *Being and Time*. Where he advances beyond the earlier analysis is in making a strong distinction between ancient and modern technology and in stating the essence of modern technology. Ancient technology reveals by means of the "bringing-forth" of art and poetry. With modern technology, by contrast, the revealing is a "challenging," a "setting-upon." In less cryptic language: where premodern technology cooperated with nature to bring forth artifacts, modern technology imposes on nature, forcing it to yield up materials and energies that are not otherwise to be found.

To clarify this characterization of modern technology as a revealing that has the special character of a "setting-upon" and "challenging-forth," Heidegger contrasts an ancient windmill or water wheel with an electric power plant. Each harnesses the energy of nature and puts it to work to serve human needs, so that at first each might be construed as a technical object in some univocal sense. But in fact the windmill and water wheel remain related to nature in a way that makes them like objects of art; they reveal the full depth of some particular part of the earth and are funda-

mentally dependent on the earth in a manner that modern technology is
not. An electric power plant, by contrast, unlocks basic physical energies
and then stores them up in an abstract, nonsensuous form. From prehistoric times until the industrial revolution the materials and forces men
worked with remained fairly constant—timber, stone, wind, falling water,
animal energy; so much so that for eighteen hundred years the construction
processes described in the writings of the Roman engineer Vitruvius could
remain authoritative for technological practice. But modern technology
proceeds in a new way to exploit the earth—extracting stored-up energy in
the form of coal, then transforming it into electricity that can be further
stored up and kept ready for distribution and use at man's will. "Unlocking,
transforming, storing, distributing, and switching are ways of revealing"
characteristic of modern technology (1954, [1977, p. 16]). Moreover, these
processes, unlike traditional techniques, never generate determinate objects.
Instead they generate a world of what Heidegger calls *Bestand*—"stock,"
"standing-reserve," things that are "in supply." The world of modern technology always stands ready and available for manipulation and consumption. *Bestand* consists of objects with no inherent value apart from human
use. Like plastic, all their character is dependent on human decisions about
what they will be used for and how they will be decorated or packaged.

What attitude toward the world makes modern techniques and the corresponding *Bestand* possible? Heidegger calls the modern attitude *Gestell*.
He admits to taking a common word, which in its normal usage means
something like "stand," "frame," or "rack," and giving it a technical philosophical meaning. The root *stell* refers back to *stellendes* ("setting-upon")
and connotes that cognitive frame of mind that conceives nature as a technologically manipulable system. The standard English translation is "en-
framing," which emphasizes the active character of *Gestell*, although perhaps an equally good translation would be "framework," keeping in mind
the English expression "framework of thought." In any case, "*Gestell* means
that way of revealing that holds sway in the essence of modern technology
and is not itself anything technological" (p. 20). *Gestell* is not another part
of technology; it is that attitude toward the world that is the foundation
for, yet wholly present in, technological activity. It is, simply, the technological attitude toward the world.

Gestell can be approached, from one point of view, as an impersonal
cognitive frame of mind. But according to Heidegger, in what is undoubtedly
one of his most provocative arguments, *Gestell* is more fundamentally an
impersonal volition. Not only does *Gestell* "set upon" and "challenge" the
world—a description that already hints at volitional elements; it also sets
upon and challenges man. Ultimately, it is not just man's personal needs
and desires that give rise to modern technology. "The essence of modern
technology starts man upon the way of that revealing through which the
real everywhere, more or less distinctly, becomes *Bestand*" (p. 24).

Gestell is thus presented as an historical destiny or fate that calls man forth to act in a particular way. It is not, however, a fate that compels in some crude sense. Heidegger suggests elsewhere that a "restoring or surmounting" of technology is possible, as "one gets over grief or pain" (1962 [1977, p. 39]). Again (1959 [1966]) he calls the process a "release," with a contemplative meaning to the term.

Heidegger's work has spawned a considerable volume of philosophical commentary on technology from what might be called a broadly "Heideggerian" perspective. The works of Magda King (1973), William Lovitt (1973), and Michael Zimmerman (1975 and 1977) can be taken as representative expositions. Important extensions of Heidegger are to be found in Kostas Axelos (1969), Reinhart Maurer (1973), Albert Borgmann (1971 and 1978), Hwa Yol Jung (1972, 1974, and with Petee Jung, 1976), and Don Ihde (1979). Finally, major developments and adaptations of the Heideggerian approach can be found in Hannah Arendt (1958), Herbert Marcuse (1964), Hans Jonas (1966, 1974, 1976, and 1979), and Hans-Georg Gadamer (1977).

Others have been strongly critical of Heidegger's approach. Engineers like Dessauer (1956), and philosophers sympathetic to engineering practice, like Moser (1958), have argued that Heidegger fails to appreciate the real nature of technology, that his analysis rests on too little concrete familiarity with the thing he proposes to understand, and that his language is needlessly complex and obscure.

For Heidegger the heart of philosophy is hermeneutics or interpretation. He thus exemplifies in a clear way the basic alternative to technological philosophy. The engineering objection, restated, is that Heidegger has overinterpreted his subject matter. Heidegger's philosophical interpretation of the meaning of technology, in turn, illustrates the primary hermeneutical problems of how to interpret without losing contact with a subject matter, and how to stay close to a subject matter without becoming enmeshed in it.

For Heidegger the problem is both simplified and complicated by the fact that technology is not just artifacts or processes or scientific theories of some particular sort, but the whole modern volitional stance toward the world. It is simplified, because modern technology is a form of practical consciousness, so it is not necessary to have detailed engineering knowledge or a close relationship with technical objects and procedures in order to interpret it; one is able to rely simply on the language of common experience and ordinary everyday consciousness if one lives in the technological milieu. The problem is complicated because our language, since it has become inherently technological, can no longer be used to talk about technology in a nontechnological way. All talk of technology tends to become just more technology. (An example of this might be the practice of technology assessment. Rather than something independent of technology, TA exhibits a strong tendency to become itself a kind of technology.) As a result,

Heidegger feels compelled, in order to describe the language of technology, to create another language. He thinks it is necessary to try to step outside the common presuppositions and framework built into our technological world-view; this explains why his language sounds somewhat obscure. Only this step outside, he implies, will enable us to move from mere descriptions of technology to its inner meaning, to move toward the essence of technology.

5. **Ancient versus Modern Technology.** The deepest philosophical analysis of technology attempts to reach that point at which technology takes on its greatest density—the point at which objects, processes, knowledge, and volition meet. Dessauer, for instance, defines technology as "real being from ideas through purposeful designing and moulding of resources given in nature" (1956, p. 234). Although this definition does not explicitly refer to technology as object, the three other modes are indicated. Rapp (1974) identifies skills, engineering science, production processes, and objects and their uses as aspects of technology—again with obvious, if not completely homogeneous, references to technology as object, process, knowledge, and volition.

This approach to the essence of technology is related to the traditional understanding of the nature of causation. Moser (1958), commenting on Dessauer, observes that the latter's definition is modeled on Aristotle's distinction between material, formal, efficient, and final causes. Natural resources constitute a material cause; ideas are the formal cause; designing and molding serve as efficient causes; and purpose is the final cause.

Conceptual clarification also leads to more substantive distinctions. At its point of greatest density, technology separates into two species conveniently designated as ancient and modern. This distinction is hinted at in various ways by Mumford's differentiation of bio- and monotechnics, Aristotle's division of the arts into those of cultivation and those of construction, and Heidegger's analysis of the differences between technology as a bringing forth and as a challenging or setting upon. Ancient technology incorporates individual objects fabricated with naturally available energies and materials on the basis of intuitive knowledge for the limited purposes of usefulness and pleasure within a variegated spectrum of activities. The ideal type for ancient or traditional technology is the handcrafted making and using of utensils; as such, a better word for it might be "technics" (from the Greek *techne*) or "art." Modern technology, by contrast, involves mass-produced objects designed to utilize abstract energies and artificial materials on the basis of scientific theories for purposes of efficiency, power, or profit. Here, the ideal type is the institutional or assembly-line production of electric devices made to function within any system that maximizes power or economic expansion.

This substantive distinction is highlighted by the Mexican philosopher Octavio Paz (1974), who observes that craftwork is "not governed by the principle of efficiency but of pleasure, which is wasteful, and for which no rules exist. . . . Its predilection for ornamentation is a violation of the principle of efficiency" (p. 21). With traditional crafts, knowledge tends to collapse into the techniques or skills of bringing forth, and the craft object stands out as a unique creation, even when it is something as mundane as a basket or a wooden bowl. With modern technology, the ideal is "the ever-increasing production of ever-more-perfect, identical objects" (p. 19). Skilled handcraft is replaced by engineering design separated from the concrete making activity, which is itself transformed into a systematic production process, with the result that artifacts no longer stand out as distinct creations but tend to disappear into their functions. Making and using are thus of two types: ancient technology or technics, and modern technology—or simply technology.

B. Ethical and Political Issues

Despite the theoretical priority of metaphysical analyses, it is ethical and political concerns that have dominated the philosophy of technology. This situation obtains in part because of the modern emphasis on practice over theory. In part, it reflects the pressing nature of the problems raised by technological progress. The commitment to practice is ironically self-confirming. Ethical-political analysis of modern technology can thus be approached through a sketch—necessarily selective—of some overlapping and more or less historically developing issues.

1. Technology and Work. Historically speaking, the original crisis in modern technological practice grew out of transformations in the nature of work during the industrial revolution. Oppression of the worker and the psychological consequences of the division of labor and of mechanization have been major themes in the discussion of technology since the eighteenth century. As poverty has ceased to be the overriding issue it once was, attention has centered more on the "problem of alienation."

Alienation is a multifaceted issue. The idea itself is grounded in a modern reflection on the complexities and ambiguities of making and using, against the background of a traditional reflection on the complexities and ambiguities of thinking and moral doing. In Platonic philosophy, alienation is a process by which a person goes out of himself to become unified with some transcendent reality. St. Augustine, for instance, speaks of *alienatio mentis a corpore* to signify that state in which the human soul is elevated above itself to become one with God. Such a separation is a positive good, is in fact the perfection of thinking. The Hebrew prophets, however, speak of immoral action and religious legalism as separating or alienating man from God. In this context, alienation is to be actively rejected by the altering of one's personal behavior.

In technology, alienation takes on a different but related character. Just as thinking does not automatically terminate in understanding, so making, instead of leading directly to appropriation and humanization of the world, involves at least a moment of alienation or estrangement. But this alienation is neither a perfection of making, nor is it the kind of separation which can be overcome by any simple change of personal behavior.

The first philosopher to deal explicitly with the issue of alienation is Georg W. F. Hegel. Hegel approaches consciousness as a kind of self-creative practical activity. He thus deepens the modern epistemological princple that to know involves being able to make. He shows how, in many instances, knowledge develops out of a process analogous to that in which a worker discovers himself and takes satisfaction in his work. The making of consciousness, in the Hegelian understanding, involves an initial self-alienation, a separation and objectification of some unconscious part of oneself. Once objectified, this element is available to be brought into consciousness; the alienation is overcome through a recognition of its true foundations in the creative self. Alienation is a process for the immanent enrichment of a creative subject through the differentiation and appropriation of its own contents.

Marx rejects the idea of alienation as a means to a higher unity within the self. He treats it as a distortion of the human essence. This is because Marx no longer sees thinking, even when understood as a kind of making, as the essence of man. The human essence is making itself; human nature is realized in work. However, this possibility is denied by the capitalist economic system. Under capitalism, work is coercive rather than spontaneous and creative; workers have little control over the work process; the products of labor are expropriated by others to be used against the worker; and the worker himself becomes a commodity in the labor market. "All these consequences result from the fact that the worker is related to the *product of his labor* as to an *alien* object" ("Estranged Labor," in *Economic and Philosophic Manuscripts of 1844*).

Subsequent to Marx, sociologists and social philosophers have enlarged the concept of alienation. The 1950s especially witnessed a rediscovery of the concept of alienation, as well as a tendency to link alienation with Romantic criticisms of technology as separating man from nature and his affective life, with the sociological categories of *anomie* (Durkheim) and "disenchantment" (Weber), and with Freud's psychological theory of repression. Automation also reinforced the importance of the problem of alienation.

Robert Blauner (1964) offers a systematic and empirical study which attempts to relate many of these aspects of the problem. For Blauner there are four dimensions of alienation: powerlessness, meaninglessness, isolation, and self-estrangement. The most visible of these dimensions is powerlessness, the opposite of freedom and control. Alienation and freedom are conceived as two poles of the experience of technology as a production process.

The ethical response to this experience involves attempts to reduce alienation and enhance freedom. Social ownership of the means of production (socialism), increased wages and fringe benefits, workers' councils, personnel enrichment programs, even automation—all these are attempts to deal with different dimensions of alienation. By means of empirical studies of a print shop, a textile mill, an automobile assembly line, and an automated chemical plant, Blauner argues that the character of alienation and freedom is affected by different kinds of technologies.

2. **Technology and War.**   There are two theories about the relationship between war and technology: one, technology makes war unnecessary, and so horrible it becomes unthinkable; two, man will always go to war, technology just makes it more horrible. Following the optimism of the Enlightenment and nearly a century of peaceful industrial expansion, there was a general European belief that the energies that men had once directed against each other would now be directed against nature for the relief and benefit of all. On this view war is the result of a scarcity of goods which abundance through technology will eliminate. As late as 1924 J. B. S. Haldane could still offer a modified defense of this position by arguing that "the tendency of applied science is to magnify injustices until they become too intolerable to be borne" (p. 85).

The trauma of World War I was not just the fact that it was the most technologically destructive war in history up to that time; it shattered the internal confidence of a social order. As a result, intellectual criticism of technological civilization turned into a pessimistic reassessment of man's tendency, contrary to all rational self-interest, to direct his immense destructive powers against himself. Yet along with recognition of the deep potentiality for evil comes awareness of a new and unqualified need to master or overcome it. Replying to Haldane, Bertrand Russell (1924) argued that "the human instincts of power and rivalry . . . will need to be artificially curbed, if industrialism is to succeed" (p. 13). Like Haldane, Russell considered world government to be the only hope, although he doubted that it could easily be achieved. Such gloomy perspectives were reinforced by World War II and the technological extermination of six million Jews, the invention and use of the atomic bomb, the subsequent Cold War spread of nuclear weapons, and the deployment of advanced means of surveillance and terror. Ideals of the brotherhood of man and the virtue of peace, which in the past could remain moral exhortations not necessarily realized in deed, have become practical demands lest man obliterate himself from the earth.

One cogent presentation of the ethical implications of modern technological weaponry is Gunther Anders' "Commandments in the Atomic Age" (1961). Anders maintains that with the possibility of nuclear apocalypse, a gap has opened between what man can make and what he can imagine making; the violence of his hands exceeds the grasp of his mind.

The primary human obligation becomes "to violently widen the narrow capacity of your imagination . . . until imagination and feeling become capable to grasp and realize the enormity of your doings" (p. 13).

This enlargement of sensibilities will lead, Anders believes, to a consequent enlargement of ethical principles. Technological objects like nuclear weapons are not neutral; they operate on principles devoted to producing definite results. Kant had formulated the fundamental principle of morals in terms of an introspective duty: "Act only on that maxim which you can simultaneously will to become a general law" (*Grundlegung zur Metaphysik der Sitten,* 1785). In light of the demanding character of modern technology, Anders reformulates this categorical imperative as: "Have and use only those things, the inherent maxims of which could become your own maxims and thus the maxims of a general law" (p. 18).

3. Technology and Culture. Different issues arise here, depending on whether "culture" is used as the value-neutral term of social anthropology, meaning the aggregate customs and institutions of a group, or as a term that refers to the educated refinement of mind and taste. Sociological studies of "cultural lag" and "future shock" use the term in the weaker sense, documenting dissonances between different components of culture—often between what could be called "institutions of doing" and those of making and using. Insofar as these studies make value judgments, they tend to affirm the primacy of making, and to argue for the adaptation of other aspects of culture.

C. P. Snow (1959), playing on the ambiguity of "culture," points out that Western intellectual life is split between two types of education or refinement—literary and scientific-technological. Literary culture, in Snow's account, is pessimistic, selfish, tied to the ideals of a pre-industrial past, whereas scientific-technological culture is optimistic, democratic, and forward looking. Because of antagonism toward industrial development (from which they nevertheless benefit), literary or "traditional" intellectuals are obstructing the liberation of humanity from hunger and disease throughout the underdeveloped world. The day will come, Snow warns, when Western culture will be pillaged for the technology it refuses to share.

Snow's essay can be read as a brief for the ethical ideals of technologists against the humanist fears articulated in criticisms of progress and anti-utopian literature over the past hundred years. This so-called "culture criticism" of technology has taken a number of forms. Mumford, for example, presents a general contrast between humanistic and technological culture. A more strictly aesthetic criticism can be traced from William Morris and John Ruskin in the Victorian period to John Julian Ryan (1972). Existentialist reactions to mass society further exemplify cultural criticism. Friedrich Georg Juenger (1949) argues that technology generates massive illusions and undermines the vitality of high culture. In the writings of

Gabriel Marcel (1952, 1954), the problem becomes the place of the person and of wisdom in a culture dominated by technique. Some discussions of the dangers of computers and the proper limitations of artificial intelligence also fall into this category, as in the thesis of the computer scientist Joseph Weizenbaum (1976), that "there are certain tasks which computers *ought* not be made to do, independent of whether computers *can* be made to do them" (p. x).

In general, the argument that the ways of modern technology take over and infect the ways of high culture and of leisure, that making and using in their modern form undermine doing as an independent realm (see Hannah Arendt, 1958), recalls pre-modern criticism of the practical arts and the acquisition of wealth as inimical to the perfection of virtue. Recently, the "counter-culture" critique of Theodore Roszak (1973) and others has interpreted culture in terms of freedom rather than virtue, with freedom understood especially in a psychological sense. In light of the commitment to material freedom detectable in technology itself, the neoromantic argument that sexual and emotional repression has been the price of technological development raises questions best left aside here.

A provocative approximation of the traditional criticism of making centers around the relationship between technology and thought. One approach is provided by the studies of Marshall McLuhan (1954) and Frederick Wilhelmsen and Jane Bret (1970). Wilhelmsen and Bret, following McLuhan, argue that machines and electronic media engender two antagonistic kinds of thinking within culture—"rationalist-analytic" and "imaginative-synthetic," respectively. Alvin Gouldner (1976) likewise focuses on thinking in moral discourse to examine a dialectical interaction of ideology and technology. Ideologies have grown up in conjunction with the development of printing technology in its capitalist form. In contrast to religions and myths, ideologies are science-like "symbol systems that serve to justify and to mobilize public projects" (p. 55) by basing commands on reports about the way things are in the material world and a belief that "life here on earth is capable of being perfected by human knowledge and effort" (p. 67). Gouldner also suggests that changes in communications technology will affect the character of ideology, and that there is the need for a "media-critical politics." In another study, Manfred Stanley (1978) stresses the problem of "linguistic technicism"—that is, the misuse of scientific and technological vocabularies with regard to human activities. Stanley's argument includes a concrete discussion of how to embody "countertechnicist" tendencies in education and open up the possibility of moral and political discourse about technological projects.

Finally, the issue of culture ultimately hinges on the question of human nature. Edward Ballard (1978), in an attempt to "measure technological culture," is forced to explicate what it means to be human. Since the human self is born, dies, and can never be identified with any of its worldly roles,

Ballard describes it as "dependent, finite, and only negatively known" (p. 146). By means of its symbols, culture should support the recognition and remembrance of such a condition and thus foster human authenticity. Ballard questions the ability of technological culture to do so. In a more positive vein, Jean Ladrière (1977) and Hans-Georg Gadamer (1977) argue for a view of culture as both "dynamical and polycentric." The essence of man is his culture, according to Gadamer; the interpretation of culture, including technology, can provide a new kind of self-knowledge. For Ladrière, man must take responsibility for enclosing technological making and the possibilities it creates within a "hermeneutic" (interpretive) doing.

   4.   Technology and Religion.   Religion is an aspect of culture often singled out for special attention. In 1968, Lynn White, Jr., published an essay, "The Historical Roots of Our Ecologic Crisis," which argued that at least in its dominant Western form Christianity was responsible for the attitude toward the world that gave rise to modern technology—and thus to the environmental crisis. The cause is the Judaeo-Christian conception of man as superior to the rest of creation, and of everything else in creation as subject to man's use and enjoyment.

   That the religious issue, prior to White's article, had not been more central to intellectual debates on technology is somewhat ironic. Max Weber, in *The Protestant Ethic and the Spirit of Capitalism* (1904), originally argued that capitalist modes of production and modern technology were born of the Protestant attempt to apply the asceticism of the monasteries to everyday life. John Nef, in the *Cultural Foundations of Industrial Civilization* (1958), maintains that Renaissance delight in the world, and Catholic attempts to infuse human relationships with a new tenderness, also played their part. Perhaps such studies did not provoke the reaction of White's article because they could easily be misunderstood as having credited Christianity with an achievement. White's article was clearly an indictment.

   The wide response to White's article has taken two forms. On one hand, his argument has been picked up and extended by many who have been concerned to develop an environmental ethics. On the other hand, White has been criticized for having too limited a view. Lewis Moncrieff (1970), for instance, objects that White ignores cultural factors that are more important than religion. René Dubos (1972) points to other religious traditions that have been as exploitative of nature as has Christianity. Dubos and Michael Foley (1977) defend neglected aspects of the Judaeo-Christian tradition. (This discussion throws new light on some of the theological rejections and affirmations of technology mentioned in the first part of this chapter.)

   Other aspects of the relationship between technology and religion concern the practical application of Christian ethics to social justice issues aggravated by modern technology, and the problematic character of religion in a secular, technological world. The former has generated a large volume of literature which has largely ignored those questions posed by White. The latter is almost intractable, precisely because those who are aware of it in any depth are also conscious of the Christian ambiguity toward the secular which is at the basis of White's argument. See also George Grant (1969) and D. J. Hall (1976).

Technology and the secular exist in evident tension with religion and the sacred. As Wilhelmsen (1975) has felicitously stated, "Religion bends the knee" but technology "makes man walk erect" (p. 85). In modern culture, the secular has become autonomous and dominates the sacred, which is at least partly because of religious ideas about the nature of the secular. Two hermeneutic attempts to re-enclose the secular within the sacred can be found in the work of William Lynch (1970) and Gabriel Vahanian (1976). Vahanian argues that, whereas originally Christianity adapted itself "to the constraints of natural religion," now it must adopt "the framework of a religious sensibility determined by technology" (p. xvi). This involves a shift in theology from an emphasis on the divine in nature to the divine in technological utopianism, and a shift in church structure from priestly to prophetic ecclesiology. For Lynch, the problem is to be met by imagining secular autonomy not as defiance but as grace.

5. Technology and Values. The original discussion of "technology and values" took place in classical political economy, and resulted in the so-called "labor theory of value." The problem was to determine the source of the monetary value of a technological object. General value theory extends what was originally an economic term to cover moral, aesthetic, religious, and other judgments, and thus suggests generalizations of the original problem. Sources of the value of technological objects evidently include labor, capital utilization, usefulness, inherent complexity, and form. So far, however, there has been little discussion, except in economics and aesthetics, of how such factors combine to create intrinsic value in artifacts.

The contemporary question of the interaction between technology and values attempts to formulate in more analytic and empirically testable terms some of the ethical issues raised by discussions of technology, culture, and religion. It rests on the typically modern distinction between objective facts and subjective values, and is related to the idea of subsuming ethics within a general theory of values. As such, the contemporary approach to technology and values has sometimes been interpreted as reflecting the influence of technological modes of thought on ethics itself.

Analyses of the technology-values relationship can stress either the way technology influences values (the Marxist thesis) or the way values impede or foster technological growth (the approach of Max Weber). Arguments about which of the two terms is primary in the relationship between technology and values have yielded to a consensus that they are interdependent.

Most studies in which these two terms are central to the discussion nevertheless focus on the effects of technology on values, as in the works by Kurt Baier and Nicholas Rescher and by Emmanuel Mesthene.

The Baier-Rescher collection (1969) stresses methodological issues, values-technology interactions, and control mechanisms. Baier's contribution develops the concept of value as a subjective assessment rather than an intrinsic property of things. Rescher outlines various processes through which values can be "upgraded" or "downgraded," proposing a cost-benefit methodology for predicting whether certain values might be altered in rank under conditions of technological change.

Mesthene's work (1970) is the best brief introduction to the technology and values field. Its argument is presented in general terms, with a substantial annotated bibliography. Mesthene's thesis is that "new technology creates new opportunities for men and societies, and it also generates new problems" (p. 26). He is not the least dismayed by this double effect. He sees the possibilities as the means of human liberation, and the problems as able to be contained either by more technology or by social change. Mesthene is a firm advocate of what Alvin Weinberg (1966) has called the "technological fix." Many social and technological problems can best be dealt with by further technological advances. For example, the best hope for solving the problem of overpopulation created by agricultural and medical innovations is contraceptive technology. Admittedly, the full implementation of technological opportunities may require transformations in social values and the enhancement of the public sphere. But even when such changes undermine traditional values, they never undermine the need for the act of valuing itself. In the final analysis it cannot be said that technology is opposed to values. Technology simply makes the intelligent exercise of values more crucial.

John McDermott, in "Technology: The Opiate of the Intellectuals" (1969), criticizes Mesthene for what he calls a *laissez-innover* attitude—the twentieth-century equivalent of *laissez-faire* economic theory. From the point of view of moral philosophy, another weakness of the Baier-Rescher and Mesthene volumes is their failure to attempt a systematic ranking of values. By contrast, Max Scheler, in the European tradition, argues for a hierarchy of values, ascending from the pleasant (versus unpleasant), to the noble (versus base), to spiritual values of beauty, right, and truth (versus ugliness, wrong, and falsity), and finally to the religious value of holiness (versus sin). In this hierarchy, Scheler places technological values in the lowest category (pleasant-unpleasant) under the concept of the useful.

6. **Technology and the Environment.** The 1960s witnessed a progressive recognition of the problems of pollution and environmental deterioration as important ethical-political issues. This focus has given rise to two new disciplines—technology assessment (TA) and environmental

ethics—plus a general reassessment of philosophical ideas in terms of their attitudes toward nature.

TA as a narrow, problem-oriented discipline was originally conceived to investigate second-order consequences of specific technological projects, in order to facilitate the decision making process. In this form, TA was restricted from dealing with value questions. In practice it has remained limited to case study research of an analytic and empirical character.

The more philosophical possibilities of TA have received considerable attention. Lynn White (1974) and Carroll Pursell (1974) have argued the need for historical perspecive. The legal theorist Laurence Tribe has argued (1973) that TA should function as a means to bridge the discontinuity between man and machines, and (1974) that it should not be based on an anthropocentric world view. Stanley Carpenter (1977) sees the main philosophical issues as being involved with theories of the nature of man, society, and labor, and thus as external to TA itself. Frederick Rossini (1979) interprets TA as leading to a new type of science in which social responsibility becomes integrated with technical activities. (Rossini's article also includes a good annotated bibliography.)

In discussions of TA, Tribe's arguments especially point to the development of an environmental ethics. An early, influential effort in this direction is the widely cited article by biologist Garrett Hardin, "The Tragedy of the Commons" (1968). In this article and subsequent books, Hardin has elaborated with ideological fervor, the notion of a "life-boat ethics" for "spaceship earth." For him, the major problem is limiting population; limited resources demand that population be limited as well.

More subtle treatments are those by philosophers John Passmore and William Blackstone. Passmore (1974) attempts a broad historico-philosophical overview of ecological problems, recognizing that at the root of much ecological thinking is the idea of a "new ethic" that would extend man's responsibilities to the natural world. Against the background of a critical analysis of Western traditions of both opposition to and cooperation with nature, Passmore examines the four major ecological problems—pollution, depletion of natural resources, destruction of species, and overpopulation. His conclusion is that the issues are more complex than is usually recognized, and that a "new ethic" should be careful about abandoning many aspects of the intellectual heritage of the West. Blackstone (1974) focuses more specifically on the implications of ecology for the theory of moral, human, and legal rights. Again, he cautions against the too-ready application of ecological concepts to ethics.

The idea that ethics needs to incorporate new forms of responsibility is nevertheless the central feature of an environmental ethic. George Sessions (1974) and Hwa Yol Jung (1976), among others, have defended such an ethic. But it is in Hans Jonas' work on the ethics of technology—undertaken as part of an influential body of work on religious thought, the philosophy of biology, and the historico-philosophical origins of modern technology—that this issue has so far received its most complete expression. Jonas (1973) begins by observing that in the past, technology could reasonably be looked upon as neutral or lacking in ethical significance; ethics could ignore nature and focus on dealings between human beings within a relatively limited spatio-temporal order. As the problems caused by modern technology come to the surface, they alter our awareness of "the critical vulnerability of nature to man's technology—unsuspected before it began to show itself in damage already done." Awareness of the damage has altered the basic character of ethics. "This discovery, whose shock led to the concept and nascent science of ecology," calls man to an unprecedented level of moral responsibility (p. 38). "It is at least not senseless anymore to ask whether the condition of extra-human nature, the biosphere as a whole and in its parts, now subject to our powers, has become a human trust and has something of a moral claim on us not only for our ulterior sake but for its own sake and in its own right" (p. 40). Moreover, the former framework of immediacy is "swept away by the spatial spread and time-span of the cause-effect trains which technological practice sets afoot" (p. 39).

This extension of moral responsibility, from the strictly human within a humanly limited place and time to the whole of nature stretching out into the future, involves a necessary "de-anthropozation" of ethical principles. Within a utilitarian framework this implies an extension of the concept of utility to include nonhuman ends. Within the human rights tradition it involves an extension of rights to nonhuman objects. Technological advances have persistently enlarged the conception of human rights—from Locke's "life, liberty and property" to the welfare state rights to food, housing, employment, and medical care. Ecological considerations now would accord some rights to animals, plants, and perhaps even to inanimate natural objects, as a basis for the recognition of legal obligations toward the nonhuman world. Along with Tribe, Christopher Stone (1975) has attempted to develop legal arguments to support this position, and John Rodman (1977) has given it serious philosophical attention.

The difficulties in responding to this new responsibility are at least twofold. First, predictive knowledge of the consequences of technological actions is severely limited. Second, without a metaphysical critique of the scientific-technological conception of the world as a force-field available to human manipulation (see Heidegger, as outlined earlier), the idea of according rights to anything other than human beings is no more than an impotent dream. Indeed, with advances in biomedical engineering, the way in which rights should be accorded to human beings is even called into question—as has been persuasively argued by C. S. Lewis (1947) and Leon Kass (1972). Given such limitations in the face of pressing needs, Jonas (1973) makes a brief for an ethics of restraint.

In a second essay on this subject, Jonas (1976) attempts a more explicit formulation of this ethics of restraint, especially insofar as it bears on future generations. Since the good is most easily perceived through its opposite (e.g., we take health for granted until we are sick), Jonas argues for a "heuristics of fear." "Moral philosophy must consult our fears prior to our wishes to learn what we really cherish" (p. 88). Given the global stakes of the modern technical project, Descartes' principle of doubt needs to be inverted and applied whenever there is the possibility of destruction. Given the awesome power of technology, we must treat the doubtful-but-possible as if it were certain.

7. **Technology and Development.**   Conceptual issues connected with what is called "technology transfer" became the subject of special interest in the late 1960s. By the early 1970s the discussion had taken on ethical-political dimensions, especially with regard to the transfer of technology from developed to underdeveloped countries. Predictions of limits to world technological growth and development have heightened the importance of these questions.

The work of Peter Berger and colleagues in *The Homeless Mind* (1973) is particularly significant. Berger offers a phenomenological description of modern consciousness, as a function of technology and bureaucracy, and then analyzes the way this consciousness is transferred to Third World countries in the process of "modernization." He is particularly interested in modernity as a "package deal," as well as in the extent to which it can or cannot be modified. For Berger, the distinguishing characteristic of the modern life-world is a segmentation or pluralization which undermines traditional social meanings and leads to a condition that can be summed up under the concept of "homelessness."

In another work Berger (1974) criticizes both capitalist and socialist ideologies of modernization; he argues that "demodernizing" tendencies must be taken seriously, not just dismissed as backward or irrational. What is needed is "a new method to deal with questions of political ethics and social change (including those of development policy)" that will integrate hard-nosed analysis with utopian imagination (p. xiv).

Studies by Denis Goulet (1971 and 1977) move in this direction. Goulet combines original field research in Latin America with a strong appreciation of thinkers like Mumford and Ellul to form a "philosophy of development." It is built on the recognition that "technology is no panacea for the ills of underdevelopment" (1977, p. 251).

A related movement growing out of work in Third World countries involves the idea of "alternative," "appropriate," or "intermediate" technology. The economist B. F. Schumacher (1973 and 1977) is a major influence here. Amory Lovins (1975, 1976, and 1977) provides sophisticated technical arguments for what he calls "soft technology" in both underdeveloped and developed countries. David Dickson (1974) gives alternative technology a radical political meaning in advanced industrial societies. For Dickson, the problems of technology "stem as much from the nature of technology as from the uses to which it is put, but . . . the former is largely determined by social and political factors, of which technology can never therefore be considered independent" (pp. 9–10). Dickson also distinguishes between alternative technology in advanced industrial societies and in the Third World; the first he calls "utopian," the second "intermediate" technology.

J. Von Brakel (1978), in a critical analysis of the concept of "appropriate technology" as applied to underdeveloped countries, concludes that to argue for appropriate technology in the abstract is merely preaching the good. That "appropriate technology is good" is either true but trivial, or masks unstated values and goals.

8. Technology and Social-Political Theory. When Machiavelli, Bacon, and Descartes argued that making is more central to human nature than the ancients thought, they laid the foundations for that modern theory of progress which sees technological change as bringing with it beneficial social change. At least since the industrial revolution, however, this theory of progress has become increasingly problematic—precisely because of such difficulties as worker alienation, war, mass society, and environmental pollution. The benefits expected from technology—increasing material welfare and decreasing labor—have brought with them unanticipated side effects. Much political theory in the modern period has concentrated on explaining how these generally undesirable consequences came about and what can be done to remedy them.

The uncertain character of technological progress introduces two kinds of questions, descriptive and prescriptive. Descriptive questions regarding technological change and its interaction with social change have become a primary focus of the history of technology (see chapter 2 in this volume) and social science studies of technology (see chapter 7). So-called "future studies" also tend to emphasize descriptive questions within a broad prescriptive framework, as do science policy studies (see chapter 9). Insofar as such disciplines rest on assumptions about what the crucial questions are and how in principle to answer them, they invite at least indirect philosophical scrutiny.

Prescriptive questions call for more direct philosophical attention. Yet questions about the proper political response to modern technology cannot be addressed without reference to some descriptive theory of the interaction between technology and society. The interdisciplinary character of social and political theory thus provides a broad field of interaction among historians, sociologists, and philosophers of technology. Some of the terms of this interaction can be indicated through a consideration of contemporary discussions centering around four topics: (1) transformations in Marxist theory; (2) the idea of "post-industrial" society; (3) liberal responses to the challenges of technology; and (4) the possibility of "autonomous technology."

Classic Marxist theory argues that the problems of technology are caused not by technology itself but by the social structure within which it is embedded. Technology is and must be used by its capitalist owners to dominate the workers. Freed from this social condition, technology will bring about the liberation for which it was originally conceived. Bernard Gendron (1977) has recently made a strong case for this view. His analysis depends on a somewhat oversimplified contrast between socialism and two equally mistaken views of technology: the "utopian," which views technology itself as producing social goods; and the "dystopian," which views technology as necessarily productive of undesirable consequences.

Two weaknesses in Gendron's defense of the socialist thesis are that he gives inadequate consideration to the predictive aspect of Marx's theory and he does not take into account the real attempts made by some countries to carry out Marx's proposals. According to Marx, the capitalist development of technology will inevitably bring about the downfall of capitalism. Also, in practice, socialist revolutions have taken place primarily in underdeveloped rather than developed countries; and in the Soviet Union and Eastern Europe, technology continues to be associated with problems such as alienation and pollution—if not war and political domination.

In response to the persistence of capitalism and the perversions of socialism, attempts have been made to revise Marxist theory. For instance, C. B. Macpherson (1967) alters the stress in Marxism from economics to philosophical anthropology. Man must no longer think of himself as a consumer but as an active producer, in order to make a fully human use of technology.

A related but more influential reformulation of Marxism has been undertaken by a group of theorists associated with the Frankfurt School of Social Research in Germany. Members of this group include Max Horkheimer, Herbert Marcuse, and Jürgen Habermas. The Frankfurt school program centers around what Horkheimer in the 1930s called "critical theory." By stressing neglected aspects of the early Marx in conjunction with insights adapted from Freud, critical theory attempts to extend Marx's critique of political economy. Where Marx sought to demystify nineteenth-century capitalist economic ideology, critical theory aims to unmask the illusions and mystifications of twentieth-century capitalist and socialist ideologies—while guarding against the tendency of critique itself to turn into a new ideology.

The fundamental argument of critical theory is that rationality is warped by domination, but that domination is not ended with the elimination of class struggle. Domination may be embodied in strictly political structures, as in the Soviet Union. It can also take on more subtle forms under advanced capitalism, some of which are essentially related to science and technology. Horkheimer's "critique of instrumental reason" attempts to expose scientific positivism as ideology. Marcuse's *One-Dimensional Man* (1964) provides the most readily accessible statement of the general argument concerning the new forms of social domination in the West and the more subtle forms of domination operating in science and technology, and in philosophy insofar as it is influenced by scientific and technological thought. In more recent work Marcuse (1972) and his student William Leiss (1972 and 1975) have made efforts to incorporate a philosophy of nature and an environmental ethic within critical theory.

Habermas has criticized Marcuse's conception of "technology and science as 'ideology'" (1968). He has also formulated what he terms a "communicative ethics" (1970) for the technological media. For Habermas, the understanding of contemporary Western society as dominated by false forms of reasoning in advertising and propaganda calls for an alternative description of the conditions of true communication. These conditions include an absence of violence and equal access to public media. In this arena, Habermas closes ranks with liberals of the John Stuart Mill tradition who argue that freedom of speech is a precondition for social justice.

The most comprehensive alternative to Marxist revisionism is the theory of "post-industrial" society. As a revision of the earlier theory of industrial society, post-industrialism integrates a number of observations about changes in advanced technological society in both capitalist and socialist forms. Prominent among these changes are what has been called the "managerial revolution," the growing importance of information and knowledge as productive forces, and the global proliferation of technology. These aspects have been emphasized in literature in which systems theory is applied to business management and social planning (see Boguslaw, 1965; and Laszlo, 1974). From another angle, Daniel Boorstin (1978) argues that technological proliferation is creating a shared public experience which democratizes the eighteenth-century Republic of Letters. He refers to this universal and homogeneous state as the Republic of Technology.

The most serious and sustained investigation of post-industrial society is that provided by Daniel Bell (1973), although his work largely ignores the universalist tendencies inherent in such a social order. According to what Bell calls a "venture in social forecasting," the centrality of theoretical knowledge is linked to an expansion of the service sector of the economy. In economic processes, telecommunications and computers (hardware), in association with the intellectual technologies of planning and decision making (software), have become more crucial than property. As a result of this change in focus, there has been a shift in the social weight of institutions such as universities, which are centers of organized research. In occupational activities, the shift appears as a rise in the economic significance of professional and technical services, paralleled by an increase in the social services of health, education, and welfare.

Bell's analysis has been subject to vigorous attack by Marxist critics.

Soviet and East European proponents of the scientific-technological revolution reject Bell's argument for a convergence of capitalist and socialist economic structures. Gendron argues that Bell thinks technology by itself can solve social problems. Bell replies that "post-industrial" refers primarily to an ideal type toward which the structure of the technological-economic order is tending; only indirectly does it refer to changes in the political and cultural realms that are independent axial structures of society as a whole.

Bell concludes with an assessment of the political problems of liberalism. In post-industrial society, one critical issue concerns equality. Against the radical left Bell defends the idea of a "meritocracy," a just social hierarchy based on technical merit. To some critics this amounts to a conservative defense of technocracy, or worse. The issue is emotionally loaded, according to Bell, because society is no longer defined by the character of the relationships between man and nature, or man and technology. Technological advances have led to the ascendancy of social relations. When this happens all relations, especially social ones, become subjects of intense political debate. In the face of resulting tendencies to fragmentation, it becomes imperative, Bell argues, "to define some coherent goals for the society as a whole and, in the process, to articulate a public philosophy which is more than the sum of what particular social groups may want" (p. 377).

Liberalism, as the reigning public philosophy of the West, is subject to considerable differentiation, and Bell represents what may be called the theory-oriented manifestation. A more practical or pragmatic-oriented manifestation of liberalism is at the basis of Paul Durbin's (1972 and 1978) arguments that philosophy of technology should be a reasoned statement of the ends that technology ought to serve and the way the technical community ought to be structured within the larger society, coupled with practical attempts to press this view in particular cases. The stress is not so much on overall theory as on practice, on attempting to deal with particular issues in a way that allows those affected to contribute to the decision making process by arguing their viewpoints.

Paul Goodman (1969) makes the same general point. He maintains that technology is a kind of practice essentially subservient to politics; it ought to be the subject of grass-roots political debate. Elting Morison, at the conclusion of a study of the history of American technology (1974), gives substantive content to this argument by explicitly rejecting general theories in favor of a piecemeal approach to particular problems under the guidance of three assumptions: (a) technology ought to fit and serve man; (b) it is manageable; and (c) it ought to be controlled democratically, because, given the opportunity, most people are the best arbiters of their own interests.

Criticisms of liberalism, especially in relation to environmental problems, have been advanced by Victor Ferkiss and Robert Heilbroner. Ferkiss (1974) rejects the liberal ideal of individual freedom in favor of what he calls an "ecological humanism" in which man recognizes his interdependence with all of nature and also perceives the immanence of a new social order. (See also Skolimowski 1975, 1977, and 1978.) Heilbroner (1974) is much less sanguine about the human prospect. His comparison of the abilities of socialist and capitalist economic structures to deal with the three basic problems of population, war, and pollution as influenced by science and technology, leads Heilbroner to grant socialist economies a slight advantage. Yet his conclusion is that " 'human nature' makes it utopian to hope that we will face the global challenge of the future" in a way which will not bring about serious diminutions in the liberal freedoms of the West (p. 113).

The possibility that technology is not subject to social control on either Marxist or liberal terms is the basic thesis behind Langdon Winner's comprehensive analysis in *Autonomous Technology* (1977). As Winner points out convincingly, the idea of technology as an autonomous force has been present at least since Mary Shelley's *Frankenstein* (1818), and has recurred in numerous works of fiction. Within the last half century it has also been given serious attention in social analyses such as those by Mumford, Giedeon, and particularly Ellul (see sections II. A. 1. and 2, above). Winner's book is an unabashed commentary on Ellul, refuting the too-easy rejection of the argument that technology has become an independent force in social and political affairs.

Winner identifies three senses in which technology can be perceived as autonomous. First, it can be seen as the basic cause of all social change, progressively transforming and incorporating the rest of society. Second, large-scale technical systems can appear to operate on their own principles, independent of human intervention. And third, the individual can appear to be overwhelmed and engulfed by technological complexity. The first perception is historiological, the second political, and the third epistemological in character. Winner's work is mainly a careful analysis of the second of these senses.

Winner justifies his concentration on the structure and functioning of technological society by criticizing theories of the philosophical origins of technology and corresponding proposals for revolutions in consciousness. Both are too abstract. In seeking to avoid "depth without direction and details without meaning," he turns to political theory (p. 134). Here he begins with a critical reformulation of the thesis of technocracy. Technological society is not determined by the dominance of technical elites so much as by the insinuation of technological ways into the political realm. Technology dominates politics not through personal agents but through the transformation of political life, turning the latter into what Winner calls "technological politics." In this view, both the problems posed and the principles upon which they are to be solved are determined primarily by technology. "To be commanded, technology must first be obeyed. But the opportunity to command seems forever to escape modern man. Perhaps more than anything else, *this* is the distinctly modern frustration" (p. 262).

Man overcomes his bondage to nature and economics only by submitting to a bondage of a different but equally powerful sort. Winner concludes by arguing that even Marxist revolutions wind up in the thralls of this technological bondage.

Two main criticisms of the "autonomous technology" thesis have been put forward. One takes the tack of accepting technological control—in effect accusing Ellul and his followers of misplaced fears. This is the viewpoint of cybernetics and systems theory, and it is closely related to ideas of technocracy and Bell's "post-industrial" theory. Winner's reply is simply that technology escapes the control even of technocrats, and that complexity inherently leads to a loss of agency.

The second line of criticism has been voiced by Daniel Callahan (1973). In concluding a serious examination of ethical-political issues raised by biomedical technologies, Callahan maintains that arguments for the autonomy of technology "make provocative bedtime reading, but little more than that, primarily because they fail to take account of the psychological reality principle of technology—that contemporary man cannot and will not live without technology." In political theory one must start from "the tangible, ineradicable fact of technology." Any ethical or political proposal "must be modest in its demands and culturally viable in its principles" (p. 260).

Winner's reply to this criticism is an admission that he has no systematic proposal to make. He implicitly takes what may be called an "anarchist" position by proposing an "epistemological Luddism," i.e., the small-scale dismantling of technology wherever possible. In a sense this position sounds suspiciously like pragmatic liberalism. The difference is that whereas Callahan and pragmatic liberals display what may be called a politics of accommodation, Winner proposes a politics of resistance. This resistance is also implicitly present in a number of conservative discussions of biomedical technology (see, for example, Lewis, 1949, and Kass, 1974).

C.   Toward a Synthesis: The Question of Humanization

The major unstated theme of ethical and political discussion concerns the question of humanization. In what ways, and to what extent, does technology promote or obstruct the realization of human nature? When asked in terms of the individual, the question is primarily an ethical one. When asked of the relation between technology and social institutions, it becomes political in character. In either case, any argument about the human meaning of technology ultimately presupposes ideas about both the nature of technology and the nature of man. The attempt to articulate these presuppositions is a central task of philosophy of technology, a task that points toward a synthesis of different aspects of the field.

With regard to ideas about the nature of technology, the question of humanization has been broadly alluded to in analyses of metaphysical issues. Technology as object raises questions about the extent to which artifacts are extensions of the human body. Technology as process explicitly broaches the subject of humanization in relation to action. Is the pursuit of technical efficiency in harmony with or opposed to realization of the nature of man? Technology as knowledge raises questions about man as a rational animal, and about the truly human forms of thinking. Technology as volition likewise addresses the issue of which practical attitudes toward the world are authentically human. The theoretical description judged most adequate for comprehending technology will undoubtedly influence the ways technology can be understood as related to man.

Theoretical analysis culminates with a distinction between two types of technology, between making and using as exercised in the traditional intuitive crafts and as found in the modern scientific processes of mass production. A similar distinction recurs in ethical and political discussions, most notably in relation to arguments concerning alternative technologies. Lovins' practical differentiation between the hard technology of nuclear power and the soft technologies of wind and solar energy at least echoes the earlier metaphysical distinction between ancient technics and modern technology. By technology that is "soft" Lovins does not mean "vague, mushy, speculative or ephemeral, but rather flexible, resilient, sustainable and benign" (1976, p. 77). On the one hand, by arguing that this type of technology exists on a more human scale and is more manageable or more ecologically sound, one could readily criticize the dehumanizing effects of modern technology. On the other hand, by arguing that modern technology exercises more power over nature or is able to generate greater material welfare, someone else could make an equally strong argument that soft technology is the more dehumanizing type. Each would be appealing, however implicitly, to a different view of the nature of man.

The question of humanization has already been suggested here, in distinctions between ancient and modern conceptions of human nature, and in the historico-philosophical speculations of Ortega and Mumford. For present purposes, the two theories of the nature of man—divided in their understandings of the relative primacy of doing and making—can also be associated with two approaches to philosophy of technology mentioned earlier. The idea of human life as essentially manifested in making can be correlated with what was called "technological philosophy"; and the idea of human life as essentially a kind of doing, with an external philosophical critique of technology. In terms of the ethical question of humanization, this contrast can be phrased as one between professional engineering ethics and what may be termed "moral philosophy of technology."

In a narrow sense, engineering ethics is comparable to medical or legal ethics. It establishes standards of professional conduct and applies generally accepted moral principles to engineering practice. Particular issues in this area have recently been highlighted in a documentary by Robert Baum and Albert Flores (1978). As Baum and Flores point out, the general principle of engineering ethics is the same as that for all professional practice: "honest service of the public interest." Problems arise over different conceptions of the public interest, as well as over how that interest is to be served. What is an engineer's responsibility if the public asks that he design something which lacks engineering integrity? In private industry, for example, there is a persistent conflict between engineering and management: the best technical solution to a problem is seldom the least expensive or the most acceptable to the labor force. Even when engineering interests are kept to a minimum, it seems likely that contemporary society is sufficiently enamored of technology that the engineering attitude is scarcely disturbed.

This possibility calls for a shift from a narrow to a broad sense of engineering ethics. Morison's (1974) historical study relates how, on one occasion, the American engineer John B. Jervis, while overlooking a section of the American wilderness, was moved to exclaim "What a place for engineering!" (p. 45). Jervis was convinced that by building as solidly as he knew how, but with determined simplicity, he was improving not only man's place in nature but nature as well. Such a belief entails a theory of the nature of man as essentially expressed in technological activity, a position most explicitly developed in some European discussions of engineering ethics. For example, the engineer-philosopher Hans Sachsse (1978) moves from a general consideration of technology in relation to human physiology, through technology and history, to an ethical argument for the technological mastery of technological problems. The engineer-philosophers Hans Lenk and Günther Ropohl (1979) have also argued for what they call a "pragmatic and interdisciplinary philosophy of technology"—that is, the articulation of a "general technology" based on systems theory, as the foundation for multidisciplinary team consideration of the practical problems of specialized technologies.

Bunge (1977) likewise argues for an enlarged understanding of the moral responsibility of the technologist leading toward "teams of experts in various fields, including applied social scientists" working "under public scrutiny and control" to establish a "global technocracy, or the rule of experts in all fields of human action" (p. 107). Ethics is an underdeveloped science that has much to learn from technology; in fact, moral rules have exactly the same logical structure as technological rules, so that "ethics could be conceived as a branch of technology" (p. 103). This argument rests on Bunge's epistemological analysis of technology as knowledge. As such, it unites theory and practice to make a persuasive case for reconstructing moral philosophy on the basis of professional engineering ethics, thus presenting a case for scientific technology as inherently humanizing.

The opposite approach, that of a "moral philosophy of technology," attempts to limit and enclose engineering ethics within a broader perspective. Though making a claim to breadth, this approach has two forms, one narrower than the other. In the first type, ethical problems associated with various technologies are interpreted as opportunities to apply and develop broader ethical principles. The difference with professional engineering ethics in the narrow sense, though subtle, is that this application is not undertaken to further the professional activity of making; neither does it in any sense assume the primacy of technology. "Moral philosophy of technology" in this sense does not extend technological ways of thinking (as with Bunge); it involves inner transformations and renewals. In its broader form, the approach culminates in a fundamental inquiry into the nature of man as the basis for understanding technology. Because it does not presume the primacy of the technological, such an inquiry often exhibits a face that is negative toward modern culture.

In some premodern thought, the realm of technics was conceived as subject to politics, and politics as subject to ethics. This is the basis of Plato's proposal for ethical-political regulation of art in the good state. All such traditional criticisms rest on the argument that doing is a more eminently human activity than making, and recognize that the products of making can adversely affect human doing. There is a traditional defense of the political realm against the incursion of attitudes and influences from the productive realm. John Rodman (1975), in a witty adaptation of arguments from Samuel Butler's *Erewhon* (1872), characterizes the work of Mumford, Ellul, and others as sharing an Erewhonian antimechanist philosophy but as refraining from drawing the obvious conclusion that technology must be limited.

The modern interaction between moral philosophy and technology typically centers around questions of responsibility, neutrality, freedom, and alienation. For instance, it is generally recognized that technology is not neutral in any meaningful sense. Particularly when interpreted as process or as volition, technology has manifold effects on psychological and social life. Furthermore, the modern commitment to technology is grounded in an attempt by man to assume responsibility for his material existence— and thereby realize his freedom. Each increase in technological power has in fact reinforced the initial commitment to technology and extended its scope, until the point is reached where man has become responsible for systems of artifacts, for even non-artificial aspects of the world of nature, indeed for future human generations (Jonas). Alienation nevertheless confronts man with barriers to the exercise of such responsibility and freedom. Alienation goes beyond worker experiences of powerlessness, meaninglessness, isolation, and self-estrangement (Blauner). In its political form it is what Winner analyzes as the idea of "autonomous technology," the public experience of technology as out of control. At the level of ethics, alienation raises a question of meaning. The original human meaning of modern technology was freedom. Alienation, rendering freedom dubious, reraises the question of humanization.

Modern ethics and politics comprise a history of proposals for recovering the original meaning of technology. Alongside the political proposals of utilitarians and Marxists, the primary ethical response has been to propose some kind of transformation of consciousness. Most common has been an argument for the greater exercise of intelligence and the expansion of technical knowledge (Mesthene, Bell). Whether this is a viable response is ultimately dependent on the extent to which alienation, in the form of predictive uncertainty, is an integral part of technological processes. A second proposal has stressed affective transformation. Romantics (following Hazlitt's characterization of Rousseau), have argued for the cultivation of an "extreme sensibility" that would enclose scientific rationality and technological powers (Anders).

Ultimately, the question of humanization as a question of consciousness becomes one of self-interpretation. Faced with the paradoxes of responsibility and alienation, reflective philosophy of technology is driven to reinterpret technological man. There have been, for instance, numerous attempts to reintegrate modern technology into the human realm by means of a transformed image of human making (Macpherson), or an enlarged understanding of human doing (Ladrière, Gadamer), or a reinterpretation of religious faith (Lynch, Vahanian). Edmund Byrne (1978) suggests that what is crucial is the goal of the man–machine relation; he distinguishes between a "cyborg" system, integrating man within the machine, and a "prosthetic" system using machines to extend human functions. The crucial issue becomes one between two theories of human nature, as essentially related to doing, and as related to making. The issue between these interpretations is at the heart of philosophy of technology as a hermeneutic enterprise; it may also be at the heart of philosophy in general.

In attempting to explore this argument, philosophers make inevitable appeals to history. Given the fact that modern history emerged in conjunction with a philosophical rejection of ancient attitudes toward the world, it is only proper that the attempt to interpret the nature and meaning of making and using should include a philosophical interpretation not only of the history of technology, but also of the history of modern philosophy.

The deepest question to be addressed in historico-philosophical studies —one which is persistently present even if seldom stated—concerns the place of philosophy itself in the technological world. Jonas argues, for instance, that although the scientific-technological revolution originated in revolutionary thought and freedom, it is now defended by the orthodox and allied with social conservatism. "What began in acts of supreme and daring freedom has set up its own necessity and proceeds on its course like a second, determinate nature—no less determinate for being man-made" (1973, p. 48). One measure of this necessity is that the question of humanization is precisely related to the questioning of technology. Another measure is that two centuries of philosophical criticism, comparable in character to the centuries of criticism that ushered in the modern period, have been ineffectual in altering the course of modern commitments in any fundamental way. Is this because such criticisms have been misguided? Is it because within the conditions that philosophy helped establish, philosophy has been deprived of its rightful powers? Or have expectations of the powers of philosophy been misconceived? In reflecting on technology, philosophy is compelled to grapple with its own nature and the extent of its powers in ways that will, ironically, influence its analysis of technology.

BIBLIOGRAPHIC INTRODUCTION

A. Nonphilosophical Source Materials

Philosophical reflection on technology requires serious acquaintance with the way men of practice (engineers, technologists, business managers) understand their own professions. It also demands some historical knowledge of technological development and its social consequences.

Engineers themselves have undertaken philosophical reflection on technology in the work of Dessauer, Florman, Lafitte, the Verein Deutsche Ingenieure, and others. For access to information about various technical aspects of engineering and business management the following can be of help:

General overviews of engineering activity are provided by: Ralph J. Smith, *Engineering as a Career,* 3d ed. (New York: McGraw-Hill, 1969); Charles Susskind, *Understanding Technology* (Baltimore: Johns Hopkins University Press, 1973); and Dustin Kemper, *The Engineer and His Profession* (New York: Holt, Rinehart and Winston, 1975). Morris Asimov, *Introduction to Design* (Englewood Cliffs, N.J.: Prentice-Hall, 1962); and Thomas T. Woodson, *Introduction to Engineering Design* (New York: McGraw Hill, 1966), give technical introductions to design as the essential engineering activity. The *McGraw-Hill Encyclopedia of Science and Technology* (1971) is an indispensable reference for clear explanations of a multitude of technical definitions and concepts. For the special area of cybernetics and information theory one should be aware of the *Scientific American* collection, *Information* (San Francisco: W. H. Freeman, 1966); and Jagjit Singh, *Great Ideas in Information Theory, Language and Cybernetics* (New York: Dover, 1966). For computers, Margaret A. Boden's *Artificial Intelligence and Natural Man* (New York: Basic Books 1977) provides an enthusiastic survey of recent research. Peter Drucker, *The Practice of Management* (New York: Harper & Row, 1954), and C. West Churchman, *Challenge to Reason* (New York: McGraw-Hill, 1968), provide basic accounts of management theory as it has come under the influence of modern science and technology. The ideas in these last two books are best interpreted against the background of the history of work as provided by Melvin Kranzberg and Joseph Gies in *By the Sweat of Thy Brow* (New York:

Putnam's, 1975). The social and technical vision of what can be called "leftist engineering" is represented in different ways in Duncan Davies, Tom Banfield, and Ray Sheahan, *The Humane Technologist* (London: Oxford University Press, 1976), and the two periodicals *CoEvolution Quarterly* (Sausalito, Calif.) and *Undercurrents* (London). Christopher Williams, *Craftsmen of Necessity* (New York: Random House, 1974), provides a perceptive if impressionistic account of traditional technologies; but it should not be read in isolation from Mircea Eliade, *The Forge and the Crucible* (New York: Harper & Row, 1962).

Bibliographies of historical and social science studies of technology can be found elsewhere in this volume; but because it is not always easy to draw hard and fast distinctions between empirical social science and philosophical social theory, the main bibliography also contains a number of references to literature in these areas. From a philosophical point of view—which is not necessarily the same as that of the historical or social science point of view—the following further references are helpful: One of the most useful historical source books is Frederick Klemm, *A History of Western Technology* (Cambridge, Mass.: M.I.T. Press; 1964); the original German title, *Technik: Eine Geschichte ihrer Probleme* [Technology: a history of its problems] (Freiburg: Alber, 1954), gives a better idea of the contents of this collection of comments by technologists, moralists, theologians, naturalists, poets, economists, and statesmen. Two basic social science background collections are Melvin Kranzberg and Carroll W. Pursell, Jr., eds., *Technology in Western Civilization,* 2 vols. (New York: Oxford University Press, 1967); and Ina Spiegel-Rösing and Derek de Solla Price, eds., *Science, Technology, and Society: A Cross-Disciplinary Perspective* (London and Beverly Hills: Sage, 1977). For complementary background on Soviet-East European work, see Frederic J. Fleron, Jr., ed., *Technology and Communist Culture: The Socio-Cultural Impact of Technology under Socialism* (New York: Praeger, 1977). Historical works that provide good background reading for the philosophy of technology include: Siegfried Giedion, *Mechanization Takes Command* (New York: Oxford University Press, 1948); John U. Nef, *Cultural Foundations of Industrial Civilization* (Cambridge: Cambridge University Press, 1958); Paolo Rossi, *Philosophy, Technology and the Arts in the Early Modern Era* (New York: Harper & Row, 1970); Cyril Stanley Smith, "Art, Technology, and Science: Notes on Their Historical Interaction," *Technology and Culture* 11 (Autumn 1970): 493–549; Lynn White, Jr., *Medieval Technology and Social Change* (New York: Oxford University Press, 1962); Jerome R. Ravetz, *Scientific Knowledge and Its Social Problems* (New York: Oxford University Press, 1971); and Arnold Pacey, *The Maze of Ingenuity: Ideas and Idealism in the Development of Technology* (Cambridge, Mass.: M.I.T. Press, 1976).

Finally, philosophical analysis of technology, especially as an ethical and political problem, should not be pursued without some awareness of

anthropological and literary studies of technology. Three good anthropological sources are: Edward H. Spicer, ed., *Human Problems in Technological Change* (London: Sage, 1952); William L. Thomas, Jr., ed., *Man's
Role in Changing the Face of the Earth*, 2 vols. (Chicago: University of
Chicago Press, 1956); and H. Russell Bernard and Pertti J. Pelto, eds.,
Technology and Social Change (New York: Macmillan, 1972). For literary
studies the most philosophically relevant are Leo Marx, *The Machine in
the Garden: Technology and the Pastoral Ideal in America* (New York:
Oxford University Perss, 1964); and Herbert L. Sussman, *Victorians and the
Machine: The Literary Response to Technology* (Cambridge, Mass.: Har-
vard University Press, 1968). Utopian and antiutopian novels, as well as
science fiction, are valuable sources for ideas about technology. Robert
Pirsig's *Zen and the Art of Motorcycle Maintenance* (New York: William
Morrow, 1974) is an autobiographical novel that deals with technology as
a philosophical problem.

B.   Philosophy of Technology Source Materials

Two useful publications that include newsletter materials, bibliographies,
and articles are the *STS Newsletter* (formerly *Humanities Perspectives on
Technology*, Lehigh University) and *Science, Technology, & Human Values*
(formerly *Public Conceptions of Science,* Harvard and MIT). There is a
Journal for the Humanities and Technology (Southern Technical Institute, Marietta, Ga.), and ongoing bibliographies are available in the *Newsletter of the Society for Philosophy & Technology* (University of Delaware).

Here we focus on bibliographies, major collections, and textbook anthologies. For bibliographies see especially Mitcham and Mackey (1973),
Mistichelli and Roysdon (1978), and Mitcham and Grote (1978). The most
important collections are by Mitcham and Mackey (1972), Lenk and Moser
(1973), Rapp (1973), and Durbin (1978, 1979). After these, the twelve items
marked with an asterisk (*) provide the most concentrated introduction
to the philosophy of technology. (It should be noted that there are some
overlaps among these volumes.)

* Anderson, Alan Ross, ed. *Minds and Machines*. Englewood Cliffs, N.J.:
   Prentice-Hall, 1964.
"Are There Any Philosophically Interesting Questions in Technology?"
   In *PSA 1976: Proceedings of the 1976 Biennial Meeting of the Philoso-
   phy of Science Association,* vol. 2, pp. 137–201. Edited by F. Suppe and
   P. Asquith. East Lansing, Mich.: Philosophy of Science Association,
   1977. Symposium including papers by Max Black, Mario Bunge, Paul
   Durbin, Ronald Giere, and Edwin Layton.
* Barbour, Ian G., ed. *Western Man and Environmental Ethics: Attitudes
   toward Nature and Technology*. Reading, Mass.: Addison-Wesley,
   1973. Best text anthology on this topic.

Bereano, Philip L., ed. *Technology as a Social and Political Phenomenon.* New York: John Wiley, 1976. Good text anthology.

Blackstone, William T., ed. *Philosophy and Environmental Crisis.* Athens, Ga.: University of Georgia Press, 1974.

* Bugliarello, George, and Doner, Dean, eds. *The History and Philosophy of Technology.* Urbana, Ill.: University of Illinois Press, 1979. Proceedings of an international symposium held in 1973.

Burke, John G., ed. *The New Technology and Human Values.* Belmont, Calif.: Wadsworth, 1966. 2d ed., enlarged 1972. Widely used text; more popular than philosophical.

* Crosson, Frederick J., ed. *Human and Artificial Intelligence.* New York: Appleton-Century-Crofts, 1970. Well-balanced collection, good text.

Crosson, Frederick J., and Sayre, Kenneth M., eds. *Philosophy and Cybernetics.* Notre Dame, Ind.: University of Notre Dame Press, 1967. Six of the eight papers are by Crosson and Sayre.

Civilization technique et humanisme [Technological civilization and humanism]. Paris: Beauchesne, 1968. Papers in French, German and English.

Dechert, Charles R., ed. *The Social Impact of Cybernetics.* New York: Simon and Schuster, 1966. From a 1964 conference at the University of Notre Dame.

Douglas, Jack D., ed. *Freedom and Tyranny: Social Problems in a Technological Society.* New York: Random House, 1970. Textbook.

Durbin, Paul T., ed. *Research in Philosophy & Technology.* Greenwich, Conn.: JAI Press, vol. 1, 1978; vol. 2, 1979. Annual publication of the Society for Philosophy & Technology; includes review and bibliography section edited by Carl Mitcham.

Feigenbaum, Edward A. and Feldman, Julian, eds. *Computers and Thought.* New York: McGraw-Hill, 1963. A classic collection of technical and intrepretative papers with a large subject-indexed bibliography.

Freyer, Hans; Papalekas, Johannes C.; and Weippert, Georg, eds. *Technik im technischen Zeitalter; Stellungnahmen tur geschichtlichen Situation* [Technology in the technological era; attitudes toward the historical situation]. Düsseldorf: Schilling, 1965. A collection of twenty papers touching many topics of an ethical and political nature.

Harvard University Program on Technology and Society, 1964–1972. In the course of its existence the Harvard program produced a series of eight "research reviews" that contain good descriptions of various fields along with substantial annotations of a select number of publications: no. 1 (Fall 1968), "Implications of Biomedical Technology"; no. 2 (Winter 1969). "Technology and Work"; no. 3 (Spring 1969), "Technology and Values"; no. 4 (Summer 1969), "Technology and the Polity"; no. 5 (1970), "Technology and the City"; no. 6 (1970), "Technology and the Individual"; no. 7 (1971), "Implications of Computer Technology"; no. 8 (1971), "Technology and Social History."

Herlitzius, Erwin. "Technik und Philosophie." *Informationsdienst Geschichte der Technik* (Dresden) 5 (1965): 1–36.

Kranzberg, Melvin, and Davenport, William H., eds. *Technology and Culture*. New York: Schocken, 1972. Selections from the first ten years of *Technology and Culture*.

Krohn, Wolfgang, Layton, Edwin T., Jr., and Weingart, Peter, eds. *The Dynamics of Science and Technology*. Boston: D. Reidel, 1978. Sociology of the sciences; a yearbook.

Lenk, Hans, ed. *Technokratie als Ideologie; sozialphilosophische Beiträge zu einem politischen Dilemma* [Technocracy as ideology; social-philosophical contributions to a political problem]. Stuttgart: W. Kohlhammer, 1973.

Lenk, Hans, and Moser, Simon, eds. *Techne, Technik, Technologie: Philosophische Perspektiven*. Pullach: Verlag Dokumentation, 1973.

Lovekin, David, and Verene, Donald Phillip, eds. *Essays in Humanity and Technology*. Dixon, Ill.: Sauk Valley College, 1978.

McLean, George F., ed. *Philosophy in a Technological Culture*. Washington, D.C.: Catholic University of America Press, 1964.

* *Man, Science, Technology: A Marxist Analysis of the Scientific and Technological Revolution*. Moscow and Prague: Academia, 1973. Also available in Russian.

"Die Marxistisch-leninistische Philosophie und die technische Revolution" [Marxist-Leninist philosophy and the technological revolution]. Special issue of *Deutsche Zeitschrift für Philosophie* 13 (1965).

Mistichelli, Judith, and Roysdon, Christine. *Beyond Technics: Humanistic Interactions with Technology: A Basic Collection Guide*. Bethlehem, Pa.: Humanities Perspectives on Technology and Lehigh University Libraries, Lehigh University, 1978. Excellently annotated bibliography in the area of general humanities reactions to technology.

Mitcham, Carl, and Grote, Jim. "Current Bibliography in the Philosophy of Technology: 1973–1974." In *Research in Philosophy & Technology*, vol. 1, pp. 313–390. Edited by P. Durbin. Greenwich, Conn.: JAI Press, 1978. Author index in vol. 2, 1979.

Mitcham, Carl, and Mackey, Robert. *Bibliography of the Philosophy of Technology*. Chicago: University of Chicago Press, 1973. Comprehensive annotated bibliography. First published as special supplement to *Technology and Culture* 14 (April 1973). Author index in *Research in Philosophy & Technology*, vol. 4. Edited by P. Durbin, Greenwich, Conn.: JAI Press, 1980.

Mitcham, Carl, and Mackey, Robert, eds. *Philosophy and Technology: Readings in the Philosophical Problems of Technology*. New York: Free Press, 1972. Essays in five categories: conceptual issues, ethical and political critiques, religious critiques, existentialist critiques, and metaphysical studies. Select bibliography.

Mohan, Robert Paul, ed. *Technology and Christian Culture*. Washington, D.C.: Catholic University of America Press, 1960.

Niblett, Roy W. *The Sciences, the Humanities and the Technological Threat*. London: University of London Press, 1975.

Proceedings of the XIVth International Congress of Philosophy; Vienna, September 2–9, 1968. Vienna: Herder, 1968. Vol. 2, colloquium 6, pp. 477–614, is on "Cybernetics and the Philosophy of Technical Science."

Proceedings of the XVth World Congress of Philosophy; Varna, Bulgaria, September 17–22, 1973, vol. 2: *Reason and Action in the Transformation of the World; Philosophy in the Process of the Scientific and Technological Revolution; Knowledge and Values in the Scientific and Technological Era; Structure and Methods of Contemporary Scientific Knowledge*. Sofia: Sofia Press Production Centre, 1973. See also vols. 3–6, 1974–1975.

Progrès technique et progrès moral [Technological progress and moral progress]. Neuchatel: La Baconniere, 1947.

* Pylyshyn, Zenon W., ed. *Perspectives on the Computer Revolution*. Englewood Cliffs, N.J.: Prentice-Hall, 1970. A comprehensive anthology with good bibliography and index.

Rapp, Friedrich. *Contributions to a Philosophy of Technology: The Structure of Thinking in the Technological Sciences*. Dordrecht and Boston: D. Reidel, 1974. Annotated bibliography.

* Richta, Radovan, ed. *Civilization at the Crossroads: Social and Human Implications of the Scientific and Technological Revolution*. Revised ed. Prague: International Arts and Sciences Press, 1969; first published as *Civilizace na rozcesti*. Prague: Svoboda, 1967.

* Sachsse, Hans, ed. *Technik und Gesellschaft* [Technology and society], vol. 1: *Literaturführer* [Leading literature]. Munich: Verlag Dokumentation, 1974. Vol. 2: *Texte: Technik in der Literatur* [Texts: technology in literature]. Munich: Verlag Dokumentation, 1976. Vol. 3: *Selbstzeugnisse der Techniker; Philosophie der Technik* [Personal testimonies of technologists; philosophy of technology]. Munich: Verlag Dokumentation, 1976. The most comprehensive sourcebook in any language.

* Sayre, Kenneth M., and Crosson, Frederick J., eds. *The Modeling of Mind: Computers and Intelligence*. Notre Dame, Ind.: University of Notre Dame Press, 1963. New York: Simon and Schuster, 1968.

"Science in a Social Context." London and Boston: Butterworths. Series of pamphlet-textbooks, including: K. Pavitt and M. Worboys, *Science, Technology and the Modern Industrial State*, 1977; K. Green and C. Morphet, *Research and Technology as Economic Activities*, 1977; E. Braun and D. Collingridge, *Technology and Survival*, 1977; E. Braun et al., *Assessment of Technological Decisions: Case Studies*, 1979; M.

Gowing and L. Arnold, *The Atomic Bomb,* 1979; J. Lipscombe and
B. Williams, *Are Science and Technology Neutral?* 1979.

Stover, Carl F., ed. *The Technological Order.* Detroit: Wayne State University Press, 1963. Proceedings of a 1962 conference first published in
Technology and Culture 3 (Fall 1962).

* "La Technique." *Les Etudes philosophiques* (April-June 1976). Includes
papers by Jacques Ellul, Sergio Cotta, and Daniel Cérézuelle.

Technology and the Frontiers of Knowledge. Garden City, N.Y.: Doubleday, 1975. Lectures by Saul Bellow, Daniel Bell, Edmund O'Gorman,
Peter Medawar, and Arthur C. Clarke.

Teich, Albert H., ed. *Technology and Man's Future.* 2d ed. New York:
St. Martin's Press, 1977. Good introductory anthology, with a section
on technology assessment.

* "Toward a Philosophy of Technology." *Technology and Culture* 7
(Summer 1966): 301–390. Seminal symposium, with papers by Lewis
Mumford, James Feibleman, Mario Bunge, Joseph Agassi, J. O.
Wisdom, Henryk Skolimowski, and I. C. Jarvie. Follow-up discus-
sions in *Technology and Culture* 7 (Summer 1976) and *Technology
and Culture* 8 (January 1967).

* "Toward a Philosophy of Technology." *Philosophy Today* 15 (Summer
1971): 75–156. Good essays on German philosophy of technology.

Tuchel, Klaus, ed. *Herausforderung der Technik; Gesellschaftliche Voraus-
setzungen und Wirkungen der technischen Entwicklung* [The challenge of technology; social prerequisites and effects of technological
development]. Bremen: Carl Schunemann Verlag, 1967.

White, Hugh C., Jr. *Christians in a Technological Era.* New York: Seabury,
1964. Papers by Margaret Mead, Michael Polanyi, Jean Ladrière,
Bernard Morel, François Russo, Jean de la Croix Kaelin, and Scott
Paradise.

Zimmerli, W. C., ed. *Technik—Oder wissen wir, was wir tun?* [Technol-
ogy—or do we know what we are doing?]. Basel and Stuttgart: Schwabe,
1976.

BIBLIOGRAPHY

Abrecht, Paul, ed. *Faith, Science, and the Future.* Philadelphia: Fortress
Press, 1979.

Anders, Günther. "Commandments in the Atomic Age." In *Burning Conscience,* pp. 11–20. Edited by Claude Eatherly and Günther Anders.
New York: Monthly Review Press, 1961. Reprinted in *Philosophy
and Technology,* pp. 130–135. Edited by C. Mitcham and R. Mackey.
New York: Free Press, 1972.

Arendt, Hannah. *The Human Condition.* Chicago: University of Chicago
Press, 1958. Doubleday Anchor, n.d.

Ashby, W. Ross. *An Introduction to Cybernetics.* New York: John Wiley,
1956.

Axelos, Kostas. *Alienation, Praxis, and Techne in the Thought of Karl Marx.* Tr. Ronald Bruzina. Austin: University of Texas Press, 1976. Translated from *Marx, penseur de la technique: De l'aliénation de l'homme à la conquête du monde.* Paris: Les Editions de Minuit, 1969.

Baier, Kurt, and Rescher, Nicholas, eds. *Values and the Future: The Impact of Technological Change on American Values.* New York: Free Press, 1969.

Bailes, Kendall E. "The Politics of Technology: Stalin and Technocratic Thinking among Soviet Engineers." *American Historical Review* 79 (April 1974): 445–469.

Ballard, Edward G. *Man and Technology.* Pittsburgh: Duquesne University Press, 1978.

Bavink, Bernard. "Philosophy of Technology." In his *The Natural Sciences,* pp. 562–574. New York: Century, 1932. Translated from the 4th German edition, 1930.

Beck, Heinrich. *Philosophie der Technik: Perspektiven zu Technik, Menschheit, Zukunft* [Philosophy of technology: perspectives on technology, mankind, future]. Trier: Spee-Verlag, 1969.

Beck, Robert N. "Technology and Idealism." *Idealistic Studies* 4 (May 1974): 181–187.

Bell, Daniel. *The Coming of Post-Industrial Society: A Venture in Social Forecasting.* New York: Basic Books, 1973. Harper Colophon, 1976.

Berger, Peter L. *Pyramids of Sacrifice: Political Ethics and Social Change.* New York: Basic Books, 1974.

Berger, Peter L.; Berger, Brigitte; and Kellner, Hansfried. *The Homeless Mind: Modernization and Consciousness.* New York: Random House, 1973.

Bergson, Henri. "Mechanics and Mysticism." In his *Two Sources of Morality and Religion.* London: Macmillan, 1935. Doubleday Anchor, n.d. Translated from *Les Deux sources de la morale et de la religion.* Paris: Alcan, 1932.

Bernanos, Georges. *La France contre les robots.* Paris: Laffont, 1947.

Bertalanffy, Ludwig Von. *Robots, Men, and Minds.* New York: Braziller, 1967.

———. *General Systems Theory: Foundations, Development, Applications.* New York: Braziller, 1968.

Blauner, Robert. *Alienation and Freedom: The Factory Worker and His Industry.* Chicago: University of Chicago Press, 1964.

Boguslaw, Robert. *The New Utopians: A Study of System Design and Social Change.* Englewood Cliffs, N.J.: Prentice-Hall, 1965.

Boirel, René. *Science et technique* [Science and technology]. Neuchatel: Editions du Griffon, 1955.

Boorstin, Daniel J. *The Republic of Technology.* New York: Harper & Row, 1978.

Borgmann, Albert. "Technology and Reality." *Man and World* 4 (February 1971): 59–69.

————. "Orientation in Technology." *Philosophy Today* 16 (Summer 1972): 135–147.

————. "Functionalism in Science and Technology," In *Proceedings of the XVth World Congress of Philosophy*, vol. 6, pp. 31–36. Sofia: Sofia Press Production Centre, 1974.

Brakel, J. van. *Chemical Technology for Appropriate Development*. Delft: Delft University Press, 1978.

Brinkmann, Donald. *Mensch und Technik; Grundzüge einer Philosophie der Technik* [Man and technology; the fundamentals of a philosophy of technology]. Bern: A. Franke, 1945. A good summary of Brinkmann's ideas can be found in his "Technology as Philosophic Problem." *Philosophy Today* 15 (Summer 1971): 122–128.

Brun, Jean. *Les Conquêtes de l'homme et la séparation ontologique* [The conquests of man and the ontological separation]. Paris: Presses Universitaires de France, 1961.

Brunner, Heinrich Emil, *Christianity and Civilization*, vol. 2: *Specific Problems*. New York: Charles Scribner's Sons, 1949.

Bunge, Mario. "Toward a Philosophy of Technology." In *Philosophy and Technology*, pp. 62–76. Edited by C. Mitcham and R. Mackey. New York: Free Press, 1972. Adapted from the author's *Scientific Research*, vol. 2: *The Search for Truth*. Berlin and New York: Springer-Verlag, 1967. A less technical version of this paper appeared as "Technology as Applied Science." *Technology and Culture* 7 (Summer 1966): 329–347; and is included in *Contributions to a Philosophy of Technology*, pp. 19–39. Edited by F. Rapp. Dordrecht: Reidel, 1974.

————. *Technologia y Filosofia* [Technology and philosophy]. Monterrey, Mexico: Autonomous University of Neuva Leon, 1976.

————. "Towards a Technoethics." *Monist* 60 (January 1977): 96–107.

————. "Philosophical Inputs and Outputs of Technology." In *The History and Philosophy of Technology*, pp. 262–281. Edited by G. Bugliarello and D. Doner, Urbana: University of Illinois Press, 1979. Similar paper published as "The Philosophical Richness of Technology," as part of the symposium "Are There Any Philosophically Interesting Questions in Technology?" in *PSA 1976*, vol. 2, pp. 153–172. Edited by F. Suppe and P. Asquith. East Lansing, Mich.: Philosophy of Science Association, 1977.

————. "The Five Buds of Technophilosophy." *Technology in Society* 1 (Spring 1979): 67–74.

Byrne, Edmund. "Humanization of Technology: Slogan or Ethical Imperative?" In *Research in Philosophy & Technology*, vol. 1, pp. 149–177. Edited by P. Durbin. Greenwich, Conn.: JAI Press, 1978.

————. "Technology and Human Existence." *Southwestern Journal of Philosophy* 10 (Spring 1979): 55–69.

Callahan, Daniel. *The Tyranny of Survival*. New York: Macmillan, 1973.

Carey, James, and Quick, John. "The Mythos of the Electronic Revolution." *American Scholar* 39 (1970): 219–241 and 395–424.

Carpenter, Stanley R. "Modes of Knowing and Technological Action." *Philosophy Today* 18 (Summer 1974): 162–168.

———. "Philosophical Issues in Technology Assessment." *Philosophy of Science* 44 (December 1977): 574–593.

Cérézuelle, Daniel. "Fear and Insight in French Philosophy of Technology." In *Research in Philosophy & Technology*, vol. 2, pp. 53–75. Edited by P. Durbin. Greenwich, Conn.: JAI Press, 1979.

Churchman, C. West. *The Design of Inquiring Systems: Basic Concepts of Systems and Organization.* New York: Basic Books, 1971.

Cooper, Julian M. "The Scientific and Technical Revolution in Soviet Theory." In *Technology and Communist Culture*, pp. 146–179. Edited by Frederic J. Fleron, Jr. New York, Praeger, 1977.

Dessauer, Friedrich, *Philosophie der Technik: Das Problem der Realisierung.* Bonn: F. Cohen, 1927. Portion translated in *Philosophy and Technology*, pp. 317–334. Edited by C. Mitcham and R. Mackey. New York: Free Press, 1972.

———. *Streit um die Technik.* Frankfurt: J. Knecht, 1956. Second edition, 1958.

Dickson, David. *The Politics of Alternative Technology.* New York: Universe, 1975. Published in England as *Alternative Technology and the Politics of Technical Change.* London: Fontana, 1974.

Diemer, Alwin. "Philosophie der Technik." In his *Grundriss der Philosophie*, vol. 2: *Die philosophischen Sonderdisziplinen*, pp. 540–549. Meissenheim: Anton Hain, 1964.

Dreyfus, Hubert L. *What Computers Can't Do: A Critique of Artificial Reason.* New York: Harper & Row, 1972.

Dubos, René. *A God Within.* New York: Charles Scribner's Sons, 1972.

Duchet, René. *Bilan de la civilisation technicienne; anéantissement ou promotion de l'homme* [Balance sheet on technical civilization; annihilation or promotion of man]. Toulouse: Privat-Didier, 1955.

Durbin, Paul T. "Technology and Values: A Philosopher's Perspective." *Technology and Culture* 13 (October 1972): 556–576.

———. "Toward a Social Philosophy of Technology." In *Research in Philosophy & Technology*, vol. 1, pp. 67–97. Edited by P. Durbin. Greenwich, Conn.: JAI Press, 1978.

Ellul, Jacques. *The Technological Society.* Tr. J. Wilkinson. New York: Alfred A. Knopf, 1964. From *La Technique, ou l'enjeu du siècle.* Paris: Colin, 1954.

———. "The Technological Order." First published in *The Technological Order.* Edited by C. Stover. Detroit: Wayne State University Press, 1963. Reprinted in *Philosophy and Technology*, pp. 86–105. Edited by C. Mitcham and R. Mackey. New York: Free Press, 1972.

———. *The New Demons.* Tr. C. Edward Hopkin. New York: Seabury, 1975. From *Les Nouveaux possédés.* Paris: A. Fayard, 1973.

———. *Le Système technicien* [The technological system]. Paris: Calmann-Levy, 1977.

Engelmeier [Engelmeyer], P. K. "Philosophie der Technik." *Atti del* 4. *Congresso internazionale di filosofia,* vol. 3, pp. 587–596. Bologna, 1911.

Espinas, Alfred Victor. *Les Origines de la Technologie* [The origins of Technology]. Paris: F. Alcan, 1897.

Ferguson, Eugene S. "The Mind's Eye: Nonverbal Thought in Technology." *Science* 197 (26 August 1977): 827–836.

Ferkiss, Victor C. *Technological Man: The Myth and the Reality.* New York: Braziller, 1969.

———. *The Future of Technological Civilization.* New York: Braziller, 1974.

Florman, Samuel C. *The Existential Pleasures of Engineering.* New York: St. Martin's Press, 1976.

Foley, Michael. "Who Cut Down the Sacred Tree?" *CoEvolution Quarterly,* no. 15 (Fall 1977), pp. 60–67.

Freyer, Hans. *Theorie des Gegenwärtigen Zeitalters* [Theory of the present age]. Stuttgart: Deutsche Verlags-Anstalt, 1955.

Friedmann, Georges. *La Crise du progrès; Esquisse d'histoire des idées, 1895–1935* [The crisis of progress; outline of the history of ideas, 1895–1935]. Paris: Gallimard, 1936.

———. *Anatomy of Work: Labor, Leisure, and the Implications of Automation.* New York: Free Press, 1961. Tr. by W. Rawson of *Le Travail en miettes; Specialisation et loisirs.* Paris: Gallimard, 1956 and 1964.

———. *Sept études sur l'homme et la technique* [Seven studies on man and technology]. Paris: Gallimard, 1966.

———. *La Puissance et la sagesse* [Power and wisdom]. Paris: Gallimard, 1970.

Gadamer, Hans-Georg. "Theory, Technology, Practice: The Task of the Science of Man." *Social Research* 44 (Autumn 1977): 529–561. This is a translated and modified version of "Theorie, Technik, Praxis: Die Aufgabe einer neuen Anthropologie," the introduction to *Neue Anthropologie,* edited by Gadamer and Paul Vogler. Stuttgart: Thieme, 1972–1975.

Gendron, Bernard. *Technology and the Human Condition.* New York: St. Martin's Press, 1977.

Gill, Eric. *Christianity and the Machine Age.* London: Sheldon, 1940. Reprinted in *Philosophy of Technology,* pp. 214–236. Edited by C. Mitcham and R. Mackey. New York: Free Press, 1972.

Goodman, Paul. "Can Technology Be Humane?" *New York Review of Books,* 13 (20 November 1969), pp. 27–34. Reprinted in *Technology and Man's Future,* pp. 207–222. Edited by A. H. Teich. New York: St. Martin's Press, 1977.

Gouldner, Alvin W. *The Dialectic of Ideology and Technology: The Origins, Grammar, and Future of Ideology.* New York: Seabury, 1976.

Goulet, Denis. *Cruel Choice: A New Concept in the Theory of Development.* New York: Atheneum, 1971.

———. *The Uncertain Promise: Value Conflicts in Technology Transfer.*

New York: I.D.O.C.–North America; Washington, D.C.: Overseas Development Council, 1977.

Grant, George P. *Technology and Empire*. Toronto: House of Anansi, 1969.

———. "Knowing and Making." *Royal Society of Canada: Proceedings and Transactions,* Fourth Series, vol. 12 (1974): 59–67.

———. "The Computer Does Not Impose on Us the Ways It Should Be Used." In *Beyond Industrial Growth,* pp. 117–131. Edited by A. Rotstein. Toronto: University of Toronto Press, 1976.

Gunderson, Keith. "Minds and Machines: A Survey." In *Contemporary Philosophy: A Survey,* vol. 2: *Philosophy of Science,* pp. 416–425. Edited by Raymond Klibansky. Florence: La Nuova Italia Editrice, 1968.

———. *Mentality and Machines*. Garden City, N.Y.: Doubleday, 1971.

Habermas, Jürgen. *Toward a Rational Society: Student Protest, Science and Politics.* Tr. J. J. Shapiro. Boston: Beacon, 1970. The last three essays are from *Technik und Wissenschaft als "Ideologie."* Frankfurt: Suhrkamp, 1968.

———. *Communication and the Evolution of Society*. Boston: Beacon, 1979.

Haldane, J. B. S. *Daedalus, or Science and the Future*. New York: Dutton, 1924.

Hardin, Garrett. "The Tragedy of the Commons." *Science* 162 (13 December 1968): 1243–1248. Reprinted in his *Exploring New Ethics for Survival*. New York: Viking, 1972.

Harrison, Andrew. *Making and Thinking: A Study of Intelligent Activities*. London: Harvester Press; Indianapolis: Hackett, 1978.

Heidegger, Martin. *Being and Time*. Tr. John Macquarrie and Edward Robinson. New York: Harper & Row, 1963. First published 1927. Page references follow the 8th German edition, 1957.

———. "The Question Concerning Technology." In *The Question Concerning Technology and Other Essays,* pp. 3–35. Tr. William Lovitt. New York: Harper & Row, 1977. First published in Heidegger's *Vorträge und Aufsätze*. Pfullingen: Neske, 1954.

———. *Discourse on Thinking*. Tr. John M. Anderson and E. Hans Freund. New York: Harper & Row, 1962. First published as *Gelassenheit*. Pfullingen: Neske, 1959.

———. "The Turning." In *The Question Concerning Technology and Other Essays,* pp. 36–49. Tr. William Lovitt. New York: Harper & Row, 1977. First published in *Die Technik und die Kehre*. Pfullingen: Neske, 1962.

———. "Only a God Can Save Us." *Philosophy Today* 20 (Winter 1976): 267–284. Another translation of this interview from *Der Spiegel* (May 31, 1976) can be found in *Graduate Faculty Philosophy Journal,* New School for Social Research, 6 (Winter 1977): 5–27.

Heilbroner, Robert L. *An Inquiry into the Human Prospect*. New York: W. W. Norton, 1974.

Hommes, Jakob. *Der Technische Eros; Das Wesen der materialistischen Geschictsauffassung* [Technological eros; the essence of the materialist interpretation of history]. Freiburg: Herder, 1955.

Horkheimer, Max. *Critique of Instrumental Reason.* New York: Seabury, 1974. Translated from *Zur Kritik der instrumentellen Vernunft.* Frankfurt: S. Fischer, 1967.

Huning, Alois. *Das Schaffen des Ingenieurs; Beiträge zu einer Philosophie der Technik* [The activity of engineers; contributions to a philosophy of technology]. Düsseldorf: VDI-Verlag, 1974.

———. "Philosophy of Technology and the Verein Deutscher Ingenieure." In *Research in Philosophy & Technology,* vol. 2, pp. 265–271. Edited by P. Durbin. Greenwich, Conn.: JAI Press, 1979.

Ihde, Don. *Technics and Praxis.* Boston Studies in the Philosophy of Science, vol. 24. Dordrecht: Reidel, 1979.

Jaspers, Karl. "Modern Technology." In *The Origin and Goal of History,* pp. 96–125. Tr. M. Bullock. New Haven: Yale University Press, 1953. From *Von Ursprung und Ziel Geschichte.* Zurich: Artemis-Verlag, 1949.

Jonas, Hans. "The Practical Uses of Theory." *Social Research* 26 (1959): 151–166. Reprinted in Jonas, *The Phenomenon of Life.* New York: Harper & Row, 1966. Also in *Philosophy and Technology,* pp. 335–346. Edited by C. Mitcham and R. Mackey. New York: Free Press, 1972.

———. "Philosophical Reflections on Experimenting with Human Subjects." *Daedalus* 98 (Spring 1969): 219–247.

———. "The Scientific and Technological Revolutions: Their History and Meaning." *Philosophy Today* 15 (Summer 1971): 76–101.

———. "Technology and Responsibility: Reflections on the New Tasks of Ethics." *Social Research* 15 (Spring 1973): 160–180.

———. *Philosophical Essays: From Ancient Creed to Technological Man.* Englewood Cliffs, N.J.: Prentice-Hall, 1974.

———. "Responsibility Today: The Ethics of an Endangered Future." *Social Research* 43 (Spring 1976): 77–97.

———. "Toward a Philosophy of Technology." *Hastings Center Report* 9 (February 1979): 34–43.

Juenger, Friedrich Georg. *The Failure of Technology.* Chicago: Regnery, 1949. From *Die Perfektion der Technik.* Frankfurt: Klostermann, 1946.

Jung, Hwa Yol. "The Ecological Crisis: A Philosophical Perspective." *Bucknell Review* 20 (Winter 1972): 25–44.

———. "The Paradox of Man and Nature: Reflections on Man's Ecological Predicament." *Centennial Review* 18 (Winter 1974): 1–28.

Jung, Hwa Yol, and Jung, Petee. "Toward a New Humanism: The Politics of Civility in a 'No-Growth' Society." *Man and World* 9 (August 1976): 283–306.

Kapp, Ernst. *Grundlinien einer Philosophie der Technik: zur Entstehungs-*

geschichte der Cultur aus neuen Gesichtspunkten [Fundamentals of a philosophy of technology; on the genesis of culture from a new point of view]. Braunschweig: Westermann, 1877. Reprint ed., Düsseldorf: Stern-Verlag Janssen, 1978.

Kass, Leon R. "The New Biology: What Price Relieving Man's Estate?" *Science* 174 (19 November 1971): 779–788.

King, Magda. "Truth and Technology." *Human Context* 5 (1973): 1–34.

Kol'man, Ernst. "Chto takoe kibernetika?" [What is cybernetics?]. *Voprosy filosofii* 9 (1955): 148–159.

Kotarbiński, Tadeusz. *Praxiology: An Introduction to the Science of Efficient Action.* Tr. O. Wojtasiewicz. New York: Pergamon, 1965.

Kuzin, A. A., and Shukhardin, S. V. *Sovremennaia naucho-teknicheskaia revoliutsiia—istoricheskoe issledovanie* [The contemporary scientific-technological revolution—an historical analysis]. Moscow. Nauka, 1967.

Ladrière, Jean. *The Challenge Presented to Cultures by Science and Technology.* Paris: UNESCO, 1977.

Lafitte, Jacques. *Reflexions sur la science des machines* [Reflections on the science of machines]. Paris: Bloud & Gay, 1932. Reprint ed., Paris: J. Vrin, 1972.

Laszlo, Ervin. *A Strategy for the Future: The Systems Approach to World Order.* New York: Braziller, 1974.

Layton, Edwin T., Jr. "Technology as Knowledge." *Technology and Culture* 15 (January 1974): 31–41.

Leiss, William. "The Social Consequences of Technological Progress: Critical Comments on Recent Theories." *Canadian Public Administration* 23 (Fall 1970): 246–262.

————. *The Domination of Nature.* New York: Braziller, 1972.

————. "The Problem of Man and Nature in the Work of the Frankfurt School." *Philosophy of the Social Sciences* 5 (1975): 163–172.

————. *The Limits to Satisfaction: An Essay on the Problem of Needs and Commodities.* Toronto: University of Toronto Press, 1976.

Lem, Stanislaw. *Summa technologiae.* Cracow: Wydawnictwo Literackie, 1964. Revised ed., 1967. German translation, same title. Frankfurt: Insel Verlag, 1976.

Lenk, Hans. *Philosophie im technologischen Zeitalter* [Philosophy in a technological age]. Stuttgart: Kohlhammer, 1971.

Lenk, Hans, and Günther Ropohl. "Toward an Interdisciplinary and Pragmatic Philosophy of Technology." In *Research in Philosophy & Technology,* vol. 2, pp. 15–52. Edited by P. Durbin, Greenwich, Conn.: JAI Press, 1979.

Leroi-Gourhan, André. *Evolution et techniques* [Evolution and techniques]. 2 vols. Paris: Albin Michel. Vol. 1: *L'Homme et la matière,* 1943. 2d ed., 1971. Vol. 2: *Milieu et techniques,* 1945. 2d ed., 1973.

————. *Le Geste et la parole* [Action and Speech]. Vol. 1: *Technique et langage*. Paris: Albin Michel, 1964.

Lewis, C. S. *The Abolition of Man*. New York: Macmillan, 1947.

Ley, Hermann. *Dämon Technik?* [Demon technology?]. Berlin: Deutscher Verlag der Wissenschaften, 1961.

Lilienfeld, Robert. *The Rise of Systems Theory: An Ideological Analysis*. New York: John Wiley, 1977. See also "Systems Theory as an Ideology." *Social Research* 42 (Winter 1975): 637–660.

Lilje, Hans. *Das technischen Zeitalter; Grundlinien einer christlichen Deutung* [The technical age; outlines of a Christian interpretation] Berlin: Furche-Verlag, 1928.

Lombroso, Gina. *The Tragedies of Progress*. Tr. C. Taylor. New York: Dutton, 1931. From *Le tragedie del progresso*. Turin: Bocca, 1930.

Lovekin, David. "Jacques Ellul and the Logic of Technology." *Man and World* 10 (1977): 251–272.

Lovins, Amory. *World Energy Strategies: Facts, Issues, and Options*. San Francisco: Friends of the Earth; Cambridge, Mass.: Ballinger, 1975.

————. "Energy Strategy: The Road Not Taken?" *Foreign Affairs* 55 (October 1976): 65–96.

————. *Soft Energy Paths: Toward a Durable Peace*. San Francisco: Friends of the Earth; Cambridge, Mass.: Ballinger, 1977. Reprint ed. New York: Harper & Row, 1978.

Lovitt, William. "A 'Gespraech' with Heidegger on Technology." *Man and World* 6 (February 1973): 44–59.

Lynch, William F., S. J. *Christ and Prometheus: A New Image of the Secular*. Notre Dame, Ind.: University of Notre Dame Press, 1970.

McDermott, John. "Technology: The Opiate of the Intellectuals." *New York Review of Books* (31 July 1969), pp. 25–35. Reprinted in *Technology and Man's Future*, pp. 180–207. Edited by A. H. Teich. New York: St. Martin's Press, 1977.

McLuhan, Herbert Marshall. *Understanding Media: The Extensions of Man*. New York: McGraw-Hill, 1964.

Macpherson, C. B. "Democratic Theory: Ontology and Technology." In *Philosophy and Technology*, pp. 161–170. Edited by C. Mitcham and R. Mackey. New York: Free Press, 1972. First published in *Political Theory and Social Change*. Edited by D. Spitz. New York: Atherton, 1967.

Marcel, Gabriel. *Man Against Mass Society*. Tr. G. S. Fraser. Chicago: Regnery, 1952.

————. *The Decline of Wisdom*. Tr. M. Harari. London: Harvill, 1954. From *Le Declin de la sagesse*, Paris: Plon, 1954.

Marcuse, Herbert. *One-Dimensional Man: Studies in the Ideology of Advanced Industrial Society*. Boston: Beacon, 1964.

————. *Counterrevolution and Revolt*. Boston: Beacon, 1972.

Marx, Karl. *Capital: A Critique of Political Economy*. Vol. 1, first published 1867. Tr. Samuel Moore and Edward Aveling. New York: International Publishers, 1967.

Maurer, Reinhart. "From Heidegger to Practical Philosophy." *Idealistic Studies* 3 (May 1973): 133–162.

Mazlish, Bruce. "The Fourth Discontinuity." *Technology and Culture* 8 (January 1967): 1–15.

Melsen, Andrew G. Van. *Science and Technology*. Pittsburgh: Duquesne University Press, 1961.

Menninger, David C. "Jacques Ellul: A Tempered Profile." *Review of Politics* 37 (April 1975): 235–246.

Mesthene, Emmanuel G. *Technological Change: Its Impact on Man and Society*. Cambridge, Mass.: Harvard University Press; New York: New American Library, 1970.

Mitcham, Carl. "Types of Technology." In *Research in Philosophy & Technology*, vol. 1, pp. 229–294. Edited by P. Durbin. Greenwich, Conn.: JAI Press, 1978.

————. "Philosophy and the History of Technology." In *The History and Philosophy of Technology*, pp. 163–201. Edited by G. Bugliarello and D. Doner. Urbana: University of Illinois Press, 1979.

Mitcham, Carl, and Mackey, Robert. "Jacques Ellul and the Technological Society." *Philosophy Today* 15 (Summer 1971): 102–121.

Moncrief, Lewis. "The Cultural Basis for Our Environmental Crisis." *Science* 170 (25 October 1970): 508–512.

Morison, Elting E. *From Know-How to Nowhere: The Development of American Technology*. New York: Basic Books, 1974. Paperback reprint, New York: New American Library, 1977.

Moser, Simon. "Toward a Metaphysics of Technology." *Philosophy Today* 15 (Summer 1971): 129–156. Translated from "Zur Metaphysik der Technik," in his *Metaphysik einst und jetzt*. Berlin: De Gruyter, 1958. Revised version, "Kritik der traditionellen Technikphilosophie," in *Techne, Technik, Technologie*, pp. 11–81. Edited by H. Lenk and S. Moser. Pullach: Verlag Dokumentation, 1973.

Mounier, Emmanuel. "The Case against the Machine." In his *Be Not Afraid: A Denunciation of Despair*. London: Rockliff, 1951; New York: Sheed and Ward, 1962. From *La Petite peur du XXᵉ siècle*. Paris: Seuil, 1948.

Mumford, Lewis. *Technics and Civilization*. New York: Harcourt Brace, 1934.

————. *Art and Technics*. New York: Columbia University Press, 1953.

————. *The Myth of the Machine*. 2 vols. New York: Harcourt Brace Jovanovich. Vol. 1: *Technics and Human Development*, 1967. Vol. 2: *The Pentagon of Power*, 1970.

Ortega y Gasset, José. "Thoughts on Technology." In *Philosophy and Technology*, pp. 290–313. Edited by C. Mitcham and R. Mackey. New York: Free Press, 1972. Translated from "Meditacion de la technica," in

Ensimismamiento y alteracion; obras completas, vol. 5. Madrid: Revista de Occidente, 1939. First published 1933.

Passmore, John. *Man's Responsibility for Nature: Ecological Problems and Western Traditions.* New York: Charles Scribner's Sons, 1974.

Paz, Octavio. "Use and Contemplation." In *In Praise of Hands: Contemporary Crafts of the World,* pp. 17–24. Greenwich, Conn.: New York Graphic Society, 1974.

Pursell, Carroll W., Jr. "Belling the Cat: A Critique of Technology Assessment"; " 'A Savage Struck by Lightning': The Idea of a Research Moratorium, 1927–1937"; " 'Who to Ask besides the Barber'—Suggestions for Alternative Assessments." *Lex et Scientia* 10 (October-December 1974): 130–177.

Rahner, Karl. "The Experiment with Man; Theological Observations on Man's Self-Manipulation." In his *Theological Investigations,* vol. 9, pp. 205–224. New York: Herder and Herder, 1972.

Reuleaux, Franz. *The Kinematics of Machinery: Outlines of a Theory of Machines.* Tr. Alexander B. W. Kennedy. New York: Dover, 1963. First German publication, 1875.

Riessen, H. van. *Filosofie en techniek* [Philosophy and technology]. Kampen: J. H. Kok, 1949.

Rodman, John. "On the Human Question, Being the Report of the Erewhonian High Commission to Evaluate Technological Society." *Inquiry* 18 (Summer 1975): 127–166.

――――. "The Liberation of Nature?" *Inquiry* 20 (Spring 1977): 83–131.

Rossini, Frederick A. "Technology Assessment: A New Type of Science?" In *Research in Philosophy & Technology,* vol. 2, pp. 341–355. Edited by P. Durbin. Greenwich, Conn.: JAI Press, 1979.

Roszak, Theodore. *Where the Wasteland Ends: Politics and Transcendence in Postindustrial Society.* Garden City, N.Y.: Doubleday, 1973.

Rotenstreich, Nathan. *Theory and Practice: An Essay in Human Intentionalities.* The Hague: Martinus Nijhoff, 1977.

Russell, Bertrand. *Icarus, or The Future of Science.* New York: Dutton, 1924.

Ryan, John Julian. *The Humanization of Man.* New York: Newman Press, 1972.

Sachsse, Hans. *Anthropologie der Technik; Ein Beitrag zur Stellung des Menschen in der Welt* [Anthropology of technology; an essay on the place of man in the world.] Braunschweig: Vieweg, 1978.

Sayre, Kenneth M. *Recognition: A Study in the Philosophy of Artificial Intelligence.* Notre Dame, Ind.: University of Notre Dame Press, 1965.

――――. *Consciousness: A Philosophic Study of Minds and Machines.* New York: Random House, 1969.

――――. *Cybernetics and the Philosophy of Mind.* Atlantic Highlands, N.J.: Humanities Press, 1976.

――――. *Moonflight: A Conversation on Determinism.* Notre Dame, Ind.: University of Notre Dame Press, 1977.

————. *Starburst: A Conversation on Man and Nature*. Notre Dame, Ind.: University of Notre Dame Press, 1977.

————, ed. *Values and the Electric Power Industry*. Notre Dame, Ind.: University of Notre Dame Press, 1977.

Schon, D. A. *Technology and Change: The New Heraclitus*. New York: Delacorte Press, 1967.

Schröter, Manfred. *Philosophie der Technik* [Philosophy of technology]. Munich: Oldenbourg, 1934.

Schroyer, Trent. *The Critique of Domination: The Origins and Development of Critical Theory*. New York: Braziller, 1973. Paperback reprint, Boston: Beacon, 1975.

Schuhl, Pierre-Maxime. *Machinisme et philosophie* [Machinism and philosophy]. Paris: F. Alcan, 1938.

Schumacher, E. F. *Small Is Beautiful: Economics as if People Mattered*. New York: Harper & Row, 1973.

————. *A Guide for the Perplexed*. New York: Harper & Row, 1977.

Schuurman, Egbert. *Technology and Deliverance: A Confrontation with Philosophical Views*. Tr. Donald Morton. Toronto: Wedge Publishing Foundation, 1980. Translated from *Techniek en Toekomst: Confrontatie met wijsgerige beschouwingen*. Assen: Van Gorcum, 1972.

————. *Reflections on the Technological Society*. Toronto: Wedge, 1977.

Sessions, George S. "Anthropocentrism and the Environmental Crisis." *Humbolt Journal of Social Relations* 2 (Fall/Winter 1974): 71–81.

Sibley, Mumford Q. *Nature and Civilization: Some Implications for Politics*. Itasca, N.Y.: F. E. Peacock, 1977.

Simon, Herbert A. *The Sciences of the Artificial*. Cambridge, Mass.: M.I.T. Press, 1969.

Simondon, Gilbert. *Du Mode d'existence des objets techniques* [The mode of existence of technological objects]. Paris: Montaigne-Aubier, 1958. 2d ed., 1969.

Skolimowski, Henryk. "The Structure of Thinking in Technology." *Technology and Culture* 7 (Summer 1966): 371–383. See also I. C. Jarvie's "The Social Character of Technological Problems: Comments on Skolimowski's Paper" in the same issue. Both papers included in *Philosophy and Technology*, pp. 42–53. Edited by C. Mitcham and R. Mackey. New York: Free Press, 1972. Also in *Contributions to a Philosophy of Technology*, pp. 72–92. Edited by F. Rapp. Dordrecht: Reidel, 1974.

————. "On the Concept of Truth in Science and Technology." In *Proceedings of the XIVth International Congress of Philosophy: September 2–9, 1968*, vol. 2, pp. 553–559. Vienna: Herder, 1968.

————. "Technology and Philosophy." In *Contemporary Philosophy: A Survey*, vol. 2: *Philosophy of Science*, pp. 426–437. Edited by Raymond Klibansky. Florence: La Nuova Italia Editrice, 1968.

————. "Extensions of Technology: From Utopia to Reality." In *Man, So-*

ciety, Technology, pp. 24–36. Edited by Linda Taxis. Washington, D.C.: American Industrial Arts Association, 1970.

———. "A Way Out of the Abyss." *Main Currents* 31 (January/February 1975): 71–76.

———. "Ecological Humanism." *Tract* (Sussex, England) nos. 19–20, pp. 1–41. Spanish version in *Folia Humanística* 15 (May 1977): 321–328; (June 1977): 417–428; and (July–August 1977): 537–549.

———. "Eco-Philosophy versus the Scientific World View." *Ecologist Quarterly* 3 (Autumn 1978): 227–248. Also published as "A Twenty-first Century Philosophy." *Alternative Futures* 1 (Fall 1978): 3–31.

Sloman, Aaron. *The Computer Revolution in Philosophy: Philosophy Science, and Models of Mind.* Atlantic Highlands, N.J.: Humanities Press, 1978.

Sobolev, S. L., Kitov, A. I.; and Liapunov, A. A. "Osnounye cherti kibernetiki" [Basic features of cybernetics]. *Voprosy filosofii* 9 (1955): 136–148.

Snow, C. P. *The Two Cultures and the Scientific Revolution.* New York: Cambridge University Press, 1959. 2d ed., enlarged, *The Two Cultures: And a Second Look.* New York: Cambridge University Press, 1963.

Spengler, Oswald. *Man and Technics: A Contribution to a Philosophy of Life.* Tr. C. F. Atkinson. New York: Alfred A. Knopf, 1932. First published 1931.

Stanley, Manfred. *The Technological Conscience: Survival and Dignity in an Age of Expertise.* New York: Free Press, 1978.

Starr, Chauncey. "Social Benefit versus Technological Risk." *Science* 165 (19 September 1969): 1232–1238.

Stone, Christopher. *Should Trees Have Standing?* New York: Avon Books, 1975.

Thomas, Donald E., Jr. "Diesel, Father and Son: Social Philosophies of Technology." *Technology and Culture* 19 (July 1978): 376–393.

Tribe, Laurence. "Technology Assessment and the Fourth Discontinuity." *Southern California Law Review* 44 (June 1973): 617–660.

———. "Ways Not To Think About Plastic Trees: New Foundations for Environmental Law." *Yale Law Journal* 83 (June 1974): 1315–1348.

Turing, A. M. "Computing Machinery and Intelligence," *Mind* 59 (1950): 433–460. Reprinted in *Minds and Machines,* pp. 4–30. Edited by A. R. Anderson. Englewood Cliffs, N.J.: Prentice-Hall, 1964.

Vahanian, Gabriel. *God and Utopia: The Church in a Technological Civilization.* New York: Seabury, 1977.

Wartofsky, Marx. "Philosophy of Technology." In *Current Research in Philosophy of Science,* pp. 171–184. Edited by Peter Asquith. East Lansing, Mich.: Philosophy of Science Association, 1979.

Weber, Max. *The Protestant Ethic and the Spirit of Capitalism.* Tr. Talcott Parsons. New York: Charles Scribner's Sons, 1930; reprint ed., 1958. First German edition, 1920.

Weinberg, Alvin M. "Can Technology Replace Social Engineering?" *Bulle-*

tin of the Atomic Scientists 22 (December 1966): 4–8. Reprinted in *Technology and Man's Future,* pp. 22–30. Edited by A. Teich. New York: St. Martin's Press, 1977.

Wiener, Norbert. *Cybernetics: Or Control and Communication in the Animal and the Machine.* New York: John Wiley, 1948. 2d ed., Cambridge, Mass.: M.I.T. Press, 1961.

———. *The Human Use of Human Beings: Cybernetics and Society.* New York: Houghton Mifflin, 1950. Garden City, N.Y.: Doubleday, 1954.

———. *God and Golem, Inc.: A Comment on Certain Points Where Cybernetics Impinges on Religion.* Cambridge, Mass.: M.I.T. Press, 1964.

Weizenbaum, Joseph. *Computer Power and Human Reason: From Calculation to Judgment.* San Francisco: W. H. Freeman, 1976.

White, Lynn Jr. "The Historical Roots of Our Ecologic Crisis." *Science* 155 (10 March 1967): 1203–1207.

———. "The Iconography of Temperantia and the Virtuousness of Technology." In *Action and Contemplation in Early Modern Europe,* pp. 197–219. Edited by T. K. Rabb and J. E. Sergel. Princeton: Princeton University Press, 1969.

———. "Technology Assessment from the Stance of a Medieval Historian." *American Historical Review* 79 (February 1974): 1–13.

———. *Medieval Technology and Religion.* Berkeley: University of California Press, 1978.

Wilhelmsen, Frederick. "Art and Religion." *Intercollegiate Review* 10 (Spring 1975): 85–94.

Wilhelmsen, Frederick D., and Bret, Jane. *The War in Man: Media and Machines.* Athens, Ga.: University of Georgia Press, 1970.

Winner, Langdon. "On Criticizing Technology." *Public Policy* 20 (Winter 1972): 35–39. Reprinted in *Technology and Man's Future,* pp. 354–375. Edited by A. H. Teich. New York: St. Martin's Press, 1977.

———. *Autonomous Technology: Technics-out-of-Control as a Theme in Political Thought.* Cambridge, Mass.: M.I.T. Press, 1977.

Zeman, J. "Cybernetics and Philosophy in Eastern Europe." In *Contemporary Philosophy: A Survey,* vol. 2: *Philosophy of Science,* pp. 407–415. Edited by Raymond Klibansky. Florence: La Nuova Italia Editrice, 1968.

Zimmerman, Michael E. "Heidegger on Nihilism and Technique." *Man and World* 8 (November 1975): 394–414.

———. "Beyond 'Humanism': Heidegger's Understanding of Technology." *Listening* 12 (Fall 1977): 74–83. See also Zimmerman's "A Brief Introduction to Heidegger's Concept of Technology." *HPT News,* Newsletter of the Lehigh University Humanities Perspectives on Technology Program, Bethlehem, Pa., no. 2 (October 1977): 10–13.

Zschimmer, Eberhard. *Philosophie der Technik; Einfuhrung in die technische Ideenwelt* [Philosophy of technology; introduction to the world of technological ideas]. 3d rev. ed. Stuttgart: F. Enke, 1933. Earlier editions appeared in 1913 and 1919 under slightly different titles.

Chapter 6
Philosophy of Medicine

H. Tristram Engelhardt, Jr., GEORGETOWN UNIVERSITY
Edmund L. Erde, UNIVERSITY OF TEXAS, GALVESTON

INTRODUCTION

Our purpose here is to provide a guide to what may appear to be a newly emerging field of philosophical study—the philosophy of medicine. The field appears new because of a marked increase in interest and productivity—an increase reflected, for example, in the development of numerous new journals. A relatively early general journal is *Perspectives in Biology and Medicine*, which was started in 1957. More recent journals include *The Hastings Center Report* (1971), *Ethics in Science and Medicine* (1975), *The Journal of Medical Ethics* (1975), *The Journal of Medicine and Philosophy* (1976), *Man and Medicine* (1975), and *Metamedicine* (begun in 1977 as *Metamed*). An *Encyclopedia of Bioethics* (1978) has in fact appeared. Earlier, interest was focused more on religion and medicine; for example, *The Linacre Quarterly*, founded in 1932. A number of recent anthologies (e.g., Feinberg, 1973; Gorovitz, 1976; Hunt and Arras, 1977; Beauchamp and Walters, 1978) and some original works of substance (E. Murphy, 1976; Feinstein, 1967; Wulff, 1976; and Foucault, 1973) also mark the emergence of this interest in the philosophy of medicine. In addition, several series of colloquia have been undertaken, and their proceedings have been published (for example, see those listed under Engelhardt and Spicker in the bibliography).

Although most of the recent interest embodied in these publications is focused on issues in biomedical ethics, interests within the philosophy of medicine are broadening. The Philosophy of Science Association in 1976 sponsored a session on epistemological issues in medicine (Engelhardt, 1977a; Grene, 1977; Wartofsky, 1977; Whitbeck, 1977a), and an issue of *Philosophy of Science* is devoted in large part to this topic (December, 1977). In short, an area of scholarship has developed that can aptly be termed "the philosophy of medicine."

Despite the appearance of novelty, however, it should be noted that philosophy and medicine have an extended record of interaction. Philosophers from Pythagoras and Plato through Spinoza and to Wittgenstein have drawn metaphors from medicine and have concerned themselves with the conceptual health of the human being. Furthermore, some of the great figures in the history of psychiatry—notably C. G. Jung—thought that their area of inquiry could best be understood as philosophy. Aristotle, whose father was a physician, seems to have believed that medicine could aid in philosophic and moral tasks to a large degree (Owens, 1977). It has also been argued (Romanell, 1974) that John Locke's philosophic work was shaped to a great extent by his medical orientation. Moreover, as Isaac Newton characterized his work on dynamics as a contribution to natural philosophy, other thinkers of Newton's day—for example, Descartes—thought that a philosophic approach to such basic sciences as physiology, as well as to clinical or applied medicine, would be highly productive. Thus, although activity in the philosophy of medicine is accelerating at a geometric rate, it is really a reemerging field, and much of our guide will be focused by historical references.

Our guide to the philosophy of medicine will consist of an overview of issues in the theory of knowledge, the philosophy of mind, and ethics, as they arise in medicine. This orientation is somewhat arbitrary, for there is a conceptual difficulty that complicates the construction of such a tour: the term "philosophy of medicine" has a troublesome vagueness about it. It does not have the precision of, for example, "philosophy of science" or "philosophy of biology." This vagueness is attributable in part to the ambiguity of the term "medicine," which has no single, univocal meaning. Consider that medicine includes surgery, internal medicine, psychoanalysis, dentistry, and public health, as well as nursing, occupational therapy, physical therapy, and other allied health sciences. Further, in each of these many domains, "medicine" refers to basic sciences (e.g., theories about the way the pancreas works); to theoretical endeavors (e.g., the development of explanatory models of health and disease, such as etiological models of the development of diabetes mellitus); and to actual practice (e.g., regimens for the control of diabetes mellitus through the use of insulin). As a result, one may have philosophical puzzles about theories of function and disease (e.g., as found in physiology and pathology); and about theories of treatment (e.g., as found in pharmacology); and about the ways in which health practitioners engage in their preventive or therapeutic activities (e.g., the ways internists make clinical judgments).

Some may object to the term "philosophy of medicine" and may wish instead to use a term such as "philosophy of health care." This objection could be motivated, for example, by a concern not to construe nursing as part of "medicine." However, "health care" and "health science" are by themselves inadequate terms, for the one does not clearly include the bio-medical sciences, and the other does not include the medical arts. One might, then, prefer a phrase such as "philosophy of health care sciences and arts." We have, instead, chosen the straightforward phrase "philosophy of medicine" because we find it less cumbersome and fairly well established. "Medicine" should thus be interpreted broadly, as it is used, for example, in referring to one of the four traditional university faculties.

Accordingly, our use of the term "medicine" will encompass health care activities as well as the sciences that undergird them. Given this use of the term, the philosopher who attends to medicine will be studying a set of undertakings gathered together under that name for various historical, social, and conceptual reasons. There is nothing in medicine approaching the conceptual unity that one expects in biology. Thus, philosophy of medicine, unlike the philosophy of biology, confronts a range of heterogeneous questions. This is the case, in part, because medicine serves diverse social goals, while biology more clearly addresses one dimension of experience.

Thus, "philosophy of medicine" brings together disparate philosophical pursuits because "medicine" has such broad meaning. We need not take sides here in the dispute over whether there is a uniqueness to the field; we need not insist that there are unique issues for the philosophy of medicine—issues unlike those found elsewhere in ethics or in the philosophy of science. It is enough if the grouping of issues proves useful and at times heuristic. In fact, that society has created such an enterprise as medicine is enough to justify such an enterprise as the philosophy of medicine, no matter how diverse its various undertakings. A student of the field must, however, be forewarned against presuming an underlying unity or systematic coherence to the philosophy of medicine.

Our overview of current work in the philosophy of medicine gathers the issues under three general rubrics. The first provides an historical introduction; although not comprehensive, it illustrates some of the roots of present discussions. (Some historical material will be presented in other sections as well.) The second section addresses basic epistemological issues; it considers issues in the language of medicine, the logic of knowledge in medicine, and issues in clinical judgment and medical decision making. Here also attention is given to the philosophy of mind in medicine, including philosophical theories of mental illness, psychosomatic illness, the development of the theory of cerebral localization, philosophical quandaries raised by the transection of the corpus callosum (i.e., split brain operations), and problems concerning psychosurgery. The third section is devoted to bioethical issues and ontological issues bearing on bioethics. It includes an examination of the definition of "persons" as this appears in issues bearing on abortion, fetal experimentation, the status of children, the senile, the mentally retarded, and the dead. Also, the rights and duties borne by persons are considered, especially in terms of the controversy between so-called "deontological" and "teleological" views concerning rights and duties with regard to health care and medicine. The standing of natural law and its bearing on bioethics are also examined.

In this fashion, we hope to acquaint the reader with the major areas of research in the philosophy of medicine and indicate where pertinent literature can be found. It is hoped as well that the reader will see how philosophy can be heuristic for science; the general analysis of concepts illuminates ways in which we can examine and understand the world.

I. EXCERPTS FROM THE HISTORY OF THE FIELD

The heterogeneity of the issues covered by the philosophy of medicine is reflected in the ongoing disputes about the term itself. Owsei Temkin, a noted historian of medicine, has remarked: "Anyone wishing to speak about the philosophy of medicine will, at the onset, encounter difficulties: the vagueness of the term and the prejudice against the subject itself" (Temkin, 1956, p. 241). Temkin, in signaling the vagueness of the phrase "philosophy of medicine," was responding in part to the efforts of a period of less cautious philosophizing than our own, for in the nineteenth century there appeared a number of works that were overly ambitious metaphysical attempts at medical theory (Risse, 1971 and 1972). That considerable literature includes Francesco Vacca Berlinghieri's *La Filosofia della Medicina* (1801); Johann Christian August Grohmann's *Philosophie der Medizin* (1808); Jean Bouillaud's *Essai sur la philosophie medicale* (1836); and Elisha Bartlett's *The Philosophy of Medical Sciences* (1844). These works offered general reflections on reasoning and attitudes in medicine. In addition, numerous works appeared under the title of "medical logic," such as Gilbert Blane's *Elements of Medical Logic* (1819); Osterlen's *Medizinische Logik* (1852); and Wladislaw Bieganski's *Logika Medyzyny* (1894). They focused on reasoning in medicine, having as their goal to show, as Blane put it, "in what medical error consists, the difficulties that have obstructed the progress of the art, and the means of obviating them" (Blane, 1819, p. 23). They were guides to the use of reason in medicine.

The philosophy of medicine engendered great expectations. In part, these expectations have persisted. This persistence is illustrated in some contemporary review articles that have attempted to indicate the scope of the philosophy of medicine. For example, in an important article Szumowski argued: "The philosophy of medicine is a science which considers medicine as a whole. It studies its position in humanity, in society, in the state, and in the medical schools. It embraces at a glance the whole of the history of medicine" (1949, p. 1138). Szumowski held that the philosophy of medicine "reveals the more general problems of the philosophy of biology. It analyzes the methodological form of medical thought, mentioning and explaining the errors of logic which are committed in medicine" (p. 1099). Temkin shares Szumowski's construal of the philosophy of medicine: "The philosophy of medicine should present us with a medical logic, medical ethics, and medical metaphysics" (Temkin, 1956, p. 244). The demands placed on the field are thus ambitious and ambiguous.

The ambiguities in the term "philosophy of medicine" have continued to cast doubt on the integrity of the enterprise. As recently as May 1974, Jerome Shaffer attacked the use of the term. In the inaugural symposium of a continuing series on philosophy and medicine, Shaffer argued that what was called "philosophy of medicine" properly belonged to the philosophy of science and moral philosophy, so that there was "nothing left for the philosophy of medicine to do" (Shaffer, 1975, p. 218). He was, however, willing to allow that those issues in the philosophy of science, the philosophy of mind, and moral philosophy that bear on medicine are in themselves worth pursuing. He was therefore disposed to tolerate the rubric "philosophy of medicine" if that phrase proved to have heuristic value in stimulating further useful study of the philosophical issues clustering around medicine.

In short, interest in medical reasoning has a considerable history, though the phrase "philosophy of medicine" has caused some anguish. As the philosophic study of medicine is now being pursued under the influence of the analytic age of philosophy, disclaimers and misgivings are giving way to an exciting new interest. The philosophy of medicine now draws concepts and methods from such contemporary philosophical enterprises as the philosophy of science, the philosophy of biology, and the philosophy of mind. These influences promise considerable progress. Further, an approach to medicine through philosophical anthropology is also making important contributions (Lain-Entralgo, 1970; Guttentag, 1953; Spicker, 1978).

The epistemological interests have been accompanied by concern for ethical problems in medicine. The roots of Western concern with ethics in medicine can be traced to reflections in the Hippocratic corpus upon questions of etiquette and morals (e.g., *On Decorum,* and the oath); Plato's comments on the proper deportment of physicians (e.g., *Laws,* IV, 720b); and Aristotle's recommendation of the use of early abortion to control population (*Politics,* VII, 16,1335b20). The reflections of these and other classical authors lead into a long tradition of oaths and codes of "medical ethics" that have for the most part been codes of medical etiquette (Etziony, 1973). They have also influenced religious and philosophical reflections upon such questions as contraception and abortion. St. Thomas, among many others, deliberates on the morality of abortion (*Summa theologiae,* I, 118, 2; among other places) and on the unnaturalness of certain sex acts (e.g., *Summa theologiae* II–II, 154, 2).

In addition to ethical thinking about contraception (Noonan, 1965) and abortion (Noonan, 1970), there has been concern about the implications of surgery, and especially transplantation surgery, with regard to moral duties to maintain a person's physical integrity. This is a point defended not only by Roman Catholic moralists under the rubric of the principle of totality (Kelly, 1958; Kenny, 1962), but by Kant (*Die Metaphysik der Sitten* 1797 [1968 ed., p. 423]). Medical ethical reflections exist as well in the Jewish religious tradition, with contributions by Maimonides and many others (see Jakobovits, 1959). In more recent times, a number of Protestant theologians have also contributed to this literature (Joseph Fletcher, 1960; Ramsey, 1970b, 1977).

Apart from such reflections within religious traditions, there has been a more secular literature, which arose in modern times under the rubric of *Medicus politicus*. Whereas this tradition had religious roots, which one finds in such works as Giovanni Codronchi's *De Christiana ac tuta medendi ratione* (1591), it later assumed a more secular character, attested to in works such as Rodrigo Castro's *Medicus politicus sive de officiis medicopoliticis* (1614). Castro, a Portuguese Jewish physician, stressed the virtues of prudence and generosity in the conduct of physicians and included analyses of truth-telling, payment for services, and abandonment of patients.

In the eighteenth century, similar concerns arose with a special focus on public health. The duties of physicians were then cast in much more social terms. As George Rosen and others have shown, there were considerable reflections on the implications of medicine for the health of the community, especially under the term "medical police" (Rosen, 1974; Cipolla, 1976; Shryock, 1976). Wolfgang Thomas Rau (1721–1772), who coined the term, contended that the physicians had moral and social duties with respect to the health of the populace. He argued that regulations needed to be developed in order, among other things, to eliminate charlatans and untrained health practitioners, to regulate the training of such practitioners, and to provide for the health of the public—including public education in health matters. In the famous synthesis of this movement, *System einer vollständigen medicinischen Polizey* (1777), Johann Peter Frank explicated the physician's duty in terms of obligations to the state (Frank, 1976). Abortion, confidentiality, and public health measures were viewed in terms of their contribution to the good of the nation. As a result, duties of confidentiality were not considered absolute; physicians were required to report the occurrence not only of infectious diseases, but also of wounds sustained during criminal acts. On the Continent, at least, the frame of moral discourse was social and even political.

In Great Britain and the United States, concern was focused more on the role of the individual practitioner and the community of practitioners. The classic treatise in this genre was John Gregory's *Lectures upon the Duties and Qualifications of a Physician* (1772). There Gregory expounded upon the way physicians should act in order to fulfill their obligations of preserving health, prolonging life, and curing disease. He discussed the character of the temper, moral disposition, and education required for the successful practice of medicine, as well as the requisite canons of medical etiquette. Within such treatises and codes one discovers a tension between attempts to develop rules of civil probity, perhaps better termed "medical etiquette," on the one hand, and moral philosophy *in sensu stricto* as applied to medicine, on the other. Much of Gregory's treatment of etiquette, however, turns on moral issues, such as the duty of physicians to upgrade their knowledge and to inform patients of the nature of their illnesses, except where disclosure would decrease the chance of cure.

American professional medical ethics developed under the influence of such men as the British physician Thomas Percival (1740–1804) and the American physician Benjamin Rush (1745–1843). Benjamin Rush contributed prominently to this discussion of issues of medical etiquette. In 1789 he delivered a paper entitled "Observations on the Duties of Physicians and the Methods of Improving Medicine" (Rush, 1789), in which he made suggestions on how to succeed not simply as a healer, but as a businessman—in a manner reminiscent of parts of the Hippocratic corpus (e.g., *On Decorum*). The greatest influence upon early American medical ethics seems to have come from Thomas Percival via his work *Medical Ethics* (1803). In that volume Percival considered physicians' hospital practice, the relation of physicians to apothecaries and the law, and their conduct in private practice. But it was his treatment of private practice that had the most profound impact upon modern American medicine. It influenced many of the early codes of medical ethics, including the first code of medical ethics of the American Medical Association, adopted May 1847 (American Medical Association, 1848; Berlant, 1975).

The interest in codes as a focus for bioethics has persisted into the twentieth century, with special consideration given in recent years to issues of experimentation (World Health Organization, 1958). As studies have shown, the interplay of economic, social, and ethical issues in the development of professional medical ethics has been extremely complex (Konold, 1962). One impetus to the study of the ethical roots of such codes has been drawn from Roman Catholic moralists who, over centuries, have developed manuals with a heavy emphasis on such ethical issues as artificial insemination, sterilization, contraception, and abortion (Noonan, 1965; Kelly, 1958; Kenny, 1962). They have been joined by an extensive Protestant (Ramsey, 1970a, 1970b, 1975, 1978; and Hauerwas 1974, 1977) and an established Jewish literature (Rosner, 1972, and issues of *Conservative Judaism*). These religious ethicists have inspired important work in their own traditions, and have also helped spark the concern of the larger community—including a reawakened interest on the part of philosophers.

II. EPISTEMOLOGICAL ISSUES IN THE PHILOSOPHY
OF MEDICINE

The epistemological concerns of the philosophy of medicine are both the most ancient and the most contemporary; the greatest current growth in scholarship lies here. This genre of the philosophy of medicine includes the following elements that were noted by Kazem Sadegh-zadeh in his introduction to the first issue of the new journal *Metamed:*

> Metamedical application(s) of the advanced methods of the philosophy of science, logic, and mathematics [to medicine]; problems of medical language, knowledge and theory construction; [the examination of] metatheory and methodology of clinical judgment and the decision process [in medicine; as well as the] study of the structure and dynamics of medical theory, and discussion and clarification of the basic concepts, conditions and problems of the medical enterprise (1977c, p. 2).

Philosophy functions here as a formal analytic tool to display the structure of reasoning in medicine, for example, by rendering explicit the ingredient logic of clinical judgment. One is, then, doing philosophy of medicine by portraying the rules of inference that guide reasoning in medicine.

In order to impose an organization on the mass of material in this domain, we will cluster the areas of current scholarship under four tasks: (1) investigations of medical language and basic concepts; (2) an analysis of the structure and dynamics of medical theories; (3) the development of a metatheory and methodology of clinical judgment and medical decision-making; and (4) problems in the philosophy of mind in medicine.

Our primary concern will be with philosophical analyses by both philosophers and biomedical practitioners, and will function on a metatheoretical level of analysis (i.e., theorizing about the theories and concepts of medicine). As yet unexplored conceptual issues in the biomedical sciences that should receive attention in the future will also be indicated.

Epistemological issues significantly affect the way we define "medicine." To decide what counts as health and disease is to shape what counts as prevention, diagnosis, treatment, cure, and medical care. These concepts are basic in that they serve to separate medical treatment from such undertakings as remedial education, political programs for the improvement of a society's living conditions, and enterprises aimed at achieving spiritual peace.

For purposes of action or policy, the distinction between medical and nonmedical is not so much to be discovered as to be defined or stipulated. The best that can be discovered is the way a particular society at a particular time has viewed the distribution of roles among the professions. What is actually at stake is the fashioning of a coherent picture of what medicine could be about. The goods of health (for example, strength) grade over into the goods of education (dietary knowledge) and social well-being (some classes have better health than others). Thus, a conceptual map of the relevant terms would indicate boundary regions, rather than boundary lines, between the concepts. In any event, no account of what medicine is in a particular society, or could be in general, is possible without a sorting out of the meanings with which one wishes to invest crucial terms such as "health," "disease," "illness," "sickness," "deformity," "disability," "dysfunction," and so on. The fact that there is no clear account of the way these terms are to function is at the root of much confusion concerning the status of both medicine and medical theory.

A.   The Language and Concepts of Medicine

The function of philosophical analysis, avowed from Plato through Peirce and Wittgenstein, defines a basic role of the philosophy of medicine. The person wanting to become clear about the philosophical foundations of the biomedical sciences will need first and foremost to understand the significance of the key terms that frame medicine's *raison d'être*. There has been, in fact, a traditional interest on the part of physicians in the analysis of the language of medicine with a view toward the removal of ambiguities in terminology and the establishment of standard usages. Such discussions, however, have often been fraught with confusion, as Sadegh-zadeh has shown with respect to attempts to clarify the language of disease. Confusion has stemmed, he argues, from a failure to distinguish among: (1) providing a stipulative definition of disease, illness, sickness, deformity, etc.; (2) indicating standards for the use of disease language within particular groups of speakers (e.g., anthropological or linguistic studies of the use of such terms as "disease," "illness," and "sickness" among particular linguistic groups); and (3) putting forward statements that announce the discovery of the true nature of disease (Sadegh-zadeh 1977b, p. 12). Uses of language in the third way often mark nonempirical attempts to disclose the true or essential nature of health and disease.

Whatever one's intentions, it is necessary to be clear about which of the three uses of language one is undertaking so as not to confuse stipulative definitions, reports of the currency of certain meanings, and metaphysical discoveries of the true nature of health and disease. The contributions by such authors as Sadegh-zadeh and Rothschuh (1977b) indicate a serious interest in analyzing the meaning and use of basic terms in the language of medicine. A great deal of future research in the philosophy of medicine is likely to occur in this area, for it is here that much effort must be undertaken in order to develop a coherent account of the nature and significance of the biomedical sciences. One should note, however, that interest in the language of medicine has a considerable lineage, including efforts to develop standard usages and terminologies. It is found, for example, in the development of nosologies by François Sauvages (1707–1767) and by William Cullen (1710–1790), whose classifications turn on various competing senses of such terms as "disease," "health," and "cause."

The contemporary discussion of the concepts of health and disease has had a particular focus on the explanatory versus evaluative meanings of these terms. The importance of such analysis should be apparent. Because medicine exists to preserve health, and to prevent, cure, or at least ameliorate the effects of disease, the focus of this social enterprise will be directed differently, depending upon the way health and disease are construed. Examples of disputes along these lines include those over the disease status of alcoholism (Szasz, 1972), homosexuality (Richard Green, 1972), menopause (Barnes, 1962), and the disabilities of aging (Engelhardt, 1977b). Depending on whether such states are diseases, the scope of health care is correspondingly expanded or contracted. As a result, many have suspected that concepts of *health* and *disease* are value laden in covert ways (Engelhardt, 1975b; King, 1954; Margolis, 1976a). These values appear to be disguised or transmuted ethical or moral values; they are expressed, for example, in beliefs about whether alcoholics should be held responsible for drinking and whether the compulsive need to drink is abnormal—that is, whether alcoholism is to be considered a disease (Trotter, 1804; Szasz, 1972).

Discussions of the force of disease and illness language have also been examined by sociologists of medicine. Here the literature on labeling theory and descriptions of the function of the sick role has been important in indicating the ways in which medicine constructs the reality of the physician-patient interchange (Siegler and Osmond, 1973; Parsons, 1951, 1957, 1958a, 1958b). Undoubtedly the work of Alfred Schutz and Thomas Luckman (1973) could be used to illuminate further the ways in which the world of health care is a social construction (see also Berger and Luckman, 1966).

If such value-laden norms inform judgments about whether a condition is functional or dysfunctional, then we may regard concepts of disease and illness as socially dependent or invented rather than as reflections of realities ingredient in nature. In studying the status of these influences, we mark one of the many places where the philosophy of medicine grades into or invades the philosophy of biology and the assessment of biological functions. For if certain psychological and physiological functions cannot be identified as natural in the sense of being normative for humans and inherent in what humans are, then disease states and illnesses are socially stipulated rather than identified as strictly natural features. Thus, they represent distinctions or evaluations that communities make with regard to nature. On the other hand, if such inherent, natural features can be identified, then definitive answers should be available as to whether drug addiction, homosexuality, alcoholism, or menopause should count as diseases.

In analyzing these issues, Christopher Boorse has employed a useful distinction between neutralist and normative ways of explaining concepts of disease (Boorse, 1975, p. 52). He has further distinguished within the latter a strong normativists' view, in which disease categories are held to be purely evaluative, and a weak normativists' view, in which disease categories would have both a descriptive and an evaluative component.

One should note that weak-normative accounts would explain cross-cultural agreements about certain states being diseases (e.g., angina) on the basis of certain nearly constant value judgments. That is, in almost any conceivable society, the pain and disability of angina would be perceived as dysfunctional and as biologically improper. Such value judgments would be made with regard to how certain physiological and/or psychological processes or states that obtain across cultures are experienced in different cultures. The accidentally constant values being given to certain widespread physiologic phenomena (e.g., coronary artery disease and myocardial eschemia) and the sensations they evoke (e.g., pain) are the kinds of phenomena the weak normativist would rely on. Strong normative accounts might appeal to the accidental constancy of such value judgments as a sufficient basis for the agreements that do in fact occur. Strong normativists, though, are likely to stress differences among cultures. However, a commitment to a normative account of disease and health does not necessarily commit one to expect great cultural divergences. Rather, one would expect a spectrum ranging from a great agreement (e.g., about angina) to considerable disagreement (e.g., about vitiligo).

Boorse has also addressed the ambiguities in natural languages of such terms as "illness," "defect," "dysfunction," "deformity," "disease," and "sickness." He has suggested a distinction between "disease"—which is used to identify physiological dysfunctions, so that concepts of disease are descriptive concepts—and "illness," which is used to identify states of disease evaluated as serious and undesired; "illness" is thus used to place individuals in sick roles (Boorse, 1975, p. 56).

One might also, to take a different model, contrast a *descriptive* sense of "illness" (meaning a complaint-charged physiological or psychological condition) with an *explanatory* sense of "disease" (meaning by the latter a way of accounting for complaints in question). The descriptive sense of "illness" would be normative in that it relies upon disvalues attributed to particular physiological or psychological states. The explanatory model would also be value laden, though only in a derivative fashion. One would know that a particular physiological state, in contrast with other particular states, is pathological because it underlies or produces states of illness. In this view, one would be required to explain diseases only because they are ways of accounting for illnesses. As an explanatory model, "disease" would ineluctably involve or refer to values because illnesses are disvalued states. (It should be remembered that both "illness" and "disease" are being used stipulatively here in order to contrast different elements of medical reasoning that are only poorly distinguished in non-medical language.) On this point, Whitbeck (1977b) argues that illnesses are intrinsic to the understanding of disease processes.

Those who have argued that attributions of illness are normative judgments have stressed the cultural relativity as well as the heterogeneity of judgments concerning illness (Engelhardt, 1977a; Fabrega, 1972; Wartofsky, 1975). Some authors (e.g., Margolis, 1976a) have emphasized that illness and disease talk can be understood only in terms of particular cultures and their ideologies. They further contend that an appeal to evolution or adaptation cannot supply a universal sense of health. After all, adaptation is always relative to a particular environment, and human environments include cultures. Thus there can be no standard definition of human health or proper function because there is no standard environment for humans or for any other species. (Against this see Boorse, 1977, pp. 557 and 563.)

In addition, thinkers like Margolis and Engelhardt can argue that intraspecies variations increase the potential adaptability of a species. Insofar as one is concerned about the long-range success or adaptation of a species, one will hope that there are some marginally adapted (i.e., "somewhat dysfunctional") individuals within the group who could help the species to survive should the environment radically change. Moreover, traits that do not directly contribute to survival may do so in recondite and obscure ways. Homosexuality, it has been contended, may be genetic and may have contributed to the overall survival of societies and thus of the species (Hutchinson, 1959; Trivers, 1974; Wilson, 1975, p. 555). In short, there may not be a univocal meaning of "normal" or "healthy."

Another question is why species survival should be valued or the results of evolution celebrated and used as a ground for holding some traits as normal and others as abnormal. For example, the existence of sickle-cell trait contributes to the general adaptability of the species, presuming, as many have contended, that it allows for better survival and reproductive efficiency in the presence of falciparum malaria and in the absence of antimalarial drugs (Livingston, 1964, p. 685). Nonetheless, one may term bearing sickle-cell trait not a state of full health—e.g., the trait may prevent one from being a jet test pilot (J. M. McKenzie, 1977). One is unlikely to accept sickle-cell trait as the price of the species' long-range success and hold the trait to be normal; whereas some descriptions of evolutionary success focus on the species, most accounts of health and disease tend to be individual- or patient-oriented. Thus it may also be possible to speak of widespread phenomena as diseases (e.g., menopause and presbyopia).

In this vein, one finds such thinkers as Peter Medawar asserting that "nature does *not* know best," but instead creates a "tale of woe [including] anaphylactic shock, allergy and hypersensitivity" (Medawar, 1959, pp. 100–101). In addition to his helpfulness in not allowing us to be romantic nature worshipers, Medawar is indicating that one can judge the value of physical states only within particular contexts.

Physical states considered out of a cultural-environmental context are neither states of health nor illnesses; they are neutral. One may think here of color blindness, which in some circumstances may be termed a defect; however, in environments where the identification of camouflage is important for survival, color blindness would be a state of special fitness (R. Post, 1962). There are no standard environments given by nature in terms of which functions can be specified as normal or abnormal. States of affairs are singled out as diseases in terms of the circumstances (environmental and cultural) of the organism as well as what a community wishes from the circumstances. Thus, due to the role of the community, a special problem will arise (and we will discuss it below) in talking about the diseases of animals (Sedgwick, 1973; Engelhardt, 1977a). All this discussion implies what Boorse would call a weak normative view of diseases, which sees them as falling along spectra with more or less descriptive aspects and thus with more or less agreement across cultures.

Thus, states of affairs may be considered *illnesses* (1) because of their thwarting of the achievement of a goal that members of a culture hold to be "normally" open in that culture; or (2) by constituting a state of pain or discomfort without leading to the achievement of some "proper" human goal—as, e.g., teething or childbirth do; or (3) because of their dysaesthetic nature, e.g., vitiligo. Given such an account of illness, "disease" would be held to refer to *explanations* of illness states and would come to be applicable to individuals even when no illness was yet expressed, but where it could be expected.

Under the preceding account of "illness," medicine would be called upon to identify either complaints or occasions for actions to prevent the need for therapy. In this way, medicine would not need the presence of a disease to legitimate its interventions. It would serve to ameliorate involuntary conditions that produce complaints by a patient or by another on behalf of a patient. That is, medicine would try to ameliorate vexations concerning dysfunctions, pain, or deformities that were physiologically or psychologically based. "Therapy" here would be used in its original breadth to indicate a service, a waiting upon, a caring for, as well as a curing.

It is against such normative accounts of health and disease that philosophers such as Christopher Boorse (Boorse, 1975, 1976a, 1976b, 1977), and Leon Kass (Kass, 1975b) have applied their neutralist views. Boorse, for example, has attempted to provide a value-neutral account of disease, specified in terms of what functions are typical for the individuals of the species of which the organism is a member. In fact, being healthy for

Boorse is being a "good specimen" of the species to which one belongs, and being diseased is a state of failing to be a good specimen (Boorse, 1975, p. 58; 1977). He can be understood to argue the following thesis: For a given class of natural organisms delineated by age and sex, the species-typical parts or processes of its members that contribute to their individual survival and reproduction are definitional of "health" as it applies to that species; diseases are states that impair health (Boorse, 1977, p. 555).

In trying to define "health" by reference to just species-typical normality and functional normality, Boorse is forced to resort to stipulating a hierarchy of goals from the cellular to the holistic level. Contrary to his thesis, this hierarchy appears value laden; at least, it does not seem possible for one to indicate its structure without building in some judgments about values—perhaps indeed in an anthropomorphic way.

In contrast to his concept of disease, Boorse's concept of illness is acknowledged as normative. Indeed, Boorse introduces illness as the normatively structured presentation of disease. However, Boorse's goal is to show that disease and health talk is not value laden. He is, in particular, writing in reaction to strong-normative accounts of disease in the psychiatric literature (e.g., Ian Gregory, 1968), though he criticizes weak-normative accounts as well (Redlich, 1952).

In Boorse's view, then, one would regard evolution as the selection of certain sorts of functions (that are necessary for particular achievements, as sight is the function of eyes), which functions are then "good" for the survival of the species.

An Aristotelian approach to accounts of health and disease in terms of internal natural standards or norms has found a number of supporters. Among these must be counted von Wright, who argues as follows:

> Because of the intrinsic connexion which holds between the goodness of organs and faculties and the good of the being to whom they belong, I propose to call the functions which are proper to the various organs and faculties *essential functions* of the being, or rather, of the kind of species of which the individual being is a member. The essentiality of the functions does not entail that every individual of the species can actually perform all those functions. But it entails that, if an individual cannot do this at the time when by nature it should —to quote Aristotle—we call it *abnormal* or *defective* or *faulty* or, sometimes, *injured*. It follows by contraposition that the essential functions of the species are functions which any *normal* individual of the species can perform. The essential functions are needed for that which could conveniently be called a *normal life* of the individual (1963, pp. 53–54).

Others have supported this mode of analysis and attempted to tie it in with consideration of both species and individuals. Thus Leon Kass has argued that health is "a natural standard or norm, . . . a state of being that reveals itself in activity as a standard of bodily excellence or fitness, relative to each species and to some extent to individuals, recognizable, if not definable, and to some extent attainable." Kass embraces an Aristotelian analysis, seeing health as "the well working of the organism as a whole, . . . an activity of the living body in accordance with its specific excellence" (1975, pp. 28–29).

In addition to accounts of general uses of "health," the language of natural norms comes into arguments like those of Gorovitz and MacIntyre (1976), who wish to speak of the need of a physician to know what it is in particular for a human to flourish. This sort of language is deeply embedded in medicine.

Other problems surround the ambiguity of "normal," a concept that has been recognized in medicine as complex and elusive. Marjorie Grene (1977, 1978) has distinguished three senses of "normal" in medicine: one is synonomous with "health"; a second is synonomous with "statistically frequent"; the third means something like "characteristic for members of the species"—the usage on which Boorse (1977) relies so heavily. Edmond Murphy further distinguishes among seven senses of normal: (1) statistical, as with reference to a Gaussian distribution; (2) average or mean; (3) typical or expectable; (4) conducive to the survival of a population; (5) innocuous or harmless, as may be a response by a physician to an inquiry by a patient concerning the significance of a particular sign or symptom; (6) commonly aspired to; and (7) the most perfect or excellent of its class. Others have found still more senses to add to this list (Sackett, 1977; Abercrombie, 1960). As the ambiguity of "normal" indicates, the importance of distinguishing senses of "normality," "disease," and "illness" is emerging as an issue in the literature, and it will likely command considerable interest and attention in the future.

The importance of analyzing the ways in which natural norms presuppose concepts of design, along with the question of the extent to which such language remains appropriate in a postcreationist, evolutionist view of health and disease, comes to the fore. As Boorse (1977) shows, notions of good functioning that surface, for example, in talk about effective heart action are clearly class terms; and since the measure of one heart is highly uninformative, we have no idea from just a single case what range of values can count as normal and what range can count as abnormal or diseased. Furthermore, variations in the ways in which subjects are selected must be taken into account. For example, we cannot use the figures from an Olympic female track star to test for the cardiac output in a forty-year-old male gym teacher or a sixty-year-old farmer. It would be highly suspect not to divide the range of standards according to age, sex, location, and even other categories. Of course, what we need is a range of values for a range of populations that we agree are healthy or diseased.

These issues remain central to ongoing research on the language of medicine. Concern with these questions has prompted efforts to display the ways in which particular concepts of health and disease are generated (Gross, 1977; Leiber, 1966; Peiffer, 1977; Rothschuh, 1977b).

Given the considerable difficulties in distinguishing evaluative from explanatory, normative from nonnormative, and culturally relative from universalist views of disease, attempts to circumvent such ways of framing the issues have emerged. Authors like Henrik Wulff (1976, p. 37) have put forward "pragmatic construals," in which particular concepts of disease would be neither true nor false, but rather more or less useful in the achievement of diagnostic and therapeutic goals (Wulff, p. 37). Whitbeck (1977b) takes a somewhat similar though perhaps more philosophical position in arguing that disease entities are in fact constructed and defined along lines of a general instrumental interest, and that causal factors are singled out as "the" cause when they are the subject of human interest from the point of view of efficacious production, control, prevention, or reversal. A similar line of reasoning led Richard Hull to argue that there may be both a conceptual and a moral problem in the label "genetic diseases" (Hull, 1978). Such approaches have the advantage of clearly identifying where the goals lie—in the social enterprise of medicine and its pursuit of certain forms of explanation (diagnosis) and intervention (treatment). But here all the issues reappear as to whether such goals are culturally relative or can be specified in terms of internal biological norms.

Issues concerning concepts of health, disease, and illness are so central that they recur when one considers questions of medical theory as well as medical decision-making in clinical judgment. It is not unexpected, therefore, that they have commanded most of the interest of philosophers. But, as already indicated through reference to Whitbeck's (1977b) discussion of causality, other key concepts that are intimately connected with defining diseases are also beginning to be analyzed.

For some time, rather naive views of causality in medicine have been said to hold sway; physicians are taken to have single-cause notions of etiology, on the model of one billiard ball causing another to move by impacting upon it. Yet fairly sophisticated distinctions among senses of "cause" were already drawn in the nineteenth century by Rudolph Virchow (1895, p. 140). Virchow, who has at times been wrongly viewed as opposing the contagion theory of disease, pointed out that infectious agents were not *the* cause of the diseases after which they were being named. One could, as Virchow demonstrated, identify infectious agents in normal carriers showing that, although they may be necessary for the development of the disease in question, they were not sufficient. One can see in Virchow's work the beginning of a distinction among necessary, sufficient, and contributory conditions for the development of human illnesses.

The argument against unique, specific causes of disease was a difficult move to make, for physicians had over a long period sought such causes

(Pagel, 1944). Moreover, the notion of specific causes for specific diseases supported hope for specific cures, thus forging a distinction between symptomatic and etiological therapy. The discovery of pathogens reinforced this view. Virchow, in contrast, was arguing for what in modern times has become a multifactorial view of the causality of diseases as expressed in such studies of multiple influences upon disease as the Framingham Study (1949) of the U.S. National Health Institute.

Modern criticisms of single-causal views of etiology are found in works by physicians like Edmond Murphy (1976, 1979), Alvan Feinstein (1967, 1969, 1973), and others who argue for multifactorial—in fact statistical—portrayals of causality in medicine. These arguments are usually conjoined with statements concerning scientific method—e.g., the importance of attending to multiple factors, of distinguishing among necessary, sufficient, and contributing factors, or of developing reliable probabilistic descriptions of events and accounts of their occurrence. One also finds such critical accounts in textbooks of epidemiology (e.g., Friedman, 1974). Such accounts offer suggestions on how to develop causal accounts within already accepted, but unanalyzed, models of medical causality.

To date, there has been little sustained analysis of the issues they raise. Authors have, instead, contrasted senses of causality in medicine with meanings of responsibility (Dagi, 1976) and have just begun to catalogue the issues (Agassi, 1976). This is not to say that such authors have failed, but rather that there is so much to be done that for a while analyses will tend to be global and somewhat unclear. Again, as Sadegh-zadeh (1977c) and others have indicated, this problem can be remedied only by careful and sustained analysis of the ways that ideas and words function in various biomedical contexts.

B.   Medical Theories and Medical Explanations

Though there has been little sustained philosophical analysis of medical theories and modes of explanation, there is a considerable history of systematic reflection on medicine. Much contemporary analysis is done in the shadow of that history. We will present an overview of the philosophical analysis of medical theory against the backdrop of historical reflection on three main areas of interest: (1) nominalist versus conceptual-realist accounts of medical nosologies (these terms will be explained below); (2) the feasibility of giving descriptive accounts of symptoms and sign complexes without commitment to more fundamental pathological or etiological theories of disease and illness; and (3) the reducibility of medical knowledge and its language and theories to the language and theories of physiology and the basic biomedical sciences. Analyses of these issues have involved presentations of various uniqueness theses—assertions that medical knowledge is unique in structure or in function or in the types of concepts upon which it is built. One of the virtues of such uniqueness theses is that, even if they do not succeed in proving the nonreducibility of medical language and theories to physiological or basic-medical-science language and theories, they are likely to disclose important characteristics of the process of framing theories and explanations in medicine.

The disputes between normative and neutralist views of disease reviewed above are mirrored in disputes about the standing and significance of medical theories and explanations; both disputes concern the extent to which disease classifications are discovered or invented. A great deal of recent literature concerns what historians of medicine have termed the dispute between "ontologists" and "physiologists of disease" (Niebyl, 1971). This can be understood, in great measure, as a dispute between a conceptual *realist's* account and a *nominalist's* account of the meaning of "health" and "disease." The conceptualist-nominalist dispute connects with the status of medical theory, since it is rarely framed in terms of the analysis of isolated concepts but is cast instead in terms of entire medical explanatory schemes or nosologies. (One of the perennial interests of theorists of medicine has been in the development of systems to classify diseases—nosologies. These have also been bound to systematic attempts to describe illnesses—nosographies.)

The terms "ontological" and "physiological" achieved currency in Broussais' (1824) discussion of the objectivity (realism) or arbitrariness (nominalism) of disease designations for constellations of signs and symptoms. The ontologists had argued that language about disease referred to real clusters of phenomena somewhat like natural species. According to them, those groupings of signs and symptoms identified by nosologies correspond to real species and distinctions existent in reality (Sydenham, 1676; Cullen, 1769)—a point described in great detail by Foucault (1973). In contrast, physiologists held that disease designations were stipulative—devised for the convenience of the nosologist—and that laws or concepts from physiology were adequate to explain diseases or illnesses (Romberg, 1909; Wunderlich, 1842).

The pattern of these controversies can be traced back to Hippocratic times to the disputes between the school of Cnidus, which, somewhat ontologically, wished to term each variation in clinical findings a "disease entity," and the school of Cos, which, in the mode of physiologists engaged in less classification, was more concerned with variables concerning pathological phenomena. Historically, the development of classifications was not motivated by purely theoretical considerations. It was bound to reflections on care in developing theories and drawing conclusions (see Hippocrates, 1923; *Precepts* I, *Nature of Man, Sacred Disease* I). Interpretations of what counted as good reasoning and proper classification, though framed for therapeutic purposes, reflect the considerable interest that even ancient thinkers had in the logic and methods of medicine (see, e.g., Galen's *Institutio Logica*).

These precedents of philosophic and formal interest in the nature of medicine provide the context for understanding the various commitments of modern writers in formulating the canons by which evidence is accepted and systematized in medicine. Moreover, they have directly influenced modern attempts in their development of classifications of disease.

To trace the historical milestones closer to the beginnings of modern medical science, we can start with Thomas Sydenham's (1624–1689) *Observationes medicae* (1676). It is a landmark in the science of nosography and nosology, for in that volume Sydenham attempted a sophisticated theory for the description and classification of diseases. His method may be characterized as "phenomenological" in that he forwarded a method of describing the signs and symptoms of diseases while discounting theoretical and causal accounts of the origin of the signs and symptoms. He attempted only to describe the natural, temporal courses of diseases—their natural histories. Sydenham thought that such an account would succeed because he believed that the world of signs and symptoms is there for all to see. He thus recommended that "all disease be reduced to definite and concrete species" (Sydenham, 1848, p. 13) by means of observation. Once they have been described, disease relationships can be exhibited just as botanists exhibit relationships among species of plants. This paradigm appeared reasonable to Sydenham, for he presupposed that signs and symptoms fell into recurring, natural, and enduring patterns. It was in this sense that he was an ontologist of diseases. In viewing natural histories of diseases as the proper and central foci of medical knowledge, Sydenham saw the science or theory of medicine as engaged in description and categorization of symptom-constellations that were like essences:

> Nature in the production of disease is uniform and constant; so much so that for the same disease in different persons the symptoms are for the most part the same; and the self-same phenomena that you would observe in the sickness of a Socrates you would observe in the sickness of a simpleton (1848, p. 15).

Others who follow Sydenham in this understanding of nosology include Carl von Linnaeus (1707–1778), in his *Genera morborum* (1763); François Boissier de Sauvages de la Croix (1707–1776), in his *Nosologia methodica sistens morborum classes juxta Sydenhami mentem et botanicorum ordinem* (1768); and William Cullen (1710–1790), in his *Synopsis nosologiae methodicae* (1769). The result of Sauvages' studies was a nosology with ten classes divided into 44 orders, 315 genera, and 2400 species in which the primary criteria for classification were descriptive. In Sauvages' nosology, causal criteria were at best of only secondary importance. Although he did retain some reference to causal factors in the development of his classification, Cullen was the most forthright of the four in stating a commitment to developing a nosology without major reliance on supposed causes of the signs and symptoms. He wished to provide a theory-neutral description of the signs and symptoms of illness—a descriptive system of disease (Bowman, 1975).

These descriptive, clinical accounts of disease were bound to a conceptual-realist or ontological view of disease entities—a view that what is real is the species of disease. (This realism of essences is unlike the naive realistic theories of disease as perhaps found in the accounts of Paracelsus, who offered another sense of an ontological view of the reality of diseases; there each case of a disease seems to have been conceived of as caused by a real, actual thing, the *ens morbi;* see Pagel, 1958, p. 137.) For the realist of essences like Sydenham, diseases were not things that invaded the body, or patho-physiological processes. They were nonetheless the ways in which symptoms existed in recognizable patterns.

In contrast to Sydenham's were approaches that did not eschew reference to physiological theories altogether. Among those others were the iatromechanistic explanations put forward by Hermann Boerhaave (1638–1738) and Friedrich Hoffman (1660–1742) (see Brooks and Cranefield, 1959; King, 1958; Lindeboom, 1968). They had more confidence than Cullen had in etiological and systematic explanations. Others engaged in even more ambitious attempts to systematize medicine in terms of various basic principles. There were also theories of an iatrochemical nature, such as those presented by van Helmont (1577–1644) and others. Some, for example John Brown (1735–1788), wished to provide medicine with the conceptual unity that Newton provided to physics (Bole, 1974; Risse, 1971). Numerous attempts were also made to generate medical theory in a speculative, *a priori* fashion. The philosopher F. W. J. Schelling (1775–1854), for example, attempted a systematic account of medical knowledge.

Older clinical-phenomenological classifications of disease in terms of fevers and fluxes seem strange to today's reader in comparison with classifications that turn on pathoanatomical and etiological considerations. This is because we see things quite differently. Michel Foucault argues in *The Birth of the Clinic* (1973) that we have undergone a "syntactical reorganization of disease" and therefore look at signs and symptoms differently, inasmuch as we "know" in a different way how those signs and symptoms should cohere. We see the world in terms of our commitments to particular etiological and pathological accounts. As J. H. van den Berg (1978) has argued concerning the psychology of discovery in medicine, the world truly appears quite different to us as our assumptions and concepts change. Our theoretical assumptions tell us what should be considered important and significant findings; prior commitments structure the psychology of discovery and in fact the logic of discovery.

Foucault among others has shown that there is a turning point in the history of medical explanation; it came in the nineteenth century with the impact of Xavier Bichat (1771–1802), Rudolph Virchow (1821–1902), and the microbiologists of the nineteenth century who successfully interpreted clinical findings in terms of anatomical-pathological structures and infectious influences. As a result, Sydenham's, Sauvages', and Cullen's attempts to group illnesses under such rubrics as fevers, fluxes, cachexias, and weaknesses came to a halt as disease began to be described in pathoanatomical and pathopsychological terms. This important conceptual move away from purely clinical observations of disorders (e.g., jaundice) to characterizations in terms of underlying pathological findings (e.g., hepatitis) continues to succeed though it may also have its shortcomings.

Etiological accounts of the origin of the pathoanatomical and physiological findings had their impact on the language of medicine, too. There was a change, for example, from simply speaking of hepatitis to speaking of viral hepatitis, toxic hepatitis, immune hepatitis, et cetera. Also, clinically diverse phenomena, such as consumption and Pott's disease, have been grouped together as manifestations of tuberculosis, while illnesses not previously distinguished, such as typhus and typhoid, have become clearly distinguished.

In the nineteenth century, the dispute between ontologists and physiologists of disease assumed its contemporary character. The claims of the ontologists were formulated in terms of cellular pathology with, for example, Rudolph Virchow holding that disease states could be correlated uniquely with cellular changes and that diseases were properly described in terms of the language and laws of cellular pathology (Virchow, 1895). The claim of these physiologists was that disease designations are arbitrary and require no special nonphysiological laws or concepts for their explanation. Broussais, for example, criticized Phillipe Pinel's (1745–1826) *Nosographie philosophique* (1789). There Pinel distinguished various diseases on the basis of an ontological typology, clustering diseases on the basis of their symptoms, not on the basis of underlying pathophysiological processes, and held their differences to be discovered, not invented. In criticizing Pinel, Broussais advanced his classic critique of ontological theories of disease:

> One has filled the nosographical framework with groups of most arbitrarily formed symptoms . . . which do not represent the affections of different organs, that is, the real diseases. These groups of symptoms are derived from entities or abstract beings, which are most completely artificial ὄντοι; these entities are false, and the resulting treatise is ontological (1824, vol. 2, p. 646).

Other thinkers (Romberg, 1909; Wunderlich, 1842) should not be interpreted simply as making the nominalist claim that only individual patients are real and disease designations are arbitrary (Temkin, 1961). Rather, they appear to be holding that the language of pathology and its explanations can be reduced to the language and explanations of physiological accounts. That is, one can recognize a disguised argument concerning the reducibility of pathology to physiology. For example, Wunderlich argued against ontological views of the concept of diseases and the medical theories that they supported, describing them as logical blunders arising from confusing abstract concepts with things by "presupposing them as actually existing and at once considering and treating them as entities . . . which contain no truly essential feature and to which we only by way of exception or by using compulsion, find an example in nature" (1842, p. ix). Romberg put his similar objection strongly. He argued that physiology laws are real and that pathology simply displays peculiar variations and substitutions for the variations described in those laws, but that no unique phenomena appear nor are unique laws required.

> And we do not regard the mere placing of the disease under this or that rubric as the final aim of diagnosis. . . . The most important thing remains the determination of the degree that the individual human is injured by his malady, and which cause has produced the momentary disorder (1909, p. 4).

Because of these nineteenth-century debates, the twentieth century arrived with a background of philosophical arguments concerning the standing and significance of classification and explanations in medicine already well in place. Though these debates largely involved physicians, they were clearly philosophical. A great deal of this literature arises from attempts by historians of medicine to unravel the disputes between ontologists and physiologists of medicine (e.g., see Cohen, 1960, p. 160). Some historians with philosophical sensitivities—for example, Peter Niebyl (1971) and Owsei Temkin—have drawn important implications. Temkin (1961) notes that an accent upon the concern for individuals rather than cases is implied in the nominalist argument that individuals are real but explanatory abstractions are not. Others attempt to diagnose certain theoretical tendencies of medical model building, such as the work of Knud Faber, who argued that ontological theories of disease indicate "the ever-recurring craving of . . . clinicians for . . . fixed categories of diseases" (1923, p. 95).

Having indicated the historical orientation, we must stress that the problems of rigor in nosology have not vanished, but have changed with the growth of medical inquiry. The epidemiologist, for instance, wishes to describe, explain, and, as a public health agent, control events and circumstances bearing on particular kinds of pathological phenomena. In order to succeed, the epidemiologist must collect relevant data that presuppose a set of categories and procedures for observing, describing, explaining, measuring, calculating, and intervening. These categories involve types of scales (e.g., nominal, ordinal, interval, and probability scales), types of rates (e.g., incidence rates, the percentage of population contracting a condition within a given time), prevalence rates (the percentage of a population suffering from a condition at a given time), and a typology for dividing the population. In addition, epidemiologists need procedures for deciding how certain observations are to be characterized. All such measures take place in terms of classifications of phenomena through nosographies and nosologies.

Nominalist critiques of concepts of disease also continue to be actively articulated. A modern nosologist writes:

> The history of the definition of the concept of disease is a story of woe. No one has yet written a definition which is illuminating and therefore everyone who needs such a definition attempts to form a new one and therefore springs anew into the abyss. The concept of disease is a fiction. The disease entity is, in contrast to the elements out of which we construct it, a true fiction in that it corresponds to nothing in reality (Koch, 1920, pp. 130–131).

In short, many would contend that one is simply binding together diverse phenomena, various signs and symptoms, into an artificial system for pragmatic reasons and then calling those clusters disease entities.

Probably some of the most philosophically interesting interpretations of these historical developments have come from French sources (Foucault, 1973; Canguilhem, 1972). As indicated, Foucault portrays conceptual developments in medicine from the phenomenological essentialism of Sydenham, Sauvages, and Cullen, through the development of the clinic and of the modern discipline of pathology. From seeing diseases as self-announced patterns, physicians came to relate symptoms to one another and to construct syndromes in the clinic, and thus to see diseases as grounded in underlying pathological processes, and finally to see them as physiological processes that are assessed as pathological under varying circumstances. For Foucault, disease is socially determined and in part a social construct. In his sketch of medicine from Thomas Sydenham's *Observationes* (1676) through Giovanni Morgagni's *De sedibus et causis morborum per anatomen indagatis* (1761) and Xavier Bichat's *Anatomie générale* (1801), Foucault is himself developing a theory for understanding medical theories, as well as a historiography and an account of the psychology of discovery in medicine. All of these efforts present significant issues for the philosophy of medicine (Guédon, 1977).

The quest for scientific categorization has continued to the present. Alvan Feinstein claims that the clinical sciences are not reducible to the language and explanations of the nonclinical basic medical sciences (Feinstein, 1967). He provides an argument for the uniqueness of the clinical sciences:

> As an organism is reduced to organ, organ to tissue, tissue to cell, cell to molecule and molecule to sub-atomic particle . . . every time the new unit of observation is changed, the forms and functions of the new unit are different from those present before (1967, p. 43).

Thus, if one can say the function of the structure called the arm is to throw a ball, one can still not say that that is the function of its muscles. The function of its muscles is to contract; the cellular bands that comprise the muscles have yet another function—to respond to the output of electrical discharge of yet another structure.

Feinstein's antireductionist stand provides leverage for the push to a science of clinical medicine. The contention is that just as laboratory science may occupy itself with discovering correlations and mechanistic conclusions as to how things work, so a clinical science may concern itself with unities and law-like generalizations, but on a different level. For clinical science, the entity under examination is a patient and not simply an organ system, an organ, or a tissue. There are also procedures peculiar to the clinical sciences. In the clinic, the subject—who is really best called a patient—to varying extents chooses the scientist—i.e., the physician. In other sciences, the general practice is for the scientist to choose his or her subject. Another difference between clinical and basic sciences is that in basic laboratory work, the researcher can assume much to be standard or typical; in clinical work, however, the practitioner can factor out only some variables and must be attentive to some that are scientifically distracting but are important to the condition in the patient. Further, the notion of "cure" turns on more subjective criteria related to the suffering and incapacities of patients, while the laboratory findings of confirmation or disconfirmation usually depend upon more objective criteria. Feinstein sees these and other characteristics as signaling the irreducibility of clinical sciences to the nonclinical sciences.

However, there has until recently been no precise attention to the ways in which a reduction of medicine or the clinical sciences could succeed or be precluded. As Kenneth Schaffner suggests, it is one thing to argue that medicine or biomedicine is not reducible to the nonbiomedical sciences (in that medicine treats of unique or special entities)—that is, to deny the reduction of a whole to its parts in vitalist fashion. It is quite different to argue that medical language, generalizations, and laws cannot be reduced without loss of meaning—that is to say, to deny nomological reduction (admittedly a form of reduction that is much less likely to succeed) (Schaffner, 1977). Nor is it clear from the writings of authors in the field whether they intend to deny one or both forms of reduction (e.g., see Cassell, 1975).

Related to the issues of the status of nosologies and reduction is the problem of how the activity of diagnosis can best be undertaken. Numerous recent articles and books have analyzed the logic and the principles of establishing diagnoses (Weed, 1970; Feinstein, 1967; E. Murphy, 1979; and D. Black, 1968). Fairly extensive attempts have also been made to understand the role and significance of medical knowledge (American Medical Association, 1971; Grene, 1977, 1978; Wartofsky, 1975). In addition, a rather novel and important approach to the analysis of the status of medical theories and knowledge has been developed by Samuel Gorovitz and Alasdair MacIntyre (1976). They have contended that the clinical sciences are the basic sciences and that the so-called basic sciences are the more mediate and derived. They point out that detailed knowledge of particular patients is necessarily historical; it depends upon getting to know the past circumstances of patients as well as upon coming to understand their internal goods. They further contend that such knowledge of particular patients cannot be derived in a deductive fashion from general biological laws. Rather, the laws depend on knowledge of particular patients. Medicine is thus characterized as a science of particulars rather than of universal laws. Each particular patient must be studied as a whole, and such wholes can be understood only by reference to what constitutes the good of each—to use their term, what it is for a whole of that sort to "flourish."

In addition to undercutting a value-neutral view of medical science, their account insists on incorrigible unknowns as inevitably involved in medical practice. According to Gorovitz and MacIntyre, the physician must know the individual's history, which can be accomplished only in part; he must have a conception of general human flourishing, in part a philosophical and in part a psychological undertaking; and, finally, he must know the individual's conception of the good—that is, he must know the patient as a person (see also MacIntyre, 1977b).

Gorovitz and MacIntyre thus use medicine as a vehicle for attacking the classical view of science, which they believe takes each particular to be fulfilling the laws of nature that govern its kind. If the kind is known well enough, so the standard view implies, then only laziness or some other fault can account for error or ignorance. In criticism of this view, Gorovitz and MacIntyre stress that particular persons have varying environments and histories that serve as elusive intervening variables and that therefore neither diligence nor complete general knowledge can suffice against the vicissitudes and idiosyncrasies of nature. Even if we knew all the law-like generalizations that are applicable to patients, or patients of a particular kind, we could never specify all the reactions and the behaviors of that kind.

Gorovitz and MacIntyre believe that this argument applies both to probabilistic and to nonprobabilistic generalizations. Basic scientists are interested in the similarities among objects of a certain sort. These similarities make generalizing possible to the degree that one can talk about *the* hydrogen atom, for example. Clinical scientists are different; they are concerned with and limited by what is distinctive about particulars. Thus, the law-like generalizations that a basic scientist employs are different in kind from the patterns of thought used by medical scientists or practitioners in making diagnoses and prognoses, and in prescribing courses of action. In taking this line of argument, Gorovitz and MacIntyre have joined the camp of those who hold the nonreducibility of clinical science to the basic sciences.

These two philosophers draw two important policy consequences from their position on medical knowledge and its fallibility. One deals with scientific policy, the other deals with legal policy. They conclude, first, that a record of error or mishap ought to be rigorously maintained in order to provide necessary data for an account of the scope and limits of medical predictive power. They also conclude that if negligence and incomplete scientific knowledge do not exhaust the categories of sources of untoward results of medical intervention, then the remaining categories will include unavoidable mishaps that may not properly be considered grounds for mal-practice suits. They contend that laws and categories of culpability ought to reflect this epistemological discovery.

For all the strengths of this account, it has not gone without powerful criticism (see Bayles and Caplan, 1978; Gorovitz, 1978; Caplan and Bayles, 1978). Michael Martin's rejoinder (1977) takes issue with the argument to establish medical knowledge as a knowledge of particulars as well as with the consequent doctrine of necessary fallibility. Martin does this by raising four problems: (1) There would, in any event, be a virtue, for purposes of research and therapy, in ignoring such fallibility even if it could be shown to exist; (2) there may be no real difference between a doctrine of necessary fallibility and the conventional view that it is difficult to apply general scientific knowledge to particulars; (3) the Gorovitz-MacIntyre contention that knowledge concerning particular patients must always be qualified by the term "characteristically and for the most part" is an empirical claim that must be substantiated—a priori arguments will not suffice; and (4) their reasoning turns upon an infinite regress: Gorovitz and MacIntyre hold that the way an individual will react in a given circumstance depends not only on the makeup and history of the individual but also on the particular makeup and history of each of the elements that are part of his or her circumstances, and these circumstances in turn cannot be completely known because completeness would require knowledge of each of the historical influences, and so on indefinitely.

Marx Wartofsky (1977) attempts to defend the uniqueness of medical knowledge by putting forward an evolutionary historical account. He con-tends that theories of science are themselves instruments of human adapta-tion and that the true value of cognitive claims must be judged in terms of their utility, their adaptive value, and their ability to satisfy our needs. Medicine constitutes a unique domain of knowledge because it addresses a peculiar or unique set of needs. Thus, Wartofsky argues, medicine is not a derivative science but is one of the fundamental ways in which humans come to terms with reality. Moreover, such coming to terms with reality is never a purely theoretical enterprise but is a form of praxis, an endeavor of doing and making.

Given this involvement in praxis, the concepts of health and disease as applied to humans are framed in terms of individual and social expectations rather than simply in terms of species goals. When applied to non-human animals, "health" and "disease" are defined in terms of the way the animals' states contribute to human goals. For example, an irritation produces pearls in oysters, but we do not describe an oyster so bothered as suffering from disease. In cases where animals serve no human good—e.g., when not pets—we tend to choose our uses of the terms "health," et cetera, on the basis of analogies with human physiology and anatomy. For example, a given condition in the heart of a fish, which is of no interest to humans, would be described as diseased if it seemed to shorten the life or interfere with the capacities of the fish taken to be an independent animal, without regard to impact on species survival. We do, of course, tend to consider states pathological with respect to their impact on species survival unless the death or loss supports the survival or general thriving of the species—for example, as the male black widow's death does in the reproductive process. Thus, it seems that "health" and "disease" have different uses when applied to humans in contrast to animals. If this is so, it lends weight to the conclusion that medical theories are not reducible to physiology or to bio-medical science generally (Engelhardt, 1977a).

This is a contested point. It reflects the disputes concerning medical language and concepts that were discussed in the previous section. Here two contrasting arguments, one from Sedgwick and the other from Kass, are illustrative. Sedgwick, for his part, holds this:

> All departments of nature below the level of mankind are exempt from both disease and from treatment. . . . The blight that strikes at corn or at potatoes is a human invention, for if man wished to cultivate parasites (rather than potatoes or corn) there would be no "blight," but simply the necessary foddering of the parasite crop. Animals do not have diseases either, prior to the presence of man in a meaningful relation with them. . . . Outside the significances that man voluntarily attaches to certain conditions, *there are no illnesses or diseases in nature.* . . . Out of his anthropocentric self-interest, man has chosen to consider as "illnesses" or "diseases" those natural circumstances which precipitate the death (or the failure to function according to certain values) of a limited number of biological species: man himself, his pets and other cherished livestock, and the plant varieties he cultivates for gain or pleasure. . . . Children and cattle may fall ill, have diseases, and seem as sick; but who has ever

imagined that spiders or lizards can be sick or diseased? . . . The medical enterprise is from its inception value-loaded; it is not simply an applied biology, but a biology applied in accordance with the dictates of social interest (1973, pp. 30–31).

An argument against this position is articulated by Kass:

Insofar as one considers only disease, there is something to be said for this position—but not much. Disease-entities may in some cases be constructs, but the departures from health and the symptoms they group together are not. Moreover, health, although certainly good, is not therefore a good whose goodness exists merely by convention or by human decree. Health, illness, and unhealth all may exist even if not discovered or attributed. That human beings don't *worry* about the health of lizards and spiders implies nothing about whether or not lizards and spiders *are* healthy, and any experienced student of spiders and lizards can discover—and not merely invent—abnormal structures and functionings of these animals. Human indifference is merely that. Deer can be healthy or full of cancer, a partially eaten butterfly escaping from a blue jay is not healthy but defective, and even the corn used to nourish parasites becomes abnormal corn, to the parasite-grower's delight (1975, p. 23).

Sedgwick's position may well lead to an irreducibility thesis insofar as human organisms have unique functions that would be captured in the concepts of health and disease as these are framed by human-centered medicine.

In summary, one finds a nascent and growing contemporary literature concerning the nature and significance of medical theories and models. The current literature would be sparse indeed if it were not viewed in terms of the historical discussion about the nature of medical explanations. Given that background, areas can be identified for the prospective or current student in the field, including analyses of the significance of medical classifications, of descriptive or clinical accounts of disease, and of the reducibility of medical explanations to those of the basic sciences.

C. Metatheory and Methodology of Clinical Judgment and Medical Decision-Making

A body of literature is emerging that outlines canons of scientific and clinical rigor, the importance and proper use of statistics, and the need to abandon hunches in favor of well-established processes in making choices of diagnosis and therapy (E. Murphy, 1976, 1979; Lusted, 1968; Wulff, 1976; D. Black, 1968). Most of this literature on clinical judgment and medical decision making is not strictly philosophical. It offers reflections on scientific method or rigor in medicine and provides raw material for philosophic refinement. Despite the fact that this field is just emerging, a number of interesting issues have been framed and discussed. We will review three main topic areas: (1) the psychology or logic of discovery; (2) the process of clinical judgment making; and (3) the possibility of giving a formula for making clinical judgments.

Clinical judgment raises, in a new context, issues that have been discussed under the title of the logic and psychology of scientific discovery. In *Patterns of Discovery* (1961), for example, Norwood Hanson argues that seeing is intrinsically organized by concepts (language) and by contextual clues; the ability to see depends upon a conceptual framework. This framework, or context, is very much like what Thomas Kuhn called a paradigm in *The Structure of Scientific Revolutions* (1962; for these references, see chapter 4 in this volume). When transferred to medicine, these contentions lead to arguments to the effect that one must be trained in seeing, hearing, feeling, and smelling patients in order to *observe* whether they are or are not in particular conditions—a point, as indicated above, that is made by Foucault and van den Berg. This is often held to involve more than the skill of discerning norms and variations from norms. It involves, rather, the development of a practical knowledge that, in the doing, forms that which is known. In other words, the activity of knowing and the presuppositions that go with it shape, at least in part, the way the thing is known.

Studies in the history of disease tend to substantiate this view. In the nineteenth century, for example, physicians who presupposed that masturbation was evil found correlations between masturbation and serious, in fact fatal, conditions (Engelhardt, 1974a). Pathologists were even able to demonstrate at autopsies that changes in the lower spinal cord correlated with the death of the masturbator and attributed these changes to the excess excitement of this pernicious activity (Jones, 1889). Similarly, slaves were found to have diseases particular to them (Cartwright, 1851; Smith, 1851–1852), and in the twentieth century there have continued to be examples of anticipations skewing clinical judgment (Engelhardt, 1976a).

Varying methods of information gathering and organizing also structure the world of signs and symptoms presented to the physician. For example, Lawrence Weed and others have argued that structuring data in a specific way would enhance medical understanding and delivery of medical care (Weed, 1970; Feinstein, 1973; Goldfinger, 1973). By clustering records around recurrent complaints and problems presented by the patient, the clinician has his or her attention focused in directions that shape the data observed.

Thomas Kuhn noted in his preface to *The Structure of Scientific Revolutions* that the gestaltist view of scientific progress, in which community and the interrelatedness of fact and theory are emphasized, was anticipated by Ludwik Fleck in his book *Entstehung und Entwicklung einer wissenschaftlichen Tatsache* (1935). Fleck himself used a medical example, namely, the development of syphilis as a clinical entity and the collective nature of research on syphilis. He argued for recognition of the importance of "thought collectives" and "styles of thought" in the appearance and development of a scientific fact, and he stressed the collective nature of scientific research, indicating how the foci of interest and the style of thought of the research community influence the character of scientific facts. One finds many of the core issues concerning the nature of the history of science and the historical nature of scientific theories displayed in this work on the history and philosophy of medicine. Fleck's work has stimulated a small but important response within the philosophy of medicine (Delkeskamp, Engelhardt, and Spicker, 1980; Schäfer, 1980; Tsouyopoulous, 1980).

Alvan Feinstein has also addressed such issues as these in indicating that nosologies or taxonomies of disease tend to skew the logic and psychology of discovery. Particular styles of medical education, record keeping, language and taxonomy, as well as statistical and empirical methodologies, develop varying portraits of health and disease. They then covertly guide clinical judgment in predetermined directions. As only one example in the heuristics of discovery, medical students are not taught to expect observer-influenced variations in observations. In fact, they rarely learn to appreciate the subtleties of testing the reliability and standardization of observations and correcting for the variations. Nor do they often appreciate that biases in the observer (the physician) lead to less helpful characterizations of patients. As an example, Feinstein indicates studies of rheumatic fever that show that patients who have arthritis have a better outcome with respect to heart function than those who do not. The two forms of rheumatic fever have two different natural histories. Thus a rheumatologist studying the disabilities associated with rheumatic heart disease would likely find a better life expectancy and lower long-term morbidity than would an internist. Each would be selecting, unknown to himself, quite different patients. To remedy such distortions, Feinstein argues, medicine should develop categories and subcategories of study that attend precisely to such differences. In particular, he contends that the differences will be recognized only if one is concerned with clinical categories, such as "pain in joints" (Feinstein, 1967).

What Feinstein is advocating thus amounts to a choice among different models of medical data-gathering and clinical judgment. He opposes what he sees as a narrow model of science that pushes physicians into laboratories for their scientific authenticity, thus distracting medicine from cultivating a scientific approach to (1) observing symptoms (the science of diagnosis at the bedside); (2) developing prognoses at the bedside; and (3) treating and managing medical complaints with a high standard of systematic clinical conduct. Instead of developing these clinical sciences, Feinstein sees the physician qua scientist as being impelled, by a narrow model of science, to study the etiology (genetic, infectious, or environmental) of various diseases, and the pathogenesis (the causal sequence of the development of diseases) apart from the clinical complaints of actual patients. Feinstein also offers a theory of discovery, outlining what he thinks would be proper clinical categories through which one could see clinical facts; he also puts forward a logic for ordering the facts; this logic (or these rules of procedure) would be scientifically rigorous, but applied to bedside and clinical data, and would lead to accurate and useful clinical judgments—bedside judgments of signs and symptoms—in contrast to laboratory determinations of etiology and pathogenesis.

Thus, Feinstein is in the tradition that aims at providing a logic of medicine, along with Blane, Oesterlen, and Biegansky, mentioned earlier. Feinstein has, moreover, inspired a resurgence of activity within that tradition. For example, while discussing at length the prospects for formalizing clinical study, he argues that a belief in all science as mathematical can become a prejudice that all data are quantitative; that in establishing a calibration for degrees of what one wishes to measure, one is doing something intrinsically different from counting quanta. However, both calibrating and counting provide methodological problems for mathematics. Feinstein formulates a solution to some of these problems. Combinations of Boolean algebra or set theory and statistics, he thinks, may allow for the formalization of descriptive concepts from clinical medicine and lend themselves to the establishment of a coherent taxonomy of clinical calculation.

The extent to which he and other contemporary proponents of the science of clinical medicine (e.g., Cassell, 1977, 1979) are in the tradition of the great clinicians Sydenham, Sauvages, and Cullen is worth stressing. Feinstein, for example, complains that because current diagnostic labels are very dependent upon morbid anatomy (from the contribution of Bichat and others), the clinician is liable to forget the importance of simple clinical observations. The use of the diagnosis "myocardial infarction," for instance, does not include reference to the amount of pain, the state of shock, the presence or absence of arrhythmias, or congestive heart failure, that may attend an infarction. Rather than signaling these important clinical diagnostic signs, such labels signal underlying pathological states. The distinction between illness and disease (drawn in Section II. A. above) thus regains our philosophic interest. Feinstein is offering a phenomenology of illness in the language of the clinician and of the clinician's world, in contrast with an emphasis on the language of pathological and etiological explanations, which is the dimension of disease models and theories.

Thus one finds Feinstein arguing for a clinical nosology of illness in many respects similar to that of William Cullen. Like Cullen's proposed nosology (see II. B. above), Feinstein's nosology is supposed to afford criteria for (1) designating each symptom as such; (2) identifying patterns or clusters of such signs and symptoms; (3) designating clinical ways of deciding which of the possible disorders is present; (4) further dividing pathological categories on the basis of clinical findings; (5) inferring which tests, if any, would confirm the diagnosis suggested by the clinical evidence; and (6) forming prognoses and treatment plans grounded on clinical data.

Allied to investigations of the nature of clinical judgment have been analyses of various implications of diagnostic and therapeutic choices. This literature has highlighted the value implications of choices among diagnoses—establishing procedures of varying costs, specificity (i.e., lowering the chance of producing false positives), and sensitivity (i.e., lowering the chance of producing false negatives). Also, this literature has offered discussions concerning the assignment of values to outcomes of diagnostic and therapeutic projects in order to decide in what direction they should be focused. For example, Edmond Murphy has argued that one should be more accepting of false positives than of false negatives, if, in the case of a serious disease, the treatment is inexpensive, relatively convenient, and has a low morbidity. Because of the seriousness of the disease, false negatives would be too important to risk; and because of harmlessness of the treatment, false positives would not be of concern. On the other hand, one would wish to avoid false positives in the case of a less serious disease where the treatment is associated with high cost, pain, and mortality. In fact, even a choice of diagnostic procedures involves value judgments, for one should avoid those procedures associated with high cost in financial terms as well as possible patient mortality and morbidity if the disease that could be diagnosed has no effective treatment. In short, the process of developing a data base and of acting upon plans of treatment is full of judgments concerning values and their realization.

Consider, for example, the choice of a highly sensitive measure for appendicitis, which avoids the evil of ruptured appendices due to delayed surgical intervention, but does so at the cost of more appendectomies when in fact the appendices are normal. Contrast this with the choice of highly specific measures for appendicitis that avoid the evil of unnecessary appendectomies at the cost of more ruptured appendices. Ways of erecting classifications and of assigning patients within particular classifications have definite implications for action and for the realization of particular sets of goods and values.

The interest in greater clarity about the nature of clinical judgment and decision-making is invoking an increasingly larger literature. There has been a great deal of recent German discussion of the process of diagnosis and the choice of treatment, including such works as Rudolph Gross's *Medizinische Diagnostik—Grundlagen und Praxis* (1969); Wolfgang Wieland's *Diagnose—Überlegungen zur Medizintheorie* (1975); and Hans Westmeyer's *Logik der Diagnostik: Grundlagen einer normativen Diagnostik* (1972). These and other individuals have examined ways of constructing formulae for decision-making in the formation of clinical judgments (e.g., Elstein, 1976, 1979; E. Murphy, 1976, 1979; Scriven, 1979; Sober, 1979). Arguments have also been developed against such attempts (see Cassell, 1979). Individuals like Hans Westmeyer (1973, 1975) have produced detailed analyses of the weighing of costs of treatment, effectiveness of treatment, and possible positive and negative results of treatment attendant upon any therapeutic choice.

A great number of the more formalistic approaches to clinical judgment and therapeutic decision-making run aground, apparently because of the obscurities of medical language, including medical nosologies. Proppe (1970) has complained, "The problem for a general program of computer diagnosis (and thus recommendations for treatment) lies in the unclarity concerning what should be diagnosed after all" (p. 131). This core difficulty for the practical goals of medicine reflects basic difficulties in the philosophical analysis of medical practice. Moreover, if the arguments succeed to the effect that there is necessary fallibility in medical knowledge (Gorovitz and MacIntyre; see II. B., above), the possibilities for formalizing clinical judgment and therapeutic decision-making will be altered to an important extent. Clinical judgment, instead of being seen as the application of generalizations to particular circumstances and instances, will be viewed as an historical process of judging individuals in part by reference to lawlike generalizations and in part by reference to the internal goods of the individuals. And it will always be, in principle, a fallible form of knowledge.

Thus, philosophical discussions of the theoretical underpinnings of clinical judgment are an important part of the field of the philosophy of medicine (Sadegh-zadeh, 1977c; Albert, 1977). Through it one is shown the conceptual structures of alternate ways of forming diagnoses with a view to treatment. But the philosophical issues implicit in such discussions have not yet been successfully addressed. Further accounts are needed of the measures of the consistency and testability of clinical descriptions, and of the extent to which clinical and pathological perspectives of maladies aim at the same phenomenon. Furthermore, rigorous discussions of computer diagnosis, its steps and nuances, are needed. Interesting recent discussions have been provided by Harry Pople, Jr., Jack Myers, and R. C. Miller (1973). They have examined the ways clinical diagnoses can be rationally reconstructed and performed by artificial intelligence. Alternate and perhaps more effective means of reaching medical diagnosis have also been explored. Just as one need not resort to feathered wings for flying planes, computer programs may prove effective aids in clinical reasoning. Programs need not mirror the reasoning processes of actual clinicians.

The literature concerning the possibility of rigorously reconstructing clinical judgment has grown both in extent and quality. Michael Scriven (1979), Arthur Elstein (1976, 1978, 1979), Elliott Sober (1979), and others (see Meehl, 1977) have addressed the extent to which the process of clinical judgment can be approximated by particular formulae for reasoning, the extent to which decision-theoretic or policy approaches help in the study of the process of data integration, and the extent to which a rational reconstruction of the process of clinical judgment can be given. Such concerns have been joined with interests in developing programs for computer diagnoses (Lange and Wagner, 1973; Rose and Mitchell, 1975). These proposals have met with criticisms focused on the inadequacy of present models of clinical reasoning, as well as on confusions between models for developing diagnoses and models for acting without sufficient data bases (McMullin, 1979; Pellegrino, 1979). These discussions have also included discussions about the usefulness of Bayes's theorem in a rational reconstruction of the process of clinical judgment (Lusted, 1968; Suppes, 1979).

The literature concerning the metatheory and methodology of clinical judgment and medical decision making thus offers a wide range of issues of philosophical interest, but only limited philosophical reflections on these points of interest. Much of philosophical interest is being discussed with little realization of the philosophical significance of the questions at hand. Examples here would be Fleck's (1935) study of the collective nature of research or Lusted's (1968) treatment of Bayes's theorem as basic to a rational reconstruction of clinical judgment. They raise numerous issues about the epistemological significance of claims to knowledge in the clinical context.

D. Philosophy of Mind in Medicine

Because of medicine's intrusion into the study of the nervous system and the psychological states of patients, a number of problems in the philosophy of mind have a special salience in the philosophy of medicine. A locus classicus for the current debates is the work of Descartes. In his *Meditations* (1641) and his *Treatise on Man* (1664), Descartes suggested classic questions in medicine and the biomedical sciences: (1) Can one study the nervous system and the psychological states of humans separately? (2) To what extent are psychological states reducible to neurophysiological states? (3) To what extent are the findings of psychology reducible to the findings of neurophysiology? (4) What are the relationships between mental states and particular organizations of the brain? (5) What is the standing of a science of mental states? These questions have been treated in the classic texts of great physicians, physiologists, and philosophers (la Mettrie, 1748; Soemmerring, 1796; Flourens, 1845; Freud, 1953; Sherrington, 1951; von Hartmann, 1869; Gaubius, 1747; Alexander Bain, 1855, 1875; Herbert Spencer, 1855; Benjamin Rush, 1789; Adolph Meyer, 1950). These issues have, for the most part, either been viewed (e.g., Toulmin, 1948; Flew, 1949) as questions concerning the status of psychology, psychiatry, and psychoanalysis as sciences (the scientific character of claims forwarded in psychiatry and psychoanalysis), or as questions concerning the mind-body relationship (questions of the reducibility of one science—e.g., psychology, psychiatry, or psychoanalysis—to another, e.g., neurophysiology) (see Feigl, 1958; Polten, 1973).

Interesting discussions of the status of the mind-body relationship have been undertaken by modern physicians and physiologists. These are noteworthy in part because they have been framed by individuals who have been instrumental in the formation of modern neurology and neurophysiology (Jackson, 1915; Ferrier, 1886; Sherrington, 1951; Penfield, 1958; Eccles, 1966, 1970). In fact, reflections of Hughlings Jackson, the father of modern English neurology, reached a strictly philosophic level. Jackson, for example, argued for adopting a mind-brain parallelism as a methodological expedient for allowing one to pursue independently the sciences of psychology and neurophysiology. He also insisted that the formulations of hypotheses and the descriptions of data in neurology and neurophysiology be free of mentalistic terms. On the other hand, he assumed that one could similarly succeed in devising a psychology that did not require reference to physiological laws or presuppositions (Engelhardt, 1975e). This view of the relationship between the neurophysiological and the psychological sciences had an important impact upon Freud's development of psychoanalysis (Stengel, 1963).

The extent of explicit philosophical activity on the part of neurologists in framing their theories is impressive. The French neuroanatomist and neurophysiologist M. J. P. Flourens (1845), for example, dedicated an important work to Descartes and saw himself as giving arguments against reductive materialists. Charles Bell (1824), Franz Josef Gall (1835), and many others have seen their investigations as having significant conceptual presuppositions. Such reflections on the neural sciences have continued unabated (Eccles, 1970) and have led to an important contemporary exchange between the philosopher Karl Popper and the neurophysiologist John Eccles (1977).

The examination of the relationship between mind and brain by neurophysiologists has produced a series of discussions concerning personal identity and its relationship to brain structure. These discussions have arisen with reference to the concept of cerebral localization and split brain operations. A central element of the debate among Flourens, Gall, and Jackson had been focused on the extent to which particular mental faculties can be located in particular brain structures (see Bynum, 1976; Benton, 1976). These discussions turned not only on issues of interaction between mind and brain, but on the intelligibility of localizing mental functions. Three factions have arisen: strict localizers, who hold that mental functions are localized in particular brain structures; moderate localizers, who suggest that mental functions can be localized in certain regions of the brain, though they are never, or rarely, localized in any particular small area and are moreover dependent on interactions with the brain as a whole; and nonlocalizers. In comparison with the other two, nonlocalizers have not commanded a significant portion of the current debate (see Magoun, 1963). In Russia, theorizing has been influenced by presuppositions of dialectical materialism that were taken to count against strict localization (Luria, 1966).

Important philosophical questions are encountered under this tripartite rubric. For example, with any attempt to localize functions there is the problem of how to designate the cerebral functions that are localized. If one uses mentalistic language, one presupposes that the brain is organized according to our mentalistic categories (Young, 1970). Basic problems in the mind-body literature have surfaced in this way.

Early in modern reflections on the significance of the cerebral localization of mental functions, a problem arose concerning the role of the two cerebral hemispheres and the implications of this duality for the unity of human consciousness. This point was raised, for example, as early as 1844 by Wigan in his book *The Duality of the Mind* and was addressed in a classic work by John Hughlings Jackson, "On the Nature of Duality of the Brain" (1915). These theoretical reflections took on a practical force when neurosurgeons began to disconnect the cerebral hemispheres surgically as a treatment for forms of intractable epilepsy. Some individuals showed signs that each hemisphere possessed an independent stream of consciousness. As a result, a series of discussions and analyses of the significance of these findings has been published. It questions whether in the bisection one creates two persons (Puccetti, 1973; R. W. Sperry, 1977; Shaffer, 1977; Gazzaniga, 1970) or even whether in some sense there are not in all of us two minds (Bogen, 1969a, b, c; Jaynes, 1977). This led to a debate over the implications of these findings for understanding personal identity and about whether, perhaps, persons can be split into two, as well as whether they could theoretically be rejoined into one.

These considerations have bearing upon general policy questions—for example, about the significance of psychosurgery as altering the embodiment of persons (National Commission Report on Psychosurgery, 1977a; Margolis, 1976b; Fodor, 1976). Similar issues of the meaning of embodiment also underlie discussions of new definitions of death and the status of the fetus as a person. It seems that, depending on what it means for a body to embody a person, different definitions (or criteria) of death will recommend themselves: (1) whole-body oriented definitions; (2) whole-brain oriented definitions; (3) neocortically oriented definitions (Engelhardt, 1975c). Different senses of embodiment and of the relation of mind and body will also recommend different notions of psychosomatic illnesses. Moreover, if one holds to a strict separation of mental problems and physical diseases, one likely presupposes a Cartesian dichotomy of mind and body so that disease states are only states of bodies (Szasz, 1978; Baruch Brody, 1978). In addition to difficulties for the concept of mental illness, difficulties follow for an adequate appreciation of psychosomatic illness (Erde, 1977). Purely somatic notions of health have, however, been offered by philosophers who acknowledge that human health can fail at what we may dualistically refer to as "the mental level." For them, conceiving the human to be a complex somatic unity does not obstruct conceiving of certain behavioral problems as problems of health (Hanna, 1970, pp. 77ff).

In addition to these philosophical considerations about theories of medicine and mind, there exists a vast literature reflecting on psychiatry and psychoanalysis as sciences and arts (Macklin, 1972, 1973; Ellis, 1956; Meehl, 1970). A preponderance of this literature involves consideration of psycho-analytic methods and the significance of the processes and capacities stipulated by psychoanalysts (Yankelovich and Barrett, 1970).

There is also a wisdom literature—reflections on how one can, through psychoanalysis or psychiatry, attain the good life, or how psychiatry and psychoanalysis indicate to us what one would mean by a good life (Fingarette, 1963; R. Schafer, 1976). Further, there are more strictly analytic endeavors: attempts to provide a rational reconstruction of the logic of explanation in psychoanalysis as well as of clinical judgments and decision-making in psychotherapy generally (Westmeyer, 1972; Sherwood, 1969; Mullane, 1971). Finally, internal disputes among proponents of different views of the meaning of psychiatry and psychology have been voluminous. Here it is enough to indicate the debate among behaviorists, biologically oriented and psychologically oriented psychologists, and psychiatrists (see J. B. Watson, 1913; Skinner, 1971; Heath, 1967).

In short, medicine has occasioned profound and wide-ranging reflections on the notion of personal identity, the relation of mind and brain, and the significance of the psychological and neurological sciences. This is enough to show that the role of the philosophy of mind in the philosophy of medicine is major. At least from the point of view of publicity, however, the most important issue concerns the status of the concept "mental illness" and the actions that turn upon that conception. (This topic alone would be sufficient to call into question the feasibility of separating fact from value as is commonly done. Issues about what ought to be done and issues about what is the case are apparently more intimately bound together here than they are on any other topic.)

A great deal of Thomas Szasz's work (e.g., 1960, 1961, 1972, 1978) has aimed at arguing that the concept of mental health is nonsensical and that public policies based on it—especially those involving involuntary civil commitment—are wrong. Szasz has drawn a great deal of criticism (e.g., Macklin, 1972, 1973), some of it aiming at his attack on the concept of mental health (Michael Moore, 1975). Nevertheless, those critical of Szasz's attack seem to have to admit that the concept is difficult to clarify and is very complex. Some researchers have displayed the existence of at least a dozen different models for what mental illness has been thought to be (Siegler and Osmond, 1974). Another side of the problem is that the use of the concept is very unreliable, even among practitioners who try to apply it appropriately and reliably (Rosenhan, 1973). Moreover, it is clearly open to abuse, as indicated by the political abuses of psychiatry in the Soviet Union.

However, the critique of psychiatry's potential for abuse need not be limited to extremely flagrant cases such as those involving the Soviets, for the standing of the psychiatrist is often ambiguous. Consider such roles as military psychiatrist, prison psychiatrist, school psychiatrist, and even private practitioner, as well as administrator for a mental institution. Each of these roles can place the psychiatrist in a position of divided loyalties and demands, so that he or she functions as a "double agent" (Burt et al., 1978). The psychiatrist is in the difficult situation of assessing the afflicted person as a danger to self and/or to others, and of using debated and often conflicting criteria in deciding on the institutional commitment of individuals (Peszke, 1975). The danger to others is debated partly on grounds of the difficulty of defining and predicting danger, but partly as a policing function improper to medicine. The danger to self is debated on the moral grounds of the standing of autonomy. With the mention of this ground, we have a direct bridge to the general field of biomedical ethics.

III. BIOETHICS

The several senses of "ethics" complicate the mapping of the field of bioethics—just as the senses of "medicine" complicate the idea of what a philosophy of medicine is in general. "Ethics" has been used to cover sentiments concerning what is right or wrong, guidelines for upholding customs or mores, including formal rules such as those of medical etiquette which give the canons of civil probity for the medical profession, as well as judgments about the propriety of conduct according to religious canons. "Ethics" has also been used to cover the philosophical enterprise of conceptual critique and argument assessment where the concepts and arguments involve the practices of praising and blaming. This level of ethical inquiry, while it involves what may be termed a first-level account of proper human conduct, also involves a second level—so-called "metaethical" accounts of how to assess the first-level accounts. Ethics, in these senses, is a rational enterprise. Moreover, it is not simply discourse about tactics or logistics (i.e., how to obtain a good); it is rather directed at the more basic level of asking why some things are taken to be moral goods or evils, or why some forms of conduct should be subject to blame and others to praise. Further, "ethics" may sometimes be used (overbroadly) to indicate concern with value judgments that are the concern of neither moral philosophy, nor religious ethics, nor medical etiquette—value judgments embedded in such concepts as health and disease. These questions were addressed in section II.

There are two kinds of approaches to ethical concerns, which we will contrast sharply for the purposes of this essay: one concerned with conse-

quences, the other with obligations not reducible to interests in consequences. "Consequential" or "teleological" ethics involves choosing goods,
ranking goods, and specifying the most efficient means of achieving the
goods to be pursued. Sometimes the consequences are defined in utilitarian
terms—maximizing pleasure and/or minimizing pain throughout a population or for the greatest number. Sometimes they are defined teleologically
without being utilitarian: pleasing God may be a desired consequence
independent of whether that gives humans pleasure. Teleological ethical
questions clearly concern efficacy, assessing maneuvers for achieving sought
ends. They often appear to be analogous to engineering questions: "How
can I (we) secure . . . ?"; and/or economic questions: "Is getting such
and such by this or that means worth the cost?" In the medical arena the
goals have to do with curing, diminishing morbidity or mortality, making
patients or clients and their families feel better, and distributing services
and the like. The means to achieve an end result (e.g., socializing medicine
in order to expand access to the poor) may require disrupting standing practices (e.g., physicians' free enterprise) and blocking alternative social pursuits (spending funds on medicine that might be spent elsewhere).

The second kind of ethical concern, which focuses on respect for persons, can be seen as Kantian or "deontological." (There are senses of deontological ethics, such as W. D. Ross's, that are non-Kantian, as well as
attempts to give a *quasi*deontological grounding for ethics, such as that of
John Rawls, 1971.) This respect for persons is not a value to be pursued, but
is a *condition for the possibility* of certain kinds of obligations. Neither is
it reducible to interests in goods and values. Rather, insofar as one can
recognize the possibility of respecting a subject solely because he or she
is a morally responsible agent, one can understand the *possibility* of a moral
community—a community bound together on the basis of mutual respect,
rather than by reference to force. In other words, actions that are attempts
to have one's will or view of the good, or the will or view of the good of
one's group, actualized by force are distinguished from those actions that
are bound by the recognition of the constraint of respecting others as
autonomous agents. Insofar as one wishes to actualize such a moral community, one is bound to reject coercion and allow only reason-giving and
peaceful and honest manipulation in agreeing upon common courses of
action. Furthermore, once one can understand the possibility of such a
moral community, one can distinguish practices of force from moral practices. Recognizing (having a sense of) obligations to others does not mean
an interest in the goods and values that accrue to one from those others.
In such a view, there are rights and duties that cannot be reduced to interests in goods and values as in the case of teleological ethics. Finally, this
deontological view is accepted not because it is itself useful, but because it
takes account of the moral life in all its richness and depth—including a
sense of obligations not reducible to interests in goods and values. It makes

explicit an intellectual possibility, a strong view of ethics and of acting ethically. In it, the freedom of others stands as a side constraint—a limiting condition, a necessary condition for the possibility of morality. It is not simply something that could be valued or disvalued (Nozick, 1974). It defines a community of moral agents founded on respect.

To put the matter in terms of a stark paradox: within the context of freedom having a standing as inviolate constraint, a rational free agent's choice to disvalue rational freedom (e.g., in submitting to a lobotomy) should be respected. It is in terms of respect for freedom as a constraint that many moral claims arise—e.g., making free and informed consent a procedural requirement of treatment or experimentation, or respecting the right of patients to refuse life-saving treatments—two examples that bear on not *imposing* a course of action on another person. Some *restrictions* may also be construed as intrusions on freedom (Szasz, 1971, 1974): e.g., refusing a person access to certain substances, such as laetrile or heroin, or refusing certain requested procedures, such as sterilization, or controlling access by requirements of prescriptions.

The deontologist, one must admit, has two special problems. First, for the deontologist there can be tragic, irresolvable conflicts between the goods of individuals and their rights. That is, the cost of respecting freedom may be the loss of many goods, especially social goods. Second, in order to decide what has been properly consented to in political programs, the deontologist must have a theory of the delegation of power and authority.

Teleological justifications have also been given for requiring free and informed consent. These justifications are grounded in the claim that a majority feels better, or is better off, when such practices exist. But despite the agreement here between the deontological and the teleological position, there is a possibility for a teleologist to waver; situations can lead to distasteful states of affairs in which all will clearly be worse off if freedom is respected. For example, certain distributions of goods or experiments—which most of us may find congenial and which may clearly be best for all individuals—could be precluded because an unconsenting minority has a right to veto the desired program. Such losses would be no less distasteful to the deontologist, but they are unrectifiable according to deontological rules. The utilitarian might, and probably would, be obliged by his principles to ignore the veto of a minority; he or she might use force to achieve the good of all.

This distinction between the teleological and deontological foci of ethical claims is meant as a heuristic device for the analysis of issues in bioethics. The heuristic is needed because the area is a thicket of complexity. For example, a claim concerning a right or a duty may be an elliptical way of making a plea for the establishment of a particular practice aimed at producing a good; claims to the right to adequate medical care might be a claim of this sort (Macklin, 1976). In contrast, other claims of rights and duties may be made in recognition of practices that seem basic to the nature of a moral community as such, and that appear to be independent of a particular community's vision of goods and values. Examples of such duties seem to include an obligation to refrain from experimenting upon humans without prior consent, or an obligation to refrain from harvesting and distributing organs from a living donor without proper consent. Such obligations produce rights on the part of the individuals—to consent freely and informedly, or to refuse treatment, experimental subject status, or donation of an organ. Finally, the community of persons specifically involved in these issues includes not merely the patients and physicians, but nurses, medical social workers, physical therapists, occupational therapists, and technicians of many sorts.

How such individuals, as witnesses and parties to actions, are to relate to physicians' authority and patients' rights and needs is an issue receiving increasing attention (see Purtilo, 1978a, 1978b; Aroskar and Veatch, 1977). Attention is also being paid to the idea that excessive talk of rights and duties may obscure the intuition that much of concrete moral life is concerned with goods, virtues, the building of moral character, and the quality of relationships (Hauerwas, 1974, 1977; Ladd, 1978). Often rights-duties language is only a rhetorical play, to remind one of the importance of different ways of living a moral life—which ways are invented as much as they are discovered.

In what follows, we will discuss basic claims of rights and duties and goods and values as they arise in medicine. First the focus will be on conflicts between physician and patient rights and duties, and among different views of the goods and values at stake in health care. (Nurses and other professionals have some issues in common with the physician, but they also have other problems of their own—especially, e.g., how to relate to questionable commands of those in authority; cf. Purtilo, 1978a.) We will then shift to talking about persons and the definition of "person"; once one has decided what rights and duties are borne by persons, or what goods and values should be provided to persons, one must go on to determine which of the objects in the world are persons and in what senses they are persons. The discussion will then turn to allocations of health-care services. Finally, some closing remarks will be made concerning arguments in natural law and from religious viewpoints regarding bioethical issues. Special aspects of problems with contraception, sterilization, and artificial insemination by a donor will be examined in this final section.

This clustering of topics will mean that in the first section we will examine issues about consent (in treatment and experimentation), truth telling, confidentiality, killing versus letting die, euthanasia, the right to refuse life-saving treatment, death with dignity, eugenics, sterilization, and genetic screening. One should note here that though much is said (in reflection of the amount and quality of the literature) about experimentation, the points raised are usually general ones. They are of importance for non-research-oriented medical practice as well. In the second section topics such as the following will be addressed: in vitro fertilization; fetal experimentation; abortion; experimentation involving incompetents, e.g., infants, the mentally retarded and the mentally ill; the role of adolescents in determining their treatment; and the definition of death. The third section, which focuses on the distribution of biomedical goods and harms, will deal with comparing health to other values; rights *to* health care, rights *in* health care; questions of rights to equality of health care; allocation of scarce resources; population control; environmental concerns (including issues concerning research with recombinant DNA); and interprofessional relations. The last section will address special questions that have been raised from natural law or religious viewpoints. In short, the topical issues will be organized in terms of philosophical considerations. This approach will lead to the separation of some issues that often are treated together—e.g., experimentation with children and adults. Such separations will, we hope, better highlight the underlying philosophical issues.

A. Rights and Duties, Goods and Values, for Physicians and Patients

Much of American medical ethics has been developed in an individualistic, contractarian mode in which the emphasis has been upon rights among physicians, patients, and various third parties defined within a marketplace for services. For example, codes of medical ethics have primarily addressed problems concerning individual physician-patient relations. These concerns, often dismissed as merely questions of etiquette, reflect background assumptions about entitlements to goods, the moral significance of what may be called "the natural lottery" (e.g., that some are born rich, others sick, while some contract terrible diseases through no fault of their own), and the moral rights and duties ingredient in particular social roles (e.g., physician, nurse, patient).

If such codes are to have moral significance, they must express moral principles or procedures for preserving such principles. Though American codes have tended to have an individualistic cast, they have also contained references to the communal aspects of medicine, such as the duty of the community to medicine as a profession, and the duty of physicians to the community (A.M.A., 1848, 1957). Views of the social significance of the medical community have also been expressed in licensing laws, regulating the ability to form physician-patient contracts (Shryock, 1967). Licensure has had as its object the protection, development, and nurturing of medicine as a responsible profession whose members will be reliable and dedicated to the goods and values held to be important to medicine. Such laws also seem to have served as an established form of restraint of trade. Indeed, the great shibboleth of the medical profession, the Hippocratic oath, speaks to the concern for the public as well as the private interests of the profession by recommending the integrity of a professional group and the restriction of medical knowledge to those who belong to the group. Further, in the discussions of medical etiquette one discovers the Hippocratic triad—of the physician, the patient, and the profession or art (see Hippocrates, 1923)— as the three interest points around which medicine is formed.

Turning to the focal topics, we may inquire into the elements of the proper physician-patient contract. Concern over the place of free and informed consent is often viewed as concern about what should count as negotiating a legitimate doctor-patient contract (Sade, 1971). Likewise, concern over confidentiality, truth telling, the right to refuse treatment, and medical paternalism are often couched in terms of what is perhaps implicit in the physician-patient agreement. Paul Ramsey, a contractarian, has argued from the metaphor of the biblical convenant as well as from an adaptation of the Kantian view of respect for persons, with a number of crucial changes to accommodate religious assumptions (Ramsey, 1970b). Similar stands have been taken by several others (John Fletcher, 1967; Campbell, 1972). Charles Fried (1974)—in a fashion reminiscent of the secular Kantian regard for the freedom of rational agents, viz., persons— argues that the physician-patient relationship should be characterized by (1) lucidity: the patient has a right to know all relevant details about the situation in which she or he is; (2) autonomy: the patient has a right to be free from force and coercion during medical care; (3) fidelity: dealings among persons create expectations of reliance and trust that should be fulfilled; and (4) humanity: individuals have a right to have their full human particularity taken into account by those who do enter into relationships with them. These four points characterize what he has elsewhere (Fried, 1975) termed rights *in* health care as opposed to rights *to* health care—rights that characterize proper physician-patient relations regardless of what rights exist or do not exist *to* health care. Under the rubric of rights *in* health care are usually placed the right to be told the truth in the physician-patient relationship, the right to confidentiality of patient records (*Personal Privacy in an Information Society*, 1977; Westin, 1977), the right to respectful treatment, and the right to consent freely and informedly to treatment.

There are numerous conflicting views. For example, Joseph Fletcher (1960, 1966) has argued that the realization of certain goods (e.g., love, *agape*) or the maximization of the greatest good for the greatest number is the crucial test in moral questions. In such arguments the realization of the overriding good (being loving) or of the greatest good of the greatest number justifies exceptions to the restraints of free and informed consent, of keeping confidence, and of truth telling.

Others have argued that patients are not in a position to judge in an informed fashion what is at stake when they attempt to consent (Ingelfinger,

1972). Some physicians have concluded from this argument that physicians have the right to do what they hold to be in the best interest of patients. Though for the most part this view has been eroded in the law (*Canterbury v. Spence*, 1972), one still finds relics of it in what has been termed "therapeutic privilege." According to this principle, it would be justifiable if, for example, a patient were suffering from cancer, to treat that individual without obtaining informed consent—i.e., without revealing the diagnosis, the prognosis, and the true reason for the types of therapy involved (see Curran, 1969).

At times this principle has been extended even to include circumstances in which research is undertaken. In the Federal Food, Drug and Cosmetic Act, Drug Amendment Acts of 1962 (see Kefauver-Harris Bill, 1962), section 505 (i) states that physicians, when engaging in therapeutic research, are relieved of the duty to require informed consent "where they deem it not feasible or, in their professional judgment, contrary to the best interests of said human being." In contrast, others have held that the difficulties in achieving consent preclude all or nearly all risk-bearing, nontherapeutic research (Jonas, 1969; Katz, 1972, p. 733).

In short, discussions of free and informed consent involve different views of the stringency of the duty to respect the freedom and autonomy of persons; the importance of achieving the goods and values of medicine; the importance of scientific progress through experimentation; the need to have patients cooperate in the teaching of medical students, interns, and residents; the ability of individuals to comprehend their circumstances, their rights to act capriciously, and the extent to which interests in the goods and values of medicine or medical science can override patient or subject autonomy.

Medical experimentation is a focal topic of bioethics in large part because of the special questions it raises regarding informed consent. If, for some patients, free and informed consent to treatment (which is presumably in their best interests, and, therefore, where failure of consent may do less damage) is reluctantly waived through a paternal fiction of proxy or presumed consent, then surely in the case of risk-laden, nontherapeutic research, action on persons without consent must be considered even more suspect or less likely to be a choice made in their best interests (Freund, 1969). Conflicts of interest between the concerns of an agent as both physician and researcher (Toulmin, 1977; Lasagna, 1975) are part of the reason. In addition, particular research designs (randomized clinical trials, the testing of placebos, and psychological research requiring deception) may fall short of the duty to respect subjects as persons by always seeking free and informed consent. Randomized clinical trials and the using of placebos are less problematic in that subjects can be informed that they are being asked to participate in double blind studies (Fried, 1974). Many psychological research protocols that involve deception (Milgram, 1965, 1974) are more problematic. Evidence of the tolerance of such deception is given by the code of ethics of the American Psychological Association (1963), which allows for deception given sufficient merits of the research, protection of the best interests of the subject, and an explanation after the experiment.

Examination of these issues has been central to the deliberations of the National Commission for the Protection of Human Subjects of Biomedical and Behavioral Research (see National Commission, 1975, 1976, 1977a, 1977b). Though the commission has attended primarily to research, the principle of informed consent is pertinent to medicine generally in that it relates to respect for autonomy, issues of beneficence, and questions of distributive justice—problems of a general ethical nature that appear in all areas of the biomedical sciences and health care.

Questions concerning the use of prisoners, children, and the mentally ill in human research reflect deeper questions about human practices and value commitments. For example, Hans Jonas (1969) has argued that "if we hold to some idea of guilt, and the supposition that our judicial system is not entirely at fault, then prisoners may be held to stand in a special debt to society and their offer to serve—for whatever motive—may be accepted with a minimum of qualms as a means of reparation" (p. 246). Jonas' remark shows the role of background assumptions in setting the frame of reference for particular questions concerning the probity of different ways of employing prisoners in research, since his remark is in an essay in which he nearly precludes the use of other humans in nontherapeutic research.

The use of prisoners in human research becomes dramatized in the modern era through the Nuremberg trials' and code's concerns with coercion and duress. Such concerns have been addressed by thinkers who are sensitively paternalistic in regard to subtle compromises of consent. One finds contentions that the environment of a prison erodes the ability of individuals to choose in free and informed fashion. For example, in one of the more influential modern cases, *Kaimowitz and Doe v. the Department of Mental Health* (1973), it was argued that the very circumstances of prison life undermine the ability of the individual to give free consent. It is out of such considerations, at least in part, that the National Commission for the Protection of Human Subjects recommended (1976) that the use of prisoners be restricted to research beneficial to prisoners or directly concerned with prison life and conditions.

These recommendations presuppose a view of human autonomy in which a coercive atmosphere may overwhelm the ability to exercise free choice. In taking its position, the commission raised issues about degrees of coercion and manipulations that stand to invalidate the authenticity of human choice. The commission's recommendations are a development of ideas raised earlier in general discussions about policies regarding the use of prisoners. The view is that the integrity of consent depends on the capacity to choose on the basis of sufficient information and without coercion (John Fletcher, 1967).

Similar concerns about respect for persons and individual freedom to pursue particular goods and values have arisen in the controversies concerning the right to refuse treatment; the right to commit suicide when death is imminent or when faced with a painful disease; the morality of assisted suicide; and the practice of euthanasia (Hook, 1975; Brandt, 1975a, 1975b; Dyck, 1975; Kohl, 1975, 1977; Downing, 1969; Veatch, 1976a). These considerations, in great measure, reflect traditional philosophical discussions of the moral probity of suicide by David Hume, Seneca and Kant (see also Bridgman, 1933). One should note that even St. Thomas More in his *Utopia* discussed euthanasia as a proper cessation of life.

The discussions of death, dying, euthanasia, suicide, and so on have focused on the nature of obligations to respect the autonomy of patients, the nature of such autonomy, characterizations of the goals and responsibilities of medicine and physicians, and general characterizations of the goods and values that society would have dying or potentially suicidal individuals embrace. The arguments concern the circumstances under which commitments to beneficence may obligate one to kill another, the extent to which respect for freedom entails not interfering in the rational choices of individuals to take their lives or to refuse life-saving treatment, and the extent to which concerns about the sanctity of life preclude the intentional taking of the life of another, as well as concerns regarding styles of dying.

The preponderance of reflections on euthanasia and letting die is drawn from the extensive literature concerning the sanctity of life and the alleged proscription against the intention to take the life of another. In particular, a great deal of this literature is situated in the framework of religious ethics (Ramsey, 1978; Kelly, 1950, 1951, 1958). Notions of the immorality of intentionally willing the death of the innocent lead to distinctions between foresight and intention, between foreseeing that an action will kill another and intending that an action should kill another. Such distinctions were developed within Roman Catholic moral theology in order to deal with such issues as the "just war" debate (the particular issues there center about killing civilians as a by-product of hostilities). Such reflections led to the theory of "double effect" or oblique intentionality, which has, in Roman Catholic circles, been broadly applied to such issues in medical ethics as contraception, sterilization, and abortion.

According to the theory of double effect, one may engage in omissions or commissions leading to a more rapid death (or other evils—see above) as long as (1) the action, considered by itself independently of its effects, is not morally evil; (2) the evil effect is not the means of producing the good effect; (3) the evil effect is sincerely not intended but merely tolerated; and (4) there is a proportionate reason for performing the actions in spite of the evil consequences (Kelly, 1951, pp. 13–14). These conditions delineate situations within which one could permissibly foresee and allow a natural evil without intending it—and thus without committing a moral evil (McCormick, 1973). The distinction between foreseeing and intending underlies the moral possibility of physicians' ceasing "extraordinary" or "heroic" measures in the care or treatment of dying patients (Pope Pius XII, 1958). The framing of such distinctions presupposes that although it is immoral to intend to kill the innocent, unintentional but foreseen killing of the innocent (killing that would not be intended and would thus not be described as a killing by the actor) is justified if it allows the achievement of sufficient good (i.e., saving resources, diminishing pain). An important criticism of the theory of double effect is provided by Elizabeth Anscombe (1969, pp. 28–49); a recent major discussion of the theory can be found in a volume edited by Richard McCormick and Paul Ramsey (1978).

Discussions concerning the moral significance of distinguishing foresight from intention (i.e., double effect, or oblique intentionality) have recently been overshadowed by secular considerations of the moral differences between acting to kill individuals and refraining from actions so as to let them die. Many of these discussions appear to take for granted that in either case one is intending the death of the individual (Rachels, 1975). Analyses in such circumstances have focused on defending or attacking the supposition that acting is morally more problematic than refraining.

Under the doctrine of double effect, however, both acting and refraining were held to be morally equivalent. What is important is intention, or not acting in order to expedite the death of another. Thus one could permissibly give a dying patient an analgesic that might, as a side effect, accelerate death. One would not be engaged in the act of murder because one would not be so intending. There would at most be an incidental homicide, never an intended killing; thus killing in order to relieve pain would not be permissible. One could also withdraw medical treatment in order not to employ medical resources uselessly—when that wasting would simply extend the processes of dying in the absence of a tolerable quality of life for the patient in his or her own terms. (For a discussion of the concept "quality of life," see Reich, 1978b.)

In contrast, the moral issues at stake in distinguishing acting from refraining are less clear. Critiques have focused on comparisons of judgments of various paradigm cases (Rachels, 1975) or attempts to dispel belief in a moral difference between acting and refraining. Jonathan Bennett (1966), for example, contends that the only essential difference between acting and omitting is that, in the case of *acting*, most things that one could do will *not* lead to the result in question—only a few would. Omissions, he maintains, may be characterized by the fact that most of the things one could do *would* prevent the result of the omission. In other words, the difference between acting and omitting turns on whether many or few things lead to the result—a difference that Bennett takes to be without intrinsic moral relevance. Choices between acting and refraining depend, so suggests a reconstruction of Bennett's position by Dinello (1971), on their consequences—a criterion reminiscent of the criterion of proportionate good used in deciding cases of double effect (McCormick, 1973). On the other hand, Bennett's claim that the distinction between acting and refraining has no moral significance is not without its critics.

The debates about action, omission, and double effect have special application to questions about euthanasia. Discussions on that topic have been pursued in other terms as well, such as consent and grounds in favor of the action in particular circumstances. In the latter view, most have argued for the moral permissibility, rather than the obligatoriness, of euthanasia under certain circumstances—though a number have contended that it is under certain circumstances praiseworthy. Marvin Kohl (1974, 1975, 1977), for example, has argued for euthanasia under conditions in which that act would be one of kindness or beneficence. This view has been criticized by those who have contended that an account of the justifiability of euthanasia requires a prior account of the moral probity of suicide (Troyer, 1977). Others have objected because of the procedures and policies they foresee being required by having an option to die (Kamisar, 1969). A philosophical literature about this is still emerging (Beauchamp, 1976; Daube, 1971; Lebacqz and Engelhardt, 1977), though, as indicated above, it builds on a considerable general literature that already exists (Downing, 1969).

Recently both in the law (Cantor, 1973; Hegland, 1965; Montange, 1974; Rabkin et al., 1976; Foreman, 1975) and in philosophy (Horowitz, 1972), there have been a number of defenses of the right to refuse treatment, even life-saving treatment. Such discussions have given a great deal of attention to the state's interests and a need to preserve a value for life. Some of these arguments have been in terms of respecting the freedom of individuals, a right to privacy, or a right to be secure against unauthorized touching. Yet others have contended that individuals are in general the best judges of their interests. Those who have argued in favor of the right to refuse treatment have given grounds, at least in part, against paternalistic judgments in medicine. Moreover, like the issues in euthanasia generally, they occasion moral questions connected with suicide.

Though physicians have been inclined to recognize increased patient self-determination in other areas of judgment, they have been more reluctant in cases of refusal of life-saving treatment. Physicians have often found such decisions by patients difficult to accept, particularly when the treatment would not simply prolong the period of dying but would actually return the patient to a state in which he or she could live out a more or less normal life for a time (White and Engelhardt, 1975; Platt, 1975). Some authors have interpreted the reluctance of physicians to accept a patient's refusal of further treatment in terms of the obligations of physicians to achieve what they take to be the proper goals of medicine, including prolongation of life:

> We know that when the physician enters the [physician-patient] relationship he acquires a responsibility for the patient that cannot be morally relieved merely by the patient's refusal to consent for treatment. But more simply, the physician could not stand aside and allow the patient to die from a disease otherwise easily treated without feeling that he, the doctor, was responsible for the death (Cassell, 1977, p. 16).

The patient also has obligations. "In giving himself into the responsibility of a physician, the patient is obligated not to injure the physician morally or legally by making it impossible for the physician to operate on the responsibility" (Cassell, 1977, p. 16).

Another position, on the surface similar to Cassell's, is emerging. It is paternalistic, but not "for medicine's sake." This updated version of Hippocratic ethics makes duty to the patient for the patient's sake a part of the role of the physician (Ladd, 1978). One could say simply that this trend is a rejection of legalism (Burt, 1978).

The subject of living wills—written instructions to physicians to withhold or withdraw life-saving treatment under particular circumstances in case one becomes incompetent—ties into those of "death with dignity" and "quality of life" discussions, i.e., controversies concerning euthanasia and the right to refuse life-saving treatment. The living will is one of the ways proposed to cope with some of the policy issues. Preferences expressed by such wills reflect antecedent judgments concerning the worth of life under varying circumstances as well as standards about how natural or appropriate it is to accept death under varying conditions. Such instruments are meant to extend the rights of a competent person into the future—to make his or her wishes binding. That is, such instruments as living wills are meant to force physicians to act in accord with the considered judgment of the patient with regard to the refusal of life-saving treatment (or else they must withdraw from the case), even after a patient has become incompetent. Many have taken these instruments to be useful ways of protecting personal freedom. Laws have been passed giving statutory support to such instruments under very restricted conditions, in the case of terminal illness (California), or under any circumstances (Alabama). (See California Natural Death Act, 1976; also Reiser, Dyck, and Curran, 1977.) Others have been critical, contending that such mechanisms will diminish both patient and physician rights (McCormick and Hellegers, 1977).

The events that have come to be known as the Karen Quinlan case have provided an example of the wide range of value and procedural issues occasioned by discussions of letting die or ceasing life-saving treatment. These discussions have included attempts to analyze talk about a "right to life," the standing of proxy consent in the refusal of life-saving treatment, the use of ethics committees, and the meaning of such terms as "ordinary" and "extraordinary care." One should note, in this regard, that there has been an increasing skepticism that the distinctions between ordinary versus extraordinary care means anything other than a contrast between obligatory versus nonobligatory care (Veatch, 1976a). In any event, the concept of extraordinary care (Pope Pius XII, 1958)—i.e., care that one would be excused from giving because of the unlikelihood of success—has received renewed examination.

Discussions of death and dying have not only focused on rights to determine the style of one's dying; they have turned as well on the goods and values that ought to be achieved in dying. These discussions have involved, in part, a revival of the sixteenth-century *ars moriendi* literature (Lipman and Marden, 1966; Bush, 1950; Miller, 1974), as well as more or less psychological and medical discussions of the ways individuals can be helped to die with "dignity" (Saunders, 1976; Craven and Wald, 1975). A great deal of this discussion consists of accounts of psychological processes or stages in adjusting to the fact of one's own impending death, e.g., stages of denial, anger, bargaining, depression, and acceptance (Kübler-Ross, 1969). Some of this work has led to claims about evidence concerning life after death (Moody, 1976). The conceptually oriented literature has attended to values associated with various circumstances and styles of dying—so-called "death with dignity" or "quality of life." These circumstances include dehumanizing situations, e.g., "tubes in every orifice," and various conditions under which sufficient goods and values cannot be realized by patients so as to make their lives worthwhile for them.

This literature presupposes what may be aptly termed an axiology of death and dying. Works in this genre have examined the question of whether death is intrinsically good, evil, or neutral. Paul Ramsey has continued the Judaeo-Christian view that sickness and death are the result of sin (Sifra 27a; I Corinthians 15:21–22; Romans 5:12). Ramsey (1974) argues, for example, that an acceptance of the value of life entails imputing a negative value to death. In contrast, Engelhardt (1975c), Kass (1974), and Morison (1974) have argued that an acceptance of death is integral to an understanding and acceptance of life, and that such an acceptance of death makes a death with dignity not only possible but in fact a good to be pursued. These discussions have included analyses of the naturalness of death and the meaning of talk about a "natural" death (Callahan, 1977). Others have addressed the values implicit in certain styles of dying (MacIntyre, 1978).

Other contemporary writers on death and dying have addressed broad cultural themes. They have asserted that modern disquietudes concerning the circumstances of death derive in part from the denial of death by our modern culture (Becker, 1973). Assertions in this vein have included the observation that there is a general unwillingness to face death, with its unpleasantness and finality; there are attempts to find a defense against death in technological intervention and to suppress mourning. This unwillingness these authors see as a shift in the response to death from the moral to the technological (Ariès, 1974; Illich, 1974).

In sum, discussions concerning euthanasia, the refusal of treatment, living wills, and styles of dying have ranged, on the one hand, from those couched in terms of rights and duties and privileges of individuals choosing the circumstances of their deaths to examinations of the goods and values to be achieved within particular styles of dying, on the other.

As with styles of dying, questions about rights and values have been raised with regard to reproduction. These questions have spurred discussions of rights to reproduce, duties to future generations, purported rights of future generations, and goods and values either to be achieved or that are put in jeopardy through varying programs of eugenics or of genetic counseling. Again, as in the case of the term "death and dying," this rubric gathers together many disparate issues, some of which are postponed for discussion later.

Recent discussions of the rights and duties and the goods and values involved in reproduction descend in part from the eugenics movement of the early twentieth century with its recommendations of forced sterilization and the use of artificial insemination, as well as from the controversies engendered by those recommendations (Holmes, 1976; Hunt and Arras, 1977; Graham, 1978). Those who favor eugenics argue that in the absence of natural selective pressures to eliminate defective members of the species, humans would become progressively dependent on medical interventions and less able to survive on their own. Such policies would lead to a general weakening of the species, with its gene pool carrying an increasing load of defective genes (Muller, 1935, 1961). In recent years it has been common to hold that there is no need to establish positive eugenics programs to improve the human gene pool (Lappé, 1972). Individuals from the fields of philosophy and theology have argued against establishing such programs, contending that they could lead to abuse—the "slippery slope" argument. Moreover, our knowledge of genetics is incomplete, and our vision of the goals worth pursuing through positive eugenics is too poorly developed, to justify such programs. Thus individuals like Paul Ramsey and Harmon Smith (Ramsey, 1970a; H. Smith, 1970) have written against positive eugenics. Others have continued to argue with qualifications in support of such programs (Glass, 1972).

Associated with discussions of positive eugenics are considerations of the moral implications of envisaged scientific technologies that could be used to prevent or treat genetic defects—techniques ranging from embryo transfer (Grossman, 1971; Ciba Foundation, 1973) to questions of genetic engineering (Hull, 1972). Genetic engineering has implications that range from the ability to treat diseases by altering genetic structure to remote, even fanciful, possibilities, such as cloning (James Watson, 1971). The issues are raised primarily because of the impact that such technology would have upon human life and lifestyles. Several writers have offered counsels of caution, given our inadequate knowledge of genetics (D. English, 1974; Hamilton, 1971).

A great deal of the controversy has centered on questions of positive eugenics and in vitro fertilization, especially in the wake of the birth of the first human produced from embryo transfer by Steptoe and Edwards in 1978 (Callahan et al., 1978).

Actual practice, however, has been predominantly concerned with negative eugenics (intervention deliberately aimed at decreasing the frequency of unwanted traits). Negative eugenics has usually taken the form of counseling—of warning prospective parents regarding the risks of producing offspring with particular genetic defects, as well as advice concerning the use of contraception, sterilization, and amniocentesis with abortion to prevent the birth of defective children. Some of the issues in genetic counseling raise common medical-ethical questions, though in their own way. There are, for example, problems of confidentiality: Should one inform a prospective spouse of the other's genetic diseases? Should one inform a couple that they need not worry about more children with a particular defect because the evidence shows that the husband was not the father of the child with the defect? There are problems of free and informed choice: How can one enable parents to understand the significance of risks expressed in percentages? (Lappé, 1971; Bergsma, 1974; Callahan, 1972a, 1973). In addition, there are problems about contraception, sterilization, and amniocentesis with abortion to eliminate defective fetuses, for those procedures take on special value as means of avoiding defective or otherwise undesired offspring.

The possibility of avoiding defective offspring raises the question of whether there is a responsibility to do so. Thus some of the questions arising from genetic counseling are singular. For example, it could be argued that parents have a duty not to burden society with the birth of defective children, if such births can be prevented. In certain cases, one would have a moral obligation to use contraception, sterilization, amniocentesis with abortion, or perhaps artificial insemination by donor (John Fletcher, 1972). The point is that reproduction has been revalued, insofar as having children is no longer simply accepting life—e.g., from God—but is viewed as an enterprise freighted with responsibilities involved in free choice. To allow possibly defective children to go to term could be regarded as blameworthy (Glass, 1972). Of course, such matters as the morality of abortion, artificial insemination by a donor, and the effects upon the nature of parenting are involved in assessing the probity of viewing reproduction as a deliberate, rational act aimed at producing "normal" children (Ramsey, 1971; John Fletcher, 1972).

The issues of reproductive responsibility have been highlighted by

legal discussions of a possible tort case for wrongful life. It has been argued
at law that one can injure another by allowing him or her to be born into an
uncomfortable or painful or markedly deprived existence; that individuals
who are responsible for such births should be liable for making up for that
injury. There are cases in which individuals have sued for having been born
with handicaps (*Williams v. New York State, Zepeda v. Zepeda*; see
Tedeschi, 1966, and Anonymous, 1977). A recent example is the case in
which a child sued physicians for not having warned its parents of the
possibilities of a pregnancy that could lead to its birth with polycystic
kidney disease (*Park v. Chessin,* 1976). These legal arguments are of particular interest, for they suggest the existence of a moral duty to future
children not to allow them to be born under certain circumstances. They
may in fact suggest a moral duty to use abortion to terminate some preg-
nancies where a fetus promises to become a defective child.

At the political level, the duty to avoid reproduction raises questions
concerning interventions by governments, physicians, and others. Until
recently, paternalistic intrusions weighed against curtailing reproductive
abilities. For example, the general practice of physicians and hospitals was
not to allow young individuals or those with few children to be sterilized
on the grounds that individuals might regret such a choice in the future,
e.g., after a divorce and remarriage (McKenzie, 1973). The current counter-
trend is to accent self-determination, if not self-accountability. Thus, one
emphasis in the literature argues for the rights of women to control their
bodies through contraception and abortion (Thomson, 1971). Another
wrinkle in the claim of a right to control one's own body arises in connection with public funding of abortion. This debated topic is obviously tied
to general issues about rights to health care and national health programs,
as recent Supreme Court decisions show (*Maher v. Roe* and *Poelker v. Doe,*
both 1977). It also has special relevance in the present context because objections to genetic screening and abortion place people who are against
those practices in a vexing political position (Annas, 1977b; Steinfels, 1977).

Yet another complex issue has to do with sterilizing retarded persons
who are not deemed competent to consent to the procedures involved.
Social and personal harms and benefits are once again in conflict, involving
paternalism and autonomy (Gaylin et al., 1978).

Finally, issues of genetic screening raise problems of falsely labeling
individuals. Labeling influences opportunities as well as outcomes of studies
of traits; indeed it may cause individuals to be psychologically injured
(Roblin, 1975). For example, one runs the risk of adversely influencing the
ability of individuals to be insured or to be accepted in competitive schools
once they have, through genetic screening, been found to have received a
deleterious trait that will express itself later in life. Even true labeling has
been deemed improper in some accounts. For example, some have argued
that individuals who have received a gene that will express itself later in

severe morbidity and early death, should not be told, even though such ignorance will mean that these individuals may reproduce and pass that trait on to their children (Hemphill, 1972).

One cannot review these issues without seeing that taking a position on them affects the patient-physician relationship. Although the tendency to argue for increasing the scope of patient responsibility and choice under the concept of informed consent is most vocally held by such nonphysicians as philosophers and lawyers, there is also a countertendency for some physicians to be conservative about the matter (Cassell, 1971). Although within the bounds of an agreed-to paternalism one could still talk of "doctor's orders," patients have become increasingly interested in establishing treatment plans with their physicians. Therapeutic actions undertaken by physicians, nurses, and other health professionals without patient consent have become increasingly unacceptable (Annas, 1975). In part this rejection of paternalism has reflected increased appreciation of patients' rights; in part it reflects acknowledgment of the plurality in our culture—acknowledgment that there no longer exists a common view of the goods and values of life. Patients are thus increasingly forced to act as their own agents (MacIntyre, 1977b). Despite professional reluctance (Pellegrino, 1977; Hellegers, 1977), such views of patient autonomy are receiving greater acceptance by the medical profession and are part of the redefinition of the relationships among physicians, patients, and other health practitioners. These questions have also been addressed in a growing literature on paternalism in medicine (Buchanan, 1978; Dworkin, 1972; Ellin, 1978; Gert and Culver, 1976; Graber, 1978).

In this context, one might note the evolution of special professional groups that have as their task the imposition of general community judgments (as through Professional Standard Review Organizations) as a result of the partial "socializing" of health care. Medicine has come to be seen as a national resource. Its research, development, and educational institutions are in large proportion supported by public monies and have thus come to be subject to direction according to socially chosen goals. In consequence, the patient has become subject to the constraints of these general social goals—e.g., in terms of the utilization of hospital beds. The background of these more general social issues will be discussed in section C., below.

B. Definitions of "Person" and "Human"; Some Ontological Problems in Medical Ethics

One result of applying the methods of analysis to medical ethics has been a rekindling of interest in basic ontological problems within the context of bioethics. In particular, there are questions about when the life of persons begins (e.g., are fetuses persons?) and when it ends (definitions of death). A conceptual distinction has been drawn between the terms "human" and

"person." Various definitions of "person" have been offered as the literature has developed, and increasing care has been given to distinguishing
between being alive, being a human, and being a person. However, confusion still remains; articles are published in which these distinctions are
not clearly marked (e.g., Joseph Fletcher, 1972, 1974b).

The distinctions may be drawn thus: being a human is being a member
of a particular zoological genus or species (i.e., either the genus *Homo,* or
a particular species, *Homo sapiens*). "Human" contrasts with "canine,"
"equine," "porcine," and "simian." In this view, being a person is being a
moral agent, an entity worthy of special regard as a member of the community of morally responsible entities. Moreover, in addressing the area of
animal research and treatment (see McCullough and Morris, 1978), there
are those (Singer, 1975; Clark, 1977) who wish to distinguish between "person" as a responsible agent and "person" as a subject whom we should not
injure or exploit. An antivivisectionist position emerges if animals are
thought to have standing as subjects of experience, and such a view would
have especially strong support if a sort of rationality could be discovered in
animals. In any event, it may not be the case that all persons are humans
(consider angels, gods, and possibly some extraterrestials who would be
persons in the sense of being moral agents), or that all humans are persons
(a brain-dead but otherwise alive human body is no longer a clear instance
of a person since his or her agency is gone).

An implication of these distinctions is that many serious moral problems—e.g., about the use of fetuses or brain-dead bodies—dissolve if we no
longer count certain humans as persons. They are not accorded the strict
protections we normally give to persons, as the development of laws for the
donation of organs shows (Capron and Kass, 1972).

Less sharp distinctions have been made that favor women over the
fetuses they bear in cases of life-and-death choices (Ramsey, 1975). The
ambivalence and ambiguity concerning the status of the fetus is reflected
in the papers presented to the National Commission for the Protection of
Human Subjects concerning fetal research (1975). The recommendations of
the commission, while apparently recognizing the existence of abortion—
which denies the standing of the fetus as a person—still accord the fetus
special protection as a human subject.

There is a considerable literature concerning the standing of fetuses
(Engelhardt, 1974c; Feinberg, 1973; Tooley, 1972; J. English, 1975). Moreover, the standing of fetuses has implications beyond their use for research,
as authors dealing with these issues have often recognized. Tooley (1972), for
example, has seen that his arguments would support infanticide while
possibly according the status of persons to higher primates. Engelhardt
(1974c) on the other hand has argued for more than one concept of person:
a strict sense to apply to self-conscious rational agents, and a social one that

imputes to nonpersonal instances of human life the standing of a person for teleological reasons.

Distinctions between humans and persons have been pursued in order that the arguments concerning abortion may be resolved in a clearer fashion (Benn, 1973). If fetuses do not count as persons, they do not require the strict safeguards of rights and interests that we usually accord to persons. (One should note that these discussions have analogues in a theological tradition that has asked when the soul enters the fetus in order to make it a person; see de Dorlodot, 1952; Donceel, 1967, 1970; Gerber, 1966.) On the other hand, if one has a firm belief that fetuses are persons, one may then feel morally obligated not to allow abortions except under very special circumstances (B. Brody, 1975).

Those who hold that fetuses are persons, or that drawing a line between being a human and being a person fails, have developed more traditional arguments concerning the weighing of conflicting claims or mutually exclusive values. An example of this genre is Judith Jarvis Thomson's (1971) defense of abortion; she does not take issue with the fetus being a person, but holds that even if it be a person, the fetus still has no strong right to the use of the mother in pregnancy. She develops the standard of the "minimally decent Samaritan," from which it would follow that many women in many circumstances would have grounds to refuse to carry a fetus to term even if fetuses are persons. Susan Nicholson (1978) has also suggested that abortion should not be described as an assault upon the fetus, but as a refusal, that is justified under some circumstances, to give assistance.

Fetal experimentation (including in vitro fertilization and embryo transfer) is problematic in many of the ways that abortion is, and for the same reasons. If fetuses are persons, then it follows that there is a *prima facie* duty not to do injurious research upon them. However, if fetuses are not persons and abortion on request is moral, it is not clear why and to what extent fetal experimentation should be forbidden. Those who would hold fetuses to be persons would clearly have strong motives against risk-laden fetal experimentation (Ramsey, 1975). There is the additional consideration that even if fetuses are not persons, women who subject their fetuses to dangerous experimentation prior to abortion may then decide not to procure an abortion. In the case of in vitro fertilization and embryo transfer, there is the special problem of causing a child to be born injured as a result of the experimental process—a process that does not include consent (Kass, 1971a). One would then be in a position of either allowing a future person to be injured or forcing the mother to submit to an abortion. In addition, others have been fearful of the impact on our sensitivities of allowing experimentation on fetuses, even if they are not persons. These sorts of arguments led to restricting nontherapeutic research on fetuses attended by significant risk, and to treating fetuses destined for abortions the same as those intended to be taken to term (National Commission, 1975). The position of the commission is at least in tension with an acceptance of the moral permissibility of abortion on request, and leaves untouched important issues about the standing of the fetus as a person.

Special difficulties arise in the case of instances of human life in which some but not all characteristics of personhood are present, i.e., very young children, the mentally retarded, and the senile. These problems for the most part have centered on issues of consent. For example, consent is problematic in cases of stopping life-saving treatment for the incompetent, especially for defective newborns (Duff and Campbell, 1973; McCormick, 1974b; Robertson and Fost, 1976). Who should give consent when proxy consent must be given? Who has a right to consent for the incompetent, on what basis, and regarding what procedures? As philosophers have indicated, it is difficult to construe most instances as truly proxy consent in the sense that one competent individual gives another a proxy to choose on his or her behalf (Abrams, 1977). One cannot be free in acting on behalf of a small child or a mentally retarded individual; one is constrained to act in his or her best interests. If the incompetent are to be treated as persons, as normal adults are, it is not clear that one is choosing in their best interests or as they would have chosen. As a debate between Freedman (1978) and McCormick (1978) indicates, it is not clear how one should understand the tests for adequate proxy consent. There may in addition be a duty to avoid forms of treatment that will do more injury than benefit to the incompetent (Engelhardt, 1975d).

Similar status questions arise with respect to nontherapeutic experimentation upon children. Some, who hold children to be persons in the same sense that normal adults are, have argued that all nontherapeutic research accompanied by pain, risk, injury, or bodily invasion is precluded because the child or incompetent has not consented to the risk. In the case of therapeutic research, where there is a positive balance of benefit over risk, it has been assumed that the incompetent would have consented (Ramsey, 1970b). Others have argued that children and other incompetents can be subjected to nontherapeutic research that has only minimal risk, because this is socially useful and is done out of recognition of what a person should have consented to as a part of a normal social duty (McCormick, 1974b). However, if incompetents are protected not because they are persons in a strict sense but out of concern for the defenseless, one may have grounds for certain nontherapeutic research with minimal risk (Engelhardt, 1978).

In all this discussion there is obviously a problem of attending to the various levels of competency in children and the mentally infirm (National Commission, 1977 and 1978). The National Commission, for example, suggests that the consent of children to nontherapeutic research be required as soon as they are able to give it. It extends this principle to require consent even for therapeutic research on older children. The problem of discerning levels of competency to refuse treatment or to assume risks in the case of teenagers is obviously vexing (Schowalter et al., 1973).

We have summarized discussions of abortion, the status of the fetus, children, and incompetents on the basis of general remarks about persons and humans. The same issues arise with regard to the definition of death. Disputes tend to center about either conceptual or operational definitions of death. The former asks for the meaning of "death"; the latter asks what are the most accurate and useful measures for testing when the conceptual definition applies to a particular case (Kass, 1971b).

Grounds for choosing a particular conceptual definition are a concern: Are the grounds discovered, or invented in order to maximize utility (Morison, 1971)? Another issue is the number of false positives (being declared dead though still alive) or false negatives (being declared alive though dead) that attach to each approach. Each kind of false finding has its cost. Since one is likely to be very cautious with regard to false positives, one may be inclined to err on the side of conservative operational definitions of death. Thus, even if one held a conceptual definition that to be a person in the world requires a minimal amount of sentience and interaction with the environment—so that this or that person should be dead—one may not wish to declare the person dead in the absence of a reliable operational definition that will not lead to an unacceptable incidence of false positives if generally applied. For example, even if one believed Karen Quinlan to have been dead at the time of the judicial considerations of her status, the lack of a reliable operational definition for certifying that judgment—out of fear for false positives—would lead one to continue to consider her alive. Veatch's *Death, Dying, and the Biological Revolution* (1976a) is very sophisticated in its treatment of these and related distinctions.

The recent discussion of the definition of death appears to turn on what amount of brain structure is requisite to being alive in the world and what tests for brain function will lead to few or no false positives (Ad Hoc Committee, 1968). This issue has led to changes in legal definitions of death throughout the U.S. (Capron and Kass, 1972), from whole-body to whole-brain criteria. It has led also to proposals that death of the neocortex be considered the basis for holding that a person is dead (Veatch, 1975; van Till, 1976). There has, in short, been an increased willingness to associate the life or death of a person with the presence of the preconditions for a minimal level of consciousness—though this trend is far from uniform (P. Black, 1978).

By forcing these issues concerning the criteria for personhood, medical ethics has contributed in important ways to theoretical reflections on the meaning of being a person. Bioethical attention to particular problems is thus shown to involve basic conceptual issues. Indeed, the issues can be tied to classic metaphysical debates—e.g., What is the soul? When is it created?—and can thus be seen to reflect questions of central and enduring philosophical interest.

C. Distributing Biomedical Goods and Harms

Discussion of the proper distribution of health care goods and values, and of harms and benefits, has played a central role in discussions of bioethics. These discussions embrace not only issues bearing on national health care, but on the distribution of general risks such as those attendant to recombinant DNA research and the population explosion. These discussions can be organized around the following issues: (1) the problem of the nature of the entitlement: Who owns which goods (e.g., health care resources, a society's monies, the services of health care professionals)?; (2) the moral significance of the results of the natural lottery (Does the fact that some individuals are born diseased, unhealthy, poor and in need of treatment impose an obligation upon the healthy and wealthy to treat them? In short, how does one draw a line between what is unfortunate and what is unjust?); (3) the extent to which one should pursue health care goods and values as opposed to other goods and values; and (4) the problem of obtaining consent from all the individuals involved in the distribution of benefits and harms.

The various positions taken regarding health care can be roughly characterized as of three kinds. The first, a contractarian model, holds that there are strong entitlements to goods—i.e., ownership implies vast liberty of control over one's possessions—and that there is not a general moral duty to make reparation for invidious outcomes of the natural lottery. Thus, if patients have the money to purchase services, and physicians have the inclination and time to sell those services, patients may purchase them (Sade, 1971). This model can be aptly described as the laissez-faire, free-market model.

The second model involves a two-tiered system of health care. One tier is seen as providing a decent minimum, with the second tier providing extra care on a more or less free market basis (Fried, 1976). In discussing the distinction between the two tiers, it becomes important to separate rights *to* health care from rights *in* health care—e.g., rights to consent to treatment, to courteous care, etc. (Fried, 1975). Such discussions have presupposed some duty of beneficence—that there is a moral duty to palliate or cure the injuries individuals receive through the natural lottery, to meet the needs of the ill (Outka, 1974; Kass, 1975). Such a view also requires some modified notion of entitlement. It may be as benign as the assertion that because health care education is subvened through public funds, therefore health care professionals have general duties to society at large. It may be more radical and assume a theory of entitlement such as John Rawls's

(1971) in which goods and values are owned only insofar as they conform to a just pattern. Adaptations of Rawls's theory, including health care among the primary social goods to be distributed in a society, have been proposed (Ronald Green, 1976). Under such a view, a two-tiered system would be countenanced as long as it redounded to the benefit of the least well-off class. However, there are basic problems in specifying the least well-off class (Nozick, 1974). How can any system of health care redound to the benefit of the truly terribly off? Another problem with the use of the concept "justice" in this context can be portrayed by the question: If one is going to try to blunt the natural lottery, should one not also distribute, in addition to services and goods, paired organs from those who have two to those who have none?

Any account that justifies the distribution of goods and services in terms of some desired good and just pattern must also give an account of why health care services are to be preferred over spending those resources on other goods—e.g., formal parks and art museums. In contrast, other accounts can accommodate the preferential pursuit of particular goods on the basis of individual choice. In fact, the former (called "end-state" accounts) require a means of distinguishing health care needs (the bases of claims to beneficence) and health care desires (wants that would not make duties of beneficence).

Attempts to make such discernments have been associated with attempts to distinguish between goods and services that are properly medical and those that are not (Kass, 1975). End-state justifications of personal health care distributions must meet further criticisms that individual health care is not as important in realizing reduced levels of morbidity and mortality as are public health measures (Dubos, 1965). Public health care allows more easily of equal distribution and is more cost-effective. Moreover, distributions based on end-state accounts may fail to take adequate account of individual responsibility in maintaining health and avoiding or preventing illness. Many difficulties with the two-tier view may be overcome by such plans as those suggested by Enthoven (1978), which aim at blending consumer choice, based upon individual judgments of the economic and health values involved in joining one of several available health service programs, with the underwriting of the cost to consumers.

The third position on access to health care is one of equal distribution. The point of those who subscribe to this position appears to be that differences in health care are too disrupting or injurious to be tolerated. Aside from the justification of such a scheme through a general contractarian theory (all individuals involved actually freely consent), one would again have to appeal to some sort of end-state or end-result account and would need to answer the questions that such accounts raise. Such end-state accounts may be either utilitarian or some adaptation of a Rawlsian view. A utilitarian might hold that the pain that would be felt throughout the moral order from knowing of the differences in distributions would be sufficient to warrant blunting those differences so as to achieve the happiness of the greatest number. Again, such accounts will have to presuppose that there is no strict entitlement to goods and/or that there is a strong duty to beneficence to the greatest number (Singer, 1972). A well-developed account of an egalitarian health care distribution system has been provided by Robert Veatch (1976b). Such theories are problematic if they are intended to forward an egalitarian distribution of health care resources, even if a two-tiered system would redound to the benefit of the least well-off class in both general goods and medical goods. As Nozick's (1974) critique suggests, such a theory requires envy as part of the characterization of the rational individual, and jealousy as a principle in the distribution of goods and services—i.e., one would rather have oneself or others be worse off than have yet others be better off.

These problems recur when one attempts to understand the problems of microallocation. Resolutions will turn in part on who owns what and what obligations of beneficence are incumbent upon the individuals concerned. The latter will, in turn, depend on the moral significance of the natural lottery as well as prior agreements among the parties concerned. As a consequence, under the rubric of microallocations, there have been extensive debates of what should count as proper procedures for distributing scarce medical resources when there are not sufficient resources to save the lives of all involved. Those, such as utilitarians, who have favored particular end-state distributions of resources have suggested utility-maximizing means of distributing health care resources that favor saving the most socially useful individuals (Rescher, 1969). In contrast, others have given special weight to the distress that individuals might feel over a nonrandom distribution and have favored a chance allocation (Gorovitz, 1966). There have been discussions as well of the advantages of altruism in the distribution of health care resources (Singer, 1973). Others have stressed particular claims that individuals may have upon others; such claims are often intended to be irreducible to interests in goods and values, as occurs in utilitarian schemes such as Rescher's (Childress, 1970). There have also been a number of criticisms of utility-maximizing systems of distributive health care services (e.g., those that give priority to socially contributing members of a community) as not acknowledging the standing of patients as persons (again see Childress, 1970). However, it is not clear whether such criticisms apply, if all participating have consented in person or through a democratic means.

These topics have also surfaced in discussions of the world population explosion. In particular, similar issues recur with regard to entitlements and to duties to be beneficent. There are also questions of whether one has a right to protect oneself and one's resources by forcing the sterilization of others as part of general social policies. As a result, a literature has developed on the subject of the ethics of population control (Berelson, 1969; Callahan, 1972b, 1973; Hardin, 1974). These discussions compare the relative values of freedom, justice, and survival in developing programs of population control; they have also weighed the implications of positive versus negative incentives, raising the issue of when physical coercion (e.g., forced sterilization) is justified. A popular designation for this discussion has been "lifeboat ethics" and has included the question of the extent to which duties of beneficence may be restricted to one's own locality—e.g., one's own nation, or one's time—thus raising the question of the rights and standing of future persons and of future generations (Lucas et al., 1976).

Considerations of allocation of goods have raised questions about allocation of risks, as for example in the recombinant DNA debate, where risks may be worldwide and harms may be catastrophic. For example, what weight in moral judgments should very unlikely but potentially catastrophic events have? Or how may consent be sought from the world at large? Does taking informed consent seriously involve the right to refuse to be a part of risky research endeavors in an environmental sense—if, as some protagonists argue, recombinant DNA research may place the world at a risk of a harm from which no one may be excluded as a conscientious objector (National Academy of Sciences, 1977)?

One response has been the development of regulations for engaging in such research as well as the beginning of a debate concerning its significance, especially its bearing upon freedom of scientific inquiry (*Recombinant DNA Research,* vols. 1 and 2, 1976, 1978; Medical Research Council: *Guidelines,* 1977). It should be clear that these issues are tied at least in part to discussions of environmental ethics generally (Potter, 1971; see also chapter 5, II. B. 6, in this volume).

In short, the problems of the micro- and macroallocation of biomedical goods and risks involve theories of entitlement to goods, duties of beneficence, the significance of the natural lottery, and the feasibility of gaining consent (out of respect for autonomy) for distribution programs of goods and exposure to harms. This latter issue is of cardinal significance, for on it hinges the status of communities, of relationships of individuals to their communities, and the standing of entitlements within communities. These broad issues in political philosophy and the theory of justice form the background of these debates.

D. Special Perspectives

Many of the important reflections on bioethics have been drawn from particular cultural or religious perspectives. Indeed, the discussion of works by thinkers within particular traditions, in the above sections, testifies to their important general contribution. In this section we will not repeat our discussions of those works, but only suggest some of their more special implications. Some scholars reflecting within particular religious traditions have drawn from general moral theories and have produced philosophical reflections on the significance of those theories. Others have attempted to work out principles within the strict confines of their own traditions. Even so, the results have rarely remained parochial. In attempting to develop fine and precious distinctions, religious thinkers have often put forward distinctions of general interest—one might think here, for example, of the doctrine of double effect, which has received sustained discussion beyond the confines of the Roman Catholic Church (Foot, 1967; McCormick and Ramsey, 1978). And finally there have been traditions that, though they are not religious in a theological or traditional sense, operate within a particular orthodoxy. An example here is the discussion of medical ethics as conducted in communist countries.

The religious discussions that hinge on prohibitions of actions by individual religious sects based simply on claims of divine revelation have tended to be least involved in general philosophical debates. Instead, they have focused on interpreting what the deity did or did not wish to forbid as evidenced in Scripture or by religious tradition. Proscription of artificial insemination by a donor in the Judeo-Christian tradition on the grounds that it is a form of adultery even when all parties consent is often presented in such a fashion. So, too, are proscriptions of euthanasia, suicide, or abortion that are founded simply on particular scriptural readings. Here also one could place proscriptions of transfusions by Jehovah's Witnesses (see *Blood, Medicine, and the Law of God*, 1961). One must note, however, that even when bioethical reflections have been framed within the confines of a particular tradition, these reflections have often produced works that not only pay close attention to argument and exegesis, but also achieve success in illuminating issues of general human concern. One must surely place here volumes by bioethicists working in the Jewish tradition (Feldman, 1968; Jakobovits, 1959).

Because of its roots in Greco-Roman philosophical traditions, the Roman Catholic Church has produced a series of reflections on bioethical issues meant as general appeals to reason concerning proscriptions based on natural law. Such discussions of bioethical issues have been derived from a general set of premises concerning the ability to discern ingredient purposes in nature, together with a notion that such purposes are good and that in fact it is obligatory not to thwart them (Kelly, 1958; Kenny, 1962). These discussions have led to proscriptions of artificial contraception, as well as of artificial insemination and sterility tests that require masturbation or coitus interruptus to procure sperm. The position is that in each of these instances the procreative act is being thwarted from its natural goal of reproduction. Some very precise distinctions have been drawn within this tradition. For example, it has been held that a sperm sample for sterility tests may be procured by using a condom with a very small perforation, allowing only the escape of a small amount of sperm, but with the retention of a sufficient amount for the sterility test (Vermeersch, 1921). Procedures like sterilization are also judged impermissible, not only because they thwart reproductive goals, but also because they involve mutilation, and mutilation is held to be immoral by reference to a principle of totality. Further, not only are sterilizations seen to be morally problematic, but also the transplantation of normal organs from a healthy living donor. Even incidental appendectomies—removing a healthy appendix to prevent the possible future need of an appendectomy while performing some other operation—are morally suspect. Although many of these arguments have fallen into disuse, they are of interest in claiming to be based on general principles of reason. In any event, they produced a large body of reflections that over the centuries have had an immense impact on what we count as natural or unnatural, moral or immoral, acts and interventions (Noonan, 1965). In particular, it is worth noting that this Roman Catholic literature has had important reciprocal relations with the work of non-Catholic thinkers as well.

Religious traditions in general have given a special cast to the moral sense of many bioethicists. As James Gustafson indicates in his volume *The Contributions of Theology to Medical Ethics* (1975), a religious perspective may strengthen respect for life and give a basis for self-criticism and caution. Religious sentiments have been seen as guards against hubris in reflections on where to set limits in the use of biomedical technology (Augenstein, 1969; Vaux, 1977). In fact, some of the most influential works in bioethics have drawn heavily upon religious views of human nature and the human condition (e.g., Ramsey, 1970b). Such borrowings have, however, met with criticism based upon skepticism as to whether religious perspectives are particularly useful or necessary, or the extent to which they can be put forward in anything but an arbitrary fashion (MacIntyre, 1977a; Delkeskamp, 1977; Ramsey, 1977). As MacIntyre indicates, the status of religious perspectives is difficult to distinguish from the status of ideologies generally.

In this regard, one must note that along with religious inspiration for views of proper conduct in medicine, there is a considerable tradition drawn from Marxism. This literature has included explorations of the way one should treat persons in a socialist society. It has stressed dedication to the community and the support of those virtues of character that allow one to attend to the interests of the sick in a general social context (Ehmann and Löther, 1975). It has also provided critical appraisals of Western medicine (Zechmeister, 1972). But it has also addressed problems of biomedical advances in ways very close to those in the Western literature (Geissler et al., 1974).

In short, discussions of bioethical issues within particular traditions tend inexorably to lead back to general concerns and thus to general philosophical and ethical reflections.

In summarizing the high points of this tour through bioethical issues, we should stress how very fast the literature in the field is growing. Attention to questions about privacy and confidentiality, research involving humans (and use of animals), informed consent, justice and distribution of goods and risks, the right to die, active and passive euthanasia, the definitions of "person" and "death," the status of abortion, sterilization, genetics and human reproduction generally, physician paternalism, and patient autonomy—all extremely "public" issues—are at the center of current discussions.

There are areas of research and treatment that appear to be problematic almost exclusively because of cultural attitudes, often religiously based. One should think here of the moral issues raised with respect to research and therapy for sexual dysfunction. Aside from the usual problem of free and informed consent or the avoidance of coercion and duress, there should be no more difficulty in doing research in reproductive physiology than in respiratory physiology. At least, there has been a considerable difficulty in specifying what the differences might be. On ethics for sex researchers and therapists, see Masters, Johnson, and Kolodny (1977, 1980). Important areas that have received less—but growing—attention include nursing ethics and ethics for allied health fields (Spicker and Gadow, 1980; Purtilo, 1978a).

No doubt, as our society evolves, both what we include in the definition of "health" and our expanding technology will teach us what should have been discussed in the literature by now and will also uncover new, even unheard of, realms for further ethical inquiry.

Acknowledgements

Many individuals have contributed generously with suggestions and criticisms concerning this essay on the philosophy of medicine. Though they are surely not responsible for its shortcomings, we are grateful to them for having provided many of its virtues. Those individuals include Jane Backlund, Deanna Barousse, James Childress, Sara Ehrman, Susan Engelhardt, Eric Juengst, John Moskop, Ruth W. Moskop, Warren Reich, LeRoy Walters, Pamela Waltrip, and Stephen Wear.

BIBLIOGRAPHIC INTRODUCTION

Our contribution to this *Guide* has dealt with fields that, though tied together now in very important ways, seem disparate. Although there are reasons to challenge the canonical fact-value distinction, that distinction has played a role in the structure of this chapter. We first addressed the more factual aspects of medicine—issues associated with the philosophy of science, clinical judgment and the language of medicine, epistemology and metaphysics (specifically mind-body problems)—then turned to the value aspects of medicine in biomedical ethics.

There are philosophical reasons for questioning the separation of these dimensions, even if they can be conceptually distinguished: For example, the definition of the term "health" seems to appeal as much to matters of fact as it does to matters of value. Or again, the definition of "health" draws on social and individual values and so presupposes value judgments as much as it does judgments of fact. Evaluation and explanation intertwine.

Our bibliography, accordingly, is not topical. However, in indicating what might be taken to be especially helpful works in the field, we will make suggestions on both sides of the putative divide.

A. Basic Sources

Because the philosophy of medicine is so rapidly emerging, long-standing or even generally acknowledged classics in the field are not readily identified.

1. Some volumes (though not all strictly philosophical works) that are useful in approaching epistemological issues in the *philosophy of medicine* are:

Buchanan, Scott. *The Doctrine of Signatures*. London: K. Paul, Trench, Trubner, 1938.

Cannon, W. B. *The Wisdom of the Body*. New York: W. W. Norton, 1939.

Elstein, Arthur S.; Shulman, Lee S.; and Sprafka, Sarah A. *Medical Problem Solving*. Cambridge, Mass.: Harvard University Press, 1978.

Engelhardt, H. T., Jr.; Spicker, Stuart F.; and Towers, Bernard, eds. *Clinical Judgment*. Dordrecht: D. Reidel, 1979.

Feinstein, Alvan. *Clinical Judgment*. Baltimore: Williams and Wilkins, 1967.

King, Lester. *The Philosophy of Medicine: The Early Eighteenth Century*. Cambridge, Mass.: Harvard University Press, 1978.

Leavitt, Judith Walzer, and Numbers, Ronald L. *Sickness and Health in America*. Madison: University of Wisconsin Press, 1978.

Wulff, Henrik. *Rational Diagnosis and Treatment*. Oxford: Blackwell Scientific, 1976.

2. For *bioethics:*

Beauchamp, Tom L., and Childress, James F. *Principles of Biomedical Ethics*. New York: Oxford University Press, 1979.

Beauchamp, Tom L., and Walters, LeRoy, eds. *Contemporary Issues in Bioethics*. Encino and Belmont, Calif.: Dickenson, 1978.

Beauchamp, Tom L., and Perlin, Seymour, eds. *Ethical Issues in Death and Dying*. Englewood Cliffs, N.J.: Prentice-Hall, 1978.

Burns, Chester, ed. *Legacies in Ethics and Medicine*. New York: Science History Publications, 1977.

Burns, Chester, ed. *Legacies in Law and Medicine*. New York: Science History Publications, 1977.

Gorovitz, Samuel, et al., eds. *Moral Problems in Medicine*. Englewood Cliffs, N.J.: Prentice-Hall, 1976.

Katz, Jay, ed. *Experimentation with Human Beings*. New York: Russell Sage Foundation, 1972.

Ladd, John, ed. *Ethical Issues Relating to Life and Death*. New York: Oxford University Press, 1979.

Ramsey, Paul. *Ethics at the Edge of Life*. New Haven: Yale University Press, 1978.

Ramsey, Paul. *Patient as Person*. New Haven: Yale University Press, 1970.

Reiser, Stanley J.; Dyck, Arthur J.; and Curran, William J., eds. *Ethics in Medicine*. Cambridge, Mass.: M.I.T. Press, 1977.

Veatch, Robert M. *Case Studies in Medical Ethics*. Cambridge, Mass.: Harvard University Press, 1977.

Veatch, Robert M. *Death, Dying, and the Biological Revolution*. New Haven: Yale University Press, 1976.

Veatch, Robert M., and Branson, Roy, eds. *Ethics and Health Policy*. Cambridge, Mass.: Ballinger, 1976.

However, this list barely scratches the surface, not merely for the obvious and common reason that there is so much to this field, but also because the field is still finding itself.

B. Reference Sources

The interested reader should seek further reading and areas of research from the following sources:

Reich, Warren T., ed., *Encyclopedia of Bioethics*. New York: Macmillan and Free Press, 1978.

Walters, LeRoy, ed. *Bibliography of Bioethics*, vols. 1–5. Detroit, Michigan: Gale, 1975–1979.

Bioethics Digest. A useful monthly annotated bibliography that began in 1975 and ceased publication in May 1978.

The Hastings Center Reports (Institute of Society Ethics, and the Life Sciences) include annual bibliographies.

Medical Ethics Film Review Project. College Park, Maryland: The Council for Philosophical Studies, 1974.

C. Journals and Series

Many medical journals, such as the *New England Journal of Medicine* and the *Journal of the American Medical Association,* frequently include important source material. The journals in the field include: *Bioethics; Ethics and Science in Medicine; The Hastings Center Reports; Journal of Medicine and Philosophy; Man and Medicine;* and *Perspectives in Biology and Medicine.*

Two series that should prove useful are listed in the bibliography under "Engelhardt and Spicker" or "Spicker and Engelhardt," and under "Engelhardt and Callahan."

For the teaching of bioethics, see:

Self, Donnie, ed. *The Role of the Humanities in Medical Education.* Norfolk, Virginia: Teagle and Little, 1978.

The Teaching of Bioethics. Report of the Commission on the Teaching of Bioethics. Hastings-on-Hudson, N.Y.: Institute of Society, Ethics, and the Life Sciences, 1976.

There is also the possibility of quick acquisition of information concerning recent publications bearing on issues in bioethics through BIO-ETHICSLINE, a computerized cross-disciplinary retrieval service of the National Library of Medicine available wherever one has a MEDLINE terminal.

BIBLIOGRAPHY

Note: We have prefaced bibliographical entries with a letter or letters to indicate whether they deal primarily with issues in the theory of knowledge (signaled with a "K"), in ethics (signaled with an "E"), or are of an historical nature (signaled with an "H").

K   Abercrombie, M. L. Johnson. *The Anatomy of Judgment.* New York: Basic Books, 1960.

E   Abrams, Natalie, "Medical Experimentation: The Consent of Prisoners and Children." In *Philosophical Medical Ethics: Its Nature and Significance,* pp. 111–124. Edited by Stuart F. Spicker and H. T. Engelhardt, Jr. Dordrecht: Reidel, 1977.

E   Addison, Philip H. "Kidney Donation and the Law." *British Medical Journal* 3 (August 18, 1973): 409.

K   Ad Hoc Committee of the Harvard Medical School to Examine the Definition of Brain Death. "A Definition of Irreversible Coma." *Journal of the American Medical Association* 205 (August 5, 1968): 337–340.

K   Agassi, Joseph. "Causality and Medicine." *The Journal of Medicine and Philosophy* 1 (1976): 301–317.

K   Albee, George W. "Emerging Concepts of Mental Illness and Models of Treatment: The Psychological Point of View." *American Journal of Psychiatry* 125 (January 1969): 870–876.

K   Albert, Daniel A. "Confirmation of a Diagnosis: A Decision-Theoretic Approach." *Metamed* 1 (1977): 213–225.

K   Alexander, Peter. "Symposium: Cause and Cure in Psychotherapy."

Aristotelian Society Proceedings, supplementary volume no. 29 (1964): 25–42.

E   American Medical Association. *Code of Medical Ethics.* New York: H. Ludwig, 1848. Reprint ed., New York: William Wood, 1877. Reprinted in the *Encyclopedia of Bioethics,* vol. 4, pp. 1738–1746. Edited by Warren Reich. New York: Macmillan and Free Press, 1978.

E   ———. *Principles of Medical Ethics.* Chicago: American Medical Association, 1957.

K   ———. *Current Medical Information and Terminology.* 4th ed. Edited by Burgess L. Gordon. Chicago: American Medical Association, 1971.

E   American Psychological Association. "Ethical Standards of Psychologists." *American Psychologist* 18 (January 1963): 56–60.

E   ———. *Ethical Principles in the Conduct of Research with Human Participants.* Washington, D.C.: American Psychological Association, 1973.

E   Annas, George J. *The Rights of Hospital Patients.* New York: Avon, 1975.

E   ———. "Allocation of Artificial Hearts in the Year 2002." *American Journal of Law and Medicine* 3 (Spring 1977a): 59–76.

E   ———. "Let Them Eat Cake." *Hastings Center Report* 7 (August 1977b): 8–9.

E   Anonymous. "A Case of Action for 'Wrongful Life.'" *Minnesota Law Review* 55 (1977): 58–81.

E   ———. *In the Matter of Karen Quinlan.* 2 vols. Arlington, Va.: University Publications of America, 1975.

E   Anscombe, G. E. M. *Intention.* 2d ed. Ithaca, N.Y.: Cornell University Press, 1969.

E   Ariès, Philippe. "Death Inside Out." In *Death Inside Out,* pp. 9–24. Edited by Peter Steinfels and Robert M. Veatch. New York: Harper & Row, 1974.

E   Aroskar, Mila, and Veatch, Robert M. "Ethics Teaching in Nursing Schools." *Hastings Center Report* 7 (August 1977): 23–26.

E   Augenstein, Leroy. *Come Let Us Play God.* New York: Harper & Row, 1969.

HK Bain, Alexander. *The Senses and the Intellect.* London: J. W. Parker, 1855.

HK ———. *Mind and Body.* New York: D. Appleton, 1873.

E   Bandman, E. L., and Bandman, B. *Bioethics and Human Rights: A Reader for Health Professionals.* Boston: Little, Brown, 1978.

K   Barnes, Allan. "Is Menopause a Disease?" *Consultant* 2 (June, 1962): 22–24.

HK   Bartlett, Elisha. *Essay on the Philosophy of Medical Science*. Philadelphia: Lea and Blanchard, 1844.

E   Bayles, M., and Caplan, A. "Medical Fallibility and Malpractice."
Journal of Medicine and Philosophy 3 (September 1978): 169–
186.

E   Bayles, M., and High, D. M. *Medical Treatment of the Dying: Moral
Issues*. Cambridge, Mass.: Schenkman, 1978.

E   Beauchamp, Tom L. "An Analysis of Hume's Essay 'On Suicide.'"
The Review of Metaphysics 30 (September 1976): 73–95.

E   Beauchamp, Tom L., and Childress, James F. *Principles of Biomedical Ethics*. New York: Oxford University Press, 1979.

E   Beauchamp, Tom L., and Walters, LeRoy, eds. *Contemporary Issues
in Bioethics*. Encino and Belmont, Calif.: Dickenson, 1978.

E   Becker, Ernest. *The Denial of Death*. New York: Free Press, 1973.

HK   Bell, Charles. *Essays on the Anatomy and Philosophy of Expression*.
2d ed., London, 1824.

E   Benn, S. I. "Abortion, Infanticide, and Respect for Persons." In *The
Problem of Abortion*, pp. 92–104. Edited by Joel Feinberg.
Belmont, Calif.: Wadsworth, 1973.

E   Bennett, Jonathan. "Whatever the Consequences." *Analysis* 26 (January 1966): 83–102.

HK   Benton, Arthur. "Historical Development of the Concept of Hemisphere Cerebral Dominance." In *Philosophical Dimensions of
the Neuro-Medical Sciences*, pp. 35–57. Edited by S. F. Spicker
and H. T. Engelhardt, Jr. Dordrecht: D. Reidel, 1976.

E   Berelson, Bernard. "Beyond Family Planning." *Science* 163 (7 February 1969): 533–543.

K   Berger, P. L., and Luckmann, Thomas. *The Social Construction of
Reality: A Treatise in the Sociology of Knowledge*. Garden City,
N.Y.: Doubleday, 1966.

E   Bergsma, Daniel, ed. *Ethical, Social, and Legal Dimensions of Screening for Human Genetic Disease*. Symposia series no. 10. March of
Dimes. Miami, Fla.: Symposia Specialists, 1974.

E   Berlant, Jeffrey L. *Profession and Monopoly: A Study of Medicine in
the United States and Great Britain*. Berkeley: University of
California Press, 1975.

HK   Berlinghieri, Francesco Vacca. *La Filosofia della Medicina*. Lucca:
aux frais de l'auteur, 1801.

HK   Bichat, Xavier. *Anatomie général appliquée à la physiologie et la
médicine*. 4 vols. Paris: Brosson, Gabon, 1801.

HK   Bieganski, W. *Logika Medyzyny*. Warazawa: Kowalewski, 1894.

K   Black, D. A. K. *The Logic of Medicine*. Edinburgh: Oliver and Boyd
1968.

E   Black, Peter. "Brain Death." *New England Journal of Medicine* 299 (August 17, 1978): 338–44.

HK   Blane, Gilbert. *Elements of Medical Logick.* London: T. and G. Underwood, 1819.

E   "Blood, Medicine, and the Law of God." Brooklyn, New York: Watch Tower Bible and Tract Society of Pennsylvania; International Bible Students Association; Watchtower Bible and Tract Society of New York, 1961.

K   Bogen, J. E. "The Other Side of the Brain I." *Bulletin of Los Angeles Neurological Societies* 34 (April 1969a): 73.

K   ———. "The Other Side of the Brain II." *Bulletin of Los Angeles Neurological Societies* 34 (July 1969b): 135.

K   ———. "The Other Side of the Brain III." *Bulletin of Los Angeles Neurological Societies* 34 (October 1969c): 191.

HK   Bole, Thomas J., III. "John Brown, Hegel, and Speculative Concepts in Medicine." *Texas Reports on Biology and Medicine* 32 (1974): 287–297.

HK   Bouillauds, Jean. *Essai sur la philosophie medicale et sur les généralités de la clinique medicale.* Paris: J. Rouvier et E. le Bouvier, 1836.

K   Boorse, Christopher. "On the Distinction Between Disease and Illness." *Philosophy and Public Affairs* 5 (Fall 1975): 49–68.

K   ———. "What a Theory of Mental Health Should Be." *Journal For the Theory of Social Behavior* 6 (1976a): 61–84.

K   ———. "Wright on Functions." *The Philosophical Review* 85 (January 1976b): 70–86.

K   ———. "Health as a Theoretical Concept." *Philosophy of Science* 44 (December 1977): 542–573.

HK   Bowman, Inci. "William Cullen (1710–90) and the Primacy of the Nervous System." Unpublished dissertation. Bloomington: Indiana University, 1975.

E   Brandt, Richard. "A Moral Principle About Killing." In *Beneficent Euthanasia,* pp. 106–114. Edited by Marvin Kohl. Buffalo: Prometheus Books, 1975a.

E   ———. "The Morality and Rationality of Suicide." In *A Handbook for the Study of Suicide,* pp. 61–76. Edited by Seymour Perlin. New York: Oxford University Press, 1975b.

E   Bridgman, P. W. "The Struggle for Intellectual Integrity." *Harper's Monthly Magazine* (December 1933), pp. 18–25.

E   Brody, Baruch A. *Abortion and the Sanctity of Human Life.* Cambridge, Mass.: M.I.T. Press, 1975.

K   ———. "Szasz on Mental Illness." In *Mental Health: Philosophical Perspectives,* pp. 251–257. Edited by H. T. Engelhardt, Jr., and S. F. Spicker. Dordrecht: D. Reidel, 1978.

E   Brody, Howard. *Ethical Decisions in Medicine*. Boston: Little, Brown, 1976.

HK   Brooks, C. McC., and Cranefield, P. F., eds. *The Historical Development of Physiological Thought*. New York: Hafner, 1959.

HK   Broussais, F. J. V. *Examen des doctrines médicales et des systèmes de nosologie*. 2 vols. Paris: Mequignon-Marvis, 1824.

K   Brunk, H. D., and Lehr, J. "An Improved Bayes' Method for Computer Diagnosis." In *Proceedings of the Conference on the Use of Computers in Radiology*. pp. B-38–B-54. St. Louis: University of Missouri Press, 1966.

E   Buchanan, Allen. "Medical Paternalism." *Philosophy & Public Affairs* 7 (Summer 1978): 370–390.

K   Burbank, P. "A Computer Diagnostic System for the Diagnosis of Prolonged Undifferentiating Liver Disease." *American Journal of Medicine* 46 (1969): 401–415.

E   Burns, Chester R., ed. *Legacies in Ethics and Medicine*. New York: Science History Publications, 1977a.

E   ———. *Legacies in Law and Medicine*. New York: Science History Publications, 1977b.

E   Burt, Robert A. "The Limits of Law in Regulating Health Care Decisions." *Hastings Center Report* 7 (December 1977): 29–32.

E   Bush, M. *The Adventure Called Death*. New York: Bond Wheelright, 1950.

HK   Bynum, William F. "Varieties of Cartesian Experience in Early Nineteenth Century Neurophysiology." In *Philosophical Dimensions of the Neuro-Medical Sciences*, pp. 15–33. Edited by S .F. Spicker and H. T. Engelhardt, Jr. Dordrecht: D. Reidel, 1976.

HK   Cabot, Richard. *Differential Diagnosis*. Philadelphia: W. B. Saunders, 1911.

E   California. *Natural Death Act. California Health and Safety Code (West)*, ch. 3.9. 1976 Cal. Stats. ch. 1439.

E   Callahan, Daniel. *Abortion: Law, Choice, Morality*. New York: Macmillan, 1970.

E   ———. "Ethics, Law, and Genetic Counseling." *Science* 176 (14 April 1972a): 197–200.

E   ———. "Ethics and Population Limitation." *Science* 175 (4 February 1972b): 487–494.

E   ———. *Population Control For and Against,* New York: Hart, 1973.

E   ———. "The Emergence of Bioethics." In *Science, Ethics, and Medicine,* vol. 1, pp. x–xxvi. Edited by H. T. Engelhardt, Jr., and Daniel Callahan. Hastings-on-Hudson, N.Y.: Institute of Society, Ethics, and the Life Sciences, 1976.

E   ———. "On Defining a 'Natural Death.'" *The Hastings Center Report* 7 (June 1977): 32–37.

E   Callahan, D.; Ramsey, P.; Toulmin, S.; Lappé, M.; and Robertson, J.
       "In Vitro Fertilization: Five Commentaries." *Hastings Center
       Report* 8 (October 1978): 7–14.

E   Campbell, A. V. *Moral Dilemmas in Medicine.* Edinburgh: Churchill
       Livingston, 1972.

K   Canguilhem, G. *Le Normal et le pathologique.* Paris: Presses Univer-
       sitaires de France, 1972.

E   *Canterbury v. Spence.* 464 F.2d 772 (D.C. Cir. 1972).

E   Cantor, Norman. "A Patient's Decision to Decline Life-Saving Medi-
       cal Treatment: Bodily Integrity Versus the Preservation of
       Life." *Rutgers Law Review* 26 (1973): 228–264.

E   Caplan, A., and Bayles, M. "A Response to Professor Gorovitz."
       Journal of Medicine and Philosophy 3 (September 1978): 192–
       195.

K   Capron, Alexander M., and Kass, Leon R. "A Statutory Definition of
       the Standards for Determining Human Death: An Appraisal
       and a Proposal." *University of Pennsylvania Law Review* 121
       (November 1972): 87–118.

HK   Cartwright, Samuel. "Report on the Diseases and Physical Peculiari-
       ties of the Negro Race." *The New Orleans Medical and Surgical
       Journal* 4 (May 1851): 691–715.

K   Cassell, Eric. "Preliminary Exploration of Thinking in Medicine."
       Ethics in Science and Medicine 2 (May 1975): 1–12.

E   ———. "The Function of Medicine." *Hastings Center Report* 7
       (December 1977): 16–19.

K   ———. "The Subjective in Clinical Judgment." In *Clinical Judgment,*
       pp. 199–215. Edited by H. T. Engelhardt, Jr., Stuart F. Spicker,
       and Bernard Towers, Dordrecht: D. Reidel, 1979.

HE   Castro, Rodericus. *Medicus-Politicus: sive de officiis medicopoliticis
       tractatus.* Hamburg: Frobeniano, 1614.

K   Cattaneo, A. D.; Luchelli, P. F.; Rocca, E.; Mattioli, F.; and
       Recchi, G. "Computer Versus Clinical Diagnosis of Bilary Tract
       Diseases." *Abdominal Surgery* 14 (1972): 71–75.

E   Childress, James F. "Who Shall Live When Not All Can Live?"
       Soundings 53 (Winter 1970): 339–355.

E   Chisholm, Roderick. "Coming into Being and Passing Away: Can the
       Metaphysican Help?" In *Philosophical Medical Ethics: Its Na-
       ture and Significance,* pp. 169–182. Edited by Stuart Spicker and
       H. T. Engelhardt, Jr. Dordrecht: D. Reidel, 1977.

E   Ciba Foundation Symposium. *Law and Ethics of A.I.D. and Embryo
       Transfer.* Amsterdam: Associated Scientific, 1973.

HK   Cipolla, Carlo. *Public Health and the Medical Profession in Renais-
       sance Italy.* New York: Cambridge University Press, 1976.

E   Clark, Stephen. *The Moral Status of Animals.* Oxford: Clarendon
       Press, 1977.

HE   Codronchi, Giovanni. *De Christiana ac tuta medendi ratione libri duo.* Ferrara, 1591.

K   Cohen, Henry. *Concepts of Medicine.* Edited by Brandon Rush. Oxford: Pergamon Press, 1960.

HE   Cook, George Wythe. "The History of Medical Ethics." *New York Medical Journal* 101 (1915): 141–146, 205–207.

E   Craven, Joan, and Wald, Florence. "Hospice Care for Dying Patients." *American Journal of Nursing* 75 (October 1975): 1816–1822.

KE   Crockett, Campbell. "Ethics, Metaphysics, and Psychoanalysis." *Inquiry* 4 (Spring 1961): 37–52.

HK   Cullen, W. *Synopsis nosologiae methodicae.* Edinburgh: William Creech, 1769.

HK   Curran, William J. "Governmental Regulation of the Use of Human Subjects in Medical Research: The Approach of Two Federal Agencies." *Daedalus* 98 (Spring 1969): 542–594.

K   Dagi, T. Forcht. "Cause and Culpability." *The Journal of Medicine and Philosophy* 1 (1976): 349–371.

KE   Daley, James W. "Psychoanalysis and the Problem of Value." *Diogenes* 59 (Fall 1967): 1–24.

E   Daube, David. "The Linguistics of Suicide." *Philosophy and Public Affairs* 1 (1971): 387–437.

HE   Davis, Nathan S. *History of Medicine with the Code of Medical Ethics.* Chicago: Cleveland Press, 1903.

HK   de Dorlodot, Canon Henry. "A Vindication of the Mediate Animation Theory." In *Theology and Evolution,* pp. 259–283. Edited by E. C. Messenger. London: Sands, 1952.

E   Delkeskamp, Corinna. "Another Response to MacIntyre, Tragedy, Reason, Religions and Ramsey." In *Knowledge, Value and Belief,* pp. 79–99. Edited by H. T. Engelhardt, Jr., and Daniel Callahan. Hastings-on-Hudson, N.Y.: Institute of Society, Ethics, and the Life Sciences, 1977.

K   Delkeskamp, C.; Engelhardt, H. T.; and Spicker, S., eds. *Technology, Science, and the Art of Medicine.* Dordrecht: Reidel, 1980.

E   Dinello, Daniel. "On Killing and Letting Die." *Analysis* 31 (April 1971): 83–86.

E   Donceel, J. "Abortion: Mediate v. Immediate Animation." *Continuum* 5 (Spring 1967): 167–71.

E   ———. "Immediate Animation and Delayed Hominization." *Theological Studies* 13 (March 1970): 76–105.

E   Downing, A. B. "Euthanasia: The Human Context." In *Euthanasia and the Right to Die,* pp. 13–29. Edited by A. B. Downing. London: Peter Owen, 1969.

E   Dubos, René. *Man Adapting.* New Haven: Yale University Press, 1965.

E   Duff, Raymond S., and Campbell, A. G. M. "Moral and Ethical

Dilemmas in the Special Care Nursery." *New England Journal of Medicine* 289 (October 25, 1973): 890–894.

HK Duffy, John. *The Healers: The Rise of the Medical Establishment.* New York: McGraw-Hill, 1976.

E Dworkin, Gerald. "Paternalism." *Monist* 56 (1972): 64–84.

E Dyck, Arthur. "Beneficient Euthanasia and Benemortasia: Alternative Views of Mercy." In *Beneficent Euthanasia,* pp. 117–129. Edited by Marvin Kohl. Buffalo: Prometheus Books, 1975.

K Eccles, Sir John, ed. *Brain and Conscious Experience.* New York: Springer-Verlag, 1966.

K ———. *Facing Reality.* New York: Springer-Verlag, 1970.

HE Edelstein, Ludwig. "The Hippocratic Oath: Text, Translation, and Interpretation." In *Ancient Medicine: Selected Papers of Ludwig Edelstein,* pp. 3–63. Edited by Owsei Temkin and C. Lilian Temkin. Baltimore: Johns Hopkins University Press, 1967.

HE ———. "The Professional Ethics of the Greek Physicians." In *Ancient Medicine: Selected Papers of Ludwig Edelstein,* pp. 319–348. Edited by Owsei Temkin and C. Lilian Temkin. Baltimore: Johns Hopkins University Press, 1967.

E Ehmann, Günter, and Löther, Rolf, eds. *Sozialismus—Medizin—Persönlichkeit.* Oberlungwitz: V.E.B. Kongressdruck, 1975.

E Ellin, J. "Comments on 'Paternalism and Health Care.'" In *Contemporary Issues in Biomedical Ethics,* pp. 245–254. Edited by J. Davis et al. Clifton, N.J.: Humana, 1978.

K Ellis, Albert. "An Operational Reformulation of Some of the Basic Principles of Psychoanalysis." In *The Foundations of Science and the Concepts of Psychology and Psychoanalysis,* pp. 131–154. Edited by H. Feigl and M. Scriven. Minneapolis: University of Minnesota Press, 1956.

K Elstein, Arthur. "Clinical Judgment: Psychological Research and Medical Practice." *Science* 194 (12 November 1976): 696–700.

K ———. "Human Factors in Clinical Judgment." In *Clinical Judgment,* pp. 17–28. Edited by H. T. Engelhardt, Jr., Stuart Spicker, and Bernard Towers. Dordrecht: D. Reidel, 1979.

K Elstein, Arthur; Shulman, Lee S.; and Sprafka, Sarah A. *Medical Problem Solving: An Analysis of Clinical Reasoning.* Cambridge, Mass.: Harvard University Press, 1978b.

HK Engelhardt, H. Tristram, Jr. "The Disease of Masturbation: Values and the Concept of Disease." *Bulletin of the History of Medicine* 48 (Summer 1974a): 234–248.

K ———. "Explanatory Models in Medicine: Facts, Theories and Values." *Texas Reports on Biology and Medicine* 32 (Spring 1974b): 225–239.

E ———. "The Ontology of Abortion." *Ethics* 84 (April 1974c): 217–234.

K ———. "The Concepts of Health and Disease." In *Evaluation and Explanation in the Biomedical Sciences,* pp. 125–141. Edited by H. T. Engelhardt, Jr., and Stuart Spicker. Dordrecht, D. Reidel, 1975a.

E ———. "The Counsels of Finitude." *Hastings Center Report* 5 (April 1975b): 29–36.

K ———. "Defining Death: A Philosophical Problem for Medicine and Law." *American Review of Respiratory Disease* 112 (1975c): 587–590.

E ———. "Ethical Issues in Aiding the Death of Young Children." In *Beneficent Euthanasia,* pp. 180–192. Edited by Marvin Kohl. Buffalo: Prometheus, 1975d.

HK ———. "John Hughlings Jackson and the Mind-Body Relation." *Bulletin of the History of Medicine* 49 (Summer 1975e): 137–151.

K ———. "Ideology and Etiology." *The Journal of Medicine and Philosophy* 1 (September 1976): 256–268.

K ———. "Is There a Philosophy of Medicine?" In *PSA 1976,* vol. 2, pp. 94–108. Edited by P. Asquith and F. Suppe. East Lansing, Mich.: Philosophy of Science Association, 1977a.

E ———. "Some Persons Are Human, Some Humans Are Persons, and the World is What We Persons Make of It." In *Philosophical Medical Ethics: Its Nature and Significance,* pp. 183–194. Edited by Stuart Spicker and H. T. Engelhardt, Jr. Dordrecht: D. Reidel, 1977b.

K ———. "Treating Aging: Restructuring the Human Condition." In *Extending the Human Life Span: Social Policy and Social Ethics,* pp. 33–40. Edited by Bernice Neugarten and Robert Havighurst. Washington, D.C.: National Science Foundation, 1977c.

E ———. "Medicine and the Concept of Person." In *Ethical Issues in Death and Dying,* pp. 271–84. Edited by T. Beauchamp and S. Perlin. Englewood Cliffs, N.J.: Prentice-Hall, 1978.

E Engelhardt, H. T., Jr., and Callahan, Daniel, eds. *Science, Ethics, and Medicine.* Hastings-on-Hudson, N.Y.: Institute of Society, Ethics, and the Life Sciences, 1976.

K Engelhardt, H. T., Jr., and Spicker, Stuart, eds. *Evaluation and Explanation in the Biomedical Sciences.* Dordrecht: D. Reidel, 1975.

K ———. *Clinical Judgment.* Dordrecht: D. Reidel, 1979.

K English, Darrel S., ed. *Genetic and Reproductive Engineering.* New York: MSS Information, 1974.

E English, Jane. "Abortion and the Concept of Person." *Canadian Journal of Philosophy* 5 (October 1975): 233–243.

E Enthoven, Alain C. "Consumer-Choice Health Plan." 2 parts. *New England Journal of Medicine* 298 (March 23, 1978), 650–658; 298 (March 30, 1978): 709–720.

K   Erde, Edmund L. "Mind-Body and Malady." *The Journal of Medi-
    cine and Philosophy* 2 (June 1977): 177–190.

E   Etziony, M. B. *The Physician's Creed.* Springfield, Ill.: Charles C.
    Thomas, 1973.

K   Faber, Knud. *Nosography in Modern Internal Medicine.* New York:
    Paul B. Hoeber, 1922, 1923. Reprinted under the title *Nosog-
    raphy: The Evolution of Clinical Medicine in Modern Times.*
    Hoeber, 1930; and A.M.S., 1978.

K   Fabrega, Horacio, Jr. "Concepts of Disease: Logical Features and
    Social Implications." *Perspectives in Biology and Medicine* 15
    (1972): 583–616.

K   Feigl, Herbert. "The 'Mental' and the 'Physical.'" In *Concepts,
    Theories, and the Mind-Body Problem,* pp. 370–497. Edited by
    H. Feigl, M. Scriven, and G. Maxwell, Minneapolis: University
    of Minnesota Press, 1958.

K   Feigl, Herbert, and Scriven, Michael, eds. *Minnesota Studies in the
    Philosophy of Science.* Minneapolis: University of Minnesota
    Press, 1956.

E   Feinberg, Joel, ed. *The Problem of Abortion.* Belmont, Calif.: Wads-
    worth, 1973.

K   Feinstein, Alvan. *Clinical Judgment.* Baltimore: Williams and Wil-
    kins, 1967.

K   ——. "Taxonomy and Logic in Clinical Data." *Annals of the New
    York Academy of Science* 161 (1969): 450–459.

K   ——. "The Problems of the 'Problem-Oriented Medical Record.'"
    Annals of Internal Medicine 78 (1973): 751–762.

E   Feldman, D. M. *Marital Relations, Birth Control, and Abortion in
    Jewish Law.* New York: Schocken Books, 1968.

HK Ferrier, David. *The Function of the Brain.* 2d ed. New York: G. P.
    Putnam's Sons, 1886.

K   Fingarette, Herbert. *The Self in Transformation.* New York: Basic
    Books, 1963.

KE Fisher, Kenneth A. "Ultimate Goals in Therapy." *Journal of Exis-
    tential Philosophy* 7 (Winter 1966–1967): 215–232.

E   Fitzgerald, P. J. "Acting and Refraining." *Analysis* 27 (March 1967):
    133–139.

K   Fleck, Ludwik. *Entstehung und Entwicklung einer wissenschaftlichen
    Tatsache.* Basel: Benno Schwabe, 1935; English version, *Genesis
    and Development of a Scientific Fact.* Edited by T. J. Trenn
    and R. K. Merton. Tr. F. Bradley and T. Trenn. Chicago:
    University of Chicago Press, 1979.

E   Fletcher, John. "Human Experimentation: Ethics in the Consent
    Situation." *Law and Contemporary Problems* 32 (Autumn, 1967):
    620–649.

E   ——. "The Brink: Parent-Child Bond in the Genetic Revolution." *Theological Studies* 53 (September 1972): 457–85.

E   Fletcher, Joseph. *Morals and Medicine*. Boston: Beacon Press, 1960.

E   ——. *Situation Ethics*. Philadelphia: Westminster Press, 1966.

E   ——. "Ethical Aspects of Genetic Controls." *New England Journal of Medicine* 285 (September 30, 1971): 776–83.

E   ——. "Indicators of Humanhood: A Tentative Profile of Man." *The Hastings Center Report* 2 (November 1972): 1–4.

E   ——. *The Ethics of Genetic Control*. Garden City, N.Y.: Doubleday Anchor, 1974a.

E   ——. "Four Indicators of Humanhood—the Enquiry Matures." *The Hastings Center Report* 4 (December 1974b): 4–7.

K   Flew, Antony. "Psycho-analytic Explanation." *Analysis* 10 (1949): 8–15.

HK   Flourens, M. J. P. *Examen de la phrénologie*. 2d ed. Paris: Paulin, 1845.

K   Fodor, Jerry A. "Psychosurgery: What's the Issue?" In *Philosophical Dimensions of the Neuro-Medical Sciences*, pp. 85–94. Edited by S. F. Spicker and H. T. Engelhardt, Jr. Dordrecht: D. Reidel, 1976.

E   Foot, Philippa. "The Problem of Abortion and the Doctrine of Double Effect." *The Oxford Review* 5 (1967): 5–15.

HE   Forbes, Robert. "A Historical Survey of Medical Ethics." *British Medical Journal*, supplement 2 (1935): 137–140.

E   Foreman, Percy. "The Physician's Criminal Liability for the Practice of Euthanasia." *Baylor Law Review* 27 (1975): 54–61.

HK   Foucault, Michel. *The Birth of the Clinic: An Archaeology of Medical Perception*. Tr. A. M. Sheridan Smith. New York: Random House, 1973.

E   *The Framingham Study*. Washington, D.C.: Public Health Service (H.E.W.) in association with American Heart Association, 1949.

E   Frank, Johann. *A System of Complete Medical Police*. Baltimore: Johns Hopkins University Press, 1976. German original, 1777.

E   Freedman, Benjamin. "On the Rights of the Voiceless." *Journal of Medicine and Philosophy* 3 (1978): 196–210.

HK   Freud, Sigmund. *An Outline of Psycho-analysis*. In *The Complete Psychological Works*, vol. 23, pp. 141–207. Tr. James Strachey. London: Hogarth Press, 1953.

HK   ——. *On Aphasia*. Tr. E. Stengel. London: Imago, 1953.

E   Freund, Paul. "Introduction to the Issue 'Ethical Aspects of Experimentation with Human Subjects.'" *Daedalus* 98 (Spring 1969): viii–xiii.

E   Fried, Charles. *Medical Experimentation*, New York: American Elsevier, 1974.

E ———. "Rights and Health Care—Beyond Equity and Efficiency." *New England Journal of Medicine* 293 (July 31, 1975): 241–245.

E ———. "Equality and Rights in Medical Care." *Hastings Center Report* 6 (February 1976): 29–34.

K Friedman, G. D. *Primer of Epidemiology*. New York: McGraw-Hill, 1974.

HE Gaeley, J. W. S. *Conferences on the Moral Philosophy of Medicine*. New York: Rebman, 1906.

HK Gall, Franz Josef. *On the Functions of the Brain and of Each of Its Parts*. 3 vols. Boston: Marsh, Capen and Lyon, 1835.

HK Gaubius, J. D. *De regimine mentis quod medicorum est*. Leyden: Balduinum van der Aa, 1747.

E Gaylin, Willard. "Genetic Screening: The Ethics of Knowing." *New England Journal of Medicine* 286 (June 22, 1972): 1361–1362.

K Gaylin, Willard; Meister, J.; and Neville, R., eds. *Operating on the Mind*. New York: Basic Books, 1974.

E Gaylin, Willard; Thompson, Travis; Neville, Robert; and Bayles, Michael. "Sterilization of the Retarded: In Whose Interest?" *Hastings Center Report* 9 (June 1978): 28–41.

K Gazzaniga, M. S. *The Bisected Brain*. New York: Appleton-Century-Crofts, 1970.

K Geissler, E.; Kosing, A.; Ley, H.; and Scheler, W., eds. *Philosophische und ethische Probleme der Molekularbiologie*. Berlin: Akademie-Verlag, 1974.

E Gerber, Rudolph J. "When is the Human Soul Infused?" *Laval Théologique et Philosophique* 22 (1966): 234–247.

E Gert, Bernard, and Culver, Charles M. "Paternalistic Behavior." *Philosophy & Public Affairs* 6 (Fall 1976): 45–57.

E Glass, Bentley. "Human Heredity and Ethical Problems." *Perspectives in Biology and Medicine* 15 (Winter 1972): 237–53.

K Goldfinger, Stephen. "The Problem-Oriented Record: A Critique from a Believer." *New England Journal of Medicine* 288 (March 22, 1973): 606–608.

E Gorovitz, Samuel. "Ethics and the Allocation of Medical Resources."
K ———. "Medical Fallibility: A Rejoinder." *Journal of Medicine and Philosophy* 3 (September 1978): 187–191.

E Gorovitz, Samuel, et al., eds. *Moral Problems in Medicine*. Englewood Cliffs, N.J.: Prentice-Hall, 1976.

K Gorovitz, Samuel, and MacIntyre, Alasdair. "Toward a Theory of Medical Fallibility." *The Journal of Medicine and Philosophy* 1 (1976): 51–71.

E Graber, G. C. "On Paternalism and Health Care." In *Contemporary*

Issues in Biomedical Ethics, pp. 233–244. Edited by J. Davis et al. Clifton, N.J.: Humana, 1978.

HE   Graham, Loren. "Attitudes Toward Eugenics and Human Heredity in Germany and the Soviet Union in the Nineteen-Twenties." In *Morals, Science, and Sociality,* pp. 119–149. Edited by H. T. Engelhardt, Jr., and Daniel Callahan. Hastings-on-Hudson, N.Y.: Institute of Ethics, Society, and the Life Sciences, 1978.

K   Green, Richard. "Homosexuality as a Mental Illness." *International Journal of Psychiatry* 10 (March 1972): 77–98.

E   Green, Ronald M. "Health Care and Justice in Contract Theory Perspectives." In *Ethics and Health Policy,* pp. 111–126. Edited by R. M. Veatch and R. Branson. Cambridge, Mass.: Ballinger, 1976.

K   Gregory, Ian. *Fundamentals of Psychiatry.* Philadelphia: Saunders, 1968.

HE   Gregory, John. *Lectures on the Duties and Qualifications of a Physician.* London: W. Strahan, 1772.

K   Grene, Marjorie. "Philosophy of Medicine: Prolegomena to a Philosophy of Science." In *PSA 1976,* vol. 2, pp. 77–93. Edited by P. Asquith and F. Suppe. East Lansing, Mich.: Philosophy of Science Association, 1977.

K   ———. "Individuals and Their Kinds: Aristotelian Foundations of Biology." In *Organism, Medicine, and Metaphysics,* pp. 121–136. Edited by Stuart Spicker. Dordrecht: D. Reidel, 1978.

HK   Grohmann, Johann Christian August. *Philosophie der Medizin.* Berlin: Schmidt, 1808.

K   Gross, Rudolf. *Medizinische Diagnostik—Grundlagen einer normativen Diagnostik.* Stuttgart: Hippokrates Verlag, 1969.

K   ———. *Zur klinischen Dimension der Medizin.* Stuttgart: Hippokrates Verlag, 1976.

K   ———. "Was ist eine Krankheit? Gibt es stabile Krankheitsbilder?" *Metamed* I (1977): 115–122.

E   Grossman, Edward. "The Obsolescent Mother." *The Atlantic* (May 1971), pp. 39–50.

HK   Guédon, Jean-Claude. "Michel Foucault: The Knowledge of Power and the Power of Knowledge." *Bulletin of the History of Medicine* 51 (Summer 1977): 245–277.

E   Gustafson, James M. *The Contributions of Theology to Medical Ethics.* Milwaukee: Marquette University, 1975.

K   Guttentag, Otto. "The Problem of Experimentation on Human Beings." *Science* 117 (27 February 1953): 207–210.

E   Hadfield, Stephen J. *Law and Ethics for Doctors.* London: Eyre and Spottiswoode, 1958.

E Hamilton, Michael, ed. *The New Genetics and the Future of Man,* Grand Rapids, Mich.: W. B. Eerdman's, 1971.

K Hanna, Thomas. *Bodies in Revolt.* New York: Holt, Rinehart, and Winston, 1970.

E Hardin, Garrett. "Living on a Lifeboat." *Bioscience* 24 (October 1974): 561–568.

K Hartmann, Heinz. "Psycho-Analysis and the Concept of Health." *The International Journal of Psycho-Analysis* 20 (1939): 308–321.

E Hauerwas, Stanley. *Vision and Virtue.* Notre Dame: Fides Publishers, 1974.

E ———. *Truthfulness and Tragedy.* Notre Dame: University of Notre Dame Press, 1977.

K Heath, R. G. "Schizophrenia as an Immunologic Disorder." 3 parts. *Archives of General Psychiatry* 16 (January 1967): 1–33.

E Hegland, Kenny F. "Unauthorized Rendering of Life-Saving Medical Treatment." *California Law Review* 53 (1965): 860–877.

E Hellegers, André. "Round Table Discussion." In *Philosophical Medical Ethics: Its Nature and Significance,* pp. 225–30. Edited by Stuart Spicker and H. T. Engelhardt, Jr. Dordrecht: D. Reidel, 1977.

K Hemphill, Michael. "Tests for Presymptomatic Huntington's Chorea." *New England Journal of Medicine* 287 (October 19, 1972): 823–824.

HKE Hippocrates. *Hippocrates and the Fragments of Heracleitus.* 4 vols. Tr. William Henry S. Jones. Cambridge, Mass.: Harvard University Press, 1923.

E Holmes, Oliver Wendell. *"Buck v. Bell."* In *Moral Problems in Medicine,* p. 128. Edited by Samuel Gorovitz et al. Englewood Cliffs, N.J.: Prentice-Hall, 1976.

E Hook, Sidney. "The Ethics of Suicide." In *Beneficient Euthanasia,* pp. 57–69. Edited by Marvin Kohl. Buffalo: Prometheus Books, 1975.

E Horowitz, Louise. "The Morality of Suicide." *The Journal of Critical Analysis* 3 (January 1972): 161–165.

K Hudson, Robert P. "The Concept of Disease." *Annals of Internal Medicine* 65 (September 1966): 595–601.

E Hull, Richard. "Changing Man's Genetic Composition: A Reply to Dr. Verle Headings." *The Humanist* 32 (September-October 1972): 13.

E ———. "Why 'Genetic Disease'?" In *Genetic Counseling: Facts, Values, and Norms,* pp. 57–70. Edited by Marc Lappé. New York: National Foundation, March of Dimes, 1979.

E Hunt, Robert, and Arras, John, eds. *Ethical Issues in Modern Medicine,* Palo Alto, Calif.: Mayfield, 1977.

E Hutchinson, G. D. "A Speculative Consideration of Certain Possible Forms of Sexual Selection in Man." *The American Naturalist* 93 (March-April 1959): 81–91.

E Illich, Ivan. "The Political Uses of Natural Death." In *Death Inside Out,* pp. 25–42. Edited by Peter Steinfels and Robert Veatch. New York: Harper & Row, 1974.

E Ingelfinger, Franz J. "Informed (But Uneducated) Consent." *New England Journal of Medicine* 287 (August 31, 1972): 465–466.

HK Jackson, John Hughlings. "On the Nature of the Duality of the Brain." *Brain* 38 (1915): 80–103.

HK ———. *Selected Writings of John Hughlings Jackson.* 2 vols. Edited by J. Taylor. London: Staples Press, 1958.

K Jacquez, John A., ed. *The Diagnostic Process.* Ann Arbor: University of Michigan Press, 1964.

E Jakobovits, Immanuel. *Jewish Medical Ethics.* New York: Bloch, 1959.

K Jaynes, Julian. *The Origins of Consciousness in the Breakdown of the Bicameral Mind.* Boston: Houghton Mifflin, 1977.

E Jonas, Hans. "Philosophical Reflections on Experimenting with Human Subjects." *Daedalus* 98 (Spring 1969): 219–247.

HK Jones, Joseph. "Diseases of the Nervous System." In *Transcript of the Louisiana Medical Society,* pp. 170–171. New Orleans: L. Graham and Son, 1889.

E Jonsen, Albert R., and Butler, Lewis H. "Public Ethics and Policy Making." *Hastings Center Report* 5 (August 1975): 19–31.

E *Kaimowitz and Doe v. Dept. of Mental Health for the State of Michigan.* Civil Action No. 2 Prison L. Rptr. 433 (1973), 73–19434–AW.

E Kamisar, Yale. "Euthanasia Legislation: Some Non-Religious Objections," In *Euthanasia and the Right to Die,* pp. 85–133. Edited by A. B. Downing. London: Peter Owen, 1969.

E Kass, Leon. "Babies by Means of *In Vitro* Fertilization: Unethical Experiments on the Unborn?" *New England Journal of Medicine* 285 (November 1971a): 1174–1179.

E ———. "Death as an Event: A Commentary on Robert Morison." *Science* 173 (20 August 1971b): 698–702.

E ———. "Averting One's Eyes, or Facing the Music?—On Dignity and Death." In *Death Inside Out,* pp. 101–114. Edited by Peter Steinfels and Robert M. Veatch. New York: Harper & Row, 1974.

K ———. "Regarding the End of Medicine and the Pursuit of Health." *The Public Interest* 40 (Summer 1975): 11–24.

E Katz, Jay. *Experimentation with Human Beings.* New York: Russell Sage Foundation, 1972.

446 PHILOSOPHY

E Katz, J., and Capron, A. M. *Catastrophic Disease: Who Decides What?* New York: Russell Sage Foundation, 1975.

E Kefauver-Harris Bill, Public Law 87–781—October 10, 1962. Food and Drug Administration, Food, Drug and Cosmetic Act, Amendments 1962 (52 stat. 1051–1055, § 505(i)).

E Kelly, Gerald, S. J. "The Duty of Using Artificial Means of Preserving Life." *Theological Studies* 11 (1950): 203–220.

E ———. "Notes—The Duty to Preserve Life." *Theological Studies* 12 (1951): 550–556.

E ———. *Medico-Moral Problems.* St. Louis: Catholic Hospital Association of the United States and Canada, 1958.

E Kenny, John P. *Principles of Medical Ethics.* Westminster, Maryland: Newman Press, 1962.

K King, Lester S. "What is Disease?" *Philosophy of Science* 21 (July 1954): 193–203.

E ———. "Development of Medical Ethics." *New England Journal of Medicine* 258 (March 6, 1958): 480–486.

HK ———. *The Philosophy of Medicine: The Early Eighteenth Century.* Cambridge, Mass.: Harvard University Press, 1977.

K Koch, R. *Die ärztliche Diagnose.* Wiesbaden: J. F. Bergman, 1920.

K Kohl, Marvin. *The Morality of Killing.* London: Peter Owen, 1974.

E ———. "Voluntary Beneficent Euthanasia." In *Beneficent Euthanasia,* pp. 130–141. Edited by Marvin Kohl. Buffalo: Prometheus Books, 1975.

E ———. "Euthanasia and the Right to Life." In *Philosophical Medical Ethics: Its Nature and Significance,* pp. 73–84. Edited by Stuart Spicker and H. T. Engelhardt, Jr. Dordrecht: D. Reidel, 1977.

E ———. *Infanticide and the Value of Life.* Buffalo, New York: Prometheus Books, 1978.

HE Konold, Donald E. *A History of American Medical Ethics 1847–1912.* Madison: University of Wisconsin Press, 1962.

E Kübler-Ross, Elisabeth. *On Death and Dying.* New York: Macmillan, 1969.

E Ladd, John. "Legalism and Medical Ethics." In *Biomedical Ethics,* pp. 1–36. Edited by John Davis. Clifton, N.J.: Humana Press, 1978.

E Ladimer, Irving. "Human Experimentation: Medicolegal Aspects." *New England Journal of Medicine* 257 (July 4, 1957): 18–24.

HE Lain-Entralgo, P. *Therapies of the Word in Classical Antiquity.* New Haven: Yale University Press, 1970.

K la Mettrie, J. O. de. *L'Homme machine.* Leyden: Zucac, 1748. Translated as *Man a Machine.* LaSalle, Ill.: Open Court, 1961.

K Lange, H. J., and Wagner, G. *Computerunterstützte ärztliche Diag-*

nostik: Bericht über die 17. Jahrestagung der GMDS. Stuttgart, 1973.

E   Lappé, Marc. "The Genetic Counselor: Responsible to Whom?" *Hastings Center Report* 1 (1971): 6–11.

E   ──────. "Moral Obligation and the Fallacies of 'Genetic Control.'" *Theological Studies* 33 (September 1972): 411–427.

E   Lasagna, Louis. *The Conflict of Interest Between Physician as Therapist and as Experimenter.* Philadelphia: Society for Health and Human Values, 1975.

K   Lazare, Aaron. "Hidden Conceptual Models in Clinical Psychiatry." *The New England Journal of Medicine* 288 (February 15, 1973): 345–351.

E   Leake, Chauncey D. "Theories of Ethics and Medical Practice." *Journal of the American Medical Association* 208 (May 5, 1969): 842–847.

E   Lebacqz, Karen, and Engelhardt, H. T., Jr. "Suicide." In *Death, Dying, and Euthanasia,* pp. 669–705. Edited by Dennis J. Horan and David Mall. Washington, D.C.: University Publications of America, 1977.

K   Ledley, R. S., and Lusted, L. B. "Reasoning Foundations of Medical Diagnosis." *Science* 130 (3 July 1959): 9–21.

K   Leiber, B. *Die klinischen Syndrome.* München: Urban and Schwerzenberg, 1966.

E   Lennane, K. Jean, and Lennane, R. John. "Alleged Psychogenic Disorders in Women—a Possible Manifestation of Sexual Prejudice." *The New England Journal of Medicine* 288 (February 8, 1973): 288–292.

HK   Lindeboom, G. A. *Herman Boerhaave.* London: Methuen and Co., Ltd., 1968.

HK   Linnaeus, Carolus. *Genera morborum, in auditorum usum.* Upsaliae: Steinert, 1763.

E   Lipman, A., and Marden, P. "Preparation for Death in Old Age." *Journal of Gerontology* 21 (July 1966): 426–31.

K   Livingston, F. B. "The Distribution of the Abnormal Hemoglobin Genes and Their Significance for Human Evolution." *Evolution* 18 (1964): 685–699.

E   Lucas, George R., Jr., and Ogletree, Thomas W., eds. *Lifeboat Ethics: The Moral Dilemmas of World Hunger.* New York: Harper & Row, 1976.

K   Luria, Aleksandr. *Higher Cortical Functions in Man.* Tr. Basil Haigh. New York: Basic Books, 1966.

K   Lusted, L. B. *Introduction to Medical Decision Making.* Springfield, Ill.: Charles C. Thomas, 1968.

E   McAllister, Joseph E. *Ethics with a Special Application to the Medical and Nursing Professions.* Philadelphia: Saunders, 1956.

E   McCormick, Richard A. *Ambiguity in Moral Choice.* Milwaukee: Marquette University Press, 1973.

E   ———. "Proxy Consent in the Experimental Situation." *Perspectives in Biology and Medicine* 18 (Autumn 1974a): 2–21.

E   ———. "To Save or Let Die: The Dilemma of Modern Medicine." *Journal of the American Medical Association* 229 (July 8, 1974b): 172–176.

E   ———. "Freedman on the Right to the Voiceless." *Journal of Medicine and Philosophy* 3 (September 1978): 211–221.

E   McCormick, Richard A., and Hellegers, André E. "Legislation and the Living Will." *America* (March 12, 1977), pp. 210–213.

E   McCormick, Richard A., and Ramsey, Paul, eds. *Doing Evil to Achieve Good.* Chicago, Ill.: Loyola University Press, 1978.

E   McCullough, L. B., and Morris, J. P., III, eds. *Implications of History and Ethics to Medicine—Veterinary and Human.* College Station, Texas: Texas A and M, 1978.

KE   MacIntyre, Alasdair. "Symposium: Cause and Cure in Psychotherapy." *Aristotelian Society Proceedings,* supplementary no. 29 (1964): 43–58.

E   ———. "Can Medicine Dispense with a Theological Perspective on Human Nature?" In *Knowledge, Value, and Belief,* pp. 25–43. Edited by H. T. Engelhardt, Jr., and Daniel Callahan. Hastings-on-Hudson, N.Y.: Institute of Society, Ethics, and the Life Sciences, 1977a.

E   ———. "Patients as Agents." In *Philosophical Medical Ethics: Its Nature and Significance,* pp. 197–212. Edited by Stuart Spicker and H. T. Engelhardt, Jr. Dordrecht: D. Reidel, 1977b.

E   ———. "The Right to Die Garrulously." In *Death and Decision,* pp. 75–84. Edited by Ernan McMullin. Boulder, Colo.: Westview Press, 1978.

K   McKenzie, J. M. "Evaluation of the Hazards of Sickle Trait in Aviation." *Aviation Space Environment Medicine* 48 (August 1977): 753–62.

E   McKenzie, James F. "Contraceptive Sterilization: The Doctor, the Patient, and the U.S. Constitution." *University of Florida Law Review* 25 (1973): 237–349.

K   McKeon, Richard. "The Concept of Mankind and Mental Health." *Ethics* 77 (October 1966): 29–37.

K   MacMahon, Brian; Pugh, Thomas; and Ipsen, Johannes. *Epidemiologic Methods.* Boston: Little, Brown, 1960.

E   McMullin, Ernan. "A Clinician's Quest for Certainty." In *Clinical*

Judgment, pp. 115–129. Edited by H. T. Engelhardt, Jr., Stuart Spicker, and Bernard Towers. Dordrecht: D. Reidel, 1979.

K   Macklin, Ruth. "Mental Health and Mental Illness: Some Problems of Definition and Concept Formation." *Philosophy of Science* 39 (September 1972): 341–365.

E   ———. "The Medical Model in Psychoanalysis and Psychotherapy." *Comprehensive Psychiatry* 14 (January-February 1973): 49–69.

E   ———. "Moral Concerns and Appeals to Rights and Duties." *Hastings Center Report* 6 (October 1976): 31–38.

K   Magoun, H. W. *The Waking Brain.* 2d ed. Springfield, Ill.: Charles C. Thomas, 1963.

E   *Maher v. Roe* 97 S. Ct. 2376 (1977).

K   Margolis, Joseph. *Psychotherapy and Morality.* New York: Random House, 1966.

K   ———. "The Concept of Disease." *The Journal of Medicine and Philosophy* 1 (1976a): 238–55.

K   ———. "Persons and Psychosurgery." In *Philosophical Dimensions of the Neuro-Medical Sciences,* pp. 71–84. Edited by S. F. Spicker and H. T. Engelhardt, Jr. Dordrecht: D. Reidel, 1976b.

K   Marmor, Judd. "Homosexuality and Cultural Values Systems." *American Journal of Psychiatry* 130 (1973): 1208.

K   Martin, Michael. "On a New Theory of Medical Fallibility: A Rejoinder." *The Journal of Medicine and Philosophy* 2 (1977): 84–88.

E   Masters, William H.; Johnson, Virginia E.; and Kolodny, Robert C., eds. *Ethical Issues in Sex Therapy and Research.* Boston: Little, Brown, vol. 1, 1977; vol. 2, 1980.

K   Mechanic, D. "Some Factors in Identifying and Defining Mental Illness." In *Mental Illness and Social Processes,* pp. 23–32. Edited by Thomas J. Scheff. New York: Harper & Row, 1967.

E   Medawar, P. *The Future of Man.* London: Methuen, 1959.

E   *Medical Ethics Film Review Project.* College Park: Institute on Moral Problems in Medicine and the Council for Philosophical Studies, University of Maryland, 1974.

E   Medical Research Council. *Guidelines for the Handling of Recombinant DNA Molecules and Animal Viruses and Cells.* Ottawa, Canada: Medical Research Council, 1977.

K   Meehl, Paul. "Specific Etiology and Other Forms of Strong Influence: Some Quantitative Meanings." *The Journal of Medicine and Philosophy* 2 (1977): 33–53.

K   ———. "Some Methodological Reflections on the Difficulties of Psychoanalytic Research." In *Analyses of Theories and Methods of Physics and Psychology,* pp. 403–416. Edited by M. Radner and

S. Winokur, Minneapolis: University of Minnesota Press, 1970.

K   Menninger, K. "Unitary Concept of Mental Illness." In *Psychopathology Today*, pp. 85–90. Edited by William S. Sahakian. Itasca, Ill.: F. E. Peacock, 1970.

K   Meyer, Adolf. *The Collected Papers of Adolph Meyer*. 4 vols. Edited by Eunice Winters. Baltimore: Johns Hopkins University Press, 1950.

E   Milgram, Stanley. "Some Conditions of Obedience and Disobedience to Authority." *Human Relations* 18 (1965): 57–75.

E   ——. *Obedience to Authority*. New York: Harper & Row, 1974.

E   Miller, R. C. *Live Until You Die*. Philadelphia: United Church Press, 1974.

E   Montange, Charles H. "Informed Consent and the Dying Patient." *The Yale Law Journal* 83 (July 1974): 1632–1664.

E   Moody, Raymond. *Life After Life*. New York: Bantam Books, 1976.

KE Moore, Asher. "Psychoanalysis, Man, and Value." *Inquiry* 4 (Spring 1961): 53–65.

K   Moore, Michael S. "Some Myths about Mental Illness." *Archives of General Psychiatry* 32 (1975): 1483–1497.

HK Morgagni, Giovanni. *De Sedibus et causis morborum pes anatomen indagatis*. Venice, 1761.

E   Morison, Robert. "Death: Process or Event?" *Science* 173 (20 August 1971): 694–98.

E   ——. "The Dignity of the Inevitable and Necessary." In *Death Inside Out*, pp. 97–100. Edited by Peter Steinfels and Robert Veatch. New York: Harper & Row, 1974.

K   Mullane, Harvey. "Psychoanalytic Explanation and Rationality." *The Journal of Philosophy* 68 (July 22, 1971): 413–426.

E   Muller, Hermann. *Out of the Night*. New York: Vanguard Press, 1935.

K   ——. "Human Evolution by Voluntary Choice of Germ Plasma." *Science* 134 (8 September 1961): 643–49.

K   Murphy, Edmond. *The Logic of Medicine*. Baltimore: Johns Hopkins University Press, 1976.

K   ——. "Classification and Its Alternatives." In *Clinical Judgment*, pp. 59–85. Edited by H. T. Engelhardt, Jr., S. Spicker, and B. Towers. Dordrecht: D. Reidel, 1979.

K   Murphy, George. "The Physician's Responsibility for Suicide." *Annals of Internal Medicine* 82 (March 1975): 301–308.

E   National Academy of Sciences. *Academy Forum: Research with Recombinant DNA*. Washington, D.C.: National Academy of Sciences, 1977.

E   National Commission for the Protection of Human Subjects of Biomedical and Behavioral Research. *Research on the Fetus*. Publ. no. (OS) 76–127, 128. Washington, D.C.: H.E.W., 1975.

E ———. *Research Involving Prisoners*. Publ. no. (OS) 76–131, 132. Washington, D.C.: H.E.W., 1976.

E ———. *Report and Recommendations on Psychosurgery*. Publ. no. (OS) 77–0001. Washington, D.C.: H.E.W., 1977a.

E ———. *Psychosurgery* (Appendix). Publ. no. (OS) 77–0002. Washington, D.C.: H.E.W., 1977b.

E ———. *Research Involving Children*. Publ. no. (OS) 77–0004, 0005. Washington, D.C.: H.E.W., 1977c.

E ———. *Research Involving Those Institutionalized as Mentally Infirm*. Publ. no. (OS) 78–0006, 0007. Washington, D.C.: H.E.W., 1978.

E National Institutes of Health. *Recombinant DNA Research: Documents Relating to "NIH Guidelines for Research Involving Recombinant DNA Molecules,"* vol. 1. Publ. no. (NIH) 78–1139. Washington, D.C.: H.E.W., 1976.

E Nicholson, Susan T. *Abortion and the Roman Catholic Church*. Studies in Religious Ethics, no. 2, *Journal of Religious Ethics* (1978).

K Niebyl, Peter. "Sennert, Van Helmont, and Medical Ontology." *Bulletin of the History of Medicine* 45 (March-April 1971): 115–137.

E Noonan, John T., Jr. *Contraception*. Cambridge, Mass.: Harvard University Press, 1965.

E ———, ed. *The Morality of Abortion: Legal and Historical Perspectives*. Cambridge, Mass.: Harvard University Press, 1970.

E Nozick, Robert. *Anarchy, State, and Utopia*. New York: Basic Books, 1974.

E Nuremberg Code. In *Trials of War Criminals Before the Nuremberg Military Tribunals under Control Council Law, No. 10*. Washington, D.C.: U.S. Government Printing Office, 1949.

HK Osterlen, F. *Medical Logic*. Edited and tr. by G. Whitney. London: Sydenham Society, 1855.

E Outka, Gene. "Social Justice and Equal Access to Health Care." *Journal of Religious Ethics* 2 (September 1974): 11–32.

E Owens, Joseph. "Aristotelian Ethics, Medicine, and the Changing Nature of Man." In *Philosophical Medical Ethics: Its Nature and Significance*, pp. 127–142. Edited by Stuart Spicker and H. T. Engelhardt, Jr. Dordrecht: D. Reidel, 1977.

HKE Pagel, Walter. *The Religious and Philosophical Aspects of van Helmont's Science and Medicine*. Baltimore: Johns Hopkins University Press, 1944.

HK ———. *Paracelsus: An Introduction to Philosophical Medicine in the Era of the Renaissance*. Basel: S. Karger, 1958.

E   *Park v. Chessin,* 400 N.Y.S.2d 110 (N.Y. Sup. Ct., App. Div., December 2, 1977).

K   Parsons, Talcott. *The Social System.* New York: Free Press, 1951.

K   ———. "The Mental Hospital as a Type of Organization." In *The Patient and the Mental Hospital,* pp. 108–29. Edited by M. Greenblatt et al. Glencoe, Ill.: Free Press, 1957.

K   ———. "Definitions of Health and Illness in the Light of American Values and Social Structure." In *Patients, Physicians and Illness,* pp. 165–187. Edited by E. G. Jaco. Glencoe, Ill.: Free Press, 1958a.

K   ———. "Illness, Therapy and the Modern Urban American Family." In *Patients, Physicians and Illness,* pp. 234–45. Edited by E. G. Jaco. Glencoe, Ill.: Free Press, 1958b.

K   Peiffer, Jargen. "Zur Zeitgebundenheit diagnostischer Krankheitsbezeichnungen." *Metamed* 1 (1977): 123–128.

E   Pellegrino, Edmund. "Moral Agency and Professional Ethics: Some Notes on Transformation of the Physician-Patient Encounter." In *Philosophical Medical Ethics: Its Nature and Significance,* pp. 213–20. Edited by H. T. Engelhardt, Jr., and Stuart Spicker. Dordrecht: D. Reidel, 1977.

K   Penfield, Wilder. *The Excitable Cortex in Conscious Man.* Springfield, Ill.: Charles C. Thomas, 1958.

K   ———. *The Mystery of the Mind.* Princeton, N.J.: Princeton University Press, 1975.

E   Percival, Thomas. *Medical Ethics.* Edited by Chauncey Leake. Baltimore: Williams and Wilkins, 1927. Reprint ed., New York: Robert E. Krieger, 1975.

E   *Personal Privacy in an Information Society: The Report of the Privacy Protection Study Commission.* Washington, D.C.: U.S. Government Printing Office, 1977.

K   Peszke, Michael Alfred. "Is Dangerousness an Issue for Physicians in Emergency Commitment?" *American Journal of Psychiatry* 132 (August 1975): 825–828.

HK   Pinel, Philippe. *Nosographie philosophique, ou la méthode de l'analyse appliqué à la médecine.* Paris: Richard, Caille, et Ravier, 1798.

E   Platt, Michael. "Commentary: On Asking to Die." *Hastings Center Report* 5 (December 1975): 9–12.

E   *Poelker v. Doe* 97 S. Ct. 2391 (1977).

K   Polten, Eric P. *Critique of the Psycho-Physical Identity Theory.* The Hague: Mouton, 1973.

E   Pope Pius XII. "The Prolongation of Life." *The Pope Speaks* 4 (1958): 393–398.

K   Pople, Harry; Myers, Jack; and Miller, R. "Dialogue: A Model of Diagnostic Logic for Internal Medicine." In *Advance Papers for the Fourth International Joint Conference on Artificial Intelligence, Tbilisi, Georgia, U.S.S.R.,* vol. 2, pp. 848–855. Cambridge, Mass.: Artificial Intelligence Laboratory, 1975.

K   Popper, Karl R., and Eccles, John C. *The Self and Its Brain.* New York: Springer-Verlag, 1977.

HE   Post, Alfred C., et al. *An Ethical Symposium: Being a Series of Papers Concerning Medical Ethics and Etiquette from the Liberal Standpoint.* New York: Putnam, 1883.

K   Post, Richard H. "Population Difference in Red and Green Color Vision Deficiency: A Review and a Query on Selection Relaxation." *Eugenics Quarterly* 9 (1962): 131–146.

E   Potter, Van R. *Bioethics: Bridge to the Future.* Englewood Cliffs, N.J.: Prentice-Hall, 1971.

K   Proppe, A. "Computer-Diagnostik." *Datenbearbeitung und Medizin.* IBM Seminar Series. Liebenwell, 1969.

K   ――――. "Notwendigkeit und Problematik einer Computer-Diagnostik." In *Computer: Werkzeug der Medizin,* pp. 127–159. Edited by C. Ehlers, N. Hollberg and A. Proppe. Berlin: Springer-Verlag, 1970.

K   Puccetti, R. "Brain Bisection and Personal Identity." *British Journal for the Philosophy of Science* 24 (1973): 339–55.

E   Purtilo, Ruth. "Ethics Teaching in Allied Health Fields." *Hastings Center Report* 8 (April 1978a): 14–16.

E   ――――. *Health Professional-Patient Interaction.* Philadelphia: W. B. Saunders, 1978b.

E   Rabkin, Mitchell T.; Gillerman, Gerald; and Rice, Nancy R. "Orders Not to Resuscitate." *New England Journal of Medicine* 295 (August 12, 1976): 364–66.

E   Rachels, James. "Active and Passive Euthanasia." *New England Journal of Medicine* 292 (January 9, 1975): 78–80.

HE   Radbill, Samuel X. "A History of Medical Ethics." *Philadelphia Medicine* 58 (1962): 873–876.

E   Ramsey, Paul. *Fabricated Man.* New Haven: Yale University Press, 1970a.

E   ――――. *The Patient as Person.* New Haven: Yale University Press, 1970b.

E   ――――. "Ethics of a Cottage Industry in an Age of Community and Research Medicine." *New England Journal of Medicine* 284 (April 1, 1971): 700–706.

E   ――――. "Shall We Reproduce? I. The Medical Ethics of *In Vitro* Fertilization." *Journal of the American Medical Association* 220 (June 5, 1972): 1345–1350.

E   ――――. "Shall We Reproduce? II. Rejoinders and Future Forecast." *Journal of the American Medical Association* 220 (June 12, 1972): 1480–1485.

E   ———. "The Indignity of 'Death with Dignity.'" In *Death Inside Out,* pp. 81–96. Edited by Peter Steinfels and Robert W. Veatch. New York: Harper & Row, 1974.

E   ———. *The Ethics of Fetal Research.* New Haven: Yale University Press, 1975.

E   ———. "Kant's Moral Theology or a Religious Ethics?" In *Knowledge, Value, and Belief,* pp. 44–74. Edited by H. T. Engelhardt, Jr. and Daniel Callahan. Hastings-on-Hudson, N.Y.: Institute of Society, Ethics, and the Life Sciences, 1977.

E   ———. *Ethics at the Edge of Life.* New Haven: Yale University Press, 1978.

E   Ramsey, Paul, and McCormick, Richard, eds. *Doing Evil to Achieve Good: Morality in Conflict Situations.* Chicago: Loyola University Press, 1979.

HE   Rau, Wolfgang T. *Gedanken von dem Nutzen und der Nothwendigkeit einer medicinischen Policeyordnung in einem Staat.* Ulm: Stettin, 1764.

E   Rawls, John. *A Theory of Justice.* Cambridge, Mass.: Harvard University Press, 1971.

E   *Recombinant DNA Research,* vol. 1: *Document Relating to "NIH Guidelines for Research Involving Recombinant DNA Molecules," Feb. 1975–June 1976.* Publ. no. (NIH) 76–1138. Washington, D.C.: H.E.W. 1976. Vol. 2: *Document Relating to "NIH Guidelines for Research Involving Recombinant DNA Molecules."* Publ. no. (NIH) 78–1139. Washington, D.C.: H.E.W., 1978.

K   Redlich, F. C. "The Concept of Normality." *American Journal of Psycho-Therapy* 6 (1952): 553.

K   ———. "Editorial Reflections on the Concepts of Health and Disease." *The Journal of Medicine and Philosophy* 1 (1976): 269–280.

E   Reich, Warren T., ed. *The Encyclopedia of Bioethics.* 4 vols. New York: Macmillan and Free Press, 1978.

E   ———. "Life: Quality of Life." In *Encyclopedia of Bioethics,* pp. 829–840. Edited by Warren T. Reich. New York: Macmillan and Free Press, 1978.

E   Reiser, Stanley J.; Dyck, Arthur J.; and Curran, William J., eds. *Ethics in Medicine: Historical Perspectives and Contemporary Concerns.* Cambridge, Mass.: M.I.T. Press, 1977.

E   Rescher, Nicholas. "The Allocation of Exotic Medical Lifesaving Therapy." *Ethics* 79 (April 1969): 173–186.

E   Resnick, Jerome H., and Schwartz, Thomas. "Ethical Standards as an Independent Variable in Psychological Research." *American Psychologist* 28 (February 1973): 134–139.

HK   Risse, Guenter B. "Kant, Schelling, and the Early Search for a Philosophical 'Science' of Medicine in Germany." *Journal of the History of Medicine and Allied Sciences* 27 (April 1972): 145–158.

HK ———. "The Quest for Certainty in Medicine: John Brown's System of Medicine in France." *Bulletin of the History of Medicine* 45 (January-February 1971): 1–12.

E   Robertson, John A., and Fost, Norman. "Passive Euthanasia of Defective Newborn Infants: Legal and Moral Considerations." *Journal of Pediatrics* 88 (1976): 883–89.

E   Roblin, Richard. "XYY Controversy in Boston." *The Hastings Center Report* 5 (August 1975): 5–8.

HK Romanell, Patrick. "Medicine as New Key to Locke." *Texas Reports on Biology and Medicine* 32 (Spring 1974): 275–285.

K   Romberg, Ernst. *Lehrbuch der Krankheiten des Herzens und der Blutgefässe.* 2d ed. Stuttgart: Verlag von Ferdinand Enke, 1909.

K   Rose, J., and Mitchell, J. H., eds. *Advances in Medical Computing.* Edinburgh: Churchill Livingston, 1975.

HE Rosen, George. *From Medical Police to Social Medicine: Essays on the History of Health Care.* New York: Science History Publications, 1974.

K   Rosenhan, D. L. "On Being Sane in Insane Places." *Science* 179 (19 January 1973): 250–258.

E   Rosner, Fred. *Modern Medicine and Jewish Law.* New York: Yeshiva University Press, 1972.

K   Rothschuh, Karl. *Prinzipien der Medizin.* München-Berlin: Urban and Schwarzenberg, 1965.

K   ———. *Konzepte der Medizin.* Stuttgart: Hippokrates Verlag, 1977a.

K   ———. "Krankheitsvorstellung, Krankheitsbegriffe, Krankheitskonzept." *Metamed* 1 (1977b): 106–114.

HE Rush, Benjamin. *Observations on the Duties of a Physician and Methods of Improving Medicine.* Philadelphia: Prichard and Hall, 1789.

K   Sackett, David. "Book Review: *The Logic of Medicine.*" *The Journal of Medicine and Philosophy* 2 (1977): 71–76.

E   Sade, Robert M. "Medical Care as a Right: A Refutation." *New England Journal of Medicine* 285 (December 2, 1971): 1288–1292.

K   Sadegh-zadeh, Kazem. "Grundlagenprobleme einer Theorie der klinischen Praxis." *Metamed* 1 (1977a): 76–102.

K   ———. "Krankheitsbegriffe und nosologische Systeme." *Metamed* 1 (1977b): 4–41.

K   ———. "The Nature and Purpose of *Metamed.*" *Metamed* 1 (1977c): 2.

HK Sauvages de la Croix, François Boissier de. *Nosologia methodica sis-*

tens morborum classes juxta Sydenhami mentem et botanicorum ordinem. 5 vols. Amsterdam: Fratrum de Tournes, 1768.

E   Saunders, Cicely. "The Problem of Euthanasia." *Nursing Times* 72 (July 1, 1976): 1003–1005.

K   Schäfer, Lothar. "Medizinische Forschung." In *Technology, Science, and the Art of Medicine*. Edited by Corinna Delkeskamp, H. T. Engelhardt, Jr., and Stuart Spicker. Dordrecht: D. Reidel, 1980.

K   Schafer, Roy. *A New Language for Psychoanalysis*. New Haven: Yale University Press, 1976.

K   Schaffner, Kenneth. "Reduction, Reductionism, Values, and Progress in the Biomedical Sciences." In *Logic, Laws, and Life*, pp. 143–171. Edited by R. Colodny. Pittsburgh: University of Pittsburgh Press, 1977.

E   Schowalter, John E.; Ferholt, Julian B.; and Mann, Nancy M. "The Adolescent Patient's Decision to Die." *Pediatrics* 51 (January 1973): 97–103.

E   Schroeder, Oliver, ed. *Physician in the Courtroom*. Cleveland: Press of Case-Western Reserve University, 1954.

K   Schutz, Alfred, and Luckmann, Thomas. *The Structures of the Life-World*. Tr. R. M. Zaner and H. T. Engelhardt, Jr. Evanston: Northwestern University Press, 1973.

K   Scriven, Michael. "Clinical Judgment." In *Clinical Judgment*, pp. 3–16. Edited by H. T. Engelhardt, Jr., Stuart Spicker, and Bernard Towers. Dordrecht: D. Reidel, 1979.

K   Sedgwick, Peter. "Illness—Mental and Otherwise." *Hastings Center Studies* 1 (1973): 19–40.

E   Self, Donnie, ed. *The Role of the Humanities in Medical Education*. Norfolk, Virginia: Teagle and Little, 1978.

K   Shaffer, Jerome. "Round Table Discussion." In *Evaluation and Explanation in the Biomedical Sciences*, pp. 215–219. Edited by H. T. Engelhardt, Jr., and Stuart Spicker. Dordrecht: D. Reidel, 1975.

K   ———. "Personal Identity: The Implications of Brain Bisection and Brain Transplants." *Journal of Medicine and Philosophy* 2 (June 1977): 147–161.

E   Shaw, Anthony. "Dilemmas of Informed Consent in Children." *New England Journal of Medicine* 289 (October 1973): 885–890.

K   Sherrington, Sir Charles. *Man on His Nature*. Cambridge: Cambridge University Press, 1951.

K   Sherwood, Michael. *The Logic of Explanation in Psychoanalysis*. New York: Academic Press, 1969.

HE   Shryock, Richard. *Medical Licensing in America, 1650–1965*. Baltimore: Johns Hopkins University Press, 1967.

HK   ———. *The Development of Modern Medicine: An Introduction to the Social and Scientific Factors Involved.* Philadelphia: University of Pennsylvania Press, 1976.

K   Siegler, Miriam, and Osmond, Humphrey. "The 'Sick Role' Revisited." *Hastings Center Studies* 1 (1973): 41–58.

K   ———. *Models of Madness, Models of Medicine.* New York: Harper & Row, 1974.

E   Singer, Peter. "Famine, Affluence, and Morality." *Philosophy and Public Affairs* 1 (Spring 1972): 229–243.

E   ———. "Altruism and Commerce: A Defense of Titmuss Against Arrow." *Philosophy and Public Affairs* 2 (Spring 1973): 312–320.

E   ———. *Animal Liberation.* New York: Avon Books, 1975.

KE   Skinner, B. F. *Beyond Freedom and Dignity.* New York: Alfred A. Knopf, 1971.

E   Smith, Harmon L. *Ethics and the New Medicine.* Nashville: Abingdon Press, 1970.

HK   Smith, James. "Review of Dr. Cartwright's Report on the Diseases and Peculiarities of the Negro Race." *The New Orleans Medical and Surgical Journal* 8 (1851–1852): 228–237.

HE   Smithies, Frank. "On the Origin and Development of Ethics in Medicine and the Influence of Ethical Formulae upon Medical Practice." *Annals of Clinical Medicine* 3 (1924–1925): 573–603.

K   Sober, Elliott. "The Art and Science of Clinical Judgment: An Informational Approach." In *Clinical Judgment,* pp. 29–44. Edited by H. T. Engelhardt, Jr., Stuart Spicker, and Bernard Towers. Dordrecht: D. Reidel, 1979.

H   Soemmerring, Samuel Thomas. *Über das Organ der Seele.* Königsberg: F. Nicolovius, 1796.

HK   Spencer, Herbert. *The Principles of Psychology.* London: Longman, Brown, Green, and Longmans, 1855.

K   Sperry, R. W. "Forebrain Commissurotomy and Conscious Awareness." *Journal of Medicine and Philosophy* 2 (June 1977): 101–126.

E   Sperry, William L. *The Ethical Basis of Medical Practice.* New York: P. B. Hoeber, 1950.

K   Spicker, Stuart F. "Commemorative Remarks in Honor of Erwin W. Straus." In *Mental Health: Philosophical Perspectives,* pp. 145–155. Edited by H. T. Engelhardt, Jr., and Stuart Spicker. Dordrecht: D. Reidel, 1978.

K   ———. "Intuition and the Process of Diagnosis: Unwarranted Epistemological Worries in the Art of Technology." In *Technology, Science, and the Art of Medicine.* Edited by Corinna Delkeskamp, H. T. Engelhardt, Jr., and Stuart Spicker. Dordrecht: D. Reidel, 1980.

K   Spicker, Stuart F., and Engelhardt, H. T., Jr., eds. *Philosophical*

Dimensions of the Neuro-Medical Sciences. Dordrecht: D. Reidel, 1976.

E   ———. *Philosophical Medical Ethics: Its Nature and Significance.* Dordrecht: D. Reidel, 1977.

E   Spicker, Stuart F., and Gadow, Sally, eds. *Nursing: Images and Ideals: Opening Dialogue with the Humanities.* New York: Springer, 1980.

E   Steinfels, Margaret, and Levine, Carol, eds. "In the Service of the State: The Psychiatrist as Double Agent." *Hastings Center Report* 8 (April 1978): 1–24.

E   Steinfels, Peter. "The Politics of Abortion." *Commonweal* (July 22, 1977), p. 456.

HK   Stengel, D. "Hughlings Jackson's Influence in Psychiatry." *British Journal of Psychiatry* 109 (1963): 348–355.

K   Suppes, Patrick. "The Logic of Clinical Judgments: Bayesian and other Approaches." In *Clinical Judgment,* pp. 145–159. Edited by H. T. Engelhardt, Jr., Stuart Spicker, and Bernard Towers. Dordrecht: D. Reidel, 1979.

K   Susser, M. *Causal Thinking in the Health Sciences: Concept Strategies of Epidemiology.* New York: Oxford University Press, 1973.

HK   Sydenham, T. *Observationes medicae circa morborum acutorum historiam et curationem.* London: G. Kettilby, 1676.

HK   ———. *The Works of Thomas Sydenham, M.D.* Edited by George Wallis, M.D. London: G. G. H. Robinson, 1848.

KE   Szasz, Thomas. "The Myth of Mental Illness." *American Psychologist* (1960): 113–118.

KE   ———. *The Myth of Mental Illness.* New York: Hoeber-Harper, 1961.

KE   ———. *The Ethics of Psychoanalysis.* New York: Basic Books, 1965.

KE   ———. "The Ethics of Addiction." *The American Journal of Psychiatry* 128 (November 1971): 541–546.

KE   ———. "Bad Habits Are Not Diseases: A Refutation of the Claim that Alcoholism is a Disease." *The Lancet* 2 (July 8, 1972): 83–84.

KE   ———. *Ceremonial Chemistry.* Garden City, N.Y.: Doubleday, 1974.

KE   ———. "The Concepts of Mental Illness: Explanation or Justification?" In *Mental Health: Philosophical Perspectives,* pp. 235–250. Edited by H. T. Engelhardt, Jr., and Stuart F. Spicker. Dordrecht: D. Reidel, 1978.

HK   Szumowski, W. "La Philosophie de la medicine, son histoire, son essence, sa denomination et sa definition." *Archives Internationales de l'Histoire des Sciences* 9 (1949): 1097–1141.

K   Taylor, F. Kraupl. "Part 1. A Logical Analysis of the Medico-Psychological Concept of Disease." *Psychological Medicine* 1 (November 1971): 356–364.

K ———. "Part 2. A Logical Analysis of the Medico-Psychological Concept of Disease." *Psychological Medicine* 2 (February 1972): 7–16.

E *The Teaching of Bioethics.* Report of the Commission on the Teaching of Bioethics. Hastings-on-Hudson, N.Y.: Institute of Society, Ethics, and the Life Sciences, 1976.

E Tedeschi, G. "On Tort Liability for 'Wrongful Life.'" *Israel Law Review* 1 (October 1966): 513–538.

HE Temkin, Owsei. "Medicine and the Problems of Moral Responsibility." *Bulletin of the History of Medicine* 23 (1949): 1–20.

K ———. "On the Interrelationship of the History and Philosophy of Medicine." *Bulletin of the History of Medicine* 30 (1956): 241–251.

HK ———. "The Scientific Approach to Disease: Specific Entity and Individual Sickness." In *Scientific Change,* pp. 629–647. Edited by A. C. Crombie, London: Heinemann, 1961.

E Thomson, Judith Jarvis. "A Defense of Abortion." *Philosophy and Public Affairs* 1 (Fall 1971): 120–139.

E Tooley, Michael. "Abortion and Infanticide." *Philosophy and Public Affairs* 2 (Fall 1972): 37–65.

K Toulmin, Stephen. "The Logical Status of Psycho-Analysis." *Analysis* 9 (1948): 23–29.

E ———. "The Meaning of Professionalism: Doctors' Ethics and Biomedical Sciences." In *Knowledge, Value, and Belief,* pp. 254–278. Edited by H. T. Engelhardt, and Daniel Callahan. Hastings-on-Hudson, N.Y.: Institute of Society, Ethics, and the Life Sciences, 1977.

E Trivers, Robert. "Parent-Offspring Conflict." *American Zoologist* 14 (1974): 249–264.

HK Trotter, Thomas. *An Essay, Medical, Philosophical, and Clinical, on Drunkenness and Its Effects on the Human Body.* 2d ed. London: Longman, Hurst, Rees, and Orine, 1804.

E Troyer, John. "Euthanasia, the Right to Life, and Moral Strictures: A Reply to Prof. Kohl." In *Philosophical Medical Ethics: Its Nature and Significance,* pp. 85–89. Edited by H. T. Engelhardt, Jr., and Stuart Spicker. Dordrecht: D. Reidel, 1977.

K Tsouyopoulos, Nelly. "The Scientific Status of Medical Research." In *Technology, Science, and the Art of Medicine.* Edited by Corinna Delkeskamp, H. T. Engelhardt, Jr., and Stuart Spicker. Dordrecht: D. Reidel, 1980.

HK van den Berg, J. H. "A Metabletic-Philosophical Evaluation of Mental Health." In *Mental Health: Philosophical Perspectives,* pp. 121–135. Edited by H. T. Engelhardt, Jr., and Stuart Spicker. Dordrecht: D. Reidel, 1978.

E   van Till, H. A. H. "Diagnosis of Death in Comatose Patients under Resuscitation Treatment: A Critical Review of the Harvard Report." *American Journal of Law and Medicine* 2 (Summer 1976): 1–40.

E   Vaux, Kenneth. *This Mortal Coil*. New York: Harper & Row, 1977.

E   Veatch, Robert M. "Drugs and Competing Drug Ethics." *Hastings Center Studies* 2 (January 1974): 68–80.

E   ———. "The Whole-Brain-Oriented Concept of Death: An Outmoded Philosophical Foundation." *Journal of Thanatology* 3 (1975): 13–30.

E   ———. *Death, Dying, and the Biological Revolution*. New Haven: Yale University Press, 1976a.

E   ———. "What Is a 'Just' Health Care Delivery?" In *Ethics and Health Policy*, pp. 127–153. Edited by R. Veatch and R. Branson, Cambridge: Ballinger, 1976b.

HE   Veith, Ilza. "Medical Ethics Throughout the Ages." *Archives of Internal Medicine* 100 (1957): 504–512.

E   Vermeersch, Arthur. *De castitate*. Rome: Gregorian University, 1921.

HK   Virchow, Rudolf. *Hundert Jahre allgemeiner Pathologie*. Berlin: Verlag von August Huschwald, 1895.

HK   ———. *Disease, Life, and Men*. Tr. Leland Rather. Stanford, Calif.: Stanford University Press, 1958.

HK   von Hartmann, E. *Die Philosophie des Unbewussten*. 3 vols. Berlin, 1869.

E   von Wright, Georg Henrik. *The Varieties of Goodness*. New York: Humanities Press, 1963.

E   Walters, LeRoy. "Some Ethical Issues in Research Involving Human Subjects." *Perspectives in Biology and Medicine* 20 (1977): 193–211.

E   ———, ed. *Bibliography of Bioethics*, vol. 1. Detroit, Mich.: Gale Research, 1975. Annual volumes thereafter.

K   Wartofsky, Marx W. "Organs, Organisms, and Disease: Human Ontology and Medical Practice." In *Evaluation and Explanation in the Biomedical Sciences*, pp. 67–83. Edited by H. T. Engelhardt, Jr., and Stuart Spicker. Dordrecht: D. Reidel, 1975.

K   ———. "How to Begin Again: Medical Therapies for the Philosophy of Science." In *PSA 1976*, vol. 2, pp. 109–122. Edited by P. Asquith and F. Suppe. East Lansing, Mich.: Philosophy of Science Association, 1977.

E   Watson, James D. "Moving Toward the Clonal Man: Is This What We Want?" *The Atlantic* (March 1971), pp. 50–53.

K   Watson, John B. "Psychology as the Behaviorist Views It." *Psychological Review* 20 (1913): 158–177.

K   Weed, Lawrence. *Medical Records, Medical Education, and Patient Care*. Chicago: Year Book Medical Publications, 1970.

E   Westin, Alan F. *Computers, Health Records, and Citizen Rights.* New York: Petrocelli Books, 1977.

K   Westmeyer, Hans. *Logik der Diagnostik—Grundlagen einer normativen Diagnostik.* Stuttgart: Kohlhammer, 1972.

K   ———. "The Diagnostic Process as a Statistical-Causal Analysis." *Theory and Decision* 6 (1975): 57–86.

K   ———. "Verhaltenstherapie: Anwendung von Verhaltenstheorien." *Metamed* 1 (1977): 55–75.

K   ———. "Technologische Modell fur die Rationalität von Therapien." In *Technology, Science, and the Art of Medicine.* Edited by Corinna Delkeskamp, H. T. Engelhardt, Jr., and Stuart Spicker. Dordrecht: D. Reidel, 1980.

K   Whitbeck, Caroline. "The Relevance of Philosophy of Medicine for the Philosophy of Science." In *PSA 1976,* vol. 2, pp. 123–138. Edited by P. Asquith and F. Suppe. East Lansing, Mich.: Philosophy of Science Association, 1977a.

K   ———. "Causation in Medicine: The Disease Entity Model." *Philosophy of Science* 44 (December 1977b): 619–637.

K   White, Robert B., and Engelhardt, H. T., Jr. "Case Studies in Bioethics: A Demand to Die." *Hastings Center Report* 5 (June 1975): 9–10.

K   Wieland, W. *Diagnose—Überlegungen zur Medizintheorie.* New York: DeGruyter, 1975.

HK   Wigan, A. L. *A New View of Insanity: The Duality of the Mind.* London: Longman, 1844.

E   *Williams v. New York State,* 260 NY 2d 953 (June 1965).

E   Wilson, E. O. *Sociobiology: The New Synthesis.* Cambridge, Mass.: Harvard University Press, 1975.

K   Wolf, Stewart. "Disease as a Way of Life: Neural Integration in Systematic Pathology." *Perspectives in Biology and Medicine* 4 (Spring 1961): 288–305.

K   World Health Organization. *Constitution of the World Health Organization,* (Preamble). In *The First Ten Years of the World Health Organization.* Geneva: World Health Organization, 1958.

K   Wulff, Henrik. *Rational Diagnosis and Treatment.* Oxford: Blackwell Scientific, 1976.

K   Wunderlich, Carl. "Einleitung." *Archiv für physiologische Heilkunde* 1 (1842): 1–24.

K   Yankelovich, Daniel, and Barrett, William. *Ego and Instinct.* New York: Random House, 1970.

HK   Young, Robert M. *Mind, Brain, and Adaptation in the Nineteenth Century.* Oxford: Clarendon Press, 1970.

K   Zechmeister, Klaus. *Arzt und Weltanschauung.* Berlin: Akademie-Verlag, 1972.

E   *Zepeda v. Zepeda,* (Ill. Ct. of Appeals) 190 N.E. 2d 849 (1963).

Part III
Sociology

Chapter 7
Sociology of Science and Technology
Jerry Gaston, SOUTHERN ILLINOIS UNIVERSITY, CARBONDALE

The sociology of science is a relatively new specialty in sociology. The state of the art, as in other parts of the social sciences, leaves much to be developed—a fortunate situation for scholars new to the field. Although young, the sociology of science has quickly become well entrenched among the research specialties of sociology, though not among the teaching specialties. The sociology of science differs considerably from the sociology of technology. "Sociology of technology" may become an area with scholars investigating systematically a set of focused problems, but the sociology of science has already become institutionalized within the larger sociological profession. Although there are many problems in technology for sociological investigation, sociologists have not embarked on that investigation in sufficient numbers to constitute an identifiable group comparable to that operating in the sociology of science.

This chapter concentrates on the state of the art, introducing the reader to the main topics through summaries of the published research. There is little critical discussion of the methods that scholars have used in collecting data, or of the researchers' interpretations, or of reasonable alternative interpretations. The chapter is primarily an introduction to the work of scholars in the field, illustrating what they have considered worthy of research. It is neither an exercise in scholarly criticism nor an attempt to convince the reader that the researchers should have been doing different types of research.

Where ethical or value questions are implicit in the research, and where there are areas in which ethical and value questions might be researched, those matters are discussed, although the bulk of this material is reserved for section IV, below. The chapter concludes with a brief bibliographic introduction to guide the reader through the literature, and a detailed bibliography.

INTRODUCTION: SOCIAL AND INTELLECTUAL
CONTEXTS OF SOCIOLOGY OF SCIENCE
AND TECHNOLOGY

It is reasonable to ask: What is the sociology of science and technology? To answer that question requires an idea of what *sociology* is. There are dozens of definitions of what sociology is or ought to be, but we will just state arbitrarily what the sociological perspective is and will not try to argue what it should be.

Sociology looks at phenomena generated in the social world and tries to explain one type of social phenomenon using another type of social phenomenon. It is interested in any sort of social behavior. Not usually concerned primarily with an individual's behavior in isolation, sociology focuses on behavior that results from group membership. Sociologists do not always observe and measure whole groups; more frequently, observation is of individual members of a particular group.

What can sociology say about science? In broadest terms, sociological inquiry is focused on three sets of questions: What social phenomena influence science? How does science influence society? What is the nature of science as a social phenomenon?

Science is an attempt to understand the natural world. Scientific knowledge is the result of a method; it is knowledge of physical or social regularities in the universe. Science is not the figment of someone's imagination. It is not a creation. Scientific knowledge may be discovered through creative imagination, but the knowledge must be more than a unique experience. How, then, can social phenomena influence science, if scientific knowledge is produced through discovery and not through creation?

By this time in the history of the sociology of science, it is clear that society can affect science. Science, on any reasonable scale, is possible only in a society that permits it to exist because science cannot exist without social support. Science cannot operate well in a hostile environment. Without an educational system that produces new recruits, for example, science would suffer. Without funds to purchase equipment, supplies, and labor, most kinds of science would be impossible. Without the institutionalized role of scientist in the occupational structure, research would be possible only by the idle, the rich, or those both idle and rich.

It is clear also that science and technology affect society. Without scientific knowledge and many technological capabilities, society would be much different. Perhaps the most vivid example is that science and technology have made it possible to support populations at a size and standard of living unknown in previous history. But the fact that science affects society at all levels, not just the material dimensions, is so well known that it needs no argument. What is equally clear, however, is that we know relatively little about the way the reciprocal relation works, and we cannot predict with much accuracy what the effect will be of particular scientific advances.

The third type of general question in the sociology of science is the nature of science as a social phenomenon. This area has received the most attention, and this chapter will reflect that research emphasis in the literature reviewed. It would be too simple to say that science is social in nature because science is comprised of human beings working in social environments producing new knowledge. Of course that is true, but it does not go far enough. We must consider the social institution of science, with its goals, norms, and systems for maintaining conformity to institutional objectives—in brief, the social system of science itself—before we can begin to understand the many dimensions that provide opportunities for sociological investigation.

If science is susceptible to social influences, in spite of a tendency of some to see science as "above" such mundane forces, it is even easier to understand how technology is influenced by social factors. Technology is not discovered, it is developed. Its development is possible only in a cultural complex that, at least minimally, nurtures it. The cultural complex has many dimensions. The basic raw materials must be present or available; the necessary labor at appropriately skilled levels must be present; the particular process or product must be known; there must be some level of demand that will justify capital investment; there must be legal, moral, or perhaps religious approval of the product; and so forth. Because of such complex sets of potential social influences, it is difficult to explain why a sociology of technology has not developed systematically.

A. Science, Technology, and the Social Perspective

Science and technology are often confused. (The general introduction to this volume, as well as the surveys in chapters 1, 2, 5, and 9, include additional discussions of the relation between science and technology.) Science differs from technology in several important respects. Science is concerned with understanding the basic natural processes in the universe; technology is concerned with developing innovative processes and products. With its task of basic understanding, science has a goal, uses a research process, is organized socially, and has peculiar norms for its participants to observe. For a complete explication, not summarized here, see Bernard Barber, *Science and the Social Order* (1952, reprinted 1978).

The tendency to confuse science and technology has two sources: a misunderstanding of the tasks of each, and the attempt to equalize the social status of the two. The nature of technology is closely related to that part of science designated as "applied science," "pure science" being its semantic opposite. Because intellectual pursuits of a "pure" nature have often been accorded higher respect, applied science has been viewed as a less strenuous intellectual endeavor. Some scholars are purposefully reticent to acknowledge differences between science and technology, in order to avoid invidious comparisons.

Brilliant researchers work in the areas of both science and technology. Sociologically, these areas of pure and applied research are separated by the differing nature of each as a social phenomenon. Researchers questioned whether their work is "pure" research could reply affirmatively if they could also answer affirmatively to the question: Are all aspects of your work publishable without constraints, and is the material available to all who might wish to read it?

What about "secret" pure research? Some scientists may work in potentially sensitive areas and not be permitted to publish their research efforts. If they would normally be able to publish their pure research, but are prevented from doing so by government contract for secret research—an unusual situation—they cannot be considered "pure" scientists. Under this definition, they are applied scientists or technologists during the period of constraint. (For a comprehensive discussion of government and secrecy in research, see Edward A. Shils's *Torment of Secrecy*, 1956.) Although mandates for secrecy may delay the publication of pure research for a time, such material cannot be protected indefinitely. Anyone attempting to keep "pure" research secret is going against the very nature of science. Under the accepted definitions of pure and applied science, any research being protected by such secretive policy must be defined as technology, not science. On the other hand, does freely publishable technological research constitute "pure" research, simply because it is freely publishable? Certainly some technology research is published without restraints, so the criterion is asymmetrical. Freedom to publish does not mandate a definition of pure research, even if it is accepted that constraint on publication necessitates defining the research as applied.

A second criterion distinguishing science from technology is the purpose of the research. Is it avowedly aimed at process or product? If so, it is technology. If it is not, and is still not directed toward fundamental knowledge, then what is it? Perhaps it is a waste of time and resources.

A difficult problem lies in dealing with researchers who claim that their work has such relevance for technology that they cannot know when they are doing science and when technology. If a new instrument or a modification is discovered during scientists' research, with descriptive papers published for others to make use of the knowledge in their work, then it might be argued that this is science. However, it merely shows that scientists are capable of making contributions to technology, just as technologists are capable of contributing to science.

B. Brief History of the Sociology of Science

This brief history focuses on the interest in science, not the chronology of ideas about sociology and science. With few exceptions (P. A. Sorokin; perhaps Thorstein Veblen), professional sociologists paid little attention to the sociology of science until about the 1950s. Robert Merton and Bernard Barber, the two main exceptions and still active in the field, did not have a substantial following until conditions were right for a professional academic specialty to develop.

Bernard Barber published the first comprehensive reviews of the state of the art in 1955 and 1956; in addition to his own *Science and the Social Order* (1952, reprinted 1978), there were not many books to mention. Because the situation changed little in the next few years, when he published an up-to-date review of the state of the art in 1959, there were not many more new references he could include. By the beginning of the 1960s, the situation began to change. According to Norman W. Storer's introduction to Merton's *Sociology of Science* (1973), the latter's "Priorities in Scientific Discovery" (delivered as his presidential address at the 1957 annual meeting of the American Sociological Association) served as the spur that was needed for the growth that began in the 1960s.

In 1960, Joseph Ben-David published two classic papers that exemplify a sociological approach to science (1960a, 1960b). In 1962, the specialty's name was first used in a book title when Barber and Hirsch published their influential *Sociology of Science*. In the same year, Norman Kaplan published a paper on the state of the art, "Sociology of Science," in the *Handbook of Modern Sociology;* in the three-year interval between Barber's 1959 review and Kaplan's survey, there had been enough work to allow many new references. The specialty was about to take off.

Merton describes the institutionalization of the sociology of science in his "The Sociology of Science: An Episodic Memoir" (1977). At the 1974 meeting of the American Sociological Association in Montreal, some eager members met to discuss the formation of a group of sociologists of science. Because many felt constrained by lack of program space at the annual A.S.A. meetings, a special section on the sociology of science was suggested, but the A.S.A. requirement of two hundred members for the establishment of a new section seemed to doom that prospect. (See chapter 8 on the A.S.A. section status of medical sociology.)

Sociologists of science have always communicated with scholars in other disciplines because other disciplines are the primary subject of their studies. There were several nonsociologists at the first meetings. Consensus established an *ad hoc* group to discuss the feasibility of a new international multidisciplinary scholarly society. Once the idea was considered feasible, the group was to prepare a charter. With Robert McGinnis as chairperson, the group discussed alternatives; by mid-1975 a charter was prepared that would establish the Society for Social Studies of Science ("4S"). At the 1975 A.S.A. meeting in San Francisco, the charter was approved and the society was founded.

Subsequently, Robert K. Merton was elected founding president; Warren Hagstrom served as second president during 1977 and 1978. The fact that sociologists have been elected as the first two presidents reflects sociology's involvement in the formation of the society as well as its role in the institutionalization of social studies of science. The third president, for 1979, was Dorothy Nelkin, a science policy specialist; this fact confirms the interdisciplinary character of the society. Membership in the elected council of the society includes scholars from history, political science, and science policy studies. Succeeding elections will most likely expand the representation of disciplines.

Jonathan Cole and Harriet Zuckerman's "Emergence of a Scientific Specialty" (1975) describes the development of the specialty, comparing the rise of the sociology of science to other specialties whose origins have been studied by sociologists of science. Their historical account presents growth rates in the literature and networks of researchers, highlighting Merton's role in the intellectual evolution of the sociology of science. Merton's "Episodic Memoir" (1977) offers a perspective on this intellectual history that goes beyond published information to provide insight into the events that helped shape the sociology of science as a professional academic discipline. One important passage gives a concise summary of a social explanation of why the sociology of science finally became institutionalized:

> By the 1960s and 1970s, public concern with science as the seeming source of social problems had become greatly enlarged, differentiated, and intensified. Developments in the social, political, economic, ecological, and technological contexts of science and in the scientific disciplines themselves led science to become increasingly problematical. Anti-science ideologies and movements emerged, in a mirrorlike reversal of popular image, ready to find science the unquestionable fount of all things evil (Merton, 1977, p. 112).

Merton then presents a long list of specific "evils" that are attributed to science and that have caused it to be declared "a social problem continually generating other social problems."

C. Brief History of the Sociology of Technology

Technology and society are obviously involved in reciprocal relationships. In many areas of society, technology has a significant influence, and various technologies are included in sets of variables in sociological research. All social institutions are affected by technology. For example, the sociology of the family includes the study of effects of technological developments—for example, the effects of automobiles and birth control pills on the institution of the family. Religion has felt both positive and negative effects of radio and television. To say that the polity has both influenced technology and been affected by it is to state the obvious. For example, the national audience of televised presidential debates is a far cry from the minuscule assemblages that attended the Lincoln-Douglas debates. And to mention the most important connection, the economy and technology are so intertwined that it seems impossible to untangle their reciprocal connections.

In spite of the importance of technology in society, now as before, the sociology of technology is still relatively undeveloped. There is no identifiable group of sociologists whose primary concern is the development of research programs and the accumulation of a body of significant research. This situation for sociology is in bold contrast to the history and philosophy of technology, which have developed scholarship to a highly sophisticated level (see chapters 2 and 5).

Although some sociologists have written about technology, much of that literature is intensely emotional. Passionate writing about the evils or the wonders of technology does not constitute a real sociology of technology. As a result of the genuine scholarship—and, to some extent, the more passionate literature—one can expect that the sociology of technology may soon develop into a visible specialty in sociology. After all, it was only when science came to be considered a social problem that a large number of scholars began systematically to develop a sociology of science. Technology is increasingly being seen as a social problem by many people; so if Merton's correct prediction about the sociology of science has sociological relevance for other areas, a systematic sociology of technology cannot be far away.

Because the sociology of technology is not now institutionalized in the sociological profession, one should not conclude that sociologists have not approached the subject, or that other scholars have not examined sociological aspects of technology. Historians' and philosophers' work often includes sociological implications, and that literature is summarized in chapters 2 and 5, respectively.

Awareness of a reciprocal relationship between society and technology is one of the first steps in identifying a sociology of technology. The social aspects of technology have been considered in the light of the way *technology affects society;* the way *society affects technology,* and the *social factors in technology,* have not been emphasized. Section II of this chapter includes a summary of issues that belong under the rubric of a sociology of technology.

D. Problems of Ethics and Values in the Sociology of Science and Technology

Ethical and value issues exist in science and technology themselves, as well as in the sociological study of science and technology. Debates have been waged for decades, by observers within and without science, over whether or not values affect science. Some have argued that values have no place in science and that values have no impact on the development of knowledge. Extremists at the other end of the spectrum claim that nothing in science is unaffected by values. Neither argument carries the weight of full truth. A more moderate view, somewhere between the two extremes, would be more tenable (see also chapter 4, IV. C. in this volume).

To define values as goals or objectives allows the inclusion of science itself as a value. Granting science as a value in itself, the next step admits the values involved in the decision to become a scientist. A further value is expressed in the choice of which discipline to pursue among the sciences. Also, continuing value choices are made in the consideration of alternative problems and methodologies for research.

One of the most notorious areas of value dispute in the sciences recently has been the debate over recombinant DNA research. (For a wide-ranging discussion of this issue, see the 1978 issue of *Daedalus*, "Limits to Inquiry," edited by Gerald Holton.) Against proponents of restraint who favor restrictions, advocates of freedom of research, who place a high value on scientific knowledge, argue that the potential benefits outweigh the potential risks. Contenders of the opposite stance propose to limit research solely on the grounds of the probability of an accident—whatever the magnitude of the probability or the danger threatened. For them, safety is more valuable than the uncertain knowledge that may be derived from such research.

This issue provides an excellent example of value conflict in science. Neither side can prove the validity of its position, for several reasons. The unknowns turn out to be the culprits in the conflict—what sort of accident might happen, and, on the other side, what the benefits might be. Different values are being preferred and weighed by each group: knowledge, safety, freedom from fear, general well-being. It remains a fascinating debate, the conclusion of which will spread its implications throughout our society.

There are also important value considerations in research on the value implications of science and technology. One primary concern is the choice of a research site; another is the choice of which value conflicts to investigate. For example, a researcher might question the importance of sociological processes in one of the following areas: where wealth and health are in conflict, as in medical services; where knowledge and health may be in conflict, as in recombinant DNA; where wealth and knowledge conflict, as in research on the moon.

One important dimension of the value problem in the sociology of science and technology is the question of which social groups stand to lose and which stand to gain. This question is exemplified in the development of the supersonic transport (SST). Its development has been encouraged in order to provide more adequate national defense as well as to bolster the economy. If it is true that the SST has serious implications for defense because it would keep contractors in a ready state for potential military programs, and that it has serious implications for the national economy because a continued positive balance of payments in the aerospace component of international trade is important, then only an extremely hard-core radical could deny that all classes in society would benefit from the SST. Adequate national defense is assumed to be of benefit to all classes in our society, while a stronger economy will obviously benefit only some classes.

Ethical and value considerations become involved with the positions of the actors in a social drama, or with those affected by the behavior of a social group. Presumably what is *ethical* can be determined by some kind of standard of ethics; certainly what is *valuable* can be determined by ranking preferred outcomes in a hierarchical arrangement. Section IV discusses some implicit and explicit problems of ethics and values inherent in the research described in sections I to III.

I. SOCIOLOGY OF SCIENCE

A. Sociological Models of Science

The work of Robert K. Merton and Thomas Kuhn is responsible for the current major orientations in the sociology of science, especially in the United States. Merton's early work, *Science, Technology, and Society in Seventeenth-Century England* (1938) did not provide the model of science that later came to be called Mertonian sociology of science; it was 1942 before the first outline of the "Merton paradigm" was presented. Then it was twenty years before Kuhn's *The Structure of Scientific Revolutions* (1962) came on the scene. Some immediately saw its import (see Merton, 1977, pp. 71–109), and it was soon diffused throughout the community of scholars, who ultimately would use and discuss it widely.

Both these models have received much attention and also much criticism. It would take the equivalent of at least a book-length monograph to deal with the various critics. It is correct to say, nonetheless, that no one has attempted to propose such a comprehensive model as these two; indeed, no one has even proposed an alternative. (For a different view of the matter, see chapter 1, II. I., in this volume.) A brief description of these models, without inclusion of all the criticism, will deny neither their existence nor their validity; it is simply a recognition of the fact that the reader will not benefit from having too many views presented in a short section.

1. The Mertonian Model. In his 1942 paper on the ethos of science, "Science and Technology in a Democratic Order," Merton described the basic normative structure of science. He described the requirements that enable science to progress most efficiently. His model assumes that the institutional goal of science is to extend certified knowledge. (*Certified* is the key element.) These requirements, or functional imperatives, are the institutional values or norms that guide individual scientists' behavior toward research and toward other scientists and their work. (The brief description that follows is elaborated in Barber's *Science and the Social Order,* 1952; Storer's *The Social System of Science,* 1966; and Gaston's *The Reward System in British and American Science,* 1978a.)

To achieve the goal of certified knowledge with maximum efficiency, the institution of science requires communality (publicly sharing information about one's research), disinterestedness (placing truth above personal gain), organized skepticism (withholding judgment on the validity of empirical or theoretical assertions regardless of the status and reputation of the source of the knowledge), and, perhaps most important, universalism (disregarding the personal and social characteristics of other scientists when evaluating the validity or importance of their research). These norms, elaborated within an institutional framework, can be developed into a model that permits comparisons with empirical observations. For instance, two cases of the systematic violation of these institutional norms—scientists under the Nazi regime in Germany (violating universalism) and Lysenkoism in the U.S.S.R. (violating organized skepticism)—are so flagrant that they can be perceived with ordinary observation, without the systematic observation techniques of empirical sociology.

Merton's model of science states that: (1) as a social institution, science has a goal; (2) to achieve the goal requires that scientists, individually and severally, conform to codes of behavior; (3) if the norms are not observed, there will be negative consequences for the development of scientific knowledge. According to Storer, in his introduction to Merton's *The Sociology of Science* (1973), this model was virtually completed when Merton presented his paper "Priorities in Scientific Discovery" (1957, reprinted in Merton, 1973), arguing that the fuel that drives the engine of scientific development is the reward system of science.

Scientists and other observers have known for a long time that discoveries are made, in many instances, nearly simultaneously by two or more scientists. The frequency of simultaneous discovery may be explained, to a large extent, as a function of the knowledge base at that point in time. However, in many cases of multiple discovery, scientists have been engaged, either personally or through the efforts of friends and colleagues, in activities designed to establish their priority. What explains these frequent battles over priority? Because of orientation and perspective, scientists might be inclined to interpret the events as evidence of individual idiosyncrasies, but a sociological interpretation—one that takes into account the fact that science is a social institution in the same way that the economy, the polity, religion, education, and the family are social institutions—gives a superior explanation.

Merton provides an account that illustrates the intrinsic improbability of alternative explanations. One alternative suggests that priority disputes stem from such factors of human nature as egotism. Merton reminds us that the history of thought is strewn with the corpses of those who have tried "to make the hazardous leap from human nature to particular forms of social conduct" (1957 [1973, p. 209]). Others have claimed that scientists as a group are particularly egotistic. Although this claim may be true for some scientists, it is certainly not empirically verified that science has more egotistical practitioners than any other group. Furthermore, even if self-selection did bring many egotistical people into science, in matters of priority disputes the "controversies often involve men of ordinarily modest disposition who act in seemingly self-assertive ways only when they come to defend their rights to intellectual property" (Merton, 1957 [1973, p. 291]). This interpretation overlooks the fact that priority disputes are more frequently pursued by the friends of scientists than by the scientists themselves. The contenders stand to gain little, if anything. Their expression of moral indignation is caused by the violation of a special norm, and they want to see a "wrong" made right. "The very fact of their entering the fray goes to show that science is a social institution with a distinctive body of norms exerting moral authority and that these norms are invoked particularly when it is felt that they are being violated. . . . Thus fights over priority . . . are not merely expressions of hot tempers, although these may of course raise the temperature of controversy; basically they constitute responses to what are taken to be violations of the institutional norms of intellectual property" (p. 293).

Plainly, scientists want due recognition for their discoveries and achievements. After all, scientists have little else to gain for their efforts, aside from a decent level of living. Apparently, peer recognition of outstanding performance is sufficient incentive to prevent capable scientists from turning their talents to more profitable endeavors.

The model of a social system of science in which scientists pursue knowledge in a social environment, hoping and expecting to receive recognition for their original contributions, provides a multitude of research questions—what has come to be called "Mertonian" sociology of science.

2. The Kuhnian Model. Kuhn's model is different from Merton's but complementary. There is a voluminous literature commenting on the varied aspects of Kuhn's ideas, and the brief description that follows is not intended to suggest that Kuhn's model is simple. He views the history of science as a series of cycles in which research efforts are directed by a shared paradigm. Through education and research apprenticeship, scientists are guided to view research problems in the light of the prevailing data and theories. They devote their research to the filling in of the various pieces of the research puzzle. Kuhn calls this stage of scientific development normal science.

This "normal" part of the cycle of science characterizes most of science most of the time. When new empirical data or theoretical insights enter the picture and call into question the accuracy of the current paradigm, then consensus on the state of certified knowledge may be threatened. When an anomaly occurs, if it occurs infrequently, it does little to disturb the consequences. If anomalies in data or calculations persist, and especially if they are replicated by different scientists, the accumulation of anomalies is an important condition that ultimately leads to a lack of consensus.

It usually takes a great deal of contradictory evidence, coupled with certain social conditions, to cause the occurrence of a "revolution." Kuhn allows that various social conditions affect the transition to a new paradigm, sometimes retarding that transition. Holding to the old paradigm—in which scientific leaders were socialized and spent their careers—retards the transition. But it will not stop it. Sooner or later the powerful retire from active work. The progress of truth—the new paradigm—will march on, even if it must march over the remains of scientists formerly recognized as outstanding researchers.

Although Kuhn is an historian of science, his model is clearly a model of the sociology of science. The processes he describes are fundamentally sociological. For example, scientific change is one form of innovation, and innovation and resistance to it have caused serious problems to arise in all social institutions, as sociologists of all kinds have documented.

Kuhn's model has been important for sociologists—probably more so than for his colleagues in history—for several reasons. First, it provides a set of concepts, including "paradigm," "normal science," and "revolution," that have potential for empirical investigation. Second, it views science as a dynamic activity operating with what many perceive to be considerable relativism. Third, it offers an alternative to the functionalist Mertonian model that is resisted by many sociologists because of the latter's association with conservatism and the status quo. Fourth, Kuhn stresses that normal science involves a community of scientists who share values as well as a body of knowledge. Because science is located in a social group and exists because of the group, the sociologist of science has a variety of questions on which to focus.

Although Kuhn's model is pregnant with sociological implications, to say it is superior to Merton's is to misunderstand both models. Kuhn's is a model of a process whereby the *content* of science changes; Merton's is a model of the *social operation* of science. It may be that Kuhn's "revolutions" do result in *momentary* disregard of some of Merton's norms of science, but this proviso means only that there may be times when institutional norms are less effective—not that the norms are irrelevant.

Because scientists often appear to resist new ideas, they seem to be contradicting the norms of science by neglecting the imperative of organized skepticism. A certain level of conservatism, however, prevents the scientific community from forever shifting itself around with the advance of every new idea. (See Barber's "Resistance by Scientists to Scientific Discovery," 1961.) Research must be more orderly and predictable than such fluctuations would indicate. On the other hand, resistance to new ideas could be construed as adherence to organized skepticism. Acceptance of a new idea is not a mandate simply because it is proposed. It must be studied, and even in Kuhn's model, it will eventually be accepted if it is consistent with an acceptable paradigm.

3. Merton, Kuhn, and Critics.

Certainly neither Merton nor Kuhn would argue that their perspectives are either comprehensive or the only viable models that could be created. The older Mertonian model describes the functional imperatives of a social institution with the manifest purpose of extending certified knowledge. Clearly, certified knowledge may become "uncertified" as new facts and new theories arise, but these in turn develop into newer certified knowledge. Merton never intended his model to predict and explain what the *content* of the certified knowledge would be. Instead, it predicts the conditions under which scientific communities will progress toward a state of certified knowledge. Kuhn's model is complementary. It depicts the social processes whereby the content of science has changed over the centuries and probably will change in the future.

Merton's model has been criticized both by sociologists of science and by scholars who have migrated into science studies from other fields. Norman Kaplan (1963b) argued that the norms are problematic because they do not appear to account for what seemed to be happening in Europe in science at that time.

Michael Mulkay (1969) makes two arguments: that the Mertonian norms cannot explain the growth of science, and that no empirical studies prove that the norms of science are peculiar to the scientific community. Mulkay does not argue against the existence of norms in science, but he believes they are technical rather than social.

S. B. Barnes and R. G. A. Dolby (1970) attack what they see as a "school" mentality among sociologists of science influenced by Merton. Their criticism is divided into three parts: (1) claims that it is not possible for skepticism, rationality, and universalism to represent "statistical norms"; (2) a description of various professed as well as statistical norms in the history of science; and (3) a criticism of Merton's idea that pairs of norms produce ambivalence in scientists.

Barnes and Dolby reach a conclusion similar to Mulkay's—that technical norms suffice to provide order and social control within the social organization of science; they conclude that science can operate without social norms. Although technical norms change frequently, they can substitute for social norms in directing the social behavior of scientists. Barnes and Dolby believe that growth and change are explained in their model, and also claim that the ethos of science does not differ from that of other institutions, or society in general.

Richard Whitley (1972) presents the most spirited critique of Mertonian sociology of science. One of his claims is that North American soci-

ologists have reduced the problem of what goes on in science to inputs and
outputs of a "black box"; only what comes out is important, not what happens in the box. His position is that sociologists of science ought to spend
their time studying how knowledge is created; in particular, they should use
a philosophical perspective on the creation of knowledge. In his efforts to
define what a true sociology of science is—a definition that has considerable
merit—Whitley jumps to the erroneous conclusion that sociologists in North
America are opposed to his views and will continue to be opposed because
they are uninterested in questions of the growth of knowledge. His only real
evidence is that sociologists, at the time of his writing, had yet to frame
questions systematically along the lines he was proposing.

Other criticisms have focused on either the content or the presumed
exclusiveness of Mertonian sociology of science. Robert Rothman (1972)
argues that, because the norms are violated, the nature of the ethos needs
reexamination. His evidence seems extremely selective, scarce, and weak.
Leslie Sklair (1972) attacks Mertonian sociology as bourgeois ideology
claiming to be scientific. I. I. Mitroff offers another criticism, basing his
observations on a longitudinal study of forty-two scientists working on moon
research (1974). Mitroff does not consider the Mertonian norms to be in-
valid, irrelevant, or unnecessary, but he is concerned with the enormous
personal commitment of some scientists to their own ideas. This commitment prevents neutrality, but he believes such violations of a norm suggest
the existence of a counternorm rather than some kind of social deviance
that ought to be explained. (See Merton's *Sociological Ambivalence*, 1976,
and Gaston's *The Reward System*, 1978a, for a detailed discussion.)

Scholars, wishing to focus the specialty on new problems as some of
the puzzles in the Mertonian paradigm have moved steadily towards solution, have criticized many aspects of Mertonian sociology. Joseph Ben-
David's paper "The Emergence of Two National Traditions" (1978), hy-
pothesizes that much of the conflict over styles of sociology of science in
Britain and North America stems from different educational systems. In
North America, sociologists of science are usually located in departments of
sociology, and are integrated into the teaching of a sociological curriculum.
The British sociologists of science are seldom located in such departments;
usually they are found in science studies units or in organizations whose
perceptions of science are much different from the traditional view of scientists and students in the universities. Indeed, science studies institutes have
a vested interest in developing a different view of science.

With few exceptions, sociologists of science today tend to favor Kuhn
over Merton. (M. D. King, 1971, is one exception; he takes issue with both
Merton and Kuhn.) The main criticisms of Kuhn come not from sociology
but from history and philosophy. (See the various papers criticizing Kuhn,
and his response to them, in Imre Lakatos and Alan Musgrave, eds., *Criticism and the Growth of Knowledge*, 1970.)

As mentioned several times, the critical difference between Kuhn's and

Merton's models is that each is a picture of a different thing. Although each model has components or logical implications that are testable, neither model has been convincingly validated or invalidated by empirical inquiry. Even with so many journal articles devoted to discussion and critique of the two models, they remain the best available. The best attitude is to take them as complementary.

B. The Scientific Community

Our concern here is the nature of science as a social phenomenon, the way science works as a social institution. This general problem area has received the lion's share of attention in the sociology of science.

Since scientists work in social environments, sociologists are interested in various aspects of the social system of science. Here we will focus on topics that involve important theoretical concepts in sociology. In all social systems, individual members perform in various roles. Social processes are monitored by control mechanisms, one consequence of which, very often, is social stratification. These two concepts are central in the literature on the internal relations of scientific communities.

Warren Hagstrom discusses these concepts extensively in *The Scientific Community* (1965), in which he offers the first integrated presentation of the so-called "exchange theory" of the reward system in science, wherein scientists exchange their research contributions for recognition.

1. Role Performance and Social Stratification in Science   The most important role of a scientist is the production of new knowledge. The priority of this role does not negate the importance of other functions, such as teaching, administration, and service to the community through committee work and consultation. However, for the scientific community these subordinate roles would be irrelevant without the production of new knowledge.

Studies of scientific productivity—predominantly publication records of scientists—consistently conclude that great variations exist in the performance of this role. Some scientists publish part of their doctoral research. Others publish a paper and are never heard from again. Some publish at a low annual rate. Others publish frequently for years and then seem to disappear. A few begin publishing early, before they receive Ph.D. degrees, and continue to publish for the rest of their working careers.

This variation, even among elite scientists, is shown in a study by Wayne Dennis of men listed in the *Biographical Memoirs* of the National Academy of Sciences for 1943–1952; the study is described in Derek Price's *Little Science, Big Science* (1963, p. 41). Of 41 scientists who lived past 70, the top producer had 768 publications, the lowest 27 papers. The group averaged more than 200 papers, and only 15 scientists published less than 100 papers.

Hagstrom (1971, 1974) has produced data on scientists with appointments in graduate departments in selected disciplines; the data show these scientists to have, on the average, a lower publication level than members of the National Academy of Science. For the five years prior to Hagstrom's questionnaire, the average number of papers was: for mathematicians, 6.3; for theoretical physicists, 9.96; for experimental physicists, 7.4; for experimental biologists, 11.6; and for chemists, 12.4. In each case the productivity of about two thirds of the scientists ranged from zero to twice the average number.

The American Council on Education conducted the largest and most comprehensive surveys of American academics, one in 1969 and one in 1972–1973 (Alan Bayer, 1970 and 1973). The surveys asked faculty to indicate their career publications. One question in the later survey asked for the number of published writings in the preceding two years. Among faculty in universities (excluding four-year colleges and other academic institutions), 37.2 percent had published no papers in the preceding two years; 45.6 percent, one to four papers; and 17.2 percent had published five or more papers. This variation over a short period of two years clearly shows that scientists and other scholars as well, at all levels of the hierarchy in American universities—National Academy members, graduate faculty, and faculty in general—differ markedly in their publication behavior.

Studies have been conducted of American scientists alone, including physicists (S. Cole and J. Cole, 1967); sociologists (Lightfield, 1971; Clemente, 1973); sociologists, psychologists, mathematicians, and historians (Babchuk and Bates, 1962); and physiologists (Meltzer, 1956). Other studies have involved several scientific disciplines (Crane, 1965; Hargens and Hagstrom, 1967), and Nobel laureates (Zuckerman, 1967 and 1977a). In each study, publication productivity is measured in a variety of ways and always varies greatly.

Variation in scientific productivity is not peculiar to the United States. Examples from British science show that even in a system of higher education much more homogeneous than that of the United States, there is variation in scientific productivity. (See Blume and Sinclair, "Chemists in British Universities," 1973; Gaston, *Originality and Competition in Science*, 1973; and Halsey and Trow, *British Academics*, 1971.) Generally the variations range from about half who have published fewer than ten papers to less than a quarter who have published twenty or thirty or more.

Although the variation in scientists' productivity has not been explained statistically by these studies, the studies are valuable nevertheless. They have examined not only the number of papers that scientists publish—a quantitative measure of role performance—but also the consequences for scientists of that role performance.

Concern about role performance is tied to its consequences, especially in terms of a stratification system. In science, social stratification is generally based on role performance, unlike the situation in the general society, where other factors, such as the position of one's parents, may be the primary determinant of one's social position. Studies of scientific productivity help to explain the relationship between role performance and placement within the social stratification system of science. A further value of these studies lies in the increase in understanding of the causes of variation in role performance. Although it may not be the actual goal of a particular scientist, the production of scientific knowledge is what allows a person the title of scientist.

Merton summarizes this state of affairs:

> On every side the scientist is reminded that it is his role to advance knowledge, and his happiest fulfillment of that role, to advance knowledge greatly. This is only to say, of course, that in the institution of science originality is at a premium. For it is through originality, in greater or smaller increments, that knowledge advances. When the institution of science works efficiently—and like other social institutions, it does not always do so—recognition and esteem accrue to those who have best fulfilled their roles, to those who have made genuinely original contributions to the common stock of knowledge. Then are found happy circumstances in which self-interest and moral obligation coincide and fuse (1973, p. 293).

Scientists produce knowledge, but they do not own it. After all the hard work, frustration, and agony, they generously "contribute" the results to their scientific peers. Their only reward—other than the usual economic security of a tenured faculty position, for which the salary is predominantly connected to teaching—is the recognition that accrues from having done a good job, of having been competent enough to produce something useful in the research of other scientists.

It is precisely this recognition that provides a large degree of the continuing motivation for scientific research. Whether or not scientists receive due recognition, based on their performance, is the subject of many studies in the sociology of science. Given a situation in which, ideally, scientists would be rewarded with recognition from peers, sociologists are naturally interested in studying the conditions under which the ideal is or is not approximated.

Because scientists working on new certified knowledge (in contrast to applied technological developments) are usually affiliated with colleges and universities, and because American colleges and universities vary both in quality and in reputation, one of the most persistent problems has been the extent of the advantage that scientists at the more prestigious universities have over their less fortunate peers.

Data on the reward system in American science have been inconsistent. Crane (1965) concluded that the prestige of institutions interferes with the universalistic reward system. Stephen Cole and Jonathan Cole, in a series of studies culminating in *Social Stratification in Science* (1973), found that universalism is widespread. They concluded, on the basis of several kinds of studies, that role performance is the single most important consideration in determining a scientist's status in the community. Harriet Zuckerman's study of Nobel laureates, *The Scientific Elite* (1977a), shows the pre-prize performance of the laureates to be much greater than that of a comparative group of other productive scientists. She could not assert that the laureates deserved the prizes more than others did, but if the basis for selection was role performance, the laureates were assured of being in the group from which the winner would be selected. Hargens and Hagstrom (1967) found that particularism was involved in the early stages of careers, but that universalism predominated later.

Studies of British scientists have shown universalism to be predominant for high energy physicists (Gaston, 1973). For chemists in all specialties, particularism seemed to interfere with the ideal operation of the reward system (Blume and Sinclair, 1973). A recent study of both British and American scientists, involving similar data from each country in the same disciplines, and using comparable methodology, concludes that the reward system in both countries operates predominantly as an ideal universalistic system (Gaston, *The Reward System*, 1978a).

The most important aspect of studies on the reward system is the understanding gained of the way this part of the social system of science works. Research remains to be done in two areas: determining how a universalistic reward system affects the speed with which new knowledge is produced, and how the production of knowledge varies with the use of universalistic and particularistic reward systems.

2. Social Processes in Science. Role performance and its subsequent rewards are among the important aspects of the social system of science. Other social aspects of the operation of scientific communities are also pertinent. Scientists doing research—and for many scientists this is almost continual—are involved in various social processes. From the sociological perspective the most significant processes are the "three Cs": communication, cooperation, and competition.

The extensive literature on communication in science is diffuse; studies range from the number of journals withdrawn from libraries to the informal network of scientists working on specific research topics (Zaltman, 1968). Just as social scientists have studied communication patterns, so have scholars in other areas—including areas related to the practical application of library policies. For a comprehensive review of important research through the early 1970s, see Crane (1971a). Because much of the research on communication has been empirical, lacking a strong theoretical perspective, interest in communication studies by sociologists has waned in the 1970s.

Sociological studies of communication in science focus especially on structures and functions of communication (Menzel, 1962, 1966). For example, on the informal and personal level, questions are asked regarding the extent of the "random" process in communication with specific scientific colleagues. Why are some people communicated with, while contact is not maintained with others who are equally "eligible" for communication? (See Crane, 1972; Mullins, 1968; and Gaston, 1973, for examples and discussion of this type of question.)

The structure of informal personal communication may or may not resemble the structure of formal written communication. A scientist may regularly be in verbal contact with a small number of colleagues who exchange information; however, when a different sort of information is required, the formal printed communications of other scientists will be consulted.

Communication between scientists is a form of cooperation, but better examples are collaboration and team work. (See Price and Beaver, "Collaboration in an Invisible College," 1966.) Indications are that collaboration in science is not a random process, but far too little research has been done on the subject. Coauthored papers are usually the best indicator of formal collaboration; throughout this century, each decade has seen an increase in the percentage of papers published by two or more authors. Especially since World War II, there has been a combined increase in coauthorships and team research (Price, *Little Science, Big Science,* 1963; and Hagstrom, 1964). These two forms of collaboration may have a common cause in the general growth of research, especially in the sophistication required for certain types of research.

Informal cooperation between scientists has not yet been studied systematically. Examples, such as the frequent sharing by scientists of unpublished data, specimens, rare cultures, et cetera, could be important areas for study, helping to counter the view that science is unreasonably competitive.

Competition generally is one of the more interesting social processes (see Reif, 1961). Scientists frequently find themselves in competition because the topics of their work are similar. Competition is keen where research, rather than being routine, holds the promise of novelty in information or ideas. Scientists do not create the competition; the nature of the social system of science is what requires originality. (The five papers in part 4 of Merton's *The Sociology of Science,* 1973, give examples and descriptions, historical and contemporary, of the nature of competition in science.)

The consequences of competition take several forms. Scientists may move out of one research area into another with fewer competitors. Scientists may work harder trying to be first. If sufficiently secure professionally and personally, scientists may worry little about the competition and decide that being "right" is worth the risk of being "second." Scientists may also respond to competition by becoming secretive about their work (see Hagstrom, 1965, 1974; Gaston, 1973), by taking greater risks with research subjects (see Sullivan, 1975; Barber et al., 1973), or by devious actions such as fraud. Fraud in science, when discovered, tends in the United States to produce headlines in *Science, Time,* and the *New York Times.*

In situations of extreme pressure to produce results for personal or professional reasons, scientists may deviate from the norms of science—but probably no more than similarly situated bank employees. Although opportunities for deviant scientific behavior are built into an institutional structure requiring originality, it may be social-psychological differences among scientists that account for deviations when they occur.

One point should be emphasized. It is easy for citizens—even for scholars who should know better—to believe that contemporary society is different, that "times have never been quite like this before." No doubt the mass media cause the impression that life is hectic, that everyone is caught up in "getting ahead," but these impressions should not lead us to believe that our society is unique in all of history.

The unhistorical character of this view can be seen in terms of scientific competition and its consequences. Merton (*The Sociology of Science,* 1973, pp. 325–342) points out that the public response to James Watson's *The Double Helix* (1969) was to conclude that scientists are human—all too human. Scientists have the emotions and passions of ordinary people. The evidence for that conclusion? It was the admission by Watson that he wanted to establish priority for discovering the structure of DNA.

What Merton points out is that this is not a unique case in history, however vividly the story may have been told by a main contender. He reiterates the undisputed historical fact that the behavior pattern discussed as the result of modern society has been with us for centuries. It is inherent in science. Although various contemporary conditions—contemporary, indeed, at any point in time—may influence the amount of attention such behavior receives, the fact is that scientific research is inherently competitive. Some kinds are more so, some kinds less so, but all science has the potential for competition.

Scientific competition may also be experienced between national systems. Ben-David (1960a, 1960b, 1968, 1968–1969, 1970, 1971) has systematically examined the rise and decline of science in various European states and the United States, as well as the organization of university systems as the predominant source of new scientific knowledge. He argues convincingly that the growth or decline of research centers is unfailingly linked to the amount of competition structured into the system. When the social structure produces a high level of competition, research productivity increases and a country is recognized as a center of world science. When competition declines, the center of world science moves on.

Although it is possible to be personally or politically opposed to competition, it is difficult to deny that science benefits from competition, despite some negative consequences. (See also Ben-David and Collins, 1966.)

3. Social Control in Science. Every social group has mechanisms for social control; science is no exception. What is unusual about science is that of all social institutions it has the least formal control mechanisms. (In *Scientific Knowledge and Its Social Problems,* 1971, Jerome Ravetz devotes chapter 10 to a thorough analysis of the problem of quality control in science, including informal mechanisms of social control.) Religion has its laws for group members; the economy, its highly codified laws on virtually every aspect of social activity. Even the family as an institution has legal requirements—though admittedly they are difficult to enforce. But science is not like these other institutions.

Scientists of course have rules to follow. By virtue of their position in universities and other organizations, scientists are bound by contractual arrangements dealing with conditions of work and so forth; but these arrangements are covered by rules pertaining to institutional spheres other than science. There are no laws, strictly speaking, about conduct as scientists —only as citizens, employees, parents, spouses. The purpose of social control is to promote conformity to the norms of a group. In this case, how could science progress if scientists were, for instance, to lie about their data or steal others' materials?

One aspect of science that distinguishes it from other social institutions, and that may explain why *formal* mechanisms for social control are unnecessary, is that every member of the scientific community is potentially a policeman. Citizens who cheat on taxes may brag about it; spouses who cheat may tell friends; physicians who err may admit their mistake to a colleague. When such a disclosure occurs, the transgressor is rarely called to task by the confidant. But science does not tolerate blatant and willful deviance; no temptation to compromise the rules—at least ideally—is strong enough to justify yielding to the temptation.

In spite of widespread publicity when scientific fraud or other serious deviance is discovered, little attention has been paid to this subject, and there exists no comprehensive study of scientific deviance. A recent analysis of what is known about deviant behavior in science shows that the technical standards and moral norms in science are connected to the goal of science in producing certified knowledge (Zuckerman, 1977b). The reward system in science generally provides enough sanctions for scientists to conform to the norms of science.

One should not conclude, therefore, that unusually honest people are recruited into science; there is no evidence for or against that idea. Scientists are caught in a structural situation that minimizes deviance: if the rewards of recognition are sufficiently attractive to risk exposure, then the problem is probably sufficiently important for others to be working on it at the same time—or for them to try to replicate the results. In either case the deviant is likely to be exposed. If, on the other hand, a problem is unimportant, if no one cares, if the rewards are so inconsequential that the risk of exposure is small—if, in other words, the problem is at the periphery of research activity—more systematic deviance can be expected. For example, in a university where research is encouraged but publications are printed in low-prestige journals, external evaluations of a scientist's work may not even be sought. Structurally, this situation can encourage deviation. However, if deviation occurs more frequently at some levels of the social stratification system in science, at the theoretically probable locations of deviance, this pattern does not confirm suspicious allegations that the famous cases of deviance in science are only the tip of the iceberg. At the upper reaches of science, where recognition rewards are high, there is little or nothing to gain from deviance.

C. Scientific Growth and Change

In the last quarter of the twentieth century, science is prospering around the world in countries whose economies can afford to support the costs of conducting modern research. Even in countries where there is no long tradition of scientific research, efforts are made to establish a research capacity, often by sending young men and women to established scientific centers for education and research experience.

It is clear that science has developed to a point, in both knowledge and size, that makes it vastly different from the beginnings of modern science in the seventeenth century. The enormous changes in those three hundred years, chronicled by historians of science, have produced changes in societies, many still to be chronicled. After World War II, as governments placed more emphasis on science, it was not surprising that a fledgling sociology of science would attempt to explain how modern science had grown and developed from its modest beginnings to its current established state.

1. Origins and Early Development of Science.   Except for Pitirim A. Sorokin's early work, in *Social and Cultural Dynamics* (1937), on the relation of science to culture—recalled in Merton's "The Sociology of Science: An Episodic Memoir" (1977, pp. 25, 116–117)—and Barber's extensive discussion in *Science and the Social Order* (1952, pp. 51–92), most historical-sociological interest in science focuses on the early experience in Europe. Ben-David (1964, 1965, 1968–1969, 1970, and 1971) and Merton, beginning with *Science, Technology, and Society in Seventeenth-Century England* (1938), have been responsible for much of our understanding of the impact of sociological processes on the origins of the Scientific Revolution (Price, 1978). Merton's classic study of Protestantism's unintentional support of the rise of science in England parallels Max Weber's *Puritanism and the Spirit of Capitalism* (1958). Merton's thesis has not escaped criticism—he describes and evaluates some of it in his 1970 preface to the latest printing of *Science, Technology and Society in Seventeenth-Century England*. Merton does not claim that Protestantism spawned science; rather, that the special nature and values of Protestantism, at that time in England, nurtured science by providing impetus to its social status as well as to its growth.

Ben-David's research has focused mainly on comparative studies. *The Scientist's Role in Society* (1971) traces the movement of scientific research centers through European countries and on to the United States after World War II. Several centers emerged, each gaining success because of specific local social factors; the main systematic factor he detects is competition. Professionalization of the scientist's role was an essential requirement for the development and flourishing of science; once science was respected as a career and lost its amateur status, it gained the approval and nurture required for growth.

A persistent growth rate has been an obvious characteristic of modern science. Derek de Solla Price, an historian with sociological interests, has been one of the most astute observers of this growth. In *Little Science, Big Science* (1963), Price charts the growth of science since the 1600s, finding that it has been exponential. The doubling period has been about every ten to fifteen years. This rate may vary according to certain historical conditions, but:

> If any sufficiently large segment of science is measured in any reasonable way, the normal mode of growth is exponential. That is to say, science grows at compound interest, multiplying by some fixed amount in equal periods of time. Mathematically, the law of exponential growth follows from the simple condition that at any time the rate of growth is proportional to the size of the population or to the total magnitude already achieved—the bigger a thing is, the faster it grows (1963, pp. 4–5).

In Price's research, there is an implicit internalist view, that science has within itself the fuel that drives the machine forward. As knowledge expands, it becomes possible for knowledge to expand further. However, the knowledge base depends on many external social and political conditions that contribute to its expansion, and even if there is an internal impetus for a time, it will inevitably cease to be the primary force. As Price accurately predicted, the growth rate of science had to slow:

> It is clear that we cannot go up another two orders of magnitude as we have climbed the last five. If we did, we should have two scientists for every man, woman, child, and dog in the population, and we should spend on them twice as much money as we had. Scientific doomsday is therefore less than a century distant (1963, p. 19).

The decreased rate of growth has given rise to a condition not previously experienced in the history of science. As the previous growth rate was attributed to science's internal combustion engine, the slowdown can be attributed to society's antipollution regulations. External influences of the 1960s have decidedly decreased the growth rate of science. Inflation and the decision to increase other government programs have resulted in a reduced level of funding and support for graduate students. With science so vulnerable to external services, a purely internalist view is scarcely tenable.

2. Disciplines, Specialties, Research Areas. A scientific discipline is generally defined as the largest research unit with which scientists identify themselves. If a nonscientist asks a scientist what his occupation is, most likely the reply will be "scientist." If one scientist asks another the same question, the answer will be more specific: "I'm a chemist." If one chemist asks another, "What do you do?" the answer is likely to be, "I'm a biochemist." The specificity grows as the questioner is assumed to be increasingly sophisticated; at the extreme, a specific research problem area will be identified as one's area.

Growth and change in the scientific community, from a sociological perspective, parallels these self-perceptions of scientists' work. In the beginning, sociologists of science did not differentiate very much among scientists; the focus was most frequently on *the* scientific community, in a sociological sense. Indeed, for most scientists it is irrelevant whether they are in contact with a great many other scientists. For some purposes, all scientists may still be viewed as a community, but this perspective is possible only at a rather high level of abstraction. Because of the location of the research activity, as well as its influence on others similarly situated—which certainly does not extend throughout the aggregate of scientists—scholars began to change their units of analysis from the whole to constituent parts.

Specific social and intellectual factors, singly and in combinations, have influenced the origin and growth of new research areas. They are not born in social isolation. Ben-David (1960a, 1960b) long ago described how both social organization and the structure of society either hinder or promote science. Does the origin of a specialty in scientific research result more from the social environment or from the state of knowledge? Ben-David and Collins (1966) answer in terms of an example: experimental psychology originated because career opportunities were negligible in one area but available in another; role hybridization occurred between philosophy and physiology, producing experimental psychology.

On the other hand, as Nicholas Mullins (1972b) argues, it was the intellectual base that created the new specialty of molecular biology. To add still a third view, Russell Meier (1976) suggests that biophysics emerged from a combination of social and intellectual factors.

Whatever the specific condition—such as the recent energy crisis, which spawned various disciplinary interests—the growth of knowledge is most visible at the problem level. Problem areas combine with other areas to make up a specialty. And, in turn, many specialties comprise a discipline. Because of its centrality in scientific growth, the specialty group has correspondingly become central for sociologists in the last decade. At first glance it would seem easy to study scientific specialties, but the problem of defining a specialty has required the prior solution of some methodological problems, as well as a multiple set of approaches, by sociologists of science.

In *Invisible Colleges* (1971), Diana Crane set out to study the growth of two research specialties: one in rural sociology, involving studies of the diffusion of innovations; and one in algebra, dealing with the theory of finite groups. She surveyed bibliographies compiled on the two subject matters, and the authors of the papers were taken to be members of the specialties. Although other questions were studied, she was interested primarily in determining whether the research conformed to a model of a rapid growth, leveling off, and decline. She found that the two specialties followed the pattern closely. Much work since Crane's pioneering effort has focused on the types of issues she raised, especially on the characteristics of groups that remain constant through time, as they first coalesce, then expand and decline as a scientific specialty. This line of research inevitably led to problems: How does one define a specialty? And what is the connection between cognitive development and social organization in a research specialty?

At first glance, it would seem relatively easy to identify a scientific specialty for sociological study. However, even the participants themselves do not always agree on the limits of a specialty; neither are they always certain about the label that should be used for their research. A variety of attempts have been made to identify a specialty. For each new technique, there is likely to be a set of criticisms advanced by others, who either misunderstand the purpose of the particular method or who claim that its proponents exaggerate their claims. For example, an approach using the *Science Citation Index* has been proposed. (Since the mid-1960s, the *SCI* has published annual compilations of all the references in a large number of journals, originally in the natural sciences but more recently also in the social sciences.) This approach has been criticized on two grounds: because of inevitable printing errors in the *SCI*, and because there are various norms, as well as deviations from norms, for giving references. Moreover, this method appears to some to be too simplistic as an approach to important intellectual developments.

(For criticisms specifically of the *Science Citation Index* data as used by various scholars, see Chubin, 1973; Chubin and Moitra, 1975; Kaplan, 1965a; Moravcsik and Murugesan, 1975; Porter, 1977; Sullivan, White, and Barboni, 1977a. It should be noted that those who use the *Science Citation Index* are not unaware of its limitations. For descriptions of the use of bibliographical sources, see Adain, 1955; Chubin, 1975a; J. Cole and S. Cole, 1973; Dieks and Chang, 1976; Garfield, 1955, 1963; Garfield, Malin, and Small, 1978; Griffith, Small, Stonehill, and Dey, 1974; MacCrae, 1969; Meadows and O'Connor, 1971; Small, 1973, 1977; Small and Griffith, 1974; Sullivan, White, and Barboni, 1971.)

Daryl Chubin (1976) discusses the problems in studying specialties; he reviews substantive findings as well as the methodological issues and problems. Chubin suggests that identifying a specialty requires more than a single method. He also feels that the specialty, because it is the locus of scientific growth, will become the research site for much sociology of science in the future.

David Edge and Michael Mulkay, in *Astronomy Transformed* (1976), give a good example of the sociology of a scientific specialty. They provide historical background on the development of radio astronomy, show how social conditions affected the development of its research problems, and emphasize the importance of research equipment. They weave together all these influences in a narrative that provides a comprehensive picture of specialty development.

II. RESEARCH IN THE SOCIOLOGY OF TECHNOLOGY

Although technology is a potentially important subject for sociology, the sociology of technology has not become an institutionalized specialty in the discipline. Even if it is clear that no sociological model of technology exists, and even if there are insufficient numbers of sociologists of technology to constitute a community of scholars, the few sociologists who have published on the subject should be recognized for their efforts. The first section here does that. The second discusses the optimistic future for a sociology of technology. A third section discusses areas the sociology of technology might pursue parallel to the primary questions in the sociology of science.

A. Sociological Research on Technology

Some early sociological treatments of technology were William F. Ogburn's *Social Change* (1922) and S. Colum Gilfillan's *The Sociology of Invention* (1935). These works did not add up to a theoretical paradigm prompting further studies on technology; they did, however, demonstrate one idea, which at that time was novel and still is in some circles: that a sociological perspective can account for aspects of invention better than an account based on the idea that individuals are the primary factor in invention.

Ogburn developed the concept of "cultural lag." This term describes the fact that technological changes are incorporated into a culture much more quickly than are ideas and values. Gilfillan tried to show that invention is caused partly by social influences, that it is not simply a process of applying knowledge about science and technology. Using a case study of ship development, he produced certain principles about social influences on invention. Although the book is useful primarily as an early historical example, it ranks as a classic because of its pioneering efforts.

Merton's *Science, Technology, and Society in Seventeenth-Century England* (1938) included chapters on the development of mining, transportation, and military techniques. Merton showed that the development of technology was fostered by the rising interest in science, and that sociological factors influenced both the development of science and the development of technology. Although his descriptions provided a limited model for a macro-sociological approach to the sociology of technology, he did not produce a comprehensive paradigm that would later develop into such a specialty in sociology.

In 1952 Isidor Thorner, following the lead of early workers in the sociology of science, published a study entitled "Ascetic Protestantism and the Development of Science and Technology." He argued that if Protestant-ism involved values that supported science and technology, then greater scientific and technological productivity should appear in Protestant socie-ties. He used data from Sorokin's *Social and Cultural Dynamics* (1937). Sorokin had argued that before the eighteenth century, Catholic countries showed greater productivity in science and technology than Protestant coun-tries. Thorner's conclusions were quite different. He found that discoveries and inventions, measured per millions of population in a country, had always favored Protestant countries, except in the case of France. Thorner argued that French scientists were not Catholics, something that would indicate further support for the effect of Protestantism on productivity in science and technology.

To assess whether the United States and other countries were experi-encing a decline in invention, Alfred Stafford (1952) examined patent sta-tistics. He concluded that patents are declining in all industrial countries and offered several speculative explanations to account for the phenomenon. He leaned strongly toward one idea, that the state of development in tech-nology had created the decline, producing different mechanisms for dealing with productivity. His view was later supported by Gilfillan (1959), who confirmed that patenting was on the decline and that the development of new organizations was largely responsible. Where in earlier years the indi-vidual inventor was likely to patent a small increment of his invention, more recently in this century the laboratory had replaced the individual inventor. In effect, a system was now likely to be patented where formerly it had been only a component of a system. Gilfillan felt that inventing had not decreased. A better method to measure inventive activity, he argued, would be to mea-sure the productivity of scientists and engineers by counting also their pub-lished papers and not simply patents. Implicit in this argument was the idea that the portion of the population formerly involved in inventing and applying for patents had now become absorbed into the pool of scientific manpower.

In early twentieth-century discussions of whether social environment or biological composition accounts for historical events, most sociologists con-cluded (and still conclude) that although it is impossible to assign a specific percentage to each influence, environment and heredity interact to produce the total personality of an individual. Interest in questions of this sort sparked studies of all kinds of occupational groups—among them one by

Sanford Winston (1939) describing the characteristics of 371 inventors. His findings cannot be generalized to cover contemporary inventors, but his paper illustrates the sort of data that might be helpful in making connections between personality types and number of inventions. Winston concluded that although inventors have considerably better social backgrounds than the population in general, their backgrounds are not as privileged as those of scientists, of scholars generally, or of members of the traditional learned professions: law, medicine, and theology.

One of the best-known contemporary commentators on technology, Jacques Ellul, calls his efforts "sociological." (See, for instance, his "The Technological Order," 1962.) Ellul's rather negative view tries mainly to describe how technology has come to dominate society. His ideas are phrased in such a way that they are virtually impossible to test empirically; hence most discussions of his work are not social-scientific. (Carl Mitcham devotes several pages to Ellul and his critics in chapter 5.) Even though his work is provocative, Ellul does not present a systematic framework for studying sociological questions about technology. In fact, his basic thesis that technology no longer exists *in* society, that today's society *is* technological, is merely postulated.

Books and papers are included in our bibliography that, by an extremely broad definition of sociology, could be considered sociology of technology. They may have "technology" in the title. They may deal explicitly with scientists in settings in which research is limited to applications of scientific knowledge. But they are not mentioned here because they do not deal with questions that would have to be included in a proper sociology of technology. Many studies on the subject of technology have some sociological implications. Indeed, much contemporary literature does. However, it would be misleading to suggest that such work represents a sociology of technology.

If sociologists have largely ignored technology, economists have not. Traditionally a social science—some would even say the "queen" of the social sciences—economics is, nevertheless, not sociology. If the reader desires a good introduction to research on economics and technology, he might consult Christopher Freeman's chapter in I. Spiegel-Rösing and D. Price, eds., *Science, Technology, and Society* (1977), and several parts of Diana Crane's chapter 9 in this book.

B. Conditions Enhancing the Future of the Sociology of Technology

In 1959 Francis Allen summarized research on technology and social change, the problem most studied by sociologists looking at the social aspects of technology. He mentioned only a few items by sociologists, including Gilfillan, Ogburn, and a book he had written with Hart, Miller, Ogburn, and Nimkoff (1957). Allen concluded that while some progress has been made:

This existing stock of knowledge does not fulfill to a satisfactory degree the scientific goal of providing systematic generalizations which are consistent with known empirical findings. Sufficient data are still unavailable concerning many subjects, and often the existing data are not theory-oriented. As a consequence some processes or theories are based on much empirical evidence, whereas others have little factual support (1959, p. 53).

Since the time that Allen presented this rather pessimistic view, scholars have provided grounds for greater optimism. In the newer literature, perspectives are emerging that require empirical research—a necessary condition for refining theoretical views. The literature consists of at least two approaches.

One type of literature deals explicitly with the *problematics* of a sociology of technology. In one example, R. D. Johnston (1972) argues that while science and technology are two largely different social phenomena, they have similarities. He shows that within technology, activities can profitably be viewed with the help of Kuhn's concept of paradigm:

A detailed examination of modern technology reveals that much of the activity of producing artifacts representing new or improved capabilities can be considered to be paradigm-bound, and many of the innovations which are commonly accepted as landmark events in technology possess the characteristics of paradigm shifts (1972, p. 123).

Others provide helpful cautionary notes relevant to this approach. Edwin Layton (1971b) shows how wrong some general views of the connection between science and technology are; the scientific and technological communities are clearly different though complementary (see also Layton, 1970 and 1977). With considerable empirical evidence, Derek Price has shown in several papers how enormous the differences are in the literatures of science and technology; because of the different reward systems of science and technology, the two cannot be equated (see Price, 1965c; 1969a; 1970a; 1970b).

Combined, these ideas provide the kind of foundation necessary for moving toward an understanding of what technology is apart from science. The next steps would involve systematic investigation and the development of a research specialty in the sociology of technology.

A second element important for the future of sociological research on technology is the scholarly response to polemical denunciatory writings that blame technology for the current condition of society. Stephen Cotgrove (1975) assesses various polemics and argues that the march of technique is not the problem. Rather, according to Cotgrove, the problem is one of value preferences that result in personal goals becoming less important than collective goals. This in itself may be a polemical view, but Cotgrove's primary point is that scholars must conduct studies of the various components of the problem. Wholesale condemnations of technology will not answer our questions.

C. Research Potential in Sociology of Technology

To suggest how the sociology of science and technology might be approached by scholars using internal or external perspectives, we might use a simple matrix:

SUBJECT	PROBLEM AREA		
	Social Effects on	*Internal Workings of*	*Effects on Society*
Science	1. External Directions	2. Scientific Community	3. Negative/Positive
Technology	4. External Directions	5. Technological Community?	6. Negative/Positive

Problem area 1 would deal with those aspects of society that produce effects on science; these aspects were discussed above as external influences. Problem area 2 would deal with the internal social workings of the scientific community: relations between individuals, groups, and disciplines. Problem area 3 involves positive and negative impacts of science on society. This is the area in which science receives the most criticism, for instance, for "dehumanizing" society. Science, through its insistence on rationality and empirical verification, is alleged to have underemphasized, if not undermined, the importance of emotions, feelings, the "unmeasurable" spiritual side of human existence. The validity of such charges is not obvious to everyone, partly because some people evaluate positively the same things that others evaluate negatively. It seems to some serious students of the subject that it is precisely here that many observers confuse science with technology.

Problem area 4 would cover aspects of technology in which need and market factors play a part. This theoretically possible research domain has not been highly developed by scholars—certainly not by sociologists. One exception is Merton's *Science, Technology, and Society in Seventeenth-Century England* (1938), in which he discusses the social influences on military and other technologies in seventeenth-century England. Although sociologists might be expected to, Jewkes et al. (1969) are not optimistic that social scientists can contribute much to the study of technology: "Up to now, their experiments in this field have not been free of serious mistakes." To support this interpretation, they cite one of the few attempts at a sociology of technology. They write: "For example, Mr. S. C. Gilfillan, who has probably made more advanced claims about the possibility of prediction than any other writer, in his book *Inventing the Ship*, 1935, speaks of the startling invention of the rotorship. He obviously imagined that there was a great future for this type of ship. But nothing has been heard of it since 1935" (1969, p. 176).

Problem area 5 is virtually unexplored. This area would parallel the sociology of the scientific community. In contrast to the sociology of the scientific community, little is known about the sociology of the technological community. One reason may be that technologists work in many different settings. Unlike the usual university base of scientists, in which a sociologist or any other observer can come and go and secrecy is at a minimum, observation of technologists is not so easy. Moreover, the term "technologists" covers many different areas of expertise; it is not as easy for sociologists to understand them as it is to understand their colleagues in scientific departments of the university. However, the most likely reason a sociology of technology has not developed is because there has been no research paradigm on the sociology of technology that produces researchable questions.

This leaves problem area 6. This is the area in which most criticism of technology has occurred. The purpose of most of this criticism has not been to understand technological development but to suggest that technology has harmed human existence, that it controls man rather than the reverse, that it may, in the end, produce the downfall of civilization. Although this sort of criticism has not led to research it may do so in the future. This point will come up later, in our concluding comments.

III. SOCIOLOGY OF SCIENCE AND TECHNOLOGY OUTSIDE THE UNITED STATES

There is important work conducted by scholars outside the United States that did not fit neatly into the categories used for the review in this chapter. It should be identified for the interested reader. The sociology of science is pursued in most of Western Europe but especially in Great Britain and Germany. In Eastern Europe what is called sociology of science, because of the nature of the governments there, is much closer to what West Europeans and Americans call science policy studies. (See chapter 9 in this book.) Several recent publications will guide the reader both to bibliographic sources and to samples of the research done. The material, of course, varies in the different countries.

In 1972, a group of sociologists of science met in London for a conference of the Sociology of Science Research Committee of the International Sociological Association. Many of the papers presented at the conference were published in a volume, edited by Richard Whitley, entitled *Social Processes of Scientific Development* (1974).

In 1974, the Institute for Advanced Studies in Vienna sponsored another international conference. Karin Knorr, Hermann Strasser, and Hans Zilian edited *Determinants and Controls of Scientific Development* (1975), made up of papers presented at that meeting.

At about the same time, an influential German journal *Kölner Zeitschrift für Soziologie und Sozialpsychologie* invited Nico Stehr and Rene

König to edit a special issue on the sociology of science. They arranged for contributions from several countries and produced a book entitled *Wissenschaftssoziologie* (1975).

More recently, an annual series has been organized. The first issue, edited by Everett Mendelsohn, Peter Weingart, and Richard Whitley, is devoted to *The Social Production of Scientific Knowledge* (1977). Later issues deal with the dynamics of science and technology (1978) and with countermovements and the sciences (1979).

Stuart Blume's *Toward a Political Sociology of Science* (1974) presents a case for expanding the sociology of science to include more explicit connections with politics. His more recent edited book, *Perspectives on the Sociology of Science* (1977), contains a variety of papers that provide additional ideas for research on the influence of social conditions in various countries on scientific growth.

A new journal, *Scientometrics,* with an international board of editors, will be devoted to publishing quantitative studies of science with special emphasis on investigations using statistical methods.

Reviews of important publications in the sociology of science for Great Britain, France, Italy, West Germany and Austria, Poland, Scandinavia, and the Soviet Union constitute the various chapters of Merton and Gaston's *The Sociology of Science in Europe* (1977). These reviews include comprehensive bibliographies, classic statements, and more recent empirical and theoretical developments. Because in some European countries technology and science are not intellectually separated as they are in the United States, some chapters, especially those on Italy and West Germany, include references to European sociology of technology. A more extensive review of sociology of science in the U.S.S.R. can be found in Linda Lubrano's *Soviet Sociology of Science* (1976), although it is presented from an outsider's perspective.

IV. SCIENCE, TECHNOLOGY, AND THE PROBLEM OF VALUES

Early in this chapter it was indicated that value concerns would be raised wherever they came up in the literature. This happened rarely because, it is clear, concern with values has been neglected in the sociology of science and technology. (For the most glaring exception see Barber et al., 1973.) Here we shall discuss some issues that scholars might be concerned with if they dealt with values in science and technology.

A. Science and Values

Opportunities for value investigations in science, from the sociological perspective, follow from two dimensions:

1. *Society values science.* Society provides financial support and an environment that encourages the pursuit of scientific research. It ranks

"scientist" very high among the occupations and professions. Although science is not the most important value for society; although support for science is not as great as scientists would prefer; although many people in society may value science for reasons that are less than ideal; and although society may even value science for erroneous reasons—all this being granted, it remains that science is valued. These observations raise a number of questions:

First, at the societal level: How do scientific values rank among other values in American society? Is the position of science changing with changes in the size, structure, and educational characteristics of the general population? What limits does public opinion place on science?

Second, at various individual levels: How do values differ among people who value science differentially? Do people with different economic interests see science as serving them or as serving others?

Third, under what conditions would society stop valuing science? Would there be changes if science could answer questions about life after death? Or, on the opposite side, what if creation could be conclusively eliminated as a possible explanation for life on earth?

Fourth, under what conditions would society raise its evaluation of science to a higher level—for example, exchange priorities between defense and science? How far would society be willing to go in valuing science?

2. *Science as a social institution values certified knowledge.* As a goal, certified knowledge is the highest value in science. However, the scientific community is made up of humans. This simple fact raises interesting questions about science and values:

First, are there conflicts between science's highest value and need for society's support? Is it ever the case that the scientific community withholds certifiable knowledge from society for fear it will not be supported?

Second, how far would the scientific community go to guarantee its opportunity to pursue knowledge of all kinds?

Third, is there value consensus within science? Do some put specialized interests above institutional interests? If so, how can value consensus be maintained? If not, what explains such value stability as there is?

Fourth, what do scientists see as the value of their research for society? Or for science itself? How much of a consideration is this in their daily work? How are perceived conflicts resolved? Do scientists believe they or others have the responsibility to resolve science-society conflicts?

B. Technology and Values

There are dimensions to the problem of technology and values that are similar to those for science and values. Although in many respects technology has a more pervasive effect on society than does science, much less is known about the value aspects of technology. Here, three dimensions may be mentioned:

1. *Society values technology.* Without necessarily knowing what it is valuing, society encourages technological advances. Nowadays, to some degree, this attitude exists because one specific technological development—television—nurtures the desire for novelties that require technological development. Although other questions would arise in less developed countries, several specific questions emerge in a society with a developed technological capacity:

First, what counts as technological sufficiency for a society? Are there limits to technological growth? For example, when the supersonic transport plane (SST) was shelved, did this rejection mean that society would place limits on faster transportation between airports in exchange for faster movement from origin to airport?

Second, does society see technological developments as a product of scientific research or of capitalistic investment? Are technological products evaluated in terms of human value or economic potential?

Third, does society see technological developments as advantageous at some times and not at others?

2. *Technology is a combination of engineering knowledge and social organization.* Unlike science, technology would not be defined by sociologists as a social institution. It is a part of other social institutions, and has important implications for them. Because it is not a social institution and has no set of institutional values and norms, technology does not raise questions similar to those about science.

3. *Technology values profits, so technologists value innovations.* Because of technology's complex connections with economic institutions, technologists must produce innovations that have potential for profit. This raises several value questions:

First, what are the values of the technologists who produce innovations in the military, consumer, health, educational, or entertainment sectors? Are there different value hierarchies in each sector, or is profit most important in all sectors?

Second, how are conflicting values resolved? If for example, profit is more important than safety or durability, are decisions stretched to just within the law?

Third, how is technology seen as a component of society? Is it viewed by technologists and administrators as a vital service, as an opportunity for profit, or as a good in itself?

Fourth, in the process of innovation, what values are sought? Does innovation involve costs to other human or social values?

C. Future Opportunities for Value Concerns

This brief list of questions offers many opportunities for future research. Most of these questions are more likely to be raised by humanistic scholars than by social scientists, but this is not to argue that value concerns are of no interest to social scientists. Nonetheless, we must be realistic, and this means that the most important value questions will have to be studied by humanists. On the other hand, generally speaking, humanists are not equipped to deal with the methodologies that social science requires. Thus, on these kinds of studies, it is important for social scientists and humanists to collaborate. And there are likely to be sufficient opportunities for such collaboration.

V. CONCLUSION

One of the main purposes of the *Guide* and its chapters is to deal explicitly with values as they are involved in research in science, technology, and medicine, or in the sociology, history, and philosophy of science, technology, and medicine. Because there is no uniformity in these vast literatures, each chapter gives a different amount of attention to values—and this chapter perhaps gives the least.

Three reasons help explain the relative lack of attention to value questions by sociologists of science:

First, although the concept of values has been around for a long time in the social sciences—studies too numerous to summarize have examined the values of college students, for instance, or tried to explain the differences in people's value preferences—the study of people's values tends to fall under social psychology. And social psychology has not been a notable feature of sociology of science.

Second, by contrast with the history or philosophy of science, sociology of science is relatively new. Merton, in the foreword to Barber's *Science and the Social Order* (1952), predicted that sociologists would begin to give increasing attention to science only when it was perceived as a social problem. Until that time, the small congregation of sociologists of science could undertake studies that simply attempted to enhance understanding of the social aspects of science, and felt little pressure to deal with value questions.

Third, even today, when there are more sociologists of science, and when science has come to be questioned, sociologists who choose the sociology of science as a specialty—more than is the case within other sociological specialties—tend to approach their research with a scientific rather than a humanistic perspective. And it is the latter perspective that is more likely to elicit explicit value questions.

The humanistic perspective on values is different from that of social scientists. For a humanist, values tend to be conceived of as *human* values; social scientists tend to conceive of values as *social* values. Since this distinction may seem arbitrary, if not ambiguous, a clarification may be in order. Humanistic concerns seem to be with those conditions of human existence that are assumed to be important to all people; knowledge of what they are is not problematic. What is problematic to the humanist is that the opportunity to experience and enjoy these values appears to have declined today.

On the other hand, the social scientist has two problems: knowing what social values are preferred, and knowing how to structure society so that the maximum number of people can obtain whatever it is that they value.

Humanistic disciplines have traditionally been interested in values. But it seems only recently that humanists have rediscovered the possibility that science and technology may be responsible for reducing the probability of conserving the ideal social and cultural environment. For instance, in this *Guide,* and in equally significant indicators of this rediscovery, humanists are giving an attention to science and technology that scholars two decades ago might have considered either a waste of time or irrelevant or both.

C. P. Snow's description of the two cultures (1959), along with his fear that the neglect of each culture by the other could have profound implications for both, becomes paradoxical. Snow was generous in placing responsibility for this neglect about equally on scientists and humanists. However, it seems demonstrable that the neglect was greater on the side of humanists. The newly developed humanistic interest in science may help to preserve humanistic values, but there is a haunting question: Is it too late?

Probably it is not too late. Sociologists of science share the natural optimism Snow ascribed to scientists, and it seems reasonable to say that even a new-found interest in the implications of science and technology for human values is better than continuing neglect. Given sufficient resources to conduct scholarly studies and to expose students and the public to various dimensions of the problem, we should be able to influence events in favor of human values.

BIBLIOGRAPHIC INTRODUCTION

In this bibliography we have attempted to include most of the important works in the sociology of science and the sociology of technology. But there are sources that could not be included because of space, and some readers may question why their favorite author is not here. Some works are included by scholars who are not sociologists, either because they provide sources for sociologists or because they deal sociologically with the topic, but our criterion has been to leave out works that have merely potential sociological implications. Such work—by historians, economists, political scientists, and others—has been left out because the sociology they contain is only implicit, and to include them would present a false picture of what the sociology of science is. There are already sufficient misconceptions about what is real sociology or what really is sociology; we do not want to add to that problem.

A. Basic Works in Sociology of Science

Our chapter has discussed most of the classic works, but that may be obvious only by the frequent reference to certain works. For colleagues in other fields, as well as for students and other interested readers, it is important to have specific directions to a sample of primary sources. We naturally risk criticism of omission, but we believe the following to be the ten books that will provide the best introduction to the sociology of science. Note: The references that follow are in abbreviated form; full information is in the main bibliography.

Barber, Bernard. *Science and the Social Order* (1952; reprinted 1978). This classic was the first to develop a systematic sociology of science. Easy and enjoyable to read.

Barber, Bernard, and Hirsch, W., eds. *The Sociology of Science* (1962; reprinted 1978). The first attempt to combine important papers from various sources into one volume on the subject.

Ben-David, Joseph. *The Scientist's Role in Society: A Comparative Study* (1971). Summarizes much of Ben-David's work on the sociohistorical development of science, and of the scientist as professional.

Crane, Diana. *Invisible Colleges* (1972). Results of studies giving data and interpretations about growth of scientific disciplines, from the perspectives developed by Derek Price.

Hagstrom, Warren O. *The Scientific Community* (1965; reprinted 1975). A classic; includes statements of the exchange theory and other ideas in the sociology of science.

Kaplan, Norman, ed. *Science and Society* (1965). A book similar to Barber and Hirsch (1962), with little overlap and interesting papers.

Merton, Robert K. "Science, Technology, and Society in Seventeenth-Century England" (1938; reprinted 1970). Merton's classic thesis in the sociology of science indicating the influence of society on science and technology.

———. *The Sociology of Science: Theoretical and Empirical Investigations* (1973). A compilation of many of Merton's important papers on the subject, representing his theoretical ideas and empirical research over three decades.

Price, Derek J. de Solla. *Little Science, Big Science* (1963). The classic text on exponential growth in science, with sociological and policy implications of such growth.

Storer, Norman W. *The Social System of Science* (1966). One of the earliest comprehensive theoretical treatments of the way the social institution of science works.

If the reader wishes to extend reading to a second ten books, we suggest the following books, which are listed in the main bibliography: Joseph Ben-David, *Fundamental Research and the Universities* (1968); Stuart Blume, *Toward a Political Sociology of Science* (1974); Jonathan Cole and Stephen Cole, *Social Stratification in Science* (1973); Yehuda Elkana et al., *Toward a Metric of Science* (1978); David Edge and Michael Mulkay, *Astronomy Transformed: The Emergence of Radio Astronomy* (1976);

Jerry Gaston, *The Reward System in British and American Science* (1978a); Lowell Hargens, *Patterns of Scientific Research: A Comparative Analysis of Research in Three Scientific Fields* (1973); James Watson, *The Double Helix* (1969); John Ziman, *Public Knowledge* (1968); and Harriet Zuckerman, *Scientific Elite: Nobel Laureates in the United States* (1977a).

Nonsociologists (Watson and Ziman) wrote two of these ten books, but they should be standard reading. Ziman's book explicitly discusses sociological ideas, and Watson's book describes events and processes in the conduct of actual research that illustrate virtually every kind of social phenomenon studied by sociologists.

B. Bibliographic Sources

There are no standard texts in the sociology of science. In addition to Merton's (1973) bibliography, extensive bibliographies and reviews are accessible in the following, which are listed in the main bibliography:

Barber, Bernard. "Sociology of Science: A Trend Report and Bibliography" (1956).

———. "The Sociology of Science" (1959).

Ben-David, Joseph, and Sullivan, Teresa A. "Sociology of Science" (1975).

Kaplan, Norman. "The Sociology of Science" (1964).

C. Basic Works in Sociology of Technology

In contrast to the sociology of science, there are few classic books in the sociology of technology, and they are dated prior to World War II. William F. Ogburn's *Social Change* (1922) is one; Gilfillan's classic *The Sociology of Invention* (1935) is another; and Merton's *Science, Technology, and Society in Seventeenth-Century England* (1938) is devoted largely to technology and society. There are of course newer works that might be viewed as sociology of technology, but because no systematic approach to the sociology of technology has been developed, these older books remain the classics.

BIBLIOGRAPHY

Adair, William C. "Citation Indexes for Scientific Literature?" *American Documentation* 6 (1955): 31–32.

Allen, Francis R. "Technology and Social Change: Current Status and Outlook." *Technology and Culture* 1 (1959): 48–59.

Allen, Francis R.; Hart, Hornell; Miller, Delbert C.; Ogburn, W. F.; and Nimkoff, M. F. *Technology and Social Change.* New York: Appleton-Century-Crofts, 1957.

Allen, Tom. *Managing the Flow of Technology.* Cambridge, Mass.: M.I.T. Press, 1977.

Allison, Paul D., and Stewart, John A. "Productivity Differences among Scientists: Evidence for Accumulative Advantage." *American Sociological Review* 39 (1974): 596–606.

Babchuck, Nicholas, and Bates, Alan P. "Professor or Producer: The Two Faces of Academic Man." *Social Forces* 40 (1962): 341–348.

Back, Hurt H. "The Behavior of Scientists: Communication and Creativity." *Sociological Inquiry* 32 (1962): 82–87.

Barber, Bernard. *Science and the Social Order.* New York: Free Press, 1952; reprint ed., New York: Free Press, 1962; 2d reprint ed., Westport, Conn: Greenwood Press, 1978.

———. "Present Status and Needs of the Sociology of Science." *Proceedings of the American Philosophical Society* 99 (1955): 338–342.

———. "Sociology of Science: A Trend Report and Bibliography." *Current Sociology* 5 (1956): 91–153.

———. "The Sociology of Science." In *Sociology Today,* pp. 215–228. Edited by R. K. Merton, L. Broom, and L. S. Cottrell. New York: Basic Books, 1959.

———. "Resistance by Scientists to Scientific Discovery." *Science* 134 (1 September 1961): 596–602. Also available in Barber and Hirsch (1962).

———. "Review of T. S. Kuhn's *The Structure of Scientific Revolutions.*" *American Sociological Review* 28 (1963): 298–299.

———. "The Functions and Dysfunctions of 'Fashion' in Science: A Case for the Study of Social Change." *Mens en Maatschappij* 43 (1968): 501–514.

Barber, Bernard, and Fox, C. "The Case of the Floppy-Eared Rabbits: An Instance of Serendipity Gained and Serendipity Lost." *American Journal of Sociology* 64 (1958): 128–136. Also available in Barber and Hirsch (1962).

Barber, Bernard, and Hirsch, W., eds. *The Sociology of Science.* New York: The Free Press of Glencoe, 1962; reprint ed., Westport, Conn.: Greenwood Press, 1978.

Barber, Bernard; Lally, John J.; Makarushka, Julia; and Sullivan, Daniel. *Research on Human Subjects: Problems of Social Control in Medical Experimentation.* New York: Russell Sage Foundation, 1973; reprint ed., New Brunswick, N.J.: Transaction Books, 1978.

Barnes, Barry, ed. *Sociology of Science.* Harmondsworth: Penquin Books, 1972.

Barnes, S. B., and Dolby, R. G. A. "The Scientific Ethos: A Deviant Viewpoint." *European Journal of Sociology* 11 (1970): 3–25.

Barnett, H. G. *Innovation, The Basis of Cultural Change.* New York: McGraw-Hill, 1953.

Bayer, Alan E. "College and University Faculty: A Statistical Description." *A.C.E. Research Reports* 5 (1970): 1–48.

———. "Teaching Faculty in Academe: 1972–73." *A.C.E. Research Reports* 8 (1973): 1–68.

Bayer, Alan E., and Folger, John. "Some Correlates of a Citation Measure of Productivity in Science." *Sociology of Education* 39 (1966): 381–390.

Beaver, Donald, deB. "Reflections on the Natural History of Eponymy and Scientific Law." *Social Studies of Science* 6 (1976): 89–98.

Ben-David, Joseph. "Roles and Innovations in Medicine." *American Journal of Sociology* 65 (1960a): 557–568.

———. "Scientific Productivity and Academic Organization in Nineteenth-Century Medicine." *American Sociological Review* 25 (1960b): 828–843. Also available in Barber and Hirsch (1962); and Kaplan (1965).

———. "Scientific Endeavor in Israel and the United States." *American Behavioral Scientist* 6 (1962): 12–16.

———. "Scientific Growth: A Sociological View." *Minerva* 3 (1964): 455–476.

———. "The Scientific Role: The Conditions of Its Establishment in Europe." *Minerva* 4 (1965): 15–54.

———. *Fundamental Research and the Universities.* Paris: Organization for Economic Cooperation and Development, 1968.

———. "The Universities and the Growth of Science in Germany and the United States." *Minerva* 7 (1968–1969): 1–35.

———. "The Rise and Decline of France as a Scientific Centre." *Minerva* 8 (1970): 160–179.

———. *The Scientist's Role in Society: A Comparative Study.* Englewood Cliffs, N.J.: Prentice-Hall, 1971.

———. *American Higher Education.* New York: McGraw-Hill, 1972a.

———. "The Profession of Science and Its Powers." *Minerva* 10 (1972b): 362–383.

———. "Organization, Social Control, and Cognitive Change in Science." In *Culture and Its Creators: Essays in Honor of Edward Shils,* pp. 244–265. Edited by J. Ben-David and T. Clark. Chicago: University of Chicago Press, 1977.

———. "Emergence of National Traditions in the Sociology of Science: The United States and Britain." In *Sociology of Science,* pp. 197–218. Edited by Jerry Gaston. San Francisco: Jossey-Bass, 1978.

Ben-David, Joseph, and Collins, Randall. "Social Factors in the Origins of a New Science: The Case of Psychology." *American Sociological Review* 31 (1966): 451–465.

Ben-David, Joseph, and Zloczower, Awraham. "Universities and Academic Systems in Modern Societies." *European Journal of Sociology* 3 (1962): 45–84. Also available in Kaplan (1965).

Ben-David, Joseph, and Sullivan, Teresa A. "Sociology of Science." In *Annual Review of Sociology,* pp. 203–222. Edited by Alex Inkeles, James Coleman, and Neil Smelser. Palo Alto: Annual Reviews, 1975.

Bennis, Warren G. "Some Barriers to Teamwork in Social Research." *Social Problems* 3 (1956a): 223–235.

———. "Values and Organization in a University Social Research Group." *American Sociological Review* 21 (1956b): 555–563.

Bereano, P. L., ed. *Technology as a Social and Political Phenomenon.* New York: John Wiley, 1976.

Bernard, Russell H., and Pelto, Pertti J., eds. *Technology and Social Change.* New York: Macmillan, 1972.

Blau, Judith. "Scientific Recognition: Academic Context and Professional Role." *Social Studies of Science* 6 (1976): 533–545.

Blissett, Marlan. *Politics in Science.* Boston: Little Brown, 1972.

Blume, Stuart. "Research Support in British Universities: The Shifting Balance of Multiple and Unitary Sources." *Minerva* 7 (1969): 649–667.

———. *Toward a Political Sociology of Science.* New York: Free Press, 1974.

———, ed. *Perspectives on the Sociology of Science.* New York and London: John Wiley, 1977.

Blume, Stuart, and Sinclair, Ruth. "Chemists in British Universities: A Study of the Reward System in Science." *American Sociological Review* 38 (1973): 126–138.

Boalt, Gunnar: *The Sociology of Research.* Carbondale, Ill.: Southern Illinois University Press, 1969.

Box, Steven, and Cotgrove, Stephen. "Scientific Identity, Occupational Selection, and Role Strain." *British Journal of Sociology* 17 (1966): 20–28.

———. "The Productivity of Scientists in Industrial Research Laboratories." *Sociology* 2 (1969): 163–172.

Broadus, R. N. "A Citation Study of Sociology." *American Sociologist* 2 (1967): 19–20.

Burton, R. E., and Kebler, R. W. "The 'Half-Life' of Some Scientific and Technical Literatures." *American Documentation* 11 (1960): 18–22.

Calhoun, D. H. *The American Civil Engineer: Origins and Conflict.* Cambridge, Mass.: Harvard University Press, 1960.

Calvert, M. A. *The Mechanical Engineer in America, 1830–1910.* Baltimore: Johns Hopkins University Press, 1967.

Caplow, Theodore, and McGee, Reece J. *The Academic Marketplace.* Garden City, N.Y.: Doubleday, 1965.

Cardwell, D. S. L. *The Organisation of Science in England: A Retrospect.* London: William Heinemann, 1957.

Carter, C. F. "The Distribution of Scientific Effort." *Minerva* 1 (1963): 172–181.

Cartter, Allan M. *An Assessment of Quality in Graduate Education.* Washington, D.C.: American Council on Education, 1966.

Chubin, Daryl E. "On the Use of the *Science Citation Index* in Sociology." *The American Sociologist* 8 (1973): 187–191.

———. "Sociological Manpower and Womanpower: Sex Differences in Two

Cohorts of American Sociologists." *The American Sociologist* 9 (1974): 83–92.

———. "The Journal as a Primary Data Source in the Sociology of Science: With Some Observations from Sociology." *Social Science Information* 14 (1975a): 157–168.

———. "Trusted Assessorship in Science: A Relation in Need of Data." *Social Studies of Science* 5 (1975b): 362–368.

———. "The Conceptualization of Scientific Specialties." *The Sociological Quarterly* 17 (1976): 448–476.

Chubin, Daryl E., and Crowley, C. J. "The Occupational Structure of Science: A Log-Linear Analysis of the Intersectoral Mobility of American Sociologists." *The Sociological Quarterly* 17 (1976): 197–217.

Chubin, Daryl E., and Moitra, Soumyo D. "Content Analysis of References: Adjunct or Alternative to Citation Countings?" *Social Studies of Science* 5 (1975): 423–441.

Clark, Terry. "Institutionalization of Innovations in Higher Education: Four Models." *Administrative Science Quarterly* 13 (1968): 1–25.

———. *Prophets and Patrons: The French University and the Emergence of the Social Sciences.* Cambridge, Mass.: Harvard University Press, 1973.

Clemente, Frank. "Early Career Determinants of Research Productivity." *American Journal of Sociology* 79 (1973): 409–419.

Clemente, Frank, and Sturgis, Richard B. "Quality of Department of Doctoral Training and Research Productivity." *Sociology of Education* 47 (1974): 287–299.

Cole, Jonathan. "Patterns of Intellectual Influence in Scientific Research." *Sociology of Education* 43 (1970): 377–403.

———. *Fair Science: Women in the Scientific Community.* New York: Free Press, 1979.

Cole, Jonathan, and Cole, Stephen. "Measuring the Quality of Sociological Research." *American Sociologist* 6 (1971): 23–29.

———. *Social Stratification in Science.* Chicago: University of Chicago Press, 1973.

Cole, Jonathan, and Zuckerman, Harriet. "The Emergence of a Scientific Specialty: The Self-Exemplifying Case of the Sociology of Science." In *The Idea of Social Structure,* pp. 139–174. Edited by Lewis A. Coser. New York: Harcourt Brace Jovanovich, 1975.

Cole, Stephen. "Professional Standing and the Reception of Scientific Discoveries." *American Journal of Sociology* 76 (1970): 286–306.

———. "Continuity and Institutionalization in Science: A Case Study of Failure." In *The Establishment of Empirical Sociology,* pp. 73–129. Edited by Anthony Oberschall. New York: Harper & Row, 1972.

———. "The Growth of Scientific Knowledge: Theories of Deviance As a Case Study." In *The Idea of Social Structure,* pp. 175–220. Edited by Lewis A. Coser. New York: Harcourt Brace Jovanovich, 1975.

Cole, Stephen, and Cole, Jonathan R. "Scientific Output and Recognition: A Study in the Operation of the Reward System in Science." *American Sociological Review* 32 (1967): 377–390.

———. "Visibility and the Structural Bases of Scientific Research." *American Sociological Review* 33 (June 1968): 397–413.

Collins, H. M., and Cox, Graham. "Recovering Relativity: Did Prophecy Fail?" *Social Studies of Science* 6 (1976); 423–470.

Collins, H. M., and Harrison, R. G. "Building a TEA Laser: The Caprices of Communication." *Social Studies of Science* 5 (1975): 441–450.

Collins, Randall. "Competition and Social Control in Science." *Sociology of Education* 41 (1968): 123–140.

Connor, Patrick E. "Scientific Research Competence: Two Forms of Collegial Judgment." *Pacific Sociological Review* 15 (1972): 355–366.

Cooley, W. W. "Attributes of Potential Scientists." *Harvard Educational Review* 28 (1958): 1–18.

Cotgrove, Stephen. "The Sociology of Science and Technology." *British Journal of Sociology* 21 (1970): 2–15.

———. "Technology, Rationality, and Domination." *Social Studies of Science* 5 (1975): 55–78.

Cournand, André. "The Code of the Scientist and Its Relationship to Ethics." *Science* 198 (18 November 1977): 699–705.

Cournand, André, and Meyer, Michael. "The Scientist's Code." *Minerva* 14 (1976): 79–96.

Cournand, André, and Zuckerman, Harriet. "The Code of Science: Analysis and Some Reflections on Its Future." *Studium Generale* 23 (1970): 941–962.

Crane, Diana. "Scientists at Major and Minor Universities: A Study of Productivity and Recognition." *American Sociological Review* 30 (1965): 699–714.

———. "The Gatekeepers of Science: Some Factors Affecting the Selection of Articles for Scientific Journals." *American Sociologist* 2 (1967): 195–201.

———. "Fashion in Science: Does It Exist?" *Social Problems* 16 (1969a): 433–441.

———. "Social Class Origin and Academic Success: The Influence of Two Stratification Systems on Academic Careers." *Sociology of Education* 42 (1969b): 1–17.

———. "The Academic Marketplace Revisited: A Study of Faculty Mobility Using the Cartter Ratings." *American Journal of Sociology* 75 (1970): 953–964.

———. "Information Needs and Uses." *Annual Review of Information Science and Technology* 6 (1971a): 3–39.

———. "Transnational Networks in Basic Science." *International Organization* 25 (1971b): 585–601.

——. *Invisible Colleges*. Chicago: The University of Chicago Press, 1972.

——. "Reward Systems in Art, Science, and Religion." *American Behavioral Scientist* 19 (1976): 719–734.

Crawford, S. "Informal Communication among Scientists in Sleep Research." *Journal of American Society of Information Science* 22 (1971): 301–310.

Dedijer, Stefan. "The Science of Science: A Programme and a Plea." *Minerva* 4 (1966): 489–504.

Dieks, Dennis, and Chang, Hans. "Differences in Impact of Scientific Publications: Some Indices Derived from a Citation Analysis." *Social Studies of Science* 6 (1976): 247–267.

Dolby, R. G. A. "Sociology of Knowledge in Natural Science." *Science Studies* 1 (1971): 3–21.

——. "What Can We Usefully Learn from the Velikovsky Affair?" *Social Studies of Science* 5 (1975): 165–175.

Downey, Kenneth J. "Sociology and the Modern Scientific Revolution." *Sociological Quarterly* 8 (1967): 239–254.

——. "The Scientific Community: Organic or Mechanical?" *Sociological Quarterly* 10 (1969): 438–458.

Edge, D. O., and Mulkay, M. J. "Case Studies of Scientific Specialties." *Kölner Zeitschrift für Soziologie und Sozialpsychologie* 18 (1974): 197–229.

——. *Astronomy Transformed: The Emergence of Radio Astronomy in Britain*. New York and London: Wiley–Interscience, 1976.

Eiduson, Bernice T. *Scientists: Their Psychological World*. New York: Basic Books, 1962.

Eiduson, Bernice T., and Beckman, Linda, eds. *Science as a Career Choice*. New York: Russell Sage Foundation, 1973.

Eliott, Clark A. "The American Scientist in Antebellum Society: A Quantitative View." *Social Studies of Science* 5 (1975): 93–108.

Elkana, Yehuda; Lederberg, Joshua; Merton, Robert K.; Thackray, Arnold; and Zuckerman, Harriet, eds. *Toward a Metric of Science: The Advent of Science Indicators*. New York: Wiley–Interscience, 1978.

Ellul, Jacques. "The Technological Order." *Technology and Culture* 3 (1962): 394–421.

Etzioni, Amitai, and Remp, Richard. *Technological Shortcuts to Social Change*. New York: Russell Sage Foundation, 1973.

Faia, Michael A. "Productivity among Scientists: A Replication and Elaboration." *American Sociological Review* 40 (1975): 825–829.

Farrall, Lyndsay A. "Controversy and Conflict in Science: A Case Study—The English Biometric School and Mendel's Laws." *Social Studies of Science* 5 (1975): 269–301.

Fisher, C. S. "The Death of Mathematical Theory: A Study in the Sociology of Knowledge." *Archives for History of Exact Sciences* 3 (1966): 137–159.

_____. "The Last Invariant Theorists." *European Journal of Sociology* 8 (1967): 216–244.

_____. "Some Social Characteristics of Mathematicians and Their Work." *American Journal of Sociology* 78 (1973): 1094–1118.

Folger, Anne, and Gordon, G. "Scientific Accomplishment and Social Organization: A Review of the Literature." *American Behavioral Scientist* 6 (1962): 51–58.

Ford, Julienne, and Box, Steven. "Sociological Theory and Occupational Choice." *Sociological Review* 15 (1967): 287–299.

Freeman, Christopher. "Economics of Research and Development." In *Science, Technology and Society*, pp. 223–275. Edited by Ina Spiegel-Rösing and Derek de Solla Price. London and Beverly Hills: Sage Publications, 1977.

Friedkin, Noah. "University Social Structure and Social Networks among Scientists." *American Journal of Sociology* 83 (1978): 1444–1465.

Friedrichs, Robert W. *A Sociology of Sociology*. New York: Free Press, 1970.

Fulton, Oliver, and Trow, Martin "Research Activity in American Higher Education." *Sociology of Education* 47 (1974): 29–73.

Garfield, Eugene. "Citation Indexes for Science." *Science* 122 (15 July 1955): 108–111.

_____. "Citation Indexes in Sociological and Historical Research." *American Documentation* 14 (1963): 289–291.

Garfield, Eugene; Malin, Morton; and Small, Henry. "Citation Data As Science Indicators." In *Toward a Metric of Science: The Advent of Science Indicators*. Edited by Y. Elkana, J. Lederberg, R. K. Merton, A. Thackray, and H. Zuckerman. New York: Wiley–Interscience, 1978.

Garvey, William D., and Griffith, Belver C. "Scientific Information Exchange in Psychology." *Science* 146 (25 December 1964): 1655–1659.

_____. "Studies of Social Innovations in Scientific Communication in Psychology." *American Psychologist* 21 (1966): 1019–1036.

_____. "Scientific Communication as a Social System." *Science* 157 (1 September 1967): 1011–1016.

_____. "Scientific Communications: Its Role in the Conduct of Research and Creation of Knowledge." *American Psychologist* 26 (1971): 349–362.

Garvey, William D.; Lin, Nan; and Nelson, C. E. "Communication in the Physical and Social Sciences." *Science* 170 (11 December 1970): 1166–1173.

Garvey, William D.; Lin, Nan; Nelson, C. E.; and Tomita, Kazuo. "Research Studies in Patterns of Scientific Communication." *Information Storage and Retrieval* 8 (1972): 111–122, 159–169, 207–221.

Garvey, William D., and Tomita, K. "Continuity of Productivity by Scientists in the Years 1969–71." *Science Studies* 2 (1972): 379–383.

Gaston, Jerry. "The Reward System in British Science." *American Sociological Review* 35 (1970): 718–732.

——. "Secretiveness and Competition for Priority of Discovery in Physics." *Minerva* 9 (1971): 472–492.

——. *Originality and Competition in Science.* Chicago: University of Chicago Press, 1973.

——. "Scientists from Rich and Poor Countries." In *Determinants and Controls of Scientific Development,* pp. 323–342. Edited by K. Knorr, H. Strasser, and H. Zilian. Dordrecht: D. Reidel, 1975a.

——. "Soziale Organisation, Kodifizierung des Wissens und das Belohnungssystem der Wissenschaft." In *Wissenschaftssoziologie,* pp. 287–303. Edited by Nico Stehr and Rene König. Opladen: Westdeutscher Verlag, 1975b.

——. *The Reward System in British and American Science.* New York: Wiley Interscience, 1978a.

——, ed. *Sociology of Science.* San Francisco: Jossey-Bass, 1978b.

Gendron, Bernard. *Technology and the Human Condition.* New York: St. Martin's Press, 1977.

Gilbert, G. Nigel. "The Transformation of Research Findings into Scientific Knowledge." *Social Studies of Science* 6 (1976): 281–306.

Gilbert, G. Nigel, and Woolgar, Steve. "The Quantitative Study of Science: An Examination of the Literature." *Science Studies* 4 (1974): 279–294.

Gilfillan, S. Colum. *The Sociology of Invention.* Chicago: Follett, 1935.

——. "An Attempt to Measure the Rise of American Inventing and the Decline of Patenting." *Technology and Culture* 1 (1959): 201–214.

Glaser, Barney G. "Variation in the Importance of Recognition in Scientists' Careers." *Social Problems* 10 (1963): 268–276

——. *Organizational Scientists: Their Professional Careers.* Indianapolis: Bobbs-Merrill, 1964.

——. "Differential Association and the Institutional Motivation of Scientists." *Administrative Science Quarterly* 10 (1965): 82–97.

Glass, Bentley, and Norwood, Sharon H. "How Scientists Actually Learn of Work Important to Them." In *Proceedings of the International Conference on Scientific Information,* vol. 1, pp. 195–197. Washington, D.C.: National Academy of Sciences-National Research Council, 1959.

Goodell, Rae. *The Visible Scientists.* Boston: Little, Brown, 1977.

Gordon, Gerald; Marquis, Sue; and Anderson, O. W. "Freedom and Control in Four Types of Scientific Settings." *American Behavioral Scientist* 6 (1962): 39–42.

——. "Freedom, Visibility of Consequences, and Scientific Innovation." *American Journal of Sociology* 72 (1966): 195–202.

Grant, Robert P. "National Biomedical Research Agencies: A Comparative Study of Fifteen Countries." *Minerva* 4 (1966): 466–488.

Greenberg, Daniel S. *The Politics of Pure Science.* New York: New American Library, 1967.

Griffith, Belver C., and Miller, A. J. "Networks of Informal Communication among Scientifically Productive Scientists." In *Communication among Scientists and Engineers*, pp. 125–140. Edited by Carnot E. Nelson and Donald K. Pollock. Lexington, Mass.: D. C. Heath, 1970.

Griffith, Belver C., and Mullins, Nicholas C. "Coherent Social Groups in Scientific Change." *Science* 177 (15 September 1972): 959–964.

Griffith, Belver C.; Small, H.; Stonehill, J. A.; and Dey, S. "The Structure of Scientific Literatures II: Toward a Macro- and Microstructure for Science." *Science Studies* 4 (1974): 339–365.

Groeneveld, Lyle; Koller, Norman; and Mullins, Nicholas C. "The Advisers of the United States National Science Foundation." *Social Studies of Science* 5 (1975): 343–354.

Gustin, Bernard. "Charisma, Recognition, and the Motivation of Scientists." *American Journal of Sociology* 78 (1973): 1118–1134.

Gvishiani, D. M.; Mikulinsky, S. R.; and Yaroshevsky, M. G. "The Sociological and Psychological Study of Scientific Activity." *Minerva* 11 (1973): 121–129.

Haberer, Joseph. *Politics and the Community of Science*. New York: Van Nostrand-Reinhold, 1969.

Hagstrom, Warren O. "Traditional and Modern Forms of Scientific Teamwork." *Administrative Science Quarterly* 9 (1964): 241–263.

———. *The Scientific Community*. New York: Basic Books, 1965; reprint ed., Carbondale, Ill.: Southern Illinois University Press, 1975.

———. "Factors Related to the Use of Different Modes of Publishing Research in Four Scientific Fields." In *Communication Among Scientists and Engineers*, pp. 85–124. Edited by Carnot E. Nelson and Donald K. Pollock. Lexington, Mass.: D. C. Heath, 1970.

———. "Inputs, Outputs, and the Prestige of American University Science Departments." *Sociology of Education* 44 (1971): 375–397.

———. "Competition in Science." *American Sociological Review* 39 (1974): 1–18.

———. "The Production of Culture in Science." In *The Production of Culture*, pp. 91–107. Edited by Richard A. Peterson. Beverly Hills: Sage Publications, 1976.

Halmos, Paul, ed. *The Sociology of Sociology*. Monograph no. 16, *The Sociological Review* Keele: University of Keele, 1970.

———, ed. *The Sociology of Science*. Monograph no. 18, *The Sociological Review* Keele: University of Keele, 1972.

Halsey, A. H., and Trow, Martin. *The British Academics*. Cambridge, Mass.: Harvard University Press, 1971.

Hargens, Lowell. "Patterns of Mobility of New Ph.D.s among American Academic Institutions." *Sociology of Education* 42 (1969): 18–37.

———. *Patterns of Scientific Research: A Comparative Analysis of Research in Three Scientific Fields*. Washington, D.C.: American Sociological Association, 1975.

——. "Theory and Method in the Sociology of Science." In *Sociology of Science*, pp. 121–139. Edited by Jerry Gaston. San Francisco: Jossey-Bass, 1978.

Hargens, Lowell, and Farr, G. "An Examination of Recent Hypotheses about Institutional Inbreeding." *American Journal of Sociology* 78 (1973): 1381–1402.

Hargens, Lowell, and Hagstrom, Warren. "Sponsored and Contest Mobility of American Academic Scientists." *Sociology of Education* 40 (1967): 24–38.

Hargens, Lowell; Reskin, Barbara; and Allison, Paul. "Problems in Estimating Measurement Error from Panel Data: An Example Involving the Measurement of Scientific Productivity." *Sociological Methods and Research* 4 (May 1976): 439–458.

Hargens, Lowell; McCann, James C.; and Reskin, Barbara. "Productivity and Reproductivity: Professional Achievement and Fertility among Research Scientists." *Social Forces* 57 (1978): 154–163.

Harmon, Lindsey R. "The High School Backgrounds of Science Doctorates." *Science* 133 (1961): 679.

——. *Profiles of Ph.D.'s in the Sciences*. Washington, D.C.: National Academy of Sciences-National Research Council, 1965.

Harwood, Jonathan. "The Race-Intelligence Controversy: A Sociological Approach I: Professional Factors." *Social Studies of Science* 6 (1976): 369–394.

——. "The Race-Intelligence Controversy: A Sociological Approach II: 'External' Factors." *Social Studies of Science* 7 (1977): 1–30.

Hirsch, Walter. "The Autonomy of Sciences in Totalitarian Societies." *Social Forces* 40 (1961): 15–22.

——. *Scientists in American Society*. New York: Random House, 1968.

Holland, J. L. "Undergraduate Origins of American Scientists." *Science* 126 (1957): 433–437.

Holton, Gerald. "Scientific Research and Scholarship: Notes toward the Design of Proper Scales." *Daedalus* 91 (1962): 362–399.

——, ed. *Limits of Scientific Inquiry*. Cambridge: American Academy of Arts and Sciences, 1978.

Hook, Sidney; Kurtz, Paul; and Todorovich, Miro, eds. *The Ethics of Teaching and Scientific Research*. Buffalo: Prometheus Books, 1977.

Inhaber, H., and Przednowek, K. "Quality of Research and the Nobel Prizes." *Social Studies of Science* 6 (1976): 33–50.

Jewkes, J.; Sawers, D.; and Stillerman, R. *The Sources of Invention*. New York: Norton, 1969.

Johnston, R. D. "The Internal Structure of Technology." In *The Sociology of Science*, pp. 117–130. Edited by Paul Halmos. Keele: University of Keele, 1972.

Jones, Robert Allen, ed. *Research in Sociology of Knowledge, Sciences, and Art*. Greenwich, Conn.: J.A.I. Press, 1978.

Kaplan, Norman. "The Role of the Research Administrator." *Administrative Science Quarterly* 4 (1959): 20–42. (Also available in Kaplan, 1965.)

——. "Research Overhead and the Universities." *Science* 132 (1960a): 400–404.

——. "Some Organizational Factors Affecting Creativity." *I.R.E. Transactions on Engineering Management* 7 (1960b): 24–30.

——. "Research Administration and the Administrator: U.S.S.R. and U.S." *Administrative Science Quarterly* 6 (1961): 51–72. Also available in Kaplan (1965).

——. "The Western European Scientific Establishment in Transition." *American Behavioral Scientist* 6 (1962): 17–21.

——. "The Relation of Creativity to Sociological Variables in Research Organizations. In *Scientific Creativity: Its Recognition and Development,* pp. 195–204. Edited by C. W. Taylor and F. Barron. New York: John Wiley, 1963a.

——. "Science and the Democratic Social Structure Revisited." Paper read at American Sociological Association, 58th annual meeting, Los Angeles, August, 1963b.

——. "The Sociology of Science." In *Handbook of Modern Sociology,* pp. 852–881. Edited by Robert E. Faris. Chicago: Rand McNally, 1964.

——. "The Norms of Citation Behavior: Prolegomena to the Footnote." *American Documentation* 16 (1965a): 179–184.

——. "Professional Scientists in Industry." *Social Problems* 13 (1965b): 88–97.

——, ed. *Science and Society.* Chicago: Rand McNally, 1965c.

Katz, Shaul, and Ben-David, Joseph. "Scientific Research and Agricultural Innovation in Israel." *Minerva* 13 (1975): 152–182.

Kidd, C. V. *American Universities and Federal Research Funds.* Cambridge, Mass.: Harvard University Press, 1959.

King, M. D. "Reason, Tradition, and the Progressiveness of Science." *History and Theory: Studies in the Philosophy of History* 10 (1971): 3–32.

Klaw, S. *The New Brahmins: Scientific Life in America.* New York: Morrow, 1968.

Klima, Rolf. "Scientific Knowledge and Social Control in Science." In *Social Processes of Scientific Development,* pp. 96–122. Edited by Richard D. Whitley. London: Routledge & Kegan Paul, 1974.

Klima, Rolf, and Viehoff, Ludger. "The Sociology of Science in West Germany and Austria." In *Sociology of Science in Europe,* pp. 145–192. Edited by Robert K. Merton and Jerry Gaston. Carbondale, Ill.: Southern Illinois University Press, 1977.

Knapp, R. H., and Goodrich, H. B. *Origins of American Scientists.* Chicago: University of Chicago Press, 1952.

Knapp, R. H., and Greenbaum, J. J. *The Younger American Scholar.*

Chicago: University of Chicago Press (for Wesleyan University, Middletown, Conn.), 1953.

Knorr, Karin D.; Strasser, Hermann; and Zilian, Hans G., eds. *Determinants and Controls of Scientific Development*. Dordrecht: D. Reidel, 1975.

Kornhauser, William. *Scientists in Industry*. Berkeley: University of California Press, 1962a.

———. "Strains and Accommodations in Industrial Research Organizations in the United States." *Minerva* 1 (1962b): 30–42.

Krohn, R. G. "The Scientists: A Changing Social Type." *American Behavioral Scientist* 6 (1962): 48–50.

———. *The Social Shaping of Science*. Westport, Conn.: Greenwood, 1971.

Kubie, Lawrence S. "Some Unsolved Problems of the Scientific Career." *American Scientist* 42 (1954): 104–112.

Kuhn, Thomas. *The Structure of Scientific Revolutions*. Chicago: University of Chicago Press, 1962; 2d ed., enlarged, 1970.

———. *The Essential Tension: Selected Studies in Scientific Tradition and Change*. Chicago: University of Chicago Press, 1977.

Lakatos, Imré, and Musgrave, Alan, eds. *Criticism and the Growth of Knowledge*. Cambridge: Cambridge University Press, 1970.

Law, J. "The Development of Specialties in Science: The Case of X-Ray Protein Crystallography." *Science Studies* 3 (1973): 275–303.

———. "Theories and Methods in the Sociology of Science." *Social Science Information* 13 (1974): 163–172.

Law, J., and French, D. "Normative and Interpretive Sociologies of Science." *The Sociological Review* 22 (1974): 581–595.

Layton, Edwin T., Jr. "Comment: The Interaction of Technology and Society." *Technology and Culture* 11 (1970): 27–31.

———. *The Revolt of the Engineers: Social Responsibility and the American Engineering Profession*. Cleveland, Ohio: Press of Case Western Reserve University, 1971a.

———. "Mirror Image Twins: The Communities of Science and Technology in Nineteenth-Century America." *Technology and Culture* 12 (1971b): 562–580.

———. "Technology as Knowledge." *Technology and Culture* 15 (1974): 31–41.

———. "American Ideologies of Science and Engineering." *Technology and Culture* 17 (1976): 688–700.

———, ed. *Technology and Social Change in America*. New York: Harper & Row, 1973.

Lazarsfeld, Paul F., and Thielans, Wagner, Jr. *The Academic Mind*. Glencoe: Free Press, 1958.

Lehman, H. C. "Men's Creative Production Rates at Different Ages and in Different Countries." *Scientific Monthly* 78 (1954): 321–326.

Libbey, Miles A., and Zaltman, Gerald. *The Role and Distribution of Written Informal Communication in Theoretical High Energy Physics.* New York: American Institute of Physics, 1967.

Light, D. "Introduction: The Structure of the Academic Professions." *Sociology of Education* 47 (1974): 2–28.

Lightfield, E. Timothy. "Output and Recognition of Sociologists." *The American Sociologist* 6 (1971): 128–133.

Lodahl, Janice B., and Gordon, Gerald. "The Structure of Scientific Fields and the Functioning of University Graduate Departments." *American Sociological Review* 37 (1972): 57–72.

Lubrano, Linda L. *Soviet Sociology of Science.* Columbus, Ohio: American Association for the Advancement of Slavic Studies, 1976.

MacCrae, Duncan. "Growth and Decay Curves in Scientific Citations." *American Sociological Review* 34 (1969): 631–635.

Magyar, George. "Typology of Research in Physics." *Social Studies of Science* 5 (1975): 79–85.

Manis, J. G. "Some Academic Influences upon Publication Productivity." *Social Forces* 29 (1951): 267–272.

Mannheim, K. *Ideology and Utopia: An Introduction to the Sociology of Knowledge.* New York: Harcourt Brace, 1936.

Mansfield, E. *Industrial Research and Technological Innovation.* New York: W. W. Norton, 1968.

Marcson, Simon. "Role Adaptation of Scientists in Industrial Research." *I.R.E. Transactions on Engineering Management,* EM–7 (1960a): 159–166.

————. *The Scientist in American Industry.* Princeton: Industrial Relations Section, Princeton University, 1960b.

————. "Decision-Making in a University Physics Department." *American Behavioral Scientist* 6 (1962): 37–38.

————. "Research Settings." In *The Social Contexts of Research,* pp. 161–191. Edited by S. Z. Nagi and R. G. Corwin. New York: Wiley–Interscience, 1972.

Masterman, Margaret. "The Nature of a Paradigm." In *Criticism and the Growth of Knowledge,* pp. 59–89. Edited by Imré Lakatos and Alan Musgrave. Cambridge: Cambridge University Press, 1970.

Mazur, Allan. "Public Confidence in Science." *Social Studies of Science* 7 (1977): 123–125.

McClelland, David C. "On the Psychodynamics of Creative Physical Scientists." In *Contemporary Approaches to Creative Thinking,* pp. 141–175. Edited by Howard E. Gruber, Glenn Terrell, and Michael Wertheimer. New York: Atherton Press, 1963.

Meadows, A. J., and O'Connor, J. G. "Bibliographic Statistics as a Guide to Growth Points in Science." *Science Studies* 1 (1971): 95–99.

Meier, R. L. "Research as a Social Process: Social Status, Specialism, and

Technological Advance in Great Britain." *British Journal of Sociology* 2 (1951): 91–104.

Meier, Russell H. *The Sociology of Scientific Growth: A Case Study*. Washington, D.C.: College and University Press, 1976.

Mellanby, Kenneth. "The Disorganization of Scientific Research." *Minerva* 12 (1974): 67–82.

Meltzer, Leo. "Scientific Productivity and Organizational Settings." *Journal of Social Issues* 12 (1956): 32–40.

Menzel, Herbert. "Planned and Unplanned Scientific Communication." In *The Sociology of Science*, pp. 417–441. Edited by Bernard Barber and Walter Hirsch. New York: Free Press, 1962.

———. "Scientific Communication: Five Sociological Themes." *American Psychologist* 21 (1966): 999–1005.

———. "Planning the Consequences of Unplanned Action in Scientific Communication." In *Communication in Science*, pp. 57–71. Edited by A. de Reuck and J. Knight. London: J. & A. Churchill, 1967.

Merton, Robert K. "Fluctuations in the Rate of Industrial Invention." *The Quarterly Journal of Economics* 49 (1935): 454–470.

———. "Puritanism, Pietism, and Science." *Sociological Review* 28 (1936): 1–30.

———. "The Sociology of Knowledge." *Isis* 27 (1937): 493–503.

———. *Science, Technology and Society in Seventeenth-Century England*. Osiris: Studies on the History and Philosophy of Sciences. Bruges, Belgium: Saint Catherine Press, 1938; with new preface, New York: Howard Fertig, 1970.

———. "Science and the Social Order." *Philosophy of Science* 5 (1938): 321–37. Reprinted in Merton (1973).

———. "Science and the Economy of Seventeenth-Century England." *Science and Society* 3 (1939): 3–27.

———. "Karl Mannheim and the Sociology of Knowledge." *Journal of Liberal Religion* 2 (1941a): 125–147.

———. "Znaniecki's *The Social Role of the Man of Knowledge*." *American Sociological Review* 6 (1941b): 111–115. Reprinted in Merton, (1973).

———. "Science and Technology in a Democratic Order." *Journal of Legal and Political Science* 1 (1942): 115–126. Reprinted in Merton (1973).

———. "The Machine, the Worker, and the Engineer." *Science* 105 (17 January 1947): 79–84.

———. *Social Theory and Social Structure*. Glencoe, Ill.: Free Press, 1949; 2d ed., rev. and enlarged. New York: Free Press, 1957; reprint ed., 1968.

———. Introduction to Bernard Barber, *Science and the Social Order*. Glencoe, Ill.: Free Press, 1952.

———. "Priorities in Scientific Discovery: A Chapter in the Sociology of

Science." *American Sociological Review* 22 (1957): 635–659. Reprinted in Barber and Hirsch (1962) and in Merton (1973).

——. "Singletons and Multiples in Scientific Discovery: A Chapter in the Sociology of Science." *Proceedings of the American Philosophical Society* 105 (1961a): 470–486. Reprinted in Merton (1973).

——. "Social Conflict over Styles of Sociological Work." *Transactions of the Fourth World Congress of Sociology* 3 (1961b): 21–36. Reprinted in Merton (1973).

——. "Resistance to the Systematic Study of Multiple Discoveries in Science." *European Journal of Sociology* 4 (1963a): 237–282. Reprinted in Merton (1973).

——. "The Ambivalence of Scientists." *Bulletin of the Johns Hopkins Hospital* 112 (1963b): 77–97. Reprinted in Kaplan (1965) and in Merton (1973).

——. *On the Shoulders of Giants: A Shandean Postscript.* New York: Free Press, 1965.

——. "The Matthew Effect in Science: The Reward and Communication Systems of Science." *Science* 159 (5 January 1968a): 55–63. Reprinted in Merton (1973).

——. "Observations on the Sociology of Science." *Japan-American Forum* 14 (1968b): 18–28.

——. "Behavior Patterns of Scientists." Copublished in *American Scientist* 57 (1969): 1–23, and *American Scholar* 38 (1969): 197–225. Reprinted in Merton (1973).

——. "Insiders and Outsiders: A Chapter in the Sociology of Knowledge." *American Journal of Sociology* 77 (1972): 9–47. Reprinted in Merton (1973).

——. *The Sociology of Science: Theoretical and Empirical Investigations.* Edited by Norman W. Storer. Chicago: University of Chicago Press, 1973.

——. *Sociological Ambivalence and Other Essays.* New York: Free Press, 1976.

——. "The Sociology of Science: An Episodic Memoir." In *The Sociology of Science in Europe,* pp. 3–141. Edited by Robert K. Merton and Jerry Gaston. Carbondale: Southern Illinois University Press, 1977.

Merton, Robert K., and Barber, Bernard. "Sorokin's Formulations in the Sociology of Science." In *P. A. Sorokin in Review,* pp. 332–368. Edited by P. J. Allen. Durham, N.C.: Duke University Press, 1963. Reprinted in Merton (1973).

Merton, Robert K., and Gaston, Jerry, eds. *The Sociology of Science in Europe.* Carbondale: Southern Illinois University Press, 1977.

Merton, Robert K., and Lewis, Richard. "The Competitive Pressures 1: The Race for Priority." *Impact of Science on Society* 21 (1971): 151–161.

Merton, Robert K., and Sorokin, Pitirim A. "Sociological Aspects of Invention, Discovery, and Scientific Theories." In *Social and Cultural Dynamics,* vol. 2, pp. 125–180, 439–476. Edited by Pitirim A. Sorokin. New York: American Book Company, 1937.

Mitroff, I. I. "Norms and Counter-Norms in a Select Group of the Apollo Moon Scientists." *American Sociological Review* 39 (1974a): 579–595.

——. *The Subjective Side of Science.* Amsterdam: Elsevier, 1974b.

Mitroff, I. I.; Jacob, Theodore; and Moore, Eileen Trauth. "On the Shoulders of the Spouses of Scientists." *Social Studies of Science* 7 (1977): 303–327.

Moravcsik, Michael J. "Technical Assistance and Fundamental Research in Underdeveloped Countries." *Minerva* 2 (1964): 197–209.

——. "Some Practical Suggestions for the Improvement of Science in Developing Countries." *Minerva* 4 (1966): 381–390.

Moravcsik, Michael J., and Murugesan, Poovanalingam. "Some Results on the Function and Quality of Citations." *Social Studies of Science* 5 (1975): 86–92.

Mulkay, Michael. "Some Aspects of Cultural Growth in the Natural Sciences." *Social Research* 36 (1969): 22–52.

——. "Conformity and Innovation in Science." In *The Sociology of Science,* pp. 5–23. Edited by P. Halmos. Keele: University of Keele, 1972a.

——. *The Social Process of Innovation: A Study in the Sociology of Science.* London: Macmillan, 1972b.

——. "Conceptual Displacement and Migration in Science: A Prefatory Paper." *Science Studies* 4 (1974a): 205–234.

——. "Methodology in the Sociology of Science: Some Reflections on the Study of Radio Astronomy." *Social Science Information* 13 (1974b): 107–119.

——. "The Mediating Role of the Scientific Elite." *Social Studies of Science* 6 (1976a): 445–470.

——. "Norms and Ideology in Science." *Social Science Information* 15 (1976b): 637–656.

——. "The Sociology of Science in Britain." In *The Sociology of Science in Europe,* pp. 224–257. Edited by Robert K. Merton and Jerry Gaston. Carbondale: Southern Illinois University Press, 1977a.

——. "Sociology of the Scientific Research Community." In *Science, Technology, and Society,* pp. 93–148. Edited by Ina Spiegel-Rösing and Derek de Solla Price. London and Beverly Hills: Sage Publications, 1977b.

——. "Consensus in Science." *Social Science Information* 17 (1978): 107–122.

Mulkay, Michael, and Edge, D. "Cognitive, Technical and Social Factors in the Growth of Radio Astronomy." *Social Science Information* 12 (1972): 25–61.

Mulkay, Michael; Gilbert, G. N.; and Woolgar, S. "Problem Areas and Research Networks in Science." *Sociology* 9 (1975): 187–204.

Mulkay, Michael, and Turner, R. S. "Over-Production of the Personnel and Innovation in Three Social Settings." *Sociology* 5 (1971): 47–61.

Mulkay, Michael, and Williams, A. T. "A Sociological Study of a Physics Department." *British Journal of Sociology* 22 (1971): 68–82.

Mullins, Nicholas. "The Distribution of Social and Cultural Properties in Informal Communication Networks among Biological Scientists." *American Sociological Review* 33 (1968): 786–797.

————. "The Structure of an Elite: The Advisory Structure of the Public Health Service." *Science Studies* 2 (1972a): 3–29.

————. "The Development of a Scientific Specialty: The Phage Group and the Origin of Molecular Biology." *Minerva* 10 (1972b): 51–82.

————. "The Development of Specialties in Social Science: The Case of Ethnomethodology." *Science Studies* 3 (1973a): 245–273.

————. *Theories and Theory Groups in Contemporary American Sociology.* New York: Harper & Row, 1973b.

Mullins, Nicholas; Hargens, Lowell L.; Hecht, Pamela K.; and Kick, Edward L. "The Group Structure of Co-Citation Clusters: A Comparative Study." *American Sociological Review* 42 (1977): 552–562.

Musson, A. E., and Robinson, Eric. *Science and Technology in the Industrial Revolution.* Manchester: University Press, 1969.

Nagi, Saad Z., and Corwin, Ronald G. *The Social Contexts of Research.* New York: John Wiley, 1972.

Oberschall, Anthony, ed. *The Establishment of Empirical Sociology.* New York: Harper & Row, 1972.

Ogburn, William F. *Social Change.* New York: Viking Press, 1922.

————. "The Great Man vs. Social Forces." *Social Forces* 5 (1926): 225–231.

Orlans, Harold. *The Effects of Federal Programs on Higher Education.* Washington, D.C.: Brookings Institution, 1962.

————. "Criteria of Choice in Social Science Research." *Minerva* 10 (1972): 571–602.

————. *Contracting for Knowledge.* San Francisco: Jossey-Bass, 1973.

Oromaner, Mark Jay. "Professional Age and the Reception of Sociological Publications: A Test of the Zuckerman-Merton Hypothesis." *Social Studies of Science* 7 (1977): 381–388.

Orth, Charles D., III. "The Optimum Climate for Industrial Research." In *Science and Society,* pp. 194–210. Edited by Norman Kaplan. Chicago: Rand McNally, 1965.

Pacey, Arnold. *The Maze of Ingenuity: Ideas and Idealism in the Development of Technology.* New York: Holmes and Meier, 1975.

Pelz, Donald, and Andrews, Frank. *Scientists in Organizations*. New York: John Wiley, 1966.

Perrucci, Robert. *Profession without Community: Engineers in American Society*. New York: Random House, 1969.

Perrucci, Robert, and Gerstl, Joel. *The Engineers and the Social System*. New York: John Wiley, 1969.

Pfetsch, Frank. "Scientific Organization and Science Policy in Imperial Germany 1871–1914: The Foundation of the Imperial Institute of Physics and Technology." *Minerva* 8 (1970): 557–580.

Platz, Arthur, and Blakelock, Edwin. "Productivity of American Psychologists: Quantity versus Quality." *American Psychologist* 15 (1960): 310–312.

Porter, Alan L. "Citation Analysis: Queries and Caveats." *Social Studies of Science* 7 (1977): 257–267.

Price, Derek J. de Solla. "Quantitative Measures of the Development of Science." *Archives Internationales d'Histoire des Sciences* 14 (1951): 85–93.

——. *Science since Babylon*. New Haven: Yale University Press, 1962.

——. *Little Science, Big Science*. New York: Columbia University Press, 1963.

——. "The Scientific Foundations of Science Policy." *Nature* 206 (1965a): 233–238.

——. "Networks of Scientific Papers." *Science* 149 (30 July 1965b): 510–515.

——. "Is Technology Historically Independent of Science? A Study in Statistical Historiography." *Technology and Culture* 6 (1965c): 553–568.

——. "Nations Can Publish or Perish." *Science and Technology* 70 (1967): 84–90.

——. "The Structures of Publication in Science and Technology." In *Factors in the Transfer of Technology*, pp. 91–104. Edited by W. H. Gruber and D. G. Marquis. Cambridge, Mass.: M.I.T. Press, 1969a.

——. "Measuring the Size of Science." *Proceedings of the Israel Academy of Sciences and Humanities* 4 (1969b): 98–111.

——. "Differences between Scientific and Technological and Non-Scientific Scholarly Communities." Paper read at World Congress of Sociology, Varna, Bulgaria, 14–19 September 1970a.

——. "Citation Measures of Hard Science, Soft Science, Technology and Nonscience." In *Communication Among Scientists and Engineers*, pp. 3–22. Edited by C. Nelson and D. Pollock. Lexington, Mass.: D. C. Heath, 1970b.

——. "A General Theory of Bibliometric and Other Cumulative Advantage Processes." *Journal of the American Society for Information of Science* 27 (1976): 292–306.

_____. "Toward a Model for Science Indicators." In *Toward a Metric of Science: The Advent of Science Indicators*, pp. 69–95. Edited by Y. Elkana, J. Lederberg, R. K. Merton, A. Thackray, and H. Zuckerman. New York: Wiley–Interscience, 1977.

_____. "Ups and Downs in the Pulse of Science and Technology." In *Sociology of Science*, pp. 162–171. Edited by Jerry Gaston. San Francisco: Jossey-Bass, 1978.

Price, Derek J. de Solla, and Beaver, Donald deB. "Collaboration in an Invisible College." *American Psychologist* 21 (1966): 1011–1018.

Rabkin, Yakov M. " 'Naukovedenie': The Study of Scientific Research in the Soviet Union." *Minerva* 14 (1976): 61–78.

Ravetz, Jerome. *Scientific Knowledge and Its Social Problems*. New York: Oxford University Press, 1971.

Reif, Frederick. "The Competitive World of the Pure Scientist." *Science* 134 (15 December 1961): 1957–1962. Reprinted in Kaplan (1965).

Reif, Frederick, and Strauss, Anselm. "The Impact of Rapid Discovery upon the Scientist's Career." *Social Problems* 12 (1965): 297–311.

Reskin, Barbara F. "Sex Differences in Status Attainment in Science: The Case of the Postdoctoral Fellowship." *American Sociological Review* 41 (1976): 597–612.

_____. "Scientific Productivity and the Reward Structure of Science." *American Sociological Review* 42 (1977): 491–504.

_____. "Scientific Productivity, Sex, and Location in the Institution of Science." *American Journal of Sociology* 83 (1978a): 1235–1243.

_____. "Sex Differentiation and Social Organization of Science." In *Sociology of Science*, pp. 6–37. Edited by Jerry Gaston. San Francsico: Jossey-Bass, 1978b.

Restivo, Sal P., and Vanderpool, Christopher K., eds. *Comparative Studies in Science and Society*. Columbus, Ohio: Charles E. Merrill, 1974.

Richter, M. N. *Science as a Cultural Process*. Cambridge, Mass.: Schenkman, 1972.

Robbins, David, and Johnston, Ron. "The Role of Cognitive and Occupational Differentiation in Scientific Controversies." *Social Studies of Science* 6 (1976): 349–368.

Robinson, D. Z. "Will the University Decline as the Center for Scientific Research?" *Daedalus* 102 (1973): 101–110.

Rogers, E. M. *Diffusion of Innovations*. New York: Free Press, 1962.

Roose, K. D., and Anderson, C. J. *A Rating of Graduate Programs*. Washington, D.C.: American Council on Education, 1970.

Rose, Hilary. "The Rejection of the WHO Research Centre: A Case Study of Decision-Making in International Scientific Collaboration." *Minerva* 5 (1967): 340–356.

Rose, Hilary, and Rose, Steven. *Science and Society*. London: Allen Lane, Penguin Press, 1969.

Rothman, Robert A. "A Dissenting View on the Scientific Ethos." *The British Journal of Sociology* 23 (1972): 102–108.

Rottenberg, Simon. "The Warrants for Basic Research." *Minerva* 5 (1966): 30–38.

Rowe, A. P. "From Scientific Idea to Practical Use." *Minerva* 2 (1964): 303–319.

Rudd, Ernest. "The Effect of Alphabetical Order of Author Listing on the Careers of Scientists." *Social Studies of Science* 7 (1977): 268–269.

Sanderson, M. *The Universities and British Industry: 1850–1970.* London: Routledge and Kegan Paul, 1972.

Schooler, Dean, Jr. *Science, Scientists, and Public Policy.* New York: Free Press, 1971.

Schwartz, Melvin. "The Conflict Between Productivity and Creativity in Modern Day Physics." *American Behavioral Scientist* 6 (1962): 35–36.

Shils, E. *The Torment of Secrecy.* New York: Free Press, 1956; reprint ed., Carbondale: Southern Illinois University Press, 1974.

———. *Criteria for Scientific Development.* Cambridge, Mass.: M.I.T. Press, 1968a.

———. "The Profession of Science." *The Advancement of Science* 24 (1968b): 469–480.

———. *The Intellectuals and the Powers.* Chicago: University of Chicago Press, 1972.

Singer, B. F. "Toward a Psychology of Science." *American Psychologist* 26 (1971): 1010–1015.

Sklair, Leslie. "The Political Sociology of Science: A Critique of Current Orthodoxies." In *The Sociology of Science,* pp. 43–59. Edited by Paul Halmos. Keele: University of Keele, 1972.

Skoie, Hans. "The Problems of a Small Scientific Community: The Norwegian Case." *Minerva* 7 (1969): 399–425.

Small, Henry. "Co-Citation in the Scientific Literature: A New Measure of the Relationship Between Two Documents." *Journal of American Society of Information Science* 24 (1973): 265–269.

———. "A Co-Citation Model of a Scientific Specialty: A Longitudinal Study of Collagen Research." *Social Studies of Science* 7 (1977): 139–166.

Small, Henry, and Griffith, Belver C. "The Structure of Scientific Literatures I: Identifying and Graphing Specialties." *Science Studies* 4 (1974): 17–40.

Snow, C. P. *The Two Cultures: And a Second Look.* Cambridge: Cambridge University Press, 1965.

Sorokin, Pitirim A., ed. *Social and Cultural Dynamics.* New York: American Book Company, 1937.

Sorokin, Pitirim A., and Merton, Robert K. "The Course of Arabian Intellectual Development, 700–1300: A Study in Method." *Isis* 22 (1935): 516–524.

Spiegel-Rösing, Ina. "Science Studies: Bibliometric and Content Analysis." *Social Studies of Science* 7 (1977): 97–113.

Spiegel-Rösing, Ina, and Price, Derek de Solla, eds. *Science, Technology, and Society*. London and Beverly Hills: Sage Publications, 1977.

Stafford, A. B. "Is the Rate of Invention Declining?" *American Journal of Sociology* 57 (1952): 539–545.

Stehr, Nico. "The Ethos of Science Revisited: The Social and Cognitive Norms of Science." In *Sociology of Science,* pp. 172–196. Edited by Jerry Gaston. San Francisco: Jossey-Bass, 1978.

Stehr, Nico, and König, René, eds. *Wissenschaftssoziologie.* Opladen: Westdeutscher Verlag, 1975.

Stehr, Nico, and Larson, Lyle E. "The Rise and Decline of Areas of Specialization." *American Sociologist* 7 (1972): 3, 5–6.

Stein, Morris L. "Creativity and the Scientist." In *The Sociology of Science,* pp. 329–343. Edited by Bernard Barber and Walter Hirsch. New York: Free Press, 1962.

Storer, Norman W. "The Coming Changes in American Science." *Science* 142 (25 October 1963): 464–467. Reprinted in Kaplan (1965).

———. *The Social System of Science.* New York: Holt, Rinehart, and Winston, 1966.

———. "The Hard Sciences and the Soft: Some Sociological Observations." *Bulletin of the Medical Library Association* 55 (1967): 75–84.

———. "Relations among Scientific Disciplines." In *The Social Contexts of Research,* pp. 229–269. Edited by Saad Z. Nagi and Ronald G. Corwin. New York: John Wiley, 1972.

———. Introduction to *The Sociology of Science,* by Robert K. Merton. Chicago: University of Chicago Press, 1973.

Strauss, Anselm L., and Rainwater, Lee. *The Professional Scientist: A Study of American Chemists.* Chicago: Aldine Press, 1962.

Studer, K. E., and Chubin, Daryl E. "Ethics and the Unintended Consequences of Social Research: A Perspective From the Sociology of Science." *Policy Sciences* 8 (1977): 111–124.

Sturgis, Richard B., and Clemente, Frank. "The Productivity of Graduates of Fifty Sociology Departments." *The American Sociologist* 8 (1973): 169–180.

Sullivan, Daniel "Competition in Bio-Medical Science: Extent, Structure, and Consequences." *Sociology of Education* 48 (1975): 223–241.

Sullivan, Daniel; White, D. Hywel; and Barboni, Edward J. "Co-Citation Analyses of Science: An Evaluation." *Social Studies of Science* 7 (1977a): 223–240.

———. "The State of a Science: Indicators in the Specialty of Weak Interactions." *Social Studies of Science* 7 (1977b): 167–200.

Swatez, Gerald M. "The Social Organization of a University Laboratory." *Minerva* 8 (1970): 36–58.

Tavis, Irene. "A Survey of Popular Attitudes toward Technology." *Technology and Culture* 13 (1972): 606–621.

Tavis, Irene, and Gerber, William. *Technology and Work*. Cambridge, Mass.: Harvard University Press, 1969.

Tavis, Irene, and Silverman, Linda. *Technology and Values*. Cambridge, Mass.: Harvard University Press, 1969.

Taylor, C. W., and Barron, F., eds. *Scientific Creativity: Its Recognition and Development*. New York: John Wiley, 1963.

Teich, Albert H., ed. *Technology and Man's Future*. 2nd ed. New York: St. Martin's Press, 1977.

Thorner, Isidor. "Ascetic Protestantism and the Development of Science and Technology." *American Journal of Sociology* 58 (1952): 25–33.

Thrall, Charles A., and Starr, Jarold M., eds. *Technology, Power and Social Change*. Lexington, Mass.: D. C. Heath, 1972.

Townes, C. H. "Differentiation and Competition between Universities and other Research Laboratories in the United States." *Daedalus* 102 (1973): 153–165.

Trow, Martin. "The Democratization of Higher Education in America." *European Journal of Sociology* 3 (1962): 231–262.

Turner, S. J., and Chubin, Daryl E. "Another Appraisal of Ortega, the Coles, and Science Policy: The Ecclesiastes Hypothesis." *Social Science Information* 15 (1976): 657–662.

Wandere, Jules J. "An Empirical Study in the Sociology of Knowledge." *Sociological Inquiry* 39 (1969): 19–26.

Waterman, A. "Integration of Science and Society." *American Behavioral Scientist* 4 (1962): 3–6.

Watson, James D. *The Double Helix*. New York: New American Library, 1969.

Weber, Max. *The Protestant Ethic and the Spirit of Capitalism*. Tr. Talcott Parsons. New York: Charles Scribner's Sons, 1958.

———. "Science as Vocation." In *From Max Weber*, pp. 134–156. Edited by H. H. Gerth and C. Wright Mills. London: Oxford University Press, 1958. Reprinted in Barber and Hirsch (1962).

Weinstock, Melvin. "Citation Indexes." In *Encyclopedia of Library and Information Science*, vol. 5, pp. 16–40. New York: Marcel Dekker, 1971.

West, S. Stewart. "The Ideology of Academic Scientists." *I.R.E. Transactions on Engineering Management*, EM-7 (1960a): 54–62.

———. "Sibling Configurations of Scientists." *American Journal of Sociology* 66 (1960b): 268–274.

Whitley, Richard. "The Formal Communication System of Science." In

The Sociology of Sociology, pp. 163–179. Edited by Paul Halmos. Keele: University of Keele, 1970a.

———. "The Operation of Science Journals: Two Case Studies in British Social Sciences." *Sociological Review* 18 (1970b): 241–258.

———. "Black Boxism and the Sociology of Science: A Discussion of the Major Developments in the Field." In *The Sociology of Science*, pp. 61–92. Edited by Paul Halmos. Keele: University of Keele, 1972.

———. "Umbrella and Polytheistic Scientific Disciplines and Their Elites." *Social Studies of Science* 6 (1976): 471–497.

———, ed. *Social Processes of Scientific Development*. London: Routledge and Kegan Paul, 1974.

Wilson, Logan. *The Academic Man*. New York: Oxford University Press, 1942.

Wilson, Sanford. "Bio-Social Characteristics of American Inventors." *American Sociological Review* 2 (1937): 837–849.

Woolf, Patricia K. "The Second Messenger: Informal Communication in Cyclic AMP Research." *Minerva* 13 (1975): 349–373.

Woolgar, S. W. "Writing an Intellectual History of Scientific Development: The Use of Discovery Accounts." *Social Studies of Science* 6 (1976): 395–422.

Wunderlich, Richard. "The Scientific Ethos: A Clarification." *British Journal of Sociology* 25 (1974): 373–377.

Wynn, Brian. "C. G. Barkla and the J. Phenomenon: A Case Study in the Treatment of Deviance in Physics." *Social Studies of Science* 6 (1976): 307–347.

Yellin, Joel. "A Model for Research Problem Allocation among Members of a Scientific Community." *Journal of Mathematical Sociology* 2 (1972): 1–36.

Zaltman, Gerald. *Scientific Recognition and Communication Behavior in High Energy Physics*. New York: American Institute of Physics, 1968.

Ziman, John. *Public Knowledge*. Cambridge: Cambridge University Press, 1968.

Zinberg, Dorothy S. "Education Through Science: The Early Stages of Career Development in Chemistry." *Social Studies of Science* 6 (1976): 215–246.

Zuckerman, Harriet A. "The Sociology of the Nobel Prize." *Scientific American* 217 (1967a): 25–33.

———. "Nobel Laureates in Science: Patterns of Productivity, Collaboration, and Authorship." *American Sociological Review* 32 (1967b): 391–403.

———. "Stratification in American Science." *Sociological Inquiry* 40 (1970): 235–257.

———. "Interviewing an Ultra-Elite." *Public Opinion Quarterly* 36 (1972): 159–175.

————. *Scientific Elite: Nobel Laureates in the United States.* New York: Free Press, 1977a.

————. "Deviant Behavior and Social Control in Science." In *Deviance and Social Change,* pp. 87–138. Edited by Edward Sargarin. Beverly Hills: Sage Publications, 1977b.

————. "Theory Choice and Problem Choice in Science." In *Sociology of Science,* pp. 65–95. Edited by Jerry Gaston. San Francisco: Jossey-Bass, 1978.

Zuckerman, Harriet A., and Cole, Jonathan R. "Women in American Science." *Minerva* 13 (1975): 82–102.

Zuckerman, Harriet A., and Merton, Robert K. "Patterns of Evaluation in Science: Institutionalization, Structure, and Functions of the Referee System." *Minerva* 9 (1971): 66–100. Reprinted in Merton (1973).

Zuckerman, Harriet A., and Merton, Robert K. "Age, Aging, and Age Structure in Science." In *A Sociology of Age Stratification,* pp. 292–356. Edited by Matilda W. Riley. New York: Russell Sage Foundation, 1972. Reprinted in Merton (1973).

Chapter 8
Medical Sociology and Science and Technology in Medicine

Linda H. Aiken, THE ROBERT WOOD JOHNSON FOUNDATION
Howard E. Freeman, UNIVERSITY OF CALIFORNIA, LOS ANGELES

INTRODUCTION

There is a popular conception that science-based medical procedures account for the drastic reductions in mortality and morbidity in industrial and postindustrial societies. Moreover, there is a vast cadre of persons engaged in policy, research, and clinical roles in medical care who are committed to new scientific discoveries and improved technology as the solutions to health problems. In many respects, the organization of medical care and delivery of health services is dominated by scientific and technological developments in medicine (Donabedian, *Benefits in Medical Care Programs,* 1976). Nonetheless, the role of science and technology in health care has received only sporadic and limited analysis from a sociological perspective. This lack is related to the way the specialty of medical sociology has evolved. Rather than giving attention to science and technology in curing and preventing illness, medical sociologists have focused on general community conditions, access to health services, preventive practices, and interpersonal relations between providers and patients as the key determinants of health status. The inattention to this aspect of science and technology by sociologists is remarkable, and from this perspective, the sociology of health has neglected perhaps the most important determinant of the quality, distribution, and impact of medical care.

This chapter has two major purposes: first, to review the development of medical sociology; second, to examine from a social policy perspective the impact of science and technology on health research, clinical practice, and the organization and delivery of health services. The review of medical sociology draws primarily on work done by sociologists—although to a lesser extent other social sciences are represented as well. On the other hand, the analysis of science and technology in medicine is based primarily on studies by health planners and policy analysts; among social scientists, only medical economists are well represented.

In one sense the two major sections that follow—on medical sociology and on science and technology in medicine—can be viewed as independent efforts. In another sense the two parts are interrelated: We draw on the health planning and policy material, and the work of the medical economists, to enhance the sociological perspective on science and technology in medicine. One hope we have is to encourage the growth of sociological research in the area.

I. THE SPECIALTY OF MEDICAL SOCIOLOGY

The very earliest of the world's great physicians realized that health and illness are social as well as biological phenomena (Rosen, 1979). In antiquity and in medieval times, as well as now, health problems are intimately related to economic, political, and social conditions (Freeman et al., eds., *Handbook of Medical Sociology,* 1979). While social scientists tend to emphasize the recency of medical sociology, both those on the frontiers of medicine and those involved with affairs of state and matters of policy—administrators, politicians, and social reformers—have continually been aware of the sociological dimensions of health and illness.

A. Development of Medical Sociology

It is difficult to single out a particular year as the formal starting date of medical sociology. McIntire, "The Importance of the Study of Medical Sociology" (1894), seems to be the first use of the term. As Rosen suggests (1979), the appearance of the term in the late nineteenth century was not accidental. This was the period when many in the health field began to clamor about the need for social rather than additional medical programs.

Studies in Britain during the last part of the nineteenth century and the first decade of the twentieth documented the need to remedy the social environment, particularly if infant and maternal mortality were to be reduced and the health status of the poor and the industrial worker improved. In Western Europe and the United States as well, it became inescapable that the major advances in medicine required a reduction of pathological social conditions. Grotjahn's classic work, published just before World War I, spurred interest in social factors in health and illness; it also set forth the need for a social science framework in order to understand health behavior, influence medical care, and improve the education of health providers. (See Grotjahn and Kriegel, *Soziale Pathologie,* 1915.)

"Medical sociology" during this period—and indeed perhaps until the 1930s—was a synonym for "social medicine," and the effort was dominated by physicians in such fields as public health, maternal and child health, and environmental medicine. Although sociologists are taught that their discipline is as old as Auguste Comte, and that modern sociology began with Emile Durkheim and Max Weber, it was not until the Great Depression of the 1930s that social research became an important input into the complex mosaic of national policy making. Moreover, until this period there were no academic research centers as we know them now in the social sciences, and graduate education was centered in only a few universities. Sociology through the 1930s was primarily a teaching discipline with the vast majority of sociologists tending to the tasks of undergraduate education.

According to a number of analyses of the makings of contemporary sociology, at least in the United States, it was the activities of social researchers during World War II that spurred the development of the field (see especially Madge, 1962). Stouffer and his famous "American soldier team," and kindred groups working for other military organizations, applied sociology to mundane problems such as whether or not front-line troops preferred bread or potatoes, as well as to major policy questions such as the most equitable scheme for releasing persons from the military after the termination of hostilities. For perhaps the first time, sociologists and other social researchers worked in fairly large teams, had to produce timely and useful products, operated with some assurance that their recommendations would be used, and were able to undertake their studies with ample funding.

It is important not to make light of the impact of applied research immediately before, during, and right after World War II on the course of sociology in general and medical sociology in particular. The cadre of persons immersed in research during this period gained not only important experience on how to undertake empirical studies but an outlook on the usefulness of their discipline that had escaped sociologists before this period.

There are both interesting parallels and differences between the development of medical sociology in the United States and in other countries. They are related to the strength of linkages found in different countries between sociology and legal, humanistic, and philosophical studies on the one hand, and life and physical science research approaches on the other. In this and subsequent sections of this chapter, the analysis clearly focuses on sociology and science and technology in North America.

Sociology boomed in the late 1940s and 1950s, and the intricacies of federal support for higher education during those years account in part for the growth of medical sociology. Subsequent to World War II, there was not only a shortage of professors to train the expanding number of university students, but there was a commitment, first at the National Science Foundation and later at the National Institute of Mental Health, to support both social research investigations and graduate social science training. Part of the commitment, of course, was related to the general policy of finding ways to keep persons off the labor market by subsidizing educational programs. Part of the commitment also represented a continuation of the trajectory in the health field, particularly in psychiatry and public health, to understand and conquer the social causes of illness and the social barriers to the delivery of health services.

Unfortunately, there is no full history of the rise in social science support by federal agencies. But the very fact that much of the responsibility for research and training support was assigned to the National Institutes of Health created an interest in medical sociology by academicians for opportunistic and entrepreneurial reasons (Freeman et al., 1975). Sociologists became interested in a wide array of problems ranging from the social correlates of disease to the socialization of health professionals. They entered medical settings and observed patients in their relations with providers of health services; they engaged in a broad range of outcome studies on social psychological determinants of health and illness and on social variables that were mediators of treatment efforts; and they focused on hospitals as organizations with complex bureaucratic arrangements.

Toward the end of the 1950s, an era of plenty began that lasted about a decade. There were literally more research funds than persons around who had creative research ideas and who could devise sound research designs. There were more positions in health science schools than competent sociologists to fill them, and there were more training opportunities for graduate students than applicants with reasonable academic qualifications. As in the case of other resource areas, we know that not only does demand increase supply but supply increases demand.

By the 1960s, medical sociology was an established specialty. Unlike the postulatings of the early great physicians, and different from the physician-dominated medical sociology of the early 1900s, medical sociology was locked into its parent discipline and benefited from the conceptual and methodological work going on in other sociology specialties. Even by the end of the 1950s, there was enough interest to merit review papers (Freeman and Reeder, "Medical Sociology: A Review of the Literature," 1957); and specialized journals appeared, such as the *Journal of Health and Social Behavior*. Medical sociology became the first and one of the largest specialized sections in the American Sociological Association.

In a sense, then, medical sociology is a very old specialty; it goes back to the observations of early physicians who noted the importance of the social milieu in understanding health and illness, and to the pioneers in such fields as public health and social medicine. In another sense, however, medical sociology is a very recent intellectual and scientific field (Barber, 1968). Prior to the 1960s, there were probably less than twenty-five tenured positions held by sociologists in schools of health sciences; only in the past two decades has medical sociology become a legitimate specialty for graduate students in sociology departments.

B. Roles of Medical Sociologists

In health science schools, sociologists participate in the training of public health workers, medical students, residents, and other health practitioners. They undertake research investigations directed at improving the delivery of health services; they also provide sociological input into studies that are predominantly biomedical in character. These activities are referred to as sociology *in* medicine (Strauss, 1957, pp. 200–204).

At the same time, medical sociologists attached to departments of sociology and social research centers undertake research or teaching that is primarily sociological in character. Instead of studying an industrial plant, they study the health organization; instead of observing college students, they take patients as the study group; and rather than examine the careers and interpersonal strengths of policemen, they examine these relationships for health personnel. These activities, guided mainly by theories and concepts of sociology and intended primarily for sociology audiences, are identified as sociology *of* medicine.

The distinction between sociology *of* and *in* medicine has never been sharp. On the one hand, studies initiated primarily in response to concerns of the health field contribute substantively and methodologically to fundamental sociological knowledge. On the other hand, work that was initiated primarily for a sociological audience, and originally characterized as sociology *of* medicine, has proved meaningful for health professionals and health science students.

During the 1950s and 1960s, the distinction between sociology *in* and *of* medicine achieved relatively wide currency. There was a pervasive belief among sociologists that to work in medicine was less respectable. In part, this is related to the demands that academia places on persons in every discipline to do studies relevant to the mainstream of thought in their field, to publish in established journals of their own discipline, and to achieve a reputation among peers in their own discipline.

Also, the *raison d'être* of the sociologist was often at variance with that of the physician. In sociology, as in all other traditional academic disciplines, the rationale persists that one is engaged fundamentally in knowledge building. The outlook in medicine and in the other health professions typically is much more pragmatic.

The situation has changed gradually. To some extent, interpersonal contacts and collaboration between sociologists and physicians have broken down some barriers. But more important, the general health scene itself has changed; it is now generally more hospitable to social research. Although

it is impossible to chart all the major changes in the health arena during
the past two decades, we can draw attention to a few:

1.   Over the last thirty years there has been a clear movement toward
some form of national health insurance, although its shape and payment
formulae are still matters of controversy.

2.   Even without national health insurance, payments for health care
have shifted markedly; the proportion provided by "third parties" is much
higher. Medicare, Medicaid, and other government programs, as well as
private health insurance, are now the payment sources for the vast majority
of encounters with the health care enterprise. No longer are the doctor
and patient alone concerned with the cost of medical care; the public and
the private insurance sector are as well.

3.   With respect to cost, medical care payments have risen faster than
the general rate of inflation. There are a number of factors that account for
the rising costs of care. They include technological innovations, entrepreneurship in medical practice, provider control over the marketplace, and
the failure of regulatory efforts.

4.   There has been a general social thrust toward greater equity, and
this trend has dramatically affected the health sector. The student composition of medical schools has changed because of public and government
pressures for affirmative action in admissions practices. The press for adequate—indeed, quality—medical care for the poor and the underserved
has created many new types of health delivery systems and affected the
selection and employment of health professionals.

5.   The deceleration of United States population growth and the shifts
in the age structure have modified the types of both physicians and medical
care facilities that are required.

6.   Finally, the climate has changed in the health professions and in
the larger community. Matters such as the right to abortion, to informed
consent, and to privacy are growing subjects of concern and discussion both
inside and outside the health professions.

The growth and development of the specialty of medical sociology
can be seen, then, as part of the shaping of the parent discipline by forces
both internal and external to it. The trend toward policy-relevant activities
in sociology and the other social sciences meshed auspiciously with governmental and public concern with matters amenable to sociological inquiry;
this convergence allowed social researchers in the health field to undertake
work that was consistent with mainstream sociology. Both the parent discipline and the specialty of medical sociology have been criticized for being
entrepreneurial in seizing opportunities to apply methods and concepts,
and for the direction of their growth.

To say that contemporary medical sociology deemphasizes humanistic concerns and ethical and philosophical issues is a reasonable criticism. As Czoniczer (1978) notes: "For the mass sociologist of this last quarter of the twentieth century, a person stricken by disease is a dropout of production: instead of producing he consumes. He has become a customer" (p. 18). Indeed, as we will discuss subsequently, the spotty contributions of sociology to the understanding of science and technology in medicine—and particularly to the emerging area of bioethics—can be accounted for by the way sociology and its subspecialties have evolved.

In both medicine and sociology, there continues to be a persistent and vocal group committed to challenging the depersonalization and dehumanization of research, planning, and practices in the health field, and the criteria of efficiency and effectiveness that dominate current policy development. In some areas, the critics have influenced sociological research directions as well as general social interests in health—including a notable surge of interest in matters of informed consent and privacy.

C. Work of Medical Sociologists

The health field is a microcosm of virtually all structural elements and social processes of the larger society. Thus, some sociologists in the health field are concerned with social-psychological problems; or with problems of social relationships of status and power; or with organization and bureaucracy; or with social, political, and economic issues. Some emerging trends in the field will be surveyed here:

1. **Epidemiological Investigations.** Even before sociologists were heavily involved in the health field, medical epidemiologists frequently included, in their field studies of the determinants of disease, measures of social characteristics. It became more necessary to incorporate social factors, as epidemiological work turned from infectious illnesses to the causes of chronic and degenerative conditions, and as interest surfaced in the etiology of disorders produced by the social and physical environment. It is clear that a full explanation for disease conditions and their origins involves an array of social and psychological characteristics of individuals as well as critical features of the social milieu. (See Graham and Reeder's discussion of chronic illness; Clausen's analytical review of research on mental disorders; Kaplan's analysis of social stress and life changes; and the discussion of environment and disease by Cohen, Glass, and Phillips—all in Freeman et al., eds., *Handbook of Medical Sociology*, 1979. In general, that is a useful source for the next seven subsections where individual citations are not given.)

2. **Orientations and Values in Health.** There have been numerous studies on community perceptions, beliefs, and outlooks on health and health care. In some areas, such as mental health, these studies have sug-

gested ways of planning and designing programs to minimize stigmatization. Other studies have sensitized medical students and practitioners to patients' reactions to procedures and regimens and to social and cultural differences in a population. (The literature on this topic and the next is too diverse to permit individual citations.)

3. **Health Behavior.** The processes involved in the individual's decisions to prevent illness, maintain good health, seek appropriate medical care, or follow health care regimens continue to be critical concerns of medical sociologists. There are two emergent issues in this area: the first is related to prevention and treatment efforts; the second involves so-called "iatrogenic" disease or induced dependency on the medical care system.

Often prevention programs also involve efforts to modify individual life styles—including such varied behaviors as smoking, diet, exercise, and alcohol consumption, as well as general orientation to major social domains of life. These are complex matters that run up against various beliefs, attitudes, and values of both individuals and cultural groups.

Health professionals and medical sociologists have also given more attention to the need to foster greater adherence to therapeutic regimens—in part to increase the effectiveness of medical intervention, in part to curtail health costs.

4. **Selection and Socialization of Health Providers.** Despite much new technology, health care still remains labor intensive. Further, education of health providers is a long and expensive enterprise. The education of health professionals has received much study, including the landmark studies by Becker et al., *Boys in White* (1961); and Merton, Reader, and Kendall, *The Student Physician* (1957). Work like Freidson's on professionalism (*Profession of Medicine*, 1975), and Mechanic's on styles of practice and physician aspirations (1979) will continue, as will studies of specialty groups and of selection and admission processes for various educational programs. However, there have been significant changes in areas of inquiry. Largely because of cost considerations, new provider groups, such as nurse practitioners, have emerged; the physician and hospital nurse are no longer the only groups to be studied. (See Reeder and Mauksch, 1979.)

As health care has moved from the individual general practitioner to specialized medical care teams and from doctors' private offices to larger organizational settings, additional research efforts have been undertaken on the desired characteristics of health providers. Among medical educators and human resource analysts, it is now widely thought that providers must be selected not only for competence in clinical medicine but also for administrative and organizational skills and willingness to locate in underserved areas.

Decisions about location of medical care centers and design of appropriate services have shifted from the individual to the organization; and health professionals are changing—in their preference for practice arrangements, in their economic expectations, and in social and political outlooks. This creates the need for additional studies to explain the selection and socialization of health providers. Some examples: Funkenstein, *Medical Students, Medical Schools, and Society During Five Eras* (1978); and Lally, "The Making of the Compassionate Physician-Investigator" (1978). The supply-and-demand problem has also shifted, from numbers of health care specialists to their distribution.

5. **Practitioner-Patient Relations.**   A long-standing area of study is the relations between providers and patients. Interest in this area is linked to optimizing the delivery of health services, to maximizing patient compliance with medical regimens, and to minimizing communication barriers that may impede understanding of patients' complaints and symptoms and the way they define their own illnesses (see Mechanic, *Medical Sociology,* 1978). Status discrepancies between practitioners of different levels and patients, as well as ethnic and cultural variations among patient groups that affect their relations with health providers, have been matters of central interest.

Studies of patient-practitioner interactions have directly affected health providers, the outstanding example being the delivery of psychotherapeutic treatment. The widely known Hollingshead and Redlich study on socioeconomic class differences in treatment for mental illness, *Social Class and Mental Illness* (1958), was the stimulus for major attention being given to the importance of social factors not only in the etiology but also in the treatment of psychiatric disorders. A large proportion of the efforts of primary care providers involves providing psychological treatment or support; and the social-psychological relations between provider and consumer tailor the former's success as a therapist and the latter's satisfaction with health care.

The nature of provider-patient relations is determined by the way health care services are structured and by the way the delivery of health services is organized. There has been sustained interest in practitioner-patient relations in various developed and less developed countries—in the way the latter, for example, are impacted by differences in structural arrangements for health care, as well as the way cultural and ideological perspectives in different communities and nations affect health care (see Gallagher, 1976).

6. **Organization of Medical Care.**   The concept of the family doctor remains an historically interesting phenomenon despite the development of family medicine as a legitimate specialty. Today, ambulatory medical practice is fragmented into a set of specialty arrangements with which patients must cope. And the hospitalized patient faces a highly technical, specialized bureaucracy. In this context, an emerging area of concern is

how to "humanize" health care in general and in organizational settings in particular; see Howard and Strauss, eds., *Humanizing Health Care* (1975).

It is also evident that the conventional private physician's office no longer is the site of care for persons in many sectors of the community, including inner cities and ethnic neighborhoods. In part this situation obtains because, in the free selection of practice sites, most physicians choose the affluent sections of metropolitan areas. Arrangements for health care for the poor continue to be inadequate. Research and planning to remedy these problems are interests of sociologists.

In recent years advocates of health maintenance organizations (HMO's) have claimed they can provide better medicine at lower cost. HMO's vary in their philosophies, organization, payment schedules, and arrangements with staff regarding salaries and fees. There are conflicting claims and commentaries as to which of these arrangements are most effective and viable. Although these are emerging objects of study, no specific citations are given here.

7. Delivery of Health Services.   There is major concern with the larger system of health care—the complex of various health providers and organizations. It would be wrong to think of the United States as having, like some other countries, a single or unified health care system. At the local level there have been attempts at health service coordination and integration, although these efforts have not been particularly successful. Efforts are emerging to control third-party costs and to control the quality and quantity of service provided. Problems of bureaucratic health organizations, as well as the need to coordinate autonomous health agencies, merit the attention of the health planner and the medical sociologist. Moreover, there is growing pressure to control the growth and rationalize the allocation of health care resources, both locally and nationally.

8. Program Evaluation.   The past few years have seen a remarkable growth in evaluation research (see Bernstein and Freeman, 1975). As organizational arrangements have become more complex, as medical care programs and practitioner actions have become more visible, as concern has mounted for better community programs to prevent illness, and as rising costs have brought pressure for greater efficiency and economy, evaluation activities have boomed.

Both public groups and private foundations have made a strong investment in evaluation programs. There is increasing use of social research techniques in "program monitoring"—the jargon phrase for studying whether programs are implemented in ways consistent with their design. Program monitoring of national health programs is an area of especially intensive development. So, too, is the use of evaluation research by public interest groups and organizations that regard consumers as their constituents.

There are no clear boundaries between the work of medical sociologists and other social scientists in the health field, especially medical anthropologists. Consistent with the different outlook of the two fields, sociologists tend to emphasize *structure* and anthropologists *culture,* but the foci of interest are often the same—for example, provider-patient relations.

In international health programs and in training in community and public health, anthropologists have had a particularly major impact. Paul's (1958) series of case studies was one of the earliest texts for social science and medicine courses. The studies cited in Colson and Seller's review of the field (1974), and the contemporary reports found in the new journal *Medical Anthropology,* are illustrative of work in anthropology closely related to the interests of medical sociologists.

Although any classification of work in a field is necessarily arbitrary, the panorama of concerns noted here clearly indicates the breadth of medical sociology. It is within this context that we wish now to consider science and technology in medicine.

II. THE SOCIOLOGICAL STUDY OF SCIENCE AND TECHNOLOGY IN MEDICINE

Our perspective on the development of medical sociology may explain the present status of work on science and technology in medicine; neither sociological colleagues nor health professionals have pressed medical sociologists to engage in research in the science and technology area. From the standpoint of the parent discipline of sociology, medical sociology has been influenced in part by creative ideas, such as Parsons's work on the sick role (*The Social System,* 1951, and "Definitions of Health and Illness," 1958), as well as empirical investigations, such as Hollingshead and Redlich's classic *Social Class and Mental Illness* (1958). But, for the most part, the work undertaken by medical sociologists has paralleled developments in other parts of the discipline, with the health system as a microcosm of the general social system, exploited as a place for study by persons utilizing virtually any theoretical outlook.

If our thesis about medical sociology—that there has been limited work on science, technology, and health—is correct, one reason is that the area of science and technology is not a particularly robust sociological field of inquiry in general. Head counts have their limitations, but it is instructive that the 1975–1976 membership directory of the American Sociological Association—based on self-declared reports of competence by members—lists only about a hundred sociologists of science. This compares with the twelve hundred A.S.A. members who claim medical sociology as a specialty and over six hundred sociologists who regard themselves as sociologists of religion (American Sociological Association, *1975–1976 Directory of Members*).

Fox, in a recent review article discussing the important area of bioethics (1976), provides a penetrating analysis of the limited attention to this area by sociologists:

> The limited contribution that sociologists have made to bioethics emanates largely from the prevailing intellectual orientation and the "weltanschauung" of present-day American sociology. Most sociologists have not chosen to concentrate on the kinds of problems with which bioethics is concerned and are unaware either that bioethics exists, or of its potential sociocultural import. This is related to sociologists' greater propensity to work in a social structural or social organization frame of analysis than in a cultural one. In part, this preference is an artifact of the traditional conceptual emphasis of sociology, especially as it distinguishes itself from social or cultural anthropology. It is also connected with the ideological conviction held by many sociologists who espouse a critical, change-oriented, or radical approach to the field. . . . [Also,] the lack of relevant interdisciplinary competence in the sociological community has been another deterrent (1976, pp. 239–240).

At the same time, within the health sector, there are persons who have been concerned with science and technology questions that have not been interpreted as sociological; predominant among these issues has been the matter of technology and costs of medical care. Sociologists are generally seen as unconcerned with the macroeconomics of health care—the level at which most attention is paid to technology and the provisions of health services.

Persons in the health sector conducting studies and developing policies regarding the use and diffusion of technology have turned to economists and medical care planners. Administrators, planners, and policy makers have focused attention on the linkages between technology and costs, with a minimal focus on the organization of health care, the delivery of services, and relations between patients and providers. Where technology is acknowledged to be important as a determinant of medical care arrangements in sociological studies, it is treated as a residual factor; such studies continue to emphasize social-psychological and social-structural dimensions.

Earlier we categorized the work of medical sociologists in terms of the way the field is presently delimited. That classification reflects the commitments of medical sociologists to substantive emphases in the parent discipline applied to interests now salient in the health care field. In order to provide an analysis of such past work as exists on science and technology in medicine, as well as a prospectus on needed work, a different categorization of areas of effort may prove helpful. The agenda for the remainder of this chapter elaborates three domains of activities—research; the organization, delivery, and costs of medical care; and clinical practice.

A. Research

Health research is a huge industry organized into a complex mosaic. One dimension of it is the range of activities from basic research to technological innovation, from seeking the causes of cancer to developing a birth control pill with fewer side effects. A second dimension includes the types of organizations undertaking the research. Health science research is undertaken in biological and physical science departments, medical schools, nonprofit research institutes, and government and commercial laboratories.

Although the image of the laboratory as the setting for medical studies prevails in the minds of the public, many medical innovations occur in nonlaboratory settings. Epidemiological investigations and field experiments are essential components of health research. There is also the broad area of clinical research. At least among faculty of medical schools, the image is perpetuated that every physician is a scientist. Clinical studies range from small-scale investigations, estimating the impact of various therapies or the identification of side effects of treatment, to large-scale national field trials. Although much of the research is on biomedical interventions, there is considerable study of sociomedical innovations, such as arrangements for care and compliance.

Much of the work undertaken is characterized by a focus on new scientific and technological tools, rather than on clinical acumen and the power of the physician's personality. As will be discussed in greater detail subsequently, there is a tradition of studying the work of research investigators. Many of the studies in the sociology of science on the way investigators undertake their tasks and the way structural arrangements relate to their activities include persons engaged in health science studies. This is perhaps the most integrated and most organized area of effort to be discussed here.

In recent years, the social consequences of medical research have been a major concern of persons identified with the growing area of bioethics. Further, use of human subjects, consequences of medical inventions, and the politics of technical health science developments are important areas of activity.

1. **Laboratory Research.** Among studies of medical research, the most systematic attention has been given to the work of laboratory scientists. Some of these investigations mirror inquiries in the general field of the sociology of science; the effort is to understand how investigators do their work. In general, there is no way to distinguish clearly between the biological scientist and what one thinks of as the "medical investigator." Some important works of persons like Hagstrom (*The Scientific Community,* 1965); Merton (1949 and 1957); Roe (1952); Kornhauser (1962); and Marcson (1966) are relevant.

These studies have produced a number of significant findings. "Research" as opposed to "development" activities tend to be conducted in settings in which there is peer rather than bureaucratic control and where there is less concern with the periods of time required for investigations: they are usually pursued by persons who receive gratification and professional standing through recognition by others in their field. There are still many unanswered questions on the way scientists do their work and reach decisions to pursue different facets of investigation. A classic article by Barber, "The Case of the Floppy Eared Rabbits" (1958), documents the pathways that lead medical scientists to seize on a serendipitous finding with regard to the development of a drug; the article counters the popular impression of luck as an important variable in the medical science discovery process.

There is limited knowledge of the way medical scientists choose careers. Indeed, although there are a variety of articles on socialization in and the selection of a scientific career, it is difficult to understand the relative importance of personality, family background, educational experience, or occupation options on the choice of a career. See, for example, Eiduson and Beckman, eds. (1973).

Research in medicine has a unique set of parameters. One issue is the appropriateness of medical education for scientific work. Despite a long tradition of practitioners being research investigators as well, there are many individuals engaged in medical research who are not trained as investigators; their work depends upon collaboration with more technically trained Ph.D.'s. Unfortunately, little is known about the comparative productivity of the two groups, and it is merely speculative at this point to argue about whether the strains between M.D.'s and Ph.D.'s—because of status differences, research opportunities, and the like—are increasing or declining.

Clearly, as research work in medicine increases in complexity, either a greater proportion of research must be done by persons with long-term technical training—generally acquired during Ph.D. graduate school years— or further education, perhaps even resocialization, of medically trained persons will be needed for activities requiring extensive technological know-how.

One of the areas in which this trend has been evident during the past two decades is research in psychiatry. Since 1957, the National Institute for Mental Health has sponsored a career development award program to create quality scientists. Most of the junior physician-awardees, supported for periods of five to twenty years under this program, use part of their time gaining increased scientific training. Most of the senior persons, supported throughout their careers, received awards because of their reputation as teachers of apprentice-scientists or for the quality of training that psychiatrists (and sometimes other behavioral scientists) can receive in their laboratories (Boothe et al., 1974). Any number of intriguing sociological questions are raised by comparisons between the scientist and the practitioner in technical training and attitude.

Another issue is whether or not it is possible for external groups such as NIMH and NIH to program the technical training of health practitioners. A more basic question arises concerning the importance of technical training in the achievement of the status of "great scientist." Answers to these questions, as well as further knowledge of how scientists work, remains an important area for study.

Since academic medicine demands that the medical school professor publish as well as treat, many physicians with aspirations to a career in academic medicine or association with the medical elite at least dabble in research. The importance and value of their contributions remain unknown. In an interesting field study—the main thrust of which was to examine intern and resident training in an elite institution, Harvard II and IV of the Internal Medicine Service at Boston City Hospital—Miller (1972) discusses the pathways into academic medicine, including the emphasis placed by both chiefs and underlings on scientific knowledge versus clinical skills.

Whether or not the practicing physician uses his scientific knowledge in his clinical practice is a questionable matter, and little is known about whether the antennae required for sensitivity to scientific issues are the same as those required for clinical practice.

There is a long tradition of the physician as a scientist. With the marked trend toward specialization that began after World War II, and with the stratification of residencies and internships into ordinary and elite, the differentiation between the physician-clinician and the physician-scientist became clear. The former is much more likely to be trained in the community hospital and the latter in the university medical center. From there on out, their career trajectories separate. The community hospital-trained physician is likely to move into the private medical sector, while the university hospital-trained physician is likely to go into academic medicine or into a setting linked to a medical school.

There are emerging questions of vital interest here. First, the emphasis on increased respectability for primary care practice raises questions regarding whether or not there will be a change and perhaps a slowdown in health research, at least as undertaken by medically trained persons. Second, predictions regarding an oversupply of physicians, and an abundance of Ph.D.'s in the sciences, together with a reduction in real dollars for research, may change markedly the types of work undertaken, the time frames for studies, and the process of career selection and research funding. In addition to examining how these emerging changes will affect science, technology, and health, it is important to learn something about their impact on various funding agencies and the research interests of different investigators.

2. Non-Laboratory Research. Most of the comments above are addressed to work of a laboratory type. However, there is also considerable effort, both in clinical research and at the community level, in epidemiology and field studies. Very little is known about this work. Epidemiology and field studies are markedly different from laboratory studies. This type of investigation is usually exceedingly expensive, particularly the social-epidemiological study. Many laboratory scientists are still shocked by the exorbitant cost of large-scale field surveys, for instance of former mental patients or of various preventive programs, such as polio immunization. Not only are such studies costly, but they differ broadly in quality, are usually inferential rather than definitive, and they require a set of interpersonal and technical qualities that differ from laboratory techniques to insure successful investigations. Many medical scientists who are laboratory investigators are still disdainful of health research in the community. Nevertheless, as health problems have moved from simple provision of drugs to the eradication of acute illness, then to comprehensive health care programs to manage chronic illness, and as efforts have been made to develop and manage large-scale preventive programs, there has been increased emphasis on epidemiology, particularly social epidemiology, and on community research.

Closely linked to epidemiology and community research, there has emerged an entirely new field, designated health services research. Health services research is generally directed at understanding participation in health care programs and at identifying and structuring efficient health care organizations. This is a field with broad boundaries drawing upon techniques and theories in economics, sociology, and social psychology, as well as knowledge and experience of health planners and health providers. It is also a controversial field, and there are questions about the contributions of health services research to the improvement of policies and programs in health delivery (see Lewis, 1977, and Mechanic, 1977).

The emphasis on social epidemiological research and health services research accounts for some of the recent growth of medical sociology. The popularity of and support for such work has also engaged the interest and attention of medically trained persons—a phenomenon similar to the case of the technical training for laboratory scientists noted earlier. Governmental and private foundations have been designing programs to increase the supply of physician investigators in these areas. Almost all questions raised about the way scientists work, about conflicts between M.D.'s and Ph.D.'s, and about related issues can also be raised about work in these areas. Indeed, we know even less here than about laboratory-type medical investigators and their relations to each other and to patients and the public.

Finally, the activities that comprise clinical research are little understood. There are, of course, vast numbers of journals that report the results of clinical studies, some describing insights developed from a sample of one patient, others documenting longitudinal investigations of carefully selected patient groups. The fact that medicine is practiced in privacy limits what is known about the way clinical scientists work, how they relate to their patient-subjects, and the consequences of clinical study on the health care of those who are selected for study.

Many of the nonglamorous inventions and discoveries in the health field are the results of clinical investigations. The motivations for the practitioner to participate in such ventures, the criteria by which "scientific truth" is formulated, and methods for the transfer of knowledge are important areas for investigation. For clinical research as well as other types of medical investigations, the question of the most compatible setting remains unanswered. Ben-David (1976) has argued that marginal settings provide greater impetus for innovation than do academic ones. Others, such as Rossi and Williams (1972), support this position. Yet many sociology of science scholars argue that the freedom of the academic world allows the individual scientist to prosper. This unanswered question requires further study, especially in the area of clinical research.

3. **Ethical Issues.**   There is another vital issue that is the subject of literally thousands of commentaries and a number of outstanding research inquiries. This is the bioethical interest in the rights and protection of human subjects. The earliest as well as the most persistent issue was the protection of subjects from irresponsible experimentation that could lead to death, permanent bodily damage, or extensive pain. In more recent times, other issues have occupied considerable attention. One is the protection of the privacy of subjects of research. Another is the issue of carrying out experiments without the informed consent of the participant (Gray, *Human Subjects in Medical Experimentation*, 1975).

There are few research scientists, policy makers, or citizens who argue against the protection of subjects. The difficult issue is setting up guidelines and policies that are satisfactory to all parties concerned—investigators, potential subjects, and moral and humanistic leaders. The specter of a return to experimentation as it occurred in Nazi concentration camps—and as it has continued in some United States jails until this decade—looms large in the minds of many engaged in the development of a framework for human experimentation. Persons imbued with strong ethical principles about individual dignity, and those emotionally affected by past events, represent a strong coalition that argues for extremely rigid and structured policies and guidelines. At the same time, some individuals, mostly those engaged in research ventures, point to the need for human experimentation and for broad flexibility in the matter of protecting human subjects; they urge flexibility to allow for the development of new treatment regimens as well as to prevent the widespread adoption of procedures that have not been tested on humans.

The roots of the concern are moral, legal, and practical. From a moral point of view, the key issue, perhaps, is whether the risks to the subjects involved are more critical than the potential gains from investigation. Sensitivity has grown about the potential psychological as well as physical harm to individuals. The concern extends the need to question not only the advisability of social experiments in such areas as attitude modification, but also the risks either of demeaning interactions with subjects or of invasions of privacy. (See, for example, the Ad Hoc Committee on Ethical Standards in Psychological Research, 1973; and Zimbardo, 1973.)

From a legal standpoint the matters are complex. As Fried (1974) indicates, one question that must be considered is whether or not experimentation is therapeutic or nontherapeutic: Is the experiment being carried out solely to obtain information of use to others, or is it done only to determine the best treatment for this one patient? In many cases, of course, the motives are mixed and the research is both therapeutic and nontherapeutic. It is in the random control of experiments or clinical trials that the law remains vague:

> Two themes appear and reappear as we consider the legal context of medical experimentation: the contrast between the law of battery and the law of negligence, and the difficulty of determining whether randomization does violence to the duty a doctor owes his patient. These two themes run together when we consider whether a doctor has an obligation to disclose the fact of randomization and whether he is guilty of battery if he does not. . . . A review of the law can do no more than raise these questions—partly because the law is incomplete, and partly because it is open to us to judge the law and to change it. So now we must consider what is right in principle (Fried, 1974, p. 43).

Efforts to legislate a set of legal and political regulations and guidelines document the inconsistencies and complexities of human subject inquiries. Neither politicians, research administrators, blue ribbon commissions, nor academic committees have solved the problem from a legalistic standpoint. Although there are a number of key legal cases in which individuals have sued investigators—and instances in which funding agencies, institutions in which research is conducted, and professional groups have taken punitive action against investigators—there is no codified body of legal or even quasilegal guidelines (Committee on Interstate and Foreign Commerce, 1974).

There are practical matters as well. These generally involve consent and privacy issues. Some investigations do require a degree of deception. For example, in numerous drug trials of psychopharmaceutical agents on disturbed mental patients, it is possible to inform them that they are participants in an experiment but it is hardly sensible to let them know whether they are of the experimental or control group. Further, how meaningful is informed consent from such patients whose emotional condition or lack of contact with reality requires long-term custodial treatment?

Other matters are more mundane. It is possible to scramble data stored on a computerized tape in order to prevent identification of subjects (Campbell et al., 1977), but how is a clerk to be prevented from obtaining private information during the very process of coding and scrambling that information? With or without the employment of certain procedures used by top secret defense organizations, there are limits to which documents, reports, and questionnaires can be protected. Rules about double-locked file cabinets help, but they do not negate the likelihood that a member of the research staff will leave confidential material on a desk during an extended coffee break in another part of the building.

Although the individual human subject is the major focus of attention, broader societal consequences of biomedical research are also of concern. For example, genetic modification procedures, if perfected, could produce inestimable social consequences. According to some antagonists, the risks of DNA research recommend either the halting, or at least the rigid controlling, of work in this area. Control of R&D activities for reasons of risks to individual subjects, or because of the lack of benefits compared with costs, is a reality; but it can be reasonably hypothesized that extensive restrictions on scientific inquiry are most likely in the cases of potential societal risks (McGill, 1977).

Although the field of bioethics, and questions that surround the use of human subjects, are within the purview of persons in many disciplines who occupy a wide range of occupational and social positions, most of the work of sociologists has centered around two areas. (Much more effort needs to be invested in an understanding of human experimentation from a sociological perspective.) One type of research can be characterized in terms of a classic investigation by Fox (*Experiment Perilous,* 1974). She describes the intertwined problems of patients and physicians on a research ward in which there is a tradeoff between the welfare of individual patients and the prospects of undertaking a well-conceived clinical experiment. The study is one of the few that provides an understanding of the stresses between patients and physicians and the evolution of social patterns of commitment to medical studies by patients and physicians.

Another recent work (Fox and Swazey, *The Courage to Fail,* 1974) extends the issues involved in experimentation and heroic treatment by means of a series of case studies. The cases document the cultural and social-structural influence of participation by patients, families, health providers, and scientists in medical research and innovative treatment programs. These studies are exceedingly important, not only because of the insights obtained, but because they suggest the critical need for systematic field research on such issues.

Although these studies focus on extreme situations, most medical experiments are trials of a much more mundane character—such as the substitution of one anesthetic for another, or the carrying out of a particular type of surgery on an in-patient or an ambulatory basis. A broader study of the sociology of science and technology in the health area would include a documentation and understanding of the cultural and social-structural determinants of such relations, as well as interpersonal decisions and strains in a wide range of different types of research situations.

Another important study is the research volume by Barber and his associates, *Research on Human Subjects* (1973). This study, using a different methodology, examines the way human subjects review committees work, and how these subjects implement guidelines in medical research settings. The authors found, among other empirical results, that ethical sensitivity tends to be high in situations in which there is clear and serious risk to subjects. In such cases diligent care is exercised in gaining informed consent and informing patients of their right to withdraw from experiments. When investigators perceive less risk in studies, they tend to be more willing to operate on what, from a moral and ethical standpoint, could be described as a selfish basis. The authors also found in their national survey that there is considerable variation in the extent to which peer review is inclusive and searching. Risk taking by investigators in institutions may be related to the extent to which the research is important to the investigator's career and to the research institution's status. The investigation of Barber and his associates is an important example of the direction medical sociology might take towards a policy-relevant issue. Their commentary on the failure of medical schools to provide ethical leaders for the establishment of controls—particularly formal peer review and the safeguarding of the life of human subjects—raises important considerations for both the academic and the medical community. They express concern over the consequences of competition in medical research, as well as over problematic relationships between persons of different status in collaborative research groups.

Barber's work certainly warrants extension, for instance to the arena of psychological and emotional research, or to studies in community medicine and public health. Further, because of the increased pressure for regulation and social control of human subject research and the changing policies within institutions, funding agencies, and regulatory groups in the federal bureaucracy, it would be useful to have time series replicates of this work.

Although our discussion of human-subjects research indicates the utility of sociological efforts in this area, it is important to note how comparatively little research has taken place. Unfortunately, most of the involvement of sociologists and other social scientists has been at the level of commentary and debate, not empirical investigation.

B. The Organization, Delivery, and Costs of Medical Care

The impact of technology on patient care, on the organization and delivery of services, and on medical care costs are of major concern now; these issues constitute perhaps the most critical emerging area of research and dialogue. The interest is due largely to skyrocketing costs of care, but political considerations and concern with individual freedom of choice are also factors. Public policy issues that surround the allocation of resources for the development and application of various technological innovations have been the subject of numerous social commentaries, regulatory hearings, and mass media discussions. With the transfer of much of the control for health policy to laymen, and with greater influence by the public and their political representatives on the allocation of health resources, there is considerable questioning of the value of technological interventions or of the desirability of further support for research and technological development.

It is not possible to unravel fully the causal linkages between scientific and technological adaptation, organizational arrangements for the provision of health services, and the economic structure of health care. The latter two have been subjected to extensive inquiry, and the results have direct relevance to the analysis of science and technology in medicine.

1. **Organization of Medical Practice.** In the early 1900s, medical education began to undergo significant changes. (See chapter 3 in this volume.) Many medical schools were closed, and the supply of physicians came under control by means of certification requirements. Most surviving medical schools were university affiliated, and a strong basic science research base was introduced. The teaching staffs, which before had been composed largely of parttime generalist physicians, were gradually replaced by specialized physician-educators with research interests. Students of medicine began to have less contact with all-purpose generalist physicians. To serve the research and clinical interests of faculties with specialist orientations, those hospitals affiliated with medical schools rapidly evolved into what are now called "tertiary care" centers. These centers are primarily for patients with unusual, complex, or multiple problems, leaving community hospitals and private practitioners to care for patients with more common illnesses.

This shift in medical education has changed the organization of medical practice as well as facilitated the continued growth of knowledge in medicine. The organization of medical education around specialty medicine, the high prestige and status within medicine associated with specialty certification, and the lack of contact between students and general practitioners, have led an ever-increasing number of graduates to choose medical and surgical specialties and subspecialties—with a corresponding steady decline in the proportion of general practitioners. Their numbers decreased from sixty-four percent of the physician population in 1949 to twenty-four percent in 1973 (Donabedian et al., *Medical Care Chartbook,* 1972, p. 142). The supply of physicians in general practice in 1990 is projected, in one report, as only six percent of the total pool of physicians (U.S. Department of Health, Education, and Welfare, 1974, p. 44).

Specialization in medicine is consistent with stratification in other professions; in medicine, as in other occupations, financial rewards and social recognition accompany unique technical knowledge. For example, the higher rank given to medical specialists by the armed forces in World War II was clear acknowledgment of the superiority of the specialist physician over his generalist colleague. This trend was reinforced by the development of strong certification boards.

Whether or not specialization in medical education influenced the growth of basic medical research, it is clear that a climate emerged in which technological innovations could be accepted and integrated rapidly. Specialists encourage scientific and technological innovations in their fields, at least partly because these innovations further distinguish them from other physicians.

2.   **Economic Structure of Health Care.**   Technological inventions and refinements have long characterized Western medical practice—including such inventions as the stethoscope, such discoveries as antibiotics, and such procedures as open-heart surgery. Contemporary innovations, at least the ones that get play in the mass media, are remarkably expensive. Computerized tomography (CT) scanners, artificial kidneys, heart transplants, and the like, all use up vast amounts of economic resources. So too do less expensive procedures that are routinely carried out on huge numbers of patients—including, for example, computerized multiphase blood analyses.

Medical practitioners and institutions adopt and use technological procedures virtually without financial risk. The extensive adoption of third-party payments, and particularly the government-sponsored programs, Medicare and Medicaid, allow the use of complex and expensive technologies in diagnosis and treatment with only limited concern for cost to the patient. Health care charges are determined mostly by groups and lobbies of medical care providers, not by the third parties who typically pay for health care. However, the third parties, as well as those directly affected by increased costs of care—including government, industry, and labor unions—have made cost containment an issue.

Hospitals and large medical practitioner groups have little to lose by increasing their dollar volume; indeed, they compete with one another for visibility in order to attract both patients and a distinguished staff. The "tertiary medical care setting"—a euphemism for the prestigious hospital usually affiliated with a medical school—indulges in the use of complex technology, sometimes with justification, at other times simply because that sets it apart from the community hospital or the ambulatory settings in which general medical care is provided. Many practitioners affiliated with these primary and secondary care settings wish to accrue the status and prestige associated with leading institutions, so they imitate and adopt technological procedures that ought economically to be treated as scarce resources.

Many technological innovations are adopted in spite of high costs and high risks to individual patients, both because of the difficulty of evaluating their utility and because physicians have been successful in maintaining a mystique of expertise. As a consequence, few patients question either the cost or the efficacy of the procedures to which they are subject. It is only in the past few years that the matter has become a public issue as it has grown obvious that the nation, collectively, cannot provide the resources necessary to support all the technologies even now available. And they are continually added to, without control (Spingarn, 1976).

The CT scanner is a case in point. This highly sophisticated X-ray machine costs anywhere from $300,000 to $700,000 to purchase, and patients may be charged up to $150 for a series of scans (Office of Technology Assessment, 1977). It is estimated that by the end of the 1970s, without external controls, the costs to patients and third-party providers could reach $1 billion. Advocates argue that the device can revolutionize medicine. Critics, in the extreme, claim it is just a toy for medical profiteering. A recent report by the National Academy of Sciences Institute of Medicine, endorsed by the Blue Cross Association, recommends that insurance companies should pay for CT scans only under limited conditions (Institute of Medicine, 1977). The authors suggest that there be rationing of scanners within communities, and that limitations be placed on their use in physicians' offices and private clinics. Similarly, other forms of diagnosis and treatment are now believed to require regulation because of high costs and excessive, unjustifiable use (see Meyer, 1977; Knous, Schroeder, and Davis, 1977; and Banta and McNeil, 1977).

Total expenditures for health in the United States in 1977 exceeded $150 billion—more than double 1969 costs. An increasing proportion of the U.S. gross national product is now spent for medical care; it is currently estimated at 8.2 percent. Much of the increase can be accounted for by increases in federal expenditures. In 1976, federal expenditures exceeded $42 billion, an inflation-adjusted increase of 60 percent since 1969 (Koleda et al., *The Federal Health Dollar*, 1977). Since health care in the United States competes vigorously with such other human service programs as income security, education, and housing and urban rehabilitation, numerous efforts have been made to control expensive technology in medicine and to reorganize health services to economize.

An increasing number of health experts and policy makers are calling for a reevaluation of national expenditures, for a better balance between new technology and traditional personal medical care. These persons actually cite a lack of curative technologies for the treatment of current diseases. Two critics, Rick Carlson and Ivan Illich, explore the limits of modern medicine in resolving the major health problems of the population—Carlson in *The End of Medicine* (1975), and Illich in *Medical Nemesis* (1976). Illich takes an extreme position, arguing that not only is modern medicine with its technologies ineffective in ameliorating the major health problems, but current practices sometimes actually injure patients, causing what is known as "iatrogenic" disease.

There is another side to the debate. McDermott (1978), citing recent declines in infant mortality, as well as in deaths from such conditions as heart disease, peptic ulcer, and cirrhosis of the liver, claims that cancer is the only major disease not showing some decline in recent years. For all other major disease conditions, measurable declines have occurred. While attributing these declines neither directly nor totally to medical care, McDermott argues that the documented declines in the major diseases do counter the pessimism that would deny medical progress. He cites evidence suggesting that health interventions may lower infant mortality in spite of continued poverty, negligent personal health habits, and adverse environmental conditions.

3. **Modifying Organizational Arrangements.** Changing patterns of disease and the movement toward more and more specialized medicine have created a strange misfit between the needs of consumers on the one hand and the commitments of providers and the organization of medical practice on the other. Patterns of disease are shaped by the state of the economy, by advances in public health, and by the development of disease intervention technologies. Although the twentieth century has seen an explosion of medical knowledge and the rapid development of diagnostic and intervention technologies, these have not sharply affected some mortality indicators. (These advances have altered some disease patterns, especially in terms of severity and in terms of the duration of required treatment.) For example, it is interesting to note that significant declines in infant mortality took place prior to 1930, before the advent of sulfanomides in the mid-1930s and the development of penicillin and other potent antimicrobial drugs in the 1940s. None of the decline in infant mortality prior to 1930 can be attributed to medical care *per se* (Fuchs, *Who Shall Live?* 1974).

Following the introduction and widespread use of antimicrobial drugs, infant mortality continued to decline while patterns of illness changed for all age groups. Infectious diseases such as influenza, pneumonia, and tuberculosis were brought under control. Physicians in pediatrics have been more willing than others to recognize the changing nature of disease patterns. New pediatric training programs provide pediatricians with skills commensurate to these changes. Haggerty and his colleagues (1975) point out

that the major health problems of children today are not those that led
to the present organization of services; they have coined a term, "the new
morbidity," to cover the influx of problems relating to children's behavioral
and learning difficulties as well as family stress reactions.

Pediatric services were organized to provide care for acute infectious
diseases and to rehabilitate children who survived the ravages, for instance,
of polio or certain congenital defects. The training of pediatricians was
hospital based. Now, with the elimination of the major infectious diseases
of childhood, the bulk of pediatric practice deals with non–life-threaten-
ing episodic illness, or with well-child care. Many of the difficulties brought
to pediatricians lie beyond the boundaries of traditional medical care, and
pediatricians are inundated with problems unrelated to their training. We
now have a situation wherein a small number of children suffer from unusual conditions that require the application of the full complement of
modern medical techniques, while the majority require a broader interdisciplinary community-based delivery system.

Recent developments in medical technology have led to new forms of
medical organization for ambulatory care, which in turn have led to the
modification of hospital structures. As a result of efforts to provide more
accessible and more equitable care—while imposing controls on rising costs
—there has been a major shift from entrepreneurial solo medical practice
to bureaucratic arrangements in which physicians and other health providers are employees of large organizations. Mechanic notes:

> In the United States, approximately 8 percent of physicians work for
> the federal government, mostly in hospital-based practices. . . . Of the
> more than 90 percent of physicians in the nonfederal sector, approxi-
> mately one-fourth are in hospital-based practice and most of the others
> have some hospital attachment. Even those primarily in office-based
> practice are dependent on their hospital affiliations to pursue their
> work, and increasingly face restrictions under the rules of the hospital
> as a social and legal entity (*Bureaucratic Medicine*, 1976, p. 49).

The emerging form of organization is called "group practice." Group
practice may take several forms, from a fee-for-service partnership between
physicians of the same specialty—who basically share offices and take calls
for one another—to large multispecialty groups that may be either fee-for-
service or prepaid-fee practices. In addition, both teaching hospitals and
municipal hospitals in urban areas provide large outpatient clinics that
serve as the primary—sometimes the only—source of care for a large portion
of the urban poor.

Given an elaborate specialized and technological system where costs
are high and where there are strong social expectations for equity, organization of medical care is necessary. The bureaucratization of medicine has

negative as well as positive effects on the delivery of health care. It is important to study the responsiveness of bureaucratic health services to the forces that bring patients into the health care system. Despite the drama of advances in modern technological medicine, most medical care is a relatively mundane activity in response to routine symptoms and illnesses. It is plausible to hypothesize that a system geared to dealing with emergencies and esoteric conditions would have difficulty providing the humanistic support that was for so long a central component of medical care (Howard and Strauss, 1975).

Technological change affects not only the organization of medical care, but also the geographical distribution of physicians. They tend to practice in locales where they can fully utilize their technological training. Physicians trained in tertiary-care centers—which are equipped with the latest technological innovations and staffed with well-trained support personnel—become dependent upon these settings and regard them as optimal loci for practice. They seek to practice in locations where technological resources abound, and thus are overrepresented in urban areas.

There is now ample evidence that what was once thought to be an inadequate supply of physicians is now adequate—at least in terms of numbers. But physicians are maldistributed geographically as well as by specialty. Despite the fact that enrollments in medical schools have doubled since 1960, there has been little change in deployment patterns (see "Undergraduate Medical Education," *Journal of the A.M.A.*, 1976, p. 296). For example, in New York there are 256 physicians per 100,000 people; in South Dakota, there are only 90 per 100,000 (Goodman and Mason, 1976). Similarly, general surgeons are in oversupply in many urban centers but continue to be a scarce resource in rural areas. The growth of technology in medicine, coupled with physicians' dependence upon it, has changed the patterns of practice. Serious inequities in the availability of resources have arisen between urban and rural communities.

The participation of physicians and other providers in complex organizations may produce conflicts in priorities between the organization and the consumer. In traditional fee-for-service practice, there were thought to be incentives for physicians to be responsive to consumers' needs. The simple one-to-one relationship between physician and patient supposedly encourages feelings of personal accountability. In some cases there is sufficient competition in the fee-for-service sector to encourage the provision of some amenities, even to encourage compliance with patient demands that are marginal to the practice of good medicine, such as the prescription of certain drugs or the providing of certain purely elective surgical procedures.

In contrast, providers working for complex organizations may feel less accountable to provide patients with convenient medical care or other amenities; they may identify more strongly with the concerns of their peers than with the concerns of their patients (Freidson, *Doctoring Together*, 1975). Presumably it is easier for an outpatient clinic to close promptly at 5:00 P.M. than for an individual physician to refuse to care for patients after hours.

The complexities of modern medicine may encourage the evolution of forms of medical practice that make possible the delegation of personnel and the application of technology in a cost-effective manner. Whereas in the business sector the demand for goods and services is tempered by the availability of resources, in the health sector the growth of public and private health insurance has, over the years, removed normal cost restraints on the seeking of medical care. In a traditional fee-for-service arrangement with third-party health insurance, neither providers nor consumers have any financial incentive to forego medical care that offers any chance of improving health. Supposed constraints imposed on the utilization of health services via insurance mechanisms—such as nonreimbursement, coinsurance, and deductibles for ambulatory care—have, for the most part, actually encouraged the use of in-hospital services and inhibited preventive programs. The general view is that insurance mechanisms, by and large, serve to support the operations of technological centers, such as hospitals, instead of promoting the optimal use of health resources by consumers. It is fair to acknowledge, however, that there is insufficient knowledge of the causal relations between the structure of third-party payments and consumer use of different types of settings and procedures.

Prepaid health insurance plans have evolved as an alternative economic arrangement for health care. The intent is to maximize the efficiency of the health care team and the cost-effective use of technology while providing quality health maintenance. Advocates of prepaid health plans do not believe it is possible to control the use of services via fee-for-service mechanisms. Their strategy is to maximize preventive care, provide early diagnosis, and stress ambulatory treatment. An equally plausible hypothesis is that they actually ration care by increasing such barriers to care as waiting time, unavailability of hospital beds, and restrictions on specialists (Mechanic, 1978).

4. **Costs and Benefits of Technology.** Hospitals are not only the setting for most technological applications in the health field; they are regarded as the villains in the rising costs of health care. Knowles (1973) illustrates the changes that give rise to this view:

In 1925 the cost of one day's hospitalization at the Massachusetts General Hospital was $3. This bill was paid directly by the patient out of his own pocket, and he stayed for 15 days for a total cost of $56.20. In 1972 a patient would stay for an average of seven days at a total cost of $1,400, and the bill would be paid by any one of a variety of third parties (p. 97).

Today's hospital provides different services than those provided in 1950, presumably as a result of the rapid diffusion of technological innovations. During the last ten years the cost of care in hospitals has risen at an annual rate of more than twelve percent. It is estimated that one half of each year's increase is due to rises in real resources—labor, supplies, equipment, and space—rather than to increases in the prices of those resources (Russell, 1976, pp. 557–580).

The factors that influence the diffusion of technology are important in our understanding of the costs of medical care. The spread of innovation in the health sector can be viewed from at least two perspectives.

The first of these perspectives would see the rapid dissemination of technology, and the difficulties in controlling its use, as attributable to the behavior of providers—physicians in particular—and to the bureaucratic organization of hospitals, which disperses responsibility for resource commitments (Greer, 1977). In the training of physicians there is little emphasis on the economics of health care or on decision making regarding the use of finite resources; also, physicians are allowed far more control over the use of technology than are technologists in other sectors of the economy. Fuchs characterizes the physician's approach to medical care as being dominated by a "technologic imperative"; medical tradition dictates that the best care technically possible be rendered at all times no matter what the cost (Fuchs, *Who Shall Live?* 1974). The only legitimate and explicit constraint on the technological imperative is the state of the art. Fuchs points out elsewhere (1972):

> All this sets medical care distinctly apart from most goods and services. Automobile makers do not, and are not expected to, produce the best cars that engineering skills permit, they are expected to weight potential improvement against potential cost. If they do not, they will soon be out of business. However, the improvements must be those as perceived by the consumer—which may be very different from those perceived by the engineer. What is true of automobiles is true of housing, clothing, food, and every other commodity (p. 66).

Freidson shares this perspective; he discusses the political monopoly that physicians have gained over the medical marketplace (*Profession of Medicine,* 1975). It is his hypothesis that physicians in bureaucratic organizations not only retain autonomy but also influence—even control—the work of others. Moreover, while patients are consumers of health care, physicians are consumers of technology. Physicians are under multiple pressures to adhere to the technological imperative. As the practice of medicine becomes more complex and more and more specialized, many physicians have resorted to "defensive medicine," performing a larger complement of diagnostic tests than is warranted by their patients' complaints.

In a study of selected conditions treated at the Palo Alto Medical Clinic from 1951 through 1971, it is reported that there was a steady and considerable increase in the use of ancillary services, both in and out of the hospital, for all conditions studied (Scitovsky and McCall, 1976). It has been estimated that each new physician in the United States accounts for $250,000 a year in national health expenditures when all ancillary services are considered (Culliton, 1978).

An increase in malpractice suits has created for physicians an additional incentive to employ available technology (Tancredi and Barondess, 1978). No one knows the extent to which the fear of legal suits has driven physicians and hospitals to order tests that may be unnecessary, but estimates of the total burden of defensive medical practices range as high as $9 billion a year (Golladay and Smith, 1975).

The rise in malpractice suits can be attributed to the growing fragmentation of care resulting from increased specialization and subspecialization. Further, it is hypothesized that, with increased bureaucratization, consumers perceive a lack of personal accountability by providers; this perception is thought to be another important factor in malpractice suits.

A second perspective for viewing innovation and the implementation of technology in the health sector is represented by the work of Navarro (1975, 1976). He rejects the thesis that emphasis on the use of technology in the health sector is a result of the prevailing values of providers. (He does not think it is a result of consumer choices either.) He postulates that the United States economy determines and maintains the present social class structure—both generally and within the health sector. The different degrees of ownership and influence that these classes hold over the means of production in the United States explain the composition, nature, and activities of the health sector. Navarro further claims that the current value system—which is alleged to be the cause of the current organization of health services and important in the diffusion of innovations in the health sector—is really a symptom of the controls and influences of the class structure.

Navarro's explanation differs significantly from conventional views in that he does not define the main conflict in health as one of providers versus consumers. He argues that the great influence of providers—especially physicians—over the health sector is based on power delegated from the corporate ruling classes. He asserts that the same financial and corporate forces dominant in shaping other parts of the United States economy also increasingly shape the health services sector. Navarro believes it is possible that the predominance of financial capital in the health sector—and specifically the power of the commercial insurance consortium—can result in the weakening of provider control over the health sector; eventually it can lead to a situation wherein doctors and other health providers would be mere employees of finance corporations. Within this framework, the diffusion of medical technology can be attributed to the vested interests of controlling corporate entities, with physicians and consumers merely responding to a set of corporate values that may or may not be in the best interests of either society or the sick. It is not necessary to accept this Marxist perspective in its entirety to recognize the importance of the interdependencies between the economy and the health sector. For further discussion of related arguments, see Sidel and Sidel (1978) and Krause (1977).

5. **Technology and Efforts at Cost Containment.** The development of medical technologies capable of producing dramatic effects in the prolongation of life has helped bring about a public awareness of the finite nature of national economic resources. There is a growing movement to develop some mechanism by which the costs and benefits of various technological interventions can be assessed and the use of technology regulated for the public good.

The growth of health insurance has increasingly insulated hospitals from the economic constraints that guide decision making in other sectors. In response to rising hospital costs, there have been increased efforts, at the national and local level, to regulate the diffusion of technology and to avoid unnecessary duplication of equipment, personnel, and programs. The National Health Planning and Resources Development Act of 1974 makes "certificate-of-need" reviews mandatory for participation in certain federal programs. An appointed board examines proposed investments that require relatively large outlays of money, such as for new hospital beds. They are charged with determining whether the expenditure is really needed.

Initial assessments indicate that regulatory mechanisms are not particularly effective, and the process may become highly politicized (Bellin, 1976). A recent study by Salkever and Bice (1976) indicates that certificate-of-need controls did not reduce the total volume of investment although they did alter its composition. Although hospital expansion in terms of new beds was curtailed, there was increasing investment in new services and equipment. The regulation of technology raises numerous questions—including how much technology is enough, and whether the necessary knowledge is available to assess the long-term social costs and benefits of technology.

A related approach to cost containment of technological innovations is regionalization. Regionalization is an ambiguous concept in health care. At one extreme it may mean allocating specialized programs, such as high-risk perinatal centers, to geographical areas with specified population densities. At the other extreme it may mean consolidating services by centralizing organizational control; for instance, having a central dispatch center for deploying ambulances across a wide area. Regionalization has proven to be neither effective nor equitable as a method of allocating resources. The glamor of many technologies restricts the motivation of health care organizations to cooperate by becoming one of the have-nots. Allocation of regional programs by public planning bodies, rather than on a voluntary basis, is often coopted by powerful political interest groups. Thus, one may find CT scanners costing hundreds of thousands of dollars located in public and private hospitals that are literally in walking distance of each other.

Efforts have been made to study the exchange of resources in the health sector (White et al., 1979), but for the most part the political processes are poorly understood. Recently there have been several in-depth efforts to provide data relevant to the understanding of cost containment and the control of technology (Ginzberg, 1977; and Saward, 1975). However, the politics of health in general, and medical politics related to technological changes in particular, require much more study.

Technological developments contribute to costs in two ways. First, there is considerable inappropriate use of technologies that, if used ideally, should lead to cost savings. For example, automated chemistry analysis began in the 1950s, offering the possibility of significant cost savings that could be passed on to the consumer. However, insurance coverage, the financial motives of doctors and hospitals, and the practice of defensive medicine all contributed to the overuse of these procedures. Physicians ordered an estimated five billion laboratory tests, costing approximately $11 billion, in 1977 (Fineberg, 1979). The increase in laboratory tests has served to reduce the dollar savings from other technologies as well. For example, Scitovsky (1976), in a study of maternity care, found that the number of laboratory tests between 1951 and 1964 almost tripled; this increase offset very nearly the entire cost savings due to a one-day decrease in length of hospital stay combined with an increase in deliveries without general anesthetic.

New technologies can be classified as substitutes or add-ons (Bennett, "Technology as a Shaping Force," 1977). A substitute technology provides a better, more efficient, or more productive way of accomplishing an existing task. An add-on technology makes possible the accomplishment of something that was previously either impossible or economically impractical. Many new medical technologies are add-ons and hence generate new costs.

A second way in which technological innovations increase the cost of medical care is through the development and widespread use of add-on technologies that benefit a very small proportion of the population but whose high costs are shared by all consumers. Thomas (1974, 1975) makes a distinction between "definitive technologies" (for the prevention, cure, and control of disease) and "half-way technologies" (for palliation, repair, or maintenance of patients with conditions for which there are no known cures). He argues that medicine does not become more costly, and therefore more inaccessible, when curative technologies are discovered. He notes that when a cure for a disease or a measure for its prevention is found, this innovation has always turned out to be very much simpler and far less costly than the technologies employed before the disease was well understood:

Compare eight weeks of hospitalization for typhoid with the cost of a
bottle of chloramphenicol, or the iron lung with the cost of polio vaccine, or sanatoria and lung surgery for tuberculosis with the cost of
chemotherapy.

   To be sure, open-heart surgery, renal dialysis, and the transplantation of vital organs are technologies of intolerably high cost, but
these are the technologies which, for the time being, we are obliged
to use for our essentially *unsolved* diseases. They are, in their way no
more decisive or conclusive than the iron lung for poliomyelitis, or
transfusions for acute leukemia (Thomas, 1975, p. 183).

   The widespread use of half-way technologies raises a number of policy
issues and research questions. How can finite resources be allocated in an
equitable way? What—if this can be measured—is the value to society of a
single life in terms of dollars? What is the quality of life of those sustained
by half-way technologies? Should resources be used primarily for targeted
research or for "basic" investigations? (For this distinction, see Comroe and
Dripps, 1976.) Should available funds, for the most part, be funnelled into
basic research—in an effort to discover definitive technologies—or should
those funds be spent in the delivery of direct services using known half-way
technologies? (See Katz and Capron, 1975; and Fox and Swazey, 1974.)

   There is an interesting illustration of these issues in legislation that
made expensive half-way technologies available to almost all persons in
the United States who have end-stage renal disease (Rettig, 1976a, 1976b).
In 1972 the Social Security Administration was mandated to finance treatment for individuals with chronic kidney disease either by hemodialysis or
kidney transplantation; this was a significant change in federal policy. The
legislation prompted a panel of the National Academy of Sciences to recommend to Congress that the coverage of discrete categories of catastrophic
disease would, in the foreseeable future, be an inappropriate course to
follow for providing expensive care in cases of universal eligibility (Institute of Medicine, 1973).

   The number of beneficiaries of end-stage renal disease treatment is
fairly small relative to the general population; in 1971 the number of reported living dialysis patients was 4,375, and the number of kidney transplants, 1,172. Renal dialysis and transplants are very expensive therapies—
the average annual charge per year for dialysis in 1972 was $30,000; the
average charge for a kidney transplant in 1973, $12,800 (Rettig, 1976a,
1976b)—requiring resources well beyond the financial means of most
Americans.

   The technological means for renal dialysis were available as early as
1960. By 1962, the mass media had publicized the dilemma created by the
existence of a life-saving therapy and the scarcity of facilities to provide it.
Treatment centers set up lay panels to assist physicians in deciding which
patients should have access to the limited number of machines (Fox and

Swazey, *The Courage to Fail,* 1974). Patients were usually evaluated to determine their social value in terms of age and sex, number of dependents, emotional stability, occupation, and future potential—in addition to medical factors.

Renal dialysis presented a direct confrontation with ethical issues about how decisions should be made regarding life and death. Although such decisions were not new to health providers, they were potentially more difficult because patients with diagnosis of end-stage renal disease were ambulatory, functioning people, for the most part, not comatose, seriously ill patients. Providers were faced with life and death decisions regarding not only the allocation of resources but also the ethical issues regarding the withdrawal of hemodialysis treatment at the patient's request. If dialysis were stopped, would doctors be legitimizing suicide, performing euthanasia, or acting in an ethically responsible way?

From a policy perspective, there has been careful scrutiny of the results of the Social Security end-stage renal program. One of the most important lessons learned is that there are neither sufficient available methodologies nor sufficient knowledge to predict the consequences of most technologies— despite the establishment by the federal government of the Office of Technology Assessment (Behney, 1976). Measuring the social impacts of new medical technologies is an emerging field (Office of Technology Assessment, *Development of Medical Technology,* 1976). In the case of renal dialysis, underestimation of costs has persisted throughout the program's history. The original cost estimates for the program were $135 million for the first year, up to approximately $1 billion by the tenth. By the end of the second year, the costs were considerably in excess of the estimates, and in 1977 the Social Security Administration projected that costs of the program in 1982 would be $1.9 billion for 56,000 patients; by 1992 it would cost $6.3 billion to care for 65,000 patients.

There are many reasons for the inaccuracy of these cost projections. It was widely thought that home dialysis—which is far less expensive than treatment in a hospital and offers the patient much more flexibility—would become the treatment of choice for most patients. However, home dialysis has not been accepted by most patients; families were either not willing or not able to assume the responsibility, and there were significant numbers of patients living alone. The greater-than-expected dependence on hospitals for long-term dialysis not only escalated total program costs but also seriously detracted from the quality of life of patients and families. It diminished the employability of patients and thus their economic contribution to society. Significant numbers of patients are maintained indefinitely by hemodialysis, including many marginal patients who are not viable candidates for transplantation. In addition, the promise of transplants from cadavers has not materialized, and enrollment in the program grows from year to year as a result of fewer transplants than were originally anticipated.

Many critics of categorical coverage for catastrophic illness are calling for a moratorium on new legislation and programs. One reason stems from the inaccuracy of the predictions about the consequences of the end-stage renal disease program. Another comes from concern over the upsurge of movements to extend federal coverage to patients with hemophilia (at an estimated annual cost of $150 million) and to patients requiring artificial hearts (annual projected cost: $1.75 billion). As Rettig (1976a, 1976b) points out, it is almost impossible for the government to eliminate an established patient-treatment program once it is launched, regardless of the cost-to-benefit ratio. Bennett (1977) calls for an open confrontation with the issue, "lest we bankrupt ourselves in a sea of artificial hearts, while other more pressing issues, such as epidemic venereal disease, are neglected" (p. 132).

The dilemmas inherent in the use of expensive half-way technologies are examined in a thought-provoking paper by Annas (1977). He examines a hypothetical Supreme Court decision. It is the year 2002, and the case concerns a national policy of restricting the number of artificial hearts to 400 per week. The rationing program allocates hearts only to persons from 15 to 70 years of age, not addicted to alcohol or drugs, who would be capable of living at least 10 additional years if the implant procedure was successful. The majority opinion rules that Congress has the right to ration a scarce resource, and that random selection among qualified candidates does not violate due process rights. It is the opinion of the majority (in this hypothetical case) that expensive medical technology such as an artificial heart can properly be seen as a luxury, even though it sustains life, and that the state is not required to provide it to all its citizens.

Given the growing demand for exotic medical care and a world of finite resources, Hiatt (1975, 1976) suggests a set of social and medical priorities for the evaluation of medical practices. He considers evaluation particularly important for three types of medical practices: (1) those that pose conflicts between the interests of the individual and of society; (2) those of no value or undetermined value; and (3) those for potentially preventable conditions. All high-cost, half-way technologies such as renal dialysis and heart transplants would fall under Hiatt's first category. The conflict between the individual and society is basically an issue of allocating economic resources. Included also in this category would be diseases for which known treatment exists but whose prevalence rate is so low that universal screening programs might be considered more detrimental than beneficial to society—if cost were the only measure, and if human life could be assigned a dollar value. (See also Frazier and Hiatt, 1978.)

Elective treatments, especially surgical procedures, represent a more difficult set of examples that might fall within this framework. The rates for tonsillectomies, which have questionable efficacy in most cases, vary tremendously from community to community; within the same state, rates may be ten times higher in some communities than in others (Wennberg and Gittelsohn, 1973). The number of hysterectomies has increased thirty percent in the last ten years, a larger increase than would be expected given the known prevalence of pathology (Gaus and Cooper, 1979). Are the benefits, in terms of the quality of life of individuals undergoing these procedures, great enough to warrant the costs to society of maintaining the required surgeons and operating facilities?

A recent study—Bunker et al., *Costs, Risks, and Benefits of Surgery* (1977)—explores the economic implications of various common surgical procedures. In a widely cited statistic, they estimate that to keep morbidity at a minimum by surgical intervention the marginal cost of saving one additional life in cases of suspected appendicitis would be $43 million per life saved. This means that so many operations would have to be performed on incorrectly diagnosed patients that the one life saved would cost $43 million and the period of convalescence would consume 2,053 person years. This study exemplifies the impracticality of a system that would try to save every life possible.

Hiatt's second category—medical practices that are of no value or of undetermined value—includes a number of procedures that were at one time practiced widely but have now been almost abandoned. These include gastric freezing for peptic ulcer, renal capsule stripping for acute renal failure, internal mammary artery ligation for coronary artery disease, lobotomy for many mental disorders, and others. Most of these practices disappeared not because better procedures came along, but because they were found to be without value. Even though some may have merited experimental introduction, no careful evaluation studies were done before these procedures were introduced on a wide scale. Furthermore, they remained in practice much too long, used far more economic resources than were warranted, and caused an unmeasurable amount of pain, discomfort, and inconvenience to individuals.

More difficult to deal with on a policy level are procedures the value of which remains undetermined. The rapid diffusion of intensive care units is an example of the way expensive but unproven technologies are widely adopted in the health care sector. In 1958 intensive care units were present in only 21 percent of the largest voluntary hospitals; by 1969, 90 percent of hospitals with over 200 beds had intensive care units. Moreover, by 1973 a significant proportion of hospitals—ranging from 20 percent of the smallest hospitals to about 90 percent of the largest—had special coronary intensive care units (Russell and Burke, 1975). Patient charges per day in intensive care units are generally more than double the average daily hospital costs. Evidence is nonconclusive on whether the long-term outlook for patients in intensive care units is any different from patients treated in less expensive settings. In fact, Cochrane (1972) found no difference in outcome between patients with acute heart attacks treated at home and those treated in a hospital coronary care unit.

A second example is coronary bypass surgery. In 1976, 80,000 patients in over 500 hospitals underwent coronary bypass surgery at an average patient cost of $12,000. Millman (1977) analyzes the issues surrounding coronary bypass surgery. According to her, the surgery benefits only a minority of patients, and even then its effects are limited to the relief of chest pains and increased tolerance for exercise and other activities. There is no evidence that bypass surgery prevents myocardial infarction or affects long-term survival rates. (See also Murphy, 1977; and Braunwald, 1977.) Millman and others assert that there was widespread diffusion of bypass surgery before the risks and limitations were known. And to compound the irony, the proliferation of cardiac teams working below capacity has caused additional unnecessary public expense. For example, of 470 hospitals doing open heart surgery, 10 percent did 9 or fewer cases per year, 20 percent did 10 to 49 cases per year, and only 15 percent did over 200 cases per year. These figures contrast with the recommendation that over 200 operations be performed in a year in order to maximize efficiency and provide sufficient experience for the cardiac team to minimize errors. The increase in coronary bypass surgery despite questions about efficacy is attributed to several factors. These include the overexpansion of community hospitals, too many thoracic surgeons, a reimbursement structure that favors surgery, and the frustration of general practitioners and patients regarding the management of chest pain.

An analysis of the rapid diffusion of technologies of undetermined value calls forth pleas for randomized trials, or at least an experimental evaluation of some nature, to try to gain a better understanding of the relative costs and benefits of new technology. The goal of public policy, as differentiated from individual need, is to adopt those innovations that yield the greatest surplus of benefits over costs (Klarman, 1972). For innovations where potential costs outweigh benefits from the societal perspective, there may be a need for regulatory mechanisms of some kind, although up to now, as noted, regulatory mechanisms have not been particularly effective.

New high cost technologies dominate the public scene, and there is increasing concern in some quarters that preventive health care gets much less attention than warranted—even though it has the potential to provide the greatest benefits relative to cost (Beauchamp, 1976; and Breslow and Somers, 1977). Some examples: Recent surveys by the Center for Disease Control of the Public Health Service indicate that one half of the children in low-income areas are unprotected against one or more of the preventable childhood diseases (Center for Disease Control, 1975, p. 405). And despite the demonstrated effectiveness of fluoride in reducing dental caries, a significant number of communities still do not have fluoridated water.

C. Clinical Practice

The study of clinical practice—i.e., the way health providers diagnose, treat, and relate to patients, and relate to each other—is central to the sociology of health. A recurrent theme here is that modern medical technologies result in impersonal and fragmentary health care. Other issues overlap with concerns about human experimentation, or with the issues involved in applying advanced technologies, including heroic surgery, life-support systems, and complex diagnostic procedures that depend upon computers and other devices. One question about such procedures concerns the desirability of imposing them on patients and the consequences—for the patients and their close friends and relatives—in terms of psychological well-being and humanistic values. A second important issue concerns the way decisions to make such technological applications are made. A third has to do with the consequences of technological innovations in terms of the life span of individuals. A further question is raised about whether or not dependence on technology increases or reduces the competence of practitioners or patients to cope with health and illness problems, as well as with general social relations; an extreme version of one such view is that computers will make practitioners obsolete within fifty years (Maxman, 1976).

Finally, there is the bioethical concern with death and dying. In part, this is related to the values involved in the use of technological devices to prolong life; in part, it is related to the issue of informed consent and whether or not patients—particularly those with serious chronic illness— are aware of the dimensions and consequences of treatment. Furthermore, diseases such as cancer, renal disorder, and the like, raise serious questions as to who is the appropriate person to make decisions regarding when to intervene, how to intervene, or when to terminate treatment.

1. **Technology and Patient Care.** Technology has had an impact on medicine in not only a general but a specific sense as well: it may help account for the fragmentation of care, for greater nonresponsiveness to human needs. Providers in many specialties—even such so-called general medical care specialties as internal medicine, pediatrics, and obstetrics— commonly refer problems to other specialists. For example, a patient with minor cuts may be sent to a specialist for suturing.

Consumer decisions may also result in fragmentary care. It is not un-usual for patients to go directly to specialists for treatment. They have been sufficiently impressed with the complexities of new diagnostic and treatment techniques to prefer their first encounters to be with specialists rather than with the "family doctor." The lack of continuity of care and the relative absence of providers who perform an integrative function lead to impersonal care, and to dissatisfaction on the part of both consumers and providers. As Mechanic has observed:

> The growth and complexity of knowledge and technology have increasingly brought into question the individual entrepreneurial tradition in medical care, and all systems of care are moving toward greater organization of health care efforts. Specialization in the absence of

aggregation of personnel and integration of services has resulted in growing fragmentation of care and poor medical care from a community point of view. Such functions as preventive care, follow-up care, health education, and wise management of patients' difficulties in the context of family and community conditions have been difficult to maintain in systems of fragmented services where no person or agency had responsibility for the whole person (1975, p. 243).

With regard to responsiveness of care, there is evidence that physicians view as trivial many of the more common problems brought to them; for example, the two most frequent problems for which women consult physicians are weight gains and headaches (National Center for Health Statistics, 1975). These problems do not lend themselves to the application of curative technologies, and physicians are frustrated at using their time to treat such complaints.

Numerous studies of patient-practitioner relations point to the substantial difference in social power between the consumer and the provider. For most medical encounters, patients and their families are minimally involved in the decision regarding what happens to them. Some would argue that there is an unnecessary monopoly control over the use of technology by the medical profession. There is some evidence that greater benefits could be achieved by providing the individual consumer with the capability of performing uncomplicated diagnostic procedures. Blood pressure equipment—which until recently was to be used only under medical supervision—is now widely available for purchase. Many similar tests are not so widely available. Katz and Clancey (1974) demonstrated that parents can perform throat cultures with accuracy comparable to that of physicians; the implications of this finding for the early treatment of infections, and reduced costs for the patient, are quite important. Nevertheless, few medical facilities provide parents with throat-culture materials or with access to laboratories without the supervision of a physician. There is also some evidence that Pap smears for the early detection of cervical cancer could be routinely accomplished by consumers.

2. Patient Rights. It is often assumed that the complexity of modern medical technology has made it more difficult for medical interventions to take place with the informed consent of the patient. Physicians point out that not only is it impractical in terms of their time to explain the nuances of risk involved in all therapeutic procedures, but patients may misinterpret the risk potential of various procedures and make decisions that are not in their best interests. Mechanic (1977) has analyzed ethical problems in the delivery of health services, and makes a series of recommendations regarding regulatory mechanisms that could be introduced to provide more consistent protection of human rights in medical care settings.

The strain between the rights of patients to be fully informed and to give consent to procedures, and the difficulties, real or attitudinal, of providing the necessary information to patients, has increased. Despite a body of literature on physician-patient relations, the informed consent issue requires considerable additional research.

There has been an experiment in one California community, for example, in which the school health program is based upon the principle that elementary school children can make, indeed have the right to make, their own decisions regarding treatment (Lewis et al., 1977).

This and similar efforts are predicated not only on an ethical belief regarding informed consent but on the view that patients' control over their health leads to increased preventive behavior and to more appropriate use of medical resources.

Despite technological gains, medical care is highly dependent upon patients' cooperation—to return for care when necessary, to adhere to the prescribed regimen, to give an accurate medical history. Poor communication between consumer and provider seriously limits the extent to which technology can be effectively applied to cure or even control the course of illness (Svarstad, 1976).

The powerlessness of consumers is a recurring theme in contemporary commentaries on medical care. And it is a fact that patients' families are often pressured, as in the case of permission for autopsies, to act in ways that have little direct benefit for the individual (Duff and Hollingshead, 1968). The general norm is to trust one's physician. As medical care organizations have grown in size and complexity, it has become increasingly difficult to protect individual rights—especially to privacy and confidentiality (Barber et al., *Research on Human Subjects,* 1973; and Gray, *Human Subjects in Medical Experimentation,* 1975). Many social scientists have described the patient's role in the hospital as one without the normal rights and privileges accorded to persons in other settings; a lack of respect for the protection of human dignity is communicated in countless ways by the hospital staff, even by the general surroundings (Taylor, 1970). As government and third-party insurers have taken a more active role in the control of costs and quality of medical care, there is reason for increasing alarm. The uncontrolled sharing of medical information should have devastating consequences on careers, on family life, even on such mundane but important facets of life as the ability to obtain a driver's license or insurance. Although issues of medical record confidentiality are sometimes highlighted, most present data storage systems are woefully inadequate in terms of protection of confidentiality (Westin, 1976).

3. Prolongation of Life. At the present time there is a major public concern about the rights of patients; the issue is sharpest in decisions either to prolong or to terminate life. Recent studies of dying have focused on the moral and ethical issues induced by technological innovations. A study by Crane of the treatment of critically ill patients (*The Sanctity of Social Life*, 1975) concludes that physicians respond to the chronically ill or terminally ill patient not simply in terms of physiological definitions of illness but also in terms of the extent to which the patient is capable of social interaction with others. Extending previous studies, Crane's data suggest that social capacity to interact with others is a more important determinant of physician behavior than the patient's value to society. Crane also suggests that certain types of problems in the treatment of the dying occur so frequently and are so highly visible to the public that norms have emerged that permit the physician to hasten death in some situations. The prescription of large doses of narcotics for terminal patients who are in great pain, as well as the termination of respirator treatment for the patient who has suffered irreversible brain damage, are examples. Crane argues that, although in the past the traditional medical norm underlying all treatment has been the sanctity of life, other counternorms are now becoming equally legitimate—e.g., the sanctity of privacy, or the humanitarian norm to alleviate suffering.

Crane's study was completed prior to the 1975 landmark court decision in which the parents of Karen Quinlan petitioned the court for guardianship of the twenty-year-old comatose woman and authorization to remove the respirator that was thought to be sustaining her life (Branson et al., 1976). The case brought to public attention moral and ethical dilemmas that practitioners and affected families had dealt with for many years. The notoriety of the Quinlan case is unusual; one hypothesis is that it reached the courts primarily because of physicians' fears of malpractice actions. The original decision went against the parents; the court sided with the power of the physician to prescribe the care of dying patients. That decision, however, was overturned by the New Jersey Supreme Court, and the parents, acting on behalf of their daughter, were recognized as having the authority to refuse extraordinary measures to sustain her life.

Since the Quinlan decision there has been increased activity among groups seeking to establish the rights of individuals to control their fate through such mechanisms as living wills (Garland, 1976). In addition, there is increasing awareness on the part of consumers that physicians may have excessive power in such decisions by controlling the information available to patients and families (Duff and Campbell, 1973). This interest in equalizing the balance of power in medical decision making has implications that extend beyond the care of the critically ill to all aspects of interaction between physicians and consumers. Women's health advocates have become some of the most outspoken constituents demanding the sharing of information, insisting on the rights of consumers to question "expert" advice.

Not much is known about the side effects of technological efforts to prolong life. Although the dramatic cases receive the most attention, vast numbers of chronically ill patients are sustained by one technique or another—including drugs, pacemakers, dialysis, and so on. Although the medical practitioner often has a single goal—maintaining life—in many cases the quality of the patient's life brings into question the social desirability of the intervention. In addition to social adjustments on the part of the patient, there may be overwhelming consequences, both economical and psychological, for the family.

Such research as has been undertaken is revealing. For example, Abram and associates (1971) documented an unusually high occurrence of suicidal behavior among renal dialysis patients. Problems of organ transplant patients have been studied by Fox and Swazey (1974), and by Simmons et al. (1977). The adjustment to myocardial infarction has been examined by Aiken (1976). Providing effective social support systems and maximizing the utility of technological interventions, however, requires much more research.

Widespread applications of technology can be viewed as beneficial or harmful, depending upon the professional or social perspective taken. The case of the mentally ill is a good illustration. The use of psychoactive drugs stimulated the development of the community mental health movement, with consequent sharp reductions in the duration and frequency of hospitalizations for functional psychoses. It has been argued that such community treatment is both more humane and cost-effective (see Bachrach, 1976; Stein et al., 1975; and Weisbrod et al., 1976). Recently, however, pressure for deinstitutionalization has been questioned because of the quality-of-life issue for patients, families, and the community.

It is important to stress that most of what is known about technology in clinical practice concerns special and unusual medical encounters. There are few systematic generalizations available about the way technology has influenced the more mundane contacts between providers and patients. Both overuse and underuse of available routine technologies represent important ethical, social, and economic issues requiring extensive research.

III. EMERGING THEMES

As is the case with other social institutions, the organization and functioning of the health sector are determined by the prevailing structure and values of the larger society. The way research problems are studied, the way innovations are developed, the way health services are organized, delivered, and financed, and the way patients are dealt with clinically, can be understood only in the context of general social-structural and social-psychological dimensions. It follows that as the outlines of the larger society change, so do research, organization, and clinical practice in medicine. Science and technology—not counting the health sector—are likewise influenced by the structure and values of the larger society; thus the health sector, as influenced by science and technology, gets a second, indirect, influence from society.

When science and technology affect the health sector indirectly in the same ways that the larger society directly shapes it, the impact is rein-forced. For example, when both society generally and science and tech-nology emphasize the prolongation of life, the results are activities within medicine that maximize the quality of physician performance—e.g., mor-tality conferences—and such technological developments as renal dialysis. Pluralism in structure and values within the larger society may result in conflicting outputs from the health sector and from the work of scientists and technologists. For example, because society emphasizes the reproduction of normal children, programs of genetic counseling have been developed; meanwhile, technological developments provide opportunities to maintain frail infants, including those with uncorrectable physical and mental defects.

The development of an agenda for future research should therefore take into account emerging social and cultural themes. We believe three such themes to have particular importance.

The first and clearest is an emphasis upon equity. There are continu-ing and intensifying efforts for equal access to and quality of health ser-vices, just as there are efforts with respect to education and employment. This theme is manifest in efforts to remove the economic barriers to care, such as Medicaid. It is also apparent in the increased emphasis on geographi-cal redistribution of physicians and other health providers, as well as the refinement of the roles of medical providers to meet the expressed health concerns of the population.

Some technologies are consistent with these objectives, but many represent counterforces. Part of the current antagonism to science and tech-nology in medicine is related to a view that the costs and complexities im-pede progress toward equity in health care. Unless the direction of scientific and technological development becomes more consistent with the societal emphasis on equity, controls on research and development—along with further regulation of the implementation of innovation—can be anticipated.

A second theme is a deepening concern with individual rights and self-determination. Involved here are such matters as confidentiality, pri-vacy, and individual choice in personal health care. As we have observed, both in research activities and in clinical encounters, the complexities of technology often seem to restrict use. As in the case of equity, an intensified emphasis on individual determinism can be expected to have a further impact on the course of science and technology.

The third theme that merits consideration is a broadening of the concept of the quality of life. There is an increasing awareness of oppor-tunities, a sense of entitlement to a fair share of community resources. The allocation of public economic resources for health care must now compete with powerful pressure from education, welfare, public security, and cul-ture, among other constituencies. As discussed above, in the face of finite resources and rising costs, funds for research and development, as well as for applications of technology in medicine, are likely to be even more carefully scrutinized in the future. There are also likely to be more restrictions on growth of support.

These views on emerging social changes are incomplete. They are indicative, however, of the array of dimensions that need to be considered in future work, including sociological work, in the undeveloped area of science and technology in medicine. At the same time that attention is paid to long-term pressures for increased equity, individual rights, and self-determination, and a broadened concept of the quality of life, the role of technology in the so-called "crisis in health care" needs continued examination.

Technology has become the persistent scapegoat for deficiencies in the health sector. Skyrocketing costs of medical care, maldistribution of health providers, discontinuities in patient care and physician-patient relations that characterize the current delivery system in the United States— all of these problems have been blamed on the widespread introduction of new technologies in medicine. Thus, control of technology is seen as the remedy for the acknowledged defects in health care.

Clearly technology has influenced the structure, organization, and economics of health care in the United States, but, as we have indicated, regulating technology is not a viable solution. Neither current regulatory activities to control the spread of existing technologies, nor efforts to reduce the growth of new ones by curtailing research and development, are likely to have a meaningful impact on health services.

It is important that science and technology not be overvillainized. Technological overabundance is often a consequence rather than a cause of existing problems in the delivery of health services. Improvements are needed in areas such as provider remuneration, pricing of health services (particularly hospital care), consumer demands, and thorough evaluation of diagnostic and treatment procedures.

Analysts have a mandate to provide a responsible perspective on the present role of science and technology in medicine, as well as to deal with long-term humanistic and sociological questions. In order to do so, it is necessary to go beyond traditional disciplinary boundaries, to identify structural and organizational changes that could provide easier access to health services of a high quality for all persons at reasonable costs.

Acknowledgment

The authors wish to acknowledge helpful critiques of earlier drafts of the paper by Bernard Barber, Harold Garfinkel, and David Mechanic.

BIBLIOGRAPHIC INTRODUCTION

In order to accomplish our two objectives—to review the field of medical sociology, and to analyze the relation between science, technology, and health from a sociological perspective—it was necessary to cite a large body of literature. The various references on which the analysis is based are reported in the main part of the bibliography.

Since the number of references is large, a small subset may help readers who desire a briefer orientation to the field. This selection is covered under three headings: (A) General Reviews, Texts, and Source Materials; (B) Studies in Medical Sociology; and (C) Policy Analyses Relevant to Science, Technology, and Medicine.

Under the first heading we include two recent texts. One is a handbook: Howard E. Freeman et al., *Handbook of Medical Sociology*, 3d ed. (1979), that provides up-to-date reviews of the field of medical sociology by leading experts. The second, David Mechanic's *Medical Sociology* (1978), is the most comprehensive integrated review of the field. The bibliographies contained in these two volumes are valuable sources for work in medical sociology. The other seven selections provide factual information on various aspects of medical care and health services.

The second section, "Studies in Medical Sociology," provides examples of work in the field. We hesitate to use the word "classics," since most of these works are of recent vintage. But they do demonstrate the range of methods, topics, and perspectives on health and illness that are utilized by researchers identified with medical sociology. Some of these studies have direct relevance for our analysis of science, technology, and medicine. Examples are: Bernard Barber et al., *Research on Human Subjects: Problems of Social Control in Medical Experimentation* (1973), and Diana Crane, *The Sanctity of Social Life: Physicians' Treatment of Critically Ill Patients* (1975). Others have been selected because of their influence on contemporary work in medical sociology—for example, August B. Hollingshead and Frederick C. Redlich, *Social Class and Mental Illness* (1958), and Talcott Parsons, *The Social System* (1951).

Under the third heading, "Policy Analyses Relevant to Science, Technology, and Medicine," some of the references are to general analyses of the sort that shape medical care and the delivery of health services. Examples: Eliot Freidson, *Profession of Medicine: A Study of the Sociology of Applied Knowledge* (1975); and Charles E. Lewis et al., *A Right to Health* (1976). Others are targeted directly on science and technology in medicine.

A theme that emerges in our chapter is the comparative recency of work in the field and the swift changes in science and technology issues in health. Nevertheless, many of the references, and particularly those identified in the three sections discussed here, contain comprehensive bibliographies that should be valuable to scholars who wish to delve more deeply into the subject matter.

Note: The references that follow are given in an abbreviated form; for full information on each item, see the main bibliography, below.

A. General Reviews, Texts, and Source Materials

Anderson, Odin W. *Health Care: Can There Be Equity? The United States, Sweden, and England* (1972).

Donabedian, Avedis. *Benefits in Medical Care Programs* (1976).

Donabedian, Avedis; Axelrod, S. J.; Swearingen, C.; and Jameson, J. *Medical Care Chartbook* (5th ed., 1972).

Freeman, Howard E.; Levine, Sol; and Reeder, Leo G., eds. *Handbook of Medical Sociology* (3d ed., 1979).

Goodman, L. J., and Mason, H. R. *Physician Distribution and Medical Licensure in the United States, 1975* (1976).

Koleda, M.; Burke, C.; and Williams, J. *The Federal Health Dollar: 1969–1976* (1977).

Mechanic, David. *Medical Sociology* (2d ed., 1978).

———, ed. *The Growth of Bureaucratic Medicine* (1976).

B. Studies in Medical Sociology

Barber, Bernard; Lally, J. J.; and Makarushka, J. M. *Research on Human Subjects: Problems of Social Control in Medical Experimentation* (1973).

Becker, Howard S.; Geer, Blanche; Hughes, Everett C.; and Strauss, Anselm L. *Boys in White: Student Culture in Medical School* (1961).

Crane, Diana. *The Sanctity of Social Life: Physicians' Treatment of Critically Ill Patients* (1975).

Fox, Renée C. *Experiment Perilous: Physicians Facing the Unknown* (1974).

Fox, Renée C., and Swazey, J. P. *The Courage to Fail: A Social View of Organ Transplants and Dialysis* (1974).

Freidson, Eliot. *Profession of Medicine: A Study of the Sociology of Applied Knowledge* (1975).

Hagstrom, Warren O. *The Scientific Community* (1965).

Hollingshead, August B., and Redlich, Frederick C. *Social Class and Mental Illness* (1958).

Merton, Robert K.; Reader, George C.; and Kendall, Patricia L., eds. *The Student Physician: Introductory Studies in the Sociology of Medical Education* (1957).

Parsons, Talcott. *The Social System* (1951).

C. Policy Analyses Relevant to Science, Technology, and Medicine

Bennett, Ivan L., Jr. "Technology as a Shaping Force" in *Daedalus* (1977).

Bunker, John P.; Barnes, Benjamin A.; and Mosteller, Frederick. *Costs, Risks, and Benefits of Surgery* (1977).

Carlson, Rick J. *The End of Medicine* (1975).

Freidson, Eliot. *Doctoring Together: A Study of Professional Social Control* (1975).

Fuchs, Victor. *Who Shall Live? Health, Economics, and Social Choice* (1974).

Gray, Bradford H. *Human Subjects in Medical Experimentation* (1975).

Illich, Ivan. *Medical Nemesis: The Expropriation of Health* (1976).

Lewis, Charles E.; Fein, Rashi; and Mechanic, David. *A Right to Health: The Problem of Access to Primary Medical Care* (1976).

Office of Technology Assessment. *Development of Medical Technology* (1976).

BIBLIOGRAPHY

Abram, Harry S.; Moore, Gordon L.; and Westervelt, Frederick B. "Suicidal Behavior in Chronic Dialysis Patients." *American Journal of Psychiatry* 127 (March 1971): 119–123.

Ad Hoc Committee on Ethical Standards in Psychological Research. *Ethical Principles in the Conduct of Research with Human Participants.* Washington, D.C.: American Psychological Association, 1973.

Aiken, Linda H. "Chronic Illness and Responsive Ambulatory Care." In *The Growth of Bureaucratic Medicine,* pp. 239–251. Edited by David Mechanic. New York: John Wiley, 1976.

American Sociological Association. *1975–1976 Directory of Members.* Washington, D.C.: American Sociological Association, 1975.

Anderson, Odin W. *Health Care: Can There Be Equity? The United States, Sweden, and England.* New York: John Wiley, 1972.

Annas, George J. "Allocation of Artificial Hearts in the Year 2002: Minerva v. National Health Agency." *American Journal of Law and Medicine* 3 (1977): 59–76.

Bachrach, Leona L. *Deinstitutionalization: An Analytical Review and Sociological Perspective.* U.S. National Institute of Mental Health (Series D, No. 4). Washington, D.C.: N.I.M.H., 1976.

Banta, David, and McNeil, Barbara. "The Costs of Medical Diagnosis: The Case of the CT Scanner." In *Health Care in the American Economy: Issues and Forecasts,* pp. 158–187. Edited by Blue Cross Association. Chicago: Health Services Foundation, 1977.

Barber, Bernard. "The Case of the Floppy-Eared Rabbits." *American Journal of Sociology* 64 (September 1958): 128–136.

——. "Theoretische beschouwingen: The Functions and Dysfunctions of Fashion in Science: A Case for the Study of Social Change." *Mens en Maatschappij* 43 (November-December 1968): 501–514.

Barber, Bernard; Lally, J.; and Makarushka, J. M. *Research on Human Subjects: Problems of Social Control in Medical Experimentation.* New York: Russell Sage Foundation, 1973.

Beauchamp, T. D. "Public Health as Social Justice." *Inquiry* 13 (March 1976): 3–14.

Becker, Howard S.; Geer, Blanche; Hughes, Everett C.; and Strauss, Anselm L. *Boys in White: Student Culture in Medical School.* Chicago: University of Chicago Press, 1961.

Behney, Clyde J. *Studies of the Implications of New Medical Technologies: The State of the Activity.* Staff paper for the National Council of Health Planning and Development, Health Resources Administration. Washington, D.C.: H.E.W., August 1976.

Bellin, Lowell E. "Politics, Not Technology, Is the Problem in Health Care Quality Control." *Man and Machine* 1 (Winter 1976): 124–125.

Ben-David, Joseph. "Scientific Productivity and Academic Organization in 19th Century Medicine." *American Sociology Review* 25 (December 1976): 828–843.

Bennett, Ivan L., Jr. "Technology as a Shaping Force." *Daedalus* (Winter 1977): 125–134.

Bernstein, Ilene N., and Freeman, Howard E. *Academic and Entrepreneurial Research.* New York: Russell Sage Foundation, 1975.

Boothe, Bert E.; Rosenfeld, Ann H.; and Walker, Edward L. *Toward a Science of Psychiatry.* Belmont, Calif.: Wadsworth, 1974.

Branson, Roy; Casebeer, Kenneth; Levine, Melvin; Oden, Thomas; Ramsey, Paul; and Capron, Alexander. "The Quinlan Decision: Five Commentaries." *Hastings Center Report* 6 (February 1976): 125–134.

Braunwald, Eugene. "Coronary-Artery Surgery at the Crossroads." *New England Journal of Medicine* 297 (September 22, 1977): 661–663.

Breslow, L., and Somers, A. R. "The Lifetime Health-Monitoring Program." *New England Journal of Medicine* 296 (March 17, 1977): 601–608.

Bunker, John P.; Barnes, Benjamin A.; and Mosteller, Frederick. *Costs, Risks, and Benefits of Surgery.* New York: Oxford University Press, 1977.

Campbell, Donald T.; Boruch, Robert F.; Schwartz, Richard D.; and Steinberg, Joseph. "Confidentiality: Preserving Modes of Access to Files and to Interfile Exchange for Useful Statistical Analysis." *Evaluation Quarterly* 1 (1977): 269–300.

Carlson, Rick J. *The End of Medicine.* New York: John Wiley, 1975.

Center for Disease Control. *Morbidity and Mortality Weekly Report no. 48.* Atlanta: Center for Disease Control, 1975.

Clausen, John A. "Sociology of Mental Disorder." In *Handbook of Medical Sociology,* 3rd ed., pp. 97–112. Edited by H. Freeman et al. Englewood Cliffs, N.J.: Prentice-Hall, 1979.

Cochrane, A. L. *Effectiveness and Efficiency: Random Reflections on Health Services.* The Nuffield Provincial Hospitals Trust, London: Burgess and Son, 1972.

Cohen, Sheldon; Glass, David; and Phillips, Susan. "Health and the Environment." In *Handbook of Medical Sociology*, pp. 134–149. Edited by H. Freeman et al. Englewood Cliffs, N.J.: Prentice-Hall, 1979.

Colson, Anthony C., and Sellers, Karen E. "Medical Anthropology." *Annual Review of Anthropology* 3 (1974): 245–262.

Committee on Interstate and Foreign Commerce, U.S. House of Representatives Subcommittee on Public Health and Environment Hearings. *Biomedical Research: Ethics and the Protection of Human Research Subjects*. Serial No. 93–79, September 27–28, 1973. Washington, D.C.: Government Printing Office, 1974.

Comroe, Julius H., and Dripps, Robert D. "Scientific Basis for the Support of Biomedical Science." *Science* 192 (9 April 1976): 105–111. For full study report, see Comroe and Dripps, *The Top Ten Clinical Advances in Cardiovascular and Pulmonary Medicine and Surgery between 1945 and 1975*. 2 vols. San Francisco: University of California Press, 1977. See also Comroe, "The Road from Research to New Diagnosis and Therapy." *Science* 200 (26 May 1978): 931–937.

Crane, Diana. *The Sanctity of Social Life: Physicians' Treatment of Critically Ill Patients*. New York: Russell Sage Foundation, 1975.

Culliton, Barbara J. "Health Care Economics: The High Cost of Getting Well." *Science* 200 (26 May 1978): 883–885.

Czoniczer, Gabor. "The Role of the Patient in Modern Medicine." *Man and Medicine* 3 (1978): 17–28.

Donabedian, Avedis. *Benefits in Medical Care Programs*. Cambridge: Harvard University Press, 1976.

Donabedian, Avedis; Axelrod, S. J.; Swearingen, C.; and Jameson, J. *Medical Care Chartbook*. 5th ed. Ann Arbor: University of Michigan Press, 1972.

Duff, Raymond S., and Campbell, A. G. M. "Moral and Ethical Dilemmas in the Special-Care Nursery." *New England Journal of Medicine* 289 (October 25, 1973): 890–894.

Duff, Raymond S., and Hollingshead, August B. *Sickness and Society*. New York: Harper & Row, 1968.

Eiduson, Bernice T., and Beckman, Linda, eds. *Science As a Career Choice*. New York: Russell Sage Foundation, 1973.

Fineberg, H. V. "Clinical Chemistries: The High Cost of Low-Cost Diagnostic Tests." In *Medical Technologies: The Culprit behind Health Care Costs?*, pp. 144–165. Edited by Stuart H. Altman and Robert J. Blendon. Washington, D.C.: H.E.W., 1979.

Fox, Renée C. *Experiment Perilous: Physicians Facing the Unknown*. Philadelphia: University of Pennsylvania Press, 1974.

———. "Advanced Medical Technology: Social and Ethical Implications." *Annual Review of Sociology* 2 (1976): 239–240.

Fox, Renée C., and Swazey, J. P. *The Courage to Fail: A Social View of*

Organ Transplants and Dialysis. Chicago: University of Chicago Press, 1974.

Frazier, Howard S., and Hiatt, Howard H. "Evaluation of Medical Practices." *Science* 200 (26 May 1978): 875–878.

Freeman, Howard E.; Borgatta, Edgar F.; and Siegel, Nathaniel H. "Remarks on the Changing Relationship between Government Support and Graduate Training." In *Social Policy and Sociology*, pp. 297–305. Edited by Nicholas J. Demerath. New York: Academic Press, 1975.

Freeman, Howard E., and Reeder, Leo G. "Medical Sociology: A Review of the Literature." *American Sociological Review* 22 (February 1957): 73–81.

Freeman, Howard E.; Levine, Sol; and Reeder, Leo G., eds. *Handbook of Medical Sociology*, 3d ed. Englewood Cliffs, N.J.: Prentice-Hall, 1979.

Freidson, Eliot. *Doctoring Together: A Study of Professional Social Control*. New York: Elsevier, 1975.

——. *Profession of Medicine: A Study of the Sociology of Applied Knowledge*. New York: Elsevier, 1975.

Fried, Charles. *Medical Experimentation: Personal Integrity and Social Policy*. New York: American Elsevier, 1974.

Fuchs, Victor R. *Essays in the Economics of Health and Medical Care*. New York: Columbia University Press, 1972.

——. *Who Shall Live? Health, Economics, and Social Choice*. New York: Basic Books, 1974.

Funkenstein, Daniel H. *Medical Students, Medical Schools, and Society during Five Eras: Factors Affecting the Career Choices of Physicians, 1958–1976*. Cambridge, Mass.: Ballinger, 1978.

Gallagher, Eugene B., ed. *The Doctor-Patient Relationship in the Changing Health Scene*. H.E.W.-N.I.H. 78–183. Washington, D.C.: H.E.W., 1976.

Garland, M. "Politics, Legislation, and Natural Death." *Hastings Center Report* 6 (October 1976): 5–6.

Gaus, C. R., and Cooper, B. S. "Controlling Health Technology." In *Medical Technologies: The Culprit behind Health Care Costs?* pp. 242–252. Edited by Stuart H. Altman and Robert J. Blendon. Washington, D.C.: H.E.W., 1979.

Ginzberg, Eli, ed. *Regionalization and Health Policy*. Washington, D.C.: U.S. Government Printing Office, 1977.

Golladay, F. C., and Smith, K. R. *Who Shall Pay? An Analysis of the Malpractice Crisis*. No. 6–75. Madison: University of Wisconsin Health Economics Research Center, 1975.

Goodman, L. J., and Mason, H. R. *Physician Distribution and Medical Licensure in the United States, 1975*. Chicago: American Medical Association, 1976.

Graham, Saxon, and Reeder, Leo G. "Social Epidemiology of Chronic Disease." In *Handbook of Medical Sociology,* 3d ed., pp. 71–96. Edited by H. Freeman et al. Englewood Cliffs, N.J.: Prentice-Hall, 1979.

Gray, Bradford H. *Human Subjects in Medical Experimentation.* New York: John Wiley, 1975.

Greer, Ann L. "Advances in the Study of Diffusion of Innovation in Health Care Organizations." *Health and Society* 55 (Fall 1977): 505–523.

Grotjahn, Alfred, and Kriegel, F. *Soziale Pathologie.* 2d ed. rev. Berlin: August Hirschwald Verlag, 1915.

Haggerty, Robert J.; Roghmann, Klaus J.; and Pless, Ivan B. *Child Health and the Community.* New York: John Wiley, 1975.

Hagstrom, Warren O. *The Scientific Community.* New York: Basic Books, 1965.

Hiatt, Howard H. "Protecting the Medical Commons: Who Is Responsible?" *New England Journal of Medicine* 293 (July 31, 1975): 235–241.

————. "Too Much Medical Technology." *Wall Street Journal* (June 24, 1976), p. 16.

Hollingshead, August B., and Redlich, Frederick C. *Social Class and Mental Illness.* New York: John Wiley, 1958.

Howard, Jan, and Strauss, Anselm, eds. *Humanizing Health Care.* New York: John Wiley, 1975.

Illich, Ivan. *Medical Nemesis: The Expropriation of Health.* New York: Pantheon, 1976.

Institute of Medicine. *Report of the Panel on Implications of a Categorical Catastrophic Approach to National Health Insurance.* Washington, D.C.: National Academy of Science, June 1973.

————. *Computed Tomographic Scanning: A Policy Statement.* Washington, D.C.: National Academy of Science, 1977.

Kaplan, Howard B. "Social Psychology of Disease." In *Handbook of Medical Sociology,* pp. 53–70. Edited by H. Freeman et al. Englewood Cliffs, N.J.: Prentice-Hall, 1979.

Katz, Harvey P., and Clancey, Robert R. "Accuracy of a Home Throat Culture Program: A Study of Parent Participation in Health Care." *Pediatrics* 53 (May 1974): 687–691.

Katz, Jay, and Capron, Alexander Morgan. *Catastrophic Diseases: Who Decides What?* New York: Russell Sage Foundation, 1975.

Klarman, H. E. "Application of Cost-Benefit Analysis to Health Systems Technology." In *Technology and Health Care Systems in the 1980s,* pp. 225–250. Edited by M. F. Collen. (HSM) 73–3016. Washington, D.C.: U.S. Government Printing Office, 1972.

Knaus, William A.; Schroeder, Steven; and Davis, David O. "Impact of New Technology: The CT Scanner." *Medical Care* 15 (July 1977): 533–542.

Knowles, John H. "The Hospital." In *Life and Death and Medicine,* p. 97ff. Edited by Kerr L. White. San Francisco: W. H. Freeman, 1973.

Koleda, Michael; Burke, C.; and Williams, J. *The Federal Health Dollar: 1969–1976*. Washington, D.C.: National Planning Association, 1977.

Kornhauser, William. *Scientists in Industry: Conflict and Accommodation.* Berkeley: University of California Press, 1962.

Krause, Elliott A. *Power and Illness: The Political Sociology of Health and Medical Care.* New York: Elsevier, 1977.

Lally, John J. "The Making of the Compassionate Physician-Investigator." *Annals of the American Academy of Political and Social Science* 437 (May 1978): 86–98.

Lewis, Charles E. "Health Services Research: Does It Make a Difference?" *New England Journal of Medicine* 297 (August 25, 1977): 423–427.

Lewis, Charles E.; Fein, Rashi; and Mechanic, David. *A Right to Health: The Problem of Access to Primary Medical Care.* New York: John Wiley, 1976.

Lewis, Charles E.; Lewis, Mary Ann; Lorimer, Ann; and Palmer, Beverly. "Child-Initiated Care: The Utilization of School Nursing Services by Children in an 'Adult Free' System." *Pediatrics* 60 (October 1977): 499–507.

McDermott, Walsh. "Medicine: The Public Good on One's Own." *Perspectives in Biology and Medicine* 21 (Winter 1978): 167–187.

McGill, William J. "Public Control of Science Is Not the Answer." *Change* 9 (December 1, 1977): 8–9.

McIntire, Charles. "The Importance of the Study of Medical Sociology." *Bulletin of the American Academy of Medicine* 1 (February 1894): 425–434.

Madge, John. *The Origins of Scientific Sociology.* New York: Free Press, 1962.

Marcson, Simon. *Scientists in Government: Some Organization Determinants of Manpower Utilization in a Government Laboratory.* New Brunswick, N.J.: Rutgers University Press, 1966.

Maxmen, Jerrold S. *The Post-Physician Era.* New York: John Wiley, 1976.

Mechanic, David. "Ideology, Medical Technology, and Health Care Organization in Modern Nations." *American Journal of Public Health* 65 (March 1975): 241–247.

——. *Ethical Problems in the Delivery of Health Services.* A Report to the National Commission for the Protection of Human Subjects of Biomedical and Behavioral Research. Research and Analytic Report Series, no. 2–77. Madison: Center for Medical Sociology and Health Services Research, University of Wisconsin, 1977.

——. *Prospects and Problems in Health Services Research.* Series no. 11–77. Madison: Center for Medical Sociology and Health Services Research, University of Wisconsin, 1977. Also to appear in David Mechanic, *The Potential for Reform,* New York: Free Press, forthcoming.

————. "Approaches to Controlling the Costs of Medical Care: Short-Range and Long-Range Alternatives." *New England Journal of Medicine* 298 (February 2, 1978): 249–254.

————. *Medical Sociology,* 2d ed. New York: Free Press, 1978.

————. "Physicians." In *Handbook of Medical Sociology,* 3d ed. pp. 177–192. Edited by H. Freeman et al. Englewood Cliffs, N.J.: Prentice-Hall, 1979.

————, ed. *The Growth of Bureaucratic Medicine.* New York: John Wiley, 1976.

Merton, Robert K. *Social Theory and Social Structure.* New York: Free Press, 1949. 2d ed., enlarged, 1957; reprint ed., 1968.

————. "Priorities in Scientific Discovery: A Chapter in the Sociology of Science." *American Sociological Review* 22 (December 1957): 635–659.

————. *On the Shoulders of Giants.* New York: Harcourt Brace Jovanovich, 1965.

Merton, Robert K.; Reader, George C.; and Kendall, Patricia L., eds. *The Student Physician: Introductory Studies in the Sociology of Medical Education.* Cambridge, Mass.: Harvard University Press, 1957.

Meyer, Lawrence. "Panel Urges Curb on Use of Costly X-Ray Device." *Washington Post* (May 3, 1977), p. A-9.

Miller, Stephen J. *Prescriptions for Excellence.* Chicago: Aldine, 1972.

Millman, Marcia. *The Unkindest Cut.* New York: William Morrow, 1977.

National Center for Disease Control. *National Ambulatory Medical Care Survey, 1973 Summary: U.S. May 1973-April 1974.* Series no. 76–1772. Rockville, Md.: National Center for Health Statistics, 1975.

Office of Technology Assessment. *Development of Medical Technology.* Washington, D.C.: U.S. Government Printing Office, August 1976.

————. *Policy Implications of the Computed Tomography Scanner.* Washington, D.C.: U.S. Government Printing Office, August 1977.

Parsons, Talcott. *The Social System.* New York: Free Press of Glencoe, 1951.

————. "Definitions of Health and Illness in the Light of American Values and Social Structure." In *Patients, Physicians, and Illness,* pp. 107–127. Edited by E. Gartly Jaco. New York: Free Press, 1958.

Paul, Benjamin D. *Health, Culture, and Community.* New York: Russell Sage Foundation, 1958.

Reeder, Sharon, and Mauksch, Hans. "Nursing: Continuing Change." In *Handbook of Medical Sociology,* pp. 209–229. Edited by H. Freeman et al. Englewood Cliffs, N.J.: Prentice-Hall, 1979.

Rettig, Richard A. *Lessons Learned from the End-Stage Renal Disease Experience.* Background paper for a November 19–20, 1976, conference: Health Care Technology and Quality of Care. Boston University, 1976a.

————.*Valuing Lives: The Policy Debate on Patient Care Financing for*

Victims of End-Stage Renal Disease. Santa Monica, Calif.: Rand Corporation, March 1976b.

Roe, Ann. "A Psychological Study of Eminent Biologists." *Genetic Psychology Monographs* 43 (May 1952): 121–239.

Rosen, George. "The Evolution of Social Medicine." In *Handbook of Medical Sociology,* pp. 23–50. Edited by H. Freeman et al. Englewood Cliffs, N.J.: Prentice-Hall, 1979.

Rossi, Peter H., and Williams, Walter, eds. *Evaluating Social Programs: Theory, Practice, and Politics.* New York: Seminar Press, 1972.

Rossi, Peter H., and Wright, Sonia R. *Doing Evaluation: A Survey of Approaches for Assessing Social Projects.* Paris: O.E.C.D., 1979.

———. *Evaluation: A Systematic Approach.* Beverly Hills, Calif.: Sage, 1979.

Russell, Louise B. "The Diffusion of New Hospital Technologies in the U.S." *International Journal of Health Services* 6 (1976): 557–580.

Russell, Louise B., and Burke, Carol S. *Technological Diffusion in the Hospital Sector.* Washington, D.C.: National Planning Association, October 1975.

Salkever, David S., and Bice, Thomas W. "The Impact of Certificate-of-Need Controls on Hospital Investment." *Health and Society* 54 (Spring 1976): 185–214.

Saward, A. W., ed. *The Regionalization of Personal Health Services.* New York: Prodist, 1975.

Scitovsky, Anne A. "Changes in the Costs of Treatment of Selected Illnesses, 1951–1965." *American Economic Review* 57, part 2 (December 1967): 1182–1195.

Scitovsky, Anne A., and McCall, N. *Changes in the Costs of Treatment of Selected Illnesses, 1951–1971.* H.R.A. no. 77–3161. Washington, D.C.: U.S. Government Printing Office, July 1976.

Sidel, Victor W., and Sidel, Ruth. *A Healthy State: An International Perspective on the Crisis in the United States Medical Care.* New York: Pantheon, 1978.

Simmons, Roberta G.; Klein, Susan D.; and Simmons, R. *The Gift of Life: The Social and Psychological Impact of Organ Transplantation.* New York: John Wiley, 1977.

Spingarn, Natalie Davis. *Heartbeat: The Politics of Health Research.* Washington: Robert B. Luce, 1976.

Stein, L. I.; Test, M. A.; and Marx, A. J. "Alternative to the Hospital: A Controlled Study." *American Journal of Psychiatry* 132 (May 1975): 417–422.

Strauss, Robert. "The Nature and Status of Medical Sociology." *American Sociological Review* 22 (April 1957): 200–204.

Svarstad, B. L. "Physician-Patient Communication and Patient Conformity with Medical Advice." In *The Growth of Bureaucratic Medicine,* pp. 220–238. Edited by D. Mechanic. New York: John Wiley, 1976.

Tancredi, Lawrence R., and Barondess, Jeremiah A. "The Problem of Defensive Medicine." *Science* 200 (26 May 1978): 879–882.

Taylor, C. *In Horizontal Orbit: Hospitals and the Cult of Efficiency.* New York: Holt, Rinehart, and Winston, 1970.

Thomas, Lewis. "Commentary: The Future Impact of Science and Technology on Medicine." *BioScience* 24 (24 February 1974): 99–105.

———. *The Influence of Science and Technology on Medicine: A Forecast.* Ciba Foundation Symposium. New York: American Elsevier, 1975.

"Undergraduate Medical Education." *Journal of the American Medical Association* 236 (December 27, 1976): 2961.

U.S. Department of Health, Education, and Welfare. *Current and Projected Supply of Health Manpower.* H.R.A. no. 75–17. Washington, D.C.: Bureau of Health Resources Development, July 1974.

Weisbrod, B. A.; Test, M. S.; and Stein, L. I. *An Alternative to the Mental Hospital: Benefits and Costs.* Institution for Social and Policy Studies, working paper 776. New Haven: Yale University Press, 1976.

Wennberg, J., and Gittlesohn, A. "Small Area Variations in Health Care Delivery." *Science* 182 (14 December 1973): 1102–1108.

Westin, Alan F. *Computers, Health Records, and Citizen Rights.* National Bureau of Standards Monograph no. 157. Washington, D.C.: U.S. Department of Commerce, December 1976.

White, Paul E.; Levin, Lowell S.; and Levine, Sol. "Community Health Organizations." In *Handbook of Medical Sociology,* 3d ed., pp. 347–368. Edited by H. Freeman et al. Englewood Cliffs, N.J.: Prentice-Hall, 1979.

Zimbardo, Philip G. "On the Ethics of Intervention in Human Psychological Research: With Special Reference to the Stanford Prison Experiment." *Cognition* 2 (1973): 243–256.

Part IV
Policy Studies

Chapter 9
Science Policy Studies

Diana Crane, UNIVERSITY OF PENNSYLVANIA

INTRODUCTION: STUDIES OF SCIENCE AND TECHNOLOGY POLICIES

A. The Subject

Studies of science and technology policies—"science policy studies" is without a doubt the most common program title, even though applied science and technology are also studied—include analyses of the nature of science and technology policies, the way they are produced, their effects upon the development of scientific and technological knowledge, and their role in dealing with the impact of science and technology upon the larger society. In other words, if this chapter is to be comprehensive, it must focus on studies relating to science and technology policies in a great variety of forms. Science will be taken to include both the natural and the social sciences, technology to include different types of technology: agricultural, military, civilian, social. Medical technologies are discussed in chapter 8.

For other definitions of the subject matter, see UNESCO, *Science Policy Research and Teaching Units* (1971); Ina Spiegel-Rösing (1977); and Ezra D. Heitowit, *Science, Technology, and Society: A Survey and Analysis of Academic Activities in the U.S.* (1977).

B. The Field

The subject of science and technology policies covers a wide range of topics, which, in turn, are studied by a number of relatively self-contained groups. To borrow a term from Richard R. Nelson and S. G. Winter (1977), who used it in referring to the literature on industrial innovation, the literature is "balkanized"; in other words, "we cannot, in general, bring together several different bodies of analysis to focus on any one question or tie together the various pieces to achieve an integrated broader perspective" (p. 47).

Perhaps most influential among these groups are the economists who have been concerned with the economic antecedents and consequences of technological innovation (see Christopher Freeman, 1977). We will touch on these aspects of the field only peripherally.

Another influential group consists of political scientists for whom the term "science policy studies" has tended to be reserved. Political scientists have studied, more than anything else, the nature and determination of policies for science and technology, although to a lesser extent they have been concerned with the effects of science and technology upon other social institutions. This group has focused mostly on basic science and civilian and military technology. The International Council of Science Policy Studies defined the field in 1971 as "the systematic investigation of scientific and technological activities and their function within society. In particular, they are concerned with policy making in scientific and technological fields, and with the interrelationship between policymaking, cultural values, and societal goals" (see Eugene B. Skolnikoff, 1973, p. 89). The council, which was created in 1971 as a section of the Division of History of the International Union for the History and Philosophy of Science, was instrumental in the preparation of the most extensive review of the science policy studies literature to date: Ina Spiegel-Rösing and Derek de Solla Price, eds., *Science, Technology and Society* (1977).

There are other groups of specialists who examine various types of technologies: agricultural, military, civilian, social, and medical. These groups are concerned with policies for the development of these technologies. Other groups of specialists are concerned with evaluating and assessing technologies or policies for the control of technologies. For example, for futures research, see Wayne I. Boucher (1977); for policy analysis, see Duncan MacRae, Jr. (1974) and Martin Trow (1973).

Science administrators and government bureaucrats, as well as eminent scientists, have written extensively about science policy issues. In addition, UNESCO and the Organization for European Cooperation and Development (OECD) have produced numerous analytical studies of science policy. UNESCO's Science Policy Division has devoted its attention largely to science policy in the Third World. OECD's Directorate for Scientific Affairs has produced a series of reviews in which the science policy organizations of particular countries are described; they are also evaluated by eminent representatives of government, industry, and the universities in other countries. For a description of this process, see Alexander King (1974). The OECD directorate has also produced reports dealing with specific issues, such as technological innovation and new directions for science policy. The Development Centre at OECD has produced studies of science and technology policies for the Third World. The Science Council of Canada, the Organization of American States, and the National Science Board of the United States National Science Foundation have also produced studies of various aspects of science policy.

Jürgen Schmandt (1978) claims that other significant bodies of research have remained virtually invisible to academic researchers in the field; he has in mind studies of the potential effects on society of automation, nuclear weapons, environmental pollution, and critical shortages of energy and natural resources.

Research specialties can be defined either in intellectual terms or in social terms—in terms of subject matter or in terms of definitions used by particular research communities. Using a subject matter definition, as we do here, emphasizes the heterogeneity of the field. Researchers come from such varied backgrounds—in terms of theoretical approaches, methodologies, and institutional affiliations—that it is unlikely they can or do communicate meaningfully with one another, either in person or on paper.

C. Approaches

A number of approaches to the study of science policy can be identified. There are historical and case studies, written for the most part by political scientists, which generally view science and technology either as controlled by bureaucratic agencies (the public administration approach) or as an interest group pushing its programs as do representatives of other social sectors. Although these approaches are well developed, their limitations are beginning to be evident. Bruce Smith (1977) comments:

> The study of the relations of government with the scientific and technological communities is a victim of its early partial success. . . . There has been no dearth of excellent specialized studies. . . . But the whole has been less than the sum of its parts. The particular studies taken together have not constituted a coherent field of study, with well-defined problems. (p. 264; see also Harvey Sapolsky, 1975; and Karl Kreilkamp, 1973).

Another approach that is becoming increasingly important utilizes quantitative indicators of scientific growth and development; this is the so-called "science indicators" approach. It received a powerful stimulus in work done by Derek de Solla Price in the early 1960s. See, for example, Price (1965). In recent years it has attracted the interest of sociologists and historians of science as well as political scientists. See National Science Board: *Science Indicators,* 1972; *Science Indicators,* 1974; *Science Indicators,* 1976; also Yehuda Elkana et al., eds., *Toward a Metric of Science* (1978); and Yakov M. Rabkin (1976a, 1976b).

Two sociological approaches have been applied to the relationships between government and science—network analysis and interorganizational analysis—and will probably become more important in the future. See Nicholas Mullins (1976) and Robert L. Hall (1972). Network analysis is a set of techniques for delineating the structure of relationships in social groups. Interorganizational analysis focuses on the nature and significance of links between organizations (see William Evan, ed., *Interorganizational Relations,* 1976; for another approach, see Stuart S. Blume, *Toward a Political Sociology of Science,* 1974).

Science policy studies can be seen as one aspect of policy studies in general; see Stuart S. Nagel, ed., *Policy Studies Review Annual,* volume 1 (1977). The dividing line between the two is not easy to draw, as Sapolsky (1975) points out. (For an inventory of findings in the science policy studies literature, see Susan Hadden, 1977.) Duncan MacRae (1974) describes the policy analysis approach as including decision criteria, such as cost-benefit analysis; statistical decision theory; social indicators, evaluation research and the measurement of desired outcomes; construction of models linking policy parameters to outcomes; and organizational politics.

For listings of science policy programs in universities and research institutes, see the UNESCO survey *Science Policy Research and Teaching Units* (1971), and Ezra Heitowit, *Science, Technology, and Society: A Survey and Analysis of Academic Activities in the U.S.* (1977).

Some reviews of national literatures on science policy studies—often included in more general reviews of science studies—are to be found in G. S. Aurora, *Administration of Science and Technology in India: A Trend Report* (1971); Linda L. Lubrano, *Soviet Sociology of Science* (1976); Robert K. Merton and Jerry Gaston, eds., *Sociology of Science in Europe* (1977); E. M. Mirsky, "Science Studies in the U.S.S.R." (1972); Ina Spiegel-Rösing, "Science Policy Studies in a Political Context: the Conceptual and Institutional Development of Science Policy Studies in the German Democratic Republic" (1973); and Bohdan Walentynowicz, "The Science of Science in Poland: Present State and Prospects of Development" (1975).

I. THE NATURE AND DETERMINATION OF POLICIES FOR SCIENCE AND TECHNOLOGY

A. The Nature of Science and Technology Policies

1. Allocation of Resources for Scientific and Technological Research. In a very crude sense, the output of science policy is money—the allocation of funds to different kinds of research activities. It is therefore appropriate to begin with some statistics concerning the relative amounts of funds allocated to each type of research, as well as some indication of the way these allocations have changed over time. Before that, the different types must be defined.

Three types of research are generally discussed in the literature on science policy: basic research, applied research, and development. Precise definitions of these categories are a subject of controversy, and the distinction between basic and applied research is particularly difficult. Useful discussions of these terms can be found in Harvey Brooks (1967); in Derek de Solla Price (1976, 1977); and in an OECD study (1970).

Although it is necessary to define these categories as if they represented discrete entities, it is probably more accurate to think of them as points on a continuum, as if there were no sharp dividing line between them. Brooks defines basic research as being influenced "almost entirely by the conceptual structure of the subject rather than by the ultimate utility of the results . . . even though the general subject may relate to possible applications" (1967, p. 23). Applied research is influenced by the ultimate utility of the results, but it may also have an important influence on theory.

Applied research is sometimes closely related to developments in basic science, sometimes not. (See Michael Fores, 1977; and Francisco R. Sagasti, 1977.) Whether related to basic science or not—though particularly in the latter case—applied research is often referred to as technology. If related to science, it is sometimes called "high technology"; if not, it is called "low technology." The word "technology" tends to be used especially when the outcome is a technique or a product. We will use the word here in a very general sense, to cover all research activities other than basic research and development.

Development is an important part of the R&D budget. The variety of activities that are involved is indicated in the following description:

> Development begins where research ends, that is, where a model or invention has been proved technically feasible and has been designated for commercial (or military) application. Development then consists in bringing the successful invention to the point where it is ready for the market; it is the stage where prototypes are evolved, where technical problems of adjustment to "scale-up" are worked out and "bugs" are eliminated; in short, it is the pilot-plant stage where, largely through a process of trial and error, research findings are translated into products and processes ready for commercial or military application (D. Hamberg, 1966, p. 11).

The institutional location in which research is conducted is an important factor in understanding science policy and the research process. Basic research is generally conducted in the university; applied research and development, generally in industrial or governmental laboratories. A fairly recent and important development is the "mission-oriented" laboratory (see Brooks, 1967, pp. 44–51). Examples of such laboratories are: federal civil service laboratories in the U.S. Department of Defense and the National Aeronautics and Space Administration; federal contract research centers operated by universities and industrial companies; and multipurpose research institutions such as the Stanford Research Institute or the A. D. Little Company.

In the following discussion, some general trends in the allocation of R&D funds are noted. Useful sources for the United States include: Willis Shapley, *Research and Development in the Federal Budget,* FY 1977 (1976); and *Funds for Research, Development, and Other Scientific Activities,* published annually by the National Science Foundation. Shapley's study discusses the United States federal research budget generally, not just for the one year.

Certain characteristics of the American research budget have remained constant throughout the post-World War II period. The majority of funds for research has always been allocated to development, with applied research the second most important category, and basic research rarely accounting for more than fifteen percent. In the 1977 budget, basic research accounted for eleven percent of the total budget (Shapley, 1976). Derek Price (1977) has argued that these statistics are an artifact of the definitions used for basic and applied research.

For the first two decades of the postwar period—and particularly during the decade 1957–1967—the amount of funds allocated for R&D by the United States federal government increased by as much as ten to twenty percent per year. After 1967, the absolute value of annual allocations continued to increase, but at a lower rate; if the decline in value of the American dollar due to inflation is taken into consideration, R&D budgets have actually been decreasing. For example, in 1967, total federal funds for R&D were $17.1 billion; by 1977 the budget had risen to $23.5 billion, but in constant 1967 dollars this was worth only $13.1 billion (Shapley, 1976).

Another generalization that has held throughout the period concerns the relative amounts of money allocated to different areas. For example, funds for research on military technology (defense) represented fifty-one percent of the total R&D budget in 1977. Among the different types of science, the natural sciences have generally been funded more heavily than the social sciences; and, until recently, the physical sciences have been more heavily funded than the biological sciences. Social and agricultural technologies are less heavily funded than engineering and health technologies.

Similar analyses cannot be provided for all the industrially advanced countries (for information on the allocation of R&D funds in some of these countries, see OECD, *Patterns and Resources Devoted to Research and Development in the OECD Area, 1963–1971,* 1974), but a study of West German science policy since the early 1960s by Otto Keck (1976) is instructive. Keck compares the German experience with that of the United States, the United Kingdom, France, the Netherlands, and Japan. In all of these countries, growth in government R&D expenditures was slower in the late 1960s than in the early 1960s. As we have seen, R&D expenditures in real terms in the United States declined rather sharply after 1966. West Germany maintained the steadiest and most rapid growth in the group, but all the European countries in the study increased the proportion of GNP devoted to R&D during this period. While reducing its expenditures, the United States still spends roughly five times as much as these countries—with the exception of the Netherlands, where R&D expenditures are considerably lower than all the rest.

There are also interesting differences among these countries in the priorities given to specific types of research. Military R&D was given a higher priority than any other area in the United States, the United Kingdom, and France. In West Germany, Japan, and the Netherlands, the "general advancement of science" received the largest proportion of R&D funds, while military R&D received the smallest. On the other hand, economic and social services had lower priority in West Germany than in any of the other countries. Agriculture has also had low priority in West Germany.

Keck's analysis shows that industrially advanced countries have utilized somewhat different priorities in their allocations of R&D expenditures. Defense and the general advancement of science played the most important roles in the six countries, but with different emphases in different countries. (For other comparative analyses of R&D funding in industrial nations see Christopher Freeman et al., 1971; K. Pavitt and W. Walker, 1976; and S. Encel, 1971.)

2. Rationales for the Allocation of Resources in R&D Budgets.

Science and technology policies are relatively recent phenomena. Although their antecedents can be discerned before World War II (see A. Hunter Dupree, *Science in the Federal Government,* 1957; and Daniel J. Kevles, *The Physicists,* 1978), it is only after, and partly as a result of, experiences during that period that such policies became a recognizable component of government policies in industrialized countries. For a review of this history, see, for example, Don E. Kash (1972); Alexander King, *Science and Policy: The International Stimulus,* 1974; and Jean-Jacques Salomon (1977).

Europe had a stronger prewar tradition of government patronage of scientific research (see Salomon, 1977; and Joseph Haberer, *Politics and the Community of Science,* 1969). Nevertheless, modern science policy begins in the United States in the postwar period. Kash, reviewing this history (1972), says that in the United States the success of particular projects conducted by academic researchers during World War II—such as the atomic bomb and radar development projects—produced favorable attitudes on the part of both the government and universities toward continued research support. With the advice of a group of statesmen scientists, a research system evolved in which, initially, a great deal of autonomy was given to the scientific community to set its own policies for basic research.

In both Europe and the United States, a complex set of organizational structures evolved in which priorities were set and funds administered. In Europe, as Alexander King (1974) shows, these developments occurred later than in the United States, spurred by Sputnik in the late 1950s and by fears of American technological superiority in the 1960s.

Salomon (1977) argues that the rationales for the support of science and technology have been similar in Western countries, although Keck's analysis (1976), reviewed above, does not entirely support this thesis. Salomon identifies three distinct phases in research support that seem to apply most accurately to the United States: the period 1945–1955, when science and technology were funded for their own sake and members of the scientific community were given considerable autonomy in the selection of goals for the system; a period of pragmatism (1956–1967), when support for research was motivated by military objectives and economic payoffs; and a period of reassessment, which began approximately in 1967.

Since 1967, attitudes of American legislators toward science and technology have become increasingly critical, perhaps owing to shifts in national goals and priorities that have continued in the 1970s. (The quite different attitude of the general public will be discussed in section III.) Each sector has been subjected to criticism (Kash, 1972). Support for military research has been attacked on the grounds that the need for it is exaggerated by industrial corporations in liaison with the defense sector of the government—the so-called "military industrial complex." Civilian technology has been criticized as being misdirected, as no longer relevant to society's needs. Basic research has lost support in relation to applied research, partly because of changing perceptions of the contribution it can make to technological development.

In general, research and development has been viewed increasingly as having negative side effects—e.g., environmental pollution. Paradoxically, however, it is still seen as providing solutions to many national problems. There has been widespread discussion of the need to reorient scientific resources, allocating more of them toward the alleviation of social problems.

B.   Determination of Science and Technology Policies

1.   **Organization of High-Level Policymaking Bodies.**   One of the central problems in the analysis of government science policies is that of the organizational structures best suited for the administration of science and technology policies. Analysts have identified as many as four models. One study (Government of Canada, Senate Special Committee on Science Policy, 1973) lists them as: (1) the *pluralistic* model, in which financial resources are assigned to different, autonomous sectors of the government—e.g., defense, agriculture, health—with each sector making independent decisions concerning the allocation of resources; (2) the *coordination* model, in which an advisory agency coordinates the individual science policies; (3) the *cen-*

tralized model, in which a strong central agency controls science activities in all sectors; and (4) the *concerted action* model, in which the individual sectors develop their own science policies but a controlling agency provides strong and effective coordination.

There are dangers in both pluralism and centralization: pluralism can lead to a lack of coordination, overcentralization to neglect of the needs of specific sectors. The Brooks Report (OECD, 1971b) calls for a blend of the two approaches. It also points out, however, that the sectoral approach works best in a period of increasing resources and consensus about social goals, whereas a centralized approach is more appropriate when the opposite conditions prevail. For a detailed analysis of the implications for science and technology policies, see Joseph Ben-David (1977).

A useful device for analyzing what is known about science policy is to distinguish three levels: the *strategic* level, defined as the level "where the over-all national problems and goals and the linkage between these goals and their research needs are discussed and general resource priorities determined" (Alexander King, *Science and Policy: The International Stimulus,* 1974, p. 59); the *research planning* level, where specialized bodies recommend broad programs of research and pass on specific projects; and the *implementation* level, where individual projects are carried out in a wide variety of institutional settings.

A major problem in both the formulation and the implementation of science policies is that of articulation among these three levels. High-level policymaking is, of necessity, somewhat remote from the research planning level, which is extremely varied and diverse. This remoteness may be a factor in the rather abrupt shifts in policy that sometimes occur at high levels, reflecting the difficulties in assessing accurately and controlling what is taking place at lower levels. Such difficulties also show up in the frequency with which high-level committees are reorganized and replaced.

American science policy is determined at many different levels of government and by many different organizations in different sectors of society. W. Henry Lambright, in *Governing Science and Technology* (1976), stresses "the fragmentation of 'the' science and technology function. . . . There is simply no one dominant center of decision-making in science and technology policy in the federal government. . . . How is such a disparate set of organizations, such an amorphous policy area, to be approached, much less understood?" (pp. 7–9).

At the highest levels, American science policy is determined by a complex political bargaining process involving a variety of organizations and committees. Congress and the Office of Management and Budget play important roles. The President's Scientific Adviser, the Office of Science and Technology, and the President's Science Advisory Committee (1957–1973) also have played important roles in this process (Katz, 1978). In 1976, the role of science adviser was revived along with an Office of Science and Tech-

nology Policy (Ronald Brickman and Arie Rip, 1977). The National
Academy of Sciences, through its Committee on Science and Public Policy
(COSPUP), and the American Association for the Advancement of Science—
in part through its journal, *Science,* which airs science policy issues—have
also had some impact on decision making (see Philip M. Boffey, *The Brain
Bank of America,* 1975).

In general, the performance of high-level scientific advisory committees
in various forms has satisfied neither the scientific community—whose views
the scientists are supposed to represent—nor the political leaders who rely
on their advice. This failure to please either group is due partly to the
inherent difficulty of the tasks that have been assigned to these committees.
As Eugene B. Skolnikoff points out in *Science, Technology, and American
Foreign Policy* (1967, p. 242), the problems assigned to such committees are
not the kind that scientific method alone can solve. Technical expertise
rapidly shades off into policy judgments—a fact that can be disconcerting
to both the scientists and their critics. This overlap of technical and policy
issues is particularly true in matters of defense policy, where scientific
advisers have found themselves making judgments that were not based
entirely on technical competence (see Sanford A. Lakoff, 1977, p. 376).

Members of these committees have also been accused of being an elite,
rather than of being truly representative of the scientific community. Stuart
Blume (*Toward a Political Sociology of Science,* 1974, pp. 198–201) shows
that, in both the United States and Great Britain, members of such committees have tended to be drawn mainly from prestigious universities and
prestigious honorary academies. Particularly in the first decades of postwar
science policy, they were drawn disproportionately from the physical sciences. As Blume points out, the combination of elite background and
political acceptability tends not only to make advisers unrepresentative of
the scientific community but to encourage politicization of the scientific
community.

Little is known about the way advisory committees have functioned
in practice, or about the political interplay between them and other
groups—legislative bodies, government agencies, scientific academies, and
professional associations. Our lack of knowledge in this area, in turn, makes
it difficult to assess how much influence advisory committees have actually
had upon the policymaking process at the highest levels. Ronald Brickman
and Arie Rip (1977) identify six areas where these bodies have played
important roles: "the promotion of science and technology as areas of
governmental support, the organization of scientific and science-policy institutions, the evaluation of scientific and technological project proposals, the
monitoring of the scientific and technological input into general policy
deliberations, forecasting and early warning of impending problems and
crises, and overview of departmental R&D in the sense of arbitrating inter-
departmental conflicts, setting general priorities and identifying lacunae"
(p. 22).

Specifically, the President's Scientific Advisory Committee is credited with the following accomplishments:

> The committee, often in liaison with the Special Assistant and his staff, played an important and sometimes crucial role: the creation of several agencies and administrative offices in the R&D sector, including the National Aeronautics and Space Administration and a high-level R&D directorate in the Department of Defense, the development and, in some cases, curtailment of major technological programs in the civil and military sectors, the control of the use of pesticides, the effective use of the sea and the 1969 ban on biological warfare (Brickman and Rip, 1977, p. 4; see also David Z. Beckler, 1974).

As far as policy for science is concerned, Don K. Price argued that, unlike others, the "scientific estate" (1965)—by which he meant basic science alone, since he defined this estate as being concerned solely with truth for its own sake—was able to obtain support from the American government without having to justify its needs. To some extent this interpretation was correct for the early decades of postwar American science policy. By the mid-1960s the situation had changed considerably. Peter Weingart (1970) claims that the scientific statesmen lost their influence by refusing to set priorities and by supporting policies that led both to the concentration of spending at prestigious universities and to the disproportionate support of research in the physical sciences. Since 1965, Congress has played an increasing role in the formulation of policy for science, insisting that criteria for the allocation of funds include needs external to science.

The United Kingdom has moved from a high degree of centralization—dating from the formation of the Department of Scientific and Industrial Research in 1916—toward increasing decentralization. The culminating point in decentralization came with Lord Rothschild's report (Central Policy Review Staff, 1971), which recommended decentralized agencies, a weak coordinating agency, and a Chief Scientific Adviser.

As a result of difficulties with respect to economic, educational, and defense policies, which were seen as being exacerbated by existing science policies, a series of innovations in the science policy system was introduced in 1965 (see Norman J. Vig, 1968; also P. J. Gummett and G. L. Price, 1977). The new system required the selection of priorities among competing projects and disciplines; according to Vig, the old scientific advisory machinery—the Advisory Council on Scientific Policy—had been unable to cope with these problems (p. 158).

This council was replaced by a Council for Scientific Policy, which also lacked the power to cope with similar issues. It in turn was replaced by a more representative committee, but one that still lacked strong powers. A Parliamentary committee on science and technology had recommended the creation of a much stronger body—to be chaired by a new Minister for

Research and Development with Cabinet status—but this arrangement was rejected by the government. It would have moved top-level scientific advisers closer to the general policymaking process. It would also have had the effect of exposing these decisions to the public view and increasing democratic control over them (see S. W. Benn, 1972; and Norman Vig, 1975).

In recent decades, the French system has developed along the lines of the *concerted-action* model. While different sectors develop their own R&D programs, centralized control is exercised by three organizations: the General Delegation on Scientific and Technical Research (DGRST); a Scientific Advisory Committee (CCRST); and an Interministerial Committee on Scientific and Technical Research. It appears that the power of the CCRST—which has been called the "committee of wise men"—to exercise a decisive influence upon the direction of French science policy has gradually diminished over the years. Robert Gilpin, in his study of French science policy in the late 'fifties and early 'sixties (1968), says of the committee: "From these natural and social scientists came most of the proposals for the reform of French scientific institutions that were implemented in the last months of the Fourth Republic and the early years of the Fifth" (p. 196). In recent years their influence has been less decisive, in part because of the increasing influence of the secretariat for the coordination of R&D—the DGRST (see Brickman and Rip, 1977).

Brickman and Rip (1977) point out that high-level science councils in the United States, France, and the Netherlands have developed along similar lines: an early period of considerable influence, followed by a period of decline in the late 'sixties and early 'seventies, with a resurgence in the mid-1970s. They argue that "the evolution of these councils can be explained in part by trends and events which are common to all three countries," and in part by "particularistic factors" that are "unique to the political and cultural situations of each country" (p. 6). Brickman and Rip suggest that the development of specialized R&D agencies made these committees less essential to the science policymaking process. Their later resurgence can be explained in part as a response to the need for "a counterweight to the regular bureaucracy" (see also G. Bruce Doern, *Science and Politics in Canada,* 1972).

A rarely used system would permit scientific decision makers to participate in the political decision-making process. The German Science Council consists of scientists and politicians, plus representatives of the federal government and of the eleven German states. Unlike similar committees in other countries, it is not limited to specific topics; instead, it can move from one topic to another, concentrating its attention upon specific issues for periods of time sufficient to make an impact. The recommendations that emerge represent an agreement between scientists and politicians. The political agreements concerning science policy are worked out there and have considerable weight as a result (see Gilbert Caty, 1972).

The science-policy advisory body representing the greatest degree of democratization in industrial nations is Japan's Science Council. It consists of 210 scientists selected by members of research institutes and university faculties. The council is divided into seven sections representing different disciplines (see T. Dixon Long, 1968, 1969). The council unfortunately does not exercise much influence upon the decisions that the government ultimately makes. Instead, it has assumed the role of critic of the administration—which really controls science policy.

It can be argued that scientist-advisers have a choice among three possible roles: critic, adviser working within the system by exercising informal influence, and politician with formal executive powers. The case of Japan's Science Council suggests that scientist-advisers who assume a critical role are likely to be cut off from effective communication with the administration. One can speculate that this system could be effective if real participation in political decision making were allowed—perhaps through the use of a model that combined elements of both the Japanese and German science councils.

Our knowledge of science policymaking at this high level is very incomplete. The conclusions of Gummett and Price concerning high-level science policymaking in Britain apply to other countries as well:

> The questions regarding the provision of government with advice which is timely and authoritative, the establishment of a balance between the need for central coordination with the traditional emphasis upon the autonomy of departments, and the assurance that scientific research would be attentive to social problems while at the same time the conditions of the growth of science would be safeguarded, remain as open today as three decades ago (1977, p. 143).

For a complete list of policymaking bodies and committees for science and technology as of June 1974 in the United Kingdom, France, Germany, and the Netherlands, see K. Pavitt and W. Walker (1976). For case studies of science policies of several European countries, see T. Dixon Long and Christopher Wright, eds., *Science Policies of Industrial Nations* (1975). For a comparison of the organization of science-policy bodies in several European countries, see G. Caty et al. (1972, 1973).

2. High-Level Policymaking: Some Major Decisions. Further insight into the process of high-level science policymaking might be obtained from an analysis of specific decisions that have accounted for major allocations of funds, but so far there have been too few detailed analyses of such cases to permit generalization. Studies of decisions involving huge sums of government money in three countries—the United Kingdom, Canada, and the United States—have all concluded that the decision making process is not wholly rational. See L. Drath et al., "The European Molecular Biology

Organization: A Case Study of Decision-Making in Science Policy" (1975); Bernard Leach, "Decision-Making in Big Science: The Development of the High Voltage Electron Microscope" (1973); Michael Gibbons, "The CERN 300 GeV Accelerator: A Case Study in the Application of the Weinberg Criteria" (1970); G. Bruce Doern, " 'Big Science', Government, and the Scientific Community in Canada: the ING Affair" (1970); and George Eads and Richard R. Nelson, "Governmental Support of Advanced Technology: Power Reactors and Supersonic Transport" (1971).

Drath et al. (1975) conclude that Britain's decision to join an inter-governmental (European) molecular biology organization was an incremental rather than a rational process. Leach (1973) notes that the British government's commitment to finance high-voltage electron microscopes was also made incrementally and that national prestige and competition with other countries were important factors in the ultimately favorable decision. Gibbons, analyzing the British decision to participate in the construction of a 300-GeV accelerator at the international high energy physics laboratory (CERN) near Geneva (1970) shows that the decision would have been more rational had the accelerator project been systematically compared with alternative projects. Gibbons says, "It is not possible . . . to approach a more rational policy for science if each proposal is considered in isolation" (p. 190). Doern (1970) concludes from an analysis of a similar decision-making process in Canada that it was less than rational—due in this case to the absence of "professional institutions in which consensus on matters of science policy can be formed" (p. 375). Finally, Eads and Nelson (1971) argue that decisions by the American government to support the development of nuclear reactors and the supersonic transport were based, not on the perception of a pressing problem or need in the civilian economy, but on pressures from within the government itself. They were made without adequate consideration of the results that would be achieved, and again there appears to have been little attempt to compare the projects with other comparable projects.

More detailed studies are needed of the factors that influence specific decisions. Much discussion of science policymaking is conducted at a level of generality not based on an intimate knowledge of factors entering into the decisions. A good analysis of an international high-level policy decision is Hilary Rose's "The Rejection of the WHO Research Centre" (1967).

C. Science and Technology Policies in Specific Sectors

Science policy on the research planning level can be analyzed in various ways. One approach is to examine the agencies that allocate funds for different types of research. W. Henry Lambright calls these organizations "technoscience agencies" and describes them as follows: "They make the day-to-day decisions, year-in, year-out, that determine who gets what, when and how in federal research and development. They play a role not only in the execution of policy but also in its formulation" (*Governing Science and Technology,* 1976). Among these agencies Lambright includes the Department of Defense (DOD), the National Aeronautics and Space Administration (NASA), the National Science Foundation (NSF), and the National Institutes of Health (NIH). As Lambright admits, the fragmentation of scientific activity is such that some technoscience agencies deal with policy for a single type of science or technology, while others handle several different types.

Another approach frequently used organizes the discussion in terms of the location where the research is performed. However, since it is frequently the case that different kinds of research are conducted in the same type of institutional setting, this can be misleading.

A third approach, and the one followed here, is to organize a discussion of government policies around the concept of a research and development system—i.e., the whole spectrum of basic and applied research associated with the production of a specific type of technology: agricultural, military, civilian, social, et cetera. These systems can also be viewed as technological delivery systems (Wenk and Kuehn, 1977). The interrelationships between the different types of organizations—governmental, academic, and industrial—that contribute to the development of science and technology are of central interest. Research and development systems for space technology, medical technology, and service-sector technology will not be discussed here. Medical technology is discussed in chapter 8; for further discussion of government policies for medical research, see Heather J. Nicholson (1977) and J. S. Murtaugh (1973). For a review of the literature on technology for the service sector, see chapter 4 of Lambright's *Governing Science and Technology* (1976); Dorothy Nelkin (1971a, 1977d); and Alvin M. Weinberg (1967). The academic research and development system, which primarily produces basic research, will also be discussed here.

For each research and development system discussed, we will examine three questions: (1) How are research and development activities organized? (2) What types of policies have been used and through what organizational arrangements have they been formulated? And (3) how do these policies affect the development of research in the field?

1. Agricultural Research and Development

i. *Organization of Agricultural R&D:* Agricultural research and development is conducted in schools, experiment stations, and extension services that are isolated from other research systems—particularly basic research in the biological sciences. There appear to be few social and intellectual links between this system and the others. As André and Jean Mayer describe it: "Intellectually and institutionally, agriculture has been and remains an island, . . . a vast, wealthy, powerful island, an island empire, if you will, but an island nevertheless. . . . The present isolation of agriculture in American academic life is a tragedy" (1974, pp. 87, 83). The agricultural research and development system has, consequently, produced its own ancillary disciplines parallel to those in the arts and sciences—e.g., agricultural demography, agricultural economics, rural sociology. Richard Levins (1973, p. 523) speaks of "two cultures in biology with very little productive interchange." The system has also developed its own scientific societies and publications. Controlling and supporting this R&D system is a complex political system, including executive departments at the federal and state levels and legislative committees in Congress. This political system also operates relatively independently of other parts of the federal government. The agricultural research and development system has a long history, beginning with the establishment of agricultural experiment stations and the emergence of agricultural research in land-grant colleges at the end of the nineteenth century. (See Charles Rosenberg, *No Other Gods: On Science and American Thought,* 1976.)

Agricultural research is also carried out within the Department of Agriculture itself—now largely within the Agricultural Research Service. (See Ernest G. Moore, *The Agricultural Research Service,* 1967.) The Cooperative State Research Service and the Forest Service support research mostly in academic settings. The agricultural research and development system is highly integrated. It has involved close collaboration between the Agricultural Research Service and state agricultural experiment stations (Moore, 1967, p. 45). Heather Nicholson (1977) says: "Most basic research in agriculture is performed in a setting which reinforces the links between government and the science it supports" (p. 42).

An unusual characteristic of this research and development system is the close relationship between the researchers, both basic and applied, and their clients. Nicholson describes these researchers as "having a close sense of identity with farmers on whose practical problems they worked" (p. 59). As Everett Rogers et al. show (1976), the system was set up in such a way as to maximize effective communication between researcher and client. Information moves in both directions in the system: from agricultural researchers to farmers, and back to researchers in the form of requests for specific types of information. Two key individuals in the process are the extension specialist and the county agent, with the former transmitting results from the researchers to the county agents and the latter transmitting them to the farmers. The clients of the system are often trained in the agricultural sciences and have confidence in their own ability to define research problems and to evaluate their significance (see Robert Merrill, 1962).

All of these factors have led to a blurring of the distinction, in the agriculture sector, between basic and applied research. Both types are supported on the basis of the likelihood of contributions to the solution of practical problems (Nicholson, 1977, p. 47). On the other hand, over time, there has been a gradual increase in the proportion of basic research performed in government and academic settings. Moore notes "a strong trend toward basic research that developed in the 'fifties in both the A.R.S. and in the state experiment stations" (1967, p. 47). Surprisingly, a third of the R&D budget in these agencies is devoted to basic research. Shaul Katz and Joseph Ben-David, in their analysis of the Israeli agricultural R&D system (1975), note a similar trend toward basic research. They attribute it to the establishment of a university faculty of agriculture and to declining motivation among researchers to work on practical agricultural projects.

In recent years, there have been important shifts in the organization of agricultural research and development. For example, there has been an increasing separation of applied research from basic research, the former being performed in industry and the latter under public support. Yujiro Hayami and Vernon Ruttan (*Agricultural Development: An International Perspective*, 1971) suggest that resources in the public sector need to be shifted toward basic research to balance the increase in applied research by industry. They are also concerned about the need for new types of institutional linkages and information channels. These links were less necessary when extension, development, applied research, and basic research were combined in a single administrative system.

ii. *Policy for Agricultural R&D:* What sort of policy has been used in this system to influence the output of basic and applied research? Nicholson (1977) argues that policy for basic research in American agriculture has emphasized accountability rather than autonomy. Basic research is defended, not for its own sake but in terms of practical benefits. In a system dominated by the principle of accountability, decisions about research support are made by full-time administrators rather than by practicing scientists. According to Nicholson:

> Final authority rests with public officials of the Department of Agriculture. Support for research is systematically planned at the upper levels by administrators of research and representatives of the clients or users. Both allocation and planning operate in categories defined by practical problems or commodities, rather than by scientific discipline (p. 47).

By contrast, in a system directed according to a policy of autonomy, basic scientists would make these kinds of decisions. It is significant that, until 1977, the research committees of the Agricultural Research Service that reviewed agricultural research and made recommendations about projects were composed mainly of laymen (Moore, 1967, pp. 83–89). Only the Committee on Agricultural Science was composed entirely of scientists.

The primary criterion for awarding funds is relevance to practical problems; scientific excellence is a secondary consideration (Nicholson, 1977, p. 42).

According to Nicholson, since 1940 the amount of money allocated to R&D for agricultural research relative to health research has been reversed. In 1940 the U.S. government was spending $25 million for agricultural R&D—more than eight times the amount spent on health. By 1975, R&D appropriations for health were nearly six times those for agriculture. On the other hand, as Nicholson shows, the budget for agricultural research has increased slowly and steadily; the A.R.S. and C.R.S. have fared well in Congress compared with the average federal bureau.

iii. *Effects of Policy on Agricultural R&D:* Economic studies of the benefits of investment in agricultural research—e.g., on particular commodities such as hybrid corn—have generally been favorable (see Yujiro Hayami and Vernon Ruttan, 1971, p. 150). But in recent years there has been increasing criticism of the research programs of the U.S. Department of Agriculture. Mayer and Mayer (1974) are critical of both basic and applied research in agriculture:

> For lack of effective outside criticism, a great deal of agricultural research has proceeded on assumptions which are very much open to question. . . . The lack of criticism of agricultural policy has been no less serious in the political and economic spheres. In Congress, only politicians identified with a farming interest have been willing to serve on the Agricultural Committees and subcommittees. This self-selection has tended to foster large-scale government programs designed to benefit narrow classes of producers without regard for consumers or even an overall production policy (p. 91).

Other critics of government R&D for agriculture have focused their attention on the quality of the basic research it supports. A study conducted by eminent academic scientists for the National Academy of Sciences concluded that basic research supported by the U.S.D.A. was of low quality and suffered from "a shocking lack of intellectual leadership" (see Nicholas Wade, 1973a, 1973b). Although the committee's criticisms have been challenged, it seems likely that basic research has not thrived in the agricultural R&D system in spite of relatively large allocations to it.

If the principle of accountability is used to evaluate research, it is important to distinguish between short-term and long-term benefits (see Merrill, 1962, p. 432). Wade suggests that short-term benefits have been stressed throughout the U.S.D.A.'s R&D system, with decision makers "tending to look for short-term results of immediate value" to specific crops (1973a, p. 392). There have been accusations recently that the entire system serves the needs of "agribusiness" corporations rather than of small farmers

(Jim Hightower, 1972). Don F. Hadwiger (1976: 157) argues that "an enduring coalition of commercial producers . . . imposed its economic interests upon practically every activity of the U.S.D.A." He shows that the percentage of U.S.D.A. spending allocated for research declined from twenty-eight percent in 1924 to three percent in 1975.

An important change took place with the passage of the Food and Agriculture Act of 1977 (Gary A. Strobel, 1978). The act created a new program of competitive grants for high-priority agricultural research, to be awarded to scientists at all colleges and universities and not just those traditionally associated with the U.S.D.A. Grant applications are reviewed by academic scientists rather than by agricultural administrators in a peer-review system like that used by the National Science Foundation. These changes introduce a policy of autonomy in at least a portion of basic research in agriculture and may help to counterbalance the predominance of "agribusiness" in the agricultural R&D system.

2. Military Research and Development.   Another relatively isolated research and development system is military R&D. Unlike the case of agricultural R&D, however, there are many intellectual links between military and civilian R&D. Both develop from the same knowledge base. As F. A. Long (1971) put it: "Military R&D is embedded in and derivative of civilian science and technology" (p. 279).

i. *Organization of Military R&D:*   Military R&D is conducted largely in organizations that specialize in that type of activity, many of which are run by the Department of Defense. According to Milton Leitenberg (1973, p. 350), DOD has close to one hundred laboratories or installations. Other military research is done in industrial laboratories that specialize in this type of work. At the present time, educational institutions play a small role, primarily in providing basic research.

Since the scope of military R&D activities varies from long-range research and exploratory development to short-range operational systems development, it requires different types of organizational settings. Carl Kaysen (1963) divides industrial firms specializing in this type of work into three groups: 1) the major weapons-systems firms that turn out finished weapons; 2) firms producing large subsystems of weapons but not the complete system, and 3) firms supplying parts and materials for the first two types. The industrial firms that work in this sector must function in ways very different from other industries. Kaysen points out that

> the typical product of this industry has never been seen before, much less produced before, when the order for it is placed. The life of a new product is short; further innovation and consequent obsolescence is continuous. Thus, the part of the total effort of the firm which is devoted to research, development and testing as opposed to production is very high (p. 224).

The military services are the sole customers for the products of this research, and the absence of alternative markets, together with the constant changes in demands made by the military, lead to a high degree of uncertainty in the industry as well as to the inapplicability of ordinary market incentives. Edmonds (1975) comments:

> The demands made by modern weapons on the resources of the arms manufacturers have dictated that they become increasingly specialized and pool resources through mergers or cross-subcontracting to ensure the necessary capital to undertake long, expensive and technically complex programs. The effect has been to see in the military-industrial sector, the emergence of huge defense-oriented corporations with high degrees of weapons specialism and virtual elimination of any semblance of competition between them (p. 155).

Kaysen argues that the style of R&D management used in these industrial companies tends to be inappropriate. He argues in favor of "a greater emphasis on R&D activities aimed at discovery and utilization of new ideas" and less emphasis on "procurement of new weapons as integrated packages" (p. 264). He favors greater separation of these activities, recommending that the former be conducted in organizations better suited to these activities—specifically, in industrial research laboratories that are more like nonprofit institutes than businesses. This separation of functions, he thinks, would lead to greater competition in the R&D process, something he sees as important in raising the level of innovation. (For another point of view, see Leonard Sayles and Margaret Chandler, *Managing Large Systems: Organizations for the Future,* 1971.)

As Long (1971) points out, the secrecy imposed on the work has important implications for the conduct of this kind of research. Secrecy inhibits the flow of information between military and civilian technologists, but the effects upon the level of innovation are unknown. Approximately three percent of research expenditures by DOD are for basic research (see Kaysen, 1963, p. 232; and Harvey Sapolsky, 1977, p. 448).

For a discussion of the European literature on military R&D, see the same article by Sapolsky; also M. Kaldor (1972); and the Stockholm International Peace Research Institute (1973).

ii. *Policy for Military R&D:* Although absolute amounts remain large, the relative share of military R&D in the total R&D budget has gradually declined both in the United States and in other industrial countries (see Sapolsky, 1977, p. 450). In the United States, the proportion declined from around sixty percent in 1955 to about thirty percent in 1970. Since then it has remained at about fifty percent of the federal R&D budget (see Willis Shapley, 1976).

Herbert York and G. A. Greb (1977) provide a detailed picture of the machinery for military science-policy decision making in the postwar period. According to them, military R&D policy was largely determined by an "action-reaction" cycle. A major event, such as Sputnik or the Soviet H-bomb, triggered a reaction by the U.S. military in the form of organizations that survived for a long time after the events that led to their creation. All this contributed to the continuing momentum of the arms race.

York and Greb trace the evolution of committees of academic statesmen-scientists to provide advice and coordination for the system. The process began during World War II; then it was handled by "a relatively small group of men working on a few interlocking boards and committees" (1977, p. 14). Later, as the cold war developed, similar boards and committees emerged, staffed partly by civilian experts and partly by high-ranking military officers. In subsequent years, these boards were continually reorganized and restructured. Without the pressures of a world war, the statesmenscientists were less able to operate effectively in their roles. For example, the Research Development Board, which existed from 1947 to 1953 and was the highest-level R&D management unit in the Pentagon, accomplished far less than its eminent members anticipated. York and Greb comment: "The reasons were the complete lack of any kind of real authority and the extreme decentralization of such responsibilities as the board did have to a very large and complex system of committees, each of which was made up of intelligent and capable but part-time people" (p. 15). They suggest that its primary influence was informal, one of fostering the exchange of information and ideas among those who held real authority.

Since the scientists believed that they could not be effective if they were working directly for the military, a Scientific Advisory Committee was created in 1951 to report directly to one of the President's assistants. In 1957, the committee's status was changed so that it reported directly to the President. The post of Special Assistant to the President for Science and Technology was created at this time. Perhaps because of the personalities of the first two special assistants, James R. Killian and George B. Kistiakowsky, the influence of the office upon military policy was substantial: "During the last years of the Eisenhower administration, Killian, Kistiakowsky and PSAC reviewed virtually every important high technology program in the Department of Defense and many of those of the AEC and the CIA as well. Few programs or ideas that did not meet their approval got very far. They were also the major factor within the government in generating real movement toward both the nuclear test ban in particular and strategic arms control in general" (York and Greb, 1977, p. 24).

After the death of President Kennedy in 1963, the influence of PSAC on military policy and technology declined under Johnson. The committee was abolished by Nixon in 1973. York and Greb see the demise of this organization as a serious loss to military R&D since it was "the only effective 'check and balance' operation in the military R&D system" (p. 26).

During those years, additional boards and committees of scientist-experts were created in the Department of Defense. York and Greb imply that they worked most effectively when PSAC was also influential. Most of the members of these committees knew each other well, a factor that contributed to their effectiveness; information and ideas spread in spite of vertical and horizontal organizational barriers.

Over time, as York and Greb point out, the importance of R&D within the military defense establishment gradually diminished; its share of the total defense budget has substantially declined since 1965. Given the history of the advisory boards sketched above, it is not surprising that there have been complaints in recent years about the absence of a rationale for military science policy (Rodney W. Nichols, 1971). In the late 'sixties, the entire R&D policy of DOD was challenged by Senators Mansfield and Fulbright, and a primary target of their critique was DOD spending on basic research. The outcome was an amendment in 1969—the so-called "Mansfield Amendment"—forbidding the Department of Defense from supporting research unless it has a direct and apparent relationship to a specific military function or operation. This policy was the culmination of a long period of questioning the value of basic military research. (Another good source on the history of military R&D is Adam Yarmolinsky, 1971.)

There is also a policy literature on the introduction of major weapons innovations (see Harvey Sapolsky, 1977; Allison and Morris, 1975; and James Kurth, 1973). These studies have been conducted primarily by political scientists using case study methods. Sapolsky summarizes the findings as follows:

> Their work . . . stresses the dominant role of the military services in determining the rate and direction of weapon innovations. Weapon innovations, in this view, gestate over many years and are often championed by bureaucratic entrepreneurs who rise up from the ranks of weapon designers or the officer corps. As the projects approach the advanced development stage, they are absorbed more fully into the politics of the sponsoring service and their fate is influenced strongly by the service's internal resolution of goals and its bargaining position at the highest levels of government (1977, p. 457).

Kurth (1973) and Karter and Thorson (1972) discuss a number of theories that attempt to account for the development and procurement of particular weapons systems. Each type of theory stresses a different element in the system: (1) strategic theories—national security requirements; (2) technocratic theories—the military R&D system itself; (3) bureaucratic theories—interaction of bureaucratic cliques; (4) democratic theories—the domestic political process; and (5) economic theories—the influence of large defense corporations.

There has been considerable discussion of economic theories—i.e., the extent to which industrial and military organizations act in concert to convince the public that a high level of defense expenditures is in the national interest. (For a series of articles on these issues, see Steven Rosen, ed., *Testing the Theory of the Military-Industrial Complex*, 1973.) According to this thesis, military contracts represent an effective device for counteracting financial difficulties experienced by large firms (see James Kurth, 1973). Rosen finds that empirical research does not support a conspiratorial interpretation of the relationship between the military and their industrial partners; instead, the relationship represents a "subtle interplay of interests and perceptions . . . that competes with other groupings, . . . agricultural, urban and educational" (p. 24). Rosen concludes that while this interest group does not always win, it tends to predominate unless there is a coalition of opposing interests.

iii. *Effects of Policy on Military R&D:* How well has military spending for R&D achieved its goals? This question can be subdivided: 1) Does the system produce military technology efficiently and economically? 2) Does the science policy machinery select appropriate goals for the system? 3) Are its expenditures for basic research useful in this context?

According to the Stockholm International Peace Research Institute (1973, p. 258), the rate of innovation in armaments is faster than in civilian technology. However, J. P. Ruina (1971) has found that the "abortion" rate of major strategic military systems—i.e., systems that have been developed but never deployed—is much higher than in other comparable areas of technology. The reasons range from unforeseen technical difficulties to the fact that the initial projection of military benefit was inaccurate.

It seems clear that the system is both inefficient and inordinately expensive:

> Studies of United States weapon developments, for example, show a persistent pattern of cost overruns, schedule slippages, and performance defects despite a declining index of technological advancement for at least certain types of weapons and much effort at managerial improvements. . . . The experience of other nations, Eastern and Western, though less accessible, appears to be similar (Harvey Sapolsky, 1977, p. 452).

J. A. Stockfisch (*Plowshares into Swords: Managing the American Defense Establishment*, 1973) argues that there tends to be such a strong commitment to specific weapons programs that they become ends in themselves, regardless of the existence of better alternatives.

Whether expenditures for basic research have been beneficial has been a subject of considerable controversy. One of the most famous studies of this problem, Project Hindsight (see Raymond S. Isenson, 1969), found that basic research had played a minimal role in the development of twenty weapons systems. Like others of its kind, this study has been criticized on the grounds that its sample was too selective and the time span too short (Karl Kreilkamp, 1971). The procedure used in Project Hindsight, as in other such studies, was to analyze events preceding a particular innovation and to assign them to categories in terms of the nature of the research— basic or applied—that produced the event. The accuracy of this procedure can be questioned, as well as the underlying assumption that all the events considered were of equal value in the development of the innovation in question.

A different kind of study by the Air Force Office of Scientific Research of science-technology interactions—although qualitative rather than quanti- tative—produced a more favorable evaluation of the role of basic research. (See William J. Price et al., 1969.) These authors point out, in analyzing the impact of "phenomenon-oriented" basic research on military technology, that it is the provision of essential background knowledge that contributes to a better understanding, rather than specific ideas that lead to applications.

One consequence of the 1969 Mansfield amendment mentioned earlier was that DOD ceased to be the federal government's prime sponsor of basic research. Another effect was to reduce the role of academic scientists in military research. Although some writers see this decline as an advantage from an ethical point of view, others (for example, Rodney Nichols, 1971) regret the loss of the special contribution that academic scientists can make. In effect, the amendment served to increase the isolation of the military R&D system from the universities (see also Adam Yarmolinsky, 1971b, and Dorothy Nelkin, 1972).

3. Academic Research

i. *Organization of Academic Research:* In this section we discuss federal science policy for research performed in faculties of arts and sciences in American universities—primarily basic research. Basic research is also performed in other parts of the university, such as the medical schools, engineering schools, and specialized institutes, but the component of applied research is generally higher in those settings (see Robert Merrill, 1962). The social organization of basic research is discussed in chapter 7 of this *Guide*.

ii. *Policy for Academic Research:* Government support for basic re- search has for the most part been oriented toward the basic sciences them- selves rather than toward social needs (see W. Henry Lambright, *Governing Science and Technology*, 1976). The principle of autonomy—rather than accountability—has predominated (see Heather Nicholson, 1977). This has been particularly true in the two agencies, the National Science Foundation and the National Institutes of Health, that are the principal supporters of academic research. An elaborate system of advisory committees has de- veloped wherein university scientists assess applications by their peers and play a major role in the distribution of funds. With some variations, this system is used to allocate funds for basic research in many industrial nations (see Stuart Blume, *Toward a Political Sociology of Science,* 1974; Gilbert Caty, 1972; and Kurt Zierold, 1968).

The advisory system in the United States has been accused of elitism. The advisers tend to be drawn from, and the majority of the funds has been allocated to, the most prestigious universities (Daniel S. Greenberg, *The Politics of Pure Science,* 1967). Congress has, from time to time, demanded that the funds be distributed on a regional basis to offset the heavy concentration of funds in prestigious universities on the East and West coasts. It can be argued, however, that the system allocates funds to those who are most qualified; if the most qualified scientists are located in the most prestigious universities, it should not be surprising that research funds are concentrated there.

A number of studies by sociologists and political scientists have attempted to assess the degree of elitism in the system. Studies of the U.S. Public Health Service—through the National Institutes of Health, a heavy funder in that area—and of the National Science Foundation show a high rate of turnover among members of advisory committees. In the PHS, close to one third of the members have moved from "positions of little power to ones of greater power" (Nicholas C. Mullins, 1972, p. 22). A study of National Science Foundation advisory committees found an overrepresentation of physical scientists and scientists from prestigious schools (Lyle Groeneveld et al., 1975; see also Nicholas Mullins, 1976; and R. F. Lipton, 1971).

Stephen Cole et al. (1977) studied the factors associated with receipt of funds in a sample of applications for NSF funds. Not surprisingly, they found that "of those applicants who obtained their degrees from the highest-ranked graduate departments, 62 percent were awarded grants, compared to 38 percent of those who were graduates from the lowest-ranking departments. Similarly, 74 percent of the applicants currently employed in the highest-ranked departments were funded, compared with 38 percent currently in either unranked departments or non-academic institutions" (p. 38). The data collected in this study show that peer review ratings are the most important factor in the agency's decisions to award grants, whereas correlations between ratings of proposals and the prestige of the scientists' past and present institutions are moderately low. The investigators conclude that the allocation of research funds by NSF is equitable. However, Mitroff and Chubin (1979) question the assumptions of this and similar studies.

Robert Hall (1972) argues that science-oriented agencies, like other federal agencies, serve a distinct clientele. He suggests that a kind of institutional symbiosis develops between a granting agency and a particular institutional sector, and that this leads to "a wide range of mutual dependencies and exchanges of resources" (p. 205). Partly as a result of exchanges of personnel, partly as a result of similarities in goals, these organizations become increasingly alike, for instance, in terms of procedures and standards for evaluating performance. They also become important elements in each other's political support (see also S. P. Strickland, 1972; and Sanford L. Weiner, 1972).

Hall comments: "With so many linkages, there is a massive amount of communication back and forth between the symbionts, resulting in sensitivity and responsiveness to needs of the other and perhaps also in a parochial viewpoint that is insensitive to other institutions" (p. 224). As a consequence, it is difficult to redirect the policies of these agencies, as is clear in attempts to make them more socially responsive (see George D. Greenberg, 1975, and Weiner, 1972).

Certain controversies over the content of policies for academic science have persisted year after year. One is the question of whether or not basic science can be directed. One point of view, strongly supported by Derek Price, is that basic science grows according to its own inherent logic; science policies have relatively little effect. Indeed, Price argues that science policy simply accommodates itself to the needs of science, rather than vice versa (1969).

On the other hand, a team of German sociologists argues that "mature" scientific disciplines are susceptible to external direction (see Gernot Böhme et al., 1976). They argue that scientific disciplines that have developed what they call "closed" or "completed" theories are "finalized." At that point, they no longer generate their own goals. Instead, external imperatives can be used to select among the many alternative possibilities for further development of knowledge.

The vagueness of the concept of "finalization" has been criticized (Ron Johnston, 1976; and John M. D. Symes, 1976). It is difficult, for one thing, to differentiate between a "finalized" discipline and applied or mission-oriented science.

Wolfgang Van den Daele et al. (1977) discuss the results of several case studies of the relationship between research policies and scientific development in Germany. They found that policies that encourage research on specific applied problems did not lead to a decline in basic research. Indeed, policy problems were translated into basic research problems because theoretical knowledge was needed to deal with the practical problems. "Scientists continued to regard disciplinary scientific communities as their primary frame of reference and to evaluate the progress of research within the theoretical and technical framework of their discipline although their research originated from external problems" (Van den Daele et al., 1977).

A related issue in the literature is the question of the contribution basic research makes to applied research and technology. A widely held view, which served for a time as the rationale for funding academic research, was the idea that applied research and technology developed from the application of discoveries in basic science. Attempts to verify this hypothesis have led to conflicting results; these inconsistencies are due largely to the enormous variety within applied science and technology itself. For an extensive review of this literature, see Keith Pavitt and W. Walker, "Government Policies toward Industrial Innovation" (1976). Some studies—e.g., that by J. Langrish et al. (1972)—have found little evidence for the role of basic research in industrial innovation; others have found the opposite—for example, the TRACES study (Illinois Institute of Technology Research Institute, 1968) and its follow-ups (Battelle Columbus Laboratories, 1973, and Samuel Globe et al., 1973).

A recent study of research citations in patent applications for a sample of major technical advances in the United States between 1950 and 1973 found that the contributions of basic research increased during the period (Philip M. Boffey, 1976). In fact, where most of the research cited in patents was formerly done in industry, it is now performed in universities. Such evidence appears to support the distinction between a high technology closely linked to developments in basic science and a low technology that is much less so. The development of low technology is almost always stimulated by economic needs, while high technology is sometimes generated by an important scientific discovery.

Much more information is needed about the relationship between basic science and technology and the ways that different types of technological knowledge develop. For some approaches to this problem, see Thomas Waldhart (1974), A. E. Cawkell (1975), and M. B. Lieberman (1978).

Another controversy concerns the issue of whether government support should be channeled to the institution through institutional grants or to the individual through project grants (see John T. Wilson, 1971; Harold Orlans, 1973; and Gilbert Caty, 1972). The scientist's desire for autonomy in this case is in conflict with the university administration's desire for control over personnel and resources. To date, institutional funding has not played an important role in American academic research (for a discussion of comparable issues in Britain, see Stuart Blume, 1969).

iii. *Effects of Science Policy on Basic Research:* Qualitative assessments of the effects of government science policy on the development of basic research have been numerous, but there have been few quantitative studies. There are enormous difficulties in quantifying the benefits of basic research. From the economic point of view, there is no generally accepted way to measure "the social rate of return" of basic research (Sanford L. Weiner, 1972, p. 229). Nevertheless, attempts have been made. H. H. Fudenberg (1972) attempts to develop economic indicators of the benefits of biomedical research by estimating the saving to society through the discovery of the polio vaccine in the period 1955–1961. The number he arrives at is $1 billion per year, and he interprets such a large saving as an indication of the value of basic research. For a different kind of attempt to measure the impact of basic research policy, see Barry Castro (1968).

Another subject of debate is whether changes in the rates of spending for basic science have influenced the output of basic science in terms of articles published. J. Davidson Frame and Francis Narin (1976) have found that levels of funding are closely associated with numbers of publications in biomedical research (see also National Science Board, *Science Indicators 1974, 1976*). A comprehensive assessment of the state of academic science by Bruce L. R. Smith and Joseph J. Karlesky (1977) argues that decreased government spending—coupled with a decline in the financial position of many universities—is leading to a gradual deterioration in the academic research environment. Evidence alleged includes declining opportunities for younger scientists, inadequately maintained facilities, and increased conservatism in the choice of research topics.

While the dominant principle used by governments in funding basic research is that of the autonomy of the basic scientist, efforts have been made from time to time to redirect basic research toward areas of social needs—in effect, to introduce the principle of accountability (Bernard R. Stein, 1973). Some notable examples in the United States include the National Science Foundation's Research Applied to National Needs (RANN), the NIH Special Virus Cancer Program in the 1960s, and the National Cancer Act of 1971 (see Constance Holden, 1973; and Richard A. Rettig, *Cancer Crusade,* 1977). On the whole, these programs have had limited success. They have been criticized on several grounds: that they are too costly in relation to results; that fundamental knowledge is insufficient to make an applied program worthwhile; and that funds have not been distributed equitably (W. Henry Lambright, *Governing Science and Technology,* 1976, pp. 152–155). The NIH Special Virus Cancer program was eventually discontinued. NSF's RANN has been renamed the Science and Engineering Directorate and drastically restructured (R. Jeffrey Smith, 1977).

Attempts in France to direct academic research toward practical ends have not been notably successful either (Ronald Brickman, 1977a). In France, these efforts have not been confined to specific programs—although these also exist, such as the Concerted Action Programs—but include a widespread effort to substitute contracts for grants in a period of declining expenditures for R&D. Brickman comments:

> The university scientific community feels more and more like a drum upon which any number of national agencies are beating and not always in unison. . . . The university scientific community has progressively lost the ability to steer its own course and has become more sensitive to the changes in objectives and procedures of a variety of national agencies and policymakers (1977a, pp. 133–134; see also Brickman, 1977b).

Lambright concludes that finding the right balance between support for science and responsiveness to social needs is one of the most difficult problems in R&D policy (1976, p. 162).

4. Civilian Research and Development. In this section we discuss civilian technology; by that we mean technological products and processes developed by industrial firms and sold to other firms or to the public at a profit. Besides being different from the sectors discussed so far, it is also distinct from nonprofit technology (technology for the service sector) developed by industrial firms under contract to government agencies and sold to state and local governments or to certain private industries, usually on a nonprofit basis.

i. *Organization of Civilian R&D:* Civilian technology is conducted in, and is largely paid for by, industry. Federal government allocations for industrial research are largely in the area of military and space technologies or atomic energy. In industry, R&D is concentrated in a few industries and in a few of the largest companies in these industries (Harold Orlans, 1973). These large companies have developed interdisciplinary R&D laboratories; Edwin Layton (1977) calls them "a major organizational innovation" in the twentieth century. Among about five thousand such laboratories in the United States, three hundred do the bulk of the research (Layton, p. 212).

In these laboratories, the R&D manager rather than the scientist is the central figure. He defends new ideas and promotes their development in the firm. Resistance from other divisions in the firm can lead to more attention being paid to the solution of short-term problems than to the exploration of more fundamental concerns (see D. Hamberg, 1966). A number of studies indicate that, although these laboratories have not always been the source of major innovations in civilian technology, they have been successful at improving and developing existing innovations. In these cases the real innovators have often been laboratories in small firms where considerable autonomy and freedom from bureaucratic constraints prevail, resembling the environment in which the basic scientist performs his work. However, recent studies show that large firms produced more innovations than small firms in the late 1960s (Philip M. Boffey, 1976). For a review of factors associated with successful innovation, see Keith Pavitt and W. Walker, "Government Policies towards Industrial Innovation" (1976).

ii. *Policy for Civilian R&D:* Civilian technology is not directed by a "technoscience agency" (Robert Hall, 1972, p. 209). Although the U.S. Department of Commerce concerns itself to some extent with industrial policy, it does not fund and set policy for R&D in the industrial sector.

One of the functions suggested for a proposed U.S. Department of Science would be to "nurture the early stages of an infant technology without its having to compete too vigorously with traditional activities of mission-oriented agencies" (Adam Yarmolinsky, 1971b, p. 299), but this is still just an idea.

W. Henry Lambright (*Governing Science and Technology,* 1976) points out that it was unusual even to speak of technology policy before 1970: "Technology policy has emerged, through no great edict from on high but rather through many small, day-to-day decisions occurring in a variety of places" (p. 187). As mentioned, the bulk of federal expenditures for industrial research are in the areas of defense, space, and atomic energy. Support for civilian technology is left to industry—although there is supposed to have been a federal contribution in the form of "spin-offs" from defense and space technology. In recent years, it has become clear that this policy has not been effective; "spin-offs" have occurred in only a few areas, such as computers, solid state electronics, and satellite communications (Harvey Brooks, 1973, p. 113). On the other hand, investment by private industry in civilian technology has not been sufficient. Brooks blames this inadequacy on excessive spending by the federal government. Federal demands for highly skilled technological and managerial manpower have raised the cost of civilian R&D to the point where it has become unprofitable for many industries. Consequently, he claims, the United States lags in many areas of civilian technology—particularly in "low-technology," market-motivated areas.

Lambright (1976) argues that civilian technology has no constituency. Technologists have remained much more dependent upon the organizations that employ them than have basic scientists. Perhaps because of their relatively subordinate status in these organizations, they have not developed strong professional organizations (Edwin Layton, 1969). Engineers appear to play less of a role in government decisions for civilian technology than basic scientists do in the formulation of science policy for basic science. (For a view of engineers and policy that stresses some exceptions to these generalizations, see David Noble, 1977.)

It is perhaps not surprising that the United States government spends relatively less on civilian R&D than do most other industrial nations (Brooks, 1973, p. 114). In any case, attempts to increase the federal role in civilian technology have encountered considerable opposition both from Congress and from private industry. The rationale is that civilian technology should be developed by private firms on the basis of what is profitable. Harold Orlans recalls that: "a small programme of aid to civilian industrial R&D proposed by the Department of Commerce in 1963 was attacked as a threat to our American way of life and defeated in Congress" (1973, p. 129).

New policies for civilian technology are currently being proposed. These include increasing the funding for basic science and greater overall direction of federal science policy by scientists. For example, Lambright argues that "central scientific and technical advice and machinery are needed" (1976, pp. 206–207). Francisco R. Sagasti (1977) stresses the necessity of distinguishing between science policy and technology policy and argues that the confusion between the two has led to many mistaken attempts at enhancing technological capabilities.

Money alone is apparently not the solution. Studies by economists indicate that, although there is a relationship between national expenditures on R&D and national output of inventions and innovations, there is no correlation between government support for high technology and economic performance (Keith Pavitt, 1976, pp. 332, 344). Expenditures, plus government direction of an appropriate kind, might be the answer.

T. Dixon Long's analysis of Japanese technology policy is suggestive: "The organizational side of technology appears settled into effective institutional forms and practices. Little activity that relates to catching-up technologically with the industrial nations of North America and Western Europe is left to chance. The Japanese approach technology as a means to status which in turn is a measure of influence. Their tactics are those of order and persistence—in short, of organization" (1975, p. 22). In other words, Japan has created strong centralized organizations for developing and administering science and technology policy. Unlike those in the United States, such organizations work effectively in the area of private industry. In addition, Japanese industry itself spends an unusually large amount for research and development, and the basic research component of this is exceptionally large. Long says statistics suggest that "qualitative differences in the status of research in the Japanese firm . . . may soon begin to show up in the emergence of original technology on a somewhat broader front" (p. 22).

iii. *Effects of Policy on Civilian R&D:* Government support for R&D in industry is often unsuccessful. Keith Pavitt argues that "a significant proportion of governmental support of research and development is concentrated in the wrong sectors and on the wrong activities, and it is directed towards the wrong objectives" (1976, pp. 330–331). Most government support for industrial research in Europe is for high technology—aircraft, space, nuclear research, computers, and automation (Pavitt and Walker, 1976, pp. 55–62). There is a risk that government support of high technology will be used to continue technological developments that are not likely to be commercially successful (J. Zysman, 1975; also George Eads and Richard R. Nelson, 1971).

According to Pavitt (1976), the French government concentrated its activities "on a few large-scale programmes in a few large firms in a few industries" but with unsatisfactory results in terms of production and sales (p. 340–341). P. Papon (1975) claims that French government policy for R&D lacks "an industrial strategy" (p. 242). By contrast, the Japanese government appears to rely more on its own R&D, coupled with strong policy directives from government agencies. The power of comparable agencies in France is limited by the excessive political power of the Ministry of Finance.

One of the severest critics of American civilian technology is Michael T. Boretsky (1975; see also Philip M. Boffey, 1971). Boretsky argues that the United States has already experienced a trade deficit with respect to low civilian technology—which he defines as manufactured products that are not technology intensive. It is on the verge of experiencing a trade deficit with respect to high civilian technology, or technology-intensive manufactured products. According to Boretsky, technological capabilities are now growing faster in other industrial countries than in the United States. Other data show that the number of both product and process patents is declining in the United States compared to foreign countries, while the proportion of United States patents granted to foreigners is steadily increasing (Boretsky, p. 79; National Science Board, 1976).

5. Social Research and Development. The concept of social technology is relatively new—though there are obvious intellectual linkages with the concept of "applied social science" and even the older "social work." The necessity for social technology in the new sense became evident in the 1960s when unsolved social problems, such as crime, poverty, and urban decay, were perceived as amenable to solutions utilizing the social sciences. Social technology differs from other technologies in that it does not consist of materials, machines, or techniques (David Donnison, 1972). It is an innovation in social policies for allocating benefits to some individuals and removing them from others. For example, Brooks (1973c) refers to such social programs as compensatory education, the negative income tax, and systems for financing health care as "social technologies." Donnison points out that "policy innovations are often much less specific than innovations in technology. A century later, historians may still dispute when, or whether, they occurred. The most important of them are *ideas,* or a way of perceiving the world, derived from related assumptions, concepts and aspirations, and carried forward by political pressures" (p. 251). Duncan MacRae, Jr. (1974) argues that an important characteristic of applied social science research is that the dependent variable is an evaluative one; it is concerned with "variables related to action that *influences* the realization of the value in question" (p. 10). Some policies maximize conflicting goals and values, as Martin Levin and Horst Dornbusch show in their analysis of policies in criminal justice and education (1973).

Although it is not clear, from these definitions, that social scientists should be involved in the development of social technology, in fact modern social technology is expected to be based on sophisticated social science methodology. This methodology may involve social experiments, as in the case of the New Jersey income maintenance experiment (David Kershaw, 1975, and Peter Rossi, 1975), or it may involve precise assessments of effects *via* evaluation techniques (Peter H. Rossi and Walter Williams, 1972). Much work by social scientists in social technology is involved with technical problems in the implementation or assessment of social policies.

i. *Organization of Social Research and Development:* The organizational settings and policies for directing social research and development are relatively underdeveloped. Applied social research is conducted in university settings, both in academic departments and in professional schools, and in government agencies and nonacademic institutes. These last, according to Harold Orlans (1971), include "competitive organizations of the most diverse character, ranging from small, short-lived consulting services to massive industrial and financial corporations" (p. 23). In a few cases, specialized nonprofit institutes for applied research have been created, such as the Bureau of Applied Social Research at Columbia University (which has been replaced by the Center for Social Science), the Poverty Institute at the University of Wisconsin, and the Urban Institute in Washington. For a description of these different types of settings in the United States, see National Academy of Sciences, *The Behavioral and Social Sciences: Outlook and Needs* (1969). Norman Perry (1976) performs the same task for the United Kingdom, Denmark, and France.

Considerable dissatisfaction with the present arrangements for performing applied social science research is expressed in the literature. Researchers in academic departments tend to use the same procedures for doing policy research as they do for basic research (Robert A. Scott and Arnold Shore, 1974), and they consequently contribute little to the formulation and implementation of social policy. Scott and Shore stress the need to identify sociological variables "that are amenable to manipulation and control in the context of an operating program" (p. 57). However, they point out that the number of such variables is rather small.

Unlike other types of technology, applied research problems in the social sciences are concerned with concrete problems that cut across many different basic research areas and disciplines (Storer, 1968). Applied problems in the natural sciences tend to be narrower and more specific; they can be broken down into a set of technical problems, and a single basic field can make a useful contribution to the solution of an applied problem. This is rarely true in applied social science.

Research institutes in universities have not functioned satisfactorily either. Although they have been more willing than researchers in academic departments to attack intractable social problems such as poverty and delinquency, they have often suffered from unstable funding, poor quality of personnel, and a tendency to deviate from their research objectives (National Academy of Sciences, *The Behavioral and Social Sciences,* 1969, p. 197; see also Albert B. Cherns, 1970).

There have been few systematic studies of nonacademic applied social science research. Norman Perry (1976) suggests that this is because they have not been considered as "contexts for serious social science research" (p. 156). Authors in these contexts rarely publish and therefore are not very visible.

A major problem in the organization of social technology is that of communication between producers and users who tend to be located in different organizational settings. Proposals have been made to create research organizations in which producers would be better informed of policy-

makers' needs. Leonard Goodwin has proposed a public research corporation to carry out experimental social research, chartered by Congress and
with a board of overseers consisting of members of Congress and eminent
social scientists elected by their peers (Leonard Goodwin, 1975). Nathan
Caplan (1975a) concludes that the production of applied sociological research that will be implemented requires the formation of an organization
of experts trained specifically to act as intermediaries between researchers
and policymakers. On the other hand, Mark Van de Vall (1975), in a study
of 120 applied social research projects in Holland, shows that applied research is more likely to be used in the formulation of policy if the researcher
is a staff member rather than an outside consultant.

Another idea is that of establishing graduate schools of applied behavioral science (see National Academy of Sciences, 1969; and Duncan
MacRae, 1974). A number of graduate schools of public policy analysis
have emerged in the United States in the last decade (Martin Trow, 1973).
It is not clear, however, how graduate schools of public policy will maintain
contacts with clients outside the university. Morris Janowitz (1972) points
out that attempts at creating applied sociological research centers in universities have often failed. Sociologists employed by these centers are unable
to establish lasting contacts with nonacademic clients and eventually adopt
the career patterns of basic researchers.

ii. *Policy for Social Research and Development:* In 1974 the United
States government spent $218 million for applied research in the social
sciences, compared to $385 million in the physical sciences, $1.5 billion in
the life sciences, and $1.8 billion in the engineering sciences (Willis Shapley,
1976). There is no centralized coordination of policies for the social sciences.
Henry Riecken (1972) comments:

> The United States presents a picture of a relatively large number of
> separate institutions both engaged in and supporting social science
> research without a single coherent coordinating body or a well-articu-
> lated set of regulatory institutions. At the same time that these enter-
> prises are large relative to social science in the rest of the world, they
> are small relative to the total body of science of the United States.
> The development of a policy for these disciplines has been marked by
> a number of ineffectual attempts to coordinate them and by the prolif-
> eration of a "non-system" of support and performance that is both
> ramified and complexly interconnected (pp. 180–181).

At least three organizations perform partial coordinative functions:
the Social Science Research Council, the Assembly for Behavioral and Social
Sciences of the National Research Council, and the Russell Sage Foundation.
Riecken calls these three organizations "important switching centers for
the traffic between the federal government and the community of academic
social scientists" (p. 186). The Assembly for Behavioral and Social Sciences

has been concerned with increasing the use of social science in policy formation (Constance Holden, 1978). A National Social Science Foundation has often been proposed, or, similarly, a Council of Social Advisers; neither has attracted sufficient support to be enacted (Harold Orlans, 1971).

In contrast to the lack of coordination in American social science, the French government attempts to allocate research funds in such a way that researchers will be required to work on practical problems (Michael Pollak, 1976). This is done by contractual financing administered by the government without participation of the researchers in the decision-making process.

In the United Kingdom, according to Norman Perry (1976), "all types of research units appear to enjoy relative freedom to select and control their substantive programme of research work" (p. 182). On the other hand, funding for basic research is so limited that academic researchers are likely to turn to alternative sources for funding (Andrew Shonfield, 1972, p. 492). Not surprisingly, Perry found that in university departments "pragmatic, policy-oriented, descriptive research far outweighs the conscious search for theoretical advance" (p. 173).

iii. *Effects of Policy on Social Research and Development:* Two major issues predominate in the literature assessing social technology. The first is the usefulness to policymakers of research on social technology; the second is the role of values in this type of technology.

It is generally argued that social technology produced in academic settings is not useful to policymakers (see N. J. Demerath et al., 1975). On the other hand, research that is useful is frequently viewed as trivial by academic researchers. Amitai Etzioni (1971) sees the social engineer or applied researcher as performing a narrow, technical role in which he accepts problems defined by clients without questioning the goals of the research. Irving Horowitz (1967b) also suggests that when the social scientist accepts problems defined by policymakers, the quality of social science deteriorates.

Research that is viewed as significant by academic researchers tends to deal with broad, long-term problems—largely in theoretical rather than empirical terms. (These two types are sometimes characterized as "engineering" and "enlightenment.") For example, Morris Janowitz (1972) claims that "the theoretical and substantive content of sociology inherently limits the development of an applied or practice specialization in sociology" (p. 107). In his view, sociological knowledge can contribute to the enlightenment of the public or policymakers, but a profession of applied sociology—distinct from basic research in the field—is unfeasible.

On the other hand, James Coleman (1972) accepts the necessity of social engineering and argues that policymakers, not social scientists, should formulate research problems; he argues that they are most familiar with the situations for which solutions are sought. Coleman complains that because applied research is generally funded by government agencies that do not have control over policy, the agencies' selection of problems is poor (see also Russell and Shore, 1976–1977).

Michael Pollak's study of French social science (1976) shows that close involvement of policymakers in the conduct of research has both advantages and disadvantages. There is, "on the credit side, the permanent contact established with research (by policymakers) and the consequent under-standing of its problems. . . . On the other hand, in certain cases, research workers may . . . complain that for political reasons they are being urged to change their conceptual framework" (p. 128).

Laurence Tribe (1972) discusses the implications for the policy analyst of accepting the values of the decision makers who have commissioned a study: "The only values that can be served will be those strongly held by persons who seek a policy analyst's aid" (pp. 103–104).

Perhaps the most notorious example of this acceptance of the decision makers' values is Project Camelot. In 1964, a group of eminent American social scientists accepted a contract from the Army's Special Operations Research Office; they were to develop a model "which would make it pos-sible to predict and influence politically significant aspects of social change in the developing nations of the world" (Irving L. Horowitz, 1967a, p. 5). However, their colleagues, both in the United States and in Latin America—where the field studies were to be conducted—did not accept the values underlying the studies. As a result of pressure on political authorities from these colleagues, the project was canceled less than a year after it was begun.

Etzioni sees need for a professional group of policy researchers who would deal with values and seek to clarify goals and the relations between them, "to prepare alternative rationales and to pry the policymaker loose from antiquated assumptions" (1971, p. 9).

It may be that the ideal situation is one in which policymakers are themselves sociologists. From the 1920s to the 1940s, rural sociologists occu-pied important positions in the U.S. Department of Agriculture. They set policy and requested appropriate studies from other rural sociologists, and a large number of useful studies of rural life were produced. Charles and Zona Loomis (1967) comment: "Never have so many rural sociologists been employed in the government service. . . . Yet in the relatively few years of heavy government employment of rural sociologists, models for future types of studies were set, and linkages between theory and practice were accomplished which were landmarks for the discipline" (pp. 675–676).

Unfortunately, the story has an unhappy ending. By the 1950s, due to cutbacks in funds, the influence of rural sociologists in the U.S.D.A. hierarchy had greatly diminished. By the late 1960s, James Copp (1972), writing of his experiences during a five-year stint with the U.S.D.A. in Washington, could describe an almost total absence of interest in socio-logical concerns among those in high administrative positions in the agency.

Another theme in the literature is that social technology is misused by policymakers. Joseph Ben-David (1973) complains that: "Typically, a social science idea or result is taken up by policymakers with much enthusiasm but little understanding, and used in a way not justified by the actual findings" (p. 43). Some writers have argued that social science knowledge is used mainly to legitimate policy decisions that have already been taken, although Karin Knorr (1976) and Nathan Caplan (1975b) found little evidence of this in their interviews with medium-to-upper-level decision makers, in Austria and in the United States respectively. (See also Michael Useem, 1976a.) Knorr found that social science studies were used to identify public reactions to political decisions before they were taken and to structure opinion with the help of market and opinion research.

At this point in its development, social science knowledge rarely takes the form of specific innovations or techniques whose adoption and diffusion can be measured precisely. Its influence is as diffuse and nonspecific as the influence of basic research on the development of other types of technology. Caplan (1975b) found that American decision makers tended to use "soft" knowledge from secondary sources subjectively integrated, rather than specific information produced directly by social scientists (p. 53).

On the other hand, the objectives of most government social programs are as imprecise as social science knowledge. Joseph Wholey et al. (1975) conclude that "the typical Federal social program cannot be managed to achieve its stated objectives because they are generally not defined . . . in such a way that progress toward objectives can be measured. . . . The proper mechanism of program intervention is not well understood or defined, or in some cases, even known" (pp. 175–176, 194).

Margaret Boeckmann's analysis (1976) also points to problems at the receiving end. Since members of Congress evaluate scientific information the same way they evaluate any other information, results of one large-scale social experiment—the New Jersey experiment in income maintenance—did not lead to more rational policymaking. Boeckmann suggests that the impact of the experiment will be, in the short run, to provide highly technical information to government administrators. In the long run, it might contribute to a change in values and opinions about the poor.

The fact that social technology deals explicitly with values means that the process of assessing policy alternatives can be politicized (Carol Weiss, 1970). Sociological evaluations of liberal social programs have frequently been negative, and this has given support to conservatives demanding their abolition. But before we discuss value issues and efforts to control the impact of science and technology, we must summarize the present discussion.

6. A Summing Up. At this point it would seem advantageous to summarize what we have seen in this lengthy discussion of the strategic and research-planning levels of science policy. Several general themes have emerged in the discussion.

At the strategic level, the unit of comparison is the nation. Certain general issues pervade science policy discussions in industrial nations, particularly the appropriate balance between coordination and pluralism and between economic-military goals and social goals. Countries have resolved these problems in different ways.

France has made the greatest effort to coordinate the activities of its researchers, both basic and applied, and to direct them toward the achievement of goals defined by policymakers. The Japanese have forged close and effective links between the government and industry—from which the universities are virtually excluded. The United Kingdom has moved toward an increasingly decentralized system in an apparent effort to emphasize applied research and development at the expense of basic research. West Germany has moved toward increasing coordination, particularly of the academic sector; it has deemphasized social goals more than most of the other major powers.

In the United States, the degree of pluralism is such that the claim is persistently made that the country has no research policy (see, for example, Bruce L. R. Smith, 1973, p. 162). Attempts at coordination of this complex system by means of presidential advisers and special offices have had varying degrees of success.

Otto Keck (1976) shows that these countries fall into two groups in terms of actual priorities for R&D: West Germany, Japan, and the Netherlands emphasize the general advancement of science; the United States, the United Kingdom, and France give greater stress to military R&D (see also Robert Gilpin, 1970).

Our discussions of policy at the research-planning level do not permit a detailed analysis of research and development systems across different countries. However, this mode of analysis has shown that similar issues predominate in the various systems—for example, efficiency, the appropriate balance between autonomy and accountability, the role of basic research in technological innovations, economic markets for innovations and the consequences of their absence, and the utilization of technology by other sectors.

Policy is expressed via organizational forms and can lead to the dominance of organizational imperatives over policy imperatives in what Robert Hall (1972) has called the "technoscience agencies." In the United States, and perhaps elsewhere, the R&D agency is a pressure group that develops a symbiotic relationship with its clientele or constituency. The level of funding which it receives is related not so much to an overall rational policy as to the success of the agency in presenting its case to Congress and the Office of Management and Budget. There has been no effective overall direction of R&D activities, balancing the interests represented by the technoscience agencies. Interests not represented by the technoscience agencies—the prime examples we have seen are civilian and social technology—fare less well.

Have studies of science and technology policies influenced the development of policies at either the strategic or the research-planning levels? There is little evidence that academic research has influenced policy at the strategic level. The impact of the OECD country reviews (1966–1973) and other such studies would be an interesting subject for empirical study. Within the various research sectors, studies have probably had most impact on the civilian and academic sectors—for example, in terms of the debate over the role of basic research in the production of technology. Jürgen Schmandt (1978) claims that "policy studies . . . of the impact of nuclear weapons on the nature of war and on relations among nations . . . had a major impact on thinking about military strategy, disarmament and international relations" in the 1950s. Studies of policies in the area of agricultural technology have not been extensive. With respect to social technology— the other area covered—overall policy has been limited, and its results, at best, ambiguous. In short, studies of science and technology policies appear to have had little impact on actual policies for R&D.

II. NATIONAL POLICIES FOR CONTROLLING THE IMPACT OF SCIENCE AND TECHNOLOGY

Policies, as well as institutions, for controlling the impact of science and technology are much less developed than the policies and institutions that produce them. Perception of the need for such institutions is relatively recent and stems from a basic shift in cultural values that occurred during the 1960s. The shift fundamentally altered public attitudes toward the quality of life and the environment.

A. Differential Impacts of Science and Technology

Analyses of the impacts of science and technology have been more sophisticated than solutions proposed for dealing with them. Each of the technology sectors discussed above has different types of social impacts and has generated a literature assessing its consequences and proposing solutions. Here is a quick overview of the different types of impacts:

   i. *Agricultural Technology:* Industrialization of food production has led to problems with the nutritional quality of food in industrial countries; this problem in turn has provoked a consumer movement to counteract the problem. The use of pesticides has had adverse effects on plants, animals, and humans (Dorothy Nelkin, 1977d, p. 413). The development of high-yield grains, in an attempt to solve the world's food problem, has led to the disappearance of many natural strains; this trend would threaten human survival in some areas if the high-yield grains were damaged.

   ii. *Military Technology:* Escalation of weapons production has led to an arms control movement (Harvey Sapolsky, 1977). Some military technologies have contributed to energy shortages or environmental pollution;

campaigns for the reduction of these effects have, accordingly, come into being. Nuclear reactors for peace-time use have been perceived as potentially dangerous sources of radiation, and there are active citizens' movements opposing them (Allan Mazur, 1975).

iii. *Academic Science:* Discoveries in some fields have been claimed to threaten the existence of the human race or at least the quality of life in industrial nations. Some examples: research involving so-called "recombinant" DNA might threaten the future of the human race, a possibility that has led to a movement to regulate research in this area; and advances in climate control could make the planet uninhabitable.

iv. *Civilian Technology:* Technological innovations have contributed to environmental pollution, energy shortages, and the exhaustion of natural resources. Recent developments in electronics technology that make national data banks possible threaten individual rights (A. F. Westin, 1967). Public response to these issues has led to an environmental movement (David Sills, 1975), as well as to numerous voluntary associations and public protests against new technological developments or the government agencies supposed to regulate them (Dorothy Nelkin, 1977d).

v. *Social Technology:* Even in its present underdeveloped state, social technology has had negative effects. One is the tendency to underscore the negative impacts—or the lack of impact—of innovative social policies (Carol Weiss, 1970).

The extent of public concern about these impacts depends on a variety of political and social, as well as purely accidental, factors. Nelkin (1977d) argues that four areas of impact have attracted the most attention: the environment, work, civil liberties, and democratic values. She argues that whether or not a technological impact will become a matter of policy debate depends on two variables—"the degree to which they provoke a public response and their relationship to organized political and economic interests" (p. 407). Harvey Brooks (1973c, pp. 252–253) argues that many of the negative effects are due to older technologies; the oldest industries are the largest sources of pollution.

B. Policies for Controlling the Impact of Science and Technology

Although policies and organizations for controlling the impact of science and technology are underdeveloped, they are gradually emerging. Nelkin (1977d) categorizes the policies as: anticipatory, reactive, or participatory. Anticipatory controls are technical procedures for predicting the impact of new scientific and technological developments; reactive controls are created to regulate existing technologies; participatory controls are efforts by citizens—including scientists—to influence particular technological developments. We discuss the first two policies in this section; the third is discussed in part 3, on value issues.

1. **Technical Controls.** The most widely discussed techniques for predicting the impact of new social and technological developments are technology assessment and technological forecasting (or futures research). François Hetman (1973) says that technology assessment tends to be defined in two ways. The first refers to a relatively narrow operational analysis of a particular technology. The impacts of alternative policies are examined, but no specific evaluations or recommendations are made. By contrast, the second version "regards control and management of technology as part of over-all planning or social engineering. In its extreme formulation, it starts by spelling out values, social policies and objectives and works down to technology assessment in order to clarify the most appropriate technical options" (p. 263). Technology assessment is a long-range process that involves: monitoring the negative side effects of existing technology; screening and selecting among new technologies—e.g., supersonic aircraft, brain manipulation; and directing research and development toward new and desirable technologies to meet social objectives.

In the United States, an Office of Technology Assessment was created by Congress in 1972 to deal with the first two of these aspects of technology assessment (Harvey Brooks, 1973c). The National Science Foundation's Research Applied to National Needs (RANN) was created in 1971 to deal with the third aspect. Harvey Brooks (1973b) says that "it could be viewed as a mechanism for translating broadly defined social problems into specific manageable technical problems and simultaneously for developing a constituency capable of attacking them" (p. 132). As noted earlier, this agency has been renamed the Science and Engineering Directorate.

Louis H. Mayo (1972) points out that the technology assessment process is highly fragmented. Most technologies develop gradually. As the impacts become more obvious, more and more interest groups become involved as operators or as receivers of benefits or as absorbers of costs. As the number of participants increases, more and more subsystems for technology assessment are created. But each assessment subsystem concerns itself with only a particular aspect of the problem. Few attempt to deal with the overall social effects of the technology.

Forecasting is a set of techniques—modeling, simulation, the Delphi method—that attempt to predict future trends on the basis of presently available data. Perhaps the most famous and controversial study using these techniques is Donella Meadows et al., *The Limits to Growth* (1972). It makes predictions about the future of industrial economies based on their present patterns of growth.

Also called "futures research," forecasting is studied by a sizeable and diverse set of people in about one hundred academic and nonacademic organizations in the United States; there are about the same number in other countries (Wayne I. Boucher, ed., 1977, p. 5). Forecasting techniques are applied in extrapolating social and economic trends and assessing the probable progress of existing technologies. In a thorough review and evaluation of these techniques, Daniel P. Harrison (1976) finds that they have limitations and need considerable improvement if social forecasting methodology is to attain scientific status. Technological forecasting appears to be more sophisticated, but neither social nor technological forecasting has as yet had any significant impact on policy (Boucher, 1977).

2. **Legislative and Regulatory Controls.** Laurence Tribe (1971) categorizes legislative and regulatory controls for technology as follows: (1) specific directives concerning the development or use of technology; (2) market incentives such as fees or penalties affecting the decision to develop or utilize a technological process; and (3) new or altered decision-making structures, by which he means changes in the composition, powers, or obligations of existing organizations, such as private corporations and professional associations, that make crucial decisions concerning technological development.

An example of a specific directive for the control of technology is the National Environmental Policy Act (NEPA) of 1969, "which requires that 'unquantified environmental amenities and values' be considered along with technological and economic or quantitative inputs to public agency decision-making on projects, permits, contracts, and other major actions when such actions are likely to result in significant environmental impacts" (Michael S. Baram, 1973, p. 467). Pollution has been the focus of a number of such control directives in recent years—including water pollution, air pollution, and noise pollution.

A new type of legislative directive to regulate academic research on recombinant DNA has been proposed, generating considerable controversy (Clifford Grobstein, 1977). Although there has been considerable resistance among scientists to the possibility of this sort of control, the American Society of Microbiologists has come out in favor of three types of controls: specific legislation, fines against individual scientists who do not comply, and a regulatory commission (Harlyn O. Halvorson, 1977).

Since a major problem with specific directives is their enforcement, such legislation often includes the creation of a regulatory agency. These agencies have a long history in the United States—for example, the Food and Drug Administration—and several new ones have been created recently. One example is the Environmental Protection Agency, which enforces the National Environmental Policy Act (Baram, 1973).

Attempts to restructure existing organizations to perform these functions may be less satisfactory. Under pressure from Congress, a number of attempts have been made to create departments within the U.S. Public Health Service to enforce pollution standards or protect the environment. These efforts were resisted by sanitary engineers as well as by physicians in the Public Health Service, and eventually the programs were replaced by the Environmental Protection Agency (George D. Greenberg, 1975).

The assessment of regulatory controls has become the domain of a growing body of policy analysts (see Duncan MacRae, Jr. 1974; and Martin Trow 1973). In fact, controlling the impact of science and technology would seem an ideal locus for the intervention of social technology— both in the form of policy innovations and techniques for their assessment— but as yet few strides have been taken in this direction.

III. VALUE ISSUES IN THE STUDY OF SCIENCE AND TECHNOLOGY POLICIES

A. An Overview

In most of the areas with which science and technology policies deal—e.g., controlling the impact of science and technology—it is not possible to make decisions entirely on the basis of empirical evidence; value judgments inevitably play a role. William A. Blanpied, in a discussion of contemporary trends in the literature on the ethical and human value implications of science and technology (1974), demonstrates the upsurge of interest in this area since the 1960s.

In the policymaking process, there are three points at which cultural values, ideologies, and ethics are particularly significant: (1) the selection of priorities for the allocation of research funds; (2) the role of the scientist as expert and political adviser; and (3) the assessment of the impact of science and technology on society. Martin Rein's definition of values in the context of public policy—as involving either normative statements of objectives or the moral assumptions that underlie practice (1976, p. 38)—is adequate for our purposes here.

1. Selection of Priorities.   Expenditures for scientific research can be justified in terms of a variety of social values. For the classic statements on these issues, see the chapters in Edward Shils, ed., *Criteria for Scientific Development* (1968); several of those contributions are discussed below. Little has been written along these lines since then, with the exception of the OECD Report *Science, Growth and Society* (1971).

The view that scientific activity does not need to be justified in terms of external criteria is expounded by Michael Polanyi in an article in *Criteria for Scientific Development* (1968). He argues that only scientists can determine scientific priorities. These priorities emerge through a continual exchange of ideas among scientists, which leads to consensus on the areas that are important at any particular time. Any attempt to have a central plan for science would be less efficient than the operation of the "scientific market."

Polanyi's ideas have no relevance to problems of setting priorities in the area in which the largest number of choices are made: applied science and development. There is some indication that even within basic science, consensus about priorities does not emerge as easily as Polanyi supposed. William W. Lowrance has reviewed the surveys of fundamental research conducted by committees working for the National Academy of Sciences' Committee on Science and Public Policy, COSPUP (1977); he suggests that these scientists found it difficult to set priorities, both within their own fields and, particularly, between fields. Lowrance claims that no ideal method has ever been developed for evaluating the relative importance of different research programs.

Alvin M. Weinberg offers a classic analysis of "Criteria for Scientific Choice" in the Shils-edited volume (1968). He argues that internal criteria alone cannot be used in selecting priorities. He proposes these external criteria: (1) scientific merit: fields that contribute most to neighboring fields should be given priority; (2) technological merit: fields should be selected for support in terms of their possible technological applications; (3) social merit: fields should be given priority if they contribute to the solution of social problems and to improvement in the quality of life. In his view, the fields that rank highest in terms of all three of these criteria should be funded.

Michael Gibbons (1970) found that Weinberg's criteria had had considerable influence on the selection of priorities by science policy bodies in the United Kingdom. He analyzes a specific decision—whether or not to fund a 300 GeV accelerator at the international high energy physics laboratory in Geneva (CERN)—and shows how the Weinberg criteria were applied. Particular difficulty was experienced in assessing social merit, as well as in assigning weights to the different criteria when the results were contradictory.

A third viewpoint is expressed by C. F. Carter in the Shils volume. Carter favors a utilitarian approach, that of selecting projects solely on the basis of economic factors—specifically, whether or not research would contribute to economic growth. Although there are undoubtedly problems here in determining the relative importance of projects in contributing to economic growth, this rationale is one that has been found effective in assessing projects in industrial technology. Its applicability to basic research, however, is questionable.

Stephen Toulmin, in two articles in *Criteria for Scientific Development*, attempts to place the problem of priorities in a broader perspective. He argues that different types of research require different types of criteria; and these apply within, and not across, categories. For example, basic research in the natural sciences should be supported either in terms of its

eventual utility (a kind of overhead charge on applied research and technology) or for cultural reasons (knowledge has intrinsic value). For other
types of research, such as speculative technology, economic criteria such
as those suggested by Carter would apply. For product-oriented research
and problem-oriented research aimed at finding solutions to practical problems, economic criteria apply, but the public, as potential customers for the
results of the research, should also play a role in the decision-making
process.

Another argument in this discussion of priority setting is that basic
research should be supported because of its contribution to applied research
(Simon Rottenberg, in the same volume). According to this view, applied
research is essentially the application of discoveries made by basic scientists.
Investment in basic science is necessary for the progress of technology.
Empirical studies that have modified this view (see above, pp. 608–609)
may have been a factor in the decline of support for basic research in recent
years.

The more recent discussion of research priorities mentioned above—
OECD, *Science, Growth and Society* (1971)—calls for a reorientation of
science and technology to include social goals as well as economic and industrial goals (see also Harold Orlans, 1972).

The philosophical rationale for the allocation of research priorities
remains incomplete. There is no objective evidence to show that any one
theory for the allocation of resources to scientific research is superior to
any other. It is interesting that although the thinkers just discussed were
residents of countries that at the time were devoting over half their expenditures for R&D to military research, the allocation of resources to military
research and development was never mentioned. Milton Leitenberg (1973)
summarizes the rationales that have been used to justify these allocations:
"The relevant military journals which deal with R&D are full of parallel
kinds of exhortative articles. One kind reiterates the need to keep ahead
of the enemy in a technological race, and makes clear that tomorrow's
weapons derive from and depend on today's research. The second kind
propounds the philosophical view that science can't be stopped, progress
can't be stopped. Sometimes these two themes are intertwined" (p. 351).

Agricultural research was also ignored in the *Minerva* debates (Shils
volume), although it has tended to be undersupported in many countries
even while half the world's population is underfed (see S. H. Wittwer,
1978).

Derek Price, in "Measuring the Size of Science" (1969), has argued
that, as long as basic science is supported adequately, the "internal ecology"
of science will reflect the output—in terms more of numbers of scientific
publications per discipline than of allocations of funds by field. In other
words, one cannot produce an excess of publications in chemistry by force-
feeding that field at the expense of physics. More recently he has argued

that his studies also show basic science to be an overhead cost, not on technology, but on a country's total GNP. Decline in support for basic research has repercussions on the adverse balance of U.S. international trade in high technology, thereby devaluing the dollar on the world exchange (Price, 1976). He proposes that the value of basic research should be considered as extrinsic, not in terms of its consequences in applications, but because participation in the basic research front provides access to the world stock of knowledge in a particular field. Without funding, researchers rapidly become obsolescent.

These discussions of priorities for allocating research funds assume a social system in which agents make rational decisions. In Section I, our examination of the organization of science policy bodies and the nature of science policy decisions indicated to what extent this assumption is justified.

2. **The Scientist as Expert and Adviser.** The role of the scientist as expert and political adviser is complicated by the absence of a set of norms governing the behavior of scientists and technologists outside their own fields. Such a set of norms may be emerging at the present time.

There has been considerable debate over the set of norms alleged to govern the behavior of basic scientists in the conduct of their research (see André Cournand, 1977; Cournand and Michael Mayer, 1976a; and Michael Mulkay, 1976b). Cournand (1977) argues, following Robert K. Merton ("The Normative Structure of Science," in his *The Sociology of Science,* 1973), that, although such a set of norms exists, it requires considerable revision in order to apply to the scientist's behavior in other contexts—specifically in the application of scientific knowledge to practical problems. Reformulating the Mertonian norms governing the conduct of basic research as honesty, objectivity, tolerance, doubt of certitude, and unselfish engagement, he suggests that scientists follow these instead of their traditional code in their dealings with nonscientists. Second, he sees a necessity for some weighting of the elements of the code in order of importance. Third, he suggests the addition of new norms—for example, an obligation of scientists to refrain from actions that would destroy science. Finally, he advocates the formulation of a new "ethic of development" for regulating the pace of cultural and technological change by encouraging the values of egalitarianism, political pluralism, and fraternalism in sociopolitical development. Cournand would link this ethic to the code of the scientist.

A former professor of medicine, Cournand does not concern himself with the question of social control over the scientist's behavior outside the scientific community. Within the scientific community, as Shils points out, the scientist's behavior is rigorously scrutinized, but "when he becomes an adviser to persons who are not scientists, he steps outside the framework which disciplines him" (Shils, 1972, p. 108). Shils sees an urgent need to develop an ethos of scientific advice. Weinberg (1972b) suggests that every scientific society should establish a committee that would evaluate the validity of scientists' claims concerning public issues stated publicly in non-scientific contexts—i.e., claims not now refereed by the scientist's peers.

3.   Impacts of Science and Technology.   Finally, values enter into assessments of the consequences or impacts of science and technology. Impacts of scientific and technological discoveries cannot be evaluated entirely objectively. What is considered a negative impact by some individuals may not seem so to others. William A. Blanpied (1974) suggests that two ethical points of view are emerging in this area: one based on the concept of distributive justice (i.e., distributing benefits as widely as possible), the other on a more traditional, individualistic attitude toward ethical problems. Stephen Elkin has pointed out that those who would maximize personal freedom are emphasizing an individualistic conception of the public interest, which tends to "restrict consideration to efficiency judgments, i.e., to judgments about getting more total wants satisfied" (1974, p. 112).

Gunther Stent sees the problem in even more fundamental terms. He argues that the traditional Western ethical system, from which science developed, is inconsistent. On the one hand, Christian ethics views morality as based on ultimate values that are desirable in themselves. On the other hand, in the "pagan" system of ethics, "there are no ultimate values, only communal purpose, and hence . . . moral judgments are relative rather than absolute" (1974, p. 43). These different ethical attitudes would have different prescriptions for scientific research. For example, the pagan ethic would favor research on hereditary differences in intelligence, where a Christian ethic of ultimate values—which still has a strong influence on contemporary society outside science—views racism as subversive to personal freedom and therefore "since there must not be any hereditary determined differences in intelligence, research that entertains the possibility of such differences is *a priori* evil" (p. 47).

Loren Graham (1978) has attempted to identify, among types of scientific and technological research that have aroused public concern, those that should be regulated and those that should not. His conclusion: for the most part, areas of basic science that have led to ethical debate should not be regulated, but most areas of technology that have evoked similar controversies should. In fact they already are being regulated to some extent. His examples include: "destructive technology," "economically exploitative research," "human subjects research," and "expensive science."

From this brief review, it is clear that the ethical issues in each area are substantially different. (For a comprehensive bibliography of this literature, see William A. Blanpied, *The Ethical and Human Value Implications of Science and Technology*, 1974; see also Wolf-Dieter Eberwein and Peter Weingart, 1977.) Though not technically science policy literature, this material constitutes an essential background for understanding the complex moral questions raised by contemporary issues concerning the direction and control of science and technology.

Two of these areas will now be taken up in greater detail: 1) value issues in assessments of the impact of science and technology policies, and 2) roles and responsibilities of scientist-advisers and laymen. The second area will require four separate subsections, focusing on the varying responsibilities and responses to the dilemmas of experts and laymen, as well as on the sources of opposition to science and technology.

B. Issues in Assessing the Impact of Science and Technology on Society

Problems of assessing and, hopefully, controlling impacts are complicated by the uncertainty of the knowledge on which decisions must be based. They are also complicated by the fact that such decisions involve choices among goals, which presuppose different values, and by the fact that many social groups are involved representing different interests and incompatible values. For example, personal freedom and environmental quality are two values that often cannot both be maximized by the same policy. For any particular type of technology, there are at least four parties involved: the government (sometimes represented by a regulatory agency); scientists and technologists; industrial producers of the particular technology; and consumers.

Consumers have tended to be the weakest party in these interactions, although in recent years the development of consumer movements has strengthened their case. As Dorothy Nelkin (1977d) points out, the courts have played an important role in strengthening the rights of consumers. At the same time, the issues on which courts are being asked to rule are qualitatively different from those with which they have traditionally dealt. In addition, regulatory agencies have sometimes been coopted by the industries they were supposed to regulate, which has reduced their effectiveness.

As noted earlier, William Blanpied (1974) suggests that the fundamental value conflict here is between those who favor the concept of distributive justice and those who favor a more traditional individualistic ethic. A distributive ethic is favored by consumer movements while industrial corporations and some scientific advisers favor an individualistic ethic.

Dorothy Nelkin and Michael Pollak (1977) suggest that the public is beginning to view such decisions as involving "values 'too important to be left to experts'" (p. 353); the public is no longer willing to accept the judgments of the experts as a basis for decision making.

The fact that technology assessment—the term has a broad as well as a narrow sense—involves decisions about social priorities and values has led to considerable concern about whose values will be maximized. B. Wynne

(1975) argues that technology assessment favors the values of a "corporate-industrial consumer society"; he says it represents an attempt to obtain widespread consensus for these values while using procedures that are only ostensibly objective. Some writers have urged the necessity of technology assessments that are not government sponsored (e.g., H. Folk, 1972; and H. P. Green, 1972), suggesting counterassessments by opposing groups or the use of technological ombudsmen—but Wynne is skeptical of such proposals on the grounds that the very definition of a technological problem is a political act that may obscure alternative courses of action (see also Irene Taviss, 1969).

C. Roles and Responsibilities of Scientists, Technologists, and Laymen

In this section, we consider the political and moral dilemmas inherent in the roles of scientist and technologist. These dilemmas are particularly evident when members of the technical community provide advice to government agencies concerning either policy for science and technology or the utilization of scientific and technological knowledge. Although most scientists are not providers of expert advice, all are perceived as having some responsibility for the direction that scientific and technological knowledge takes, as well as for the impact of their own discoveries. Increasingly, laymen are also assuming roles and responsibilities in this area. Some are concerned about certain consequences of science while others are ambivalent or even hostile. Opposition to science and technology comes not only from relatively unsophisticated laymen but also from scholars and intellectuals.

1. Political and Moral Dilemmas in the Roles of Expert and Adviser

i. *Political Dilemmas:* Acceptance of the advisory role entails acceptance of certain governmental policies, such as the cold war. Rejection of the advisory role, in most instances, means losing the opportunity to influence governmental policies. By and large, scientists accepted the military policies of the American government until the Vietnam war led to a shift in their attitudes and to a drastic reassessment of advisory roles (see Harvey Brooks, 1976).

Another dilemma in the advisory role is that the adviser has only limited control over the way his advice is used. For example, expertise may be used to legitimize a policy rather than to evaluate it; or it may be misrepresented. As Dorothy Nelkin comments: "A delicately calibrated scale of probabilities may be replaced by such general terms as 'little risk' and 'high cost' " (1974, p. 35). (See also Nelkin, 1975; and Frank Von Hippel and Joel Primack, 1972.)

The adviser's prestige or eminence may affect the way in which politicians and the public respond to his or her advice; the advice may not be evaluated entirely on its own grounds (MacRae, 1976, p. 174). MacRae points out that the policy pronouncements of basic scientists are likely to receive more attention than those of applied scientists. Even experts who are not eminent may be accused of elitism; their expertise perforce makes them members of an elite whose interests they may be accused of supporting (Stuart S. Blume, *Toward a Political Sociology of Science,* 1974).

    ii. *Moral Dilemmas:* The advisory role often entails a conflict: Should the advice concern the selection of goals or only of means? If the adviser leans toward the latter view, he or she may be held responsible for the selection of goals. Giving advice on the selection of goals, according to some, means going beyond the limits of scientific knowledge. Alvin Weinberg says:

> I propose the term trans-scientific for these questions since, though they are, epistemologically speaking, questions of fact and can be stated in the language of science, they are unanswerable by science; they transcend science. . . . The role of the scientist in contributing to the promulgation of such policy must be different from his role when the issues can be unambiguously answered by science (1972a, p. 209).

    Among these "trans-scientific" issues, Weinberg includes the problem of establishing priorities as well as making moral and aesthetic judgments. Scientific problems for which technical evidence is incomplete are also "trans-scientific." Conflicts between experts are likely in such cases (see Dorothy Nelkin, 1974 and 1975). Harvey Brooks (1976) suggests that scientists may "exploit uncertainty of evidence in an incompletely researched field to provide different interpretations of evidence consistent with their policy predilections" (p. 239).

    There is constant tension, in the advisory role, between advice and advocacy. Advocacy of a particular policy alternative may entail bias in the interpretation of evidence and even suppression of evidence. Knowledge is replaced by ideology (see R. V. Jones, 1972; and Lauriston King and Philip Melanson, 1972).

    Advocacy can also involve overselling one's own knowledge and one's belief in what it can accomplish in the service of government. King and Melanson cite as examples the Mohole Project, in which earth scientists convinced the National Science Foundation to pay for a deep-drilling project in the absence of appropriate technology for doing so, and Project Camelot, in which social scientists convinced policymakers that they could develop a model for predicting insurgency in less developed countries.

    There is always a suspicion of advocacy when disputes between experts occur (Allen Mazur, 1975). Duncan MacRae, Jr., suggests that: "the advice given by applied scientists conforms more with the norm of 'adversary science,' in which rival advisers are satisfied to argue in partisan fashion, as lawyers in a courtroom, for their organizations or their own practical political preferences" (1976, p. 175).

On the other hand, some disputes between experts may reflect genuine scientific—and not ideological—differences (David Robbins and Ron Johnston, 1976). Different specialty backgrounds and institutional settings within which scientists work often lead to differing norms of behavior that may affect the nature of advice and lead to conflicts between experts coming at an issue from different situations. If scientists from different specialties are using different concepts and terminologies, they may simply fail to communicate with one another.

Whether or not the role of expert and political adviser differs in each type of science and technology is an unresolved issue. Are any of these problems more characteristic of advisory roles in some types of science and technology than in others? Brooks (1976) suggests that social scientists are more prone to advocacy because the tradition of expert autonomy is less well established in these fields. However, Nelkin (1975) found the roles to be very similar among experts from both the hard and soft sciences who were advisers on highly controversial issues.

All of these problems arise when scientists as a group undertake to give scientific advice. The National Academy of Sciences has been called "the largest consulting firm in the United States," serving primarily in the United States government. Philip Boffey (*The Brain Bank of America,* 1975) concludes that many of the academy's reports are "mediocre or flawed by bias or subservience to the funding agencies" (p. 245).

2. Responsibilities of Scientists and Technologists for the Impacts of Scientific and Technological Innovations.   If most scientists are not asked to advise their governments on scientific and policy issues, what are the responsibilities of these scientists for the ways in which their work is used? Does their responsibility extend to consequences that they are unable to predict? Are they responsible for the negative consequences of scientific and technological discoveries made by others?

It seems that relatively few scientists are willing to accept responsibilities of this kind. David Nichols (1976) estimates that in 1971 about 35,000 persons—including scientists, doctors, engineers, and students—were members of scientific political interest groups. Through these groups, members of scientific occupations challenge public and private institutional policies. The number increased rapidly during the 1960s but, as Nichols stresses, "the total number of members is but a small fraction of the membership of scientific occupations" (p. 162).

There are several major and numerous minor political interest groups that deal with science policy—most of them moderate rather than radical in their political orientation. Their concerns range from relatively specific issues, such as nuclear arms control or the supersonic transport, to a general interest in the ethical responsibilities of scientists— as individuals, as members of scientific societies, as workers. For example, the Federation of American Scientists has concentrated its activities largely upon attempting to influence Congress on such specific issues as nuclear arms control and unemployment among scientists. By contrast, the Society for Social Responsibility in Science aims to provide the scientific community with "the sort of information upon which essentially moral judgments may be made by scientists" (Stuart Blume, 1974, p. 159; see also Hilary and Steven Rose, 1972; and Dorothy Nelkin, 1972).

Scientific societies generally ignored their responsibilities in this area until the late 1960s, when, under pressure both from younger members and from the general ethos of the period, some societies took steps in this direction. Blume comments: "Responses of major scientific institutions to this growing awareness of the obligations of science and scientists to society are few, but they are to be found. Moreover an increasing number of individuals are attempting to stimulate them" (p. 126).

Blume points out that Nader's call to scientists and engineers—to report malpractices, such as "industrial dumping of mercury and other toxic materials, suppressed occupational disease data, . . . and adverse and undisclosed effects of pesticides and insecticides"—was ignored by scientific societies.

Charles Ruttenberg (1972), in an extensive analysis of responses by scientific societies, political interest groups, and individual scientists to the issue of chemical and biological warfare, shows that, although some professional associations did take stands on the issue, there were important differences by discipline. Biologists, for instance, were more likely to oppose chemical and biological warfare than were chemists. The activities of individuals were generally motivated by a sense of social responsibility rather than by economic self-interest. Ruttenberg notes a tendency for scientists to view these issues in two quite different ways. Those who were less likely to take a political stand tended to view the issues from a narrow technical perspective, whereas scientists with a broader perspective saw the issues in terms of consequences with respect to a broad range of social as well as scientific factors.

In some cases, scientific societies have been drawn into political debates in which their subject matter—sometimes unexpectedly—is perceived as relevant to the issues. Dorothy Nelkin (1977a) documents the effects of the environmental movement upon the small academic discipline of ecology. At first, ecologists were gratified by the popularity of ecological issues and contributed to the development of federal environmental legislation. Later, they were overwhelmed—by demands made upon their expertise by corporations, government agencies, and citizens' groups; by misuse of their expertise; and by threats to their autonomy as a professional group.

On the other hand, some writers have argued that scientists and technologists have a responsibility to provide advice and information to outsiders, particularly to Congress and citizens' groups. Frank Von Hippel and Joel S. Primack (1972) advocate "public interest science" (see also Primack

and Von Hippel, *Advice and Dissent: Scientists in the Political Arena,*
1974). This responsibility is especially important, they think, when government agencies try to keep controversial advice secret; they claim that the
activities of a very few scientists have been effective in bringing to public
attention such issues as "the negative aspects of supersonic transport, the
decision to deploy the Sentinel and Safeguard anti-ballistic missile systems,
the program of crop destruction and defoliation in South Vietnam, and the
regulation of pesticides" (Von Hippel and Primack, 1972, p. 1169).

Funding is a major problem for scientists who wish to play such a role.
Von Hippel and Primack argue that funding for these activities should come
from the scientific community itself—from scientific societies or from the
universities. (For a British version of "public interest science"—called
"critical science"—see David Dickson, 1972.)

Yaron Ezrahi (1976–1977), in a critique of the views of Von Hippel
and Primack, points out that their position involves a conception of experts
accountable to the public only in a very abstract sense, whose behavior is
not subject to the usual social controls that limit elected officeholders.

In a somewhat broader perspective, the various political roles of the
scientist can be understood in terms of Seymour Martin Lipset's categorization (1975) of the political roles of intellectuals. Experts and consultants
who advise the government on the implementation of public policies, in
his view, are *caretakers*. Scientists who remain outside the government perform conservative roles as *preservers*. *Gatekeepers*—his next category—are
spokesmen for innovative ideas that may involve the central values of the
society. Finally, *moralists* are brilliant innovative critics who are not specialists and who "hold up society to scorn for failing to fulfill basic agreed-upon
values. They challenge those running the society with the crime of heresy"
(p. 451). Exactly how scientists fit these categories might make an interesting
study.

**3.   Roles and Responsibilities of Laymen in the Determination of
Science and Technology Policies.**   The advice of laymen on matters affecting scientific and technological policy is rarely sought. However, in recent
years, more and more citizens have attempted to express their views, to
exert an influence. "Participatory technology," according to James Carroll
(1971), involves public participation in developing, implementing, and
regulating technology when the interests of the public are substantially
affected by it. Citizen participation can take several forms: law suits con-
testing specific uses of technology; involvement in technology assessment
proceedings; and "sporadic activities of loose coalitions of individuals and
groups energized by particular situations and issues" (p. 651). Dorothy
Nelkin (1977d) asserts that citizens' groups have proliferated in the United
States: "Some are temporary coalitions formed to protect local interests by
opposing projects such as power plants or airports. Other groups are or-

ganized on a more permanent basis, often as branches of a national organization such as the Sierra Club or Friends of the Earth" (p. 418).

In 1973, a study of volunteers and voluntary associations in the environment and conservation field in the United States found twenty thousand primary associations, plus another twenty thousand in related fields. Most of these organizations are small—less than five hundred members. However, the same study estimated that the environmental movement as a whole in the early 1970s involved between five and ten million persons, drawn primarily from the upper middle class (see David L. Sills, 1975; see also Barbara J. Culliton, 1978).

Allan Mazur (1975) has shown that movements opposing different technological innovations—such as nuclear power plants and fluoridation of water supplies—actually have many similarities. He suggests that such controversies are only apparently concerned with specific issues. Larger questions, such as compulsory medicine and anxiety about physical survival, as well as political and theological beliefs, are actually at stake.

In some countries, governments deliberately attempt to inform their citizens about decisions concerning controversial technological issues, such as nuclear power plants; this disclosure is intended to be a means of obtaining broad acceptance for controversial government policies. In some cases, however, the programs have had the opposite effect, arousing more concern than acquiescence. Dorothy Nelkin and Michael Pollak, in a study of Sweden, the Netherlands, and Austria (1977), point out that there is no clear constituency for many national technological issues, so that it is unclear who represents the public interest. At the same time, the public is increasingly unwilling to permit the government and its experts to make decisions.

Helmut Krauch (1971–1972) has demonstrated that the general public can have a perspective on priorities for science policy strikingly different from that reflected in a national R&D budget. He used public opinion polls to assess the preferences of the West German public with respect to government expenditures on research and development. While funds for research and development by the Federal Republic of Germany were spent primarily on defense, nuclear research, and space research, a representative sample of the general public ranked medicine, nutrition, and pollution as top priorities for R&D. A recent study of the American public yielded similar results (National Science Board, 1976).

A barrier to public involvement in decision making about science and technology is the low level of public knowledge about highly esoteric subjects. Some of the scientific political interest groups have concentrated their activities upon providing information for the layman—e.g., Scientists' Institutes for Public Information, the Center for Science in the Public Interest, the Center for Concerned Engineers, and the Union of Concerned Scientists (see Blume, *Toward a Political Sociology of Science*, 1974, chapter 7; and David Nichols, 1976).

Stuart Blume discusses the effectiveness of the mass media in communicating information about scientists to the general public. He considers both the science writer and the scientist himself as disseminators of scientific information. Blume documents the limitations of the mass media as sources of information about science for the layman and notes: "Different newspapers and programs have different interests in science; for one it is a source of education, for another of news, for a third of horror or fantasy" (1974, p. 238; see also Luc Boltanski and Pascale Maldidier, 1970).

Other approaches to scientific education for the layman include science-technology centers (Lee Kimche, 1978), which have proliferated in the last decade. The enormous public demand for these facilities—"among the most rapidly developing institutions of learning in contemporary society" (p. 270)—suggests that there may be a desire for knowledge about contemporary developments in science and technology that is not satisfied by the mass media. By way of exhibits and workshops that appeal to all ages, these centers aim to present information objectively. Kimche says that their goal is to provide enough facts so that people can make intelligent decisions by themselves (p. 271).

4. **Sources and Effects of Opposition to Science and Technology.** Opposition to science and technology is found among intellectuals and scholars as well as among the general public—although, in both groups, it appears to be a (fairly visible) minority position. There is a small literature—by such authors as Jacques Ellul (*The Technological Society*, 1965); Jürgen Habermas (*Toward a Rational Society*, 1971); Herbert Marcuse (*One-Dimensional Man*, 1964); and Theodore Roszak (*The Making of a Counter-Culture*, 1968)—that appears to be widely read in certain social groups (see Richard L. Meier, 1971; and Langdon Winner, *Autonomous Technology*, 1977).

In reviews of the antiscience literature, Leslie Sklair (1971) and Stephen Cotgrove (1975) are critical of the conclusions of these authors and of the assumptions on which their thinking is based. According to Cotgrove, these writers see technology as threatening freedom, as limiting the scope for human choice (p. 60). While criticizing the mixture of dogma, polemic, and speculation in the antitechnology literature, as well as its lack of empirical evidence, Cotgrove emphasizes that the issues raised by these authors are central to an attempt to understand the characteristics of postindustrial society (1975, pp. 61–62; for another interpretation, see Paul T. Durbin, 1972; further discussion can be found in chapter 5 of this volume).

The antiscience movement argues that the values of the individual become subordinate to material goods, but it is unclear whether the general public would agree. Cotgrove, in any case, concludes that the debate is less about the negative consequences of science and technology than about the factors that have produced the postindustrial society—factors that are as yet poorly understood.

The emergence of intellectual currents opposing science and technology has been explained in a variety of ways. For example, Gerald Holton (1973) has argued that one reason behind the antiscience movement is the increasing separation between scientists and nonscientific intellectuals. The modern intellectual is not educated in scientific subjects and is unable to understand the increasingly esoteric knowledge that scientists produce. The intellectual is no longer able to act as an informed mediator between science and the public. Holton notes that, in the past, intellectuals like Hobbes, Voltaire, Jefferson, or Franklin were able to play such a role. In the past, amateurs took a substantial interest in science. This interest seems to have flagged in the twentieth century; or, rather, it takes different forms—such as technological hobbies. Science as an institution is increasingly isolated from other cultural institutions.

There are few empirical studies of popular attitudes toward technology. Although some authors have argued that distrust of science and technology in the United States is increasing, the evidence from national opinion polls is mixed. Allan Mazur (1977b), analyzing responses to a question concerning public confidence in the national leadership of such institutions as science, medicine, and education, found that confidence in the leadership of all institutions had declined since 1966. However, the relative ranking of science had moved up from fourth place in 1966 to second place in 1975. Similarly, Todd La Porte and Daniel Metlay's survey of the attitudes of the public toward science and technology (1975a) did not support the speculative statements by scholars to the effect that the public is either mindlessly enamored of science and technology (e.g., Ellul) or hostile and alienated by it. La Porte and Metlay found more support for basic science than for technology; the implementation of technology is perceived as a problem needing regulation. Technical experts were highly rated and were seen as exercising legitimately a great deal of influence over decisions (see also Wade, 1978). By contrast, government and business leaders were rated much lower. Respondents saw the public as having the right to be involved in such issues, but as being given the least opportunity (see also La Porte and Metlay, 1975; and Irene Taviss, 1972).

It seems likely that these studies tap the rational components of the public's assessments of science and technology; underlying these assessments, there is some indication of a widespread ambivalence toward science and technology that is mirrored, on a much higher intellectual level, in the scholarly literature discussed above. George Basalla (1976), studying the depiction of scientists in such forms of popular culture as comic strips and

TV programs, finds that the scientist is generally portrayed as "a distorted individual bent upon disrupting the lives of peaceful citizens"; he is seen "as a dangerous figure who tended toward mental instability and social irresponsibility" (p. 263).

One example of a popular movement that challenges basic tenets of modern science is the creationists' challenge to Darwinism. Creationists argue that the biblical description of creation is *scientifically* accurate (see Dorothy Nelkin, 1977c, 1977e; and John A. Moore, 1976). Creationists have attempted to exert an influence over the content of biology textbooks and the teaching of evolution—and in some cases have succeeded. One interesting aspect of this movement is the high level of education of its members; in some organizations, all the members are required to have at least a master of science degree. These organizations function like scientific societies; they present papers at annual meetings and design projects to evaluate their ideas. They also, of course, try to influence educational policy by working through local school boards, curriculum committees, and state legislatures.

IV: SCIENCE, TECHNOLOGY, AND INTERNATIONAL RELATIONS

International policies for the development and control of science and technology represent an entirely different sort of value issue for policy analysis. Although there are numerous specific issues, there is an overriding issue that runs through all discussions of the matter. Many scientists, technologists, and agency officials are convinced that there is a moral imperative to apply science and technology to the solution of world problems; on the other side, there are widespread criticisms of specific technological developments, particularly in the Third World countries.

International policies in this area are gradually emerging, while external influences upon national policies for the development and control of science and technology are increasing. These include pressures from other nations, from international organizations—both governmental and nongovernmental—and from multinational corporations.

Factors that make it increasingly necessary to consider these external influences are enumerated by Eugene Skolnikoff (1977). They include: (1) the emergence of global technologies—"technological systems that are global in operation or application, such as space technologies, transportation, communication" (p. 511)—that have required joint policies for their development and control; (2) the impact of one nation's science and technology upon other countries, for example with respect to environment and climate; (3) the role of technology in economic growth—successful technologies are rapidly diffused from one country to another (John E. Tilton, 1971) and play an important role in international economic competition; (4) the role of technology in modern warfare—military technology also diffuses rapidly from one country to another and plays an obvious role in international politics; and (5) the emergence of new international agents, governmental and nongovernmental organizations and multinational corporations, whose influence upon international scientific cooperation and competition is increasing.

Because of space limitations, we briefly outline only some of the major concerns in this area.

A. Emergence of Regional and International Science and Technology Policies

While international policies have developed most rapidly for the *control* of science and technology, regional policies tend to be concerned with the *development* of science and technology. We will point to examples in each of these areas.

1. Regional and International Policies for the Development and Utilization of Science and Technology at the National Level.   Although European countries have recognized the need for the development of regional science policies, progress in this direction has been slow (Pierre Piganiol, 1968). The scope of European scientific and technical collaboration is suggested by Manfredo Macioti:

> European scientific and technical cooperation is marked today by the activities of more than 50 intergovernmental and over 400 international non-governmental organizations, by some 160 *ad hoc* governmental agreements (mostly bilateral), by over 400 international meetings per year and last but not least, by a network of private arrangements established by a variety of industrial companies (many of them "multinationals") (1975, p. 296).

In spite of all this activity—which includes a European Science Foundation to foster cooperation in basic research, as well as international committees concerned with applied research and technology—Macioti's conclusions are mixed. Cooperation in academic science—e.g., European laboratories such as CERN and EMBO—and in civilian technology has been reasonably successful; however, the results of intergovernmental cooperation in the development of high technology—nuclear, computer, and aircraft industries—have been disappointing (in addition to Macioti, see Henry R. Nau, 1974 and 1975, for Western Europe; and Lloyd Jordan, 1970, for Eastern Europe).

International policies for the development of scientific and technological knowledge have encouraged international research programs as well as conferences and committees; for a partial inventory, see Margaret Galey (1977). The International Geophysical Year (1957–1958, which involved sixty countries), the International Hydrological Decade, and the International Biological Program required the cooperation of numerous national science organizations as well as international committees of nongovernmental organizations. Often it is an intergovernmental organization that coordinates these programs. UNESCO, for example, coordinated the activities of the International Hydrological Decade, but assisting UNESCO were other United Nations agencies and the International Council of Scientific Unions (Diana Crane, 1971, p. 594).

2. **Regional and International Policies for the Development and Utilization of Science and Technology at the Global Level.** Global technologies are developed to deal with global resources—environment, climate, oceans, space—and controlling the impact of these technologies almost always involves the creation of international organizations. John G. Ruggie speaks of "international regimes," which he defines as

> a set of mutual expectations, rules and regulations, plans, organizational energies and financial commitments, which have been accepted by a group of states. One example of a regime, as we are using the term, is the international system of safeguarding nuclear materials, involving national and international materials, accounting rules and practices, regulations about inspection, and obligations about submitting specified aspects of national behavior to the regime (1975, pp. 570–571).

Ruggie says the three basic purposes of these regimes are: (1) the acquisition of new capabilities; (2) making effective use of them; and (3) coping with the consequences of the use of these new capabilities (p. 571). He argues that international organizations or programs are accepted and supported by nations only when they are capable of "so transforming the activities of national services that they would come to constitute a global system" (p. 576). (See also Ernest B. Haas, 1975.)

Seyom Brown and Larry L. Fabian (1975) claim that the existing regimes for the oceans, outer space, and weather are no longer appropriate; by the mid-1980s, a comprehensive ocean authority, an outer-space projects agency, a global weather and climate organization, and an international scientific commission on global resources and ecologies should be created. Eugene Skolnikoff (1972) also calls for strong international regimes to handle global issues. He argues that, in the development of new international machinery, the worldwide scientific and engineering communities should play a major role (p. 165).

Who will control these regimes? Michael Brenner (1975) shows that, in one case, the Intergovernmental Oceanographic Committee was created "by a group of Anglo-American scientists who conceived of a global body to promote scientific studies of the oceans and to coordinate cooperative ventures in advancing this goal" (p. 774). The I.O.C. is a highly complex, decentralized organization in which coordination is provided by scientists who perform many different and overlapping roles (Brenner, 1975, p. 781). However, in the area of international fisheries' management, according to Barbara Johnson (1975), it is bureaucrats, usually with scientific training, who make national and international policy. The conjunction of national commercial interests with the need for international regulation may in part explain these different patterns. (See also Ernest B. Haas et al., 1978.)

B. Influence of External Pressures on National Science and Technology Policies

Science policies of nations are not developed in isolation from external influences. The industrial nations, by choice, follow each other's lead in these matters—in part because they foresee economic advantages.

The countries that are most vulnerable to external pressures are the Third World countries. These are the countries whose systems for producing science and technology are least developed (Derek Price, 1967). J. Davidson Frame et al. (1977) comment that "Third World science is almost inconsequential when measured against world science *in toto*" (p. 507). Some countries, like Japan in an earlier period and Spain now, do not support research and development but depend entirely on imported technology (Philip H. Abelson, 1973). Others, such as India, have developed enormous R&D systems of their own (G. S. Aurora, 1971). Success in meeting the needs of these countries depends in part upon the characteristics of these R&D systems; for a review of this literature, see Diana Crane (1977). However, it also depends upon the extent to which multinational organizations succeed in dominating the industrial systems of these countries, thereby diminishing the market for indigenous technology (Abdul Said and L. Simmons, eds., 1974). An emphasis upon appropriate, low-cost, intermediate or "soft" technology represents an attempt to utilize indigenous technological skills and resources to meet a country's technological needs—largely outside formal R&D systems (see Nicolas Jéquier, 1976).

In some cases, multinational corporations play an important role in the transfer of technology to Third World countries (Dimitri Germidis, 1977). Germidis suggests that one solution that might strengthen the Third World countries' abilities to benefit from external influence is regional technological cooperation. He argues that multinational firms would behave differently toward Third World countries if these countries "resolved to club together or simply to cooperate seriously in a regional grouping decided on *by themselves*" (p. 30). Regional associations would create units comparable in size to markets in the developed countries. (On "appropriate" technology" and technology transfer, see also chapter 5, II. B., in this volume.)

In some Third World countries, attempts are being made to exercise control over external influences (Francisco R. Sagasti, 1977). Regional agreements for the development and utilization of technology already exist in Bolivia, Colombia, Chile, Ecuador, Peru, and Venezuela, which are members of the Andean Pact (see Junta del Acuerdo de Cartagena, 1976a, 1976b; also Ziauddin Sardar and Dawud G. Rosser-Owen, 1977; and David B. Lewis and F. A. Long, 1977).

The value issues here—indeed the paradoxes—are translucently obvious. But to treat these issues of the future in adequate detail would require another chapter as long as this one.

BIBLIOGRAPHIC INTRODUCTION

A. Introductions to the Field

Although it is difficult to speak of "classics" in a literature as diffuse and diverse as this one, the following general works provide a useful introduction to many aspects of the field.

Blume, Stuart S. *Toward a Political Sociology of Science.* New York: Free Press, 1974.

> Examination of the interactions between science and government, drawing on perspectives from both political science and sociology.

Boucher, Wayne I., ed. *The Study of the Future: An Agenda for Research.* Washington, D.C.: Government Printing Office, 1977.

> A collection of articles reviewing the current status of "futures" research.

Greenberg, Daniel S. *The Politics of Pure Science.* New York: New American Library, 1967.

> A muckraking study, by a science reporter, of the organization and policies for funding basic science in the 1960s.

Haberer, Joseph, ed. *Science and Technology Policy.* Lexington, Mass.: Lexington Books, D. C. Heath, 1977.

> A collection of recent papers on American science and technology policy.

Holton, Gerald, and Blanpied, William A., eds. *Science and Its Public: The Changing Relationship.* Boston Studies in the Philosophy of Science, vol. 33. Dordrecht: D. Reidel, 1976.

> A collection of papers on popular and scholarly attitudes toward science.

Lambright, W. Henry. *Governing Science and Technology.* New York: Oxford University Press, 1976.

> Excellent analysis of the role of technoscience agencies in the American government in the conduct and support of science and technology.

Long, T. Dixon, and Wright, Christopher, eds. *Science Policies of Industrial Nations.* New York: Praeger, 1975.

A collection of case studies of the recent history of science policy in the United States, Soviet Union, United Kingdom, France, Japan, and Sweden.

Lyons, Gene M., ed. *Social Research and Public Policy: The Dartmouth/OECD Conference.* Hanover, N.H.: Public Affairs Center, Dartmouth College, 1975.

A collection of articles on applied social research and its impact on policy.

Nagi, Saad Z., and Corwin, Ronald G., eds. *The Social Contexts of Research.* New York: John Wiley, 1972.

A collection of articles on various aspects of the relationship between science, government, and society.

Shils, Edward, ed. *Criteria for Scientific Development.* Cambridge, Mass.: M.I.T. Press, 1968.

A collection of "classic" papers on the selection of priorities for scientific research.

B. Reviews of the Literature

The following articles and books *review* various aspects of the literature:

Cotgrove, Stephen. "Technology, Rationality and Domination." *Social Studies of Science* 5 (1975): 55–78.

Crane, Diana. "Technological Innovation in Developing Countries: A Review of the Literature." *Research Policy* 6 (1977): 374–395.

Pavitt, Keith, and Walker, W. "Government Policies towards Industrial Innovation: A Review." *Research Policy* 5 (1976): 11–97.

Sapolsky, Harvey. "Science Policy." In *Handbook of Political Science,* vol. 6: *Policies and Policymaking,* pp. 79–110. Edited by Fred I. Greenstein and Nelson W. Polsby. Reading, Mass.: Addison-Wesley, 1975.

Spiegel-Rösing, Ina, and Price, Derek, eds. *Science, Technology and Society: A Cross-Disciplinary Perspective.* Beverly Hills, Calif.: Sage Publications, 1977.

C. Bibliographic Information

Blanpied, William A. "Selected Bibliography of Recent Works." In *The Ethical and Human Value Implications of Science and Technology: A Preliminary Directory Reviewing Contemporary Activity.* Edited by William A. Blanpied. Newsletter 8 of the Program on Public Conceptions of Science. Cambridge, Mass.: Harvard University Press, June 1974.

Boucher, Wayne I. "A Bibliography of Research on Futures Research." In *The Study of the Future: An Agenda for Research,* pp. 283–316. Edited by Wayne I. Boucher. Washington, D.C.: U.S. Government Printing Office, 1977.

Busch, Lawrence, and Sachs, Carolyn. *The Social and Economic Organization of the Agricultural Sciences: A Preliminary Bibliography.* Lexington: University of Kentucky Agricultural Experiment Station, 1977.

Hill, P. *Technological Change in Less Developed Countries: A Guide to Selected Literature.* Ames: Iowa State University, Department of Electrical Engineering, Engineering Research Institute, 1979.

Spiegel-Rösing, Ina. *Wissenschaftsentwicklung und Wissenschaftssteuerung.* Frankfurt: Atheneum Verlag, 1973.

D. Statistics

Statistics are an important source for studies of science and technology policies. The following references provide an introduction to these materials.

National Science Board. *Science Indicators—1976.* Washington, D.C.: U.S. Government Printing Office, 1977.

National Science Foundation. *Funds for Research, Development and Other Scientific Activities.* Washington, D.C.: U.S. Government Printing Office, annually.

Organisation for Economic Cooperation and Development. *Patterns and Resources Devoted to Research and Development in the OECD Area, 1963–71.* Document SPT (74) 12. Paris: OECD, 1974.

Shapley, Willis. *Research and Development in the Federal Budget, FY 1977.* Washington, D.C.: American Association for the Advancement of Science, 1976.

E. Journals

The following journals publish articles related to studies of science and technology policies: *Daedalus, Minerva, Public Policy, Research Policy, Science, Science and Public Policy, Social Studies of Science,* and *The Public Interest.* The emphasis here and in the following bibliography on current works is upon recent literature—i.e., works published since 1970.

BIBLIOGRAPHY

Abelson, Philip H. "Spain—Another Japan?" *Science* 180 (4 May 1973): 449.

Allison, G. T., and Morris, F. A. "Armaments and Arms Control: Exploring the Determinants of Military Weapons." *Daedalus* 104 (1975): 99–129.

Archibald, Kathleen A. "Alternative Orientations to Social Science Utilization." *Social Science Information* 9 (April 1970): 7–34.

Ashby, Sir Eric; Orlans, Harold; Ziman, John; and Wynne-Edwards, V.C. "The Choice and Formulation of Research Problems: Four Comments on the Rothschild Report." *Minerva* 10 (April 1972): 192–208.

Aurora, G. S. *Administration of Science and Technology in India: A Trend Report*. Hyderabad: Administrative Staff College of India, 1971.

Baram, Michael S. "Technology Assessment and Social Control." *Science* 180 (4 May 1973): 465–473.

Basalla, George. "Pop Science: The Depiction of Science in Popular Culture." In *Science and Its Public: The Changing Relationship*, pp. 261–278. Boston Studies in the Philosophy of Science, vol. 33. Dordrecht: D. Reidel, 1976.

Battelle Columbus Laboratories. *Science, Technology and Innovation*. Columbus, Ohio: Battelle Columbus Laboratories, 1973.

Beckler, David Z. "The Precarious Life of Science in the White House." *Daedalus* 104 (1974): 115–134.

Bell, Daniel. *The Coming of Post-Industrial Society*. New York: Basic Books, 1973.

Ben-David, Joseph. "How to Organize Research in the Social Sciences." *Daedalus* 102 (Spring 1973): 39–51.

————. "The Central Planning of Science." *Minerva* 15 (Autumn–Winter 1977): 539–553.

Benn, A. W. "For Science, Open Government Has Arrived." *New Scientist* (11 May 1972): 314–317.

Black, Guy. "The Effect of Government Funding on Commercial R&D." In *Factors in the Transfer of Technology*, pp. 202–218. Edited by William H. Gruber and Donald G. Marquis. Cambridge, Mass.: M.I.T. Press, 1969.

Blanpied, William A. "Subjective Impressions Regarding Contemporary Concerns and Trends." In *The Ethical and Human Value Implications of Science and Technology: A Preliminary Directory Reviewing Contemporary Activity*, pp. 136–156 (see next entry).

————, ed. *The Ethical and Human Value Implications of Science and Technology: A Preliminary Directory Reviewing Contemporary Activity*. Newsletter of the Program on Public Conceptions of Science. Cambridge, Mass.: Harvard University, June 1974.

Blanpied, William A., and Weisman-Dermer, Wendy, eds. *Interdisciplinary Workshop on the Interrelationships between Science and Technology and Ethics and Values*. A.A.A.S. Miscellaneous Publication 75–8. Washington, D.C.: American Association for the Advancement of Science, 1975.

Blume, Stuart S. "Research Support in British Universities: The Shifting Balance of Multiple and Unitary Sources." *Minerva* 7 (Summer 1969): 649–667.

————. *Toward a Political Sociology of Science*. New York: Free Press, 1974.

Boeckmann, Margaret E. "Policy Impacts of the New Jersey Income Maintenance Experiment." *Policy Sciences* 7 (1976): 53–96.

Boffey, Philip M. "Technology and World Trade: Is There Cause for Alarm?" *Science* 172 (2 April 1971: 37–41.

———. "Technology Incentives Program: Success or a Phony Hard Sell?" *Science* 189 (26 September 1975): 1066–1067.

———. *The Brain Bank of America.* New York: McGraw-Hill, 1975.

———. "Science Indicators: New Report Finds U.S. Performance Weakening." *Science* 191 (12 March 1976): 1031–1033.

Böhme, Gernot; van den Daele, Wolfgang; and Krohn, Wolfgang. "Finalization in Science." *Social Science Information* 15 (1976): 307–330.

Boltanski, Luc, and Maldidier, Pascale. "Carrière scientifique, morale scientifique et vulgarisation." *Social Science Information* 9 (1970): 99–118.

Boretsky, Michael. "Trends in U.S. Technology: A Political Economist's View." *American Scientist* 63 (1975): 70–82.

Boucher, Wayne I., ed. *The Study of the Future: An Agenda for Research.* Washington, D.C.: Government Printing Office, 1977.

Brenner, Michael. "The Intergovernmental Oceanographic Commission and the Stockholm Conference: A Case of Institutional Non-Adaptation." *International Organization* 29 (1975): 771–804.

Brickman, Ronald. "French Science Policy and the Changing Role of the University." *Research Policy* 6 (1977a): 128–151.

———. "The Promotion of Mobility of Scientists: A Problem of French Science Policy." *Minerva* 15 (Spring 1977b): 62–82.

Brickman, Ronald, and Rip, Arie. "Science Policy Advisory Councils in France, the Netherlands, and the U.S., 1957–1977: A Comparative Study." *Social Studies of Science* 9 (May 1979): 167–198.

Bronowski, J. *Science and Human Values.* New York: Harper Torchbooks, 1959.

Brooks, Harvey. "Basic Science and Agency Missions." *In Research in the Service of National Purpose.* Edited by J. Weyl. Washington, D.C.: Government Printing Office, 1966a.

———. "National Science Policy and Technology Transfer." In *Proceedings of a Conference on Technology Transfer and Innovation,* May 15, 1966, pp. 21–56. NSF 67–5. Washington, D.C.: National Science Foundation, 1966b.

———. "Applied Research Definitions, Concepts, Themes." In *Applied Science and Technological Progress,* pp. 21–56. A Report to the Committee on Science and Astronautics, U.S. House of Representatives, by the National Academy of Sciences. Washington, D.C.: Government Printing Office, 1967.

———. "Physical Sciences: Bellwether of Science Policy." In *Science and the Evolution of Public Policy,* pp. 105–134. Edited by J. A. Shannon. New York: Rockefeller University Press, 1973a.

———. "Knowledge and Action: Dilemma of Science Policy in the 1970s." *Daedalus* 102 (Spring 1973b): 125–144.

———. "Technology Assessment As a Process." *International Social Science Journal* 25 (1973c): 247–256.

———. "The Federal Government and the Autonomy of Scholarship." In

Controversies and Decisions: The Social Sciences and Public Policy,
pp. 235–258. Edited by Charles Frankel. New York: Russell Sage Foundation, 1976.

Brooks, Harvey, and Bowers, Raymond. "The Assessment of Technology."
Scientific American 222 (February 1970): 13–21.

Brown, Seyom, and Fabian, Larry L. "Toward Mutual Accountability in
the Nonterrestrial Realms." *International Organization* 29 (1975):
877–892.

Busch, Lawrence, and Sacks, Carolyn. *The Social and Economic Organization of the Agricultural Sciences: A Preliminary Bibliography.* Lexington: University of Kentucky Agricultural Experiment Station, 1977.

Cahn, Anne H. *Eggheads and Warheads: Scientists and the A.B.M.* Cambridge, Mass.: M.I.T. Press, 1971.

Caplan, Nathan. "A Minimal Set of Conditions for the Utilization of Social
Science Knowledge in Policy Formulation at the National Level."
Presented at the conference on Social Values and Social Engineering,
International Sociological Association, Warsaw, April 18–19, 1975a.

———. "The Use of Social Science Information by Federal Executives." In
Social Research and Public Policy, pp. 46–67. Edited by Gene M. Lyons.
Hanover, N.J.: Dartmouth College Public Affairs Center, 1975b.

Carroll, James D. "Participatory Technology." *Science* 171 (19 February
1971): 647–653.

Carter, C. F. "The Distribution of Scientific Effort." In *Criteria for Scientific Development,* pp. 34–43. Edited by Edward Shils. Cambridge,
Mass.: M.I.T. Press, 1968.

Castro, Barry. "The Scientific Opportunities Foregone Because of More
Readily Available Federal Support for Research in Experimental than
Theoretical Physics." *Journal of Political Economy* 76 (1968): 601–614.

Caty, Gilbert. "The Financing and Organisation of Research in the Universities and in the Peripheral System." In *The Research System,* vol. 1:
France, Germany and the United Kingdom. By G. Caty, G. Drilhon,
G. Ferné, and S. Wald. Paris: Organization for Economic Cooperation
and Development, 1972.

Caty, G.; Drilhon, G.; Ferné, G.; and Wald, S. *The Research System,* vols.
1, 2, and 3. Paris: Organization for Economic Cooperation and Development, 1972–1974.

Cawkell, A. E. "Connections between Engineering and Science." *I.E.E.E.
Transactions on Professional Communication,* PC–18 (1975): 71–73.

Central Policy Review Staff. "The Organisation and Management of Government R&D." In *A Framework for Government Research and Development,* pp. 1–25. Cmnd. 4814. London: Her Majesty's Stationery
Office, 1971.

Cherns, Albert B. "Relations between Research Institutions and Users of
Research." *International Social Science Journal* 22 (1970): 226–242.

————, ed. *Social Science and Government: Policies and Problems.* London: Tavistock Publications, 1972.

————, ed. *Social Science Organization and Policy.* Paris: UNESCO; and The Hague: Mouton, 1974.

Cole, Stephen; Rubin, Leonard; and Cole, Jonathan R. "Peer Review and the Support of Science." *Scientific American* 237 (October 1977): 34–41.

————. *Peer Review in the NSF: Phase One of a Study.* Washington, D.C.: National Academy of Science, 1978.

Coleman, James S. *Policy Research in the Social Sciences.* Morristown, N.J.: General Learning Press, 1972.

Copp, James H. "Rural Sociology and Rural Development." *Rural Sociology* 37 (1972): 515–533.

Cotgrove, Stephen. "Technology, Rationality, and Domination." *Social Studies of Science* 5 (1975): 55–78.

Cournand, André. "The Code of the Scientist and Its Relationship to Ethics." *Science* 198 (18 November 1977): 699–705.

Cournand, André, and Meyer, Michael. "The Scientist's Code." *Minerva* 14 (Spring 1976): 79–96.

Crane, Diana. "Transnational Networks in Basic Science." *International Organization* 25 (Summer 1971): 585–601.

————. "Technological Innovation in Developing Countries: A Review of the Literature." *Research Policy* 6 (1977): 374–395.

Culliton, Barbara J. "Science's Restive Public." *Daedalus* 107 (Spring 1978): 147–156.

Demerath, N. J. III; Larsen, Otto; and Schuessler, Karl F., eds. *Social Policy and Sociology.* New York: Academic Press, 1975.

Dickson, David. "Science to Help the People." *New Scientist* 54 (4 May 1972): 277–278.

Doctors, Samuel I. *The Role of Federal Agencies in Technology Transfer.* Cambridge, Mass.: M.I.T. Press, 1969.

————. *The NASA Technology Transfer Program: An Evaluation of the Dissemination System.* New York: Praeger, 1971.

Doern, G. Bruce. " 'Big Science,' Government and the Scientific Community in Canada: The ING Affair." *Minerva* 8 (July 1970): 357–375.

————. *Science and Politics in Canada.* Montreal: McGill-Queen's University Press, 1972.

Donnison, David. "Research for Policy." *Minerva* 10 (October 1972): 519–536.

Drath, L.; Gibbons, Michael; and Ronayne, J. "The European Molecular Biology Organization: A Case Study of Decision-Making in Science Policy." *Research Policy* 4 (1975): 56–78.

Dupree, A. Hunter. *Science in the Federal Government: A History of Policies and Activities to 1940.* Cambridge, Mass.: Harvard University Press, 1957.

Durbin, Paul T. "Technology and Values: A Philosopher's Perspective." *Technology and Culture* 13 (1972): 556–576.

Eads, George, and Nelson, Richard R. "Governmental Support of Advanced Technology: Power Reactors and Supersonic Transport." *Public Policy* 19 (1971): 405–428.

Eberwein, Wolf-Dieter, and Weingart, Peter. "Science and Ethics from the German Perspective: An Annotated Bibliography, 1965–1976." *Newsletter on Science, Technology and Human Values* 20 (June 1977): 25–38.

Edmonds, Martin. "Accountability and the Military-Industrial Complex." In *The New Political Economy: The Public Use of the Private Sector,* pp. 154–165. Edited by Bruce L. R. Smith. New York: Wiley, 1975.

Elkana, Yehuda; Lederberg, Joshua; Merton, Robert K.; Thackray, Arnold; and Zuckerman, Harriet, eds. *Toward a Metric of Science: The Advent of Science Indicators.* New York: Wiley Interscience, 1978.

Elkin, Stephen L. "Political Science and the Analysis of Public Policy." *Public Policy* 22 (1974): 399–422.

Ellul, Jacques. *The Technological Society.* London: Cape, 1965.

Encel, S. "The Support of Science without Science Policy in Australia." *Minerva* 9 (July 1971): 349–360.

Etzioni, Amitai. "Policy Research." *American Sociologist* 6 (June 1971): 8–12.

Evan, William M., ed. *Interorganizational Relations.* New York: Penguin, 1976.

Evenson, Robert E. "Technology Generation in Agriculture." In *Agriculture in Development Theory,* pp. 192–233. Edited by Lloyd Reynolds. New Haven: Yale University Press, 1975.

Evenson, Robert E., and Kislev, Yoav. *Agricultural Research and Productivity.* New Haven: Yale University Press, 1975.

Ezrahi, Yaron. "The Political Resources of American Science." *Science Studies* 1 (April 1971): 117–134.

————. "The Authority of Science in Politics." In *Science and Values,* pp. 215–251. Edited by Arnold Thackray and Everett Mendelsohn. New York: Humanities Press, 1974.

————. "The Antinomies of Scientific Advice." *Minerva* 14 (1976–1977): 577–585.

Ezrahi, Yaron, and Tal, E. *Science Policy and Development: The Case of Israel.* London: Gordon, 1973.

Folk, H. "The Role of Technology Assessment in Public Policy." In *Technology and Man's Future,* pp. 246–253. Edited by Albert H. Teich. New York: St. Martin's Press, 1972.

Fores, Michael. "Technical Change and Innovation." In *Industrial Efficiency and the Role of Government,* pp. 102–117. Edited by Colette Bowe. London: Her Majesty's Printing Office, 1977.

Frame, J. Davidson, and Narin, Francis. "NIH Funding and Biomedical Publication Output." *Federation Proceedings* 35 (1976): 2529–2532.

Frame, J. Davidson; Narin, Francis; and Carpenter, Mark P. "The Distribution of World Science." *Social Studies of Science* 7 (1977): 501–516.

Freeman, Christopher. "Economics of Research and Development." In *Science, Technology and Society: A Cross-Disciplinary Perspective,* pp. 223–275. Edited by Ina Spiegel-Rösing and Derek de Solla Price. Beverly Hills: Sage Publications, 1977.

Freeman, Christopher; Codham, C. H. G.; Cooper, C. M.; Sinclair, T. C.; and Achilladelis, B. C. "The Goals of R&D in the 1970s." *Science Studies* 1 (October 1971): 357–406.

Freeman, Howard E. "Evaluation Research and Public Policies." In *Social Research and Public Policy,* pp. 141–174. Edited by Gene M. Lyons. Hanover, N.H.: Dartmouth College Public Affairs Center, 1975.

Fudenberg, H. H. "The Dollar Benefits of Biomedical Research: A Cost Analysis." *The Journal of Laboratory and Clinical Medicine* 79 (March 1972): 353–362.

Galey, Margaret E. "Trends and Dimensions in International Science Policy and Organization." In *Science and Technology Policy,* pp. 109–128. Edited by Joseph Haberer. Lexington, Mass.: Lexington Books, D. C. Heath, 1977.

Gans, Herbert J. "Social Science for Social Policy." In *The Use and Abuse of Social Science,* pp. 13–33. Edited by Irving L. Horowitz. New Brunswick, N.J.: Transaction Books, 1971.

Germidis, Dimitri, ed. *Transfer of Technology by Multinational Corporations,* vol. 2. Paris: Development Centre of OECD, 1977.

Gibbons, Michael. "The CERN 300 GeV Accelerator: A Case Study in the Application of the Weinberg Criteria." *Minerva* 8 (April 1970): 181–191.

Gibbons, Michael, and Johnston, Ronald. "The Role of Science in Technological Innovation." *Research Policy* 3 (1974): 220–242.

Gilpin, Robert. *France in the Age of the Scientific State.* Princeton, N.J.: Princeton University Press, 1968.

———. "Technological Strategies and National Purpose." *Science* 169 (31 July 1970): 441–448.

Globe, Samuel; Levy, Girard W.; and Schwartz, Charles M. "Key Factors and Events in the Innovation Process." *Research Management* 16 (July 1973): 8–15.

Goodwin, Leonard. *Can Social Science Help Resolve National Problems? Welfare: A Case in Point.* N.Y.: Free Press, 1975.

Government of Canada, Senate Special Committee on Science Policy. *A Science Policy for Canada,* vol. 3: *A Government Organization for the Seventies.* Ottawa: Information Canada, 1973.

Graham, Loren R. "Concerns about Science and Attempts to Regulate Inquiry." *Daedalus* 107 (1978): 1–21.

Green, H. P. "The Adversary Process in Technology Assessment." In *Technology and Man's Future,* pp. 254–262. Edited by Albert H. Teich. New York: St. Martin's Press, 1972.

Greenberg, Daniel S. *The Politics of Pure Science.* New York: New American Library, 1967.

———. *Science & Government Report: International Almanac 1977.* Washington, D.C.: Science and Government Report, 1977.

Greenberg, George D. "Reorganization Reconsidered: The U.S. Public Health Service, 1960–1973." *Public Policy* 23 (1975): 483–522.

Grobstein, Clifford. "The Recombinant DNA Debate." *Scientific American* 237 (July 1977): 22–23.

Groeneveld, Lyle; Koller, Norman; and Mullins, Nicholas C. "The Advisers of the United States National Science Foundation." *Social Studies of Science* 5 (August 1975): 343–364.

Gummett, P. J., and Price, G. L. "An Approach to the Central Planning of British Science: The Formation of the Advisory Council on Scientific Policy." *Minerva* 15 (Summer 1977): 119–143.

Haas, Ernest B. "Is There a Hole in the Whole? Knowledge, Technology, Interdependence and the Construction of International Regimes." *International Organization* 29 (Summer 1975): 827–876.

Haas, Ernest B.; Williams, Mary Pat; and Babai, Don. *Scientists and World Order: The Uses of Technical Knowledge in International Organizations.* Berkeley: University of California Press, 1978.

Haberer, Joseph. *Politics and the Community of Science.* New York: Van Nostrand-Reinhold, 1969.

Habermas, Jürgen. *Toward a Rational Society.* London: Heinemann, 1971.

Hadden, Susan. "Technical Advice in Policy-Making: A Propositional Inventory." In *Science and Technology Policy,* pp. 81–98. Edited by Joseph Haberer. Lexington, Mass.: Lexington Books, D. C. Heath, 1977.

Hadwiger, Don F. "The Old, the New, and the Emerging United States Department of Agriculture." *Public Administration Review* 36 (1976): 155–165.

Hall, Robert L. "Agencies of Research Support: Some Sociological Perspectives." In *The Social Contexts of Research,* pp. 193–227. Edited by Saad Z. Nagi and Ronald G. Corwin. New York: Wiley Interscience, 1972.

Halvorson, Harlyn O. "Recombinant DNA Legislation—What Next?" *Science* 198 (28 October 1977): 357.

Hamberg, D. *R&D Essays on the Economics of R&D.* New York: Random House, 1966.

Harrison, Daniel P. *Social Forecasting Methodology: Suggestions for Research.* New York: Russell Sage Foundation, 1976.

Harvard University Program on Technology and Society. *Technology and Values*. Research Review no. 3. Cambridge, Mass.: Harvard University Press, 1969.

———. "Technology and the Polity." *Research Review* 4 (Summer 1969).

Hayami, Yujiro, and Ruttan, Vernon W. *Agricultural Development: An International Perspective*. Baltimore: Johns Hopkins University Press, 1971.

Heitowit, Ezra. *Science, Technology, and Society: A Survey and Analysis of Academic Activities in the U.S.* Ithaca, N.Y.: Cornell University Program on Science, Technology, and Society, 1977.

Hendricks, Sterling B. "The Transition from Research to Useful Products in Agriculture." In *Applied Science and Technological Progress*, pp. 153–170. A Report to the Committee on Science and Astronautics, U.S. House of Representatives, by the National Academy of Sciences. Washington, D.C.: Government Printing Office, 1967.

Hetman, François. "Steps in Technology Assessment." *International Social Science Journal* 25 (1973): 257–272.

Hightower, Jim. *Hard Tomatoes: Hard Times*. Washington, D.C.: Task Force on the Land-Grant College Complex, Agribusiness Accountability Project, 1972.

Holden, Constance. "RANN Symposium: NSF Puts Its Brainchild on Display." *Science* 182 (7 December 1973): 1006.

———. "ABASS: Social Sciences Carving a Niche at the Academy." *Science* 199 (17 March 1978): 1183–1187.

Holton, Gerald. "Modern Science and the Intellectual Tradition." In *Thematic Origins of Scientific Thought: Kepler to Einstein*. Cambridge, Mass.: Harvard University Press, 1973.

Horowitz, Irving L. "The Rise and Fall of Project Camelot." In *The Rise and Fall of Project Camelot*, pp. 3–44. Edited by Irving L. Horowitz. Cambridge, Mass.: M.I.T. Press, 1967a.

———. "Social Science and Public Policy: Implications of Modern Research." In *The Rise and Fall of Project Camelot*, pp. 339–376. Edited by Irving L. Horowitz. Cambridge, Mass.: M.I.T. Press, 1967b.

———. "The Academy and the Polity. Interaction between Social Scientists and Federal Administrators." *Journal of Applied Behavioral Science* 5 (1969): 309–335.

———, ed. *The Rise and Fall of Project Camelot*. Cambridge, Mass.: M.I.T. Press, 1967c.

Illinois Institute of Technology Research Institute. *Technology in Retrospect and Critical Events in Science*, vol. 1. Prepared for the National Science Foundation under contract NSF–C535, 15 December 1968.

Isenson, Raymond S. "Project Hindsight: An Empirical Study of the Sources of Ideas Utilized in Operational Weapons Systems." In *Factors in the Transfer of Technology*, pp. 155–176. Edited by William H. Gruber and Donald G. Marquis. Cambridge, Mass.: M.I.T. Press, 1969.

Janowitz, Morris. "Professionalization of Sociology." *American Journal of Sociology* 78 (1972): 105–135.

Jéquier, Nicolas. *Appropriate Technology: Problems and Promises.* Paris: Development Centre, Organisation for Economic Cooperation and Development, 1978.

Johnson, Barbara. "Technocrats and the Management of International Fisheries." *International Organization* 29 (1975): 745–770.

Johnston, Ron. "Finalization: A Start for Science Policy?" *Social Science Information* 15 (1976): 331–336.

――――. "Contextual Knowledge: A Model for the Overthrow of the Internal-External Dichotomy in Science." *Australia and New Zealand Journal of Sociology* 12 (October 1976): 193–203.

Johnston, Ron, and Gillespie, Brendan. "Scientific Advice and Risk Assessment: A Sociological Analysis." Paper presented at the second annual meeting of the Society for Social Studies of Science, Harvard University, October 14–16, 1977.

Jones, R. V. "Temptations and Risks of the Scientific Advisor." *Minerva* 10 (July 1972): 441–451.

Jordan, Lloyd. "Scientific and Technical Relations Among Eastern European Communist Countries." *Minerva* 8 (July 1970): 376–395.

Junta del Acuerdo de Cartagena. *Andean Pact Technology Policies.* Ottawa: International Development Research Centre, 1976a.

――――. *Technology Policy and Economic Development.* Ottawa: International Development Research Centre, 1976b.

Kaldor, M. *European Defense Industries: National and International Implications.* Sussex, England: Institute of International Organization, University of Sussex, 1972.

Kapitza, Peter L. "Scientific Policy in the U.S.S.R.: Problems of Soviet Scientific Policy." *Minerva* 4 (Spring 1966): 391–397.

――――. "Basic Factors in the Organization of Science and How They Are Handled in the U.S.S.R." *Daedalus* 102 (Spring 1973): 167–176.

Karter, A., and Thorson, S. J. "The Weapons Procurement Process: Choosing Among Competing Theories." *Public Policy* 20 (1972): 479–524.

Kash, Don E. "Politics and Research." In *The Social Contexts of Research,* pp. 97–128. Edited by Saad Z. Nagi and Ronald G. Corwin. New York: Wiley Interscience, 1972.

Katz, James E. *Presidential Politics and Science Policy.* New York: Praeger, 1978.

Katz, Shaul, and Ben-David, Joseph. "Scientific Research and Agricultural Innovation in Israel." *Minerva* 13 (Summer 1975): 152–182.

Kaysen, Carl. "Improving the Efficiency of Military Research and Development." *Public Policy* 12 (1963): 219–273.

Keck, Otto. "West German Science Policy since the Early 1960s: Trends and Objectives." *Research Policy* 5 (April 1976): 116–157.

Kendrick, J. B. "What Is Agricultural Research?" *Science* 191 (27 February 1976): 4220.

Kershaw, David N. "The New Jersey Negative Income Tax Experiment: A Summary of the Design, Operations, and Results of the First Large-Scale Social Experiment." In *Social Research and Public Policy: The Dartmouth-OECD Conference,* pp. 87–116. Edited by Gene M. Lyons. Hanover, N.H.: Public Affairs Center of Dartmouth College, 1975.

Kevles, Daniel J. *The Physicists.* New York: Alfred A. Knopf, 1978.

Kimche, Lee. "Science Centers: A Potential for Learning." *Science* 199 (20 January 1978): 270–273.

King, Alexander. *Science and Policy: The International Stimulus.* London: Oxford University Press, 1974.

King, Lauriston R., and Melanson, Philip H. "Knowledge and Politics: Some Experiences from the 1960s." *Public Policy* 20 (1972): 83–101.

Kishida, Junnosuke. "Technology Assessment in Japan." *International Social Science Journal* 25 (1973): 326–335.

Knorr, Karin. "Policymakers' Use of Social Science Knowledge: Symbolic or Instrumental." Paper presented at the annual meeting of the Society for Social Studies of Science, Cornell University, October 1976.

Komarovsky, Mirra, ed. *Sociology and Public Policy.* New York: Elsevier, 1975.

Krauch, Helmut. "Priorities for Research and Technological Development." *Research Policy* 1 (1971–1972): 28–39.

Kreilkamp, Karl. "Hindsight and the Real World of Science Policy." *Science Studies* 1 (January 1971): 43–66.

———. "Towards a Theory of Science Policy." *Science Studies* 3 (January 1973): 3–30.

Kurth, James. "Aerospace Production Lines and American Defense Spending." In *Testing the Theory of the Military-Industrial Complex,* pp. 135–156. Edited by Steven Rosen. Lexington, Mass.: Lexington Books, D. C. Heath, 1973.

Lakoff, Sanford A. "The Vicissitudes of American Science Policy at Home and Abroad." *Minerva* 11 (April 1973): 175–190.

———. "Scientists, Technologists, and Political Power." In *Science, Technology and Society: A Cross-Disciplinary Perspective,* pp. 335–392. Edited by Ina Spiegel-Rösing and Derek de Solla Price. Beverly Hills, Calif.: Sage Publications, 1977.

Lambright, W. Henry. *Governing Science and Technology.* New York: Oxford University Press, 1976.

Lambright, W. Henry, and Sapolsky, Harvey. "Terminating R&D." *Policy Sciences* 7 (Spring 1976): 199–213.

Lang, Kurt. "Pitfalls and Politics in Commissioned Policy Research." In *The Use and Abuse of Social Science,* pp. 212–233. Edited by Irving L. Horowitz. New Brunswick, N.J.: Transaction Books, 1971.

Langrish, J.; Gibbons, M.; Evans, W. G.; and Jevons, F. R. *Wealth from Knowledge: A Study of Innovation in Industry*. London: Macmillan, 1972.

La Porte, Todd R., and Metlay, Daniel. "Technology Observed: Attitudes of a Wary Public." *Science* 188 (11 April 1975a): 121–127.

La Porte, Todd R., and Metlay, Daniel. "Public Attitudes Toward Present and Future Technologies: Satisfactions and Apprehensions." *Social Studies of Science* 5 (November 1975b): 373–398.

Layton, Edwin. "Science, Business, and the American Engineer." In *The Engineers and the Social System*, pp. 51–72. Edited by Robert Perrucci and Joel Gerstl. New York: John Wiley, 1969.

———. "Conditions for Technological Development." In *Science, Technology, and Society: A Cross-Disciplinary Perspective*, pp. 197–222. Edited by Ina Spiegel-Rösing and Derek de Solla Price. Beverly Hills, Calif.: Sage Publications, 1977.

Leach, Bernard. "Decision-Making in Big Science: The Development of the High Voltage Electron Microscope." *Research Policy* 2 (1973).

Leitenberg, Milton. "The Dynamics of Military Technology Today." *International Social Science Journal* 25 (1973): 336–357.

Levin, Martin A., and Dornbusch, Horst D. "Pure and Policy Social Science: Evaluation of Policies in Criminal Justice and Education." *Public Policy* 21 (1973): 383–423.

Levins, Richard. "Fundamental and Applied Research in Agriculture." *Science* 181 (10 August 1973): 523–524.

Lewis, David B., and Long, F. A. "Policy and Research Strategies for Science and Technology in Developing Nations." In *Science and Technology Policy*, pp. 129–139. Edited by Joseph Haberer. Lexington, Mass.: Lexington Books, D. C. Heath, 1977.

Lieberman, Marvin B. "A Literature Citation Study of Science-Technology Coupling in Electronics." *Proceedings of the IEEE* 66 (January 1978): 1–13.

Lipset, Seymour M. "Intellectual Types and Political Roles." In *The Idea of Social Structure*, pp. 433–470. Edited by Lewis A. Coser. New York: Harcourt Brace Jovanovich, 1975.

Lipton, R. F. "Decision and Representation in the National Science Foundation." Ph.D. dissertation, New York University, 1971.

Long, F. A. "Growth Characteristics of Military Research and Development." In *Impact of New Technologies on the Arms Race*, pp. 271–303. Edited by B. Feld, T. Greenwood, G. W. Rathjens, and S. Weinberg. Cambridge, Mass.: M.I.T. Press, 1971.

Long, T. Dixon. "Science Policy in Postwar Japan." Ph.D. dissertation, Columbia University, 1968.

———. "Policy and Politics in Japanese Science: The Persistence of a Tradition." *Minerva* 7 (Spring 1969): 426–453.

———. "Japanese Technology Policy: Achievements and Perspectives." *Research Policy* (1975): 2–26.

Long, T. Dixon, and Wright, Christopher, eds. *Science Policies of Industrial Nations.* New York: Praeger, 1975.

Loomis, Charles P., and Loomis, Zona K. "Rural Sociology." In *The Uses of Sociology,* pp. 655–691. Edited by Paul F. Lazarsfeld, William H. Sewell, and H. S. Wilensky. New York: Basic Books, 1967.

Lowrance, William W. "The NAS Surveys of Fundamental Research, 1962–1974—In Retrospect." *Science* 197 (23 September 1977): 1254–1260.

Lubrano, Linda L. *Soviet Sociology of Science.* Columbus, Ohio: American Association for the Advancement of Slavic Studies, 1976.

Macioti, Manfredo. "Science and Technology in the Common Market: A Progress Report." *Research Policy* 4 (1975): 290–310.

MacRae, Duncan, Jr. "Science and the Formation of Policy in a Democracy." *Minerva* 11 (April 1973): 228–242.

———. "Policy Analysis As an Applied Social Science Discipline." Paper presented at the annual meeting of the American Sociological Association, Montreal, 1974.

———. "Technical Communities and Political Choice." *Minerva* 14 (Summer 1976): 169–190.

Marcuse, Herbert. *One-Dimensional Man.* Boston: Beacon, 1964.

Mayer, André, and Mayer, Jean. "Agriculture, the Island Empire." *Daedalus* 103 (1974): 83–95.

Mayo, Louis H. "The Management of Technology Assessment." In *Technology Assessment,* pp. 71–116. Edited by R. Kaspar. New York: Praeger, 1972.

Mazur, Allan. "Disputes between Experts." *Minerva* 11 (April 1973): 243–262.

———. "Opposition to Technological Innovation." *Minerva* 13 (Spring 1975): 58–81.

———. "Science Courts." *Minerva* 15 (Spring 1977a): 1–14.

———. "Public Confidence in Science." *Social Studies of Science* 7 (February 1977b): 123–125.

Meadows, Donella; Meadows, Dennis; Randers, Jorgen; and Behrens, William III. *The Limits to Growth.* New York: Universe Books, 1972.

Meier, Richard L. "On Living during the Reformation of Science: Comment." *Comparative Studies in Society and History* 13 (1971): 236–239.

Merrill, Robert. "Some Society-Wide Research and Development Institutions." In *The Rate and Direction of Inventive Activity,* pp. 409–434. National Bureau Committee for Economic Research. Princeton, N.J.: Princeton University Press, 1962.

Merton, Robert K. "The Normative Structure of Science." In *The Sociology of Science,* pp. 267–278. Edited by Norman W. Storer. Chicago: University of Chicago Press, 1973.

Merton, Robert K., and Gaston, Jerry, eds. *Sociology of Science in Europe.* Carbondale: Southern Illinois University Press, 1977.

Mesthene, Emmanuel G. *Technological Change: Its Impact on Man and Society.* Cambridge, Mass.: Harvard University Press, 1970.

Mirsky, E. M. "Science Studies in the U.S.S.R.: History, Problems and Prospects." *Science Studies* 2 (July 1972): 281–294.

Mitroff, I. I., and Chubin, D. E. "Peer Review at the NSF: A Dialectical Policy Analysis." *Social Studies of Science* 9 (May 1979): 199–232.

Moore, Ernest C. *The Agricultural Research Service.* New York: Praeger, 1967.

Moore, John A. "Creationism in California." In *Science and Its Public: The Changing Relationship,* pp. 191–227. Boston Studies in the Philosophy of Science, vol. 33. Dordrecht: D. Reidel, 1976.

Mulkay, Michael. "The Mediating Role of the Scientific Elite." *Social Studies of Science* 6 (September 1976a): 445–470.

———. "Norms and Ideology in Science." *Social Science Information* 15 (1976b): 637–656.

Mullins, Nicholas C. "The Structure of an Elite: The Advisory Structure of the United States Public Health Service." *Science Studies* 2 (January 1972): 3–29.

———. "Power, Social Structure, and Advice in American Science: Federal Science Advisory System from 1950 to 1972." Unpublished MS, Indiana University, 1976.

Murtaugh, Joseph S. "Biomedical Sciences." In *Science and the Evolution of Public Policy,* pp. 157–187. Edited by J. A. Shannon. New York: Rockefeller University Press, 1973.

Nagel, Stuart S., ed. *Policy Studies Review Annual,* vol. 1. Beverly Hills, Calif.: Sage Publications, 1977.

National Academy of Sciences. *The Behavioral and Social Sciences: Outlook and Needs.* Englewood Cliffs, N.J.: Prentice-Hall, 1969.

National Science Board. *Science Indicators 1972.* Washington, D.C.: Government Printing Office, 1973.

———. *Science Indicators 1974.* Washington, D.C.: Government Printing Office, 1975.

———. *Science Indicators 1976.* Washington, D.C.: Government Printing Office, 1977.

National Science Foundation. *Funds for Research, Development, and Other Scientific Activities.* Washington, D.C.: Government Office, annually.

———. *National Patterns of R&D Resources.* Washington, D.C.: Government Printing Office, 1972.

Nau, Henry R. *National Politics and International Technology: Nuclear*

Reactor Development in Western Europe. Baltimore: Johns Hopkins University Press, 1974.

———. "Collective Responses to R&D Problems in Western Europe, 1955–1958 and 1968–1973." *International Organization* 29 (Summer 1975): 617–653.

Nelkin, Dorothy. *The Politics of Housing Innovation.* Ithaca, N.Y.: Cornell University Press, 1971a.

———. "Scientists in an Environmental Controversy." *Science Studies* 1 (October 1971b): 245–262.

———. *The University and Military Research.* Ithaca, N.Y.: Cornell University Press, 1972.

———. "The Role of Experts in a Nuclear Siting Controversy." *Bulletin of Atomic Scientists* 30 (November 1974): 29–36.

———. "The Political Impact of Technical Expertise." *Social Studies of Science* 5 (February 1975): 35–54.

———. "Science or Scripture: The Politics of 'Equal Time.'" In *Science and Its Public: The Changing Relationship,* pp. 209–227. Edited by Gerald Holton and William A. Blanpied. Boston Studies in the Philosophy of Science, vol. 33. Dordrecht: D. Reidel, 1976.

———. "Scientists and Professional Responsibility: The Experience of American Ecologists." *Social Studies of Science* 7 (February 1977a): 75–95.

———. *Technical Decisions and Democracy: European Experiments in Public Participation.* Beverly Hills, Calif.: Sage Publications, 1977b.

———. "Creation vs. Evolution: The Politics of Science Education." In *The Social Production of Scientific Knowledge,* pp. 265–287. Edited by Everett Mendelsohn, Peter Weingart, and Richard Whitley. Dordrecht: D. Reidel, 1977c.

———. "Technology and Public Policy." In *Science, Technology and Society: A Cross-Disciplinary Perspective,* pp. 393–442. Edited by Ina Spiegel-Rösing and Derek de Solla Price. Beverly Hills, Calif.: Sage Publications, 1977d.

———. *Science Textbook Controversies and the Politics of Equal Time.* Cambridge, Mass.: M.I.T. Press, 1977e.

———. "Threats and Promises: Negotiating the Control of Research." *Daedalus* 107 (Spring, 1978): 191–210.

Nelkin, Dorothy, and Pollak, Michael. "The Politics of Participation and the Nuclear Debate in Sweden, the Netherlands, and Austria." *Public Policy* 25 (Summer 1977): 334–357.

Nelson, Richard R., and Winter, S. G. "In Search of a Useful Theory of Innovation." *Research Policy* 6 (January 1977): 36–77.

Nichols, David. "The Associational Interest Groups of American Science." In *Scientists and Public Affairs,* pp. 123–170. Edited by Albert Teich. Cambridge, Mass.: M.I.T. Press, 1976.

Nichols, Rodney W. "Mission-Oriented R&D." *Science* 172 (2 April 1971): 29–36.

——. "Some Practical Problems of Scientist-Advisers." *Minerva* 10 (October 1972): 603–613.

Nicholson, Heather J. "Autonomy and Accountability in Basic Research." *Minerva* 15 (Spring 1977): 32–61.

Noble, David F. *America by Design: Science, Technology, and the Rise of Corporate Capitalism.* New York: Alfred A. Knopf, 1977.

Organization for Economic Cooperation and Development (OECD). *Reviews of National Science Policy:* Belgium, 1966; France, 1967; United Kingdom-Germany, 1967; Japan, 1967; United States, 1968; Italy, 1969; Norway, 1971; Austria, 1971; Spain, 1971; Switzerland, 1972; Iceland, 1973; Netherlands, 1973. Paris: Organization for Economic Cooperation and Development, 1966–1973.

——. *The Measurement of Scientific and Technical Activities: Proposed Standard Practice for Surveys of Research and Experimental Development.* Paris: Organization for Economic Cooperation and Development, 1970.

——. *The Conditions for Success in Technological Innovation.* Paris: Organization for Economic Cooperation and Development, 1971a.

——. *Science, Growth and Society: A New Perspective.* Paris: Organization for Economic Cooperation and Development, 1971b.

——. *Patterns and Resources Devoted to Research and Development in the OECD Area, 1963–1971.* Document SPT (74) 12. Paris: Organization for Economic Cooperation and Development, 1974.

——. *Social Science Policy:* France, 1975; Norway, 1976; Japan, 1977. Paris: Organization for Economic Cooperation and Development, 1975–1977.

Orlans, Harold. "Social Science Research Policies in the United States." *Minerva* 9 (January 1971): 7–31.

——. "Criteria of Choice in Social Science Research." *Minerva* 10 (October 1972): 571–602.

——. " 'D&R' Allocations in the United States." *Science Studies* 3 (April 1973): 119–159.

Papon, P. "Research Planning in French Science Policy: An Assessment." *Research Policy* 2 (October 1973).

——. "The State and Technological Competition in France, or Colbertism in the 20th Century." *Research Policy* 4 (1975): 214–244.

Pavitt, Keith. "Governmental Support for Industrial Research and Development in France: Theory and Practice." *Minerva* 14 (Autumn 1976): 330–354.

Pavitt, Keith, and Walker, W. "Government Policies towards Industrial Innovation: A Review." *Research Policy* 5 (1976): 11–97.

Perry, Norman. "Research Settings in the Social Sciences: A Re-Examination." In *Demands for Social Knowledge: The Role of Research Organizations,* pp. 137–190. Edited by Elisabeth Crawford and Norman Perry. London: Sage Publications, 1976.

Piganiol, Pierre. "Scientific Policy and the European Community." *Minerva* 6 (Spring 1968): 354–365.

Polanyi, Michael. "The Republic of Science: Its Political and Economic Theory." In *Criteria for Scientific Development,* pp. 1–20. Edited by Edward Shils. Cambridge, Mass.: M.I.T. Press, 1968.

Pollak, Michael. "Organizational Diversification and Methods of Financing as Influences on the Development of Social Research in France." In *Demands for Social Knowledge: The Role of Research Organizations,* pp. 115–134. Edited by Elisabeth Crawford and Norman Perry. London: Sage Publications, 1976.

Price, Derek de Solla. "The Scientific Foundations of Science Policy." *Nature* 206 (15 April 1965): 233–238.

———. "Nations Can Publish Or Perish." *Science and Technology* 70 (1967): 84–90.

———. "Measuring the Size of Science." *Proceedings of the Israel Academy of Sciences and Humanities* 4 (1969): 98–111.

———. "The State of the Art in Science Policy Studies." Copenhagen lecture, 23 March, 1972.

———. "An Extrinsic Value Theory for Basic and 'Applied' Research." *Policy Studies Journal* 5 (Winter 1976): 160–168.

———. "Toward a Model for Science Indicators." In *Toward a Metric for Science: The Advent of Science Indicators,* pp. 69–95. Edited by Y. Elkana, J. Lederberg, R. K. Merton, A. Thackray, and H. Zuckerman. New York: Wiley Interscience, 1977.

Price, Don K. *The Scientific Estate.* New York: Oxford University Press, 1965.

Price, William J.; Ashley, William G.; and Martino, Joseph P. "The Effect of Government Funding on Commercial R&D." In *Factors in the Transfer of Technology,* pp. 117–136. Edited by William J. Gruber and Donald G. Marquis. Cambridge, Mass.: M.I.T. Press, 1969.

Primack, Joel, and von Hippel, Frank. *Advice and Dissent: Scientists in the Political Arena.* New York: Basic Books, 1974.

Rabkin, Yakov M. "Scientometric Studies in Chemistry." *Social Studies of Science* 6 (February 1976a): 128–132.

———. "Measuring Science in the U.S.S.R.: Uses and Expectations." *Survey* 2 (Spring 1976b): 69–80.

———. "'Naukovedenie': The Study of Scientific Research in the Soviet Union." *Minerva* 14 (Spring 1976c): 61–78.

Reddy, Amulya Kumar N. "Alternative Technology: A Viewpoint from India." *Social Studies of Science* 5 (August 1975): 331–342.

Rein, Martin. *Social Science and Public Policy.* New York: Penguin, 1976.

Rettig, Richard A. *Cancer Crusade: The Story of the National Cancer Act of 1971*. Princeton, N.J.: Princeton University Press, 1977.

Riecken, Henry W. "The Federal Government and Social Science Policy in the United States." In *Social Science and Government: Policies and Problems*, pp. 173–190. Edited by Albert B. Cherns, Ruth Sinclair, and W. I. Jenkins. London: Tavistock Publications, 1972.

Robbins, David, and Johnston, Ron. "The Role of Cognitive and Occupational Differentiation in Scientific Controversies." *Social Studies of Science* 6 (September 1976): 349–368.

Rogers, Everett M.; Eucland, J. D.; and Bean, Alden. *The Agricultural Extension Model and Its Extension to Other Innovation-Diffusion Problems*. Palo Alto, Calif.: Department of Communication, Stanford University, 1976.

Rose, Hilary. "The Rejection of the WHO Research Centre: A Case Study of Decision-Making in International Scientific Collaboration." *Minerva* 5 (Spring 1967): 340–356.

Rose, Hilary, and Rose, Steven. "The Radicalisation of Science." *Socialist Register* (1972): 105–133.

Rosen, Steven. "Testing the Theory of the Military-Industrial Complex." In *Testing the Theory of the Military-Industrial Complex*, pp. 1–25. Edited by Steven Rosen. Lexington, Mass.: Lexington Press, 1973.

Rosenberg, Charles E. *No Other Gods: On Science and American Thought*. Baltimore: Johns Hopkins University Press, 1976.

Rossi, Peter H. "Field Experiments in Social Programs: Problems and Prospects." In *Social Research and Public Policy: The Dartmouth-OECD Conference*, pp. 87–116. Edited by Gene M. Lyons. Hanover, N.H.: Public Affairs Center, Dartmouth College, 1975.

Rossi, Peter H., and Williams, Walter, eds. *Evaluating Social Programs: Theory, Practice, and Politics*. New York: Academic Press, 1972.

Roszak, Theodore. *The Making of a Counter-Culture: Reflections on the Technocratic Society and Its Youthful Opposition*. London: Faber, 1968.

Rottenberg, Simon. "The Warrants for Basic Research." In *Criteria for Scientific Development*, pp. 134–142. Edited by Edward Shils. Cambridge, Mass.: M.I.T. Press, 1968.

Ruggie, J. G. "International Responses to Technology: Concepts and Trends." *International Organization* 29 (Summer 1975): 557–583.

Ruina, J. P. "Aborted Military Systems." In *Impact of New Technologies on the Arms Race*, pp. 304–340. Edited by B. T. Feld, T. Greenwood, G. W. Rathjens, and S. Weinberg. Cambridge, Mass.: M.I.T. Press, 1971.

Russell, Beverly, and Shore, Arnold. "Limitations on the Governmental Use of Social Science in the United States." *Minerva* 14 (Winter 1976–1977): 475–495.

Ruttenberg, Charles L. "Political Behavior of American Scientists: The Movement Against Chemical and Biological Warfare." Ph.D. dissertation, New York University, 1972.

Sagasti, Francisco R. "Underdevelopment, Science, and Technology: The Point of View of the Underdeveloped Countries." *Science Studies* 3 (January 1973): 47–66.

———. "Guidelines for Technology Policies." *Science and Public Policy* 4 (February 1977): 2–15.

Said, Abdul, and Simmons, L., eds. *The New Sovereigns: Multinational Corporations as World Powers.* New York: Prentice-Hall, 1974.

Salomon, Jean-Jacques. "Europe and the Technological Gap." *International Studies Quarterly* 15 (March 1971): 5–31.

———. "Science Policy and Its Myths: The Allocation of Resources." *Public Policy* 20 (Winter 1972): 1–33.

———. *Science and Politics.* London: Macmillan, 1973.

———. "Science Policy Studies and the Development of Science Policy." In *Science, Technology and Society: A Cross-Disciplinary Perspective,* pp. 43–70. Edited by Ina Spiegel-Rösing and Derek de Solla Price. Beverly Hills, Calif.: Sage Publications, 1977.

Sapolsky, Harvey. *The Polaris System Development: Bureaucratic and Programmatic Success in Government.* Cambridge, Mass.: Harvard University Press, 1972.

———. "Science Policy." In *Handbook of Political Science,* vol. 6: *Policies and Policymaking,* pp. 79–110. Edited by Fred I. Greenstein and Nelson W. Polsby. Reading, Mass.: Addison-Wesley, 1975.

———. "Science, Technology, and Military Policy." In *Science, Technology and Society: A Cross-Disciplinary Perspective,* pp. 443–472. Edited by Ina Spiegel-Rösing and Derek de Solla Price. Beverly Hills, Calif.: Sage Publications, 1977.

Sardar, Ziauddin, and Rosser-Owen, Dawud. "Science Policy and Developing Countries." In *Science, Technology and Society: A Cross-Disciplinary Perspective,* pp. 535–575. Edited by Ina Spiegel-Rösing and Derek de Solla Price. Beverly Hills, Calif.: Sage Publications, 1977.

Sayles, Leonard, and Chandler, Margaret. *Managing Large Systems: Organizations for the Future.* New York: Harper & Row, 1971.

Schmandt, Jurgen. "United States Science Policy in Transition." In *Science Policies of Industrial Nations,* pp. 191–212. Edited by T. Dixon Long and Christopher Wright. New York: Praeger, 1975.

———. "Science and Technology as Objects of Social Science Research." *Society for Social Studies of Science Newsletter* 3 (1978): 14–22.

Scott, Robert A., and Shore, Arnold. "Sociology and Policy Analysis." *American Sociologist* 9 (1974): 51–59.

———. *Why Sociology Does Not Apply.* New York: Elsevier North Holland, 1979.

Seidel, R. N. *Toward an Andean Common Market for Science and Technology*. Ithaca, N.Y.: Program on Policies for Science and Technology in Developing Countries, Cornell University, 1974.

Shapley, Willis. *Research and Development in the Federal Budget, FY 1977*. Washington, D.C.: American Association for the Advancement of Science, 1976.

Shils, Edward. "The Obligations of Scientists as Counsellors: Introductory Note." *Minerva* 10 (January 1972): 107–110.

———, ed. *Criteria for Scientific Development*. Cambridge, Mass.: M.I.T. Press, 1968.

Shonfield, Andrew. "The Social Sciences in the Great Debate on Science Policy." *Minerva* 10 (July 1972): 426–438.

Sills, David. "The Environmental Movement and Its Critics." *Human Ecology* 3 (1975): 1–41.

Sklair, Leslie. "The Sociology of Opposition to Science and Technology: With Special Reference to the Work of J. Ellul." *Comparative Studies in Society and History* 13 (1971): 217–235.

Skolnikoff, Eugene B. *Science, Technology, and American Foreign Policy*. Cambridge, Mass.: M.I.T. Press, 1967.

———. *International Imperatives of Technology: Technological Development and the International System*. Research series no. 16. Berkeley, Calif.: Institute of International Studies, University of California, 1972.

———. "International Commission for Science Policy Studies." *Science Studies* 3 (January 1973): 89–90.

———. "Science, Technology, and the International System." In *Science, Technology and Society: A Cross-Disciplinary Perspective*, pp. 507–534. Edited by Ina Spiegel-Rösing and Derek de Solla Price. Beverly Hills, Calif.: Sage Publications, 1977.

Smith, Bruce L. R. "A New Science Policy in the United States." *Minerva* 11 (April 1973): 162–174.

———. "The Strengths and Limitations of the Present Tradition in the Study of Science Policy." Review of W. Henry Lambright's *Governing Science and Technology*. *Minerva* 15 (Summer 1977): 264–269.

Smith, Bruce L. R., and Karlesky, Joseph J. *The State of Academic Science: The Universities in the Nation's Research Effort*. New York: Change Magazine Press, 1977.

Smith, R. Jeffery. "More Fingers in the RANN Pie?" *Science* 197 (30 September 1977): 1347.

Spiegel-Rösing, Ina. *Wissenshaftsentwicklung und Wissenschaftssteuerung*. Frankfurt: Athenaeum Verlag, 1973.

———. "Science Policy Studies in a Political Context: The Conceptual and Institutional Development of Science Policy Studies in the German Democratic Republic." *Science Studies* 3 (October 1973): 393–413.

————. "The International Council for Science Policy Studies." *Social Studies of Science* 6 (February 1976): 133–135.

————. "Science Studies: Bibliometric and Content Analysis." *Social Studies of Science* 7 (February 1977): 97–112.

————. "The Study of Science, Technology, and Society (SSTS): Recent Trends and Future Challenges." In *Science, Technology and Society: A Cross-Disciplinary Perspective,* pp. 7–42. Edited by Ina Spiegel-Rösing and Derek de Solla Price (see next entry).

Spiegel-Rösing, Ina, and Price, Derek de Solla, eds. *Science, Technology, and Society: A Cross-Disciplinary Perspective.* Beverly Hills, Calif.: Sage Publications, 1977.

Stein, Bernard R. "Public Accountability and the Project-Grant Mechanism." *Research Policy* 2 (1973): 2–16.

Stent, Gunther S. "The Dilemma of Science and Morals." *Genetics* 78 (1974): 41–51.

Stockfisch, J. A. *Plowshares into Swords: Managing the American Defense Establishment.* New York: Mason and Lipscomb, 1973.

Stockholm International Peace Research Institute. *World Armaments and Disarmament.* SIPRI Yearbook, 1973. Stockholm: Almquist and Wiksell, 1973.

Storer, Norman. "The Organization and Differentiation of the Scientific Community: Basic Disciplines, Applied Research, and Conjunctive Domains." Prepared for a colloquium on Improving the Social and Communication Mechanisms of Education Research, sponsored by the American Educational Research Association, Washington, D.C., November 1968.

Strickland, S. P. *Politics, Science, and Dread Disease: A Short History of United States Medical Research Policy.* Cambridge, Mass.: Harvard University Press, 1972.

Strobel, Gary A. "A New Grants Program in Agriculture." *Science* 199 (3 March 1978): 935.

Symes, John M. D. "Policy and Maturity in Science." *Social Science Information* 15 (1976): 337–347.

Taviss, Irene. "Futurology and the Problem of Values." *International Social Science Journal* 21 (1969): 574–584.

————. "A Survey of Popular Attitudes toward Technology." *Technology and Culture* 13 (October 1972): 606–621.

Tilton, John E. *International Diffusion of Technology: The Case of Semiconductors.* Washington, D.C.: Brookings Institution, 1971.

Toulmin, Stephen. "The Complexity of Scientific Choice: A Stocktaking." In *Criteria for Scientific Development,* pp. 63–79. Edited by Edward Shils. Cambridge, Mass.: M.I.T. Press, 1968.

————. "The Complexity of Scientific Choice II: Culture, Overheads, or

Tertiary Industry?" In *Criteria for Scientific Development*, pp. 119–133. Edited by Edward Shils. Cambridge, Mass.: M.I.T. Press, 1968.

Tribe, Laurence H. "Legal Frameworks for the Assessment and Control of Technology." *Minerva* 9 (April 1971): 243–255.

———. "Policy Science: Analysis or Ideology." *Philosophy and Public Affairs* 2 (Fall 1972): 66–110.

Trow, Martin. "Public Policy Schools Attuned to Modern Complexities." *Policy Studies Journal* 1 (1973): 250–252.

Uliassi, Pio D. "The Prince's Counselors: Notes on Government-Sponsored Research on International and Foreign Affairs." In *The Use and Abuse of Social Science*, pp. 309–342. Edited by Irving L. Horowitz. New Brunswick, N.J.: Transaction Books, 1971.

UNESCO. *Science Policy Studies and Documents* (by country). Paris: UNESCO, 1965–1977.

———. *Science Policy Research and Teaching Units*. Science Policy Studies and Documents, no. 28. Paris: UNESCO, 1971.

U.S. House of Representatives. Committee on Science and Technology. *Special Oversight Review of Agricultural Research and Development*. Washington, D.C.: Government Printing Office, 1976.

Useem, Michael. "Government Influence on the Social Science Paradigm." *The Sociological Quarterly* 17 (1976a): 146–161.

———. "State Production of Sociological Knowledge: Patterns in Government Financing of Academic Sociological Research." *American Sociological Review* 41 (August 1976b): 613–629.

Van de Vall, Mark. "A Theoretical Framework for Applied Social Research." *International Journal of Mental Health* 2 (1973): 6–25.

———. "Utilization and Methodology of Applied Social Research: Four Complementary Models." *Journal of Applied Behavioral Science* 11 (1975): 14–38.

Van den Daele, Wolfgang, and Weingart, Peter. *The Utilization of the Social Sciences in the Federal Republic of Germany: An Analysis of Factors of Resistance and Receptivity of Science to External Direction*. Bielefeld: Science Studies Unit, n.d.

Van den Daele, Wolfgang; Krohn, Wolfgang; and Weingart, Peter. "The Political Direction of Scientific Development." In *The Social Production of Scientific Knowledge*, pp. 219–242. Edited by Everett Mendelsohn, Peter Weingart, and Richard Whitley. Dordrecht: D. Reidel, 1977.

Vig, Norman J. *Science and Technology in British Politics*. Oxford: Pergamon, 1968.

———. "Policies for Science and Technology in Great Britain: Postwar Development and Reassessment." In *Science Policies of Industrial Nations*, pp. 59–108. Edited by T. Dixon Long and Christopher Wright. New York: Praeger, 1975.

Von Hippel, Frank, and Primack, Joel. "Public Interest Science." *Science* 177 (29 September 1972): 1166–1171.

Wade, Nicholas. "Agriculture: Critics Find Basic Research Stunted and Wilting." *Science* 180 (27 April 1973a): 390–393.

———. "Agriculture: Signs of Dead Wood in Forestry and Environmental Research." *Science* 180 (4 May 1973b): 474–477.

———. "Contrary to Fears, Public Is High on Science." *Science* 199 (31 March 1978): 1420–1421.

Waldhart, Thomas. "Utility of Scientific Research: The Engineer's Use of the Products of Science." *I.E.E.E. Transactions on Professional Communication* 17 (June 1974): 33–35.

Walentynowicz, Bohdan. "The Science of Science in Poland: Present State and Prospects of Development." *Social Studies of Science* 5 (May 1975): 213–222.

Walsh, John. "Electric Power Research Institute: A New Formula for Industry R&D." *Science* 182 (19 October 1973): 263–265.

———. "Social Studies of Science: Society Crosses Disciplinary Lines." *Science* 198 (18 November 1977): 706–707.

Weinberg, Alvin M. "Scientific Choice and the Scientific Muckrakers." *Minerva* 7 (Autumn-Winter 1968–1969): 52–63.

———. "Social Problems and National Socio-Technical Institutes." In *Applied Science and Technological Progress*. Report to the Committee on Science and Astronautics, U.S. House of Representatives, by the National Academy of Sciences, pp. 415–434. Washington, D.C.: Government Printing Office, 1967.

———. "Criteria for Scientific Choice." In *Criteria for Scientific Development*, pp. 21–33. Edited by Edward Shils. Cambridge, Mass.: M.I.T. Press, 1968.

———. "Science and Trans-Science." *Minerva* 10 (April 1972a): 209–222.

———. "A Useful Institution of the Republic of Science." *Minerva* 10 (July 1972b): 439–440.

Weiner, Sanford L. "Resource Allocation in Basic Research and Organizational Design." *Public Policy* 20 (1972): 227–255.

Weingart, Peter. *Die Amerikanische Wissenschaftslobby*. Düsseldorf: Bertelsmann Universitäts-Verlag, 1970.

Weiss, Carol H. "The Politicization of Evaluation Research." *Journal of Social Issues* 26 (1970) 57–68.

———. *Evaluation Research*. Englewood Cliffs, N.J.: Prentice-Hall, 1972.

Wenk, Edward, Jr., and Kuehn, Thomas J. "Interinstitutional Networks in Technological Delivery Systems." In *Science and Technology Policy*, pp. 153–165. Edited by Joseph Haberer. Lexington, Mass.: Lexington Books, D. C. Heath, 1977.

Westin, A. F. *Privacy and Freedom*. New York: Atheneum, 1967.

Wholey, Joseph S.; Nay, Joe N.; Scarlon, John W.; and Schmidt, Richard E.

"If You Don't Care Where You Get To, Then It Doesn't Matter Which Way You Go." In *Social Research and Public Policy*, pp. 175–197. Edited by Gene M. Lyons. Hanover, N.H.: Public Affairs Center, Dartmouth College, 1975.

Williams, Roger. "Some Political Aspects of the Rothschild Affair." *Science Studies* 3 (January 1973): 31–46.

————. *European Technology: The Politics of Collaboration*. London: Croem Helm, 1973.

Williams, Walter. *Social Policy Research and Analysis: The Experience in the Federal Social Agencies*. New York: American Elsevier, 1971.

Wilson, John T. "A Dilemma of American Science and Higher Education Policy." *Minerva* 9 (April 1971): 171–196.

Winner, Langdon. *Autonomous Technology: Technics out of Control as a Theme in Political Thought*. Cambridge, Mass.: M.I.T. Press, 1977.

Wittwer, S. H. "The Next Generation of Agricultural Research." *Science* 199 (27 January 1978): 375.

Wynne, B. "The Rhetoric of Consensus Politics: A Critical Review of Technology Assessment." *Research Policy* 4 (1975): 108–158.

Yarmolinsky, Adam. "The Policy Researcher: His Habitat, Care, and Feeding." In *The Use and Abuse of Social Science*, pp. 196–211. Edited by Irving L. Horowitz. New Brunswick, N.J.: Transaction Books, 1971a.

————. *The Military Establishment*. New York: Harper & Row, 1971b.

York, Herbert F. *The Advisors: Oppenheimer, Teller, and the Superbomb*. San Francisco: W. H. Freeman, 1976.

York, Herbert F., and Greb, G. A. "Military Research and Development: A Postwar History." *Bulletin of Atomic Scientists* 33 (January 1977): 12–27.

Zierold, Kurt. *Forschungsförderung in drei Epochen: Deutsche Forschungsgemeinschaft: Geschichte, Arbeitsweise, Kommentare*. Weisbaden: Franz Steiner Verlag, 1968.

Zysman, J. "Between the Market and the State: Dilemmas of French Policy for the Electronics Industry." *Research Policy* 3 (1975): 312.

Index

669